SELECTED WORKS OF YAKOV BORISOVICH ZELDOVICH

Ya. B. Zeldovich, ca. 1956

SELECTED WORKS OF YAKOV BORISOVICH ZELDOVICH

Editor of English Edition
J. P. Ostriker

Coeditors of English Edition
G. I. Barenblatt
R. A. Sunyaev

Technical Supervisor of English Edition
E. Jackson

Translators of English Edition
A. Granik
E. Jackson

VOLUME II
PARTICLES, NUCLEI, AND THE UNIVERSE

PRINCETON UNIVERSITY PRESS
PRINCETON, NEW JERSEY

Published by Princeton University Press, 41 William Street,
Princeton, New Jersey 08540
In the United Kingdom: Princeton University Press, Chichester, West Sussex

Library of Congress Cataloging-in-Publication Data

Zel'dovich, IA. B. (IAkov Borisovich)
[Chastitsy, iadra, vselennaia. English]
Particles, nuclei, and the universe / editor of English edition J. P. Ostriker; coeditors of English edition G. I. Barenblatt, R. A. Sunyaev; technical supervisor of English edition E. Jackson; translators of English edition A. Granik, E. Jackson.
p. cm. -- (Selected works of Yakov Borisovich Zeldovich; v. 2)
Translation of: Chastitsy, iadra, vselennaia.
Includes bibliographical references.
ISBN 0-691-08742-3
1. Nuclear physics. 2. Particles (Nuclear physics). 3. Nuclear astrophysics. 4. Cosmology. I. Ostriker, J. P. II. Barenblatt, G. I. III. Sunyaev, R. A. IV. Title. V. Series: Zel'dovich, IA. B. (IAkov Borisovich). Selections. English. 1992 ; v. 2
QC3.Z44213 1993 vol. 2
[QC776]
530 s--dc20
[539.7'2] 91-14814

Selected Works of Yakov Borisovich Zeldovich, Volume I: *Chemical Physics and Hydrodynamics*, was published in 1992

Both Volumes I and II were composed in Computer Modern Roman using the TEX typesetting system

Princeton University Press books are printed on acid-free paper, and meet the guidelines for permanence and durability of the Committee on Production Guidelines for Book Longevity of the Council on Library Resources

Printed in the United States of America

10 9 8 7 6 5 4 3 2 1

Contents

III. *Atomic Physics and Radiation*

Part Two

Astrophysics and Cosmology

IV. *Elementary Particles and Cosmology*

V. *General Theory of Relativity and Astrophysics*

VI. Neutron Stars and Black Holes. Accretion.

VII. Interaction of Matter and Radiation in the Universe

VIII. Formation of the Large-Scale Structure of the Universe

IX. Observational Effects in Cosmology

Part Three
The History of Physics. Personalia

Preface to the English Edition of the Selected Works of Ya. B. Zeldovich

There has been no physical scientist in the second half of the twentieth century whose work shows the scope and depth of the late Yakov Borisovich Zeldovich. Born in Minsk in 1914, he was the author of over 20 books and over 500 scientific articles on subjects ranging from chemical catalysis to large-scale cosmic structure, with major contributions to the theory of combustion and hydrodynamics of explosive phenomena. His passing in Moscow in December of 1987 was mourned by scientists everywhere. To quote Professor John Bahcall of the Institute for Advanced Study: "We were enriched in Princeton as in the rest of the world by his insightful mastery of physical phenomena on all scales. All of us were his students, even those of us who never met him." In his range and productivity, Zeldovich was the modern equivalent of the English physicist Raleigh (1842-1919) whose name is associated with phenomena ranging from optics to engineering.

The breadth of Zeldovich's genius (characterized as "probably unique" by the Soviet physicist Andrei Sakharov) was alternately intimidating or enthralling to other scientists. A letter sent to Zeldovich by the Cambridge physicist Steven Hawking, after a first meeting in Moscow, compares Zeldovich to a famous school of pre-war mathematicians who wrote under a single fictitious pseudonym: "Now I know that you are a real person, and not a group of scientists like Bourbaki."

No selection from an opus of such scope can capture its full range and vigor. While basing ourselves primarily on the Russian edition, published by the Soviet Academy of Sciences in 1984–1985, we were delayed repeatedly as important and hitherto untranslated (but frequently cited) papers were brought to our attention as clearly warranting inclusion. Zeldovich played a major role in re-editing the Russian edition before translation and in choosing additional material for the present work. All told, this edition is approximately 15% longer than the Russian edition and the second volume contains one largely new section: *The History of Physics. Personalia*, including impressions of Einstein and Landau, and ending with *An Autobiographical Afterword.*

Because he wrote in Russian during a period when relations between that culture and the western world were at an historically low ebb, international recognition for Zeldovich's achievements were slower to arrive than merited. Within the Soviet Union his accomplishments were very well recognized, in part, due to his major contributions to secret wartime work. As the world's leading expert on combustion and detonation, he had naturally been drafted early into the effort for national survival. He had written entirely prescient papers in 1939 and 1940 (included in this volume) on the theoretical possibility of chain reactions among certain isotopes of uranium. The physicist Andrei Sakharov wrote that "from the very beginning of Soviet work on the atomic (and later the thermonuclear) problem, Zeldovich was at the very epicenter of events. His role there was completely exceptional." Zeldovich was intensely proud of his contributions to the wartime Soviet scientific effort and was the most decorated Soviet scientist. His awards include the Lenin Prize, four State Prizes, and three Gold Stars.

As a corollary to internal recognition, of course, Zeldovich's scientific work was burdened by the enormous handicaps of isolation, secrecy and bureaucracy in a closed society, made more extreme for him by restrictions due to defense work. He was not permitted to attend conferences outside of the Soviet Block until August 1982 at age 68, when he delivered an invited discourse "Remarks on the Structure of the Universe" to the International Astronomical Union in Patras, Greece. When asked then by this Editor when he was last out of the Soviet Union, he answered without hesitation "sixty eight years ago," i.e., only in a prior life. Previous to that meeting, his access to preprints, normal correspondence, all of the human interchange of normal scientific life, were severely circumscribed with contacts increasing as he moved out of defense work. Then, as international relations improved, international acclaim followed. Elected in 1979 as a foreign associate of the U. S. National Academy of Sciences, he had already been made a member of the Royal Society of London and other national scientific academies. Despite having turned relatively late in his scientific career to astrophysics, his accumulated achievements in that area, rewarded with the Robertson Prize of the U. S. National Academy of Sciences for advances in cosmology, put him among the world's leading theoretical astrophysicists.

The science is of course more interesting than the honors it wins for the scientist. Let me note just two items from astrophysics, my own specialty, where Zeldovich showed extraordinary vision and imagination. He argued shortly after their discovery that quasars were accreting black holes, and that the universe was likely to have a large-scale porous structure, anticipating in both cases the standard paradigms for interpreting these cosmic phenomena. In addition, he was among the first to realize that the early universe could be used as our laboratory for very high energy physics, leaving as fossils strange particles and cosmic microwave background fluctuations.

If the matter is more important than the recognition, it was also true, for Zeldovich in particular, that the manner was as significant as the matter. He always proceeded by a direct intuitive *physical* approach to problems. Even in areas where his ultimate accomplishment was a mathematical formulation adopted by others such as the "Zeldovich number" in combustion theory or the "Zeldovich spectrum" and the "Zeldovich approximation" to linear perturbations in cosmology, the reasoning and approach are initially and ultimately physical and intuitive. His view was that if you cannot explain an idea to a bright high school student, then you do not understand it. He backed up this conviction, and his interest in the education of young scientists, with the book *Higher Mathematics for Beginners*, which presented in a clear and intuitive way the elementary mathematical tools needed for modern science. Here again Zeldovich was in good company; from Einstein to Feynman, the greatest physicists have felt that they could and *should* make clear to anyone who cared to listen, the excitement of modern science.

The value of Zeldovich's papers, unlike those of most scientists, has outlived the novelty of his results. But, inevitably, one must question the logic of republishing scientific papers. Is not all valid scientific work included in and superseded by later work. Of course there is a value in collecting, for the record, in one place the major works of a truly great scientist. The fact that we include with each paper, commentaries (often revised from the Soviet edition) by the author on the significance of these papers will further enhance their value to historians and philosophers of science. But Zeldovich was almost above all else the teacher, the founder of a school of today's world famous scientists and author of widely read texts at all levels. He had strong views on *how* science should be done and how it should be taught. To him, the "how" of the scientific method, of his own scientific method, was central; it was what he most wanted to communicate in making his work available to a broader audience.

We are happy to be able to provide a complete enlarged edition of the works of this great scientist for the English-speaking world. We would like to thank the Academy of Sciences of the USSR for permission to utilize (a) *Izbrannye Trudy: I. Khimicheskaĭa Fizika i Gidrodinamika* and (b) *II. Chastitsy, Iadra, Vselennaĭa*, but especially offer our thanks to Professors G. I. Barenblatt and R. A. Sunyaev for their dedication and expertise in closely reading (Volumes One and Two, respectively) the entire manuscript in its English edition.

J. P. Ostriker
19 January 1990

Acknowledgements

The editors acknowledge with thanks permission from the following owners of copyrights, both of translations, some of which have been modified at the request of the scientific editors of this edition, and of original English-language publications.

Acta Physicochimica

Akusticheskiĭ journal

American Institute of Physics

Annual Review of Fluid Mechanics

Astronomy and Astrophysics

Astrophysics and Space Science

Blackwell Scientific Publications, Ltd., *Royal Astronomical Society of London Monthly Notices*

Cambridge University Press

Combustion and Flame

Communications in Mathematical Physics

Editrice Compositori

Elsevier Science Publishing Co., Inc.

Fizika goreniĭa i vzryva

W. H. Freeman and Co.

Gordon and Breach Science Publishers, Inc.

Journal of Fluid Mechanics

National Academy of Sciences, *Proceedings of the National Academy of Sciences*

National Advisory Comm. Aeronautics, Techn. Mem.

North-Holland Physics Publishing

Professor E. S. Phinney

Physikalische Zeitschrift der Sowjetunion

Reidel Publishing Company, "The Theory of Large-Scale Structure of the Universe" by Ya. B. Zeldovich, in *Large-Scale Structure of the Universe*, D. Reidel, Dordrecht, pp. 409–421. Copyright 1978 by International Astronomical Union. Reproduced by permission of Kluwer Academic Publishers.

Second international colloquium on explosion and reacting systems gasdynamics

Società Italiana di Fisica

Springer-Verlag

Supplemento al Nuevo Cimento

Я. Б. ЗЕЛЬДОВИЧ

Избранные труды

ЧАСТИЦЫ, ЯДРА, ВСЕЛЕННАЯ

Под редакцией
академика Ю. Б. ХАРИТОНА

МОСКВА
«НАУКА»
1985

PART ONE

PARTICLES AND NUCLEI

I

Nuclear Physics

1

On the Problem of the Chain Decay of the Main Uranium Isotope*

With Yu. B. Khariton

In this work we consider the problem of the moderation of neutrons which form in uranium decay and of the conditions necessary for the chain decay of uranium.

For the chain decay of the main uranium isotope to be possible it is essential that the neutrons which form in the fission of the uranium atom manage with sufficient probability to induce the next decay event, not only before they leave the mass of the uranium involved in the decay [1], but also before they are slowed to an energy below which they are no longer able to induce decay of the main isotope.

In the present note we consider precisely this last problem. If we find a probability γ that neutrons forming with energy E_0 without absorption accompanying the act of decay are slowed to the energy E_k, below which the decay of the main isotope can no longer be induced, then under optimal conditions of a maximal mass of uranium the probability of a chain reaction will be determined by the inequality

$$\nu(1-\gamma) > 1, \tag{1}$$

where ν is the (average) number of neutrons arising for one neutron captured in the energy interval $E_0 - E_k$, and γ is the probability that a neutron is slowed without being absorbed (equivalent to breaking of the chain).

*Zhurnal eksperimentalnoĭ i teoreticheskoĭ fiziki **9** 12, 1425–1427 (1939).

Below we carry out a calculation which accounts only for elastic neutron scattering.

The conditions found under this assumption are necessary for an explosion to occur, but may not be sufficient due to the presence of inelastic collisions.

Let us find the quantity γ. To do this we simultaneously consider the equation determining the variation in the mean energy of the particles in scattering and the equation for the variation in the number of particles due to their absorption.

Restricting ourselves, as we have said, to elastic collisions, we write

$$\frac{dE}{dt} = -Eu \sum \sigma_{s_i} c_i \lambda_i, \tag{2a}$$

$$\frac{dN}{dt} = -Nu \sum \sigma_{c_i} c_i, \tag{2b}$$

where E and N are the mean energy and the number of particles, respectively, u is the velocity of the neutrons, σ_{s_i} and σ_{c_i} are the scattering and capture cross-sections by the nuclei of atoms of the i-th sort, c_i is the concentration, $\lambda_i = 2m_i/(m_i + 1)^2$, where m_i is the mass of the i-th nucleus, expressed in neutron masses. Hence

$$\frac{dN}{dt} = \frac{N \sum \sigma_{c_i} c_i}{E \sum \sigma_{s_i} c_i \lambda_i} = \frac{N\psi}{E}. \tag{3}$$

If ψ is independent of the energy

$$\gamma = \frac{N_k}{N_0} = \left(\frac{E_k}{E_0}\right)^{\psi} \tag{4}$$

and in the general case

$$\gamma = \frac{N_k}{N_0} = \exp\left(\int_{E_0}^{E_k} \psi(E)\, d\ln E\right). \tag{4a}$$

In constructing the criterion of feasibility of chain reaction (1) it should be kept in mind that the number of neutrons ν entering into it is taken with respect to one neutron captured in the energy interval $E_0 - E_k$. Thus, if the number of neutrons produced in a single decay event is ν_f, then we obtain the number ν entering into (1) by the formula

$$\nu = \frac{\nu_f \sigma_f c_U}{\sum \sigma_{c_i} c_i}, \tag{5}$$

where the summation in the denominator includes, as in formulas (2b) and (3), the term $\sigma_f c_U$ as well.

Let us do a concrete calculation applied to the proposed use of uranium oxide (see, for example, [1]).

We take the following values: $\nu = 1.5, 2, 3$ [2–4]; E_0 in the two versions of the calculation is equal to 3 and 2 MeV [5]; $E_k = 1.5$ MeV [6, 7]; $\sigma_{SO} = 2 \cdot 10^{-24}$ cm^2; $\sigma_{SU} = 6 \cdot 10^{-24}$ cm^2; $\sigma_{CO} = 0$, $\sigma_{CU} = 0.5 \cdot 10^{-24}$ cm^2 [7]; $c_O : c_U = 8 : 3$ for the composition of U_3O_8.

Substituting formula (1) into (4) we obtain the final table for the quantities $\nu(1-\gamma)$.

E_0, MeV	ν		
	1.5	2	3
3	0.63	0.84	1.26
2	0.3	0.4	0.6

As we see from the table, the presence of oxygen, if not completely eliminates, then strongly inhibits the chain decay of uranium.

The situation for pure uranium is completely different: in this case, in all versions, we obtain negligible chain breaking, not exceeding $5 \cdot 10^{-3}$ (in the absence of inelastic scattering) and the feasibility of explosion.

It is clear that these considerations may also be applied to the question of the feasibility of decreasing the critical mass of uranium by surrounding it with material which slows the diffusion of neutrons to the outside [1]. As a result of neutron moderation for a large number of collisions (note that the subsequent fate of the slowed neutrons is unimportant for the chain reaction), the effectively working thickness of the neutron-isolation layer is of order $\lambda\sqrt{M/\sigma}$, where λ is the free path and M is the mass of a nucleus of the isolating material.

This last remark is related to the problem of the controllability of decay by the effect of fast neutrons: in the immediate vicinity of the explosion limit (critical conditions for development of the chain) a change by even a very small additional number of neutrons arising from the fission of the 235 isotope under the action of slow neutrons can affect the behavior of such a very sensitive system. Thus, in principle, it is possible to regulate the decay of the main isotope using the decay of the 235 isotope under the action of slow neutrons in conditions when this latter decay cannot possibly lead to explosion.

All of the above calculations were carried out under the assumption that the system is of unlimited extent, i.e., that there are no additional losses of neutrons carried from the system by diffusion.

Comparison with Perrin's calculations allows us to conclude that with the decrease of $\nu(1-\gamma)$ approaching unity, the critical size of the system grows as $[\nu(1-\gamma)-1]^{-1/2}$ and the volume as $[\nu(1-\gamma)-1]^{-3/2}$.

In contrast, for $\nu(1-\gamma) < 1$ critical conditions for chain branching cannot be achieved for any size of the system.

We note, finally, that in light of the above individual experiments which observe an increase in the number of source neutrons by 10–20% [8] in the presence of uranium cannot yet be considered proof of the realizability of uranium chain decay. Such proof can be provided only by an increase of 5–10 times, corresponding to multiple chain branching, which requires using uranium mass of the same order as the critical mass.

Institute of Chemical Physics
Leningrad

Received
October 7, 1939

Note added in proof. On the basis of the theory of N. Bohr and J. Wheeler [*Bohr N., Wheeler J.*—Phys. Rev. **56**, 299 (1939)], published while the present article was in press, we carried out a calculation of chain breaking related to inelastic neutron scattering. Due to the absence of data on the energy levels of uranium 238, the calculation was done for Th C, which would appear not to introduce significant errors. The results of the calculation lead to the conclusion that even in the case of pure metallic uranium, no chain reaction apparently takes place.

REFERENCES

1. *Perrin F.*—C. r. Acad. Sci. **208**, 1394 (1939).
2. *Anderson H. L., Fermi E., Szillard L.*—Phys. Rev. **56**, 284 (1939).
3. *Anderson H. L., Fermi E., Hanstein H. B.*—Phys. Rev. **55**, 797 (1939).
4. *Halban H., Joliot F., Kowarski L.*—Nature **143**, 680 (1939).
5. *Halban H., Joliot F., Kowarski L.*—Nature **143**, 939 (1939).
6. *Roberts R. B., Mayer R. C., Hafstad L. R.*—Phys. Rev. **55**, 416 (1939).
7. *Ladenburg R., Kanner M. N., Barshall H., Van Voorhis C. C.*—Phys. Rev. **56**, 168 (1939).
8. *Haenny C., Rosenberg A.*—C. r. Acad. Sci. **208**, 898 (1939).

2
On the Chain Decay of Uranium Under the Action of Slow Neutrons*

With Yu. B. Khariton

In realizing the chain decay of uranium under the action of slow neutrons [1], it is necessary for continuation of the chain to slow neutrons created in the act of decay with an energy of several million volts to a velocity close to the thermal velocity at which they are sufficiently likely to cause the next act of decay of the isotope with atomic weight 235. In the interval between the energy of the neutrons formed and the region in which they cause decay (continue the chain) there is a region near 25 eV of resonant absorption of neutrons[1] by the basic isotope 238; this absorption does not lead to the appearance of new neutrons and is, consequently, a break in the chain just as is the absorption of neutrons by any admixtures to uranium present in the system.

However, quantitatively there is a significant difference between these two types of chain breaking. The capture cross-sections of slow neutrons by various atoms, including the cross-section for capture by the uranium nucleus which leads to its decay, vary identically with the energy of the neutrons (inversely proportional to the velocity, i.e., as $E^{-1/2}$). The distribution of neutrons among the various possible processes—absorption by various nuclei, absorption by uranium with its subsequent decay—does not depend on the energy, and therefore on the velocity, of the moderated neutrons. The probabilities of the different processes are in a constant ratio; specifically, they are proportional to the products of the number density of the nuclei participating in the process with the capture cross-section, measured (bearing in mind its dependence on the energy, i.e., as $E^{-1/2}$) for all processes at a single energy, for example at room temperature. Thus, the number of neutrons captured in one or another particular way is easily found by a formula of the form

$$N_i = \frac{N\sigma_i c_i}{\sum_l \sigma_l c_l}. \tag{1}$$

*Zhurnal eksperimentalnoĭ i teoreticheskoĭ fiziki **10** 1, 29–36 (1940).

[1]The action of slow neutrons causes decay of isotope 235 whose content in uranium is 0.7%.

The situation is quite different with resonant capture. Qualitatively it is clear that, due to the sharply different dependence of the cross-section on the energy of the neutrons (compared with the capture of slow neutrons according to the law $E^{-1/2}$), the simple formula (1), which contains neither the dependence on the initial neutron energy nor their rate of moderation, which determines the time that they spend in the dangerous zone of resonant capture, cannot correctly describe the dependence of the amount of resonance-absorbed neutrons on the number density.[2] By analogy, while all processes with cross-section $\sim E^{-1/2}$ are included in parallel, resonant capture in the region of 25 eV, where the above processes in practical terms have not yet begun, is included in series before them during the course of slowing of the neutrons.[3]

In the present note we shall attempt to establish the patterns of resonant capture and their consequences relating to chain decay on slow neutrons.

Preliminary calculations using the methods of our previous note showed that for uranium oxide or, even more so for pure uranium, the resonant absorption is extremely large and completely eliminates the possibility of a chain reaction on slow neutrons.

In order to accomplish this reaction strong slowing of the neutrons is necessary, which may be practically accomplished by the addition of a significant amount of hydrogen. Thus the conditions for the reactions on slow and fast neutrons turn out to be significantly different so that their simultaneous calculation is necessary in the closest vicinity to the critical conditions for one of the reactions, where the system is so sensitive that even minimal participation of the second reaction can change its behavior (see our note [2]).

Let us note the peculiarities of the forthcoming calculations.

1. The quite significant radius of scattering of neutrons by protons and the favorable ratio of the masses make calculation of the slowing of neutrons in scattering by the other nuclei unnecessary.

2. In each collision of a neutron with a proton the energy of the neutron after scattering varies throughout the interval from 0 to the initial energy before the collision. A neutron with an energy of 35–50 eV has a greater probability of being immediately slowed to an energy less than 25 eV, after which it cannot be resonance-absorbed. For such a strong exchange of energy the method of the previous article, in which the change in energy after collisions was considered to be a continuous process, is inapplicable.

3. Absorption (even very little) of thermal neutrons by protons restricts the possible dilution of uranium with hydrogen with the aim of accelerating

[2]Under constant conditions (above all, composition) resonant absorption can of course be approximately described by the introduction of an equivalent capture cross-section of neutrons at room temperature (Joliot [5]). However, this fictive cross-section is an unknown function of all the parameters of the system.

[3]Near 25 eV cross-sections proportional to $E^{-1/2}$ are negligibly small.

the energy exchange and decreasing resonant absorption near 25 eV. We shall have to find the optimal proportions of uranium and hydrogen taking both effects into account.

4. Finally, for concrete calculations the form of the energy distribution function of neutrons with given initial energy after one collision with a proton is extremely important.

As may be shown in a general form, the wave function of scattered particles possesses spherical symmetry;** the other wave functions vanish in the center and therefore enter with coefficients $\sim r/\lambda$, where r is the radius of the nucleus and λ is the wavelength of the neutron, so that at the energies much less than 10^6 eV of interest to us they may be disregarded. The equal probability of all directions, independent of the angle of scattering, just as in the collision of two elastic spheres in classical mechanics, leads in the calculation of the conservation laws to a very simple energy distribution; specifically, it leads to equal probabilities of all values of the energy less than the initial value:

$$\begin{aligned} a\,(E)\,dE &= \frac{dE}{E_0}, \qquad 0 < E < E_0, \\ a\,(E)\,dE &= 0, \qquad E > E_0. \end{aligned} \tag{2}$$

In the derivation of the formula we disregard the thermal energy of the scattering protons since it is quite small compared to the energy in the resonance region.

The cross-section of resonant capture in the presence of one level obeys the Breit-Wigner formula:

$$\sigma_E = \sigma_r \sqrt{\frac{E_r}{E}} \frac{(\Gamma/2)^2}{(E - E_r)^2 + (\Gamma/2)^2}. \tag{3}$$

In order to close this part of the task, we have extracted from expression (3) the term σ_2, which behaves at small E as $E^{-1/2}$, so that the corresponding cross-section may be accounted for with the others in the form (1):

$$\left.\begin{aligned} &\sigma = \sigma_1 + \sigma_2;\ \sigma_1 = \sigma_r \sqrt{\frac{E_r}{E}} \left[\frac{(\Gamma/2)^2}{(E - E_r)^2 + (\Gamma/2)^2} - \frac{(\Gamma/2)^2}{E_r^2 + (\Gamma/2)^2}\right]; \\ &\sigma_2 = \sigma_r \sqrt{\frac{E_r}{E}} \frac{(\Gamma/2)^2}{E_r^2 + (\Gamma/2)^2}. \end{aligned}\right\} \tag{3a}$$

The remaining function vanishes at $E = 0$, and from now one we will consider resonant absorption to mean only this part of the capture function.

The advantage of such a definition lies in the fact that we may now rigorously pose the question of the probability that neutrons will be slowed without suffering absorption at the resonant level, and this quantity is no longer tied to the question of the fate of thermal neutrons, as it would have

**This refers to the symmetry in the system of the center of inertia of a proton at rest and a moving neutron—*Editor's note.*

been had we not excluded from the resonance curve the capture at small energies.

The probabilities that a neutron will be resonantly captured in the next collision by uranium or scattered by hydrogen are in the ratio $\sigma_{cU}c_U : \sigma_{sH}c_H$ (we do not consider all the other processes). Let us introduce the normalized probability of capture in the next collision:

$$W = W(E) = \frac{\sigma_{cU}(E)c_U}{\sigma_{cU}(E)c_U + \sigma_{sH}c_H}. \tag{4}$$

We construct the equation for the joint probability $\varphi(E)$ that we seek for a neutron having energy E to be slowed without being captured in the resonant region: in the first collision the neutron has a probability W of being captured, and with probability $(1 - W)$ may be scattered with a uniform distribution of energy. Hence we obtain the integral equation

$$\varphi(E) = \frac{1 - W(E)}{E} \int_0^E \varphi(\varepsilon)\, d\varepsilon. \tag{5}$$

The equation is integrated in quadratures; to do this we take the derivative φ' and express the integral entering into it again in terms of the quantity φ. Thus we arrive at the differential equation

$$\varphi' = \frac{-W\varphi}{E} - \frac{W'\varphi}{1 - W}, \tag{6}$$

and finally find

$$\varphi = (1 - W) \exp\left(- \int_0^E W\, d\ln E \right). \tag{7}$$

At an energy E which is greater than the resonant energy, beginning from the value at which W may be disregarded, φ no longer depends on E so that the desired limiting value is

$$\varphi = \exp\left(- \int_0^\infty W\, d\ln E \right). \tag{8}$$

For the simplest form of the dependence of the cross-section of resonant capture on the energy

$$\sigma_{cU}(E) = \infty, \quad E_1 < E < E_2; \qquad \sigma_{cU} = 0, \quad E < E_1 \text{ and } E > E_2, \tag{9}$$

formula (8) will yield

$$\varphi = \frac{E_1}{E_2}. \tag{10}$$

We again emphasize that our arguments and the results (7), (8) and (10) refer exclusively to slowing by protons for which each scattering collision leads to a uniform distribution in the energy interval from the initial value to zero.

For calculations with a resonance curve of the form (3) we note that for very large values of the capture cross-section W approaches unity and then changes very little; therefore, (3) may be replaced by the expression

$$\sigma_{cU} = \text{const} \cdot (E - E_r)^{-2}. \tag{11}$$

Now it is easy to write the concrete form of the function, although we do not need it:

$$W = W_0[c_U(E - E_r)^{-2}/c_H], \tag{12}$$

$$\int W\, d\ln E = \text{const} \cdot \sqrt{c_U/c_H}, \tag{13}$$

or, denoting $c_H/c_U = \eta$, we obtain

$$\varphi = e^{-\alpha\sqrt{c_U/c_H}} = e^{-\alpha\eta^{-1/2}}. \tag{14}$$

Let us turn to the consideration of uranium chain decay itself. We denote by N the total number of fast neutrons appearing in the system in unit time both from the source (N_0) and from the fission of uranium nuclei by slow neutrons (N_1) so that $N = N_0 + N_1$. The number of neutrons which arise for each slow neutron captured (in any way) by uranium we denote by ν, and the probability that a neutron (already slowed to an energy much less than $E_r = 25$ eV) will be captured by uranium rather than hydrogen calculated from a formula of the form (1), $-\theta$:

$$\theta = \frac{\sigma_{cU} c_U}{\sigma_{cU} c_H + \sigma_{cH} c_H} = \frac{1}{1 + \beta c_H/c_U} = \frac{1}{1 + \beta\eta}. \tag{15}$$

We note that if we had wanted to use the number of neutrons ν_f arising for each decay of a uranium atom, then, bearing in mind the possibility of capture of neutron by uranium without decay, we would have had to introduce in place of the general probability of capture by uranium θ the probability of capture with decay:

$$\theta_f = \frac{\sigma_f c_U}{\sigma_{cU} c_U + \sigma_{cH} c_H} = \frac{\gamma}{1 + \beta\eta}, \tag{16}$$

where

$$\gamma = \frac{\sigma_f}{\sigma_{cU}} = 1 - \frac{\sigma_1}{\sigma_{cU}},$$

so that identically

$$\nu_f \theta_f = \nu\theta = \frac{\nu}{1 + \beta\eta}. \tag{17}$$

Of the overall number of fast neutrons N arising in unit time, the number which are slowed and pass successfully through the resonant level is φN; these, causing decay, lead to $N_1 = \nu\theta\varphi N$ new neutrons in unit time. By definition

$$N = N_0 + N_1 = N_0 + \nu\theta\varphi N, \tag{18}$$

$$N = \frac{N_0}{1 - \nu\theta\varphi}, \tag{19}$$

whence the critical condition [6]

$$\nu\theta\varphi = 1. \tag{20}$$

Thus, the determination of optimal conditions for branching of the nuclear chain reduces to finding the maximum of $\theta\varphi$ as a function of $\eta = c_H/c_U$, i.e., to finding the maximum of the function

$$\frac{\exp(-\alpha\eta^{-1/2})}{1+\beta\eta}. \tag{21}$$

Turning to practical calculations, it should be noted that while the quantity β, equal to the ratio of the cross-sections, is known comparatively well, data on resonant capture in contrast have been insufficiently determined.

A numerical calculation of the quantity φ directly from formula (7), where the capture cross-section was taken in the form (3) with the constants $\sigma_2 = 3000 \cdot 10^{-24}$, $\Gamma = 0.2$ and a scattering cross-section by hydrogen as $\sigma_{sH} = 20 \cdot 10^{-24}$ leads for $\eta = 1$ to the quantity $\varphi = 0.844$, which corresponds in the interpolation formula (14) to $\alpha = 0.168$.

A direct experiment by Halban, Kowarski and Savitch [3] gives practically the same quantity 0.84 at $\eta = 62$ (see note below), which corresponds to significantly greater capture under equal conditions and a corresponding $\alpha = 1.36$.

There have been indications [4] that the simple formula (3) with a single level is not applicable at all.[4]

Until the problem is clarified experimentally we have nothing but to perform a dual calculation with the two values:

$$\alpha = 0.168 \quad \text{and} \quad \alpha = 1.36.$$

In the following we take $\sigma_{cH} = 0.27 \cdot 10^{-24}$, $\sigma_{fU} = 2 \cdot 10^{-24}$, and the cross-section of idle capture by uranium as $1.2 \cdot 10^{-24}$.

From the numerical calculations we obtain respectively the position and height of the maximum of the quantity $\theta\varphi$ and the minimum value of ν for which the critical inequality (20) holds. We show the detailed calculation.

From the relations between the cross-sections we find in formulae (17), (16), (15):

$$\frac{\nu}{\nu_f} = \gamma = \frac{\sigma_f}{\sigma_{cU}} = 0.625; \qquad \beta = \frac{\sigma_H}{\sigma_{cU}} = 0.0845.$$

The function whose maximum we seek in two variants has the following forms:

$$\theta\varphi_1 = \frac{10^{-0.0745\eta^{-1/2}}}{1+0.0845\eta}; \qquad \theta\varphi_2 = \frac{10^{-0.603\eta^{-1/2}}}{1+0.0845\eta}. \tag{22}$$

[4] It should be noted that when a nucleus is strongly excited due to capture of a neutron, it is natural to expect the presence of a series of resonant levels which differ from one another by several dozen volts. This, however, will have little effect on the form of formula (14). Thus recalculation of the experimental data using (14) will be completely legitimate even when a series of levels is present.

We compile a table of both functions:

η	62	17	8	4	2	1	1/2	1/4
θ	0.160	0.410	0.597	0.748	0.855	0.922	0.960	0.980
φ_1	0.980	0.960	0.942	0.918	0.885	0.840	0.785	0.710
φ_2	0.840	0.716	0.613	0.501	0.377	0.521	—	—
$\theta\varphi_1$	0.157	0.384	0.562	0.686	0.757	0.775	0.752	0.696
$\theta\varphi_2$	0.134	0.284	0.366	0.374	0.331	0.231	—	—

At $\alpha = 0.168$, $\eta_{\max} = 1$, $\theta\varphi_{\max} = 0.775$, $\nu_{\min} = 1.29$.
At $\alpha = 1.36$, $\eta_{\max} = 4$, $\theta\varphi_{\max} = 0.374$, $\nu_{\min} = 2.64$.

The difference in the results of the calculations for the two variants decreases if we take into account the fact that the calculation of ν itself from experiments like those of Joliot and Fermi must also be consistently carried out in two variants. Here the greater value of α, which is less favorable for chain decay (yielding a smaller ν), obviously leads in processing the experimental data to an increase in the neutron output ν calculated from the observed experimental data. Thus, the quantity $\nu\theta\varphi$ in which we are ultimately interested varies much less with the choice of one or another α.

Let us do a detailed calculation from Joliot's experiment [5]. In order to avoid introducing the new concept of neutron lifetime, we will now consider the number of neutrons N_H which are absorbed by hydrogen of the solution in both the presence and absence of uranium salt in the solution. Since the neutron detector used by Joliot absorbs neutrons also with a probability proportional to $E^{-1/2}$, the absorption of neutrons by hydrogen is exactly proportional to the product of the detector indication and the hydrogen concentration.

Integrating over the entire volume we obtain for the spherically symmetric problem

$$N_H \sim \int c_H I r^2 \, dr, \tag{23}$$

where I is the detector reading at the given point. In a solution with a constant concentration of the dissolved substance and, consequently, of hydrogen

$$N_H \sim c_H \int I r^2 \, dr. \tag{24}$$

We find the quantity c_H in the two solutions used by Joliot, taking as 1 the concentration of hydrogen in pure water. For this we supplement the data on the relative density of solutions of ammonium nitrate and uranyl nitrate as a function of percentage of dissolved substance z taken from the physical-chemical tables of Landolt, Bernstein and Roth with the following quantities: the molarity of the solution according to the formula $\mu = 1000\, dz/100M$, where M is the molecular weight of the compound; the water content in a unit volume of the solution $\varepsilon_1 = d(1 - z/100)$; and the hydrogen content (with respect to pure water) in the solution ε, to which in the case of ammonium nitrate is added the hydrogen content of the salt itself. Finally we find the desired result by graphic interpolation in the coordinates $\mu - \varepsilon$. At $\mu = 1.6$ we find the hydrogen content in a 1.6-molar solution of ammonium nitrate to be 0.982, and in the same solution of uranyl nitrate—0.893.

We have discussed these elementary calculations in such detail because in Joliot's note one finds the assertion that the concentrations of hydrogen in the two solutions used by him differ by not more than 2%. These results, which contradict our own, can be obtained either by forgetting about the decrease in the hydrogen concentration in dissolving uranyl nitrate in water and comparing the ϵ of the solution NH_4NO_3 with $\epsilon = 1$ for pure water, or by forgetting the hydrogen content in the ammonium nitrate itself and having equal water contents (or hydrogen only in the form of water) of the two solutions. Both assumptions are obviously unfounded.

The integral on the right side of (24) is nothing other than the area under the curve Ir^2, whose variation was determined by Joliot. According to his data, this area increases in the ratio 1 : 1.05 when ammonium nitrate is replaced by uranyl nitrate.

Thus, the total number of neutrons absorbed by hydrogen varies in the ratio

$$\frac{0.893}{0.982} \cdot 1.05 = 0.955.$$

Consequently, when uranium is introduced, as our calculation shows, the number of neutrons absorbed by hydrogen in fact falls.

This still does not preclude the formation of more than one neutron for each thermal neutron absorbed by a uranium nucleus since when uranium is introduced an immediate consequence is the quite noticeable absorption of fast neutrons of the source as they are slowed to thermal velocity on the resonant capture level at 25 eV.

We shall show this numerically. In the notations introduced earlier, of the total number N of fast neutrons arising in unit time, φN neutrons will be slowed without absorption on the resonant level; the slowed neutrons are distributed between uranium and hydrogen as $\theta : (1 - \theta)$ so that, finally, from (19) the number of neutrons absorbed by hydrogen is

$$N_H = N_0(1 - \theta)\varphi/(1 - \nu\theta\varphi), \tag{25}$$

rather than N_0 without uranium.

Equating

$$(1 - \theta)\varphi/(1 - \nu\theta\varphi) = 0.955, \tag{26}$$

we find φ at $\eta = 62$, $\theta = 0.160$ in two variants: $\varphi_1 = 0.98$ (extrapolated by calculations based on data on the capture curve) and $\varphi_2 = 0.840$ (directly measured by Savitch, Halban, and Kowarski). Finally we obtain

$$\nu_1 = 0.88, \qquad \nu_2 = 1.95. \tag{27}$$

The corresponding quantities taken with respect to a single event of uranium decay, under the relations taken between capture cross-sections with decay and idle capture of slow neutrons:

$$\nu_f = 3.2\nu/2.0, \qquad \nu_{1f} = 1.41, \quad \nu_{2f} = 3.12. \tag{28}$$

The last number, 3.12, was calculated under the same assumptions as Joliot used to obtain the output $\nu_f = 3.5$. Thus, refining the hydrogen content in the solution and refining the calculation (Joliot considered all effects related to the introduction of uranium as small and systematically discarded terms of second order) have changed the final value relatively little.

Returning to the question of interest, we find the magnitude of the criterion for explosion $\nu\theta\varphi$; in the two consistently performed versions of the calculation it turns out to be equal:

$$(\nu\theta\varphi)_{1\,\max} = 0.88 \cdot 0.775 = 0.68 \quad \text{at } \eta = 1,$$
$$(\nu\theta\varphi)_{2\,\max} = 1.95 \cdot 0.374 = 0.73 \quad \text{at } \eta = 4,$$

This corresponds to a maximum intensity of the source due to an increase in neutrons from uranium decay by not more than 3–4 times under the optimal choice of uranium–water ratio.

Thus Joliot's experimental data give a value for the product $\nu\theta\varphi$ which is almost independent of the choice of α and is insufficient for chain decay to occur.

A calculation from Fermi's experiment is difficult due to the separate distribution of uranium and water in his instrument. In any case, it would not provide any more consoling result, and our conclusion based on experimental data about the impossibility of powerful chain decay in a uranium-water system turns out to be related in the final analysis only to the law chosen for variation of capture on the resonant level as a function of the ratio of hydrogen to uranium in the form (14); over a broad range it is independent of the value of the coefficient α, as is clear from comparison of the final results of the two variants of the calculation with two widely different values of α.

From this it follows that in order to realize conditions for the chain explosion of uranium it is necessary to use for neutron moderation heavy hydrogen or, perhaps, heavy water, or some other substance which ensures a sufficiently small capture cross-section. The significantly smaller scattering cross-section compared to hydrogen and the somewhat lower effectiveness of energy exchange may be compensated by the negligibly small capture cross-section of neutrons and the related possibility of extreme dilution of uranium (large η).

Another possibility lies in the enrichment of uranium with the isotope 235.

If water (hydrogen) is used as the solvent, the quantity $\nu\theta\varphi$ becomes equal to one when the uranium 235 content is increased by 1.9 times (from 0.7 to 1.3%) at an optimal value $\eta \sim 8$.

All that has been said above refers to a solution of uranium and hydrogen (or water solution of uranium salt, since the effects of other nuclei may be neglected) of infinite extent. Taking account of the finite size of the volume occupied by the mixture (solution) leads to lowering of the effective value of $\theta\varphi$ due to diffusion of neutrons to the outside.

Near the critical conditions of explosion it may be shown that the following relation holds:

$$(\theta\varphi)_{\text{eff}} = \theta\varphi(1 - A/d^2), \tag{29}$$

where d is the characteristic size of the system, and A is a quantity which depends on the free path.

It is clear that the achievement of critical conditions here is made more difficult. Conversely, the greater is the quantity $\nu\theta\varphi - 1$, the smaller the critical size of the system may be.

We would like to take this opportunity to express our gratitude to I. I. Gurevich, I. V. Kurchatov and I. Ya. Pomeranchuk for a number of valuable comments in discussions of this work.

Institute of Chemical Physics
Leningrad

Received
October 22, 1939

REFERENCES

1. *Perrin F.*—C. r. Acad. Sci. **208**, 1573 (1939).
2. *Zeldovich Ya. B., Khariton Yu. B.*—ZhETF **9**, 1425 (1939).
3. *Halban H., Kowarski L., Savitch P.*—C. r. Acad. Sci. **208**, 1396 (1939).
4. *Anderson H. L., Fermi E., Szillard L.*—Phys. Rev. **56**, 284 (1939).
5. *Halban H., Joliot F., Kowarski L.*—Nature **143**, 470, 680 (1939).
6. *Semenov N. N.* Tsepnye reaktsii [Chain Reactions]. Leningrad: Goskhimtekhizdat, 555 p. (1934).

3

Kinetics of the Chain Decay of Uranium*

With Yu. B. Khariton

We consider the development of a chain nuclear reaction in a mass of uranium in the transition across the critical mass. It is shown that thermal expansion is a powerful regulating factor which makes the transition across the limit—if such exists—completely safe. For a critical mass of 1 ton heating to 1000° can be accomplished by adding only $\sim$ 50 kg above the critical mass. A gradual increase in mass above the critical value leads to an oscillatory reaction regime whose period is inversely proportional to the square root of the rate of uranium supply. Delayed neutrons significantly increase the oscillation period of the reaction rate.

1. *Introduction*

In our previous papers [1, 2] we considered the question of the possibility, in principle, of realizing a chain decay reaction of uranium on fast and slow neutrons, without taking into account diffusional evacuation of neutrons, i.e., in essence, the calculations related to an infinitely extended mass of uranium or solution of a uranium compound in water.

It would appear (the lack of experimental data precludes any categorical assertions) that by applying some technique, creating a large mass of metallic uranium either by mixing uranium with substances possessing a small capture cross-section (e.g., with heavy water) or by enriching the uranium with the U^{235} isotope, which is thought to decay under the action of slow neutrons—it will be possible to establish conditions for the chain decay of uranium by branching chains in which an arbitrarily weak radiation by neutrons will lead to powerful development of a nuclear reaction and macroscopic effects. Such a process would be of much interest since the molar heat of the nuclear fission reaction of uranium exceeds by $5 \cdot 10^7$ times the heating capacity of coal. The abundance and cost of uranium would certainly allow the realization of some applications of uranium.

Therefore, despite the difficulties and unreliability of the directions indicated, we may expect in the near future attempts to realize the process.

In this paper we investigate the details of the behavior of a system in which conditions for branching of chains of nuclear reaction have somehow

*Zhurnal eksperimentalnoĭ i teoreticheskoĭ fiziki **10** 5, 477–482 (1940).

been achieved. In an infinitely extended system the neutron number density and reaction intensity then grow exponentially (see, for example, Flügge [3]):

$$\frac{dn}{dt} = bn \tag{1}$$

until a significant portion of the substance has already reacted. If the probability of branching is 0.1, i.e., the quantity $\alpha = \nu_f(1-\gamma) - 1 = 0.1$ (see our paper [1]), or $\alpha = \nu\theta\varphi - 1 = 0.1$ (see our paper [2]), then the inverse time of relaxation b of equation (1) for fast neutrons turns out to be of order $10^7\ \mathrm{s}^{-1}$, for slow neutrons it is around $10^4\ \mathrm{s}^{-1}$. In general form [4]

$$b = \alpha u \sum n_i \sigma_{c_i}, \tag{2}$$

where α is defined above, u is the neutron velocity, n_i is the number of particles of the i-th sort in a unit volume, σ_{c_i} is the capture cross-section of particles of the i-th sort.

In the case of a system of finite size, the evacuation of neutrons into the surrounding space is of course equivalent to breaking of chains. From these considerations, analyzing the diffusion of neutrons, Perrin [5] found the critical dimensions beginning from which a branching chain of reactions is possible. Perrin's calculation was generalized by Peierls [6]; he confirmed the critical condition for the existence of a steady regime, found by Perrin, and also analyzed the practically unimportant case of high probability of branching and large decay cross-section in which the critical dimensions of the system are small compared with the free path and the diffusion equation cannot be written.

Restricting ourselves to the only interesting case when the critical dimensions are significantly larger than the free path, we construct the equation of the variation in the number density of neutrons in the absence of an outside source:

$$\frac{\partial n}{\partial t} = bn + D\Delta n, \tag{3}$$

where the coefficient of diffusion is

$$D = \frac{1}{3}\lambda u = \frac{u}{3\sum n_i \sigma_{s_i}}. \tag{4}$$

The general solution of (3) may be found in the form of a sum:

$$n = n(x, y, z, t) = \sum \psi_i(x, y, z) e^{(b-g_i)t}, \tag{5}$$

where ψ_i and g_i are the eigenfunctions and eigenvalues, respectively, of the equation

$$D\Delta\psi + g\psi = 0 \tag{6}$$

with boundary condition $\psi = 0$ on the surface of the body.

From dimensionality considerations it is clear that

$$g_i = \frac{k_i D}{d^2}, \tag{7}$$

where k_i is a dimensionless coefficient depending only on the form of the body, d is the linear dimension of the body. In the case of a spherical form, equating d with the diameter of the sphere we find

$$k_i = 4\pi^2 i^2, \qquad k_1 = 4\pi^2 \cong 40. \tag{8}$$

The critical condition is

$$b - g_1 = 0; \quad b = \frac{k_1 D}{d^2}, \quad d_{\rm cr} = 2\pi\sqrt{\frac{D}{b}}, \tag{9}$$

which coincides with Perrin's result [5].

Substituting (2) and (4), we find at the limit

$$d_{\rm cr}\sqrt{\sum n_i \sigma_{c_i} \sum n_i \sigma_{s_i}} = \frac{2\pi}{\sqrt{3\alpha}}. \tag{10}$$

In a mixture of constant composition in which all n_i are in a constant ratio, at the limit

$$d_{\rm cr} n_i = \text{const}, \tag{11}$$

and the critical mass,

$$M_{\rm cr} = d_{\rm cr}^3 n_i \sim n_i^{-2}, \tag{12}$$

decreases as the density increases.

2. The Kinetics of the Decay

This calculation obviously is not sufficient to give a macroscopic description of the process under realistic conditions.

As is clear from equation (1), far from the critical conditions, when diffusional evacuation is small, the neutron number density grows exponentially at a huge rate, increasing by e times in a time of order 10^{-7} s for decay on fast neutrons, 10^{-3} s for decay on slow neutrons. Given such rapid development of the chain decay we can no longer put off consideration of the creation of the supercritical conditions which are uniquely necessary for chain decay to occur.

The time of occurrence of processes which bring about the transition to critical conditions, e.g., the time of approach of two uranium masses, each of which separately is in the subcritical region with respect to chain decay, is hardly likely to be even comparable with the time required for the reaction to get going. We may expect, therefore, that in reality in all cases we shall have to deal with conditions which are quite close to critical. On the one hand, it is necessary to consider the start-up and acceleration of the reaction not under given conditions (of unknown origin), but in the gradual transition of critical conditions corresponding to some concrete setup of the experiment, the approach of two uranium masses, addition of uranium powder, etc. On the other hand, in the immediate vicinity of the critical conditions the behavior of the system is extremely sensitive to factors whose effect could be neglected far from the limit. As examples of such factors which need investigation we may note uranium consumption and the appearance of new nuclei capable of capturing neutrons in decay, the thermal expansion of the uranium mass being used as a result of the release of decay energy; the release of some small ($\sim 10^{-2}$) amount of all neutrons with a delay of about 10 s after decay. The effect of all these factors on the critical conditions, in themselves insignificant, turns out to be decisive in the case when the system is so close to the critical conditions that the effect, for example, of thermal expansion or the release of neutrons which have been delayed by a half-period of 10 s, can carry the system from the supercritical to the subcritical region or *vice versa*.

The kinetics of the development of chain decay are decisive in judging one or another path for practical energy or explosive use of uranium decay. Hasty conclusions made without regard for the considerations above [3], for example on the extreme danger of experiments with large masses of uranium and the catastrophic consequences of such experiments (counting on complete decay of all the uranium nuclei) do not correspond to reality. Also unnecessary, it appears, are special additives such as cadmium to control the process [7, 8]. In all the works cited the specifics of the reaction, its extreme sensitivity near the limit, were ignored.

Let us turn to setting up the equations. It is important for us to note that in the general formula (5) directly below the limit and in the supercritical region the coefficient of the first eigenfunction (with the smallest characteristic number) is incomparably larger than all the other coefficients. Disregarding the latter, we come to the conclusion that, in practical terms, throughout the region of interest to us the spatial distribution of neutrons remains self-similar and is described by the first eigenfunction of the Poisson equation of our problem (6). Because of this in what follows we do not need to consider the dependence of the neutron number density on both coordinates and time, which would at best lead to an equation in partial derivatives. Instead, in our investigation of the kinetics of the reaction we will limit ourselves to consideration of the dependence on one variable—time—of the coefficient of the first eigenfunction or of the total number of neutrons in the system.

For constant external conditions the exponential growth (or decrease) of the total number of neutrons (proportional to the coefficient of the first eigenfunction) with time in the absence of an external source,

$$N = \int n\,dv, \quad n = c_0 e^{pt}\psi(x,y,z), \quad N = N_0 e^{pt} \tag{13}$$

corresponds to the differential equations

$$\frac{\partial n}{\partial t} = pn \quad \text{or} \quad \frac{dN}{dt} = pN, \tag{14}$$

$$p = b - \frac{k_i D}{d^2} = \alpha u \sum n_i \sigma_{c_i} - \frac{k_1 u}{3d^2 \sum n_i \sigma_{s_i}}. \tag{15}$$

The supply of neutrons by an external source is introduced with a coefficient β which depends on the position of the source; this last, however, even in the least favorable case is not much smaller than the ratio of the free path to the dimensions of the system, i.e., in any case is not less than several one hundredths:

$$\frac{dN}{dt} = pN + \beta m, \tag{16}$$

where m is the intensity of the source—the number of neutrons per second, β is the already-mentioned coefficient. Together with this we introduce into the analysis the rate of change of the quantity p itself, which characterizes the distance from the limit: $p < 0$ in the subcritical region, $p > 0$ in the supercritical region:

$$\frac{dp}{dt} = c - \alpha N, \tag{17}$$

here c characterizes the rate at which uranium is added, the approach of two masses of uranium, or another process by means of which we carry the system through the critical conditions. In contrast, the coefficient α describes the self-regulation of the system, its departure from the limit as a result of the consequences of uranium decay due to consumption of the material, thermal expansion of the system in connection with the release of energy in decay. The numerical values of c and α under given experimental conditions are easily found from the definition of p in formula (15) which reveals the dependence of p on the dimensions and form of the system, the uranium concentration, and so on.

We introduce, finally—for the first time in our paper—consideration of delayed neutrons.

The observed half-period ~ 10 s is, apparently, the half-period of the process of β-transformation of one of the fragments which form in the decay; the evaporation of a neutron from a nucleus which has gained sufficient energy as a result of the β-transformation occurs, according to existing conceptions, in a time not exceeding 10^{-13} s. Denoting by l the number of nuclei capable after β-transformation of discarding one neutron, we write the equation

$$\frac{dl}{dt} = \zeta g N - fl, \tag{18}$$

where f is the probability of the β-decay of interest, 10^{-1} s^{-1}, according to what we have said, gN is the number of decay events occurring in unit time, ζ is the probability (dimensionless) of formation in the decay event of the neutron-active nucleus of interest.

In equation (16) an additional term appears

$$\frac{dN}{dt} = pN + \beta m + \xi\zeta\eta fl, \tag{19}$$

which results precisely from the "delayed" neutrons.

The number of delayed neutrons arising in a single β-decay event is denoted by η. The product $\zeta\eta$ is the experimentally determined ratio of the delayed neutron output to the number of decays which have occurred ($\sim 10^{-2}$).

The factor ξ has been introduced to account for the fact that delayed neutrons have a different energy distribution and therefore are equivalent with respect to causing further decay by the primary neutrons which formed in the decay process with a delay of 10^{-13} s. The magnitude of ξ does not differ from 1 in working on slow neutrons, and is not smaller than 10^{-2} in working on fast neutrons due to the presence of a concentration 10^{-2} of isotope U^{235}.

Let us first consider the system (18) and (19) assuming constant p. We find the solution in the form

$$N = Ae^{\gamma t} + B, \qquad l = Ce^{\gamma t} + E.$$

For the quantity γ we obtain the quadratic equation

$$(\gamma - p)(\gamma + f) - \zeta\eta\xi f g = 0. \tag{20}$$

The critical condition $\gamma = 0$ in which we are interested will be reached at

$$p = -\xi\eta\zeta g, \tag{21}$$

i.e., earlier than the limit is reached in the absence of delayed neutrons, $p = 0$.

Substituting the expression for (15) and recalling that $b = \nu g$ we obtain

$$p = \nu g - \frac{H_0 D}{r^2} = -\xi\eta\zeta g. \tag{22}$$

It is clear that the critical radius changes by no more than 1% in accordance with the small output of delayed neutrons.

We further write

$$\left(\frac{d\gamma}{dp}\right)_{\gamma=0} = \frac{1}{1 + \zeta\eta\xi g/f}. \tag{23}$$

This quantity turns out to be significantly smaller than unity in decay on both fast and slow neutrons. Physically this means that in the region where delayed neutrons are necessary for the realization of a branching chain, i.e., $\zeta\eta\xi f < p < 0$, a process at an equal distance from the limit develops more slowly—in the absence of delayed neutrons, obviously, it would be

$$\gamma = p, \qquad \frac{d\gamma}{dp} = 1.$$

Calculations of particular cases in which integration of the equations is relatively simple (for example, a steady regime, small oscillations, etc.) have convinced us of the feasibility of the following approximate interpretation of the influence of delayed neutrons: equation (19) is replaced by

$$\frac{dN}{dt} = \gamma N + \beta m,$$

with

$$\frac{d\gamma}{dt} = \left(\frac{d\gamma}{dt}\right)_{\gamma=0}, \quad \frac{dp}{dt} = \frac{c}{1+\zeta\eta\xi g/f} - \frac{\alpha N}{1+\zeta\eta\xi g/f}$$

in accord with (17) and (23).

Let us consider several particular solutions which illustrate the properties of the system.

Disregarding the release of neutrons by the external source, which is quite small for any macroscopic process, we find the stationary state

$$\gamma = 0, \quad c = \alpha N, \quad N_{\text{stationary}} = \frac{c}{\alpha}$$

The stationary number of neutrons is such that $\gamma = 0$ is maintained despite the supply of uranium (the term c). At constant density and form in the stationary state, the amount of uranium decaying in unit time, accurate to within a numerical factor close to one, is equal to the amount of uranium supplied.

It turns out, however, that, for example, for a solid mass of uranium weighing 1 ton the decay of 10^{-3} g of uranium will heat the uranium to a temperature of about 1000°, which corresponds to an expansion of about 1%. This expansion resulting from the decay of 10^{-3} g of uranium compensates for the effect on the limit of the addition of 50 kg of uranium so that in this temperature interval natural regulation through the density leads to burning of a $1.5{\cdot}10^{-8}$ part of the amounts supplied. Conversely, when heat is removed from a mass of uranium which has heated to 1000°, its temperature falls to room temperature only after 10^{12} kcal has been removed, i.e., after ~ 50 kg of the uranium has burned (in fact it occurs earlier due to the effect of other regulating factors).

An analysis of small oscillations about the stationary state gives us the period of these oscillations:

$$\tau = 2\pi\sqrt{\frac{1+\zeta\eta\xi f/g}{c}},$$

which characterizes the relaxation time of the system. In the absence of a supply of neutrons by an external source these oscillations turn out not to decay. The equation is integrated in the variables $\gamma - N$ by separation of variables even at large amplitudes:

$$N_{\text{max}} = N_{\text{stationary}} \ln N_{\text{stationary}} \, / \, N_{\text{min}}.$$

The period here of the oscillations varies only logarithmically. The order of magnitude of the period of oscillation and of the relaxation time of the system, e.g., at a critical mass of 10^6 g and supply of 10 g/s, are around 0.1 s (for fast neutrons). It is not difficult to estimate the initial amount of neutrons from which oscillations will begin (cycles in the $\gamma - N$ plane) when the limit is reached: in order of magnitude this amount is

$$N_{\min} = \beta m \tau = \beta m \sqrt{\frac{1 + \zeta\eta\xi f/g}{c}}$$

equal to the product of the supply rate (source intensity) and the relaxation time. Accordingly

$$N_{\max} = \frac{c}{2a} \ln \frac{\beta^2 a^2 m^2}{c^3} \cdot \left(1 + \frac{\zeta\eta\xi f}{g}\right).$$

It turns out here that taking account of the neutrons of the source leads to gradual decay of the oscillations.

When the uranium supply is abruptly cut off, or two uranium masses suddenly stop their approach, the amount of uranium which burns "by inertia" as the neutron number density falls as a result of the departure from the limit—this amount is equal to the average amount which burns over the relaxation time in the stationary supply regime.

Let us summarize the results of this last part of the work.

A chain disintegration** of uranium, unlike the combustion of explosives and other similar processes, practically instantaneously stops when the system moves back from the super- to the sub-critical region without affecting the remaining amount of uranium, which is quite close to the critical value.

When the process runs isothermically the amount of uranium which decays in unit time is equal to the amount supplied.

In the adiabatic process as a result of thermal expansion the amount burned is $\sim 10^8$ times less than that supplied.

The relaxation time of the process, inversely proportional to the square root of the rate of uranium supply, of order 10^2 s for a supply $\sim$ 50 kg/hr and at a critical mass of about 1 ton, is approximately 10^3 times larger than that which would result in the absence of delayed neutrons. These numbers refer to chain decay on fast neutrons. The formulas obtained are of course applicable to decay on slow neutrons as well. Such properties of the system (above all the regulation *via* thermal expansion) make experimental investigation and energy production use of uranium decay safe. Explosive use of chain decay requires special devices for a very fast and deep transition to the supercritical region and decrease in the natural thermal regulation.

Institute of Chemical Physics
Leningrad

Received
March 7, 1940

REFERENCES

1. *Zeldovich Ya. B., Khariton Yu. B.*—ZhETF **9**, 1425 (1939).
2. *Zeldovich Ya. B., Khariton Yu. B.*—ZhETF **10**, 29 (1940).
3. *Flügge S.*—Naturwissenschaften **27**, 402 (1939).

**Today the term "disintegration" is not used; it has been replaced by the term "uranium fission."—*Editor's note*

4. *Semenov N. N.* Tsepnye reaktsii [Chain Reactions]. Leningrad: Goskhimtekhizdat, 555 p. (1934).
5. *Perrin F.*—C. r. Acad. Sci. **208**, 1934 (1939).
6. *Peierls R.*—Proc. Cambridge Philos. Soc. **35**, 610 (1939).
7. *Perrin F.*—C. r. Acad. Sci. **208**, 1537 (1939).
8. *Halban H.*—Nature **143**, 793 (1939).

Commentary

The discovery in 1938–1939 of the fission of uranium nuclei, which led eventually to the development of nuclear energy, heralded a new, extraordinarily fruitful stage in Ya. B.'s scientific activity. His interests became concentrated on the study of the mechanism of fission of heavy nuclei and, what proved especially important, on the development of a theory of the chain reaction of uranium fission. On these subjects during two years (1939–1940) Ya. B., in collaboration with Yu. B. Khariton, performed three basic studies which are of enormous, fundamental value. The papers of this cycle are the foundation of the modern physics of reactors and nuclear power, they are widely known and do not require special commentary—just a short summary of the results is eloquent enough.

Ya. B.'s interest in problems of nuclear physics, and then in the physics of elementary particles, was also stimulated by the discovery of the phenomenon of fission of heavy nuclei and by practical work in nuclear power.

We shall comment on each paper in turn.

Paper 1. The conditions for the appearance of a chain fission reaction of the main isotope of uranium are considered, taking into account the slowing of neutrons below the threshold of U^{238}. The basic text of the article considers only elastic moderation of neutrons. At the same time the authors, taking plausible values for the number of secondary neutrons from fission, conclude that chain reaction, practically speaking, is impossible in uranium oxide, and possible in pure uranium. However, in a note added in proof the authors report that a calculation carried out by them on the basis of the Bohr-Wheeler theory of inelastic scattering (moderation) of neutrons shows that even in the case of metallic uranium a chain reaction is impossible. Thus the impossibility of a chain nuclear reaction on fast neutrons in natural uranium was shown.

In this same paper, for the first time, an estimate is given for the thickness of the reflector for fast neutrons in the case of threshold fission.

Paper 2 is pioneering and classical in the highest sense of the word. The basic content of the work may be summarized as follows.

a) The clear introduction of resonant absorption of U^{238} as one of the determining factors in the coefficient of multiplication in systems on slow neutrons (uranium + moderator).

b) The formulation of the history of one generation of a neutron and the derivation of the famous expression for the multiplication coefficient in an infinite medium: $K_\infty = \nu\varphi\theta$, where ν is the number of secondary neutrons per event of capture of a thermal neutron by uranium, φ is the probability of avoiding resonant capture in the process of moderation, and θ is the coefficient of consumption of thermal neutrons.

c) The derivation of an equation for φ which relates it with the resonant cross-sections of U^{238} for a hydrogen moderator. The equation for φ is easily generalized for a non-hydrogen moderator ($A > 1$), but the solution is more complicated since it leads to equations in finite differences. The paper does not do this.

d) The introduction of the effect of self-shielding of uranium atoms. A square root law is obtained for the dependence of the resonance absorption on the concentration. The later-understood possibility of reducing resonance absorption by heterogeneous (block-wise) placement of uranium in the moderator is based on the existence of two effects, one of which is the self-shielding of uranium atoms discovered in the present paper (internal regions of a block do not participate in the absorption). The second effect, not considered in this paper, is that only resonance neutrons formed directly near the block are absorbed. The remaining neutrons in the process of moderation will exit the dangerous resonant zone before they reach the block.

e. A detailed recalculation of Joliot's experiments is performed, with the conclusion that it is not possible to obtain a self-supporting chain fission reaction in a system of natural uranium + light water at any concentration. A homogeneous mixture is of course assumed. On this basis the paper makes the very important assertion, later fully justified, that new moderators must be used and proposed the use of heavy water and graphite due to the smallness of the absorption cross-sections of thermal neutrons by deuterium, carbon and oxygen (see below).

Regarding the points d) and e), more detailed explanations should be given.

1. In a number of isotopes, including Pu^{239}, there is a deviation from the $1/v$ law in the thermal region, however at the time this was unknown. For U^{235} which is considered in the paper the $1/v$ law holds.

2. Accounting for reaction on fast neutrons, which is mentioned in the paper but not carried out, becomes more important in a heterogeneous (block) system. The corresponding coefficient μ attains $1.03 - 1.04$ and is quite significant for reactors with graphite and water cooling moderation.

3. Formula (11) is valid only for large uranium concentrations (omission of the factor $\Gamma^2/4$ in the denominator of the Breit-Wigner formula). An exact solution of equation (8) gives

$$
\begin{aligned}
-\ln\varphi &= \frac{\pi}{2}\sum_i \frac{\Gamma_i}{E_{0i}}\left[\frac{\eta\sigma_s(1+\eta\sigma_s/\sigma_{0i})}{\sigma_{0i}}\right]^{-1/2} \\
&\cong \frac{\pi}{2}\sum_i \frac{\Gamma_i}{E_{0i}}\sqrt{\frac{\sigma_{0i}}{\eta\sigma_s}} - \sigma_{0i} \gg \eta\sigma_s \qquad \text{(I)}
\end{aligned}
$$

in the case of strong levels (self-shielding), and

$$
\ln\varphi \simeq \frac{\pi}{2}\sum_i \frac{\Gamma_i}{E_{0i}}\frac{\sigma_{0i}}{\eta\sigma_s} - \sigma_{0i} \ll \eta\sigma_s \qquad \text{(II)}
$$

in the case of weak levels.

The context of the paper corresponds to taking only strong (self-shielding) levels of resonance absorption into account.

For hydrogen at concentrations $c_{\mathrm{H}}/c_v = 62$ and even 17 levels in the region 100–200 eV should be considered using the exact formula.

4. Subsequent measurements significantly changed the values of the constants. For current values of the constants $\varphi \simeq 0.6$ at $c_H/c_v = 1$. However, the estimate taken in the article is much more realistic than the completely incorrect estimate of A. Khalban and L. Kovarskiĭ at which even a heavy-water reactor could not work on natural uranium.

The ratio of the cross-sections of fission and capture in uranium on thermal neutrons is taken close to the current value. But σ_a itself for natural U is 2.4 times smaller than the true value (3.2 barn instead of 7.5 barn), while for hydrogen it is only 20% lower. Therefore sharply lowered values of θ are obtained. Now the optimum is close to $c_H/c_v = 3$ and $\{\varphi\theta\}_{\max} = 0.69$, i.e., the minimum enrichment for a homogeneous system at $\mu = 1.02$ corresponds to $\nu_{ef} = 1.43$ for $K_\infty = 1$, i.e., enrichment of order 1%:

$$\theta = \frac{8.96}{8.96 + 0.96} = 0.9; \qquad \varphi = 0.77;$$

$$\mu\varphi\theta = 0.707, \qquad K_\infty = 1.01.$$

Finally, we note that self-shielding of resonance levels of capture leads to the advantageousness of using uranium in the form of bodies (blocks) several centimeters in size. The block-effect was discovered in the USSR by I. I. Gurevich and I. Yu. Pomeranchuk in 1945, after the paper by Ya. B. and Yu. B. Khariton. The block-effect noticeably increases K_∞ and is extremely important for work with reactors which use natural unenriched uranium. At the same time, the introduction of the block-effect has not changed the very important qualitative conclusions of the paper here. Even when the size of the blocks is made optimal, natural (unenriched) uranium with ordinary (light) water in an infinite system does not attain criticality.

We note that in a review article in *UFN*[1] Ya. B. and Yu. B. Khariton indicated substances which should be investigated as moderators, including helium, heavy water (D_2O) and carbon. As we know, heavy-water and graphite reactors are in practical use.

Paper 3, like paper 2, is classic and pioneering. For the first time the kinetics of the chain decay of uranium were considered in detail in the transition to the supercritical state. We note here the most important results.

a) Most important in the paper is the consideration of the role of delayed neutrons in the kinetics of the chain reaction. In the interval of effective multiplication coefficients (given by $K = K_\infty P$, where P is the probability that a neutron will be absorbed rather than leave the system),

$$1 < K < 1 + \beta, \tag{III}$$

where β is the fraction of delayed neutrons, the kinetics become very soft and are primarily determined by the periods of the delayed neutrons. The authors produce a complete system of kinetic equations with delayed neutrons from which, in particular, follows the equation of the so-called "inverse clock" which characterizes the rate of acceleration of the reactor; this last was not directly obtained in the paper. This brilliant idea explains the fact that a nuclear reactor proved to be an easily regulated system, which in turn was one of the basic factors ensuring the success of atomic energy (all reactors work in the interval (III)).

[1] *Zeldovich Ya. B., Khariton Yu. B.*—UFN **23**, 329–357 (1940).

b) Also prophetic in the paper is the statement of the possibility of the appearance of new, strongly absorbing nuclei. The well-known phenomenon of the "iodine well" in reactors is related to the accumulation of an isotope of xenon which absorbs thermal neutrons at a record rate (Xe^{135}: $\sigma_c = \pi \lambda\!\!\!^{-2}/10 = 3 \cdot 10^6$ barn).

c) The effect of heating of the uranium on the kinetics is considered in detail and it is shown that thermal expansion is an effective regulating factor.

d) Fluctuations near the equilibrium position are studied which, in the absence of external sources, prove to be non-decaying.

e) Finally, the conditions for generation of a strong explosion follow directly from the paper—significant supercriticality in the initial state and multiplication on fast neutrons. These conclusions, not explicitly formulated, were fully used by the authors in subsequent work.

Let us note that a correct estimate of the critical mass of U^{235} was given by the authors together with I. I. Gurevich as early as 1941. It was also noted then that in the distant past the content of U^{235} was greater than now, which ensured the appearance of the chain reaction. This is contained in the second part of the review,[2] submitted in 1941 but published only in 1983 in connection with the eightieth birthday of I. V. Kurchatov. As is known, signs of a chain reaction which occurred 2 billion years ago were discovered in a uranium deposit in Oklo (Gabon, Africa).

Overall, the papers 1–3 are unique in world literature. Similar papers in other countries were not published until the Geneva conference in 1955.

An introduction to the subsequent development of the ideas presented in these articles may be found in the books *The Physical Theory of Neutron Chain Reactors*[3] and *Theoretical Foundations and Calculation Methods for Nuclear Power Reactors.*[4]

[2] *Zeldovich Ya. B., Khariton Yu. B.*—UFN, Part 1 **25**, 381–405 (1941); Part 2 **139**, 501–527 (1983).

[3] *Weinberg A., Wigner E. The Physical Theory of Neutron Chain Reactors.* Chicago: Univ. of Chicago Press (1958).

[4] *Vartolomeĭ G. G., Bat' G. A., Baibakov V. D., Altukhov M. S. Osnovy teorii i metody rascheta ĭadernykh energeticheskikh reaktorov [Theoretical Foundations and Computational Methods for Nuclear Power Reactors].* Moscow: Energiĭa, 511 p. (1982).

4

On the Theory of Disintegration of Nuclei*

With Yu. A. Zysin

The possible state of a nucleus at the moment of its disintegration into two approximately equal nuclei is considered. A calculation of the energy of two ellipsoids of rotation in contact refutes Ya. I. Frenkel's arguments in favor of the existence of significantly non-spherical nuclei. The order of magnitude of the energy of ellipsoids found allows us to satisfactorily describe the observed formation of several fast neutrons for each disintegration event as evaporation of these neutrons by fragments excited in the process of fission.

Bohr's theory describes a nucleus as a drop of liquid with uniform charge density which gives rise to the electrostatic energy. The short-range attractive forces of nuclear particles specify their particular evaporation heat and also the surface tension of the drop.

In 1939 a very important success of the theory was the description of the fission discovered by L. Meitner and O. Frisch [1] of heavy nuclei under neutron bombardment into two approximately equal fragments with the release of huge—even for radioactive processes—amounts of energy (100–200 MeV) and the formation of several neutrons ("neutron dust") for each disintegration event. This last peculiarity is of particular interest since it opens the possibility in principle of the chain decay of macroscopic amounts of uranium [2]. The theory of the decay, which has been especially thoroughly developed by three physicists—N. Bohr (Denmark), J. Wheeler (USA) and Ya. I. Frenkel (USSR) [3], considers the stability of a spherical uniformly charged drop of incompressible fluid possessing a specific surface tension.

As it turns out, the spherical form becomes instable with respect to small deformations when the ratio of the electrostatic energy E to the surface energy O is

$$E/O \geq 2. \tag{1}$$

All three of the above authors then consider two contacting spherical nuclei which have resulted from the division of the original nuclei. It is not difficult to find that the energy of two contacting spheres of half the volume is equal to or less than the energy of the original sphere if for the

*Zhurnal eksperimentalnoĭ i teoreticheskoĭ fiziki **10** (8), 831–834 (1940).

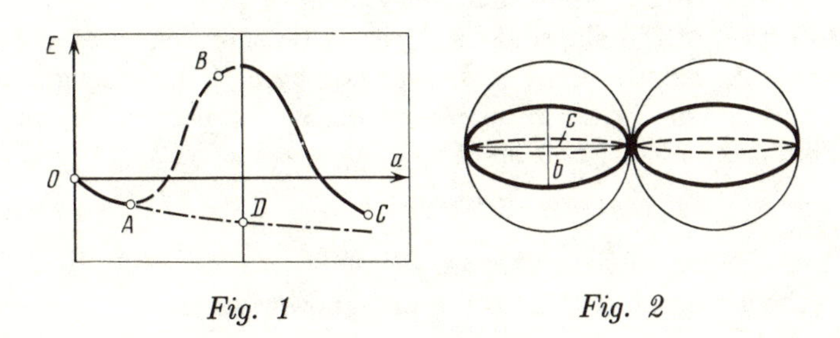

Fig. 1 *Fig. 2*

latter[1]

$$E/O \geq 2.42. \tag{2}$$

Let us introduce some parameter a which describes the process of division; for convenience in the graphical representation we choose a such that for a spherical nucleus $a = 0$. At the moment when the two nuclei which have formed are in contact at a point (which is a necessary stage of the division), $a = 1$; finally when the two nuclei have separated by an infinite distance, $a = 2$.

For the value

$$2 < E/O < 2.42 \tag{3}$$

the above calculations lead to the form of the curve of the energy variation during decay shown in Fig. 1 with solid lines.

The left segment OA is the result of the calculation of small deformations of the drop at $E/O > 2$, the right, BC—of a calculation of the energy of the two spheres as a function of the distance. By comparing the solid lines shown in Fig. 1 two substantially different conclusions may be made.

1. Connecting the two solid segments with a smooth curve (the dashed line $OABC$), we necessarily obtain the minimum energy at $a < 1$. Physically this means that heavy nuclei have a stable non-spherical form (Frenkel).

2. The other possibility is decay through a form which differs from two equal contacting spheres—"a tidally perturbed form" (Bohr and Wheeler). If this form corresponds to a sufficiently small energy (point D in Fig. 1), then the basis for conceptions of stable nonspherical forms disappears.

The calculations of the present paper relate precisely to the energy at the moment of decay, i.e., to finding the ordinate of the point D, $a = 1$.

It is easy to see that pear-shaped forms possess the minimum energy for a given charge and volume at the moment of contact.

[1]The number 2.17 cited by Frenkel is the result of an arithmetic error since the corresponding formula in his article

$$\Delta W = E(1 - 2^{1/3}/2 - 5 \cdot 2^{1/3}/24) - u(2^{1/3} - 1)$$

is written correctly. Cf. also the graph of the function $f^*(x)$ (the above-cited work by Bohr and Wheeler, Fig. 4): $f^*(x) = 0$ at $x = 1.2$, where $x = E/2O$ is Bohr's parameter.

However, even a calculation for ellipsoids of revolution elongated along the line of centers (Fig. 2) gave sufficiently definite qualitative results.

A cross-section of the ellipsoids is depicted by the solid lines in Fig. 2. Let us denote the length of the large semiaxis by c, and the length of the smaller one by b. For a given ratio c/b, we easily find each of the semiaxes from the condition of conservation of volume:

$$2cb^2 = r_0^3, \tag{4}$$

where r_0 is the radius of the original nuclei.

We find the surface energy from the well-known formula

$$O = 2\pi ab\left(\frac{b}{c} + \frac{1}{\varepsilon}\arcsin\varepsilon\right), \tag{5}$$

where $\varepsilon = \sqrt{1 - b^2/c^2}$.

We determine the electrostatic energy of an individual ellipsoid by the formula

$$F = \frac{3}{10}\frac{(e/2)^2}{c\varepsilon}\ln\frac{(1+\varepsilon)}{(1-\varepsilon)}. \tag{6}$$

It is somewhat more complicated to find the mutual energy of the ellipsoids. An exact analytical calculation for the mutual energy of two ellipsoids, carried out on the basis of the method proposed by Laguerre for the particular case of two uniform elongated ellipsoids of rotation, led to the formula

$$E_{1,2} = \frac{9}{16}\frac{(e/2)^2}{\pi}\int_{-1}^{+1}\int_{-1}^{+1}\int_{0}^{\pi}\frac{dt\,dt_0\,d\varphi\,(1-t^2)(1-t_0^2)}{2c - t\sqrt{c^2-b^2} - t_0\sqrt{c^2-b^2}}. \tag{7}$$

The evaluation of this formula led to a quite cumbersome expression containing more than 100 terms. The calculation, which was carried out for $c/b = 2$, agreed within 3% with the corresponding data of the approximate method given below, which was used for all the calculations.

If one also takes into account that the mutual energy comprises only a part of the total energy, then the possible error in the calculation of the total energy will be of the order of a fraction of a percent. For given charge $(e/2)$ of each ellipsoid and given distance between the centers $2c$, the electrostatic energy of interaction is easily found from the two limiting cases:

at $b = c$ spheres, the thin lines in Fig. 2,

$$E_{1,2} = \frac{(e/2)^2}{2c}; \tag{8}$$

at $b \to 0$, "sticks", the dashed lines in Fig. 2,

$$E_{1,2} = \frac{(e/2)^2}{1.74c}. \tag{9}$$

In the interval of interest,

$$0 < b < c \tag{10}$$

E/O	c/b				
	1	2	3	4	5
2.60	3.581	3.414	3.318	3.312	3.493
2.40	3.402	3.254	3.118	3.180	3.293
2.20	3.231	3.095	3.038	3.050	3.05
2.00	3.047	2.939	2.898	2.918	2.929
1.80	2.869	2.779	2.755	2.789	2.807
1.70	2.778	2.70	2.686	2.723	2.747
1.65	2.734	2.659	2.650	2.69	2.718
1.60	2.689	2.620	2.614	2.658	2.687
1.40	2.511	2.462	2.473	2.527	2.565

we interpolate according to the formula

$$E_{1,2} = \frac{(e/2)^2}{\sqrt{3.04c^2 + 0.96b^2}}. \tag{11}$$

The form of formula (11) reflects the very simple conceptions of the dependence of the energy $E_{1,2}$ on b; the coefficients in (11) are chosen such that both limiting expressions (8) and (9) are satisfied.

The results of the calculations are summarized in the table. For various ratios E/O of the original nucleus (Column 1) we show the energies with respect to the surface energy of the original nucleus of two contacting ellipsoids resulting from it for various values of $c/b = 1, 2, 3, 4, 5$.

As is clear from the table, the minimum energy is attained in the interval of E/O under consideration at

$$3 < \frac{c}{b} < 4. \tag{12}$$

This energy is less than the energy of the original nucleus at $E/O > 1.65$.

A nucleus for which relation (3) holds cannot be in the form of a sphere. It also cannot decay through the form of two contacting spheres. However, our calculation shows that decay through two elongated contacting ellipsoids is not prohibited.

For the interval

$$1.65 < E/O < 2 \tag{13}$$

our data on the change in energy in the process of decay are shown in Fig. 3 where all notations are taken from Fig. 1.

The segment OMA (where M is the maximum of the energy) is borrowed from Bohr. The position of the point D, which corresponds to two ellipsoids, is taken from our data. Since in the interval (13) D is located below B, there is no basis for assuming the presence of any additional maxima in the interval. If we restricted our attention to the spheres (segment BC), we would reach completely different conclusions having no connection with reality.

If we consider non-symmetric pear-shaped forms rather than ellipsoid we will undoubtedly decrease the lower boundary (13).

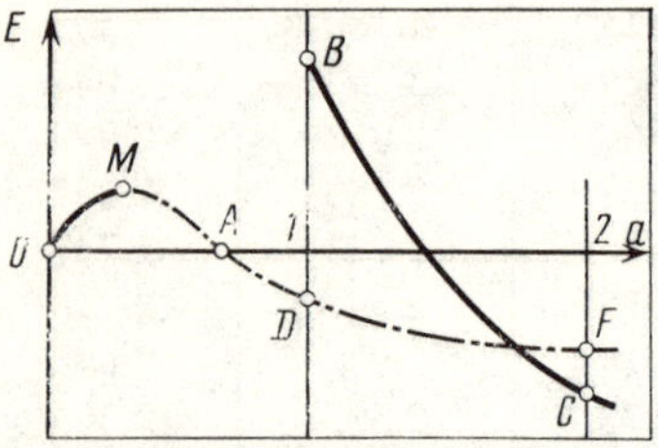

Fig. 3

Finally, comparing the energy at the point D with the energy at the point M, but not at the point O (Fig. 3) we must obtain $E_D < E_M$ for any value of E/O.

It is interesting that after separation by a large distance the energy of two ellipsoids, naturally, proves larger than the energy of two separated spheres (see the positions of the points F and C at $a = 2$ in Figs. 1 and 3).

The energy difference at $c/b \sim 3.5$, $E/O \sim 1.8$ attains about $0.078\,O$ (O is the surface energy of the original nucleus), i.e., around 42 MeV for each nucleus that forms.

The excitation energy of the nucleus will first of all be directed to evaporation of neutrons. For a comparatively small binding energy of neutrons in the nuclei of the fragments with an anomalous ratio of the charge to the mass one could thus explain the release of a large number of neutrons per fission event, as well as the observed, sometimes very large energies of the neutrons [4].

In fact, even in the case when the fission occurs through the form of two contacting ellipsoids, their form changes as they move away from one another. Calculations of the part of the energy of deformation of the fragments which goes to kinetic energy and the part which in the form of excitation energy may be used for the evaporation of neutrons, is the task of nuclear fluid dynamics, an area which is completely undeveloped.

Our elementary calculations have one meaning—they indicate the order of magnitude of the possible energy of excitation.

In any case, evaporation of neutrons by excited fragments seems more likely to us than the mechanism proposed by N. Bohr and J. Wheeler. They note that in the division of one droplet into two, there usually form several small droplets at the point where the connection breaks and they identify neutrons precisely with these small droplets.

Institute of Chemical Physics
Leningrad

Received
June 22, 1940

REFERENCES

1. *Bohr N.*—Phys. Rev. **56**, 426 (1939) [see Bibliography therein].
2. *Zeldovich Ya. B., Khariton Yu. B.*—ZhETF **9**, 1425 (1939); **10**, 475 (1940) [and earlier references therein].
3. *Frenkel J.*—Phys., USSR **1** (1939).
4. *Joliot F.*—C. r. Acad. Sci. **208**, 341 (1939).

Commentary

A calculation is performed of the energy of two fragments in contact at the moment of decay of a nucleus undergoing fission. The form of the splitting nucleus is modeled using two ellipsoids in contact. An estimate is obtained of the difference in the energy between nonspherical contacting fragments and spherical fragments flying apart which is transformed into excitation energy (thermal energy) of the fragments ($\sim$ 40 MeV). The magnitudes of this energy are sufficient to explain the release of fission neutrons by evaporation from the excited fragments. Thus, the authors propose a new mechanism for the emission of secondary fission neutrons which differs from that considered by N. Bohr and J. Wheeler, who assumed that the neutrons are emitted by the "neck" at the moment of fission.

Even the first experiments by J. Fraser and J. Milton[1] on the angular distribution of neutrons (the authors studied the fission of U^{233} by thermal neutrons) showed a clear correlation between the direction of neutron emission and the direction of motion of the fragments and in this way completely confirmed the validity of the Zeldovich-Zysin mechanism.

Subsequent experiments and theoretical calculations continued and deepened the picture of emission of fission neutrons. Thus J. Terrell,[2] recalculating the distribution curves of the energy and masses of the fragments for U^{233}, U^{235} and Cf^{252}, estimated the number of secondary neutrons ν as a function of the fragment mass and showed that the ν of a light fragment is larger than that of a heavy fragment. An investigation by H. Bowman, S. Thompson, J. Milton and W. Swiatecki[3] was devoted to the study of the dependence of the number of secondary neutrons on the mass of the fragment, its critical energy, and the angle of emission for the spontaneous fission of Cf^{252}. Using the measurement of the flight time of the neutrons and fragments, they found a strong correlation between the directions of motion of the fragment and neutron and measured the ratios of the numbers of neutrons emitted from a light fragment, a heavy fragment and at an angle of 90° to the direction of motion of the fragments. These ratios turned out to be equal to 9 : 5 : 1. In a paper by R. Vanden Bosch[4] it was shown that the experimentally observed kinetic energy and number of fission neutrons require for their explanation the introduction of a shell-like dependence of the rigidity of the fragments with respect to quadrupole deformation (rigidity is defined to be the co-

[1] *Fraser J. S. A., Milton J. C. D.*—Phys. Rev. **93**, 818–824 (1954).

[2] *Terrell J.*—Phys. Rev. **127**, 880–904 (1962).

[3] *Bowman H., Thompson S., Milton J., Swiatecki W.*—Phys. Rev. **129**, 2120–2147 (1963).

[4] *Vanden Bosch R.*—Nucl. Phys. **46**, 129 (1963).

efficient c in the expression for the quadrupole energy of deformation: $E = c\alpha^2/2$). Subsequent calculations have usually used the method of Strutinskiĭ to account for shell corrections.

Thus, this paper by Ya. B. and Yu. A. Zysin gave the correct mechanism for the emission of secondary fission neutrons and stimulated a large number of experimental and theoretical studies on the physics of fission (see the collection *Achievements of the Physics of Nuclear Fission*).[5]

We should remark separately on the statement at the beginning of the article about "refuting Ya. I. Frenkel's arguments in favor of the existence of significantly non-spherical nuclei." Today it is well known that nonspherical nuclei exist and that, in fact, Ya. I. Frenkel was right. It is also true, however, that this is related to shell rather than electrostatic effects, and also that nuclear fission is not a proof of nonsphericality.

[5] *Uspekhi fiziki deleniĭa ĭader [Achievements of the Physics of Nuclear Fission].* Collection of Papers. Moscow: Atomizdat, 307 p. (1965).

5
Storage of Cold Neutrons*

The idea of retaining slow neutrons has been mentioned many times, but the corresponding experiments have not yet been performed, and the literature does not contain even rough estimates pertaining to this problem.

It is known that slow neutrons experience total internal reflection in glancing incidence on the surface of most substances. At sufficiently low velocities, the neutrons cannot penetrate in such a substance even under normal incidence. Thus, for carbon with a density ~ 2 g/cm^3 the critical neutron velocity is close to 5 m/s, for beryllium it is approximately 7 m/s. Let us place neutrons in a cavity surrounded on all sides by graphite. The neutrons of speed higher than critical will rapidly leave the cavity, but neutrons of less than critical speed are blocked in the cavity and vanish only as they decay, with a half-life of approximately 12 minutes. Such slow neutrons will penetrate into the wall only a depth on the order of their wavelength; taking into account dimensionless factors, the depth is $\sim 10^{-6}$ cm. Therefore if the cavity has a considerable volume, the fraction of the time that the neutrons stay in the material of the shell is quite small; for a one-cubic-meter cavity this fraction is $\sim 10^{-7}$.

The capture cross section of carbon (4.5×10^{-27} cm^2 at $v = 2.2\times10^5$ cm/s) obeys the $1/v$ law and corresponds to a neutron lifetime in carbon of ~ 0.01 s regardless of its velocity. For neutrons in a cavity we obtain an absorption time of $0.01/10^{-7}$s $= 10^5$ s $=$ 1 day. Slow neutrons will also be lost, as they acquire energy by collision; obviously, however, this process is greatly suppressed, because the neutrons are for the most time in the cavity and not in the material of the shell.

The most difficult feat is to obtain a sufficient number of such neutrons. For a Maxwellian distribution at room temperature, the fraction of such neutrons is on the order of 10^{-8}.

It is advisable first to cool the neutrons in a volume filled with liquid helium, and then the fraction of the necessary neutrons increases to 10^{-5}. As a result of the long life of the slow neutrons in the cavity, their number density after a few seconds becomes equal to the Maxwellian equilibrium density. The principal difficulty is connected with the need for having a large volume of liquid helium, because of the long range of the neutrons in helium (50 cm).

*Zhurnal eksperimentalnoĭ i teoreticheskoĭ fiziki **36** (6), 1952–1953 (1959).

With a fully moderated neutron flux of 10^{12} $cm^{-2} s^{-1}$ from a reactor, the flux of neutrons emitted with a temperature of 3°K can amount to 10^{11} $cm^{-2} s^{-1}$, which corresponds at an average velocity on the order of 2×10^4 cm/s to a density of 5×10^6 cm^{-3} of thermal neutrons, including 50 cm^{-3} slow ones (with velocity less than 500 cm/s). Thus, under the most favorable assumptions, it is possible to accumulate up to 5×10^7 slow neutrons in a cavity 1 m^3 in volume.

By placing a graphite partition over the opening that joins the cavity with the liquid helium it is possible to remove the cavity with slow neutrons from the reactor and make the measurements at a small background.

It may prove advantageous to use a palliative variant without helium, by cooling neutrons, say, to 70°K and accumulating up to 10^5 neutrons. We note that the index of refraction of the moderator should be less than the index of refraction of the cavity material, or else the moderator will not admit necessary neutrons in from the vacuum, and consequently will not let any out.

An experiment of this type is quite difficult, but it seems that it can give experimenters a valuable method of investigating the interaction of slow neutrons with substances introduced into the cavity. By introducing an (n, γ) absorber of neutrons into the cavity, it is easy to measure the number of neutrons left intact at the instant of observation.

We note that the neutrons in the cavity can be effectively heated to a speed above critical by mechanical displacement of the graphite surfaces at a speed of several meters per second.

The theory of the coefficient of refraction and the total internal reflection of neutrons is well known; we note only that it remains valid also at those small energies, at which the absorption cross section, following the $1/v$ law, becomes equal to or greater than the scattering cross section. It is easy to verify that the imaginary part of the pseudo-potential, the part describing the absorption, is small compared with the real part, which describes the scattering. Their ratio is equal to $\sqrt{\pi\sigma_s}/\lambda_1$, where λ_1 is the wavelength of the neutron for which $\sigma_s = \sigma_a$. Consequently, in the case of total internal reflection, absorption does not change the exponential law of damping of the wave function of the neutron in the medium.

Received
April 3, 1959

REFERENCES

1. *Hughes D. J. Pile Neutron Research.* Cambridge, Mass.: Addison-Wesley (1953).
2. *Vlasov N. V. Neĭtrony [Neutrons].* Gostekhizdat, 427 p. (1955).

Commentary

A method is proposed for containment of very slow neutrons in a cavity, based on the fact that, beginning from some critical energy (velocity) the refraction index of the neutrons attains a zero value and complete internal reflection of neutrons occurs at any angle of incidence. "Zeldovich's nuclear bottle" opened an area of neutron physics which is now undergoing rapid development—the physics of ultra-cold neutrons (UCN).[1] Using UCN measurements are made of the electrical dipole moment of a neutron (to date only an estimate of its upper bound has been achieved) in connection with the violation of symmetry in time reversal, the lifetime of the neutron, related states of the neutron in matter and much more. All of this has its roots in this ground-laying paper by Ya. B. We note also another method—the magnetic method of neutron containment,[2] proposed by V. V. Vladimirskiĭ shortly after the paper by Ya. B. Vladimirskiĭ's method may prove more convenient for the exact determination of the probability of beta-decay of a neutron. However, for certain important physical experiments a magnetic field, which interacts with a neutron, is inadmissible. This relates to the measurement of the electrical dipole moment of a neutron and to the detection of neutron-antineutron oscillations.

[1] *Shapiro F. L.*—In: *Nuclear Structure Study with Neutrons.* Plenum Press, 259 p. (1974).

[2] *Vladimirskiĭ V. V.*—ZhETF **39**, 1062–1070 (1960).

6

Quasistable States with Large Isotopic Spin in Light Nuclei*

We consider an odd nucleus A with one excess neutron, with a minimum value of isotopic spin $T = 1/2$ in the ground state, and with a neutron binding energy Q. The excited states of the nucleus A^* with excitation energy $E > Q$ have as a rule a rather large probability of neutron emission, i.e., a large width Γ_n of the process $A^* \to B+n$, where B is an even nucleus.

Let the ground state of the nucleus B have $T = 0$, and let the state B^* with $T = 1$ have an excitation energy Δ. We assume that the nucleus A has an excited state A_3^* with $T = 3/2$ and excitation energy E_3 such that $Q < E_3 < Q + \Delta$. The decay of A_3^* to $B^* + n$ is energetically impossible, while the decay of A_3^* into $B + n$ proceeds via a change in isotopic spin and should therefore have an anomalously small width Γ_n. The state A_3^* is quasistable and should appear in a unique manner in the scattering of neutrons by nuclei B, and also in the photoeffect $A + \gamma = B + n$.

When n is scattered by B the isotopic spin of the system in the initial state is $T = 1/2$, and it is usually assumed that states with $T = 3/2$ should make only a small contribution to the scattering cross section. However, if a quasistate exists, then sharp scattering resonance takes place at a neutron energy $E_n = E_3 - Q$, with a maximum cross section

$$4\pi\lambda^2 \frac{2J+1}{2S+1}.$$

The low probability of the process, connected with the disturbance of the isotopic spin, manifests itself not in a reduction in the scattering cross section, but in a reduction of the width of the resonance scattering. Therefore observation of resonance is quite possible if the neutrons are sufficiently monochromatic.

At resonance the increase in the scattering cross section will be accompanied by an increased probability of the process $B(n,\gamma)A$, since $\sigma_{n,\gamma}/\sigma_{sc} =$

*Zhurnal eksperimentalnoĭ i teoreticheskoĭ fiziki **38** (1), 278–280 (1960).

Γ_γ/Γ_n and an anomalously small Γ_n should give[1] an anomalously large Γ_γ/Γ_n. Incidentally, the inequality $\Gamma_\gamma/\Gamma_n \ll 1$ remains in force, since $\Gamma_n \sim e^2$, when the isotopic spin is disturbed by the Coulomb interaction, like Γ_γ, which contains, however, other small factors, $(v/c)^u$, $(R/\lambdabar)^v$, and $\hbar/Mc\lambdabar)^w$ in degrees that depend on the type of transition (for $E1$: $u = 2$, $v = 2$, and $w = 0$; for $M1$: $u = v = 0$, and $w = 2$, etc.).

The existence of a quasistable A_3^* should lead to a narrow resonance in the reverse process[2] $A(\gamma, n)B$ and also to resonant scattering of γ by A. Incidentally, owing to the inequality $\Gamma_\gamma/\Gamma_n \ll 1$, the latter process can apparently not be observed.

The state A_3^* forms an isotopic multiplet with the ground state of the nucleus with three excess neutrons, and, by introducing a known Coulomb correction, it is possible to determine the expected position of the quasistable level. Thus, knowing the masses [1] of the boron isotopes B^{12} and B^{13}, it is possible to determine the energies of the corresponding states of C_2^{12*} ($T = 1$) and C_3^{13*} ($T = 3/2$). The result (in our notation) is $E_3 = 11.2$ MeV at $Q = 4.95$ MeV and $\Delta = 11.54$ MeV. Consequently, the level C_3^{13*} should be quasistable, since its energy is insufficient for decay into $C_2^{12*} + \mathrm{n}$.

One should expect a narrow resonance in the scattering of n by C^{12} at $E_n = 11.2 - 4.95 = 6.25$ MeV, corresponding to a neutron energy of 7.20 MeV in the laboratory system.

From the similarity between C^{12} in the state with $T = 3/2$ and the ground state of B^{13} one expects C_3^{13*} to be in the state $3/2^-$, which leads to a scatter of neutrons in the state $P_{3/2}$ on C^{12}, with a cross section

$$4\pi\lambdabar^2 \frac{2J+1}{2S+1} = 0.8 \text{ barn}$$

A relatively narrow resonance was observed experimentally [2] at $E_n = 6.30$ MeV, along with a superposition of two resonances at $E_n = 7.4$ and 8.7 MeV.

The state of interest to us can be investigated by studying the angular distribution and polarization of the scattered neutrons. On the other hand, at least in principle, there is a possibility of ascertaining the existence of the unknown isobars by resonance in the scattering of neutrons by stable nuclei. Thus, narrow resonance in neutron scattering on Be^{10} or C^{14} could denote the existence of stable (with respect to emission of neutrons) nuclei Li^{11} or B^{15}.

[1] $A_3^* \to A + \gamma$ is allowed, Γ_γ has a normal value.

[2] It is possible that the best method of observing the quasistable level is to let the reaction proceed against the continuous spectrum of bremsstrahlung and to determine the maxima in the spectrum of the emitted neutrons from the time of flight, using a pulsed γ source.

I take this opportunity to express my gratitude to V. I. Goldanskiĭ for discussions.

Institute of Theoretical
and Experimental Physics
USSR Academy of Sciences. Moscow

Received
September 24, 1959

REFERENCES

1. *Ajzenberg-Selove F., Lauritsen T.*—Nucl. Phys. **11**, 5 (1959).
2. *Bondelid R., Dunning K. L., Talbott F. L.*—Phys. Rev. **105**, 193 (1957).

Commentary

This paper was the first to indicate that since the isotopic spin in light nuclei is a sufficiently good quantum number, the decay of a level with $T = 3/2$ per nucleus to a state with $T = 0$ and a neutron should be forbidden.

Generalizing the idea of the paper, one may say that it was the first to point out the possibility of the existence of peculiar "isospin isomers"—narrow nuclear levels which are higher than the bonding energy of a nucleon, but which decay not through strong, but through electromagnetic interaction since the release of a nucleon from these levels is possible only through a change in the isotopic spin ($\Delta T = 1$).

Unfortunately, the paper proposes far from the best means of populating such narrow levels—through neutron scattering or the photo-effect. This means has still not been realized.

However, as was first shown by V. I. Goldanskiĭ, "isospin isomers" should be clearly evident in processes of release of β^+-delayed protons[1] and proton pairs[2] in the form of so-called analog states, populated at superallowed ($\Delta T = 0$) β^+-decay of the mother-nuclei, for example, $^{33}\mathrm{Ar} \longrightarrow {}^{33}\mathrm{Cl}^* \longrightarrow p + {}^{32}\mathrm{S}$ or $\underset{T=2}{^{22}\mathrm{Al}} \xrightarrow{\beta^+} \underset{T=2}{^{22}\mathrm{Mg}^*} \longrightarrow 2p + \underset{T=0}{^{20}\mathrm{Ne}}$. Many such examples have indeed been observed in experiments. The paper also predicts the existence of nuclei $^{11}\mathrm{Li}$ and $^{15}\mathrm{B}$ which are stable with respect to neutron decay; these were later discovered in experiments.

[1] *Goldanskiĭ V. I.*—Dokl. AN SSSR **146**, 1309–1311 (1962).
[2] *Goldanskiĭ V. I.*—Pisma v ZhETF **32**, 572–574 (1980).

7
New Isotopes of Light Nuclei and the Equation of State of Neutrons*

The limits of stability (relative to nucleon emission) of light nuclei are considered. The existence (in the sense of stability against decay with emission of a nucleon) of the following nuclei is predicted: He^8, Be^{12}, O^{13}, $B^{15,17,19}$, C^{16-20}, N^{18-21}, Mg^{20}. The problem of the possibility of existence of heavy nuclei composed of neutrons only is considered. The problem is reduced to that of a Fermi gas with a resonance interaction between the particles. The energy of such a gas is proportional to $\omega^{2/3}$, where ω is its density. The accuracy of the calculations is not sufficient to determine the sign of the energy and answer the question as to the existence of neutron nuclei.

The problem of the possible isotopes has been treated by Nemirovskiĭ [1, 2] for $8 \leq Z \leq 84$, and by Baz [3] for the region $17 \leq A \leq 40$. The former uses the one-particle approximation, with an attempt to find the dependence of the parameters of the well on the numbers of neutrons and protons. For nuclei with an excess of protons Baz bases his discussion on the experimental data on the mirror nuclei (with excess of neutrons) and on the well-known expression for the Coulomb energy. For nuclei with an excess of neutrons he extrapolates the binding energy in a series of nuclei with constant isotopic spin.

These papers predict the existence of many as yet unknown β-active isotopes. In the table given below the isotopes so predicted are enclosed in dashed-line squares. One of them (O^{20}) has very recently been observed experimentally [4].

In the present paper (Sec. 1) we make additional predictions in the region of the lightest nuclei; the isotopes so predicted are enclosed in solid-line squares in the table. We point out particularly the conclusion that there is a large probability that He^8 exists. For nuclei with an excess of neutrons the writer has tried to take the effect of shells and the pair interaction of neutrons into account as accurately as possible.

In Sec. 2 the question is raised of the existence of nuclei composed solely of neutrons. In the limiting case of a large number of neutrons, by using the data on resonance in the ^1S scattering, one can find the general form of the dependence of the energy on the density of the nuclear matter, but the

*Zhurnal eksperimentalnoĭ i teoreticheskoĭ fiziki **38** (4), 1123–1131 (1960).

accuracy of the first approximation obtained in this paper is insufficient to give a definite answer to the question of the existence of such nuclei.

1. Light Nuclei

Following the method of Baz [3], one easily convinces oneself that there should exist a nucleus O^{13} with a proton binding energy not smaller than 1.2 MeV and with β^+-decay energy 16 to 17 MeV. Using the data [4] on the mass of O^{20}, we conclude that the mirror nucleus Mg^{20} should exist with proton binding energy not less than 2.7 MeV and β^+-decay energy about 7 MeV. The existence of O^{12}, Ne^{16}, and Mg^{19} is not excluded (empty spaces in the table);[1] the corresponding mirror isotopes Be^{12}, C^{16}, and N^{19} are predicted in this paper (see later argument), but their energies cannot be predicted with enough accuracy to give a definite conclusion about O^{12}, Ne^{16}, and Mg^{19}. The isotopes Ne^{17}, Na^{19}, Mg^{21}, and Mg^{22} are predicted by Baz.

Regarding all the other nuclei in the upper right-hand part of the table we can assert with assurance that they are unstable against emission of a proton, i.e., they do not exist, which is shown in the table by the minus signs in all the upper cells.

Let us turn to the nuclei with an excess of neutrons. A nucleus with an excess of neutrons does not exist in the case in which all the discrete levels are already filled up with neutrons. An important point here is that the nuclear forces fall off rapidly with distance, and therefore the number of levels in the field of the nuclear forces is limited (in contradistinction, for example, to the case of the Coulomb field). With the spin taken into account the number of levels is always even; therefore if a nucleus exists containing an odd number of neutrons $(2n+1)$, then there is also a place for a subsequent $(2n+2)$-nd neutron. On account of the mutual attraction of a pair of neutrons the binding energy of the $(2n+2)$-nd neutron is always larger than that of the preceding $(2n+1)$-st neutron.

In each cell of the table that corresponds to an experimentally known isotope there is written the binding energy of the last neutron. It is easily verified that in all cases $E_{2n+2} > E_{2n+1}$. Therefore the existence of the nuclei Be^{12} and C^{16} definitely follows from the existence of Be^{11} and C^{15}. As a rough estimate, the binding energy of a neutron in Be^{12} is about 2–3 MeV, and the β-decay energy is 12–13 MeV; for C^{16} these values are 3–4 MeV and 8–9 MeV, respectively.

It is much harder to settle the existence of other isotopes. Extrapolation

[1] These nuclei may be unstable with respect to the emission of two protons at once. On the other hand, at the limit of stability the expression for the Coulomb energy of the last proton, $1.2(Z-1)A^{-1/3}$, gives too large a result; for example, in the pair Li^8–B^8 we have for Li^8 the binding energy $Q_n = 2$ MeV and for B^8 the value $Q_p = 0.2$ MeV, so that the difference is 1.8 MeV, whereas by the formula we would get $1.2 \times 4 \times 7^{-1/3} = 2.5$ MeV.

	p^{1}	—	—	—	—	—	—	—	—	—	—	—
N^{1}	D^{2} 2,23	He^{3}	—	—	—	—	—	—	—	—	—	—
—	T^{3} 6,26	He^{4} 20,58	—	—	—	—	—	—	—	—	—	—
—	—	—	Li^{6}	Be^{7}	B^{8}	C^{9}	—	—	—	—	—	—
	?H^{5}	He^{6} 0,94	Li^{7} 7,25	Be^{8} 18,10	B^{9} 18,58	C^{10} 21,9	—		—	—	—	—
		—	Li^{8} 2,04	Be^{9} 1,67	B^{10} 8,44	C^{11} 13,3	N^{12}	O^{13}	—	—	—	—
		He^{8}	Li^{9} 3,58	Be^{10} 6,81	B^{11} 11,50	C^{12} 18,73	N^{13} 20,9	O^{14}	—		—	—
		—	—	Be^{11} 0,55	B^{12} 3,36	C^{13} 4,95	N^{14} 10,55	O^{15} 13,23	—	Ne^{17}	—	
		—		Be^{12}	B^{13} 4,89	C^{14} 8,17	N^{15} 10,84	O^{16} 15,66	F^{17} 17,01	Ne^{18}	Na^{19}	Mg^{20}
				—		C^{15} 1,23	N^{16} 2,50	O^{17} 4,15	F^{18} 9,17	Ne^{19} 11,39	Na^{20}	Mg^{21}
					B^{15}	C^{16}	N^{17} 5,84	O^{18} 8,07	F^{19} 10,42	Ne^{20} 16,92	Na^{21} 18,57	Mg^{22}
				—		C^{17}	N^{18}	O^{19} 3,96	F^{20} 6,61	Ne^{21} 6,76	Na^{22} 11,05	Mg^{23}
					B^{17}	C^{18}	N^{19}	O^{20} 7,65	F^{21} 8,18	Ne^{22} 10,36	Na^{23} 12,42	Mg^{24} 16,4
				—		C^{19}	N^{20}	O^{21}	F^{22}	Ne^{23} 5,20	Na^{24} 6,97	Mg^{25} 7,34
					B^{19}	C^{20}	N^{21}	O^{22}	F^{23}	Ne^{24} 8,90	Na^{25} 9,15	Mg^{26} 11,12
								O^{23}	F^{24}	Ne^{25}	Na^{26}	Mg^{27} 6,44
								O^{24}	F^{25}	Ne^{26}	Na^{27}	Mg^{28} 8,56
								O^{25}	F^{26}	Ne^{27}	Na^{28}	Mg^{29}
								O^{26}	F^{27}	Ne^{28}	Na^{29}	Mg^{30}
								O^{27}	F^{28}	Ne^{29}	Na^{30}	Mg^{31}
								O^{28}	F^{29}	Ne^{30}	Na^{31}	Mg^{32}

for fixed isospin [3] T is not reliable, since it involves comparison of neutrons that are in different shells.

For the lightest nuclei the idea of a smooth dependence of the parameters of the well on N and Z [1, 2] does not take sufficient account of the individual peculiarities of the shells. We shall try to make maximum use of the experimental data. It is known from the scattering of neutrons by He^4 that for the partial wave $P_{3/2}$ there is a resonance at the energy +1.0 MeV (i.e., in the continuous spectrum) with width 0.55 MeV (which corresponds to an He^5 lifetime of 10^{-21} s). The nucleus He^5 does not exist, and consequently there is no discrete bound state of a neutron in the field of He^4.

In the same sense, the dineutron does not exist, since from experiments on the scattering of neutrons by protons it is known that in the ^1S state, which is allowed for two neutrons by the Pauli principle, the attraction is not sufficient for the formation of a bound state. Therefore the He^6 nucleus is a remarkable system of three particles $(n+n+He^4)$, which are not bound together in pairs, but all three together form a bound system. Quite crudely we can imagine that He^6 consists of two neutrons in the state $(P_{3/2})^2$ in the field of He^4. The energy of interaction between the two neutrons (about −3 MeV) is more than enough to compensate for the positive energy of each neutron in the state $P_{3/2}$ (+1 MeV) in the field of He^4.

The $P_{3/2}$ shell has four places in all. Therefore we can raise the question of the possibility of He^7 and He^8. According to Kurath [5], in the limit of small range of the forces and weakly bound nucleons, and for large radius of the orbits of the shell ($r_0 \ll r_1$, $L = 3K$, in his notation), one gets a simple result: if the energy of interaction of two neutrons is B, then the energy of the interaction of three neutrons is also B, and the energy of the interaction of four neutrons is $2B$, i.e., the neutrons combine in pairs, as it were. From this there follows the conclusion that He^7 does not exist, but He^8 exists; the expected binding energy of a neutron is 0.5–0.8 MeV, and the β=decay energy is about 12 MeV. It would be extremely desirable to verify the existence of He^8 experimentally and determine its binding energy.

How accurately the rule of the combining of neutrons in pairs in a single shell around a doubly magic (closed) core holds experimentally can be seen from two examples.[2]

1) The filling up of the $d_{5/2}$ shell on the closed O^{16} (see table). We quote the binding energies (in megavolts). The subscript on E is the number of neutrons in the $d_{5/2}$ shell (the upper index is the atomic weight):

$$E_1^{17} = 4.15, \quad E_2^{18} = 8.07, \quad E_3^{19} = 3.96, \quad E_4^{20} = 7.65.$$

There are no data on E_5 and E_6, which finish the filling of the shell; the nuclei O^{21} and O^{22} have not yet been observed.

2) The filling up of the $f_{7/2}$ shell on Ca^{40}, which has closed shells (this

[2]The mass data are taken from review articles [6–8].

example has been treated partially by Nemirovskiĭ [2]). The binding energies are:

E_1^{41}	E_2^{42}	E_3^{43}	E_4^{44}	E_5^{45}	E_6^{46}	E_7^{47}	E_8^{48}
8.3	11.4	8.0	11.4	7.4	11.0(?)[3]	6.8?[3]	10.8

At the end of the filling-up of the $f_{7/2}$ shell the binding energy E falls sharply: $E_9^{49} = 5.1$. Since He^4 is a closed doubly magic nucleus (and an even more stable one than O^{16} and Ca^{40}), these examples speak convincingly for the existence of He^8.

If the proton shell is not filled, then E drops off extremely sharply within the range of the given neutron shell; we may imagine that the first neutrons unite in pairs with the "free" protons (those outside the closed shells), and later neutrons can no longer do this. As an example let us consider the $d_{5/2}$ shell of Ne^{18}—a nucleus with two protons beyond O^{16}. We have:

E_1^{19}	E_2^{20}	E_3^{21}	E_4^{22}	E_5^{23}	E_6^{24}
11.4	16.9	6.8	10.4	5.2	8.9

If the proton shell falls short of being closed by one, two, or three protons, the binding energy of the neutrons is decreased as compared with the binding to a closed shell (cf. C^{15}, N^{16}, and O^{17} in the table). But within the limits of a given neutron shell (on a core with holes in the proton shell) E varies little, in contradistinction to the case in which excess protons are present.

We give examples of the filling of the $f_{7/2}$ shell with neutrons in nuclei with unfilled proton shells:

Nucleus K_{19}^{39}:	$E_1^{40} = 7.9$	$E_2^{41} = 10.0$	$E_3^{42} = 7.4$	$E_4^{43} = 10.8$
Nucleus Ar_{18}^{38}:	$E_1^{39} = 6.7$	$E_2^{40} = 9.7$	$E_3^{41} = 6.1$	

Thus we can formulate the rule that on nuclei with closed proton shells and with holes in the proton shell (but not on nuclei with excess protons), the binding energies of the odd neutrons are practically constant within the limits of a given neutron shell. The binding energies of the even neutrons are also constant within the limits of a given shell, but are larger by the amount of the pairing energy. Carrying this rule over to the $d_{5/2}$ shell, we come to the conclusion that the experimental fact of the existence of bound $d_{5/2}$ states in the nuclei C^{15} and N^{16}, N^{17} guarantees the possibility of filling up the entire $d_{5/2}$ shell, to C^{20} and N^{21}, respectively.

An examination of the binding energies of neutrons in the table reveals a regular increase of E in each row, with increase of the number of protons (the single exception is the pair Li^8–Be^9, which is due to the special structure of B^8). Extrapolation of E to the left along the rows makes probable the

[3]The nucleus Ca^{46} has not been studied, so that one knows experimentally only the sum $E_6^{46} + E_7^{47} = 17.8$; the separate terms in the table are obtained by interpolation.

existence of B^{15}, and from this—by the principle of the constancy of the binding energy in a shell—of B^{17} and B^{19}. The existence of the nuclei with odd numbers of neutrons, B^{14}, B^{16}, B^{18}, remains questionable. With considerable assurance we can assert that the odd (in n) nuclei Be^{13}, Be^{15}, Be^{17}, Li^{10} do not exist.

On the whole, however, the assertions that can be made reliably about nuclei with excess neutrons not known to exist are extremely weak. From studies of scattering only the nonexistence of n^2 and He^5 is quite accurately proved. From principles of the pair interaction of neutrons it is obvious that n^3 and He^7 do not exist. There is no longer such certainty regarding H^5 (H^5 is entered in the table with a question mark), and the hypothesis that it exists has been suggested [9]. We note that if n^4 and H^5 existed, then there would be isotopically similar quasi-stable systems H^4 with $T = 2$ and He^5 with $T = 3/2$, which would manifest themselves in the scattering of n by T and of n by He^4; this situation has been examined in detail in a separate note [10]. At present there are no experimental data in the required range of neutron energies.

Unlike the upper right-hand part of the table, which is almost solidly filled with minus signs ("does not exist"), in most of the cells of the lower left-hand part we can put neither a minus nor the symbol of a nucleus ("exists"). The obscurity of the problem of the limits of existence of isotopes with excess neutrons is a consequence of the fact that the limiting case is not clear; it is not known whether a heavy nucleus composed solely of neutrons could exist.

2. The Neutron Liquid

The problem of the limiting number of neutrons that can adhere to a heavy nucleus has been considered by Wheeler [11]; he came to the conclusion that for $Z \sim 90$–100 the maximum mass number is $A_{max} \sim 500$–600. Wheeler used the Weizsäcker formula; Nemirovskiĭ [2] correctly criticizes this formula near the limits of existence, and therefore, Wheeler's conclusions are not reliable.

Let us consider the extreme case of a very large nucleus consisting of neutrons alone. If it does exist, it surely does so only with a density much smaller than that of ordinary nuclei. Let us first examine the properties of a neutron liquid of small density; these properties are determined by the pair interactions of the neutrons at small energies (up to a few MeV). In this region only the interaction of pairs of neutrons in the ^1S state is of importance, and here this interaction is completely determined by the scattering length[4] (cf., e.g., [12]):

$$a = -\left(\frac{d\ln\varphi}{dr}\right)^{-1} = -19\cdot 10^{-13}\,\mathrm{cm};$$

the sign corresponds to the absence of a bound state, and the quantity a corresponds to the so-called energy of a virtual level (μ is the reduced mass, equal to $M/2$):

$$E_v = \frac{\hbar^2}{2\mu a^2} = 0.11\,\mathrm{MeV}.$$

We cite here the well-known calculation [13, 14] of the energy of interaction of particles in the continuous spectrum, confining ourselves at once to the S wave. As usual, we consider first a spherical box for $\mathbf{r} = \mathbf{r}_1 - \mathbf{r}_2$, where $\mathbf{r}_1$ and $\mathbf{r}_2$ are the coordinates of the two particles, i.e., we set $\psi(r) = 0$ at $|\mathbf{r}| = R$. Without interaction the normalized S-wave function in such a box is

$$\psi = \frac{\sin(n\pi r)}{r\sqrt{2\pi R}}.$$

With an interaction corresponding to scattering with the phase shift α we have

$$\psi = \frac{\sin[\alpha + R^{-1}(n-\alpha/\pi)\pi r]}{r\sqrt{2\pi R}},$$

which corresponds to a change of the energy of the n-th state given by

$$\Delta E_n = -\frac{\hbar^2 n\pi\alpha}{\mu R^2}.$$

Let us eliminate the auxiliary quantities R and n from the expression for ΔE_n. The state under consideration is characterized by the momentum of the relative motion

$$p_n = \frac{\hbar n\pi}{R}$$

and the density at the coordinate origin in the unperturbed motion

$$\rho_n(0) = \psi^2(0) = \frac{\pi n^2}{2R^3}.$$

Let us express ΔE_n in terms of p and $\rho(0)$; after this we can set $R \to \infty$, $n \to \infty$, and forget about n. We get

$$\Delta E = -\frac{2\pi\hbar^3\alpha\rho(0)}{\mu p}. \tag{1}$$

We express the phase in terms of the scattering length:

$$\alpha = -\tan^{-1}\left(\frac{ap}{\hbar}\right).$$

[4]For pp-scattering $a = -17.2$, and for np-scattering $a = -23.7$; we assume that a depends linearly on the product of the magnetic moments.

For $E \ll E_v$, $ap \ll \hbar$ we have

$$\alpha = \frac{ap}{\hbar}, \qquad \Delta E = -\frac{2\pi\hbar^2 a\rho(0)}{\mu}; \tag{2}$$

for $E \gg E_v$, $ap \gg \hbar$ we get

$$\alpha = \frac{\pi}{2}, \qquad \Delta E = -\frac{\pi^2\hbar^3\rho(0)}{p\mu}. \tag{3}$$

Let us apply the expressions (2) and (3) to a Fermi gas consisting of neutrons only with mean density ω. We single out one neutron with a definite spin direction. At the point where this neutron is located, the density of other neutrons with the same spin direction is zero by the Pauli principle; if there were no interaction, the density of the other neutrons with antiparallel spins would not differ from that of those with parallel spins on the average over all space; that is, $\omega(0) = \omega/2$. We recall that ω is the total density of neutrons with both spin directions and that the formula for ΔE contains just the density in the state without interaction.[5]

We still have to take into account the fact that the change of energy ΔE relates to a system of two particles; in order not to include the interaction of each pair twice, we recall that the decrease of the energy of one particle is $\Delta E/2$. We finally find that if for a pair of particles in the ^1S state $\Delta E = k\rho(0)$, where k is a coefficient that depends on the momentum, then the change of the energy of all the gas in unit volume on account of the interaction is

$$U = \frac{\omega^2\overline{k}}{4}; \tag{4}$$

here k is averaged over the Fermi distribution.

The Fermi distribution is characterized by the boundary momentum p_f, the boundary energy E_f, and the total kinetic energy $\mathcal{E}$ of all the gas in unit volume; as is well known,

$$\mathcal{E} = \omega\overline{E} = \frac{3}{5}\omega E_f, \qquad E_f = \frac{p_f^2}{2M},$$

$$\omega = \frac{p_f^3}{3\pi^2\hbar^3}, \qquad \mathcal{E} = \frac{p_f^5}{10\pi^2\hbar^3 M}. \tag{5}$$

When we average k we get a result which depends on the ratio of E_f to the energy E_v of the virtual level. For $E_f < E_v$ the quantity k is constant

[5]Another possible approach is based on the fact that the statistical weights of the triplet and singlet are in the ratio 3 : 1; a given neutron interacts with only 1/4 of the others. But in the singlet state without scattering the density at the coordinate origin is twice as large as the average density throughout the volume, since in the singlet state only even angular momenta l are possible, and therefore the S state, the only one that contributes to $\rho(0)$, makes up twice as large a fraction of all singlet states as in the case of different particles. We finally find (1 is the index for the singlet) $\omega(0) = 2\overline{\omega}_1 = 2(\omega/4) = \omega/2$, which agrees with the result obtained in the text.

and ($\mu = M/2$)

$$U = -\frac{\pi\hbar^2 \alpha\omega^2}{2\mu}. \tag{6}$$

In the limiting case $E_f \gg E_v$ we must average over the Fermi distribution p^{-1}, where p is the momentum of the relative motion of two particles. We have

$$\mathbf{p} = \mu(\mathbf{v}_1 - \mathbf{v}_2) = \frac{1}{2}M(\mathbf{v}_1 - \mathbf{v}_2) = \frac{1}{2}(\mathbf{p}_1 - \mathbf{p}_2). \tag{7}$$

Using the electrostatic analogy[6] we easily find

$$\overline{|\mathbf{p}_1 - \mathbf{p}_2|^{-1}} = \frac{6}{5p_f}, \qquad \overline{p^{-1}} = \frac{12}{5p_f}, \tag{8}$$

and finally,

$$U = -\frac{3\pi^2\hbar^3\omega^2}{5\mu p_f} = -\frac{2p^5}{15\pi^2\hbar^3 M} = -\frac{4\mathcal{E}}{3}. \tag{9}$$

This is a remarkable result: the interaction energy is a constant multiple of the kinetic energy.

If we take these results literally, we get the following physical conclusions about the dependence on the density of the average energy of a neutron, $E_1(\omega) = (\mathcal{E} + U)/\omega$: at small density, in the limit

$$E_1 = \frac{3}{5}E_m > 0, \qquad E_1 \sim \omega^{2/3}, \tag{10}$$

the interaction is proportional to a higher power of ω (first) at the density ω_0 that corresponds to $E_f = 5E_v$, the energy E_1 goes to zero, and then changes sign and at larger densities

$$E_1 = -\frac{1}{3}E_m < 0, \qquad E_1 \sim \omega^{2/3}. \tag{11}$$

This expression holds for[7] $\omega > \omega_0 \approx a^{-3}$. From this it follows that a nucleus can exist that consists of neutrons only, with a binding energy given by $-E_1$.

This treatment does not give the equilibrium density, since according to (11) as the density increases E_1 continues to decrease (E_1 is negative and its absolute value increases). To find the equilibrium density and the binding energy at this density we must bring in the effective range of nuclear forces and the interaction in states with $l \neq 0$. Qualitatively, however, the fact of the existence of neutron nuclei itself follows just from the change of sign of E_1, which is obtained from a calculation at the density $\omega_0 = a^{-3}$. Since a is extremely large, we have $\omega_0 \approx 0.001\omega_n$, where ω_n is the density of ordinary nuclei. In a state corresponding to the density ω_0 for which $E_1 = 0$

[6] For any body, $\overline{r_{12}^{-1}} = \int\int r_{12}^{-1}\, dv_1\, dv_2 = \int \varphi_1\, dv_1 = \overline{\varphi}$, where φ is the potential for unit charge density, which satisfies the equation $\Delta\varphi = -4\pi$ inside the body and $\Delta\varphi = 0$ outside the body.

[7] A consistent calculation on the assumptions made above gives a value of the coefficient very close to unity.

the boundary kinetic energy E_f is about 0.5 MeV, so that the contribution from $l \neq 0$ and the influence of the effective range are negligible; thus the assumptions about the interaction of the neutrons that were the basis for the calculation are very well satisfied at $\omega = \omega_0$. We note that if the existence of a range of values of ω in which $E_1 < 0$ is confirmed, then the surface tension of the neutron liquid will give a definite critical size of the neutron droplet, i.e., a minimum number of neutrons for which the existence of a neutron nucleus is possible. Therefore if it is proved that bound states n^4, n^6 or n^8 do not exist, this does not by itself exclude the existence of the heavier neutron nuclei.

Nevertheless the main result—the change of sign of E_1—is by no means to be regarded as established, since only the pair interaction of the neutrons has been considered and no account has been taken of the influence of the other neutrons on the wave functions of the interacting pair. The result is doubly unreliable because for $\omega > \omega_0$ the desired quantity E_1 is the small difference of two nearly equal quantities:

$$E_1 = \overline{E} + U_1, \quad \overline{E} = \frac{3}{5}E_f, \quad U_1 = -\frac{4}{5}E_f = -\frac{4}{3}\overline{E}. \tag{12}$$

For $\omega > \omega_0$, $E_f \gg E_v$, the scattering does not depend on the length a, and we can set $a = \infty$, $a^{-1} = 0$, i.e., consider resonance scattering. Then the problem contains no dimensionless parameters. From dimensional considerations it follows that in this region

$$E_1 \sim U_1 \sim \overline{E} \sim E_f \sim \omega^{2/3}. \tag{13}$$

The formula (11) for E_1 is in agreement with this requirement. But then the correction to E_1 because of the influence of a third neutron on the wave functions of a given pair is also proportional to E_f, i.e., depends on the same power of the density and can differ from E_f and E_1 only by a numerical coefficient. This case is not like the usual one; in the Fermi gas at absolute zero with resonance scattering one cannot expand in a series of powers of the density.

We have not found the corrections for the interactions of three and more particles; it is quite possible that they will change the sign of E_1 in the region $\omega < \omega_0$. We know that $E_1 > 0$ for $\omega < \omega_0$. On the other hand, for values of ω approaching the density of ordinary nuclei it is to be expected that the energy will lie above that calculated from the resonance S scattering.[8] Therefore, if from an exact solution of the problem of the Fermi gas with resonance interaction it is found that $E_1 > 0$, this will mean that the existence of nuclei composed of neutrons only is impossible.

We note that the expression (11) for E_1 found by using the pair interaction is not the mathematical expectation of the energy, calculated with

[8] By the method described above we would get for nuclear matter consisting of equal numbers of neutrons and protons, with the Coulomb interaction neglected, the result $U_1 = -4\overline{E}$; for the ordinary nuclear density this would give a binding energy ~ 60 MeV, many times the experimental value.

the unperturbed functions of the problem without interaction (otherwise we could assert that the true E_1 could only be lower than that so found); in the calculation of the interaction the change of the wave functions was taken into account from the very start (see beginning of Sec. 2). Actually the calculation of the energy of the pair interaction includes within itself the change of the wave function at the origin. We recall that $\rho(0)$ is the density that would exist in the absence of interaction; in the presence of the interaction we get for small r

$$\psi \sim r^{-1}, \quad \rho = \frac{1}{4}\pi^2\left(\frac{h^2}{pr}\right)^2 \rho(0).$$

It is obvious that the change of the density and the wave function (and consequently also of the momentum spectrum) affects the interaction of the pair under consideration with other particles. We note that with a finite change of the total energy in this way of treating the pair interaction the mathematical expectations of the kinetic and potential energies are infinite and of opposite signs.

Resonance scattering with a singular potential that is nonvanishing in a small region gives in the limit zero interaction in the first order, second order, and so on, in perturbation theory; a finite result is given only by the sum of an infinite number of terms (for details see [15]). The expression for E_1 given above is not the first approximation of perturbation theory for a Fermi gas with pair interaction between the neutrons. E_1 is the result of including in a definite way a chosen infinite succession of the terms of the perturbation-theory series, and therefore it is not clear what is the sign of the correction to E_1. The assertion of Yang and Lee [16] that not only in a Bose gas, but also in a Fermi gas any attraction always leads to a condensation seems not to be well founded.

I take this occasion to express my gratitude to A. I. Baz, V. I. Goldanskiĭ, L. D. Landau, A. B. Migdal, and P. E. Nemirovskiĭ for discussions, and to D. V. Grigor'ev for help in the preparation of this article.

Received
October 22, 1959

REFERENCES

1. *Nemirovskiĭ P. E.*—ZhETF **36**, 883 (1959).
2. *Nemirovskiĭ P. E.*—ZhETF **33**, 746 (1957).
3. *Baz A. I.*—Atom. Energiĭa **6**, 664 (1959).
4. *Jarmie N., Silbert M. G.*—Phys. Rev. Lett. **3**, 50 (1959).
5. *Kurath D.*—Phys. Rev. **88**, 804 (1952).
6. *Ajzenberg-Selove F., Lauritsen T.*—Nucl. Phys. **11**, 1 (1959).
7. *Endt P. M., Kluyver J. C.*—Rev. Mod. Phys. **26**, 95 (1954).
8. *VanLatter D. M., Whaling W.*—Rev. Mod. Phys. **26**, 402 (1954).
9. *Blanchard M., Winter R. G.*—Phys. Rev. **107**, 744 (1957).
10. *Zeldovich Ya. B.*—ZhETF **38**, 278 (1960).

11. *Wheeler J. A.*—in: *Niels Bohr and the Development of Physics.* New York: McGraw-Hill (1955).
12. *Bethe H., Morrison P. Elementary Nuclear Theory.* New York: Wiley (1956).
13. *Landau L. D., Lifshitz E. M. Statistical Physics.* Reading, MA: Addison-Wesley (1969).
14. *Lee T. D., Yang C. N.*—Phys. Rev. **105**, 767 (1957).
15. *Zeldovich Ya. B.*—ZhETF **38**, 278 (1960).
16. *Lee T. D., Yang C. N.*—Phys. Rev. **105**, 1119 (1957).

Commentary

This paper considers the boundaries of stability of light nuclei with respect to the release of nucleons. For nuclei with excess neutrons the author took into account shell effects and pair interaction of neutrons. An important result is the prediction of the possibility of existence of stable light nuclei which are strongly enriched by neutrons—neutron-rich nuclei. In particular, specific predictions were made regarding the possibility of existence of He^8 (whereas He^7 is obviously unstable). This prediction of the existence of He^8 was experimentally confirmed at the Joint Institute of Nuclear Research in Dubna.

The observation of He^8 was the beginning of the experimental discovery of light neutron-rich nuclei. Today we know 16 light stable neutron-rich nuclei:[1,2] $He^8, Li^9, Li^{11}, Be^{12}, Be^{14}, B^{15}, B^{17}, C^{18}, C^{19}, C^{20}, N^{19}, N^{20}, N^{21}, O^{22}, O^{23}, O^{24}$.

The second part of the paper considers the question of the equation of state of a neutron gas with interaction and of the possibility of existence of heavy nuclei consisting solely of neutrons.

There are ideas about the similarity of neutrons and atoms of He^3. In both cases the attraction of a pair of particles is insufficient for the formation of a connected system, of a nucleus of dineutron in the first case and of a molecule consisting of two He^3 atoms in the second. Nevertheless, liquid He^3 exists at zero temperature and its vapor pressure is equal to zero. Knowing the surface tension of He^3 we may determine the critical size of a droplet with negative energy—it consists of approximately 30 atoms of He^3.

By analogy we may expect the existence of a liquid neutron phase as well and a critical droplet size, i.e., the existence of some sufficiently heavy nucleus of neutrons alone, subject to β-decay.

The non-existence of the dineutron and of a nucleus of 4 neutrons is confirmed by the negative results of meticulous experimental search.[3]

Heavy neutron nuclei, if they exist, could with low probability arise in fission. They would undoubtedly be unstable with respect to β-decay, however, their lifetime should be sufficiently long that they are able to reach the shell of the reactor and cause unusual nuclear reactions there. We note finally that the present paper

[1] *Baz A. I., Goldanskiĭ V. I., Goldberg V. Z., Zeldovich Ya. B. Legkie i promezhutochnye ĭadra vblizi granits nuklonnoĭ stabilnosti [Light and Intermediate Nuclei Near the Boundary of Nucleon Stability].* Moscow: Nauka, 172 p. (1972).

[2] *Karnaukhov V. A., Petrov L. A. Ĭadra, udalennye ot linii beta-stabilnosti [Nuclei Far from the Line of Beta-Stability].* Moscow: Energoizdat, 200 p. (1981).

[3] *Baz A. I., et al.*

obtains a new asymptote of the equation of state of a degenerate fermi-gas arising from particles whose interaction is characterized by a large cross-section length a for small radius r_1 of the potential hole. In the limit of small density n the virial correction is proportional to n^2, as it should be, but in the region where the wavelength of particles on the Fermi boundary lies between a and r, $a > \lambda_f > r_1$, the energy of interaction varies as $n^{5/3}$.

This set of problems was later investigated in greater detail in connection with the theory of neutron stars. According to the latest papers of Bethe and his colleagues, the energy of a neutron liquid at any density is greater than the energy of a rarified gas. If they are correct, the heavy neutron nuclei do not exist.[4]

[4] *Pandharinande A.*—Nucl. Phys. **129**, 1141 (1971).

8
Vortex Isomers of Nuclei*

If nuclear matter is a superfluid liquid, then a state corresponding to a drop of this liquid is possible, i.e., a nucleus with a quantized vortex[1] passing along the axis of the drop.

The circulation of the velocity along the contour surrounding the vortex, as is well known, equals $\hbar/m$, where m is the mass of the bosons making up the superfluid liquid. This means that each such boson makes a contribution equal to $\hbar$ to the angular momentum. Consequently the total angular momentum of the nucleus in the vortical state is equal to $n\hbar = z\hbar/2$. It is assumed that the role of the bosons, whose number equals n, is played by α particles.

Since the rotation is not similar to rotation with constant angular velocity ($\omega \sim 1/r^2$ in the presence of a vortex), the equilibrium shape of the drop has the form shown in Fig. 1, with a dip on the axis. The greatest interest attaches to the minimum energy E_m of the nucleus as a function of I.

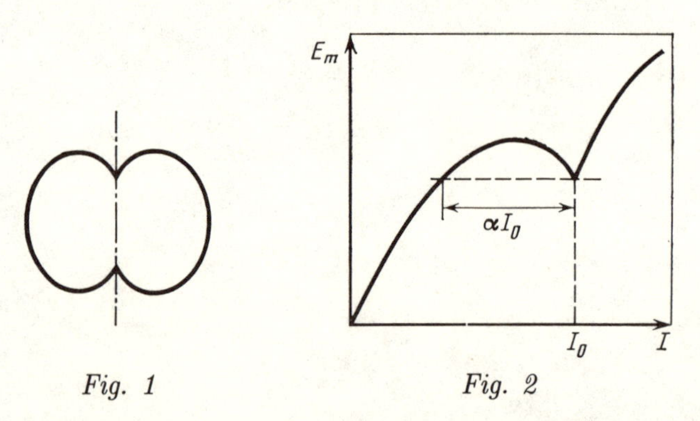

Fig. 1 *Fig. 2*

A qualitative plot of $E_m(I)$ for the case of superfluidity is shown in Fig. 2. The vortical state with $I = I_0 = n\hbar = z\hbar/2$ is a sharp minimum of the $E_m(I)$ curve with a discontinuity of the derivative $\partial E_m/\partial I$, since the increment of the energy depends linearly on the modulus of the momentum, $\Delta E_m \approx k\Delta I$, for the excitations imposed on the vortical state.

It follows from the form of the curve in Fig. 2 that the vortical state can be regarded as isomeric: its energy exceeds that of the ground state, but

*Pisma v Zhurnal eksperimentalnoĭ i teoreticheskoĭ fiziki **4** (3), 78–80 (1966).

[1] *Feynman R. P.*—Proc. First Int. Conf. Low Temperatures (1958).

the energy can be decreased by emission of a photon or of some particle only by changing simultaneously the momentum by an amount equal to not less than a certain fraction αI_0 (see Fig. 2). Therefore the transition to the ground state, direct or cascade, has low probability and the vortical state of the nucleus is expected to be long-lived.

It is of interest also to realize experimentally a vortical state of drops of liquid superfluid helium. In this case the superfluidity and the existence of quantum vortices are well known in themselves. Nonetheless, the behavior of the droplets can have curious peculiarities: it is interesting to observe the Magnus effect during falling, the change of the direction of the momentum while its magnitude is conserved, and the evaporation of a vortex drop. When it comes to the nucleus, the question may be raised regarding the smallest number of bosons above which the notions of superfluidity and quantum vortex can be employed. We note in this connection that in the single-particle treatment of the non-interacting bosons in a spherical self-consistent field the function $E_m(I)$ also has a kink, i.e., a break in the derivative, at $I = n\hbar$, $I = 2n\hbar$ ....[2] This property remains also when an interaction between bosons is turned on. This leaves only the quantitative question whether

$$\left.\frac{\partial E}{\partial I}\right|_{I = n\hbar - 0} < 0,$$

in the system, as shown in Fig. 2.

Consequently, an isomer state with $I_0 = z\hbar/2$ can be realized in principle also in a relatively light nucleus.

The preparation of such isomeric states is apparently most probable by collision of bulky particles, but not by the action of photons, protons, or neutrons.

Received
May 31, 1966

Commentary

This paper predicts a new type of isomeric states of nuclei which should arise in the collision of heavy ions. This type of isometry appears when one looks at the nuclear excitations of a boson liquid. It turns out that the minimum energy of the vortical state then depends on the total angular momentum of the nucleus, with a minimum at a particular value of the angular momentum and a discontinuity in the first derivative. Such a dependence makes transition into the basic state unlikely.

Experiments carried out so far on the collisions of heavy ions with nuclei have not yet given a basis for answering the question of the existence of vortical isomers. A determined search in this direction seems quite desirable.

[2] An exception is the degenerate case of the harmonic-oscillator potential $U = ar^2$.

II

Theory of Elementary Particles

9

On the Theory of Elementary Particles. Conservation of the Nuclear Charge and a Possible New Type of V-Particles*

In recent years the existence of neutral particles which decay into a proton and a π-meson, the so-called V-particles, has been established with certainty. It is possible that in the decay of V-particles a third uncharged particle is emitted as well [1–3].

The V-particle is the first particle of this kind to yield a nucleon when it decays. The law of conservation of nucleons should be generalized to account for the existence of V-particles.

As is known, a spin of $1/2$ for nucleons—protons and neutrons—implies the existence of an antiproton and antineutron which are capable of annihilating ordinary nucleons.

Free antinucleons have still not been observed directly, however, an indirect confirmation of their existence is the agreement with experiment of theoretical predictions of the decay of a neutral π-meson into two photons: these predictions and the calculation of the probability of decay were based on the concept of formation of virtual proton-antiproton pairs.

From the symmetry of all laws of nature with respect to particles and antiparticles it follows that together with the V-particles there must exist anti-V-particles which yield an antiproton when they decay. This conclusion

*Doklady Akademii Nauk SSSR **86** (3), 505–508 (1952).

does not depend on whether the V-particle has an integer or half-integer spin. Below we give in more detail the considerations which lead to this conclusion, and study the possibility of distinguishing the V and anti-V properties; we also give very rough considerations of the ratio of the number of V and anti-V which form under the influence of cosmic rays.[1]

Let us digress briefly from the formation and decay of V-particles. In all other nuclear processes observed so far there occurs generation of π- and μ-mesons, electrons and positrons, neutrinos and gamma-photon and transformation of neutrons into protons or *vice versa*, but with the sum of protons p and neutrons n remaining constant:

$$p + n = \text{const} = Y.$$

By analogy with the electric charge it seems reasonable to introduce the concept of nuclear charge and to ascribe to the proton and neutron an identical nuclear charge $+y$. The nuclear charge of π- and μ-mesons, electrons, neutrinos and photons is equal to zero. The fact that only transformation of protons into neutrons and back occurs, without any change in their total number, is formulated as the "law of conservation of nuclear charge." The other conservation laws (of energy, spin, electric charge) do not prohibit the K-capture of an electron by a proton with complete annihilation of an atom of hydrogen and formation of two photons, or the transformation of a proton into a positron or a neutron into a neutrino with the release of gamma-photon; they also do not prohibit processes where there are no special reasons to expect a low probability: annihilation of two neutrons with the formation of two photons or two π-mesons. At the same time, the most trivial fact of all nuclear physics is that in nuclei neutrons exist indefinitely without annihilating. In terms of the nuclear charge such annihilation is impossible because the nuclear charge of an atom of hydrogen is equal to $+y$, while the nuclear charge of a pair of neutrons is equal to $+2y$.

In the formation of proton-antiproton pairs (the number of protons is p, the number of antiprotons is $\tilde{p}$) and neutron-antineutron pairs $(n, \tilde{n})$, as is well known, the quantity $p - \tilde{p} + n - \tilde{n} = \text{const}$ is conserved. In order to equate it to the nuclear charge Y it is necessary to ascribe to antinucleons a unit negative nuclear charge $-y$. Assigning to each particle, both stable ones and those that spontaneously decay or annihilate with other nuclear particles, a definite nuclear charge we may ensure the impossibility of spontaneous annihilation of a pair of neutrons in any process at all, with any virtual particles.

Since all of our experience shows that such a process is absolutely prohibited, the law of conservation of nuclear charge is an exact law like the law of conservation of electric charge, and not an approximate law like the

[1] L. Schiff [5] recently reached similar conclusions. Since his note does not study the problem of the spin of V-particles or the possibility of the existence of antiparticles at zero spin, we consider it appropriate to give here a detailed consideration of the problem.

conservation of isotopic spin. It is obvious that the above considerations contain nothing essentially new and are given only as a convenient form for representing generally known facts.

Let us turn to V-particles and consider them in light of the symmetry of particles and antiparticles. Without using the concept of nuclear charge we could assume that neutral V-particles may equally often decay into a proton and antiproton:

$$V = p + \pi^-, \qquad V = \tilde{p} + \pi^+.$$

However, in this case we could use the virtual V-particles to construct a process of annihilation of two neutrons

$$n + n = p + \pi^- + n = V + n = \tilde{p} + \pi^+ + n = \tilde{p} + p = 2\gamma$$

or $2\pi^0$ or $\pi^+ + \pi^-$. Since such a process is not observed it is clear that we must prohibit the decay of V into $\tilde{p} + \pi^+$. Such a prohibition will obtain automatically if on the basis of the first reaction of decay of V into $p + \pi^-$ we assign to V a positive nuclear charge[2] $+y$. From the symmetry of particles and antiparticles it follows that together with the neutral V-particle with nuclear charge $+y$ a neutral anti-V-particle $\tilde{V}$ with a nuclear charge of $-y$ must exist which decays according to the equation

$$\tilde{V} = \tilde{p} + \pi^+.$$

If V indeed has a half-integer spin,[3] the existence of two different neutral particles V and $\tilde{V}$ is quite natural; a well known example is the neutron–antineutron neutral pair. By analogy we may expect, for example, that V will have a magnetic moment, with V and $\tilde{V}$ differing by the sign of the magnetic moment.

Let us now assume that the V-particles decay, emitting a neutrino:

$$V = p + \pi^- + \nu,$$

so that V can have a 0 spin. In this case again we should expect the existence of two different particles, V and $\tilde{V}$, with the decay equations

$$V = p + \pi^- + \nu, \qquad \tilde{V} = \tilde{p} + \pi^+ + \tilde{\nu}.$$

The basic purpose of the present note is in fact to point out that, in addition to mass, spin, electrical charge, magnetic moment and parity there is yet another intrinsic coordinate of a particle—its nuclear charge. Due to the presence of such a coordinate we have the possibility of existence of two particles with the same mass, zero spin, and identical electrical charge and magnetic moment, which nevertheless are strongly different, differing in the sign of the nuclear charge, i.e., differing in their decay products.

From the decay equations of V and $\tilde{V}$ under the assumption of emission of a neutrino also follows their identical parity: indeed, both theory and

[2]Here, as should be the case, the decay of V into a neutron and neutral meson is allowed, but the decay of V into several π-mesons and lighter particles is prohibited.

[3]A half-integer spin is obtained in the decay to $p + \pi^-$ or to $p + \pi^- + \pi^0$.

experiment show that π^+ and π^- have the same parity (are pseudoscalar). At the same time, in the case of spin 1/2 (the Dirac equation) particles and antiparticles $\bar{p}$ and p, ν and $\tilde{\nu}$) have opposite parity, as was first shown by V. B. Berestetskiĭ [4] using the example of positron and electron.

The fundamental possibility of existence of two particles which differ only in the sign of their nuclear charge may be confirmed by considering complex (composite) particles. Let us examine an atom of hydrogen H (consisting of p and e^-) in the S-state with total angular momentum $F = 0$, and an analogous object $\tilde{\mathrm{H}}$ made from an antiproton and a positron in the same state. From a fundamental point of view it is unimportant that the binding energy of H (or $\tilde{\mathrm{H}}$) is quite small (13.5 eV) or that an even smaller interaction by many orders of magnitude (hyperfine structure!) would transform H or $\tilde{\mathrm{H}}$ into the state $F = 1$ where they will differ in their magnetic moments. There are no strict criteria for distinguishing a composite particle and an elementary particle. Considered as elementary particles, H and $\tilde{\mathrm{H}}$ in the S-state with $F = 0$ have the same parity, the same mass, the same zero electrical charge, spin and magnetic moment; however, it is obvious that they differ: when put in contact with ordinary matter H is stable, while $\tilde{\mathrm{H}}$ annihilates, releasing about $2 \cdot 10^9$ eV.

Let us compare the energy needed for the formation of V and $\tilde{V}$ in the collision of two nucleons. In the first case according to the reaction $n + p = V + p$ the minimum energy needed in the center of mass frame is $M_V c^2 - M_n c^2 = (2200 - 1836)mc^2 = 364mc^2$; in the second case, with the reaction $n + p = \tilde{V} + p + 2n$, $M_\nu c^2 + M_n c^2 = (2200 + 1836)mc^2 = 4036mc^2$. The number of neutrons in the right-hand side of the equation describing the formation of V is determined by the condition of conservation of nuclear charge. Moving to a laboratory coordinate system, in which one of the nucleons is at rest, we find the threshold of formation (kinetic energy of the colliding nucleon) of $760mc^2 = 390$ MeV for V and $12550mc^2 = 6400$ MeV for $\tilde{V}$.

At the same time the number of $\tilde{V}$ which form in cosmic rays should be smaller than the number of V, not only because the threshold is higher and there are fewer primary particles capable of creating $\tilde{V}$, but also because, at a large collision energy which exceeds the threshold of $\tilde{V}$ formation, statistically more efficient processes of multiple π-meson formation compete strongly with the formation of $\tilde{V}$.

The ratio of the numbers of V and $\tilde{V}$ will be somewhat closer to 1 if these particles are not primary, but are formed in the decay of some very short-lived, even heavier primary particles.

Conclusions

1. We propose the concept of "nuclear charge" which allows us to conveniently formulate the well-known principle of the impossibility of nucleon annihilation.

2. In addition to the well-known neutral V-particle decaying with the formation of a proton, there must exist an anti-V-particle which decays with the formation of an antiproton; this is independent of whether the spin of the V-particle is integer or half-integer.

Note added in proof. A. Pais [6] cites a remark by I. R. Oppenheimer on the necessity of generalizing the law of conservation of the number of nucleons to nucleons and heavy V-particles. However, A. Pais subsequently assumes that heavy V-particles are fermions and does not consider the possibility noted by us of the existence of antiparticles for neutral Bose-particles.

Institute of Chemical Physics
USSR Academy of Sciences, Moscow

Received
July 14, 1952

REFERENCES

1. *Fretter W. B.*—Phys. Rev **85**, 773 (1952).
2. *Fretter W. B.*—Phys. Rev **83**, 1053 (1951).
3. *Armenteros R., Barket K. H.*, et al.—Philos. Mag. **42**, 1113 (1951).
4. *Berestetskiĭ V. B.*—ZhETF **21**, 93 (1951).
5. *Schiff L. I.*—Phys. Rev. **85**, 374 (1952).
6. *Pais A.*—Phys. Rev. **86**, 663 (1952).

Commentary

The nuclear charge is now called baryonic charge. For nucleons the baryonic charge was introduced by Stueckelberg, who called it the heavy charge and compared the conservation of this charge to a special global phase transformation of nucleons and antinucleons.[1] The baryonic charge was later discussed by E. Wigner.[2,3]

Conservation of baryonic charge was extended to hyperons in 1952 independently by Ya. B. in the present article and by A. Pais.[4] (It is the latter who coined the term baryon which combines nucleons and hyperons, as well as the term baryonic charge.)

The V-particles discussed in the article are Λ-hyperons. The hyperons with integer spin discussed in the article have not yet been discovered, but searches for

[1] *Stueckelberg E. C. G.*—Helv. Phys. Acta **11**, 299–328 (1938).

[2] *Wigner E. P.*—Proc. Amer. Philos. Soc. **93**, 521 (1949).

[3] *Wigner E. P.*—Proc. Nat. Acad. Sci. U.S. Phys. Sci. **38**, 449 (1952).

[4] *Pais A.*—Phys. Rev. **86**, 663–672 (1952).

these particles are being carried out (these particles should contain a new type of quarks—quarks with integer spin). The extension of the law of conservation of baryonic charge to hyperons played an important role in the creation of modern classification of hadrons. The question of possible non-conservation of baryonic charge has attracted particular attention in connection with the development of grand unification models which predict instability of the proton. Experimental investigations of proton decay have so far not revealed this phenomenon and have established a lower bound for the lifetime of a proton of the order of 10^{32} years.

The chief flaw of the present article, as in the works cited, consists in their failure to note the fundamental difference between electric charge on the one hand, and baryon or neutrino charges. However, this flaw is partially corrected in the next article (**10**). The electric charge characterizes the interaction with the electromagnetic field, and Maxwell's equations ensure charge conservation. Corresponding fields interacting with baryonic or neutrino charge do not exist. The non-conservation of electric charge could exist only if the electromagnetic field had a nonzero mass (the rest mass of photons) as examined in the paper by Ya. B. and L. B. Okun, "Paradoxes of the Unstable Electron."[5]

[5] *Okun L. B., Zeldovich Ya. B.*—Phys. Lett. B. **78**, 597–600 (1978).

10

On the Neutrino Charge of Elementary Particles*

At present proofs are gradually accumulating of the following suppositions:

1. Double β-decay with emission of two electrons and without emission of two neutrinos does not occur [1].
2. The spectrum of positrons which form in the decay of μ^+-mesons does not contradict the assumption that the two neutral particles which form in this decay are identical and obey the Pauli principle [2, 3].
3. Decay of μ^+ into e^+ and photons without emission of two neutral particles with spin 1/2 does not occur [4].

None of these suppositions taken separately can in any way be considered completely proved, and further experimental work is very desirable; at the same time, if the enumerated suppositions follow from a single principle, then it is quite timely to clearly formulate this principle and its consequences.

As will be shown, elementary particles may be characterized by a "neutrino charge"—a quantity which is conserved in all processes of mutual transformations of particles just as the electrical charge and the number of neutrons are conserved. The neutrino charges of μ^--meson and electron are apparently different, which leads to special, absolutely rigorous selection rules if the initial assumptions are correct.

It is commonly known that the absence of double β-decay indicates the necessity to distinguish neutrino (ν) and antineutrino ($\tilde{\nu}$), i.e., that the theory of Majorana in which real wave functions are possible is not applicable to the neutrino and, consequently, the neutrino is described by complex wave functions.

In the theory of a complex particle, naturally, birth or annihilation (without a change in the number of other particles participating in the process) can occur only for a particle-antiparticle pair ($\nu + \tilde{\nu}$), not pairs of identical particles ($\nu + \nu$) or ($\tilde{\nu} + \tilde{\nu}$). This assertion is exact and valid for processes of any order. A formal proof of the assertion consists in investigating the Lagrangian of the system, taking into account the transformation of one set of particles into another. Let there be processes of the

*Doklady Akademii Nauk SSSR **91** (6), 1317–1320 (1953).

type

$$a_1'A + b_1'B + \ldots = n_1\nu + a_1''A + b_1''B + \ldots,$$
$$a_2'A + b_2'B + \ldots = n_2\nu + a_2''A + b_2''B + \ldots, \tag{1}$$
$$\ldots\ldots\ldots\ldots\ldots\ldots\ldots\ldots\ldots$$

Here A, B are particle designations, a_1', a_1'', b_1', b_1'' are the number of particles of a given type participating in the reaction. If there is some number of such processes which are considered as original ones, then in a higher order of perturbation theory any processes are possible whose equations are linear combinations of the original processes. The assertion consists in the statement that no combination of processes will give $0 = n\nu$ ($n \neq 0$) or, equivalently, that it is impossible to realize the process

$$a_kA + b_kB + \ldots = n\nu + a_kA + b_kB + \ldots \tag{2}$$

with identical $a_k, b_k, \ldots$ in the left and right sides. Indeed, the Lagrangian of free neutrinos consists of terms each of which contains ψ_ν and ψ_ν^*, and is invariant under multiplication of ψ_ν by $e^{i\alpha}$.

The original processes (1) yield terms in the Lagrangian of the form

$$(\psi_A^*)^{a_1'}(\psi_B^*)^{b_1'}(\psi_\nu)^{n_1}(\psi_A)^{a_1''}(\psi_B)^{b_1''} + \text{complex conjug.} \tag{3}$$

Such terms prove invariant under multiplication of ψ_ν by $e^{i\alpha}$ if we simultaneously multiply ψ_A by $e^{i\alpha n_a}$, ψ_B by $e^{i\alpha n_b}$, etc.

The numbers n_a, n_b must satisfy the equation

$$a_1'n_a + b_1'n_b + \ldots = n + a_1''n_a + b_1''n_b + \ldots \tag{4}$$

If some linear combination were to give a process of type (2), then in the corresponding approximation there would appear a term of the form

$$(\psi_\nu)^n(\psi_A^*\psi_A)^{a_k}(\psi_B^*\psi_B)^{b_k}\ldots + \text{complex conjug.} \tag{5}$$

which cannot in any way be made invariant with respect to the multiplication of ψ_ν by $e^{i\alpha}$. From this, in fact, it is clear that the natural requirement of invariance of the Lagrangian with respect to multiplication of the complex ψ_ν by $e^{i\alpha}$ leads to exclusion of processes of type (2). The condition that no linear combination of the original processes (1) give a process of type (2) algebraically is the compatibility condition of the algebraic equations (4) obtained from (1). In other words, we may introduce the concept of neutrino charge (n-charge for short), with the n-charge of the particles A and B given by the numbers n_a, n_b; the n-charge is conserved in any elementary processes. We may introduce the density of the n-charge, calculated as the sum of the density of the neutrinos and the densities of the particles A and B taken with the coefficients n_a, n_b; the n-charge current may be defined analogously. The n-density and n-current exactly (in any approximation) satisfy the equation of continuity. Taking into account the conservation of charge and nuclear charge (number of nucleons) [5–8] the most general gauge group, with respect to which the Lagrangian is invariant, depends on three parameters: α, β, γ and consists in the multiplication of each ψ_A by $\exp(in_a\alpha + iz_a\beta + iy_a\gamma)$, where z_a, y_a and n_a are the electric, nuclear and neutrino charges of the particle A.

The above is a necessary and essentially obvious consequence of the assertion of exclusion of double β-decay.

Let us turn to exposing the n-charge of a μ-meson from the data on its decay. The most likely explanation of the absence of direct decay, $\mu^\pm = e^\pm + \gamma$, is the assumption that the n-charge of a μ^+-meson differs by two units from the n-charge of e^+, which in fact leads to the absolute prohibition of the unobserved process of decay without a neutrino; the same applies to μ^- and e^-. The assumption of a difference in the neutrino charges of μ^+ and e^+ is less certain than the very assumption of the existence and conservation of the neutrino charge. Moreover, it is difficult even to propose a practically realizable experiment which would confirm this assumption. If we assume identical n-charges for μ^+ and e^+ and we do not introduce any new neutral particles (μ°), the ordinary decay of μ^+ should be written as $\mu^+ = e^+ + \nu + \tilde{\nu}$; the process $\mu^+ = e^+ + \gamma$ is possible, but there is no means to calculate or compare with experiment the probability of decay to $e^+ + \gamma$ from the data on ordinary decay (see the diverging expression in [9]). Nevertheless, in what follows we shall assume that the n-charges of μ^+ and e^+ are different and shall investigate the consequences of this assumption.

The solution of equations (4), which determine the n-charges of the particles, is not unique. If we find some solution—the system of n-charges $n'_a, n'_b, \ldots$ which satisfy the conditions of n-charge conservation in reactions, then any linear combination of these n-charges with the electrical and nuclear charges of the particles

$$n''_a = n'_a + kz_a + by_a \tag{6}$$

will also be a quantity which is conserved in reactions. Consequently, the quantities n''_a may be as reasonably called n-charges as n'_a.

By definition, the n-charge of a neutrino $n_\nu \equiv 1$. We obtain one system if we take a neutron as having only a nuclear charge $y_N = 1$, $z_N = 0$, $n_N = 0$, and an electron as having only electrical charge $y_e = 0$, $z_e = -1$, $n_e = 0$. However, this system leads to a non-zero n-charge of p and $\pi^\pm$ and to an n-charge 2 of the μ-meson and is therefore inconvenient. The most convenient system is obtained if we take for the electron $n_{e^-} = +1$; here $n_{\mu^-} = -1$. For antiparticles the n-charge changes sign: the n-charges of P, N, $\pi^\pm$ are zero.

In the analogous processes of K-capture of an electron and K-capture of a μ^--meson different particles ν and $\tilde{\nu}$ should be released:

$$e^- + P = N + \nu, \tag{7}$$

$$\mu^- + P = N + \tilde{\nu}. \tag{8}$$

Another curious conclusion lies in the fact that, for example, a hypothetical particle (neutral) with an integer spin which would decay to $P + e^-$ could not decay to $P + \mu^-$; otherwise, the neutrino charge would not be conserved. In the same way the decay of one and the same neutral particle with a half-integer spin into two channels to $\pi^+ + \mu^-$ and to $\pi^+ + e^-$ is forbidden. However, decay of such a particle into the channels $\pi^+ + \mu^-$, $\pi^- + e^+$, $\pi^0 + \tilde{\nu}$, $\gamma + \tilde{\nu}$ is permitted by the laws of conservation of charges.

It should be particularly noted that the well-known experimental fact of decay of a charged π-meson with formation of a μ-meson and the absence of decay of π with formation of an electron is not related to prohibition by the n-charge; conservation of the n-charge allows both reactions:

$$\pi^+ = e^+ + \nu; \qquad \pi^- = e^- + \tilde{\nu}; \tag{9}$$

$$\pi^+ = \mu^+ + \tilde{\nu}; \qquad \pi^- = \mu^- + \nu; \tag{10}$$

The possibility of (9) on the basis of n-charge is directly established by the combination of the process $N = P + \pi^-$ and the β-process [9].

It is known that those variants of β-interaction [9]—vector V and tensor T—which are presently considered most likely [10, 11] lead to exclusion of β-decay of a pseudoscalar π-meson via polarization of the nucleon vacuum; this exclusion is related to properties of π with respect to spatial transformations (spin, parity). Therefore the exclusion may be overcome with the release of additional gamma-photons [12] with a corresponding decrease in the probability of the process. From the fact of decay to π and μ it definitely follows that the interaction (8) should take place along the pseudoscalar (P) and/or the pseudovector (A) variants.[1]

The operators V and T on the one hand, and P, A, S, y on the other belong to different groups with respect to charge conjugation (transition to antiparticles). It is possible that the membership of the two interactions (7) and (8), in which the particle and antiparticle, ν and $\tilde{\nu}$, respectively, are generated, to different groups is not accidental.

In conclusion we note that the concepts of neutrino and nuclear charge differ significantly from the electric charge. The electric charge is not only a quantity which is conserved in elementary processes (which unites it with nuclear and neutrino charge). The equations of the electromagnetic field are such that the electric charge may be measured without studying the atomic structure of the body. The law of conservation of electric charge was

[1]If we consider the original process to be (10) then for (8) we obtain directly, applying perturbation theory, the variant P. If we consider (8) to be the original then in order for the matrix element (10) to be nonzero, it is necessary that the interaction (8) contain P and/or A.

established by macroscopic experiments and entered into Maxwell's theory long before the creation of the electron theory.

The nuclear and neutrino charges cannot be measured macroscopically. In particular, the mass of bodies, due to the Einsteinian relation between mass and energy, is only an approximate (under ordinary conditions, in a substance which does not contain antinucleons, accurate to within about 1%) measure of the nuclear charge of a body.[2] The laws of conservation of nuclear and neutrino charge may only be established by studying elementary processes.

Are there still other charges in existence which are conserved, like the electric, nuclear, and neutrino charges? If any elementary particle may in the final analysis decay into some number of particles of three sorts—protons, electrons, and neutrinos (or the corresponding antiparticles)—then there is no place for a fourth conservation law. A fourth conservation law would lead to the existence of a fourth sort of stable elementary particles. Together with the majority of physicists we assumed that the neutral particles which form in the decay of π or the decay and capture of μ are the same neutrino and antineutrino which participate in the β-process, rejecting the μ^0-meson due to a complete lack of evidence. This assumption leads to the conclusion that there are only three conserved charges.

I take this opportunity to express my gratitude for discussions of the work to N. N. Bogolïubov, V. B. Berestetskiĭ, I. E. Tamm, and L. P. Feoktistov.

Institute of Chemical Physics
USSR Academy of Sciences, Moscow

Received
July 2, 1953

REFERENCES

1. *Fireman E. L., Schwarzer C.*—Phys. Rev. **86**, 451 (1952).
2. *Tiomno J., Wheeler J. A., Rau R. R.*—Rev. Mod. Phys. **21**, 144 (1949).
3. *Sagane R., Gardner W. L., Hubbard H. W.*—Phys. Rev. **82**, 557 (1951).
4. *Hincks E. P., Pontecorvo B.*—Phys. Rev. **73**, 257 (1948).
5. *Jordan P.*—Ztschr. Naturforsch. A **7**, 78, 701 (1952).
6. *Wigner E. P.*—Proc. Nat. Acad. Sci. U.S. Phys. Sci. **38**, 449 (1952).
7. *Zeldovich Ya. B.*—Dokl. AN SSSR **86**, 505 (1952).
8. *Oneda S.*—Progr. Theor. Phys. **8**, 255, 568 (1952).
9. *Vortuba V., Muzikar C.*—Phys. Rev. **82**, 99 (1951).
10. *Wu C. S.*—Physica **18**, 989 (1952).
11. *De Groot S. R., Tolhoek H. A.*—Physica **16**, 456 (1950).
12. *Ruderman M.*—Phys. Rev. **85**, 157 (1952).

[2]The relation of the nuclear charge to the meson field (6) is not necessary, but if it exists the meson field would not permit macroscopic measurement of the nuclear charge.

Commentary

The nuclear charge discussed in the article is currently called the leptonic charge. The law of conservation of the leptonic charge was formulated in 1953 independently of Ya. B. by E. Konopinskiĭ and H. M. Mahmoud[1] and by G. Marx[2] and played an important role in establishing the modern classification of elementary particles and interactions.

Today the statement of the problem has changed. The existence of not one, but three leptonic charges is assumed. The comment that there are no analogues to the electromagnetic field which would make conservation absolute remains valid. In new grand unification and supersymmetry theories the probability of processes which change leptonic charges is discussed.

The question whether in nature the law of conservation of leptonic charge is satisfied is today one of the central questions of elementary particle physics. Not a single case of violation of this law has been observed. However, as became clear after the discovery of nonconservation of parity, in a number of cases the appropriate exclusions may be caused not by the conservation of leptonic charge, but by longitudinal polarization of neutrinos. Searches for nonconservation of leptonic charge are being actively carried out in two directions: first, in the search for a neutrino-free double β-decay; second, in the search for the mass of neutrinos and the so-called neutrino oscillations whose possible existence was indicated by B. M. Pontecorvo in 1957.

[1] *Konopinskiĭ E., Mahmoud H. M.*—Phys. Rev. **92**, 1045–1049 (1953).
[2] *Marx G.*—Acta Phys. Acad. Sci. Hung. **3**, 55 (1953).

11

On the Decay of Charged π-Mesons*

Considering the π-meson as a composite particle [1] (nucleus), it is obvious that the interactions $N+\nu = P+e^-$ and $N+\overline{\nu} = P+\mu^-$ should be considered as primary (see [2, 3] on the insignificant difference here between ν and $\overline{\nu}$).

The decay, for example, of π^+ to $e^+ + \nu$ or $\mu^+ + \overline{\nu}$ in the terminology accepted in the theory of β-processes is classified as a 0–0-transition with a parity change (0–0, yes), since from an odd nucleus π^+ (besides light fermions) a vacuum is obtained, which is even by definition. The decay $\pi^+ = \pi^0 + e^+ + \nu$ is a 0–0-transition without a change in parity (0–0, no) between similar states with identical isotopic spin $T = 1$.

From this there immediately follows the well-known result [4, 5] that the S-, V-, and T-variants of β-interaction do not yield the decay $\pi^+ = e^+ + \nu$; $\pi^+ = \mu^+ + \overline{\nu}$. Decay is allowed only in the A- and P-variants. For electron β-interaction (unlike μ-meson) the variant A is presently excluded experimentally. However, recently it has been felt that the variant P enters into a linear combination of variants which describe β-decay [6, 7]. For this variant we give a calculation of the probability of decay below. Comparison of this calculation with experiment permits us to draw conclusions about the possible magnitude of the coefficient with which the variant P enters into the linear combination describing β-interaction of nucleons with light fermions.

For $\pi^+ = \pi^0 + e^+ + \nu$-decay by analogy with ordinary β-decay of nuclei we conclude that for the V-variant the square of the matrix element per nucleus $\mathcal{M}^2 = \langle 1 \rangle^2 = T(T+1) - T_z(T_z - 1) = 2$ so that the time cited for such a process $f\tau$ is the same as in other 0–0-transitions, e.g., $O_{14} \to N^*_{14}$ or $Cl_{34} \to S_{34}$, and is equal to $\sim 2600\,$s.

Substituting the energy of decay $\pi^+ \to \pi^0$, equal [8] to $10.6\, m_e c^2$, we find the maximum angular momentum of an electron $p_m = m_e c\sqrt{(10.6)^2 - 1} = 10.6\, m_e c$ (the correction for π^0 recoil does not exceed 2%) and by the expression f for large energies $f = 1/30(p_m/m_e c^5$ we obtain that $\tau^{-1}_{\pi^+\pi^0} = 1.7\,\mathrm{s}^{-1}$. This quantity should be compared with the probability of π–μ-decay [9] $\tau^{-1}_{\pi\mu} = 3.4 \cdot 10^7\,\mathrm{s}^{-1}$.

The decay process in which we are interested, $\pi^+ = \pi^0 + e^+ + \nu$, will occur on the average once in $2 \cdot 10^7$ ordinary decays $\pi^+ = \mu^+ + \overline{\nu}$. However, the small probability does not exclude the possibility of observation of the decay

*Doklady Akademii Nauk SSSR **97** (3), 421–424 (1954).

$\pi^+ = \pi^0 + e^+ + \nu$ in work with a powerful beam of artificial mesons; indeed the decay $\pi^+ = \pi^0 + e^+ + \nu$ in a moderator block (in which π^+ and e^+ are stopped) will give a quite characteristic picture: π^0 decays into 2 γ-quanta, each with an energy around 65 MeV, and e^+ will stop and annihilate, also giving 2 γ-quanta with 0.5 MeV each. In each pair the γ-quanta fly in opposite directions.

In ordinary nuclei in the non-relativistic approximation the variants S and V, which yield nucleon matrix elements $\langle\beta\rangle$ and $\langle 1\rangle$, respectively, give coincident results for allowed transitions. The π-meson cannot be studied non-relativistically since it consists of a nucleon and an anti-nucleon.

The generalized theorem of Farri shows that the decay $\pi^+ = \pi^0 + e^+ + \nu$ via vacuum polarization is allowed in the V-variant and forbidden in the S-variant. In calculating the number of odd operators one should account for the operator τ_3 in the expression relating a neutral meson with nucleons; therefore the odd operator V gives a non-zero matrix element, while the even S gives a zero one.

Above we have gathered together conclusions pertaining to meson decay which arise from spatial and charge symmetry. The considerations below require calculation of divergent integrals which in the first approximation of perturbation theory describe the decay of a π-meson in a pseudoscalar variant of the β-interaction (Fig. 1):

$$A = \frac{gg_p}{\pi i}\int[\gamma_5(\hat{p}+\hat{q}-M)^{-1}\gamma_5(\hat{p}-M)^{-1}]\,d^4p \tag{1}$$

and the decay $\pi^+ = \pi^0 + e^+ + \nu$ in the vector variant (Figs. 2 and 3):

$$B = B_1 + B_2,$$

$$B_1 = \frac{g^2 g_V}{\pi i}\int \tau_-\gamma_5(\hat{p}+\hat{q}-M)^{-1}\tau_+\hat{b}(\hat{p}+\hat{q}-M)^{-1}\tau_3\gamma_5(\hat{p}-M)^{-1}\,d^4p, \tag{2}$$

$$B_2 = \frac{g^2 g_V}{\pi i}\int \tau_-\gamma_5(\hat{p}+\hat{q}-M)^{-1}\tau_3\gamma_5(\hat{p}-M)^{-1}\tau_+\hat{b}(\hat{p}+\hat{q}-M)^{-1}\,d^4p.$$

In the last case B_1 corresponds to Fig. 2, B_2 to Fig. 3; the operators τ act on the isotopic variables of nucleons, $\hat{b} = (\psi^+_{e^+}\gamma_\rho\psi_\nu)\gamma_\rho$.

Regarding the first process the literature has the following comments: for identical values of the integral (1), the ratio of the probability of the decay $\pi^+ \rightleftharpoons e^+ + \nu$ to the probability of the decay $\pi^+ = \mu^+ + \nu$ is

$$\frac{\tau_{\pi e}^{-1}}{\tau_{\pi\mu}^{-1}} = 6\left(\frac{g_{P_e}}{g_{P_\mu}}\right)^2 \tag{3}$$

The first index on g denotes the pseudoscalar variant of the interaction, the second (the index e or μ)—to which particles ($PNe\nu$ or $PN\mu\nu$) the constant g applies. The factor 6 depends on the greater release of energy in e^+-decay, which gives a larger phase volume, and on the somewhat larger

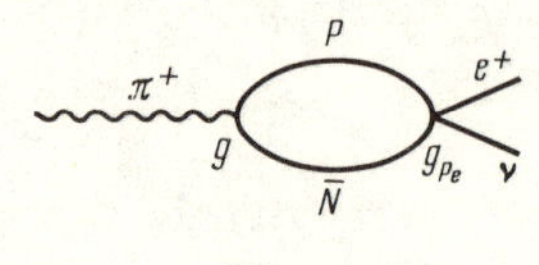

Fig. 1

Fig. 2 *Fig. 3*

matrix element for light particles. Since in experiment [10] $\tau_{\pi e}^{-1}/\tau_{\pi\mu}^{-1} \leq 7 \cdot 10^{-4}$, then $g_{P_e} \leq 0.01 g_{P_\mu}$.

Consequently, the idea of "universal fermi-interaction" cannot be understood simply to mean equality of all coefficients in all invariants for $PN\mu\nu$- and $PNe\nu$-interactions.

For the absolute calculation J. Steinberger [4] carried out the regularization formally by constructing analogously to A expressions with auxiliary M_i and coefficients C_i together with a number of conditions: $\sum C_i M_i = 0$, $\sum C_i M_i^2 = 0$, etc. Then he obtains (taking $g_P = g_{T_e}$) the decay probability for π–μ as $\tau_{\pi\mu}^{-1} = 0.2g^2\mathrm{s}^{-1}$ and of π–e decay as $\tau_{\pi e}^{-1} = 0.7g^2\mathrm{s}^{-1}$. The first value, even for $g^2 = 200$ is 10 times smaller than the experimental value [11]. Hence J. Steinberger concludes that π–μ-decay is not described by vacuum polarization and the $PN\mu\nu$-interaction of 4 fermions. He considers the interaction $\pi = \mu + \nu$ to be primary, and the interaction $PN\mu\nu$ as occurring through the virtual generation of π-mesons, for example by the scheme: $\mu^- + P = \pi^- + \nu + P = N + \nu$. Such a conclusions in even greater measure contradicts the tempting idea of a universal interaction of 4 fermions [13] since the electron β-process is certainly not described by schemes with virtual π-mesons.

We apply to the calculation of A the definition of the integral I, given in [14, formulas (1) and (2)]. In essence, by so doing the assumption is made that the removal of the divergence and the obtaining of the equality $I = m_\pi^2$ (2) occurs due to a change in the meaning of d^4p or the propagation factors, but not due to a change in the operator γ_5 in the interaction of the nucleon with the meson; otherwise the analogy between the expressions A

and I would be violated.

For the probability of electron decay we obtain (the expression is independent, practically speaking, of m_e since the electron is relativistic)

$$\tau_{\pi e}^{-1} = 0.04 m_\pi^5 \hbar^{-7} c^{-4} g_{P_e}^2 (\hbar c/g^2) \tag{4}$$

and for β-decay (the ratio m_μ/m_π has entered into the numerical coefficient)

$$\tau_{\pi\mu}^{-1} = 0.007 m_\pi^5 \hbar^{-7} c^{-4} g_{P_\mu} (\hbar c/g^2). \tag{5}$$

For comparison we write out the probability of ordinary β-decay:

$$\tau_\beta^{-1} = 0.016 m_e^5 \hbar^{-7} c^{-4} g_\beta^2 \mathcal{M}^2 f, \tag{6}$$

where $\mathcal{M}$ is the matrix element taken for the nucleus and $f\tau$ is the effective time.

The maximum value of g, which gives the minimum probability of decay of π, encountered in the literature [11] is: $g^2/4\pi\hbar c = 15$, $g^2\hbar c = 180$. The upper limit established by experiment [10] is: $\tau_{\pi e}^{-1} \leq 7 \cdot 10^{-4} \tau_{\pi\mu}^{-1} = 7 \cdot 10^{-4} \cdot 3.4 \cdot 10^7 = 2.5 \cdot 10^4 \mathrm{s}^{-1}$, whence by formula (4)

$$g_{P_e} < 0.08 g_\beta = 0.08 g_{Te}. \tag{7}$$

The result contradicts the estimates [7, 12], which are based on an analysis of the Ra E spectrum (the transition $0-0$, yes),[1] according to which $g_{Pe} \cong g_\beta$.

For the interaction of μ-mesons with nucleons, assuming that the interaction is described by the same variants as ordinary β-interaction, we obtain, as is well-known, $g_\mu \cong g_\beta$, where $g_\mu = g_{S\mu}$ or $g_{V\mu}$, or $g_{T\mu}$, or $g_{A\mu}$.

Substituting the experimental value of $\tau_{\pi\mu}$ into (5), we find that $g_{P\mu} = 9g_\mu$. If the effective g of the interaction of π-mesons with nucleons is smaller, the $g_{P\mu}$ is accordingly also smaller.

If only the P-variant acts in the interaction of μ with nucleons, then the chief role is played by simply forbidden transitions. For K-capture of a μ-meson at rest:

$$g_{P_\mu}(\psi_\nu^* \beta\gamma_5 \psi_{\mu^-}) \int e^{i\mathbf{q}_\nu \mathbf{r}} \psi_N^* \beta\gamma_5 \psi_P dV = g_{P_\mu} \frac{\mathbf{q}_\nu}{2Mc} (\psi_\nu^* \beta\gamma_5 \psi_{\mu^-}) \int \psi_N^* \sigma \psi_P dV \tag{8}$$

(cf. proofs of the equivalence of pseudoscalar and pseudovector interaction).

The matrix element $(\psi_\nu \beta\gamma_5 \psi_{\mu^-})$ is of the same order as the matrix elements for light particles in other variants.

Consequently, in the pure P-variant of $PN\mu\nu$-interaction the role of the effective constant is played by $g_{P\mu} q_\nu/2Mc \cong 0.05 g_{P\mu}$. Equating it to $g_\mu = g_\beta$, we obtain (for $g^2/4\pi\hbar c = 15$) the value calculated from (5), $\tau_{\pi\mu} = 2 \cdot 10^8$, which is 3–4 times greater than the experimental value. Therefore, the decay $\pi = \mu + \nu$ may be described *via* vacuum polarization and $PN\mu\nu$-interaction, contrary to J. Steinberger's opinion.

[1] *Note in proof*: Lately this proposition has been subject to doubt (see: *King R. W., Peaslee D. C.*—Phys. Rev. **94**, 795 (1954).)

The decay $\pi^+ = \pi^0 + e^+ + \nu$ has not been investigated in the literature. In order to obtain the selection rules associated with charge symmetry and the properties of the isotopic operator τ_3, we shall make use of formulas (2), in which all the isotopic operators are written out (for $\pi = e + \nu$ or $\pi = \mu + \nu$, writing them out yields only trivial results and so they are not included).

First performing the summation over isotopic variables, we note that if in B_1: $\tau_-\tau_+\tau_3 = +1/\sqrt{2}$, then in B_2: $\tau_-\tau_3\tau_+ = -1/\sqrt{2}$. Consequently, the spatial parts of B_1 and B_2 enter with opposite signs; in compiling the expressions B_1 and B_2 we have already neglected in advance the difference in the masses of P and N and also (in the propagation factors) the 4-momentum of the positron–neutrino pair, which is equal to the difference of the 4-momentum of π^+ and of π^0, i.e., $m_{\pi^+} - m_{\pi^0}$. This difference, on which the possibility of the decay $\pi^+ = \pi^0 + e^+ + \nu$ depends, is accounted for in the calculation of the phase volume so that the probability $\tau^{-1}_{\pi^+\pi^0} \sim (m_{\pi^+} - m_{\pi^0})^{-5}$.

In the vector variant of β-interaction the expression entering into B_1 is exactly similar to the expression for interaction with an electromagnetic field (formulas (3)–(5) of [14]), so that we find

$$B_1 = g_{Ve}\sqrt{2}(bq), \tag{9}$$

where b is the 4-vector of light particles with components $b_\rho = \overline{\psi}_e\gamma_\rho\psi_\nu$ and q is the angular momentum of the decaying meson. We further note that by replacement of the integration variable $p' = p + q$ we may establish the equality

$$B_2(b, q, M) = B_1(b, -q, M), \tag{10}$$

whence

$$B_1 = -B_2, \qquad B = g_{Ve}2\sqrt{2}(bq). \tag{11}$$

Taking into account the normalization of wave functions of the bosons π^+ and π^0 we obtain finally the square of the transition matrix element:

$$\mathcal{M}_V^2 = 2g_{Ve}^2|\overline{\psi}_{e^+}\gamma_4\psi_\nu|^2 = 2g_{Ve}^2|\psi^*_{e^+}\psi_\nu|^2, \tag{12}$$

which precisely corresponds to the assumption stated above by analogy with mirror nuclei that $\mathcal{M}^2 = 2$.

In the case of scalar interaction in B_1 and B_2 instead of the operator $\hat{b}$ there is a number (scalar)

$$c = (\overline{\psi}_{e^+} \cdot 1 \cdot \psi_\nu) = (\psi^*_{e^+}\beta\psi_\nu). \tag{13}$$

Considering $B_1 = B_2(c, q, M)$ and $B_2(c, q, M) = B_2(c, -q, M)$ by symmetry we see that $B_1 = B_2$, $B = B_1 - B_2 = 0$, which is indeed what follows from Farri's theorem.

I take this opportunity to express my gratitude to G. M. Gandelman, L. D. Landau and L. P. Feoktistov for discussions and help in this work.

Received
May 26, 1954

REFERENCES

1. *Fermi E., Yang C. N.*—Phys. Rev. **76**, 1739 (1949).
2. *Reines F., Cowan C. L.*—Phys. Rev. **92**, 830 (1953).
3. *Zeldovich Ya. B.*—Dokl. AN SSSR **91**, 1317 (1953).
4. *Steinberger J.*—Phys. Rev. **76**, 1180 (1949).
5. *Ruderman M., Finkelstein R.*—Phys. Rev. **76**, 1458 (1949).
6. *Marshak R. E., Petschek A. G.*—Phys. Rev. **85**, 698 (1952).
7. *Peaslee D. C.*—Phys. Rev. **91**, 1447 (1953).
8. *Panofsky W. K. H., Aamodt R. L., Hadley J.*—Phys. Rev. **81**, 565 (1951).
9. *Lederman L. M., Booth E. T., Byfield H., Kessler J.*—Phys. Rev. **83**, 685 (1951).
10. *Friedman H. L., Rainwater J.*—Phys. Rev. **84**, 684 (1951).
11. *Brueckner K. A., Watson K. M.*—Phys. Rev. **92**, 1023 (1953).
12. *Ruderman M.*—Phys. Rev. **89**, 1227 (1953).
13. *Finkelstein R., Kaus P.*—Phys. Rev. **92**, 1316 (1953).
14. *Zeldovich Ya. B.*—Dokl. AN SSSR **97** 2 (1954).

Commentary

When reading this article it should be kept in mind that until 1957 it was believed that the V- and A-variants in β-decay were experimentally excluded. (The experiments on which this opinion was based, as it subsequently turned out, had been erroneous). Thus the very consideration of a vector weak current in the paper is a good example of theoretical intuition.

The article gives for the first time the value of the square of the matrix element of β-decay of a π^+-meson, $\pi^+ \to \pi^0 e^+ \nu$, for the vector variant:

$$T(T+1) - T_z(T_z - 1) = 2$$

and investigates how this number is reproduced in the composite model of a π-meson proposed by Fermi and Yang. In this model π-mesons are considered to be bound states of a nucleon and antinucleon. It should be noted that not long before this article, Ya. B. Zeldovich used the Fermi–Yang model to estimate the difference in masses of π^+- and π^0-mesons[1]. The basic idea of this estimate was that the Coulomb attraction of a proton and antiproton is the reason for the decrease in the mass of π^0 compared to π^- and π^+. In the modern quark language the same result is obtain due to the fact that in π^0-meson quarks and antiquarks have charges of opposite sign, whereas in the π^+-meson they are of the same sign.

The problem of β-decay of a π^+-meson acquired particular importance after the development in 1957 by R. Feynman, M. Gell-Man, G. Marshak, and E. Sudarshan of the universal $V-A$ theory of weak interaction (see the Commentary to the next article).

The first attempt to measure the λ-dependent probability of the decay $\pi^+ \to \pi^0 e^+ \nu$ was undertaken in Dubna, in the Laboratory of Nuclear Problems of the

[1] *Zeldovich Ya. B.*—Dokl. AN SSSR **97** 2, 225–228 (1954).

Joint Institute of Nuclear Research. One event, $\lambda < 7 \cdot 10^{-8}$ (1961); two events, $\lambda < 1.5 \cdot 10^{-8}$ (1962); four events, $\lambda = (1^{+1.5}_{-0.5}) \cdot 10^{-8}$ (1962); 43 events, $\lambda = (1.1+0.2) \cdot 10^{-8}$ (1964).[2] A CERN group reported a measurement of the probability of the decay $\pi^+ \to \pi^0 e^+ \nu$ (14±2 events, $\lambda = (1.7 \pm 0.5) \cdot 10^{-8}$) in 1962.[3] The last work was on the measurement of the decay[4] $\pi^+ \to \overline{\pi}^0 e^+ \nu$. The result is based on the observation of 1235 ± 36 events, $\lambda = (1.036 \pm 0.039) \cdot 10^{-8}$. The agreement between theory and experiment is characterized by the ratio $(\lambda_{\text{theor}} - \lambda_{\text{exp}})/\lambda_{\text{theor}} = (1.2 \pm 3.7)\%$.

According to the well-known tables in *Particle Properties* (1982), the relative probability of the decay $\pi^+ \to \pi^0 e^+ \nu$ is $(1.02 \pm 0.07) \cdot 10^{-8}$ in complete agreement with the theory. The difference between this number and the numerical estimate contained in this article ($\lambda = 5 \cdot 10^{-8}$) is related to the fact that in the article the mass difference of π^+- and π^0-mesons was taken to be the quantity known at the time, 5.4 MeV; the current value, however, is 4.6 MeV. Also changed are the experimental values of the lifetime of a π^+-meson and the quantity f^+ for the nucleus of ^{14}O. In addition, the phase volume was not completely correctly estimated and a ln 2 was not accounted for in the translation from the half-life to the lifetime.

[2] *Dunaĭtsev A. F., Petrukhin V. I., Prokoshkin Yu. D., Rykalin V. I.*—ZhETF **42**, 632–637, 1421–1424 (1962); Phys. Lett. **1**, 138–140 (1962).

[3] *Depommier P., Heintze J., Mukhin A.*, et al.—In: Proc. 1962 Int. Conf. on High-Energy Phys. at CERN. Geneva, p. 441 (1962).

[4] *Farlana W. K. M., Auerbach L. B., Gaille F. C.*, et al.—Phys. Rev. Lett. **51**, 249–252 (1983)

12

Meson Correction in the Theory of Beta Decay*

With S. S. Gershtein

In a recent note, Finkelstein and Moszkowski [1] discuss the effect of strong coupling between nucleons and pions on the β-decay of nucleons.

Using the language of Feynman diagrams the authors of [1] consider, in addition to the fundamental process (Fig. 1*a*), another process involving the virtual emission of one π^0 meson (Fig. 1*b*). The calculation is carried out on the basis of the hypotheses under which Chew [2] discusses nuclear forces and the creation and scattering of mesons, and Friedman [3] discusses the anomalous magnetic moment of the nucleon: the nucleon is assumed to be infinitely heavy, and integrals over the momenta of virtual mesons are cut off at a specified value $p_{\max}$. From comparison with experiment it is found that $p_{\max}$ is close to Mc. A system with charge symmetry is considered so that the operator τ_3 enters into the expression for the coupling of nucleons to π^0.

Our notation will be very similar to that of Sachs [4]. Let P_1 be the probability that there is a virtual π^0 meson around the nucleon, compared with the probability that the nucleon is "bare," i.e., has no mesons around it.[1] The beta-decay coupling constants of a bare nucleon are denoted by g'_F (Fermi interaction with S and V interaction types) and g'_T (Gamow-Teller interaction with T and A interaction types). In the case of a real nucleon surrounded by a meson cloud, the same constants as obtained experimentally are denoted by g_F and g_T, and the ratio g_T^2/g_F^2 equals R; from experiment [5] $R = 1.75$, i.e., $R > 1$. The results of Finkelstein and Moszkowski [1] then appear as follows:

$$g_F = g'_F(1 - P_1); \qquad g_T = g'_T\left(1 + \frac{1}{3}P_1\right).$$

$$\frac{g'_T}{g'_F} = \frac{1 - P_1}{1 + \frac{1}{3}P_1}\frac{g_T}{g_F} = \frac{1 - P_1}{1 + \frac{1}{3}P_1}\sqrt{R} \cong 1.$$

While agreeing with their principal conclusion, which is that $g'_T/g'_F \cong 1$, we wish to make a few comments regarding the calculation.

1. The calculation does not account for renormalization of the nucleon wave function as a result of the possibility of creating virtual mesons. By

*Zhurnal eksperimentalnoĭ i teoreticheskoĭ fiziki **29** (5), 698–699 (1955).

[1] In the notation of [1] $P_1 = 3\delta$.

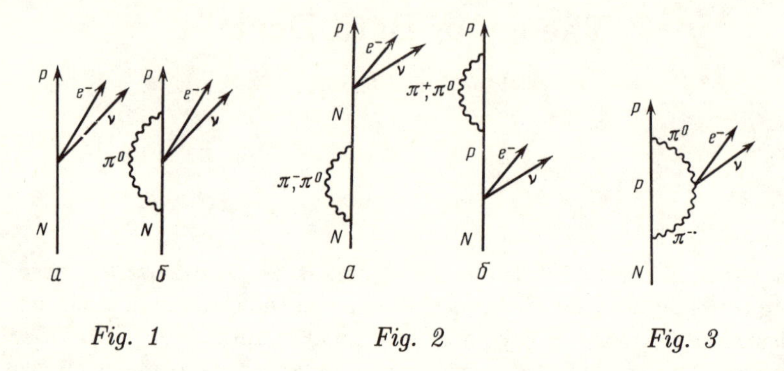

Fig. 1 *Fig. 2* *Fig. 3*

Feynman's [6] method the renormalization is made by adding the self-energy contribution to the free ends of the diagrams (Figs. 2*a* and 2*b*). In these diagrams it is necessary to consider not only neutral but also charged pions (a charged meson in the vertex part of Fig. 1*b* obviously gives a vanishing result). It is easily seen that on the basis of the hypotheses which were adopted by Finkelstein and Moszkowski, and taking renormalization into account, the correct result is

$$g_F = g'_F \frac{1 - P_1}{1 + 3P_1}; \qquad g_T = g'_T \frac{1 + \frac{1}{3}P_1}{1 + 3P_1}.$$

Thus the correction does not affect the ratio g'_T/g'_F. However, the correction may be significant when comparing the absolute value of $g_\beta = g'_F = g'_T$ with the coupling constant g_μ that determines the probability of $\mu^\pm = e^\pm + 2\nu$ decay; Finkelstein and Moszkowski [1] had this comparison in mind. A numerical result in accordance with both the latter authors and Chew [2] is that $1 + 3P_1 = 1.7$.

2. Finkelstein and Moszkowski [1] do not consider the possibility of β-conversion of mesons, i.e., processes such as $\pi^\pm = \pi^0 + e^\pm + \nu$, which may be represented by diagrams like Fig. 3.

If the probability of $\pi^\pm \to \pi^0$ β-decay is the same as for mirror nuclei, the contribution from the process represented by the diagram of Fig. 3 is of the same order as the meson correction in Fig. 1*b*; the absence of experimental indication of the $\pi^\pm = \pi^0 + e^\pm + \nu$ process does not contradict the hypothesis, since as a result of strong competition from $\pi^\pm = \mu^\pm + \nu$ decay, the former process may occur in only the small fraction 10^{-7} of the decays of free charged pions.

The question of the β-decay of pions has been thoroughly examined by one of the authors of the present note [7]. It was shown that the S and T beta-interaction types do not give β-decay of a pion in the approximation to which the theory of isotopic invariance is valid. Thus the possibility of neglecting diagrams such as Fig. 3, as is done by Finkelstein and Moszkowski [1], is actually associated with the representation of the Fermi beta interaction by the scalar S rather than by the vector type V, which is in accordance with the latest experimental findings [8].

It is of no practical significance but only of methodological interest that in the case of the vector interaction type V we should expect the equality

$$g_{F(V)} \equiv g'_{F(V)}$$

to any order of the meson-nucleon coupling constant, taking nucleon recoil into account and allowing also for interaction of the nucleon with, e.g., the electromagnetic field, etc. This result might be foreseen by analogy with Ward's identity for the interaction of a charged particle with the electromagnetic field; in this case virtual processes involving particles (self-energy and vertex parts) do not lead to charge renormalization of the particle.

3. We have calculated the meson corrections by invariant perturbation theory, using pseudoscalar coupling between pion and nucleon (coupling constant g).

In the expression for the self-energy and vertex parts a convergence factor $C(k^2)$ was introduced, where k is the momentum 4-vector of a virtual meson:

$$\Sigma = \int \tau_i \gamma_5 (\hat{p} - \hat{k} - m)^{-1} \gamma_5 \tau_i (k^2 - \mu^2)^{-1} C(k^2) d^4k,$$

$$\Gamma_0 = \int \tau_i \gamma_5 (\hat{p} - \hat{k} - m)^{-1} \tau_+ \hat{O} (\hat{p} - \hat{k} - m)^{-1} \gamma_5 \tau_i (k^2 - \mu^2)^{-1} C(k^2) d^4k,$$

$$C(k^2) = \frac{\lambda^2}{\lambda^2 - k^2}.$$

In addition to integration over momentum space (d^4k), a summation was carried out over the index i of meson isotopic spin. The beta process operator was represented as the product of the operator τ_+ which transforms a neutron into a proton, and the operator $\hat{O}$ which consists of the γ matrix ($\hat{O} = 1$ for S; $\hat{O} = \gamma_i \gamma_k$ for the T interaction type).

The nucleon mass renormalization term was calculated from Σ in the usual way: m is the mass of the nucleon, μ is the mass of the meson, and terms of the order μ/m are neglected. Taking renormalization of the wave functions into account, the result becomes

$$g_{F(S)} = g'_{F(S)} \left[1 - \frac{g^2}{32\pi^2} \left(5 \ln \left(\frac{\lambda^2}{m^2} \right) - \frac{1}{2} \right) \right],$$

$$g_{GT(T)} = g'_{GT(T)} \left[1 - \frac{g^2}{32\pi^2} \left(3 \ln \left(\frac{\lambda^2}{m^2} \right) + \frac{1}{2} \right) \right].$$

For small g and large λ a relativistic calculation also gives a decrease of g'_{GT}/g'_F compared with g_{GT}/g_F.

In the present state of the theory of interactions of pions with nucleons one cannot give preference to a relativistic perturbation theory calculation over the calculations of Finkelstein and Moszkowski [1], who employ coupling constants derived from experimental data.

Received
June 8, 1955

REFERENCES

1. *Finkelstein R. J., Moszkowski S. A.*—Phys. Rev. **95**, 1695 (1954).
2. *Chew G. F.*—Phys. Rev. **95**, 285 1669 (1954).
3. *Friedman M. H.*—Phys. Rev. **97**, 1123 (1955).
4. *Sachs R. G.*—Phys. Rev. **87**, 1100 (1952).
5. *Gerhardt R.*—Phys. Rev. **95**, 288 (1954).
6. *Feynman R. P.*—In: Symposium on Recent Developments in Quantum Electrodynamics (1954).
7. *Zeldovich Ya. B.*—Dokl. AN SSSR **97**, 421 (1954).
8. *Alforda P., Hamilton D.*—Phys. Rev. **95**, 1354 (1954).

Commentary

This paper formulates the important idea that a weak charged vector hadron current should be conserved, and that as a result of this conservation the effective vector constant in the β-decay of a neutron does not change under the effect of virtual strong interactions. The constants of β-decay of a "bare" nucleon and of a nucleon interacting with a π-meson field should be equal. The relation between this equality and the value of the amplitude of β-decay of a π-meson is investigated (see the Commentary to the preceding article). An analogy is drawn between the conservation of weak vector current and the conservation of electromagnetic current.

The ideas of this and the preceding articles were newly discovered by Feynman and Gell-Mann in their article on the universal $V - A$-interaction.[1] This article does not contain a reference to S. S. Gershtein and Ya. B.'s work, but in the next paper by M. Gell-Mann on the verification of the nature of the vector interaction in β-decay[2] there is such a reference. Since that time, in the literature on the physics of elementary particles the hypothesis of conserved vector current is firmly tied to the names of S. S. Gershtein and Ya. B.

The hypothesis of conserved vector current and the analogy between the weak and electromagnetic currents has played an important role in the development of the modern conception of the weak interaction, and of the unified gauge theory of the electromagnetic and weak interactions in particular. In essence, it is precisely the conservation of vector current that forced the theoreticians to take up the description of the weak interactions on the basis of the Yang–Mills theory.

[1] *Feynman R. P., Gell-Mann M.*—Phys. Rev. **109**, 193–198 (1958).
[2] *Gell-Mann M.*—Phys. Rev. **111**, 362–365 (1958).

13

Determination of the Limits of Applicability of Quantum Electrodynamics by Measurement of the Electron Magnetic Moment*

With G. M. Gandelman

It is presently becoming clear that modern quantum electrodynamics with point-like interactions, including renormalization of mass and charge, is still not a completely closed theory.

L. D. Landau and I. Ya. Pomeranchuk [1] approach the point interaction as the limit of a spread-out interaction with a small radius a; to the radius of spreading a there corresponds a momentum $\Lambda = \hbar a^{-1}$—the spreading out indicates a change (weakening) of the interaction in the region of the radius a around an electron and a decrease of the probability of all processes related to the transfer of momentum on the order of or greater than Λ. In the natural system of units ($\hbar = c = 1$) Λ is expressed in units of mass; the spreading is characterized by a dimensionless quantity $(am_e)^{-1} = \Lambda/m_e = l$, where m_e is the mass of the electron.

Consideration of the electrodynamics leads to the inequality

$$l < \exp\left(\frac{\hbar c}{e^2}\frac{3\pi}{2\nu}\right),$$

where ν is the number of different types of elementary particles. For $\nu = 1$ we obtain $l < 10^{260}$; for $\nu = 10$, $l < 10^{26}$.

The renormalization method allows us to prove that for momentum $p \ll lm_e$ all physical results expressed in terms of the experimentally determined mass m_e and electron charge e contain a dependence on l only in the form of additional terms l^{-1}, l^{-2}, $\frac{p}{m_e}l^{-1}$, $\frac{p}{m_e}l^{-2}$ and, consequently, in the limit of large l, are independent of l.

Since Landau's considerations show that we may not expect unlimited applicability of electrodynamics, the question arises, what in fact is known about the lower bound on l, i.e., about the region of momenta and lengths in which the applicability of electrodynamics has been established experimentally, and about the methods of determining this bound.

*Doklady Akademii Nauk SSSR **105** (3), 445–447 (1955).

At first glance the most natural way appears to be a direct investigation of interactions of energetic γ-photons with electrons. However, in the relativistic region it is known that in the collision of a quantum having momentum Am_ec and energy $Am_ec^2 = 0.5$ MeV, $A \gg 1$ with an electron at rest, in the frame of reference in which the total momentum is zero, the momenta of the electron and quantum are equal to $m_ec\sqrt{A/2}$. In order for the electron and quantum to have oppositely directed momenta, equal to lm_ec in a system in which the total momentum is zero, it is necessary in the laboratory system where the electron is at rest to take a quantum with energy $2l^2m_ec^2$, which makes investigation of the region $l > 100$ quite difficult. The same applies to the interaction of a fast electron or positron with an electron at rest. Colliding beam experiments are practically impossible because of the low beam intensity.

In principle, the conservation laws allow large transfers of momentum in the collision of photons with heavy nuclei. However, as noted by L. D. Landau and I. Ya. Pomeranchuk [1], in fact the pair production by hard photons in the field of a nucleus in the vast majority of cases occurs far from the nucleus, with small momentum transfer (on the order of m_ec) in the rest frame of the pair.

In considering those rare events when the momentum transfer is large, the difference between the coulomb field of the nucleus and the point charge field is crucial.

A change in the interaction at small distances should manifest itself in a displacement of levels. Let us consider the S-level of a hydrogen atom. If we replace the coulomb potential of the nucleus by a constant (or double it) at $r < \hbar/lm_ec$, the level displacement will be of order $\mathrm{Ry}\,(\alpha^2/l^2) \cong 10^6/l^2$ MHz. More likely, however, is the following procedure. We expand the coulomb potential of the nucleus in a Fourier integral, then in this integral we delete components with a wave vector larger than $lm_ec/\hbar$, and find the change in energy of the electron for this change in potential, calculated from the fixed wave function.[1] We obtain $\Delta E = \mathrm{Ry}\,(\alpha^3/l^3) \approx 10^4/l^3$ MHz. For $l = 100$ this correction is 10^{-2} MHz, which is within the limits of experimental accuracy (limited by the width of the level $2P_{1/2}$). The influence of the structure of the proton itself, i.e., of effects related to virtual generation of π-mesons [2], also gives a displacement of the S-level by 10^{-1}–10^{-2} MHz; this quantity cannot, obviously, be calculated exactly.

In our opinion, the most promising method of determining the limit of

[1]For the calculation it is also convenient to make the Fourier expansion of $\rho(r) = |\psi(r)|^2$ and to replace the integration over space $W = \int V(r)\rho(r)\,d\omega$, where V is the potential, by integration over the wave vector of the Fourier-transform. Then for Coulomb's law $W = e^2 \int_0^\infty \frac{dk}{(1+4a_0^2k^2)^2}$, where a_0 is the radius of the Bohr orbit, and for a modified law of interaction $W = e^2 \int_0^\Lambda \frac{dk}{(1+4a_0^2k^2)^2}$ the energy change is equal to $\Delta W = e^2 \int_\Lambda^\infty \frac{dk}{(1+4a_0^2k^2)^2}$. All expressions are given with accuracy to within a numerical coefficient.

applicability of electrodynamics is the exact determination of the magnetic moment of an electron.

A calculation by perturbation theory up to fourth order yields

$$\mu = \mu_0(1 + \alpha/2\pi - 2.973\alpha^2/\pi^2) = 1.001145356 \pm 0.000000013$$

(the numerical value and error bars are cited from [3]).

The first term $\alpha/2\pi$ has the simple meaning of a correction which arises from the fact that the electron emits, then absorbs virtual photons. A change in the magnetic moment is expressed as a sum over all values of the momentum of the virtual photon k; this sum converges. The assumption of the existence of a limit of applicability of electrodynamics means that one only need sum over photons with momentum less than $\Lambda = lm_e c$. Such a calculation was performed by us in practice using the Feynman convergence factor, i.e., by replacing in the integral of the photon propagation function $1/k^2$ by the function $(1/k^2)[\Lambda^2/(\Lambda^2 - k^2)]$, which ensures Lorentz-invariance of the result. Unlike the usual calculation, we do not neglect terms which contain Λ in the denominator.

The result has the form

$$\mu = \mu_0 \left\{ 1 + \frac{\alpha}{2\pi}\left(1 + \frac{3}{l^2}\ln l^2\right) + \dots \right\}.$$

At present the best experimental value [4] is $\mu = 1.001148 \pm 0.000006$. Comparing this value with the formula, we obtain $l > 70$. The accuracy with which the theoretical value of μ is presently known allows us to discover the effect of l up to 500–700. Further increase in l requires not only greater accuracy in experiment (for $l = 1000$ the correction is equal to $5 \cdot 10^{-8}\mu_0$), but also accounting of the subsequent terms l^6, l^8 (i.e., α^3, α^4) in the theoretical expression for μ/μ_0.

The essential limitation of the method is related to the fact that the electron interacts not only with the electromagnetic field, but also with the field of nucleons, neutrinos and μ-mesons (this interaction is manifested in the β-decay of a neutron and μ-meson). The β-interaction–dependent correction in the magnetic moment diverges in the existing theory. Natural assumptions give by order of magnitude

$$(\Delta\mu)_\beta = \mu_0 g_\beta^2 \frac{M^4c^2}{\hbar^6} \sim 10^{-10}\mu_0 \quad \text{or} \quad (\Delta\mu)_\beta = \mu_0 g_\beta^2 (lm_e)^4 \frac{c^2}{\hbar^6}$$

depending on whether we cut off the β-interaction at a momentum of the order of the nucleon mass M or at the fundamental length corresponding to a momentum $\Lambda = lm_e c$. From the condition that the correction found above should be greater than the poorly known β-correction, we obtain $l < 10^4$ or $l < 4 \cdot 10^3$, respectively, in the two cases mentioned.

Thus, the method of measuring the magnetic moment allows us to extend the limits of applicability of electrodynamics no further than to a momentum of several nucleon masses, and to a length not less than 10^{-14} cm.

Compared with the position at present such an extension would be quite important and would allow us to shed light on the possible role of the fundamental length in meson theory.

We take this opportunity to extend our gratitude to V. B. Adamskiĭ and L. P. Feoktistov for discussions and help with the calculations.

Institute of Chemical Physics
USSR Academy of Sciences. Moscow

Received
June 10, 1955

REFERENCES

1. *Landau L. D., Pomeranchuk I. Ya.*—Dokl. AN SSSR **102** 3 (1955).
2. *Ivanenko D. D.*—In: *The Latest Developments in Quantum Electrodynamics.* Moscow: Izd-vo Inostr. Lit. (1954).
3. *Karplus R., Kroll N. M.*—Phys. Rev. **77**, 536 (1950).
4. *Beringer R., Heald M. A.*—Phys. Rev. **95**, 1474 (1954).
5. *Louisell W. H., Pidd R. W., Crane H. R.*—Phys. Rev. **94**, 7 (1954).
6. *Vladimirskiĭ V. V.*—Dokl. AN SSSR **102** 6 (1955).

Commentary

This paper was the first to pose the question of verification of the applicability of electrodynamics at small distances via measurement of the magnetic moments of leptons. Subsequently, in an article by V. B. Berestetskiĭ, O. N. Krokhin, and A. K. Khlebnikov,[1] the corrected formula for the magnetic moment was obtained:

$$\frac{\mu}{\mu_0} = 1 + \frac{\alpha}{2\pi}\left(1 - \frac{2}{3l^2}\right)$$

and note was made of the fact that in the case of the muon magnetic moment it is possible with the same experimental accuracy to verify the applicability of electrodynamics at distances which are m_μ/m_e times smaller.

We note that quantum electrodynamics in order α^2 gives the following expressions:

for the magnetic moment of an electron:

$$\mu/\mu_0 = 1 + \alpha/2\pi - 0.328\alpha^2/\pi^2$$

and for the magnetic moment of a muon,

$$\mu/\mu_0 = 1 + \alpha/2\pi - 0.77\alpha^2/\pi^2.$$

To date calculations of μ/μ_0 have been carried out to terms of order α^4 inclusively. Theoretical and experimental values of μ/μ_0 agree with one another with an accuracy of order 10^{-9} for the electron and 10^{-8} for the muon, which corresponds to distances of order 10^{-16}–10^{-17} cm.

[1] *Berestetskiĭ V. B., Krokhin O. N., Khlebnikov A. K.*—ZhETF **30**, 788–789 (1956).

14
The Heavy Neutral Meson: Decay and Method of Observation*

The classification of the elementary particles admits the existence of a meson with strangeness 0 and isotopic spin 0 (see, e.g., the review by Okun [1]). Such an hypothesis has been advanced repeatedly [2, 3]. Evidently such a meson (let us call it ρ) should be neutral and interact strongly with nuclei; in particular, it may be produced singly in collisions of two nucleons.

We assume that the ρ meson differs from the neutral pion only in mass and isotopic spin, but that the space spin and parity of the ρ meson are the same as for the neutral pion (pseudoscalar). That means the ρ meson is equivalent to a nucleon-antinucleon pair in the state $0^{S1}S_0$, whereas the neutral pion is equivalent to a pair in the state $1^{S1}S_0$, $T_Z = 0$ according to the classification of Bethe and Hamilton [4] (see also [5]). These two pair states differ by the relative phase of $\overline{P}P$ and $\overline{N}N$.

In this Letter we consider the possible decay modes of the ρ meson and the method of observing it in an experiment. It is obvious that the mass of the ρ meson is greater than that of the neutral pion; otherwise the ρ meson would have been discovered in experiments on neutral pion production. The transformations $\rho \to 2\pi^0$ and $\rho \to \pi^+ + \pi^-$ are not possible as they would not conserve parity: two pions in a state with $L = 0$ are an even system, whereas the ρ meson is odd.

By applying the operator CT (C is charge conjugation, i.e., conversion of particles into antiparticles; T is charge symmetry, i.e., conversion of protons into neutrons), Bethe and Hamilton have shown that three-pion annihilation cannot occur in a 0^S state. Hence the decay modes $\rho \to 3\pi^0$ and $\rho \to \pi^+ + \pi^- + \pi^0$ are forbidden. This applies to any odd number of pions.

To treat the decay into four pions, we separate them into two pairs and denote the isotopic spin of the first pair by t_1, its orbital angular momentum by l_1, the corresponding quantities of the second pair by t_2 and l_2 and the angular momentum of the center of mass of the first pair relative to the other by L. From the assumption that the ρ meson is pseudoscalar and has $T = 0$, it follows that $t_1 = t_2$, $L = |\mathbf{l}_1 + \mathbf{l}_2|$, and that $L + \mathbf{l}_1 + \mathbf{l}_2$ is odd. If t_1 is even, then $\mathbf{l}_1$ and $\mathbf{l}_2$ are even; if t_1 is odd, then $\mathbf{l}_1$ and $\mathbf{l}_2$ are also odd.

If $l_1 = l_2$, then both pairs can be regarded as identical bosons, and the wave function must be symmetric with respect to their exchange.

*Zhurnal eksperimentalnoĭ i teoreticheskoĭ fiziki **34** (6), 1644–1646 (1958).

The lowest values of the momenta that satisfy all these conditions are $l_1 = l_2 = 2$, $L = 1$, $t_1 = t_2 = 0$, or $t_1 = t_2 = 2$.

For $l_1 \neq l_2$, such a state is $l_1 = 1$, $l_2 = 3$, $L = 3$, and $t_1 = t_2 = 1$. The need for large orbital momenta can reduce substantially the probability of the $\rho \to 4\pi$ decay.

The decay $\rho \to \pi^0 + \gamma$ is forbidden, since radiative 0–0 transitions are forbidden. The decay $\rho \to \pi^+ + \pi^- + \gamma$ is allowed, and the pion pair is then in a state with $L = 1$. Also allowed is the decay $\rho \to 2\gamma$, which is analogous to the decay $\pi^0 \to 2\gamma$. If $m_\rho > 2m_\pi$, one can expect the single photon decay to be more probable.

The expected time of decay is 10^{-18} to 10^{-20} s.

It will be extremely difficult to identify the decay $\rho \to \pi^+ + \pi^- + \gamma$ in the presence of photon background from the $\pi^0 \to 2\gamma$ decay.

We propose below a method for detecting events of single production of ρ mesons in interactions of charged particles by energy-momentum balance. Consider the reaction $p_1 + p_2 \to p_3 + p_4 + \rho$, where p_1 is a proton from the accelerator, p_2 is a proton at rest, and p_3 and p_4 are also protons.

This process is followed by the decay of the ρ meson, but we do not record the decay products. The energies and momenta of the protons p_1, p_3, and p_4 must be measured with great accuracy. Let us form the expression

$$A = \left[(E_1 + Mc^2 - E_3 - E_4)^2 - c^2(\mathbf{p}_1 - \mathbf{p}_3 - \mathbf{p}_4)^2\right].$$

For single production of ρ mesons we have $A = m^2\rho c^4$. In the case of an arbitrary process with production of two or more pions, we have a continuous spectrum of A values.

If it is observed in an experiment that there is a sufficiently narrow line (whose width must correspond to the accuracy of measurement of the magnitudes and directions of p_3 and p_4) in the distribution of A, the existence of a neutral meson with a strong nuclear interaction will have been demonstrated and its mass will have been determined.

I am grateful to V. B. Berestetskiĭ and L. B. Okun for their valuable advice.

Institute of Chemical Physics
USSR Academy of Sciences. Moscow

Received
March 9, 1958

REFERENCES

1. *Okun L. B.*—UFN **56**, 535 (1957).
2. *Teller E.*—Sci. News Lett. **71**, 195 (1957).
3. *Okun L. B.*—ZhETF **34**, 469 (1958).
4. *Bethe H. A., Hamilton J.*—Nuovo Cim. **4**, 1 (1956).
5. *Lee T. D., Yang C. N.*—Nuovo Cim. **3**, 749 (1956).

Commentary

The neutral meson whose search is discussed in this article was discovered in 1961[1] in a bubble chamber filled with liquid deuterium and exposed on a beam of π^+-mesons from the accelerator at a laboratory at Berkeley (USA). The meson, called a η-meson by the physicists who discovered it, was found in the reaction

$$\pi^+ + d \rightarrow p + p + \eta^0, \quad \eta^0 \rightarrow \pi^+ + \pi^- + \pi^0.$$

In this experiment the π^0-meson was not registered, while its momentum-energy and mass were determined by the energy-momentum balance, as was discussed in the present article by Ya. B. Then the spectrum of invariant mass of a system of three π-mesons was considered, and in this spectrum a peak was discovered which corresponded to the mass of the η-meson ($m_\eta \simeq 550$ MeV).

The experimentalists who discovered the η-meson apparently were unaware of Ya. B.'s article. The kinematic method proposed by Ya. B. and used in the discovery of the η-meson received in the literature the name missing mass method. It has been and is widely used in the discovery and study of numerous hadron resonances. The strong decay of a η-meson into 4 π-mesons, discussed by Ya. B., does not occur since the mass of the η-meson is insufficiently large. There occurs the decay $\eta \rightarrow 3\pi$, which is forbidden by the isotopic invariance of the strong interaction. The decay $\eta \rightarrow 3\pi$ occurs due to the fact that isotopic invariance is not a strict symmetry.

[1] *Pevsner A., Kraemer R., Nussbaum M.,* et al.—Phys. Rev. Lett. **7**, 421–423 (1961).

15

Parity Nonconservation in the First Order in the Weak-Interaction Constant in Electron Scattering and Other Effects*

We assume that besides the weak interaction that causes beta decay,

$$g(\overline{P}ON)(\overline{e}^{-}O\nu) + \text{Herm. conj.}, \tag{1}$$

there exists an interaction

$$g(\overline{P}OP)(\overline{e}^{-}Oe^{-}) \tag{2}$$

with $g \approx 10^{-49}$ and the operator $O = \gamma_{\mu}(1 + i\gamma_5)$ characteristic [1] of processes in which parity is not conserved.[1]

Then in the scattering of electrons by protons the interaction (2) will interfere with the Coulomb scattering, and the nonconservation of parity will appear in the terms of first order in the small quantity g. Owing to this it becomes possible to test the hypothesis used here experimentally and to determine the sign of g.

The matrix element of the Coulomb scattering is of the order of magnitude of e^2/k^2, where k is the momentum transferred ($\hbar = c = 1$). Consequently, the ratio of the interference term to the Coulomb term is of the order of gk^2/e^2. Substituting $g = 10^{-5}/M^2$, where M is the mass of the nucleon, we find that for $k \sim M$ the parity nonconservation effects can be of the order of 0.1 to 0.01 percent.

In the large-angle scattering of fast ($\sim 10^9$ eV) longitudinally polarized electrons by unpolarized target nuclei it can be expected that the cross-sections for right-handed and left-handed electrons (i.e., for electrons with $\sigma \cdot \mathbf{p} > 0$ and $\sigma \cdot \mathbf{p} < 0$) can differ by 0.1 to 0.01 percent. Such an effect is a specific test for an interaction not conserving parity.

A magnetized iron plate can serve as a source of polarized electrons [6]. When electrons are ejected from it by ion or electron impacts (dynatron effect), photoelectric effect [6], or as delta electrons, one can expect a polarization even larger than that corresponding to the ratio of the number of

*Zhurnal eksperimentalnoĭ i teoreticheskoĭ fiziki **36**, 964–966 (1959).

[1] Such an interaction has been repeatedly discussed in the past in connection with the problem of the isotope shift of electron levels (I. E. Tamm). On an analogous interaction between the neutron and the electron, see [4, 5]. New experimental possibilities arise in connection with the conservation of parity in the interaction (2).

magnetization electrons to the total number of electrons in the iron, since the inner electrons do not take much part in these processes. From thermoelectric emission or field emission one cannot expect an appreciable polarization of the emerging electrons, since the chemical potential of the electrons with spins parallel and antiparallel to the magnetization is evidently the same.

The interaction (2) leads to a shift of the electron levels of different parities in an atom.

In the hydrogen atom the probability of the metastable transition $2S_{1/2} \to 1S_{1/2}$, which appears on account of the admixture of $2P_{1/2}$ to the $2S_{1/2}$, still turns out to be even smaller than the transition probability on account of the magnetic moment of the electron, and is less than the probability of the two-quantum transition $2S \to 1S$ by a factor of more than 10^7. Finally, the interaction (2) leads to a rotation of the plane of polarization of visible light by any substance not containing molecules optically active in the ordinary sense of the words. The rotation of the plane of polarization also occurs because the weak interaction mixes atomic electronic states of different parity. A calculation of the effect gives an expression of the form

$$|n_{\text{right}} - n_{\text{left}}| \sim \frac{N_0 \frac{a^4}{\lambda} g |\psi_s(0)||\psi_P(0)|}{E_P - E_S}, \tag{3}$$

where n is the index of refraction for circularly polarized light; $N_0 \sim a^{-3}$ is the number density of the atoms; a is the linear dimension of an atom; λ is the light wavelength; $|\psi_S(0)| \sim 1/a^{3/2}$; in $|\psi_P(0)|$ there are nonvanishing "small components" χ, given by $\chi \sim (\hbar/2mc)\sigma \operatorname{grad} \varphi$, where φ are the "large components"; $|\psi_P(0)| \sim (\hbar/mc)a^{-5/2}$, so that

$$|n_{\text{right}} - n_{\text{left}}| \sim \frac{g}{a^3 \Delta E_{SP}} \frac{\hbar}{mc\lambda} \sim 10^{-20}. \tag{4}$$

Rotation of the plane of polarization by 1 radian occurs in a length of the order $\lambda/10^{-20} = 10^{15}$ cm, so that even in the first order in g the effect obviously cannot be observed.

How plausible is the assumption that the interaction (2) exists? Let us regard νe^- as a doublet in isotopic space and denote by l_1 the two-component quantity ν, e^-. We denote by B (baryons) the two-component quantity P, N. The interaction that causes β decay is written

$$g(\overline{B}\tau_+ OB)(\bar{l}_1\tau_- Ol_1) + g(\overline{B}\tau_- OB)(\bar{l}_1\tau_+ Ol_1) \equiv$$
$$\equiv 2g(\overline{B}\tau_x OB)(\bar{l}_1\tau_x Ol_1) + 2g(\overline{B}\tau_y OB)(\bar{l}_1\tau_y Ol_1). \tag{5}$$

It is natural to add to this formulation a term that complements the expression (5) to make the scalar product $\tau_B \tau_{l_1}$:

$$2g(\overline{B}\tau_z OB)(\bar{l}_1\tau_z Ol_1) = 2g[(\overline{P}OP) - (\overline{N}ON)][(\overline{\nu}O\nu) - (\overline{e}^- Oe^-)], \tag{6}$$

and this term contains the direct interaction in which we are interested here. In addition the formalism leads to the conclusion that the sign of the interaction will be different for the proton and the neutron. For the μ-meson interactions one will have to introduce another two-component quantity l_2, consisting of ν, μ^-. Besides, in the scalar product $(\bar{l}_1 \tau l_1)(\bar{l}_2 \tau l_2)$ there are not terms that would give the decay $\mu^- \to 2e^- + e^+$, and the objection of Gell-Mann and Feynman [1] disappears.

Assumptions have been proposed for a direct interaction, $g(\overline{e}^- O\nu)(\overline{\nu} O e^-)$, which would lead to a scattering of neutrinos by electrons [1], and also for a weak interaction of four nucleons [2], which leads to parity nonconservation in first order in g in nuclear reactions and the stationary states of nuclei. The four-nucleon interaction has as a consequence that odd nuclei (spin $\neq 0$) will have an "anapole" moment [3] proportional to g. The electromagnetic interaction of the electron with the anapole moment leads to parity nonconservation in the order ge^2. Thus in the absence of a direct weak interaction of electrons with nucleons the effects considered above, caused by mixtures of atomic electron levels of different parities, do not vanish, but are weakened by a factor of about 100.

I take this occasion to express my gratitude to G. M. Gandel'man, A. S. Kompaneets, L. D. Landau, L. B. Okun, I. Ya. Pomeranchuk, and Ya. A. Smorodinskiĭ for valuable comments and discussions.

Received
December 25, 1958

REFERENCES

1. *Gell-Mann M., Feynman R. P.*—Phys. Rev. **109**, 193 (1958).
2. *Roberts*—Proc. of the 8th Int. Conf. on High Energy Nuclear Physics. Geneva: CERN (1958).
3. *Zeldovich Ya. B.*—ZhETF **33**, 1531 (1957).
4. *Foldy L. L.*—Phys. Rev. **87**, 693 (1952).
5. *Lopes J. L.*—Nucl. Phys. **8**, 234 (1958).
6. *Tolhoek H. A.*—Rev. Mod. Phys. **28**, 277 (1956).

Commentary

This article contains one of the earliest discussions of the problem of weak neutral currents which violate spatial parity. Particularly notable are the use of the isospin in weak interactions and the conclusions reached on this basis that weak neutral currents must be diagonal and therefore should not lead to decays of the type $\mu^- \to e^- e^- e^+$. These ideas were advanced by Ya. B. independently of S. Bludman, who not long before had studied a model of isotopically invariant weak interaction.[1]

[1] *Bludman S.*—Nuovo Cim. **9**, 433–445 (1958).

This article by Ya. B. was the first to pose the question of the rotation of "a polarization plane of visible light by any substance which does not contain optically active (in the usual sense) molecules." Subsequent theoretical work (M. A. Bush'ĭa, I. B. Khriplovich and others) showed that Ya. B.'s conclusion that "the effect, apparently, cannot be observed," was too pessimistic and that in heavy atoms the effect was significantly greater than in hydrogen. Rotation of light polarization in vapors of an atomic substance (bismuth) was first observed by L. M. Barkov and M. S. Zolotarev.[2] This phenomenon was then observed in a few other experiments.[3]

The experiment to measure the difference in cross-sections of left- and right-polarized electrons on nucleons proposed in the article was carried out at Stanford.[4]

As regards reactions caused by a neutrino neutral current, these were discovered in 1973, several years before the discovery of electron neutral current.

The P-odd interaction of leptons with nucleons discussed in the article is an essential part of the modern unified theory of the electromagnetic and weak interactions.

[2] *Barkov L. M., Zolotarev M. S.*—Pisma v ZhETF **27**, 379–383 (1978); **28**, 544–547 (1978).

[3] *Emmons T. P., Reeves J. M., Fortson E. N.*—Phys. Rev. Lett. **51**, 2089–2092 (1983).

[4] *Prescott C. Y., Atwood W. B., Cottrell R. L.*, et al.—Phys. Lett. B **77**, 347–352 (1978).

16
The Relation Between Decay Asymmetry and Dipole Moment of Elementary Particles*

The assumption that parity is not conserved in weak interactions has led Lee and Yang [1] to new conclusions with respect to the behavior of elementary particles possessing spin.

1) Asymmetry of decay is possible, in which the emitted particles are directed primarily along or against the angular momentum of the decaying particle.

2) An elementary particle may have a dipole moment, and this is also parallel (or antiparallel) to its angular momentum.

As is well known, the first conclusion of Lee and Yang was brilliantly verified in experiments on the β-decay of oriented nuclei [2] and on μ mesons [3]. The order of magnitude of the dipole moment they predict, however, is too small for experimental observation. Landau [4] has given a complete theory relating parity nonconservation in charged-particle decay with space reflections.[1]

One of Landau's conclusions is that the dipole moment of elementary particles vanishes identically. It would seem at first that a decay asymmetry would lead necessarily to a dipole moment. Let us consider, for instance, a polarized neutron whose angular momentum is directed vertically upward. We may consider as established the fact that such a neutron decays by emitting electrons primarily in the upward direction. Let us now consider such a polarized neutron in the spherically symmetric field of a nucleus in which the energy relations are such that the neutron is stable and cannot decay. It then becomes possible and necessary for the neutron to undergo virtual decay, emitting an electron and capturing it again instantaneously. We may speak of a cloud of virtual electrons around the nucleus.

It would seem that asymmetry in real decay should correspond to a similar asymmetry in virtual decay. This would lead to asymmetry in the

*Zhurnal eksperimentalnoĭ i teoreticheskoĭ fiziki **33** (6), 1488–1496 (1957). Due to limited space, only the introductory part of the article is printed here.

[1]Lee and Yang [5] have also independently indicated the possibility of combining space reflections with the transitions to the antiparticle.

virtual electron cloud, and therefore to a dipole moment. Landau's work shows that such simple concepts are, in general, mistaken.[2] Ioffe [6] (who has kindly communicated his work to the present author before its publication) has recently shown that the decay asymmetry and dipole moment depend on whether or not the theory is invariant with respect to time reversal. Essentially the matter reduces to the following. A linear relation between the momentum of the emitted particle and the direction of polarization (spin direction) of the decaying particle is consistent with time reversal, since both quantities change sign. A static dipole **d**, however, or the analogous static center-of-mass **r** of a virtual-particle cloud, does not change sign under time reversal. Therefore the spin **s** can be related to the static quantities **d** and **r** only in a theory which is not invariant under time reversal.

Section 1 of the present article contains a coordinate-representation treatment of the emission of a nonrelativistic particle in the transformation of a spin-$\frac{1}{2}$ particle, assuming parity nonconservation. This example shows clearly the dependence of the decay asymmetry of a polarized spin-$\frac{1}{2}$ particle on the phase of the coupling constants in the expression for the interaction leading to the decay.

It is shown that in the first approximation the decay asymmetry depends on the imaginary part of the vector coupling constant. It is shown how the interaction potential can lead to asymmetric decay in the case of a real coupling constant, i.e., in the Ioffe-Rudik-Okun theory [7] (see assumption II of Ioffe [6]).

In Landau's theory charged particles are "odd." In molecular physics two types of odd phenomena are known. These are enanthiomorphic molecules (such as right and left tartaric acid) and diatomic molecules with Λ-doubling (such as a nitric oxide molecule with the projection of the electron angular momentum on the axis directed from the N to the O, or the same molecule with the angular momentum directed in the opposite way). Section 2 explains the properties of such molecules with respect to their decay asymmetry and dipole moment. It is shown that according to Landau's theory elementary particles are similar to enanthiomorphic molecules, rather than to molecules with Λ-doubling.

In Sec. 1 it is found convenient to consider decay by treating the wave function of the produced particle in configuration space, rather than in momentum space. In Sec. 3 we show why it seems to us that this approach leads more simply and directly to the known formulas for the decay probability, without calculating the level density in phase space.

[2] *Note added in proof* (November 25, 1957). In the absence of a dipole moment, parity nonconservation leads to certain specific magnetic properties, as is shown by the author in another paper in this present issue (ZhETF **33** 6, 1531–1533 (1957)).

Received
July 1, 1957

REFERENCES

1. *Lee T. D., Yang C. N.*—Phys. Rev. **104**, 254 (1956).
2. *Wuet C. S.* et al.—Phys. Rev. **105**, 1413 (1957).
3. *Garwin R., Ledermann L., Weinrich W.*—Phys. Rev. **105**, 1415 (1957).
4. *Landau L. D.*—ZhETF **32**, 407 (1957).
5. *Yang C. N.*—Rev. Mod. Phys. **29**, 231 (1957).
6. *Ioffe B. L.*—ZhETF **32**, 1246 (1957).
7. *Ioffe B. L., Okun L. B., Rudik A. P.*—ZhETF **32**, 396 (1957).

17
Interpretation of Electrodynamics as a Consequence of Quantum Theory*

In classical theory, action is the sum of three parts pertaining to the particles S_p, their interaction with the field S_i, and to the field itself S_f:

$$S = S_p + S_i + S_f = -mc\int dS - \frac{e}{c}\int A_i\,dx^i - \frac{1}{8\pi c}\int (E^2 - H^2)\,d^4x. \quad (1)$$

In the quantum theory of fermions we have

$$S_p = -\int \overline{\psi}\hat{p}\psi\,d^4x, \qquad S_i = -\frac{e}{c}\int \overline{\psi}\hat{A}\psi d^4x. \quad (2)$$

Expressing E and H in terms of A, we write $S_f = L(A_{\mu,\nu})$, where L is a quadratic expression.

In quantum theory, interaction between field and vacuum leads to renormalization of the charge; this means that a new term appears in the action, of the form [1]

$$S_v = ne^2 L(A_{\mu,\nu}), \qquad n = \frac{2\nu}{3\pi\hbar c}\ln\frac{\Lambda}{mc}. \quad (3)$$

Combining S_v with S_p and denoting the renormalized quantities by primes, we have

$$S'_f = S_f + S_v = (1+ne^2)L(A_{\mu,\nu}) = L(A_{\mu,\nu}\sqrt{1+ne^2}) = L(A'_{\mu,\nu}),$$

$$A'_{\mu,\nu} = A_{\mu,\nu}\sqrt{1+ne^2}, \quad eA_\mu = e'A'_\mu, \quad e' = e\frac{A_\mu}{A'_\mu} = \frac{e}{\sqrt{1+ne^2}}. \quad (4)$$

Landau and Pomeranchuk [1] note that when $ne^2 \gg l$, as $e'^2 \to n^{-1}$, it becomes possible to neglect the effect of the free electromagnetic field in the Lagrangian. See also Fradkin [2]. Following this remark, let us consider a theory in which we start from the expression

$$\mathbf{S} = \mathbf{S}_p + \mathbf{S}_i \quad (5)$$

and the ideas related to this theory.

Physically, equation (5) is equivalent to assuming that there exists a "field" $A(x,y,z,t)$ acting on the motion of the charged particles, as can be verified by varying the particle trajectory or wave function in (5). In the

*Zhurnal eksperimentalnoĭ i teoreticheskoĭ fiziki **6** (10), 922–925 (1967). Due to limited space, only the introductory part of the article is printed here.

non-quantum theory that follows from (5), there is no action of the field, there is no field energy, it is impossible to obtain Maxwell's equations or wave propagation, etc.

In quantum theory, on the other hand, it follows from (5) that the field A acts not only on real particles, but also on the vacuum. The theory is such that when $A = 0$ the energy and action of the vacuum are also identically equal to zero. In the presence of a "field" acting on the particles, and in the absence of free particles, virtual electron-positron pairs are created in vacuum, a nonzero vacuum energy appears (which we call field energy), and a contribution to the action appears.

Institute of Applied Mathematics
USSR Academy of Sciences

Received
September 26, 1967

REFERENCES

1. *Landau L. D., Pomeranchuk I. Ya.*—Dokl. AN SSSR **102**, 489 (1955).
2. *Fradkin E. S.*—ZhETF **29**, 258 (1955).

18

The Maximum Charge for Given Mass of a Bound State*

With V. N. Gribov, A. M. Perelomov

If a system of two particles A and B is capable of transforming into a stable particle D, the interaction $A+B \rightleftharpoons D$ is characterized by a "charge" g. The mass of D determines the position of the pole in the amplitude for scattering of A by B.

From the definition of the physical charge g, the pole term is

$$A = -\frac{mg^2}{2\pi(E_{AB} - m_D c^2)} = -\frac{g^2}{E'_{AB} + Q}\frac{m_A m_B}{2\pi(m_A + m_B)}, \tag{1.1}$$

where E'_{AB} is the energy neglecting the rest mass, and Q is the binding energy of D. On the other hand, as was first shown by Heisenberg [1], the residue at the pole of a bound state can be expressed in terms of the constant in the asymptotic form of the normalized wave function:

$$\psi(r) \to \frac{Ce^{-\kappa r}}{2\pi\sqrt{2}r} \tag{1.2}$$

for $r \to \infty$, where $\kappa = \sqrt{2mQ}$,

$$\text{Res}\, A = -\frac{|C|^2}{4\pi m}. \tag{1.3}$$

This relation holds also when the state of the physical particle D is a superposition of the "bare" particle D_0 and the "cloud" $A+B$, which for the case of an S state is described by the function $\psi(r)$.

If the interaction is local, expression (1.1) holds for any $r > 0$, and the normalization condition (including the amplitude for the bare particle D_0) gives:

$$\int |\psi|^2\, dv = \frac{|C|^2}{4\pi\kappa} \le 1. \tag{1.4}$$

By using (1.3) and (1.1), Ruderman and Gasiorowicz [2] then obtained the inequalities

$$-\text{Res}\, A \le \frac{\kappa}{m} = \sqrt{\frac{2Q}{m}}, \qquad g^2 \le \frac{2\pi\kappa}{m^2} = 2\pi\sqrt{\frac{2Q}{m^3}} \tag{1.5}$$

*Zhurnal eksperimentalnoĭ i teoreticheskoĭ fiziki **40** (4), 1190–1198 (1961). Due to limited space, only the introductory part of the article is printed here.

(we give their formulas for the simplest case of an S wave and zero range of interaction).

Exact equality in (1.5), upon which Landau [3] insists, corresponds to the case where the particle D consists entirely of $A + B$ (so that $\int |\psi|^2 \, dv = 1$), i.e., D is a composite particle with local interaction with A and B. There do exist in nature weak interactions, whose charge is several orders of magnitude less than its maximum value. At the present time, theory is not in a position to predict that there exists no interaction intermediate in charge value between weak and electromagnetic interactions, on the one hand, and the strong interactions (π- and K-mesonic interactions) on the other. Thus Landau's statement should be regarded not as a theoretical derivation, but rather as a hypothesis.

The inequalities (1.5) can be derived by means of dispersion relations, without using any pictorial representations of the cloud $A + B$.

In Section 3 we treat the properties of a system with the maximum residue. In the real case of two particles whose interaction is described by an attractive potential $U(r)$, the inequality (1.5) is violated. The residue A is greater than κ/m, for example in the case of the deuteron the residue is 1.5 times greater than κ/m.

A proof of this and an explanation of the violation in the language of dispersion relations is given in Section 4.

Finally, in Section 5 we give the results of a computation in a nonrelativistic field model with three elementary particles A, B and D; for a given physical (renormalized) mass of D, we give the expression for the physical (renormalized) charge g satisfying inequality (1.5). For the case of a single channel $D \rightleftharpoons A+B$, and for increase to infinity of the bare (unrenormalized) charge, g^2 tends to its upper limit. The presence of several channels does not change this conclusion.

In Appendix I we present the computations whose results are given in Section 5. In Appendix II we consider the case of an unstable particle D; in this case, when the bare charge goes to infinity, the pole (which is on the second sheet of the complex energy plane) goes out to infinity, and the scattering amplitude tends to zero; in the limit of strong interaction, unstable particles should not be considered.

Received
November 16, 1960

REFERENCES

1. *Hu Ning*—Phys. Rev. **74**, 131 (1948).
2. *Ruderman M. A., Gasiorowicz*—Nuovo Cim. **8**, 861 (1958).
3. *Landau L. D.*—ZhETF **39**, 1856 (1960).

19
Electromagnetic Interaction with Parity Violation*

Until the discovery of parity violation it was assumed that the interaction of an elementary particle with spin 1/2 with a weak electromagnetic field was completely described by three terms in the energy

$$q\varphi,\ \mu(\sigma\mathbf{H}),\ a\,\mathrm{div}\,\mathbf{E} = 4\pi a\rho,$$

where σ is the spin, $\mathbf{q}$ is the charge, μ is the magnetic moment and the constant a characterizes the field of a spherical capacitor, equal to zero outside but interacting with a charge ρ inside [1].

In their well known article about nonconservation of parity, Lee and Yang [2] indicate the possibility of an electrical dipole moment, i.e., an interaction $d(\sigma \cdot \mathbf{E})$. However, if, with parity violation, there is invariance with respect to combined inversion (and consequently, also with respect to reflection in time) then, as Landau [3] has shown, no dipole moment is possible. It is easiest to see this by noting that under time reversal σ changes sign but $\mathbf{E}$ does not.

What kind of electromagnetic interactions, forbidden in the case of parity conservation, are possible in a theory invariant with respect to combined inversion only?

The interaction $(\sigma \cdot \mathbf{A})$ is not allowed because of gauge invariance. There remains

$$b(\sigma\Delta\mathbf{A}) = b(\sigma\,\mathrm{curl}\,\mathbf{H}) = \frac{4\pi b}{c}(\sigma\mathbf{j}).$$

Here $\mathbf{j} = \rho\mathbf{v}$ is the current density which produces the magnetic field $\mathbf{H}$. Under parity transformation such a term would be a pseudoscalar (σ is a pseudovector, $\mathbf{j}$ a vector) and could not occur in the expression for the energy if parity is conserved. On the other hand, both σ and $\mathbf{j}$ change sign under time reflection. The moment of force corresponding to such an interaction energy is $\mathbf{M} = (4\pi b/c)[\sigma \times \mathbf{j}]$.

Such an interaction is directly obtained from a model of virtual decay of a spin-$\frac{1}{2}$ particle A into a particle B of spin 0 and a particle C of spin $\frac{1}{2}$ [4]. If this decay depends on a weak interaction which does not conserve parity, then two particles can simultaneously be produced in either S- or in P-states. Invariance of the theory relative to combined inversion corresponds to the

*Zhurnal eksperimentalnoĭ i teoreticheskoĭ fiziki **33** (6), 1531–1533 (1957).

relation of phases of the S- and P-waves which is such that the probabilities of finding the particle C above and below the equatorial plane (perpendicular to the direction of spin of the particle A) are equal, so that there is no electrical dipole moment. However, the virtual particles C in the equatorial plane have a transverse polarization; their spin σ_C has a component directed along $[\mathbf{r} \times \sigma_A]$. Thus, around the spin axis of A there is a ring of elementary magnets—virtual particles C with spin along the equator. The magnetic interaction, proportional to $(\sigma_C \cdot \mathbf{H})$ gives a term proportional to the integral $\oint \mathbf{H}\,dl$, taken along the equator. Such an integral can be expressed in terms of curl $\mathbf{H}$. It is relevant to remember Ioffe's work [5]: for a real decay, combined inversion gives an asymmetry in the direction of emission and a longitudinal polarization, whereas the second variant of Ioffe (invariance relative to charge conjugation) gives a symmetrical emission and transverse polarization of the emitted particles. In virtual decays, the correlation is the other way around; as noted by Ioffe, in the latter variant there can be a dipole moment, whereas in combined inversion there is no dipole moment, but, as can be seen from [4] and this note, there is a transverse polarization leading to the interaction $(\sigma \cdot \operatorname{curl} \mathbf{H})$. From the point of view of classification of magnetic properties of the particle, the interaction $(\sigma \cdot \operatorname{curl} \mathbf{H})$, obviously, does not correspond to any magnetic multipole (dipole, quadrupole, etc.); we will call it an "anapole." For an understanding of the anapole, transformation of the energy to the form $(\sigma \cdot \mathbf{j})$ is essential; the anapole interacts only with the current which flows into the point at which the particle is to be found. Consequently, the external field of the anapole is identically zero (more accurately—falls off just as the probability density in the cloud of virtual particles C falls off).

In this connection, the anapole is analogous to a spherical capacitor in which the field differs from zero only inside the capacitor. The difference from a capacitor comes from the fact that the anapole is a vector, has a definite direction (along the spin of the particle A considered), whereas a spherical capacitor is characterized by a scalar quantity.

A classical model of the anapole can be represented as a wire helix (solenoid) bent in a ring (toroid). The current flowing through the helix creates a magnetic field only inside the toroid. If the toroid is rigid, then no magnetic field produced by external currents can act on the toroid as a whole. However, if this wire toroid is immersed in an electrolyte which fills also the space inside the solenoid and a current is passed through the electrolyte, then a moment of force, proportional to the sine of the angle between the toroid and the direction of the current in the electrolyte, will act on the toroid. This corresponds to an interaction energy $(\sigma \cdot \mathbf{j})$, since the axis of the toroid is directed along σ.

The anapole moment of elementary particles, i.e., the constant b, can be estimated by multiplying the magnetic moment by the Compton wavelength

$\hbar/mc$ and by the square if the dimensionless constant of weak interaction f^2, i.e., is of order $10^{-26}\mu$, insofar as only the interaction in which parity is not conserved gives the anapole. It is possible that, for experimental observation, the interaction of the anapole, not with the current but with a varying electric field giving a moment of force $\mathbf{M} = (4\pi b/c)[\sigma \times \dot{\mathbf{E}}]$, is more important. However, in view of the smallness of b, at the present time such an experiment would not seem to be possible.

The anapole interaction is an example which directly refutes the assertion of V. G. Solov'ev [6], according to which combined inversion and gauge invariance lead to conservation of parity in electrodynamics.

After finishing this work, I learned that V. G. Vaks obtained independently results analogous to ours.

I should like to use this opportunity to thank B. L. Ioffe, A. S. Kompaneets (who proposed the name "anapole"), L. D. Landau, and Ya. A. Smorodinskiĭ for discussions.

Received
September 26, 1957

REFERENCES

1. *Fermi E., Marshall L.*—Phys. Rev. **72**, 1139 (1947).
2. *Lee T. D., Yang C. N.*—Phys. Rev. **104**, 254 (1954).
3. *Landau L. D.*—ZhETF **32**, 405 (1957).
4. *Zeldovich Ya. B.*—ZhETF **33**, 1488 (1958).
5. *Ioffe B. L.*—ZhETF **32**, 1246 (1957).
6. *Solov'ev V. G.*—ZhETF **33**, 1957 (1957).

Commentary

Articles 16–19, which have been presented in abbreviated form, have a common direction. Their theme is the search for an adequate theory to describe the complex and/or compound structure of interacting particles; a parallel theme is the visual description of virtual particles. In the quantum theory of interacting particles their transformation is possible, as is the generation of some particles and the destruction of others. At one stage many physicists considered desirable the investigation only of the initial and final states, i.e., particles entering from infinity and particles going away to infinity. In contrast to this, article 16 looks in detail at the wave function of exiting particles throughout the space, particularly near the source, i.e., a decaying particle.

A case is possible where there is interaction which could cause decay, however, this process is not possible from energy considerations. In this case particles which are generated will form a cloud whose amplitude decays exponentially with distance. One speaks of a cloud of "virtual" particles.

A stable system spends part of the time in the state of the initial particle, and part in the modified form of a cloud of virtual particles.

Article 16 is closely associated with the discovery of non-conservation of spatial parity.

In beta-decay a strong correlation has been discovered between the spin direction of the initial particle and the direction of escape of the electron or positron. It would appear that the same symmetry should be found for a cloud of virtual electrons about a stable nucleus. However, in the approximation of T-parity conservation (time-reversal), the electric dipole moment of a stable nucleus is identically equal to zero. The solution of this paradox is given in article 16. It later turned out that CP-parity and T-parity are also not conserved, and this sparked new interest in the search for the electric dipole moment of particles.

Another aspect of the fact that a particle A spends part of the time in the form of a virtual cloud $B + C$ is related to the definition of the "charge" characterizing the transformation $A \Leftrightarrow B + C$ (article 18).

The charge should be defined as the effective (renormalized) interaction of the observed particle A. It turns out that the limit of maximally strong interaction is determined by the closeness to unity of the time spent by the particle in the cloud state. In other words, in this limit we are dealing with a composite particle. The renormalized charge then takes a specific value.

In paper 17 an attempt is made to take a new approach to the description of the electromagnetic field as a perturbation of a vacuum of electrons and positrons (and all other charged particles as well). For this the vector-potential and its interaction with charges is introduced, but the energy of the field itself, $(E^2 + H^2)/8\pi$, and the corresponding contribution to the Lagrangian, $(H^2 - E^2)/8\pi$, are assumed to be dependent on the vacuum polarization. This approach, in which all virtual particles participate, seems more elegant than attempts to describe a photon as a pair consisting of an electron and a positron.

There are other papers which are related to those above, but which did not enter the book due to insufficient space. In particular: the moment of unstable elementary particles,[1] "Unstable Particles in the Lie Model,"[2] and part of the monograph of Baz, Perelomov and Zeldovich,[3] "Scattering, Reactions, and Fission in Non-Relativistic Quantum Mechanics."

It is notable that paper 18 was discussed with L. D. Landau. It is interesting to compare it with one of his last papers.[4]

Finally, article 19 is closely related to the problem of violation of spatial parity. Together with the direct interaction investigated earlier (see article 15 and the commentary to it), a change is also possible in the electromagnetic field of a particle or system (nucleus) related to giving up parity.

The article proposes a new type of electromagnetic interaction which is called "anapole." Today intensive attempts at experimental methods of measuring the anapole moment of heavy nuclei are being carried out.[5]

[1] *Zeldovich Ya. B.*—ZhETF **39**, 1483–1485 (1960).

[2] *Zeldovich Ya. B.*—ZhETF **40**, 1155–1159 (1961).

[3] *Baz A. I., Zeldovich Ya. B., Perelomov A. M.*—Rasseǐanie, reaktsii i raspady v nereliativistskoǐ kvantovoǐ mekhanike [Scattering, Reactions, and Fission in Non-Relativistic Quantum Mechanics]. Moscow: Nauka, 544 p. (1971); translation available through U.S. Dept. of Commerce, Clearinghouse for Federal Scientific and Technical Information, Springfield, VA.

[4] *Landau L. D.*—ZhETF **39**, 1856–1859 (1960).

[5] *Dubovik V. M., Chesnov A. A.*—Elementarnye chastitsy i atomnoe ǐadro [Elementary Particles and the Atomic Nucleus] **5**, 791 (1974).

III

Atomic Physics and Radiation

20

Energy Levels in a Distorted Coulomb Field*

We shall consider the energy levels in the field of a potential W that for $r > r_0$ is the same as the Coulomb potential $V = -(e^2/\varepsilon r)$. When $r < r_0$, the potential W can differ greatly from the Coulomb potential, and, in particular, the potential W can be such that there will be several bound states with wave functions that are localized at $r < r_0$. The boundary point r_0 is assumed to be small in comparison with the radius of the Bohr orbit.

The problem is an idealization of the problem of the energy levels of an electron in an impurity semiconductor.

The motion of an electron that has broken away from the donor is considered; in this case, the donor becomes a positively charged ion; its field is at a large distance the field of a single charge in a medium with a dielectric constant ε.

The radius of the Bohr orbit in such a field is $a = \varepsilon(m/m_{\text{eff}})a_0$, where a_0 is the Bohr radius of the hydrogen atom, m is the mass of the electron, m_{eff} is its effective mass on the lattice. Thus, when the value of ε is large, the radius of the orbit a can exceed a_0 many times. The radius r_0 of the ion is of the order a_0, so that $r_0 \ll a$.

When $r < r_0$, it is obvious that inside donor ion W has nothing in common with the expression given for V, in particular, the field of the ion is such that $(z-1)$ electrons are retained in the inner orbits.

*Fizika tverdogo tela **1** (11), 1637–1641 (1959).

In the Coulomb field $V_C = -(e^2/\varepsilon r)$ the energy levels are given by the expression $E_n = -(m_{\text{eff}}/2\varepsilon^2 n^2 m)$ (in atomic units). However, in the present case, this equation is not applicable due to the fact that near the origin the potential W is very much different from V_C. Moreover, the expression for E_n in quantum mechanics is obtained from two conditions that are imposed on the wave function: a decrease as $r \to \infty$ and regular behavior at the origin. Therefore, it is not clear whether the expression for E_n is preserved when the potential and wave function are greatly distorted for small values of r.

It is clear that it is possible, given any value of the energy $E < 0$, to specify the wave function with proper behavior at $r \to \infty$. By integrating the Schrödinger equation in the direction of diminishing r, we obtain a solution ψ_E for all values of $r \geq r_0$. It is always possible to choose the potential W in such a way that for $r \leq r_0$ a wave function that is regular for $r = 0$ has to match up with ψ_E when $r = r_0$.

Consequently, in the present problem, it is possible, in principle, to have any values for E, and the Bohr energy levels E_n can be separated only as the most probable when the behavior of V is given and the variation in W is arbitrary and independent. The problem posed in this way differs in concept from the ordinary problem of finding the energy eigenvalue E in a given potential.

Let us consider a wave function in a Coulomb potential with $r > r_0$.

In the new atomic units and for $\varphi = r\psi$, the equation for the s term has the form

$$-\frac{1}{2}\frac{d^2\varphi}{dr^2} - \frac{\varphi}{r} = E\varphi.$$

With the boundary condition $\varphi(0) = 0$, this equation has the eigenvalues $E_n = -\frac{1}{2}n^2$, $E_1 = -\frac{1}{2}$. When $n = 1$, the eigenfunction, normalized by the condition $\int_0^\infty \varphi_1^2\, dr = 1$, has the form $\varphi_1 = 2re^{-r}$.

Let us find the wave function φ for $E = E_1 + \Delta$ that will satisfy the condition $\varphi(\infty) = 0$. For this we apply the Lagrange method:[1] we shall find the irregular solution χ_1 of the equation for $E = E_1$:

$$\chi_1 = \varphi_1 \int^r \frac{1}{\varphi_1^2}\, dr.$$

We shall choose χ_1 in such a way that $D = \varphi_1\, d\chi_1/dr - \chi_1\, d\varphi_1/dr \equiv 1$.

For small values of r we have $\varphi_1 = 2r$, $\chi_1 = 2r\int^r (1/4r^2)\, dr = -1/2 + Kr + \ldots$. For large values of r we have $\chi_1 \sim e^{2r}$.

We shall seek the solution for $E = E_1 + \Delta$ in the form

$$\varphi(r) = a(r)\varphi_1 + b(r)\chi_1.$$

Instead of the set of two equations cited in footnote 1 for $a(r)$ and $b(r)$,

[1] *Zeldovich Ya. B.*—ZhETF **31**, 1101 (1956).

it is convenient to transform to a single equation for

$$z(r) = \frac{b(r)}{a(r)}; \qquad \frac{dz}{dr} = -2\Delta(\varphi_1 + z\chi_1)^2.$$

This equation is an exact one. We shall look for its solution by specifying $z(\infty) = 0$. For small values of Δ, we can restrict ourselves to the first approximation:

$$\frac{dz_1}{dr} = -2\Delta\varphi_1^2, \quad z(r) = 2\Delta\int_r^\infty \varphi_1^2\,dr, \quad z|0| = 2\Delta\int_0^\infty \varphi_1^2\,dr = 2\Delta.$$

Let us substitute this expression for $z(r)$ in the first approximation into the right-hand side of the exact equation:

$$\frac{dz}{dr} = -2\Delta\varphi_1^2 - 8\Delta^2\varphi_1\chi_1\int_r^\infty \varphi_1^2\,dr - 8\Delta^3\chi_1^2\left(\int_r^\infty \varphi_1^2\,dr\right)^2.$$

We can easily see that the corrections to $z(0)$ in the first approximation are of a higher order in Δ (they are proportional to Δ^2 and Δ^3), and have no diverging factors. In fact, for large values of r we have $\varphi_1 \sim e^{-r}$, $\chi_1 \sim e^r$, $\int_r^\infty \varphi_1^2\,dr \sim e^{-2r}$, so that both the terms with Δ^2 and Δ^3 are proportional to e^{-2r}. For small values of r, we have $\chi_1 \sim -1/2$, and the term with Δ^3 remains finite. Consequently, the validity of the expression for $z(0)$ when Δ is small has been demonstrated. We note that dz/dr is finite when $r = r_0$ (while in the first and higher approximations, $dz/dr = 0$ everywhere), so that the difference $z(r_0) - z(0)$ is small and, in what follows, we shall replace $z(r_0)$ by $z(0)$.

Let us determine the logarithmic derivative of φ for small values of $r = r_0$, applying the Lagrange condition; we shall denote it by h.

$$h = \left.\frac{d\ln\varphi}{d\ln r}\right|_{r_0} = \left.\frac{r_0}{\varphi(r_0)}\frac{d\varphi}{dr}\right|_{r_0} = r_0\frac{a(r_0)\frac{d\varphi_1}{dr} + b(r_0)\frac{d\chi_1}{dr}}{a(r_0)\varphi_1(r_0) + b(r_0)\chi_1(r_0)}$$

$$= r_0\frac{\frac{d\varphi_1}{dr} + z(r_0)\frac{d\chi_1}{dr}}{\varphi_1(r_0) + z(r_0)\chi_1(r_0)} = r_0\frac{\varphi_1'(r_0) + z(0)\chi_1'(r_0)}{\varphi_1(r_0) + z(0)\chi_1(r_0)}.$$

Here $z(0) \ll 1$, $\varphi_1'(r_0) \sim \chi_1'(r_0)$, so that, in the numerator, $z\chi_1'$ can be neglected. However, in the denominator, when r_0 is small, $\varphi_1 = 2r_0$, $\chi_1 = -1/2$, $|\chi_1| \gg \varphi_1$, and the term with $z(0)$ is retained.

Finally, we obtain

$$h = \frac{1}{1 - \Delta/2r_0}.$$

This value of h must necessarily be "matched up" with the value h_i obtained in the solution of the equation inside the potential well. Substituting h_i, we find the shift in level

$$\Delta = -2r_0\frac{1 - h_i}{h_i}.$$

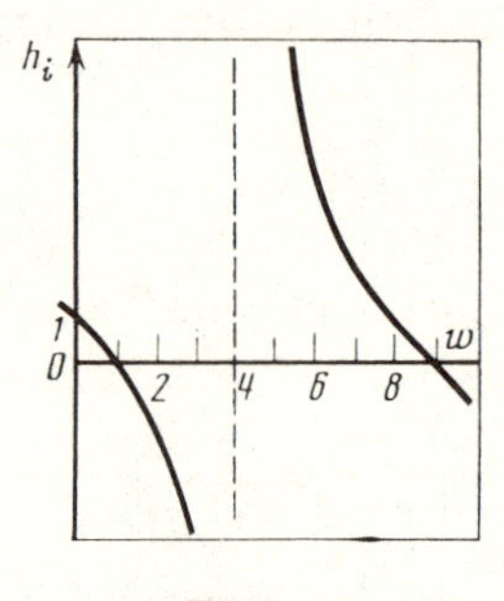

Fig. 1

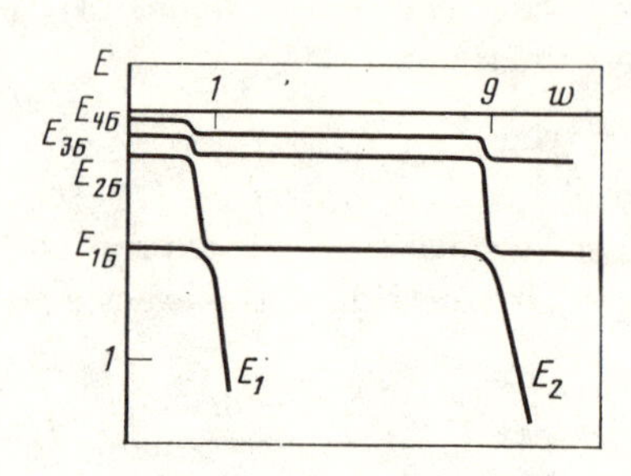

Fig. 2

In considering the internal problem, there exist no special reasons for h_i to be particularly near 0. In the general case, $|h_i| \approx 1$, $|1 - h_i| \approx 1$, we have

$$\Delta = 2gr_0, \quad \text{where} \quad g = \frac{1 - h_i}{h_i} \approx 1.$$

Consequently, it has been proved that the presence of a strong perturbation in the potential at a small radius r_0, as a rule, leads to a small perturbation in the energy of the Bohr level (in dimensional units) $4gE_1r_0/a$, where a is the radius of the Bohr orbit for given values of ε and m_{eff}, and g is of the order 1.

For the nth Bohr level, as is known, $\varphi_n(0) = 2r/n^{3/2}$. Performing a similar calculation, we find

$$\chi_n(0) = -\frac{n^{3/2}}{2} \quad \text{and} \quad \Delta_n = \frac{2r_0}{n^3}\frac{1 - h_i}{h_i} \sim |\psi_n(0)|^2.$$

Consequently, although the theory of small perturbations is inapplicable to a potential well when the values of r are small, the perturbation in the energy level is proportional to the density in the region where there is a perturbation $\psi_n^2(0)$ (we recall that $\varphi = r\psi$, while the volume density of the particles is just ψ^2), and where this is chosen depending on the unperturbed function. This result could have been predicted by imagining an equivalent potential that would create the same change in phase (scattering) as the well considered, but that would satisfy the condition that the theory of small perturbations be applicable. This point of view was adopted by Fermi in the theory of neutron scattering.

Taking into account the perturbation, we can write

$$E_n' = E_n + \Delta_n = -\frac{1}{2n^2} + \frac{p}{n^3} \simeq \frac{1}{2(n+p)^2}$$

in correspondence with the serial equation written for the hydrogen-like spectrum.

The conclusions made above are based on the fact that there exist no particular reasons for h_i to be small.

We shall illustrate this assumption by considering the special case of a rectangular well $W = \text{const} \ll 0$ for $r < r_0$; then, when $|E| \ll |W|$ inside the well, the solution is

$$\varphi \sim \sin kr; \qquad k = \frac{1}{\hbar}\sqrt{-2mW}$$

and at the edge with $r = r_0$,

$$h_i = kr_0 \operatorname{ctg} kr_0.$$

The dependence of h_i on W is shown in Fig. 1 where the quantity h_i is plotted on the vertical, while on the horizontal,

$$w = -\frac{8mr_0^2}{\pi^2\hbar^2}W.$$

The value $w = 1$, for which $h_i = 0$, corresponds to the condition that the first bound level appears in the well in the absence of a Coulomb potential; when $w = 1$, resonance takes place in the scattering of slow electrons off the well (near zero a real level with $w = 1 + \varepsilon$ and a virtual level with $w = 1 - \varepsilon$). When $w = 9$, the second level appears, with $|E| \ll |W|$, and once again resonance takes place in the scattering of the slow electrons. Here the first level already has an energy equal to $E_1 \simeq 0.7W$, i.e., it is very deep-lying. When $w = 25$, the third bound level arises, etc.

Consequently, it is possible for the well to have a substantial influence on the energy level in the Coulomb field only in the presence of resonance scattering of slow particles in the well; in this case, it is necessary that the energy of the real or virtual levels be of the same order in absolute value as the Bohr energy—just as one would expect from the general physical ideas involved.

This result can be presented in another way by means of the scattering length l. The scattering length is the coordinate of the point at which the wave function goes to zero with $E = 0$ in the field of W; it is connected with h by the following relation: $\varphi(l) = \varphi(r_0) + (l - r_0)\left.\frac{d\varphi}{dr}\right|_{r_0} = 0$,

$$l = \frac{r_0 - q}{q'} = r_0\left(1 - \frac{q}{r_0 q'}\right) = r_0\left(1 - \frac{1}{h_i}\right) = r_0\frac{h_i - 1}{h_i},$$

so that in the absence of resonance (i.e., when $|h_i| \sim 1$), $|l| \sim r_0$. The unperturbed Coulomb functions satisfy the condition $\varphi(0) = 0$. In the presence of the perturbing potential W, this condition must be replaced by $\varphi(l) = 0$. The corresponding variation in the energy when l is small is also small, being of the order l/a.

The general pattern of the variation in the spectrum for our problem with the variation of the depth of the well is shown in Fig. 2. The energy scale along the horizontal axis is the new atomic energy unit, and it is many times smaller than the energy units that we have used in scaling the depth of the well:

$$\frac{me^4}{\varepsilon^2\hbar^2} \ll \frac{\pi\hbar^2}{8mr_0^2}.$$

This inequality follows from the condition $r_0 \ll a = h^2\varepsilon/ml^2$. Figure 2 shows how everywhere except in those narrow regions around the resonance depth of the well the normal Bohr spectrum is preserved. The size of these narrow regions in the scale w is of the order $r_0/a \ll 1$.

Let us consider, for example, the region $9 < w < 25$. As is evident from the figure, in this region, the energy E_{1B} is ascribed to a level that is obtained adiabatically from the level with energy E_{3B} of the unperturbed Coulomb problem. Consequently, the level with energy E_{1B} has two nodes when $9 < w < 25$, the wave function has two nodes, i.e., reverts twice to zero [without considering $\varphi(0) = 0$] and changes sign. But when $9 < w < 25$, there are two strongly bound levels with energy $|E_{1,2}| \simeq |W|$ with functions having no nodes and with a single node. The function that corresponds to the energy E_{1B} is orthogonal to it; consequently, both of its nodes lie in the region of the well with $r \simeq r_0$. When $r \simeq a \gg r_0$, the function corresponding to E_{1B} has no nodes and hardly differs at all from the function of the fundamental state ($E = E_{1B}$ of the unperturbed Coulomb problem).

The present problem arose in the course of a discussion on a paper by L. Keldysh at a seminar held by L. Landau. E. Rabinovich participated in this discussion. I wish to take this opportunity to express my gratitude to them.

Summary

It was shown that, given a Coulomb potential everywhere except for a small region in the neighborhood of the origin, the spectrum, as a rule, differs little from the normal Bohr spectrum of the hydrogen atom. It is possible to have a marked distortion in the spectrum only when the perturbing potential has a resonance in the scattering of the low-energy particles. Without applying perturbation theory, it was shown that the variation in the energy of the Coulomb levels is proportional to the particle density at the origin in the unperturbed solution.

Institute of Theoretical and
Experimental Physics
USSR Academy of Sciences. Moscow

Received
February 6, 1959

Commentary

This paper gives an original treatment of the problem of energy levels in a potential which is a Coulomb potential up to distances much smaller than the radius of the first Bohr orbit, while at smaller distances it differs strongly from a Coulomb potential. Here, in principle, a situation is possible in which the energy levels with energies of the order of the Bohr energies differ significantly from their "purely Coulomb" values. It is shown in the article, however, that this situation is, in a certain sense of the word, improbable. It arises only in the case when the potential at small distances has a resonance for scattering particles of small energy, for which there is no basis in the general case. Much more probable, therefore, is a small shift of the Coulomb levels proportional to the density of particles at the coordinate origin in the unperturbed Coulomb problem and to the scattering length on the perturbing potential.

Recently this result was applied to a proton-antiproton system (the analog of the positronium); see papers by V. S. Popov *et al.*[1,2]

[1] *Kudrǐavtsev A. E., Popov V. S.*—Pisma v ZhETF **29**, 311–316 (1979).

[2] *Popov V. S., Kudrǐavtsev A. E., Lisin V. I., Mur V. D.*—ZhETF **80**, 1271–1287 (1981).

21

On the Theory of Unstable States*

A perturbation theory is developed and an expression is given for the amplitude (corresponding to a given initial state) of a state that decays exponentially with time. A final expression is obtained which plays the role of the norm of such a state.

An exponentially decaying state that describes, for example, the phenomenon of α decay, is characterized by a complex value of the energy, the imaginary part of the energy giving the decay probability. The wave function of this state increases exponentially in absolute value at large distances, and therefore the usual methods of normalization, of perturbation theory, and of expansion in terms of eigenfunctions do not apply to this state. We develop here a perturbation theory which gives an expression in terms of a quadrature for the changes of the mean energy and of the decay probability corresponding to an arbitrarily small change of the potential.

If the state is initially described by a certain wave function, then for a long time interval thereafter the wave function is close to an exponentially decaying function with a definite amplitude. This amplitude is also calculated in terms of quadratures.

The solutions of both problems—that of the energy and that of the amplitude of the exponentially decaying state—involve a quantity that plays the role of the norm of this state:

$$\lim_{\alpha\to 0}\int_0^\infty \chi^2 e^{-\alpha r^2} r^2\, dr.$$

For the calculation of this quantity we shall give a direct method which enables us to avoid the limiting process $\alpha \to 0$.

1. Let us consider a particle moving in a spherical potential with a barrier, i.e., moving like the α particle in the Gamow theory of α decay.

Let the corresponding Schrödinger equation have the formal solution

$$\psi(r,t) = e^{-iE't}\chi(r)$$

with the complex value $E' = E_0 - i\gamma$. The discrete value E' is obtained from the condition that at large distances $\chi(r)$ contains only an outgoing wave:

$$\chi(r) \approx Cr^{-1}e^{ikr}, \quad r\to\infty, \quad k=\pm\sqrt{2E'}, \quad \hbar = m = 1.$$

This solution is of interest not only as a description of an unstable state; the corresponding eigenvalue E' is a singular point—a pole—(in the complex

*Zhurnal eksperimentalnoĭ i teoreticheskoĭ fiziki **39** (3), 776–780 (1960).

plane) of the matrix for the scattering of a particle by the potential.

As is well known, $|\chi(r)|$ increases exponentially for $r \to \infty$; the function χ cannot be normalized, and in particular cannot be regarded as a wave function in the usual sense: it does not belong to the complete system of eigenfunctions ψ_n of the Hamiltonian operator. We cannot apply to χ the usual formulas of perturbation theory, for example,

$$\delta E_n = \frac{\int \psi^* \delta H \cdot \psi_n \, dr}{\int \psi_n^* \psi_n \, dr},$$

and the expansion of the function of an arbitrary state in terms of eigenfunctions:

$$\varphi(r) = A_n \psi_n, \qquad A_n = \frac{\int \psi_n^* \varphi \, dr}{\int \psi_n^* \psi_n \, dr}.$$

We shall find the expressions that replace these well known formulas in the case of the function χ.

Let us begin with the perturbation theory. In the simplest case of an S wave and a potential such that $V(r) = 0$ for $r > R$, by using the methods developed in [1, 2] we get without difficulty the expression

$$\delta E' = \frac{\int \chi^2(r) \delta V(r) \, dr}{\int [\chi^2(r) - (Cr^{-1}e^{ikr})^2] \, dr - C^2/2ik}, \tag{1}$$

where C is the coefficient in the asymptotic formula for the unperturbed solution χ: $\chi(r) \approx Cr^{-1}e^{ikr}$ as $r \to \infty$. The two integrals—in the numerator and in the denominator—can be thought of as taken over all space: in the numerator the region of integration is fixed by the region of the perturbation $\delta V(r)$, and in the denominator the integrand is zero for $r > R$.

We note that the integrands do not contain the square of the absolute value, but the complex quantity χ^2, and therefore $\delta E'$ is complex. The expression (1) gives not only the change of the energy E_0, but also the change of the decay probability $w = 2\gamma$.

To derive formula (1) we introduce the variable

$$y = \frac{d \ln \chi}{dr}; \qquad \chi(r) = \exp\left\{\int_0^r y(q) \, dq\right\}. \tag{2}$$

Schrödinger's equation then takes the form

$$\frac{dy}{dr} = -y^2 - 2[E' - V(r)], \tag{3}$$

and the equation for the perturbation of y is

$$\frac{d\delta y}{dr} = -2y\delta y + 2[\delta E' - \delta V(r)]. \tag{4}$$

The condition of regularity of χ at $r = 0$ uniquely determines $y(0)$, so that $\delta y(0) = 0$, and from this we have

$$\delta y(r) = \exp\left\{-2\int_0^r y \, dr\right\} \int_0^r [\delta V(q) - \delta E'] \exp\left\{2\int_0^q y \, dq\right\} dq$$

$$= \frac{2}{\chi^2(r)} \int_0^r [\delta V(q) - \delta E']\chi^2(q)\, dq. \tag{5}$$

The boundary condition for the perturbed problem for $r > R$ is

$$\frac{d \ln \chi'}{dr} = y + \delta y = i\sqrt{2(E' + \delta E')} = i\sqrt{2E'} + \frac{i\delta E'}{2E'}.$$

$$\delta y = \frac{i\delta E'}{\sqrt{2E'}} = \frac{i\delta E'}{k}. \tag{6}$$

Comparing (6) and (5), we now get (1) by an elementary calculation.

If we prescribe $\delta V = \varepsilon =$ const in the entire infinite volume, we must obviously get $\delta E' \equiv \varepsilon$. Therefore the finite expression in the denominator of (1) can be regarded as the definition of the diverging integral $\int \chi^2\, dr$. This latter integral does not have any unambiguous meaning because of the fact that

$$|\chi| \sim e^{\gamma r/\sqrt{2E_0}} \to \infty \quad \text{for} \quad r \to \infty$$

and does not become convergent if we multiply the integrand by $e^{-\alpha r}$ and subsequently take the limit $\alpha \to 0$. Convergence can be achieved by multiplication by $e^{-\alpha r^2}$:

$$\int \chi^2\, dr \equiv \lim_{\alpha \to 0} \int \chi^2 e^{-\alpha r^2}\, dr = J = \int [\chi^2 - (Cr^{-1}e^{ikr})^2]\, dr - \frac{C^2}{2ik}. \tag{7}$$

Equation (1) can then be written in the form

$$\delta E' = \frac{\int \chi^2 \delta V\, dr}{\int \chi^2\, dr}. \tag{1a}$$

2. Let us now consider the nonstationary problem. Suppose that at the initial time the wave function

$$\psi(r, t = 0) = \varphi(r)$$

is prescribed. It is well known that the asymptotic form of the solution is

$$\psi(r, t) = Ae^{-iE't}\chi(r) + \mathrm{O}(r, t), \tag{8}$$

where $\mathrm{O}(r,t)$ falls off like $t^{-3/2}$ for small r (for further details about $\mathrm{O}(r,t)$ see the paper by Khalfin [3]).

Despite the fact that the first term decreases exponentially as $e^{-\gamma t}$ and the second only by a power law, the separation of the first term is justified over a wide range of values of t for $\gamma \ll E_0$. Drukarev [4] has shown that the approach of $\psi(r,t)$ to the asymptotic expression (8) occurs nonuniformly at small r ($r < vt$, where v is the speed of the particle corresponding to the energy E_0). As has been shown by Fok and Krylov [5], the coefficient A in the first term of (8) is proportional to the residue (at the pole $E = E'$) of the spectral density of the initial state $\varphi(r)$ when it is expanded in terms of the continuous-spectrum eigenfunctions $\psi(E, r)$ that correspond to real values of E.

The coefficient A can be expressed in terms of $\varphi(r)$ and $\chi(r)$ by a simple quadrature:

$$A = \frac{\int \varphi\chi\, dr}{\int \chi^2\, dr}, \tag{9}$$

where $\int \chi^2\, dr$ is defined by (7).

To verify this we introduce, following N. A. Dmitriev, a function $\psi(r, s)$ defined by the formula

$$\psi(r, s) = -i \int_0^\infty \psi(r, t) e^{ist}\, dt, \tag{10}$$

for those values of s for which the integral converges. In the region where the integral diverges, we define $\psi(r, s)$ as the analytic continuation of the function defined by the integral (1).

The Schrödinger equation gives

$$-s\psi(r, s) - \frac{1}{2}\Delta\psi(r, s) + V(r)\psi(r, s) = -\varphi(r). \tag{11}$$

In the region $r > R$, where $V(r) = 0$ and $\varphi(r) = 0$, the solution is of the form

$$\psi(r, s) = \frac{f(s)e^{ir\sqrt{2s}} + f_1(s)e^{ir\sqrt{2s}}}{r}, \tag{12}$$

where f and f_1 are arbitrary functions.

Considering the region $\operatorname{Im} s > 0$, where $\psi(r, s)$ is given by a convergent integral, we convince ourselves that $f_1(r, s) \equiv 0$, since $|\psi(r, s)|$ cannot increase for $r \to \infty$. The condition that $\psi(r, s)$ is a diverging wave for large r is extended by the analytic continuation to arbitrary values of s.

The function $\chi(r)$ that describes the decaying state satisfies the same condition for $r \to \infty$ and an equation analogous to (11) but without the right member:

$$-E'\chi(r) - \frac{1}{2}\Delta\chi(r) + V(r)\chi(r) = 0. \tag{13}$$

It follows from this that the solution of equation (11) with nonzero right-hand side has a pole at $s = E'$(with $\operatorname{Im} s = -\gamma < 0$):

$$\psi(r, s) = \frac{a\chi(r)}{s - E'} + \psi_1(r, s), \tag{14}$$

where $\psi_1(r, E')$ is regular.

To determine a we multiply (11) by $\chi(r)$ and (13) by $\psi(r, s)$ and subtract one equation from the other. We then integrate over the volume $0 < r < R$ and substitute the expression for $\psi(r, s)$ in the form (14). Then finally for $s \to E'$ we get an expression for a that coincides with the expression (9) for A.

Inverting the relation (1), we find that the pole term in (14) gives the exponential term in (8) with $A = a$, and this completes the derivation of (9).

3. The formulas are easily extended to the case of states with $l \neq 0$. In this case all formulas contain instead of χ^2 the product $\chi(\mathbf{r})\tilde{\chi}^*(\mathbf{r})$, where $\tilde{\chi}(\mathbf{r})$ is the solution of the adjoint equation (cf. [6]). In the present case, since the operator H is Hermitian, the taking of the adjoint reduces to changing the sign of i in the boundary condition

$$\frac{\partial \ln r\chi}{\partial r} = +i\sqrt{2E'}, \qquad \frac{\partial \ln r\tilde{\chi}}{\partial r} = -i\sqrt{2E'^*}$$

$$(r \to \infty).$$

After separating off the angular factor in $\chi(\mathbf{r}) = P(\theta, \varphi)z(r)$, we get

$$\tilde{\chi} = \tilde{P}\tilde{z}, \qquad \tilde{P} = P, \qquad \tilde{z} = z^*,$$

so that finally

$$\delta E' = \frac{\int \tilde{\chi}^* \chi \delta V \, dr}{\int \tilde{\chi}^* \chi \, dr}, \tag{15}$$

$$\psi(r,t) = Ae^{-iE't}\chi(\mathbf{r}) + \mathrm{O}(r,t), \tag{16}$$

$$A = \frac{\int \tilde{\chi}^*(\mathbf{r})\varphi(\mathbf{r}) \, dr}{\int \tilde{\chi}^*(\mathbf{r})\chi(\mathbf{r}) \, dr}. \tag{17}$$

In the equation for the radial function $z(r)$ the effective potential $U(r)$ includes the centrifugal potential:

$$U(r) = V(r) + r^{-2}l(l+1),$$

and therefore in the region $r > R$, where $V(r) = 0$, the functions $z(r)$ can be expressed in terms of a Hankel function of half-integral order of the complex argument kr (k is complex when E' is complex).

For $r \to \infty$ we also have $|\chi| \to \infty$, and therefore to give a definite meaning to the integral that plays the role of the normalization we must again either multiply the integrand by $e^{-\alpha r^2}$ and then let $\alpha \to 0$ or else use a finite expression of the type of (1), which does not require the limiting procedure:

$$\int_0^\infty z^2 r^2 \, dr = \int_0^r z^2 r^2 \, dr + r^2 z^2 \frac{\partial^2}{\partial E' \partial r} \ln(rz). \tag{18}$$

For $r > R$ we get into the region where $z(r)$ can be expressed in terms of a Hankel function, and the derivatives in the second term can be taken in an elementary way. Furthermore, it is easily verified that in virtue of the equation satisfied by $z(r)$ the right-hand side of (18) does not depend on r. The problem is solved in a similar way for the Coulomb potential, the only difference being that for $r > R$ the quantity z is expressed by a hypergeometric function.

Finally, in the case of a $V(r)$ that contains, besides the Coulomb and centrifugal potentials, another part that is everywhere different from zero but that decreases sufficiently rapidly (exponentially), we must bring into the treatment, along with the solution $z(r)$ of the complete equation, another function $z_1(r)$ that is a solution of the equation with $V(r) = 0$ and coincides with $z(r)$ in the limit $r \to \infty$ $[z(\infty) = z_1(\infty)]$. The function z_1 can be expressed in terms of known (Hankel or hypergeometric) functions:

$$\int_0^\infty z^2 r^2\, dr = \int_0^\rho z^2 r^2\, dr + \int_0^\infty (z^2 - z_1^2) r^2\, dr + z_1^2 \frac{\partial^2}{\partial E' \partial r} \ln(r z_1)\Bigg|_{r=\rho}. \quad (19)$$

For $l \neq 0$ we must treat separately a neighborhood of the origin, $0 < r < \rho$, because of the fact that the Hankel function has a nonintegrable singularity at zero. It is assumed that after we have separated out from $V(r)$ the terms of orders $1/r$ and $1/r^2$, $V(r)$ falls off in such a way that $z^2 - z_1^2$ is a function that is integrable for $r \to \infty$.

In the case of a potential of complicated form, for which the integrals can only be calculated numerically, the advantage of the expression (19) as compared with (18) is that in (19) one takes the derivative of known functions.

I take occasion to note with gratitude that G. A. Drukarev, A. B. Migdal, V. A. Fok, and L. A. Khalfin have taken part in the discussion of this work. I must note particularly the important participation and assistance of N. A. Dmitriev, who provided formal proofs of a number of assertions contained in this paper.

Received
April 16, 1960

REFERENCES

1. *Zeldovich Ya. B.*—ZhETF **31**, 1101 (1956).
2. *Los' F. S.*—ZhETF **33**, 273 (1957).
3. *Khalfin L. A.*—ZhETF **33**, 1371 (1957).
4. *Drukarev G. A.*—ZhETF **21**, 59 (1951).
5. *Krylov N. S., Fok V. A.*—ZhETF **17**, 93 (1947).
6. *Kapur L., Peierls R.*—Proc. Roy. Soc. London A **166**, 277 (1938).

Commentary

This paper is devoted to the consistent development of the perturbation theory for "quasistationary" states of quantum mechanics. As is well known, the wave functions of these states do not enter into the complete set of eigenfunctions of the Hamiltonian of the system: they must be considered as a superposition of wave functions of a continuous spectrum. The usual perturbation theory is inapplicable for them because of the divergence of the normalization integrals. On the other hand, it is clear that, if the life-time of a state is sufficiently large, it is physically in many respects similar to states of a discrete spectrum and a means of describing it should exist in which this similarity is apparent. It is this method of description that is developed in the present article.

The author begins from the Schrödinger equation written in the form of a first-order nonlinear equation. The application of perturbation theory to this equation automatically leads to convergent expressions. At the same time an expression is obtained for the normalization integral, which does not contain an artificial limit transition.

We note that recent papers have shown the effectiveness of applying perturbation theory to a first-order equation for the usual problems of quantum mechanics as well.[1,2]

Consideration of the nonstationary problem with initial conditions led to formulas, which generalize to the case of quasistationary states of nonstationary quantum mechanics problems, and which are solved by expansion of the original function in functions of stationary states.

The problem is discussed in more detail in the monograph by Ya. B., A. N. Baz and V. M. Perelomov, "Scattering, Reactions, and Fission in Non-Relativistic Quantum Mechanics," Moscow: Nauka, 544 p. (1971).[3]

[1] *Dolgov A. D., Popov V. S.*—ZhETF **75**, 2010–2026 (1978).
[2] *Dolgov A. D., Eletskiĭ V. L., Popov V. S.*—ZhETF **79**, 1704–1718 (1980).
[3] Cf. Art. 19 for translation reference.

22
Scattering and Emission of a Quantum System in a Strong Electromagnetic Wave*

1. Introduction

Interaction of a strong monochromatic electromagnetic wave with a quantum-mechanical system (atom or molecule) is of particular interest in connection with the development of masers and lasers.

Coherent scattering of a monochromatic wave, the dependence of the scattering cross section and of the refractive index on the amplitude, coherent generation of harmonics, incoherent emission at the eigenfrequencies of the atom, change of these frequencies—this is but a brief and incomplete list of the questions raised. These effects are considered in a large number of papers, of which we shall cite the most interesting ones [1], in which references to earlier works are also given. Various methods were used: perturbation theory (with time-dependent potential), density-matrix (with allowance for dissipation), and others. A detailed and rigorous review of nonstationary perturbation theory was given recently by Langhoff *et al.* [2]. Problems involved in the emission of perturbed states are not considered in [2].

However, the most consistent, systematic, and at the same time simplest method is the use of quasienergy and quasienergy states (QES). Therefore by way of perfecting the method, and without claiming new results, we shall describe the quasienergy method. In this method, the influence of a classical electromagnetic wave on the atom is taken into account rigorously (at least in principle), and the spontaneous emission is regarded as a small dissipative perturbation. This method is exactly equivalent to the standard theory of the free atom, in which one first obtains exact solutions of the Schrödinger equation without taking the spontaneous emission into account, and a set of energy levels and (stationary) eigenstates is obtained. The spontaneous emission is regarded as a small perturbation of the system. In principle, when radiation is taken into account, only the lower state is exactly stationary. It is remarkable that in the presence of a strong wave, in the same approximation with allowance for the spontaneous processes, all the states are qualitatively identical in the sense that none of the QES is rigorously stationary. Let us recall the history of the quasienergy concept.

*Uspekhi fizicheskikh nauk **110** (1), 139–151 (1973). Reported at the Conference on strong electromagnetic waves (Balaton, Hungary, September 1972).

It is universally known that an electron situated in the spatially periodic field of a crystal lattice possesses a conserved quasimomentum p, $\psi(x+a) = e^{ipa}\psi(x)$, where a is the lattice constant and we have put $\hbar = 1$. Considering an electron (relativistic, obeying the Dirac equation) in the field of a strong wave, Nikishov and Ritus [3] introduced the concept of four-dimensional quasimomentum. Its fourth component was called quasienergy.

Two papers [4, 5], published practically simultaneously in 1966, applied the quasienergy concept to an atomic system in the field of a wave (see also [6, 7]). Ritus [4] considered a concrete method of obtaining the wave functions of QES. By definition, these functions satisfy the condition

$$\psi_k(t+T) = e^{-iF_kT}\psi_k(t). \tag{1}$$

If we separate the harmonic factor, we can write

$$\psi_k(t) = e^{-iF_kt}\varphi_k(t), \tag{2}$$

where $\varphi_k(t+T) = \varphi_k(t)$, so that $\varphi_k(t)$ is a strictly periodic (but not harmonic) function of the time.[1] ψ_k and φ_k also depend on x.

The function ψ_i can be expanded in a Fourier series:

$$\psi_i = \sum_{n=-\infty,\, k=0}^{n=\infty,\, k=\infty} c_{ink}\varphi_k e^{-i(F_i+n\omega)t}, \qquad \omega = \frac{2\pi}{T}. \tag{3}$$

Ritus has constructed equations for the determination of the coefficients c_{ink} and of the quasienergy F_i itself.

Zeldovich [5] also introduced the concept of quasienergy and QES, did not consider the concrete method of calculating F_i and ψ_i, but instead considered in detail the question of the radiation by the system. It has been noted, in particular, that F_i is defined in modulo ω, i.e., $F_i = F_i \pm \omega = F_i \pm 2\omega$, so that we cannot say, for example, that $F_2 > F_1$, since it is always possible to choose[2] integers (not necessarily positive) m and n such that

$$F_2 + n\omega < F_1 + m\omega.$$

We thus have a democracy—all the QES spontaneously go over into one another and there are no energy exclusions, since the strong wave is a reservoir of energy. All that remain are exclusions of the type of parity. A transition between two specified states 2 and 1 gives not a single line, but a series, in accordance with the fact that the quasienergy is defined in modulo ω.

[1] We call attention to the fact that F_kT is a number in the exponential of (1) but F_kT is a linear function of the time in the exponential of (2).

[2] However, such a change in the definition of F_i simultaneously changes the corresponding function φ_i, but F_i nevertheless remains periodic. On the other hand, we shall note below the possibility of obtaining QES by a smooth transition from a rigorous stationary state for a time-independent Hamiltonian $\mathcal{H}_0$. In the latter case, the energy of the k-th state E_k is uniquely defined. During the course of a smooth transition one can see which of the values $F_k + n\omega$ goes over into E_k; this value of F_k can naturally be called the principal value, and the remaining, which differ by an integer multiple of ω, can be called satellites.

However, since this modulus is the same for both states, the series is of the one-parameter type, and the frequency depends on $(n - m)$ but not on n and m separately.

Obviously, individual terms of the Fourier expansion give results that do not differ from the standard perturbation theory. The concept of quasienergy is useful but not essential. It would be difficult, however, to consider strong nonlinear effects and ignore the quasienergy.

Near resonance, i.e., at $E_2 - E_1 - n\omega \ll \omega$, the QES differ strongly from the stationary eigenstates of the unperturbed atom even at the relatively weak electromagnetic field of a maser or laser wave. It is precisely in this situation that the departure from the framework of perturbation theory, realized with the aid of the QES theory, is particularly fruitful. In a number of cases, the energy differences within a definite group of levels (two levels or more, but not a continuum!) are small, and the mixing of these levels with one another by a relatively weak field is of importance; remote levels and the ionization continuum are unaffected.

To the contrary, in the case of optical transitions in an atom far from resonance, there appears, simultaneously with nonlinear effects, a strong ionization of the atom, and the observation of nonlinear effects becomes difficult. In this case the quasienergy spectrum turns out to be continuous, and the discrete states have a complex quasienergy the imaginary part of which characterizes the probability of ionization of the atom in a given state by the wave. Among the cases in which one can hope for a useful application of quasienergy are nearby levels of atoms with molecules resulting from spin-orbit or hyperfine splitting; the system of degenerate levels of the hydrogen atom and the almost degenerate hydrogen-like high levels of atoms and ions; rotational levels of dipolar molecules, particularly those split by the magnetic field; level pairs produced in the presence of two equivalent states with low probability of spontaneous transition between them.[3] Some of the foregoing examples are considered in greater detail at the end of the article.

2. Evolution of the System

We consider in greater detail the analogy between the stationary states of an unperturbed system and the QES of a system in a periodic field, and the application of these concepts to the problem of the evolution of a system.

In both cases, the Schrödinger equation holds:

$$i\frac{\partial\psi}{\partial t} = \mathcal{H}\psi \tag{4}$$

and in the unperturbed case $\mathcal{H} = \mathcal{H}_0(x, \partial/\partial x)$ does not depend on the time. Thus, we have before us a linear partial differential equation. No one,

[3] We note also the application of quasienergy in the analysis of an electron acted upon by two fields, a constant magnetic field and a wave field [8].

however, even in possession of a superpowerful electronic computer, will start to solve this equation numerically,[4] by a difference method, by finding the increments

$$\psi(x, t + \Delta t) = \psi(x, t) - i\mathcal{H}_0\psi(x, t)\Delta t. \tag{5}$$

Instead, the solution is broken up into several stages:

1) We find the eigenstates, i.e., particular solutions of the type

$$\psi_k = \varphi_k(x)e^{-iE_k t}; \tag{6}$$

2) We represent the arbitrary initial state $\psi(x, t_0)$ as a superposition of eigenstates, i.e., we find c_k in the expression

$$\psi(x, t_0) = \sum_k c_k \varphi_k(x) e^{-iE_k t_0}; \tag{7}$$

3) The solution of the evolution problem, i.e., of calculating the value of $\psi(x, t)$ at an arbitrary instant of time, is then written out immediately:

$$\psi(x, t) = \sum_k c_k \varphi_k(x) e^{-iE_k t}. \tag{8}$$

This procedure is universally known. It is presented here only to demonstrate the complete analogy with quasienergy theory. Let $\mathcal{H} = \mathcal{H}(x, \partial/\partial x, t)$ contain the time t in explicit form. Then there are no solutions of the type (6). However, in the case of a periodic dependence $\psi(t + T) = \psi(t)$ we can find QES, i.e., solutions that reproduce themselves periodically. Thus, in the case of a periodic $\mathcal{H}(t)$, the first stage consists of finding solutions of the type

$$\psi_k = \varphi_k(x, t)e^{-iF_k t}, \tag{9}$$

where φ_k are not constant but depend on the time periodically. The second stage consists in expanding an arbitrary function $\psi(x, t_0)$, specified at the instant of time t_0:

$$\psi(x, t_0) = \sum_k c_k \varphi_k(x, t_0) e^{-iF_k t_0}. \tag{10}$$

Obtaining the coefficient c_k, we construct a solution that satisfies both the Schrödinger equation and the initial condition at the instant t_0:

$$\psi(x, t) = \sum_k c_k \varphi_k(x, t) e^{-iF_k t} = \sum_k c_k \psi_k. \tag{11}$$

The analogy is thus complete. The coefficients of the various ψ_k are strictly constant and the general character of the solution is directly evident.

A feature common to the solution with the time-independent Hamiltonian $\mathcal{H}_0$ and with the periodic Hamiltonian $\mathcal{H}(t)$ is that the spontaneous emission (transitions from one state to another) are disregarded for the time

[4]The discussion that follows pertains to a situation in which the principal role is played by discrete states. In a continuous spectrum, particularly for a free electron, direct methods can be effective.

being. The principal basis in the case of $\mathcal{H}_0$ is the existence of a complete orthonormal set of functions $\varphi_k(x)$, which are solutions of the eigenvalue equation

$$\mathcal{H}_0\varphi_k(x) = E_k\varphi_k(x).$$

In the case of a periodic potential, the mathematical formulation of the problem of finding the QES functions $\varphi_k(x,t)$ is at first glance entirely different, and it cannot be reduced to the problem with time eliminated. It turns out, however, that the QES also form a complete orthonormal system.

In practice, as a rule, a periodic potential consists of a time-independent Hamiltonian $\mathcal{H}_0$ and a periodic part $\mathcal{H}'(t)$, $\mathcal{H} = \mathcal{H}_0 + \mathcal{H}'$. The problem, however, is to determine the *exact* solutions of the Schrödinger equation for $\mathcal{H}(t)$, unlike in perturbation theory, in which $\mathcal{H}'$ would be regarded as small. Nonetheless, we can trace the correspondence between the solutions for $\mathcal{H}_0$ and the solutions for $\mathcal{H}(t)$. We introduce formally for this purpose a parameter α, define $\mathcal{H} = \mathcal{H}_0 + \alpha\mathcal{H}'$, and trace the variation of the solutions when α varies from 0 to 1. Each eigenfunction $\mathcal{H}_0$ turns out to be set in correspondence to one quasienergy solution. If we consider a closed group of states $\mathcal{H}_0$, which are "intermixed" with one another by the "perturbation" $\alpha\mathcal{H}'$, then it is clear that the number of QES does not differ from the initial number of states $\mathcal{H}_0$.

In this case we can also indicate an explicit algorithm for finding the wave functions of the QES and the values of the quasienergy. We shall consider below concretely the simplest example of a group consisting of two states.

We proceed from the mathematical problem of states that depend on the parameter α in the Hamiltonian $\mathcal{H} = \mathcal{H}_0 + \alpha\mathcal{H}'(t)$ to the physical problem of smoothly turning on a periodic potential. Let $\alpha = \alpha(t)$ increase slowly (during a time much longer than the period T, which is assumed constant) from 0 to 1. If the system is in a specified k-th state of $\mathcal{H}_0$ at the initial instant of time and at $\alpha = 0$, then when α grows slowly the system goes over into a pure k-th QES state.[5]

Thus, the QES have a clear-cut physical meaning, that of states obtained when a periodic perturbation $\mathcal{H}'$ is turned on smoothly. Moreover, we have obtained a non-obvious theorem: from an eigenstate of $\mathcal{H}_0$ we obtain, by smoothly turning on $\mathcal{H}'$, a state that has a periodicity property with period T equal to the period of $\mathcal{H}'$ (see the review [2]). If we use this property, we can obtain the desired solution of the Hamiltonian $\mathcal{H} = \mathcal{H}_0 + \mathcal{H}'$ even without considering the process of turning on $\mathcal{H}'$.

These statements are undoubtedly correct in the case of a finite group of states and under the additional condition that there be no exact resonance: $F_n - F_k \neq m\omega$, where m is an integer. When the continuum plays a role, an additional investigation is necessary. Exact resonance exists apparently

[5]In the case of instantaneous switching ($\alpha = 0$, $t < t_1$, $\alpha = 1$, $t > t_1$) we would have to expand $\varphi_k(x)$ in terms of $\varphi_n(x, t_1)$, and then the coefficients c_{nk} with $n \neq k$ are not equal to zero in principle.

only for noninteracting levels, just as terms of the same symmetry intersect in the stationary theory.

3. Radiation of the Eigenstates of $\mathcal{H}(t)$

In the preceding section we have emphasized the similarity of the QES to the ordinary eigenstates of $\mathcal{H}_0$, and in particular the complete analogy in the solution of the problem of the evolution of an arbitrary initial state. One should see, however, not only the similarity but also the difference between the QES and stationary states. This difference is most strikingly pronounced when one considers the radiation of the system itself, radiation not included in the Hamiltonians $\mathcal{H}_0$ and $\mathcal{H}'$.

In the case of stationary states, radiation takes place only on going from an upper state to a lower one. Yet a QES radiates by itself (a given n-th QES, without transition to others, $n \pm 1$, $n \pm 2, \ldots$). The quantity $\int \psi_n^* x \psi_n \, dV$ is not constant, but varies in time with the same period as the wave. From the condition of periodicity of $\psi_n(x,t)$ it follows that an electromagnetic wave is radiated and has the same period as the exciting wave. The radiated wave, however, is not harmonic! Its Fourier expansion contains a principal component with frequency ω and small harmonic components with frequencies $n\omega$ that are multiples of ω.

Obviously, this radiation must be classified as coherent scattering; the radiation of a given QES (without transition to another!) contributes to the real part of the refractive index of a medium filled with the considered atoms. We can calculate the nonlinear polarizability of the medium and the change in the waveform of the wave as a result of the harmonics.

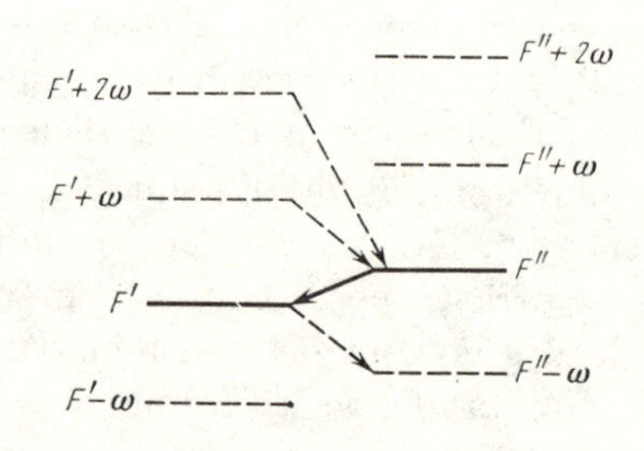

Transitions between different QES are similar to transitions from an excited stationary state to the ground state (or from one excited state to another). Such transitions occur in accordance with the laws of probability and are accompanied by emission of photons with a frequency characteristic of the atom, i.e., different from the wave frequency.

We have already noted in Sec. 1 (Introduction) that in the presence of two QES ψ_n and ψ_k it is impossible to say which of them has the larger and which the smaller quasienergy (see the figure). Spontaneous transitions go in both directions: $\psi_n \to \psi_k$ with emission $\omega' = F_n - F_k + p\omega$, and $\psi_k \to \psi_n$ with emission $\omega'' = F_k - F_n + q\omega$, where p and q are integers[6] such that $\omega' > 0$ and $\omega'' > 0$. In its aggregate, the transition $n \to k \to n$ is accompanied by emission of two photons, such that $\omega' + \omega'' = (p+q)\omega$, so that the energy taken from the field of the strong wave (from $\mathcal{H}'(t)$) is equal to an integer multiple of the frequency ω of the strong wave, as expected.

We assume that the transition probability is small and we disregard the natural widths of the quasienergy levels. However, the probability ratio of the transitions $\psi_n \to \psi_k$ and $\psi_k \to \psi_n$ is significant, since it governs the stationary number of atoms in different states,[7] and consequently also the characteristics of the radiation and the refractive index, averaged over the ensemble of the atoms.

Of particular interest is the question of inversion. In a static field ($\mathcal{H}_0$) it is possible to choose an inverted initial state, i.e., to specify initial number densities $N_k > N_n$ at $E_k > E_n$. By creating a resonator tuned to the frequency $\omega_{kn} = E_k - E_n$, we obtain a coherent ("laser") pulse. In the absence of the resonator, we obtain a spontaneous emission of frequency ω_{kn}. In the course of time, however, all the atoms go over by radiation to the lower energy state. In the field $\mathcal{H}_0$ without energy pumping, the excited states, and particularly the inverted state of an aggregate of atoms exist only for a limited time.

An entirely different case is that of a system situated in a periodic field $\mathcal{H}(t)$ in the quasienergy situation. It was noted above that in such a system there are, in principle, spontaneous transitions for all states into all states. As $t \to \infty$, a definite distribution of the atoms with respect to the QES is established and is characterized by a set of numbers N_k, where k is the number of the QES. All these numbers differ from zero at $t \to \infty$.

However, if there are spontaneous optical transitions between each pair of levels in both directions, $k \rightleftharpoons n$, and furthermore with different frequencies ω' and ω'', then the situation is reversed for one of these transitions, say for $k \to n$, $\omega'' = F_k - F_n + q\omega$ (see above), if $N_k > N_n$. Let us recall the meaning of inversion. We have considered so far only spontaneous emission at the frequency ω''. However, if there is spontaneous emission, then there also exists a proportional induced emission, and there is also absorption of the radiation in the inverse transition. The density relation $N_k > N_n$ causes the induced emission to be stronger than the absorption. Thus, although the spontaneous emission at the frequency ω'' may turn out to be weaker than the spontaneous emission of the other lines, say ω', nevertheless the emission at ω'' can be enhanced to the level of a laser pulse by means of a suitable resonator. In practice one should nevertheless expect the lowest

[6] Not necessarily positive. It is also possible for one of them to vanish.

[7] Without taking the spontaneous transitions into account, this quantity will depend on the initial conditions and on the manner in which the wave is turned on.

QES to be the most populated in the stationary state. More accurately speaking, in accordance with all the rules of the game, this QES should be given the expanded name "QES obtained adiabatically from the lower level of the time-independent $\mathcal{H}_0$."

If $\mathcal{H}'(t)$ is small, then the high-frequency components in this state (remote satellites with large p) are correspondingly small, the probability of spontaneous transition of the "former lower" state into one of the "former upper" states is small, and consequently high and difficult requirements are imposed on the Q of the resonator at the frequency ω'' that ensures generation.

We have not considered here a large number of questions of importance in the practical calculation of the generation (for example, the Doppler line broadening, the kinetics of accumulation of the inverted population, etc.). Within the framework of a theoretical article on quantum-mechanical systems in a periodic field, we must confine ourselves to the indication that generation of coherent radiation at a frequency different from the pump frequency is possible in principle.

The greater part of the general considerations given above has been advanced earlier [1, 5].

4. Two-Level System

Let us show, following [4], how quasienergy considerations are realized in the simplest case of a two-level system. We disregard spontaneous emission in this section. The wave function and the evolution of this system are determined completely by two amplitudes, $a(t)$ and $b(t)$:

$$\psi(x,t) = a(t)\varphi(x) + b(t)\chi(x), \tag{12}$$

where φ and χ are the spatial part of the stationary states of the unperturbed system with energies A and B. Thus, for an unperturbed system we have

$$\psi_0 = a_0 e^{-iAt}\varphi(x) + b_0 e^{-iBt}\chi(x). \tag{13}$$

We assume that φ and χ have different parity; in the dipole approximation, the interaction with the total energy E is given by the matrix element

$$M = eE \int \varphi(x) x \chi(x)\, dv, \tag{14}$$

by virtue of which, the final Schrödinger equation in the presence of a field is

$$i\frac{da}{dt} = Aa + 2V\cos(\omega t)b, \qquad i\frac{db}{dt} = 2V\cos(\omega t)a + Bb; \tag{15}$$

we have put here $\hbar = 1$, V is proportional to the amplitude of the wave and to the matrix element, the number 2 has been separated for convenience, and ω is the wave frequency.

We write down directly the solution with quasienergy F in the form of Fourier series, i.e.,

$$a(t) = \alpha_0 e^{-iFt} + \alpha_2 e^{-i(F+2\omega)t} + \alpha_{-2} e^{-i(F-2\omega)t} + \dots,$$
$$b(t) = \beta_1 e^{-i(F+\omega)t} + \beta_{-1} e^{-i(F-\omega)t} + \beta_3 e^{-i(F+3\omega)t} + \dots \qquad (16)$$

Obviously, these series satisfy identically the condition $a(t+T) = e^{-iFT}a(t)$, and analogously for b, since $T = 2\pi/\omega$. It is convenient to write $2\cos\omega t = e^{i\omega t} + e^{-i\omega t}$, and this makes obvious the rules of parity of the numbers for the Fourier series for a and b.

Substituting the series into the equation, we obtain an infinite system of coupled equations for the coefficients. Equating the determinant to zero, we obtain two eigenvalues F' and F'', and then the coefficients α'_n, β'_n, α''_n, and β''_n corresponding to these two solutions. We write down this system, or more accurately its first two equations:

$$F\alpha_0 = A\alpha_0 + V(\beta_1 + \beta_{-1}), \qquad (F+\omega)\beta_1 = B\beta_1 + V(\alpha_0 + \alpha_2). \qquad (17)$$

Let

$$B > A, \quad \omega - (B - A) = \delta \quad (\delta < \omega); \qquad V < B - A.$$

Then a reasonable iteration yields $F = A + \varepsilon$ $(\varepsilon < \omega)$,

$$\varepsilon\alpha_0 = V\beta_1, \quad (\delta+\varepsilon)\beta_1 = V\alpha_0, \quad \varepsilon(\delta+\varepsilon) = V^2,$$
$$\varepsilon', \varepsilon'' = -\frac{\delta}{2} \pm \sqrt{\frac{\delta^2}{4} + V^2}. \qquad (18)$$

When the wave is adiabatically turned on (by increasing V from $V = 0$ at $\omega = \text{const}$), the first state, corresponding to the solution $\varepsilon'(+)$ is obtained from the lower state:

$$V \to 0, \quad F' = A + \frac{V^2}{\delta}, \quad \alpha'_0 = 1 - \frac{v^2}{2\delta^2}, \quad \beta'_1 = \frac{V}{\delta}. \qquad (19)$$

The second state ε'' also pertains to the upper unperturbed state:

$$V \to 0, \quad F'' = A - \delta - \frac{V^2}{\delta} = B - \omega - \frac{V^2}{\delta}, \qquad (20)$$
$$\alpha''_0 = -\frac{V}{\delta}, \qquad \beta''_1 = 1 - \frac{V^2}{2\delta^2}.$$

At first glance it seems strange that the quasienergy does not tend to the energy of the unperturbed upper state B as $V \to 0$. But we see here precisely a manifestation of the fact that F is defined in modulo ω. In the second solution, the principal term at $V \to 0$ is

$$\psi'' = \beta''_1 e^{-i(F+\omega)t}\chi \to \chi e^{-iBt}$$

as expected.

At $V \sim \delta$, the coefficients α_0 and β_1 are of the same order; in particular, at $V \gg \delta$ we have

$$F' = A + V, \quad \psi' = e^{-i(A+V)t}(\varphi + e^{-i\omega t}\chi),$$

$$F'' = A - V, \quad \psi'' = e^{-i(B-V)t}(\varphi e^{i\omega t} - \chi). \tag{21}$$

In a typical "resonant" case, the ratio of V and δ can be arbitrary, $V \lesseqgtr \delta$, but both V and δ are small in comparison with $B - A$. It is then easy to verify that the discarded terms of the series $(\alpha_{\pm 2,\pm 4}, \beta_{-1,\pm 3}, \ldots)$ constitute an expansion in powers of the perturbation, for example,

$$\alpha'_{\pm 2} \sim \frac{V^2}{(B-A)^2}; \quad \beta'_{-1}, \beta'_{+3} \sim \frac{V^3}{(B-A)^3}; \quad \alpha'_{\pm 4} \sim \frac{V^4}{(B-A)^4} \cdots$$

At fixed V/δ, we can also speak of expansion in powers of the "degree of resonance" δ/ω. The denominators of the expressions for the higher-order terms will contain $B - A + \omega$, $B - A + 2\omega$, ...

The iteration and the entire quasienergy approach are good in the case when the natural width of the upper stationary state γ_0 is small (the probability of the spontaneous transition $B \to A$ is small): $\gamma_0 = W_{BA} \ll B - A$; then there is a region

$$B - A \sim \omega \gg \delta \gg \gamma_0.$$

Even the first approximation described above (with α_0 and β_1 retained) contains nontrivial results. Each of the two states F' and F'' has a variable dipole moment proportional to $\alpha'_0\beta'_1$ and $\alpha''_0\beta''_1$, respectively, varying with the time at the frequency ω. When $V < \delta$, the coherent radiation is proportional to V^2/δ^2, corresponding to a scattering cross section that does not depend on the amplitude and depends in Lorentz fashion on the difference $\delta = \omega - (B - A) = \omega - \omega_0$. When $V > \delta$, saturation sets in: the radiation tends to a constant limit, by virtue of which the cross section decreases like V^{-2}. Thus, V plays the role of the width in the shape of the scattering resonance; the stronger the wave, the larger the frequency interval in which the cross section has a plateau.

Let us pause briefly on the history of the problem. By the method of adiabatic perturbation theory, without explicitly using the concepts of quasienergy and QES, the two-level system was considered in a paper by Popular [9]. In terms of the quasienergy, the question was considered in [10], where both scattering and harmonics were considered. For an account of the quantized character of the electromagnetic field in this case, see [11].

5. Spontaneous Transitions

We turn now to spontaneous transitions. The probability of the transition $F'' \to F'$ is proportional to $|\alpha'_0\beta''_1|^2$. As $V \to 0$ we obtain the maximum probability, equal to $\gamma_0 = W'_{BA}$, of the unperturbed system. The probability of the inverse "unnatural" transition $F' \to F''$ is proportional to $|\beta'_1\alpha''_0|^2 \sim V^4/\delta^4$ when $V < \delta$. The radiated frequencies are

$$\omega''{}' = F'' - F' + \omega = B - A - \frac{2V^2}{\delta} = \omega - \delta - \frac{2V^2}{\delta},$$

$$\omega',''= F' - F'' + \omega = 2\omega - (B - A) + \frac{2V^2}{\delta} = \omega + \delta + \frac{2V^2}{\delta}. \qquad (22)$$

The complete cycle $F' \to F'' \to F'$ reduces to the radiation of $\omega','' $ and $\omega'','$ with taking of two quanta of energy 2ω from the classical field of the wave. We note that the width of the spectral line of the spontaneous transitions does not become larger than γ_0 for any value of V, regardless of the broadening of the coherent-scattering resonance. When it comes to spontaneous emission, it is the line shift and not the line width that depends on the wave amplitude (on V).

In the stationary state, the ratio of the number of atoms in F'' to the number of F' is the inverse of the ratio of the probabilities. At $V < \delta$ we have $n''/n' \sim V^4/\delta^4$. When $V > \delta$, however, n' and n'' become comparable. These two states make contributions of opposite sign to the real part of the forward scattering amplitude, i.e., to the refractive index. Therefore with increasing V the refractive index decreases like V^{-4}, i.e., more rapidly than the scattering cross section. An examination of the next terms of the expansion $\alpha_{\pm 2,\pm 4}, \beta_{-1,\pm 3}, \ldots$ makes it possible to calculate the coherent emission of the harmonics by each of the states. It is easy to verify here that only odd harmonics 3ω and 5ω, proportional to V^6 and V^{10} at small V are produced. In addition, it is possible to find the probability of the spontaneous emission $\pm(F'' - F') + n\omega$ with large n in transitions from one QES to another.

The system of equations written out above can also be used to determine the probability of overtone excitation. Let us specify, for example, $3\omega = B - A - \delta$ with $\delta \ll \omega$; we then obtain a resonant solution with large α_0, β_1, α_2, and β_3, and with a resonance condition $\delta \sim V^3/\omega^2$.

It is probable that the method developed above will also be useful for multilevel atomic-molecular systems. If the unperturbed system has a continuous spectrum (in addition to a discrete one), the quasienergy becomes complex, and this describes multiphoton ionization. Finally, in addition to spontaneous emission at frequencies different from ω, we can also consider the induced process, i.e., generation of frequencies ω' and ω'' such that $\omega' + \omega'' = n\omega$.

6. Quasienergy and Linear Stark Effect

Kovarskiĭ and Perel'man [10] applied the quasienergy concept to a consideration of excited states of hydrogen. The "random" degeneracy, for example of $2S$ and $2P$ levels, makes the excited states of the hydrogen atom with $n \geq 2$ particularly sensitive to the action of an electric field.[8] As is well known, it is precisely in such atoms that the linear Stark effect takes

[8] The influence of random degeneracy in the problem of multiphoton ionization of the atom was considered by Keldysh [12].

place. From the levels $2S$ and $2P$ with $m = 0$ we can construct linear combination with definite values of the dipole moment $\varphi_+ = (2S + 2P)/\sqrt{2}$ and $\varphi_- = (2S - 2P)/\sqrt{2}$. In this basis, the electric field yields diagonal matrix elements; in other words, the field does not cause transitions from φ_+ to φ_- or from φ_- to φ_+. The QES wave functions are therefore particularly simple in this case:

$$\psi_+(x,t) = \frac{1}{\sqrt{2}} e^{i\gamma \cos \omega t}(2S + 2P),$$

$$\psi_-(x,t) = \frac{1}{\sqrt{2}} e^{-i\gamma \cos \omega t}(2S - 2P). \tag{23}$$

The expansion of the function $\exp(i\mu \cos \omega t)$ in a Fourier series is universally known, namely, the coefficient of $\cos(n\omega t)$ is the Bessel function $J_n(\mu)$. We consider now the absorption spectrum of a normal hydrogen atom near the line Ly_α, i.e., the transition $1S \to (2S, 2P)$. If the atoms are in the field of a strong wave, the Ly_α line splits and satellites $\omega_n = \omega_0 \pm n\omega$ appear, where ω_0 is the frequency of the unperturbed Ly_α line. The amplitude of the satellites depends on the amplitude of the wave in which the atoms are situated, like $J_n^2(\mu)$, where $\mu = eEa_0 \cdot \sqrt{3}/\omega$, E is the field of the wave, and $ea_0 \cdot \sqrt{3}$ is the dipole moment of the hydrogen atom in the state $2S \pm 2P$. It is curious, therefore, that the amplitude of the satellite depends periodically on the amplitude of the wave. The higher the number n of the satellite, the larger the field necessary to obtain the maximum amplitude of the given satellite (the functions J_n behave in this manner).

This example demonstrates not only the strength but also the difficulty of the quasienergy approach. The point is that two QES are degenerate here for precisely the same reason that made it so much easier to find these states. Therefore an exact analysis of the absorption probability in the vicinity of Ly_α as a function of the angle, of the polarization, and of the satellite number calls for allowance for the phase relations between two QES. This analysis is beyond the scope of the present article.

The problem of the excitation of the atom was considered earlier [1, 13] without the use of the quasienergy concept. It is of interest to consider also the rotational levels of a dipolar molecule of the HCl type or an atomic system with nonzero angular momentum, but without random degeneracy. At $l \neq 0$ there are $2l + 1$ sublevels. From this we can make up linear combinations with a definite quadrupole moment that interacts with the gradient of the electric field of the strong wave. On the other hand, in $l \to l'$ transitions with different parity, there is a dipole matrix element, and when the strong-wave frequency is close to resonance, the sublevels of two systems, l and l', are mixed in the QES. The rules for mixing (selection) depend on the polarization of the strong wave, for example $m' = m \pm 1$ for plane polarization and $m' = m + 1$ for circular polarization.[9]

In the case of a diatomic amplitude with constant moment of inertia, there are whole-number relations between the energies of the successive levels, and the resonance for one transition $(0-1)$ coincides with the resonance of the higher orders for the next transitions, accurate to within the centrifugal deformation of the molecule.

The situation with a magnetic dipole in a magnetic field, where the equidistance and the final number of levels combine, was considered in detail in the theory of magnetic resonance. Even this short list demonstrates the large size of the class of phenomena to which the quasienergy theory can be applied.

7. Numerical Estimates

At what value of the field and at what power of the laser beam do the characteristic phenomena described above become manifest? Phenomena in weak fields can be described with the aid of quasienergy, but they are also successfully described by stationary perturbation theory—a circumstance noted on the very first page of this review. The advantages of the quasienergy approach appear in fields such that the first nonvanishing approximation of perturbation theory is not adequate. Let us estimate the required field.

We start with the simplest case (see Sec. 6 above and [10]), of the system of $(2S-2P)$ levels of the hydrogen atom. In this case: a) the unperturbed Hamiltonian $\mathcal{H}_0$ is degenerate, b) the perturbation is factorized, $\mathcal{H}' = V(x)\cos\omega t$, c) it is possible to select the eigenstates of $\mathcal{H}_0$ and $V(x)$ simultaneously, d) the QES are also factorized as a result, i.e., they take the form $\varphi(x)f(t)$ where, however, $f(t)$ is not a harmonic function:

$$f \sim \exp(-i\gamma\omega \int \cos\omega t\, dt). \tag{24}$$

Obviously, the condition that the result be nontrivial is $\gamma \gtrsim 1$. The dimensionless criterion γ is equal in this case to

$$\gamma = \frac{M_{2S,2P}}{\hbar\omega} = \frac{\sqrt{3}ea_0E}{\hbar\omega}, \tag{25}$$

where M is the matrix element of the transition in the field E, and a_0 is the Bohr radius $(a_0 = h^2/me^2)$.

The condition $\gamma = 1$ gives the field amplitude $E = me\omega/\hbar\sqrt{3}$. Let us find the corresponding power of an ideally focused laser beam. We specify the focal-spot area λ^2, where λ is the wavelength; we obtain

$$W = \frac{E^2\lambda^2 c}{8\pi} = \frac{\pi}{6}mc^2\frac{mc^2}{\hbar}\frac{e^2}{\hbar c} = 2\cdot 10^{12}\ \text{erg/s} = 2\cdot 10^5\ \text{W}.$$

This value can be regarded as rather modest in comparison with the presently attainable powers. It should be borne in mind, however, that it pertains to

[9]The case of an arbitrary number of degenerate states was considered by Kovarskiĭ and Perelman [14].

a focused beam. Spectroscopic experiments need a field that fills a sufficient volume. However, consideration of experimental devices is not the task of the present article.

If we are dealing with electron spin flip, then the matrix element of the interaction is equal to $M = \mu H$; it is smaller than in the preceding case by a factor $(\hbar c/e^2)\sqrt{3} \approx 240$. Accordingly, the required power is 6×10^4 times larger, $\sim 10^{10}$ W. Realization of experiments with this power is a difficult task.

Finally, a typical task is the excitation of a nondegenerate state by radiation close to resonance. In this case we cannot confine ourselves to the first term of the perturbation-theory series, when

$$\gamma' \sim \frac{M}{\hbar}|\omega - \omega_0| \sim 1. \tag{26}$$

For an allowed transition we have $M \sim ea_0E$, and the condition $\gamma' = 1$ gives for a beam with cross-section area S a power value

$$W_S = 6 \cdot 10^5 \frac{S}{\lambda^2} \left| \frac{\omega - \omega_0}{\omega} \right|^2 \text{ W}.$$

8. *Concluding Remarks*

It must be emphasized that the majority of the results were obtained earlier by other methods. We note below the corresponding investigations and the results. The quasienergy method developed above, however, seems to be the most adequate and economical even where the results are known. New ways of obtaining these results will be of methodological interest.

The task of the present methodological article is to demonstrate, with a very simple example, the method of operating with quasienergy and the benefits of this concept, especially in the simplest resonant case. The described simple methods can be useful in problems connected with astrophysical maser radiation, and in the theory of light transmission through rarefied media (see the note in [15]). The author did not intend to review fully the entire literature. Concerning more complicated problems, we confine ourselves to indicating the references [10, 13, 16, 17]. We shall discuss briefly only the principal question, namely, do states with definite real quasienergy always exist for a system situated in a periodic field?

Any action on the quantum mechanical system can be regarded as a unitary operator. We consider the action during one period. The quasienergy is the result of diagonalization of the operator of single-period action. In a finite-dimensional system (for example, in a two-level system), diagonalization is always possible. The unitarity of the operator ensures reality of the quasienergy. In an infinite-dimensional system, however, diagonalization is not always possible. Perelomov and Popov (see [17]) have constructed an interesting example by considering a harmonic oscillator with frequency ω_0 and a perturbation of the type $Vx^2 \cos \omega t$. At values of ω close to parametric resonance ($\omega \approx 2\omega_0$), the energy of the oscillator increases without limit for any initial state, meaning that there are no QES that repeat with a period $2\pi/\omega$.

Thus, the very existence of quasienergies is not trivial. In real systems, the quasienergy concept is approximate; it exists to the extent to which one can neglect multiphoton ionization. But one need not fear the approximateness once the character of the approximation is understood. It is appropriate to refer here to the remarkable article by Fok [18] concerning approximate solutions in physics.

REFERENCES

1. *Rautian S. G., Sobelman I. I.*—ZhETF **41**, 456 (1961); **44**, 934 (1963); *Kuznetsova T. I., Rautian S. G.*—ZhETF **49**, 1605 (1965).
2. *Langhoff P. W., Epstein S. T., Karplus M.*—Rev. Mod. Phys. **44**, 602 (1972).
3. *Nikishov A. I., Ritus V. I.*—ZhETF **46**, 776 (1964).
4. *Ritus V. I.*—ZhETF **51**, 1544 (1966).
5. *Zeldovich Ya. B.*—ZhETF **51**, 1492 (1966).
6. *Shirley J. N.*—Phys. Rev. **B138**, 595 (1965).
7. *Young R. H., Deal W. J., Jr., Kestner N. R.*—Mol. Phys. **17**, 369 (1969).
8. *Oleinik V. I.*—ZhETF **52**, 1049; **53**, 1997 (1967).
9. *Popular R.*—C. r. Acad. Sci. **B265**, 595 (1967).
10. *Kovarskiĭ V. A., Perelman N. F.*—ZhETF **60**, 509; **61**, 1389 (1971); Fiz. tverd. tela **13**, 1888 (1971).
11. *Kovarskiĭ V. A.*—ZhETF **57**, 1217 (1969).
12. *Keldysh L. V.*—ZhETF **47**, 1945 (1964).
13. *Karplus R., Schwinger J.*—Phys. Rev. **73**, 1020 (1948); *Basov N. G., Prokhorov A. M.*—UFN **57**, 485 (1955); *Javan A.*—Phys. Rev. **107**, 1579 (1970).
14. *Kovarskiĭ V. A., Perelman N. F.*—ZhETF **61**, 1389 (1971); **63**, 831 (1972).
15. *Varshalovich D. A.*—UFN **101**, 369 (1970).
16. *Adamov M. N., Balmakov M. D.*—Vestn. LGU (fizika, khimiĭa) **4**, 83 (1971); **1**, 29 (1972).
17. *Perelomov A. M., Popov V. S.*—ZhETF **57**, 1684 (1969); *Baz A. I., Zeldovich Ya. B., Perelomov A. M.—Rasseĭanie reaktsii i raspady v nereliativistskoĭ kvantovoĭ mekhanike [Scattering, Reactions, and Decays in Nonrelativistic Quantum Mechanics]*, 2nd Ed. Moscow: Nauka (1971); cf. Art. 19 Commentary for translation.
18. *Fok V. A.*—UFN **16**, 1070 (1936).

Commentary

The creation of laser light sources capable, in principle, of creating a field with intensities of atomic order has significantly changed the character of problems which present interest for the quantum theory of interaction of radiation with matter. Before this, theory was built in the majority of cases as a theory of electromagnetic field perturbations. The various processes of interaction with the field were then considered as processes of transition from one set of stationary (in the absence of the field) states with a certain energy to another set. In a strong field this method is inapplicable since the classification of states according to their energy values no longer makes sense. The paper above is devoted to a presentation of the mathematical apparatus which may be used for the description of phenomena in strong time-period fields. It is based on a classification of states according to the values of their "quasienergy," which, when given, determines the behavior of the wave function with respect to a time-shift by the field period. The concept of quasienergy was introduced practically simultaneously and independently by Ya. B. and a number of other authors. In the present paper it is shown what information on the behavior of the system may be obtained from the very fact of the existence of quasienergy, without any more detailed calculations. Of particular significance is the fact that the quasienergy is determined with accuracy only of an integer multiple of the field frequency, which automatically takes into account the possibility of absorption and emission of an arbitrary number of field quanta. Further applications of the concept of quasienergy may be found in a number of monographs.[1,2,3]

In is particularly noted in this article as well that in a stationary state in the system there is always inversion and a negative coefficient of absorption (which allows the realization of production) at certain frequencies.

Later, the ideas associated with quasienergy served as a basis for Ya. B.'s short article[4] on a possible method for obtaining low temperatures.

To date there is no information on experimental realizations of these proposals. We note also that the calculation of quasienergy states is simplified in the case when the electromagnetic wave is circularly polarized; the energy and angular momentum of the system prove to be related.[5]

To this day in the literature other methods of analysis of the behavior of a system in a periodic field are frequently used;[6] these methods are inferior to the quasienergy method in being both less graphic and less fundamentally simple.

[1] *Delone N. B., Krainov V. P.* Atom v silnom elektromagnitnom pole [The Atom in a Strong Electromagnetic Field]. Moscow: Atomizdat, 288 p. (1978).

[2] *Rapoport L. P., Zon B. A., Manakov N. L.* Teoriĭa mnogofotonnykh protsessov v atomakh [Theory of Multi-Photon Processes in Atoms] Moscow: Atomizdat, 182 p. (1978).

[3] *Kovarskiĭ V. A., Perelman N. F., Averbukh I. Sh.*, et al. Neadiabaticheskie perekhody v silnom elektromagnitnom pole [Non-Adiabatic Transitions in a Strong Electromagnetic Field]. Kishinev: Shtiintsa, 176 p. (1980).

[4] *Zeldovich Ya. B.*—Pisma v ZhETF **19**, 120–123 (1974).

[5] *Zeldovich Ya. B., Manakov N. L., Rapoport L. P.*—UFN **117**, 563–565 (1975).

[6] *Allen L., Aberley G.* Optical Resonance and Two-Level Atoms. New York: Wiley (1975).

23
Shock Wave Structure of the Radiation Spectrum During Bose Condensation of Photons*

With R. A. Sunyaev

The spectrum of photons interacting with electrons via the induced Compton effect is considered. Assuming weak energy transfer per collision, we previously predicted that a discontinuity may arise in the dependence of intensity on frequency ("shock wave" in phase space). In the present paper it is shown that if the finite temperature of the electrons is taken into account, the slope of the spectrum increases only up to a certain limit, after which an oscillatory dependence of intensity on frequency arises on the shortwave side of the spectrum. The structure of the shock wave is thus found to be more complex than previously assumed. It is similar to a collisionless wave in a plasma and not to a viscous wave in a neutral gas.

1. Introduction

The discovery of powerful compact sources of low frequency radiation in astronomical objects has attracted attention to the interaction between intense electromagnetic waves having a brightness temperature $kT_b \gg m_e c^2$ and free electrons.

Let us consider a spatially-homogeneous and isotropic situation, wherein the radiation spectrum is specified by the spectral energy density $\mathcal{E}_\nu$ of the radiation or by the occupation number $n(\nu) = (c^3/8\pi h\nu^3)\mathcal{E}_\nu$. The space is filled with electrons with density N_e and temperature T_e, which we assume specified and constant, although the results remain unchanged for a variable, and particularly equilibrium temperature (which the electrons acquire in the given radiation field). The scattering of the radiation by the free electrons leads to a redistribution of the energy and of the intensity over the spectrum; thus, the object of the investigation is the function $\mathcal{E}_\nu(t,\nu)$, where t is the time. Instead of using $\mathcal{E}_\nu$, we can characterize the radiation by the brightness temperature T_b, which is connected with $\mathcal{E}_\nu$ (we are considering long-wave

*Zhurnal eksperimentalnoĭ i teoreticheskoĭ fiziki **62** 1, 153–160 (1972).

unpolarized radiation) by the Rayleigh-Jeans relation

$$\mathcal{E}_\nu = \frac{8\pi k T_b}{c\lambda^2} = \frac{8\pi k T_b \nu^2}{c^3}.$$

Since $\mathcal{E}_\nu$ is in the general case not in equilibrium, $T_b(\nu, t)$ is also a function of the frequency. We are investigating the case of long-wave radiation of high-intensity, so that $T_b \gg T_e$ in a wide frequency range. As is well known [1, 2], the interaction of the radiation having the higher brightness temperature with the colder electrons is then accompanied by drawing of energy from the radiation,[1] as a result of which the radiation spectrum is altered in the low-frequency region [3].[2] If, furthermore, $kT_b \gg m_e c^2 \gg kT_e$, then (i) the induced scattering is stronger than the spontaneous scattering and (ii) the integral equation for the realignment of the spectrum can be transformed into a differential equation at a spectrum width $\Delta\nu \gg \Delta\nu_D = \nu\sqrt{2kT_e/m_e c^2}$. Here $\Delta\nu_D$ is the Doppler width of the spectrum and corresponds to the thermal velocities of the electrons.

Such a transformation was first performed by Kompaneets [5], who obtained the equation

$$\frac{\partial n}{\partial t} = \frac{\sigma_T N_e h}{m_e c} \frac{1}{\nu^2} \frac{\partial}{\partial \nu} \nu^4 \left(n^2 + n + \frac{kT_e}{h} \frac{\partial n}{\partial \nu} \right). \tag{1}$$

As applied to our problem, this equation can be written in simpler form

$$\frac{\partial g}{\partial \tau} = g \frac{\partial g}{\partial \nu}, \tag{2}$$

where

$$g = \nu^2 n = \frac{c^3 \mathcal{E}_\nu}{8\pi h \nu}; \quad d\tau = \frac{2\sigma_T N_e h}{m_e c} dt; \quad \sigma_T = \frac{8\pi}{3} \left(\frac{e^2}{m_e c^2} \right)^2.$$

This nonlinear equation was studied by Levich and one of the authors [3]; its characteristics are the lines

$$\frac{d\nu}{d\tau} = -g, \tag{3}$$

corresponding to a decrease of frequency at a rate proportional to the spectral density of the radiation energy (more accurately, proportional to the quantity g, which is connected with this density).

Under definite initial conditions, the spectrum evolution in accordance with equation (2) leads in the course of time to the formation of an infinite derivative $\partial g/\partial \nu$. For this purpose it is necessary and sufficient that there exist a point of inflection on the low-frequency side of the $g(\nu, 0)$ curve, i.e., the derivative $dg(\nu, 0)/d\nu$ should have a maximum at definite values $\nu = \nu_0$

[1] To realize this case it is necessary that the electrons lose energy in some manner that does not depend on the considered interaction with the low frequency radiation.

[2] This effect was apparently observed in experiment [4].

and $g = g_0$:

$$\frac{dg}{d\nu}(\nu_0, 0) > 0, \quad \frac{d^2g}{d\nu^2}(\nu_0, 0) = 0, \quad \frac{d^3g}{d\nu^3}(\nu_0, 0) < 0.$$

The situation is mathematically similar to nonlinear propagation of an acoustic wave in a gas, wherein the dependence of the wave velocity on the amplitude gives rise first to an infinite derivative $\partial\rho/\partial x \to \infty$ and $\partial p/\partial x \to \infty$, and then to a shock wave. In analogy, the formation of a "shock wave" was predicted in [3] also for the spectrum of the electromagnetic radiation under the conditions described above.

It should be noted that such situations were considered much earlier as applied to plasma oscillations. A nonlinear equation similar to (2) was derived in [6] for longitudinal plasma waves and it was noted that the evolution leads to a narrowing of the wave front. In its idea, this reference (see also [7]) anticipates the results of [3].

We note also that Kompaneets [5] and the workers that followed him [1–3] have used the quantum language in that they have considered Compton scattering of photons and the corresponding transfer of the momentum $h\nu/c$ and of the energy, with account taken of both spontaneous and induced processes, as is indicated in equation (1) by the factor $(1+n)$, which is characteristic of Bose particles (photons). Planck's constant h, however, is canceled out everywhere, so that actually the problem in question is classical, and the quantum language only makes the description more convenient.[3]

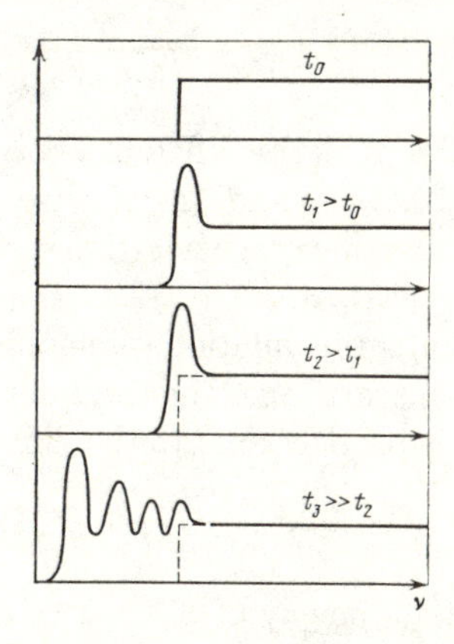

Let us turn from the history to the gist of the problem. The occurrence of an infinite derivative $\partial g/\partial\nu$ during the course of the evolution means violation of the condition necessary for changing over from the integral scattering equation to Kompaneets' differential equation (1), namely that g be smooth. An analysis of the integral equation shows that instead of a shock wave moving as proposed in [3] along the energy axis towards lower frequencies, there occurs an oscillatory frequency dependence of the radiation intensity. The resultant spectrum is represented by a set of narrow spectral lines of width $\Delta\nu \sim \Delta\nu_D$, spaced a distance $\Delta\nu \sim \Delta\nu_D$ apart and moving towards the lower frequencies, as may be seen in the figure above depicting the evolution

[3]Compare the ideas of Tsytovich [8] (and also Zeldovich [9]) concerning the number of quanta as an invariant of the classical field, and also those of Paradoksov [10] concerning the usefulness of quantum language. We note that spontaneous scattering by free electrons is also classical.

of $g(\nu)$ from an initial θ-function form due to the Compton interaction with thermal electrons. The movement of photons along the axis to the right of the discontinuity is from the induced Compton effect, to the left, by the spontaneous effect.

2. Form of the Integral Equation

The present paper is devoted to an analysis of the situation arising when the spectrum abruptly becomes steep. For such an analysis it is necessary to return to an integral equation of the form

$$\frac{\partial g(\nu,t)}{\partial t} = Ag(\nu,t)\int K(\nu,\mu)g(\mu,t)\,d\mu. \tag{4}$$

Here $A = 2\sigma_T N_e h/m_e c = \tau/t$. The effective width of the kernel $K(\nu,\mu)$ of equation (4) corresponds to the average change of the frequency following a single scattering by the moving electrons:

$$K \neq 0 \quad \text{for} \quad \left|\ln\frac{\mu}{\nu}\right| = \frac{|\mu-\nu|}{\nu} \leq \frac{v}{c} \sim \sqrt{\frac{kT_e}{m_e c^2}}.$$

The kernel K is the difference between the probability of the photons moving from all μ to ν and the inverse process of scattering from ν to all μ. It is therefore natural that the kernel K alternates in sign and moreover is antisymmetrical, $K(\nu,\mu) \equiv -K(\mu,\nu)$.

The integral equation (4) should correspond to a number of conditions. In Compton scattering, the total number of photons in the system is conserved. Since g is proportional to the number of photons per unit frequency, we have

$$0 = \frac{dN_\nu}{dt} = \frac{\partial}{\partial t}\int_0^\infty g(\nu)\,d\nu = \int\!\!\int K(\nu,\mu)g(\nu)g(\mu)\,d\nu\,d\mu. \tag{5}$$

The antisymmetry of K ensures identical satisfaction of this condition. Additional information concerning K can be obtained by using the fact that when the number of photons in the system is sufficient the joint action of the induced and spontaneous Compton processes should lead to a spectrum of the Rayleigh-Jeans type at low frequencies. The determination of the exact form of the kernel K is not part of our problem, however.

In the limit as the electron temperature $T_e \to 0$ and K is correspondingly narrow, the kernel can be replaced by the singular derivative $\delta'(\nu-\mu)$ of the Dirac function. In this approximation, the integral equation is transformed into the differential equation (2) written out above. Our task is to analyze the integral equation at low but finite T_e. We note that as $T_e \to 0$ the differential equation leads to a discontinuity, so that the analysis of the structure of the front requires allowance for $T_e \neq 0$ at any T_e.

The first seemingly natural assumption is that the solution contains instead of the discontinuity $g(\nu_0+0) \neq g(\nu_0-0)$ a transition region of width

$\Delta\nu \sim \nu_0 v/c \sim \nu_0\sqrt{kT_e/m_e c^2}$, which moves downward along the frequency axis as a viscosity-smeared shock wave with velocity proportional to the average value $\frac{1}{2}[g(\nu_0+0)+g(\nu_0-0)]$.

However, even without a detailed mathematical analysis it can be seen that the solution is more complicated. In fact, the use of a simple "step" ($g = g_1$ when $\nu < \nu_0$ and $g = g_2$ when $\nu > \nu_0$ with $g_1 < g_2$) as the initial condition causes the growth to be slower to the left of the discontinuity than to the right (since the integral in (4) is continuous, it follows that $\partial g/\partial t \sim g$). Thus, the discontinuity tends not to be smoothed out but to increase, and a sharp maximum $g > g_2$ is produced to the right of the discontinuity. The situation is clearly illustrated by the limiting case when $g_1 = 0$, i.e., when the initial spectrum is described by a step function. Then at $\nu > \nu_0$ we have predominant induced Compton scattering and motion of the photons towards lower frequencies. In the region $\nu < \nu_0$ there are no photons, and therefore there is no induced process and the photons can enter this region only with the aid of spontaneous Compton scattering. As a result, the photons should accumulate and a narrow spectral region with width $\Delta\nu \sim \Delta\nu_D$ should be produced near the discontinuity. The influence of the spontaneous Compton effect on the parameters of this line will be considered in Sec. 4.

In the general case $g_2 > g_1 \neq 0$ it turns out that at $\nu \sim \nu_0$ the single smooth discontinuity gives way to a complicated periodic dependence of g on ν. Simultaneously with the general displacement of the singularity towards lower frequencies we have a growth of the amplitude of the oscillations, whose period in the frequency scale is of the order of $\nu v/c$.

The question of possible realization and observation of such a characteristic spectrum under astrophysical conditions is beyond the scope of the present paper. In Sec. 5 below we shall give only an example of a situation in which the described picture can be realized. To answer this question it is first necessary to solve the problem in a realistic geometry and with allowance for the back-reaction of the radiation on the electrons.[4]

We confine ourselves here to an idealized formulation of the problem, for the purpose of disclosing under very simple assumptions the structure of the wave in pure form. In numerical calculations, on the other hand, we can use the following form of the kernel:

$$K(\nu,\mu) = \frac{d}{d\mu}[a\exp\{-a^2(\nu-\mu)^2\}], \qquad a \sim \sqrt{\frac{m_e c^2}{kT_e}\frac{1}{\nu}}. \tag{6}$$

In Sec. 3 below we shall determine the analytic properties of (4).

[4]It is curious that when $\mathcal{E}_\nu$ has a periodic dependence on ν one can expect a particularly strong increase of the induced radiation pressure on the electrons [11].

3. Analytic Properties of Equation (4) and Limiting Cases

We choose a scale, changing over to $x = \nu/a$, $y = \mu/a$, $t = b\tau$, such that the kernel of the equation

$$\frac{\partial g(x,t)}{\partial t} = g(x,t) \int K(x-y) g(y,t)\, dy \tag{4'}$$

has the following properties ($z = x - y$):

$$a) \int K(z)\, dz = 0, \quad b) \int zK(z)\, dz = 1, \quad c) \int z^3 K(z)\, dz = 1,$$

$$d)\, K(z) = \frac{d\Phi(z)}{dz}, \quad e)\, \Phi(+z) = \Phi(-z), \quad f) \int \bar{\Phi}(z)\, dz = 1. \tag{7}$$

Conditions (a) and (e) reflect the antisymmetry of the kernel $K(z)$, condition (d) is a definition of the symmetrical function $\Phi(z)$ (which can be considered to be Gaussian), and (b), (c), and (f) are normalization conditions.

Equation (4) can be rewritten in the form

$$\frac{\partial g}{\partial t} = g \int \Phi \frac{\partial g}{\partial y}\, dy, \tag{8}$$

which yields for a slowly varying g

$$\frac{\partial g}{\partial t} = g \frac{\partial g}{\partial x}. \tag{2'}$$

According to (2), one can visualize the photons as "moving" with velocity $g/2$. We note that the flow velocity is not equal to the perturbation-propagation velocity, but is one-half the latter. The differential equation (2) has two obvious "conservation laws" that are essential in what follows:

$$\frac{d}{dt} \int_{x_1}^{x_2} g\, dx = -\frac{g^2(x_1)}{2} + \frac{g^2(x_2)}{2}, \tag{9}$$

$$\frac{d}{dt} \int_{x_1}^{x_2} \ln g\, dx = -g(x_1) + g(x_2). \tag{10}$$

It is easy to verify that both laws are valid also for the integral equation, if the function g in the vicinities of x_1 and x_2 is constant and equal respectively to g_1 and g_2 over several units of the chosen scale (we recall that the region of influence of the kernel $K(x-y)$ is of the order of unity).

Let us attempt to construct a solution using a transition region of arbitrary shape moving in stationary manner to the left with velocity u:

$$g(x,t) = \varphi(x + ut), \qquad \varphi(-\infty) = g_1 \quad \varphi(+\infty) = g_2. \tag{11}$$

For such a solution, regardless of the form of φ, the derivatives of the integrals in (9) and (10) have perfectly defined values:

$$\frac{d}{dt} \int_{-\infty}^{\infty} g\, dx = u(g_2 - g_1) = \frac{1}{2}(g_2^2 - g_1^2),$$

$$\frac{d}{dt}\int_{-\infty}^{\infty} \ln g\, dx = u \ln\left(\frac{g_2}{g_1}\right) = g_2 - g_1,$$

from which we obtain two different values of u:

$$u' = \frac{g_1 - g_2}{2}, \tag{12}$$

$$u'' = \frac{g_2 - g_1}{\ln g_2 - \ln g_1}, \tag{13}$$

which coincide only in the limit as

$$g_2 \to g_1 \to g_0, \qquad u' \to u'' \to g_0.$$

This means that when $g_2 \neq g_1$ no stationary solution is possible!

The difference between u' and u'' is small: putting $g_2 = g_0 + \alpha$ and $g_1 = g_0 - \alpha$, we obtain $u' = g_0$ and $u'' = g_0 - \alpha^2/3g_0 + \ldots$.

Let us attempt to determine the possible quasistationary solution when $\alpha \ll g_0$, neglecting the difference between u' and u''. To this end we seek the asymptotic forms of a solution of the type $\varphi(x + g_0 t)$ to the left and to the right of the discontinuity

$$\varphi = g_1 + \beta(x + g_0 t), \tag{14}$$

$$\varphi = g_2 + \gamma(x + g_0 t). \tag{15}$$

Assuming β to be small in (14) and γ to be small in (15), we seek a solution in exponential form

$$\beta(\eta) = Be^{p\eta}; \quad \gamma(\eta) = Ce^{-q\eta}; \quad \eta = x + g_0 t. \tag{16}$$

Substituting such β and γ in (4), we get

$$g_0 = g_1 e^{p^2/4} = g_2 e^{q^2/4}. \tag{17}$$

Since $g_1 < g_0 < g_2$, it follows from (17) that p is real but q is imaginary. This means that in the quasistationary solution the front of the wave clings exponentially to the low-frequency region of x_1, g_1, but the high-frequency asymptotic form $g \to g_2$ is reached in an oscillatory manner.

The completely linearized case $\delta = g_2 - g_1 \ll g_0$ does not contain the important property of the nonlinear problem—the steepening of $g(x,t)$, i.e., the growth of $|\partial g/\partial x|_{\max}$ when the initial function $g(x,0)$ is slowly varying. Nonetheless, the linear case is not devoid of interest. The equation

$$\frac{\partial \delta(x,t)}{\partial t} = \int K\delta(y,t)\, dy \tag{18}$$

for the Fourier components

$$\delta = \int \delta_k e^{i\omega t + ikx}\, dk \tag{19}$$

yields the dispersion equation

$$\omega(k) = ke^{-k^2/4}. \tag{20}$$

At small k we have

$$\omega = k - 0.25k^3. \tag{21}$$

At large k,

$$\omega = ke^{-k^2/4}. \tag{22}$$

We see therefore that the slowly varying δ moves (as $k \to 0$) as a unit with unit velocity, so that

$$i\omega t + ikx = ik(t + x). \tag{23}$$

To the contrary, the discontinuity, or any other singularity characterizing the asymptotic form as $k \to \infty$, stands still, since at large k the dependence on t disappears in the limit as $\omega \to 0$.

4. Discussion

A detailed clarification of the picture calls for a numerical calculation. However, even the presented considerations show that in induced Compton interaction between a high-intensity radiation whose spectrum has an inflection point on the low-frequency side and thermal electrons one should expect a unique resultant spectrum with several minima and maxima. Spontaneous scattering may smooth them out on the high-frequency side.

The very appearance of intensity maxima at frequencies not corresponding to any resonances in the system is so curious that it is worthwhile to discuss the situation even before the entire picture is quantitatively explained.

The structure and evolution of the radiation spectrum can easily be explained qualitatively in the simplest example in which the initial spectrum is chosen in the form of a step function: $g(\nu > \nu_0) = \text{const}$, $g(\nu < \nu_0) = 0$. Then at $\nu > \nu_0$ the motion of the photons along the frequency axis is determined by the induced Compton effect and has a velocity

$$\left.\frac{d\nu}{dt}\right|_{\text{ind}} = -Ag = -\frac{\sigma_T N_e h\nu^2}{m_e c} n. \tag{3'}$$

When $\nu < \nu_0$ there are no photons, the induced process does not take place, and the photons can enter in this region only with the aid of the spontaneous Compton scattering. The probability of this spontaneous scattering is $w = \sigma_T N_e c$ and the average change of frequency in one act is $|\Delta\nu_{\text{sp}}| \sim v\nu/c$. Comparing $|\Delta\nu_{\text{sp}}|$ with the induce frequency shift

$$|\Delta\nu_{\text{ind}}| = \frac{1}{w_{\text{sp}}} \left|\frac{d\nu}{dt}\right|_{\text{ind}}$$

during the same time $1/w_{\text{sp}}$ we see that at

$$kT_b = nh\nu > m_e cv = \sqrt{m_e c^2 kT_e} \tag{24}$$

the motion of the photons along the frequency axis as a result of the induced processes is faster than that due to the spontaneous processes, and that in the region $\nu \sim \nu_0$ there should occur an accumulation of the photons. Obviously in the zone with $\Delta\nu/\nu_0 \sim v/c$ the number of accumulated photons should be such that their outflow towards lower frequencies as a result of spontaneous scattering should equal the influx from the higher frequencies, due to the induced processes. As a result a narrow line is produced near ν_0 (see the figure). Subsequently, since the flow of photons towards the lower frequencies is conserved, this line should move with velocity (3′), but the spectrum can already acquire a complex oscillatory structure. It is easy to make a rough estimate (which is obviously exaggerated in view of the neglect of the role of the induced processes in the formation of the line) of the stationary height of the line:

$$\frac{T_b(\Lambda)}{T_b} = \frac{n_\Lambda}{n} = \frac{g_\Lambda}{g} \sim \frac{|\partial\nu/\partial t|_{\mathrm{ind}}(g)}{|\Delta\nu|_{\mathrm{sp}}(\Lambda)w_{\mathrm{sp}}} \sim \frac{h\nu_0 n}{m_e c v} = \frac{kT_b\sqrt{m_e c^2/kT_e}}{m_e c^2}. \tag{25}$$

5. *Possibility of Astrophysical Applications*

It was noted in [12] that induced Compton interaction of low-frequency radiation with a thermal plasma can greatly distort the spectra of quasars, galactic nuclei, and pulsars in the frequency region where the brightness temperature of the radiation is $kT_b > m_e c^2/\tau_T$. Here $\tau_T = \sigma_T N_e l$ is the optical thickness with respect to the Thomson scattering of the radiation region with characteristic dimension l. Since the brightness temperatures of the radiation of compact radio sources in quasars and galactic nuclei reach $T_b \sim 10^{13}\mathrm{K} \sim 10^3 - 10^4\, m_e c^2$ and in pulsars even $T_b \sim 10^{25}\mathrm{K} \sim 10^{15}\, m_e c^2$, it follows that one can hope to observe a narrow intense line in the spectra of sources whose initial spectra have inflection points on the low-frequency side. We note that in this case we have a stationary problem wherein the photons emerge from a spatially limited region in which the photons are produced and scattered.

In order for the inhomogeneity of the source (and the differences between the emission spectra in spatially separated regions of the source) to cause no smearing of the lines in question, it would be of particular interest to consider the following geometrical arrangement of the scattering electrons. An electron cloud of optical thickness τ_T and linear dimension l is located between a source of radius R and the observer, the distance r between the cloud and the source being much larger than either R or l. Then, if $kT_b > (m_e c^2/\tau_T)(r/R)^3$ [12], a narrow intense emission line having $T_b(\Lambda) > T_b$ in the continuous spectrum should be produced in the spectrum.

We take the opportunity to thank G. I. Barenblatt, Yu. A. Berezin, A. A. Galeev, E. V. Levich and R. Z. Sagdeev for numerous discussions. We are particularly grateful to A. Kh. Rakhmatulina for performing the preliminary numerical calculations of the spectrum distortions.

Institute of Applied Mathematics
USSR Academy of Sciences. Moscow

Received
July 28, 1971

REFERENCES

1. *Peyraud P. I.*—J. Phys. **29**, 88, 306, 872 (1968).
2. *Zeldovich Ya. B., Levich E. V.*—Pisma v ZhETF **11**, 497 (1970).
3. *Zeldovich Ya. B., Levich E. V.*—ZhETF **55**, 2423 (1968).
4. *Krasyuk I. K., Pashinin P. P., Prokhorov A. M.*—Pisma v ZhETF **12**, 439 (1970).
5. *Kompaneets A. S.*—ZhETF **31**, 876 (1956).
6. *Galeev A. A., Karpman V. I., Sagdeev R. Z.*—Dokl. AN SSSR **157**, 1088 (1964).
7. *Drummond L.*—Prepr. (1967).
8. *Tsytovich V. I.* Nonlinear effects in plasma. New York: Plenum (1970).
9. *Zeldovich Ya. B.*—Dokl. AN SSSR **163**, 1359 (1965).
10. *Paradoksov P.*—UFN **89**, 707 (1966).
11. *Levich E. V.*—ZhETF **61**, 112 (1971).
12. *Sunyaev R. A.*—Astron. Zh. **48**, 244 (1970).

Commentary

In thermodynamic equilibrium the "chemical potential" of photons is identically equal to zero since photons may be generated and absorbed. Therefore, true equilibrium Bose-condensation of photons is impossible.

However, kinetic Bose-condensation is possible in the situation when radiation and absorption may be disregarded and only scattering with a frequency change occurs, but with conservation of the number of photons. It is just this situation that is considered in the article using A. S. Kompaneets' kinetic equation.

The article is of particular interest in connection with the predicted possibility of formation of narrow spectral lines in the spectra of cosmic radio-sources with a high brightness temperature of radiation. It is notable that the lines formed do not bear any relation to any quantum-mechanical system with discrete energy levels, but are a consequence of the induced Compton-effect.

The ideas expressed in the article were developed in Ya. B.'s review,[1] and in connection with processes in plasma in a number of articles.[2,3,4,5]

[1] *Zeldovich Ya. B.*—UFN **115**, 161–197 (1975).

[2] *Galeev A. A., Sunyaev R. A.*—ZhETF **63**, 1266–1281 (1972).

[3] *Galeev A. A., Sagdeev R. Z.*—In: Voprosy fiziki plazmy [Problems of Plasma Physics]. Moscow: Atomizdat **7**, 3–145 (1973).

[4] *Galeev A. A., Sagdeev R. Z.*—In: Osnovy fiziki plazmy [Foundations of Plasma Physics]. Moscow: Energoatomizdat **1**, 590–638 (1983).

[5] *Zeldovich Ya. B., Levich E. V., Sunyaev R. A.*—ZhETF **62**, 1392-1408 (1971).

24

Equilibrium Concentration of Positrons in an Optically Thin Relativistic Plasma*

With G. S. Bisnovatyĭ-Kogan, R. A. Sunyaev

In a plasma that is opaque to radiation, positron production begins at temperatures $kT \sim 0.1\,mc^2$. Very rapidly (at $kT \sim 0.4\,mc^2$, where m is the electron mass), the pair pressure becomes equal to the radiation pressure, and can greatly exceed the pressure of the initial electrons [1]**. In a low-density plasma, when the radiation goes off freely, the positron number density is determined by collisions of e^- and e^+ with the nuclei and by e^+ e^- pairs' annihilation (with emission of photons). No detailed equilibrium takes place and the equilibrium thermodynamic formulas are not valid. We present in this note the physical picture of the processes in such a plasma.

The main result is the absence of an equilibrium state at temperatures exceeding 20 MeV, thus establishing the upper limit of the temperature of an optically thin relativistic plasma.

Annihilation is a process of second order in the charge, and its cross section is of the order of

$$\left(\frac{\hbar}{mc}\right)^2 \alpha^2 g\left(\frac{E}{mc^2}\right) = r_0^2 g\left(\frac{E}{mc^2}\right),$$

where $\alpha = e^2/\hbar c = 1/137$ is the fine-structure constant, $r_0 = e^2/mc^2 = 2.8 \cdot 10^{-13}$ cm is the classical electron radius, and E is the pair energy in the c. m. s. When $E \gg mc^2$ we have $g \sim E^{-2}$.

The number of annihilations per unit volume and per unit time, obtained by integrating over the Maxwellian distributions of the electrons and positrons, is

$$A = \pi n_+ n_- c r_0^2 \psi(\theta), \qquad \theta = \frac{kT}{mc^2},$$

*Pisma v Zhurnal eksperimentalnoĭ i teoreticheskoĭ fiziki **12** (2), 64–66 (1970).

**Here is meant the situation in a hot universe where, before generation of pairs, the number density of electrons is 10^9 times less than that of photons (Editor's note).

where

$$\psi(0) = 1 \quad \text{and} \quad \psi(\theta) \sim \theta^{-2} \qquad \text{for} \quad \theta \gg 1.$$

Pair production in collision of charged particles is of fourth order in the charge, so that its cross section is of the order of

$$\left(\frac{\hbar}{mc}\right)^2 \alpha^4 f\left(\frac{E}{mc^2}\right) = r_0^2 \alpha^2 f\left(\frac{E}{mc^2}\right).$$

After integrating over the Maxwellian distribution, we obtain the number of pair productions

$$B = \pi(n_p + n_+ + n_-)(n_+ + n_-)cr_0^2\alpha^2\varphi(\theta).$$

$$\varphi(\theta) \sim e^{-2/\theta} \quad \text{for} \quad \theta < 1 \quad \text{and} \quad \varphi \sim \text{const} \quad \text{for} \quad 1 \ll \theta < M/m,$$

so that the nuclei are nonrelativistic. We disregard here the slowly varying factors that are logarithmic in θ and the difference between the numerical coefficients for the pe^-, pe^+, e^-e^-, and e^-e^+ collisions. Electroneutrality yields $n_p + n_+ = n_-$, so that finally we have

$$\frac{dn_+}{dt} = B - A = \pi c r_0^2 n_- n_+ \left[2\alpha^2\varphi(\theta)\left(1 + \frac{n_-}{n_+}\right) - \psi(\theta)\right].$$

We see therefore that in the stationary state to each temperature corresponds a definite ratio of the positrons to the electrons

$$\left.\frac{n_+}{n_-}\right|_{\text{st}} = \frac{2\alpha^2\varphi(\theta)}{\psi(0) - 2\alpha^2\varphi(\theta)}.$$

An analogous situation arises in the determination of the degree of ionization of a low-density plasma in which the ionization is produced by electron impact and radiative recombination occurs (Elwert's formula).

Exact calculations using the formulas given in [2] yield the following expression for the positron number density ($\theta > 1$, $\psi > 2\alpha^2\varphi$, $n_+/n_- < 1$)

$$n_+ = \frac{56 n_p}{27\pi^2}(\alpha^2\theta^2 \ln^2(1+\theta)) \approx 10^{-5} n_p \theta^2 \ln^2(1+\theta).$$

We note the difference from the case of complete statistical equilibrium, where the product $n_+ n_-$ is a constant on the order of $(mc/\hbar)^6\chi(\theta)$.

When $n \ll (mc/\hbar)^3 = 10^{33}\,\text{cm}^{-3}$, in an optically thin and low-density plasma such as is encountered in astrophysics, the stationary positron concentration calculated above is quite small compared with the statistic-equilibrium positron concentration at the same temperature. New phenomena arise, in which $\alpha^2\varphi/(\psi - \alpha^2\varphi)$ becomes equal to unity. The equality $n_+ = n_-$ sets in, with both n_+ and n_- tending to infinity, so that there is no stationary solution. This takes place at $T \simeq 40\,mc^2 \approx 20$ MeV. In the case of a cascade-like growth of the number of pairs $n_+ = n_-$ the kinetic equation takes the form

$$\frac{dn_+}{dt} = D(\theta)n_+^2, \quad D = \pi c r_0^2[2\alpha^2\varphi(\theta) - \psi(\theta)].$$

This equation would have n_+ go off to infinity within a finite time. An obvious limitation is the statistical equilibrium which would set in if the plasma were to become optically thick also with respect to bremsstrahlung absorption of the photons, and not only with respect to the Compton scattering and pair production by the photons. In the astrophysics of discrete radio sources, however, we are always exceedingly far from equilibrium, which would require a gigantic emission of energy.

Actually, the condition that the temperature be constant is violated, and the temperature adjusts itself to produce $D \equiv 0$, while the positron number density corresponds to the energy pumping power.

We note finally that energywise pair production is always much less (by at least a factor α) than the bremsstrahlung. The unique role of pair production lies in the fact that in the presence of a magnetic field the positrons remain within the limits of the considered region of space, whereas the photons go off from the optically thin region ($\tau < 1$ with respect to Compton scattering). At an ultrarelativistic temperature, the bremsstrahlung photons cause pair photoproduction with a cross section smaller by a factor α than the photon-scattering cross section.

This factor α cancels out the ratio α^{-1} of the bremsstrahlung to pair production in collisions. The condition $\tau < 1$ suffices therefore for our results to be valid. When $\tau > 1$, so long as the bremsstrahlung absorption is small and there is no thermodynamic equilibrium, there also exists a temperature at which the stationarity vanishes; it is lower than 40 mc^2 and is close to mc^2 when $\tau \sim 10$.

Under astrophysical conditions, a relativistic plasma can consist for a long time of the same particles, and the energy loss is offset by pumping by radiation, shock waves, or alternating magnetic fields. In particular, such a situation can take place near radio-emitting regions of quasars and pulsars [3]. Another case is also possible, wherein fast particles are injected in the given region of space, and leave the region simultaneously with the energy loss.

Our calculations pertain only to a plasma with a lifetime larger than the time required to establish equilibrium. Observation of a plasma having an electron temperature (or effective energy) much higher than critical indicates that the plasma has stayed in such a state for a short time. A detailed article will be published in "Astronomicheskii zhurnal."

We are grateful to A. Z. Dolginov for the remark that stimulated this research.

Institute of Applied Mathematics
USSR Academy of Sciences. Moscow

Received
June 5, 1970

REFERENCES

1. *Landau L. D., Lifshitz E. M.* Statistical Physics. Reading, MA: Addison-Wesley (1969).
2. *Akhiezer A. I., Berestetskiĭ V. B.* Quantum Electrodynamics. New York: InterScience Publ. (1965).
3. *Levich E. V., Syunyaev R. A.*—Radiofizika **13**, 1873–1878 (1978).

Commentary

Now, after the discovery of amazingly powerful source in the annihilation line of positronium in the region of the Galactic Center;[1] after finding annihilation lines in the spectra of cosmic gamma-bursts;[2] and after the triumph of the Ruderman-Sutherland model[3] which relates observed properties of radiopulsars with the generation of electron-positron pairs, their acceleration and radiation—after this it is not surprising that the above article, the first to point out the possibility of catastrophically strong generation of pairs in high-temperature plasma, is widely cited and discussed. Its conclusions are being discussed, modified, and developed.[4,5,6]

We note that the basic citations are to a more detailed article by the authors[7] where they touch upon the problem of gamma-radiation of a high-temperature electron-positron plasma.

At the same time it must be mentioned that the posing of the question in the present article contains a far-reaching idealization. In the astrophysical situation one must consider the specific mechanism of supply of energy to the plasma and the spatio-temporal picture of the phenomenon.

[1] *Riegler G. R., Ling J. C., Mahoney W. A., et al.*—Astrophys. J. Lett. **248**, L13–L16 (1981).

[2] *Mazets E. P., Golenetskiĭ S. V., Aptekar R. L.*—Nature **290**, 378–382 (1981).

[3] *Manchester R., Taylor G.* Pulsars. San Francisco: W. H. Freeman (1977).

[4] *Cavallo G., Rees M.*—Month. Not. RAS **183**, 357–365 (1978).

[5] *Aharonian F. A., Atoyan A. M., Sunyaev R. A.*—Astrophys. Space Sci. **93**, 229–245 (1983).

[6] *Svensson R.*—Astrophys. J. **258**, 335–348 (1982).

[7] *Bisnovatyĭ-Kogan G. S., Zeldovich Ya. B., Sunyaev R. A.*—Astron. Zh. **46**, 24–31 (1971).

PART TWO

ASTROPHYSICS AND COSMOLOGY

IV

Elementary Particles and Cosmology

25
Quarks: Astrophysical and Physicochemical Aspects*

With L. B. Okun, S. B. Pikelner

1. Introduction

The preceding issue of Uspekhi contained a number of predictions about future directions and problems in elementary particle physics. In connection with this discussion it is worthwhile to explore another possible direction for investigation which has developed recently alongside studies using accelerators and cosmic rays.

We refer to attempts to find new rare types of stable particles in nature. Thus, in studying elementary particles one can use, in addition to the traditional methods (but of course not in place of them) physicochemical methods of enrichment and investigation. We shall discuss various aspects of this approach.

There has recently been discussion of the possible existence of new types of particles that are heavier than the proton and have fractional charge. The classification of the known strongly interacting particles (SU_3 or SU_6 symmetry) leads naturally to the assumption that there exist three particles (quarks) with charges $+\frac{2}{3}e$, $-\frac{1}{3}e$, $-\frac{1}{3}e$ [1, 2] (cf. the popular summary [48]).

There is still also the possibility that new types of particles with integral charge exist, both within the framework of the SU_3 symmetry [3, 8][1] and also without any connection to the SU_3 symmetry [9].

*Uspekhi fizicheskikh nauk **87** (1), 113–124 (1965).

[1]Schemes that combine the particle classification according to SU_3 or SU_6 with the assumption of the existence of fundamental particles with integral, instead of fractional, charges, are more complicated than the quark scheme, and contain more fundamental particles.

Among the quarks, these particles with fractional charge, there is one, the lightest, that should be stable not only in vacuum but, because of its fractional charge, also stable in contact with ordinary matter (nuclei, electrons). The particles with integral charge may be unstable, but there may be selection rules that cause these particles to be stable. These selection rules may be connected with the conservation of a new quantum number that is like the charge—the supercharge.[2] Under strict conservation of supercharge the creation and annihilation are possible only in pairs—particle and anti-particle. The selection rules may also be related to the fact that the particles have unusual combinations of the usual quantum numbers. In this case their numbers should not be conserved absolutely, but, as for the quarks, modulo some number. For example, baryons with integer spin can be created in pairs in nucleon collisions (conservation modulo 2). We note that the quarks are conserved modulo 3, i.e., in reactions only 3, 6, 9, etc., quarks can be created or destroyed; for example, 7 quarks cannot convert completely into ordinary matter—there still remains one free quark.

2. Formation of Quarks by High Energy Particles

High energy experiments have so far not led to the discovery of new stable particles and have shown that the masses of such particles cannot be small. Experiments at accelerators [10–15] have detected no quarks up to masses of 3–5 GeV with production cross-section larger than 10^{-34}–10^{-32} cm^2.

A cosmic-ray experiment [21] at a height of 2.5 km gives an upper limit of $I_{1/3} < 1.6 \cdot 10^{-8} \mathrm{cm}^{-2}\mathrm{s}^{-1}\mathrm{sr}^{-1}$ for the flux of relativistic quarks with charge $\frac{1}{3}e$ (a sea level experiment [22] gives the flux limit $20 \times 10^{-8}\ \mathrm{cm}^{-2}\mathrm{s}^{-1}\mathrm{sr}^{-1}$). If we assume that the slowing down of quarks in the atmosphere is approximately the same as for nucleons, and that the cross section for creating them is $\sim$ 0.01 mb, the upper limits given show that $m_q > 7$ GeV. In this estimate we have used the integral spectrum of the primary cosmic radiation

$$N(E) = 0.9E^{-1.5}\ \mathrm{cm}^{-2}\mathrm{s}^{-1}\mathrm{sr}^{-1}, \tag{1}$$

where E is in GeV.

Searches for long-lived particles with integral charge at the accelerators at Brookhaven and CERN [16–20] have given a negative result up to a mass of 4 GeV, and have shown that if such particles are formed their numbers are approximately three orders of magnitude smaller than the number of antiprotons.

Aside from a search for new stable particles in experiments on collisions at high energies, there is another procedure—a search for particles that were formed long ago and have taken the temperature of their surroundings. If

[2]By the supercharge we mean three times the average charge of an SU_3 supermultiplet. If SU_3 triplets with integral charge (like the p, n, Λ) exist, their supercharge is 1. For the quarks the supercharge is zero.

one assumes that quarks can be created by the primary cosmic ray nucleons with $E > 300$ GeV, then, for the parameters cited above, during 5×10^9 years there should have been 10^{11} cm^{-2} created in the atmosphere, and this is approximately 10^8 quarks per gram of absorbing layer of atmosphere. The quarks diffuse from the upper to the lower layers of the atmosphere, there serve as nuclei for condensation of drops, fall with precipitation onto the Earth's surface, and are mixed into the oceans. We then get approximately 10^5 quarks per gram of water. But if the precipitation is collected in outdoor reservoirs which are then allowed to evaporate, the number density may be 1–2 orders of magnitude higher.

Because of the absence of mixing, the number density of quarks formed by cosmic rays may be of the order of 10^9 per gram in meteorites. But the size of the meteorite must be very large (~ 30 cm) to retain the quarks at the time they are created.

Under special circumstances the number density of cosmic rays may be much higher than average. On the Sun, at the time of chromospheric flares, many cosmic rays are produced, albeit of relatively low energy. Much more powerful sources are the variable stars like τ Tauri which are in the process of gravitational collapse: in these stars there is strong convection which results in varying magnetic fields and acceleration of particles. Thus the stars are the source of the "proper cosmic rays." This is indicated by the anomalously large content of Li and certain other elements that are fragments of heavier nuclei. There are stars with an anomalously high content of He^3 which, according to [23], is formed from He^4 by cosmic rays. The increased deuterium and Li content of the Earth is also related to the period of formation of the solar system, when the Sun was a star like τ Tauri and irradiated the planetary matter. This energy spectrum is not known, but if it were sufficiently hard, quarks would have been formed along with the Li.

The most powerful sources of cosmic rays, superstars or quasars [24], also must produce quarks. Finally quarks may be produced in small-scale explosions occurring in galactic cores. But the relatively short duration of the explosion process and the small mass of gas leads one to think that the main contribution in the Galaxy comes from ordinary cosmic rays.

3. Creation and Burnout of Quarks in the Initial Period of Expansion of the Metagalaxy

If the hypothetically stable particles (not quarks) possess some particular strictly conserved quantum number, like the baryonic charge, their minimum number density is a universal constant like the total baryonic charge of our portion of the Universe, and can in principle be arbitrary. If, however, these particles are conserved modulo some base, their number density will depend on their history, i.e., on the earlier physical condition of the material and on the processes that lead to the annihilation of the particles. For example, let us consider annihilation of quarks.

Since quarks are heavier than nucleons we have the process

$$q_1 + q_1 \to q_3 + q_{-1} \tag{2}$$

followed by $q_1 + q_{-1} \to nq_0$. Here the subscript indicates the number of quarks contained in the particle, the minus sign denoting the antiparticle. Thus q_1 denotes single quarks, q_2 pairs of quarks coupled by the strong interaction, q_3 ordinary baryons made up of three quarks, and q_0 mesons. Because of relation (2), annihilation of the quarks can occur in a series of pair collisions instead of via the much rarer triple collisions.

The process (2) does not consider the possibility of the existence of a q_2. But including q_2 does not alter the conclusion about the role of pair collisions. If $m_2 > m_1+m_3$, the reaction $q_2 \to q_3+q_{-1}$ goes; but if $m_2 < m_1-m_3$, the free quarks are unstable: $q_1 \to q_3 + q_{-2}$. In this case the particles with fractional charge that can exist in nature are the diquarks. If, finally, m_2 lies in an interval that guarantees the simultaneous stability of q_2 and q_1, both types of particles are annihilated in all variants with double collisions. One can treat q_4, q_5, ..., similarly with same conclusions.

From the point of view of possible creation and burnout of quarks, the most important period is the initial expansion of the metagalaxy, if this expansion proceeded from a singular state. We must start from some definite cosmological hypothesis. Here we shall assume that the Friedmann model of a homogeneous isotropic universe (cf., for example, review [25]) is applicable with sufficient accuracy up to $t \lesssim 10^{-7}$ s. Deviations from homogeneity and isotropy, which may be significant at an early stage [26, 27], can of course significantly change the results. The choice between open and closed models is entirely unimportant at the early stage. On the other hand the Friedmann theory leaves free the values of the thermodynamic parameters, the specific entropy and specific leptonic charge of unit rest mass of the matter.

Let us assume following Gamow [28] that in the singular state at infinite density the specific entropy was large ("hot model").[3] Then in the early stages of the expansion the densities of photons and of all types of particle-anti-particle pairs greatly exceeded the excess density of baryons, corresponding to a charge asymmetry in our vicinity, which we extrapolate to the whole Universe. For $t \to 0$, $\rho \to \infty$, $T \to \infty$. Fixing the quark mass m, one can easily find the relative equilibrium number density of quarks for $T \leq m$:[4] this number density is $n \sim e^{-m/T}$. There are no conditions at the present time under which a significant number of quarks would be in equilibrium. We must therefore determine the moment when, in the course of the expansion, the actual quark number density ceased following the equilib-

[3]This hypothesis has also been supported recently by Hoyle and Tuler [44].

[4]We work in units with $\hbar = c = k = 1$. In these units, mass, energy and temperature have the same dimensions, and can be in degrees, MeV, proton masses, etc.

rium behavior (the moment of "freezing" of the equilibrium). The analogous problem was solved earlier for the freezing in of antinucleons [29].

From the equations of general relativity for the Friedmann solution and from thermodynamics

$$\rho = ahT^4 = \frac{3}{32\pi G t^2}, \qquad s = \frac{4ahT^3}{3\rho_b}, \tag{3}$$

here $a = 4\sigma/c = \pi^2/15$, where σ is the Stefan-Boltzmann constant, ρ_b the density of rest mass of excess baryons, s the specific entropy per unit of this mass, h a dimensionless number that takes account of the presence of other particles, that are in equilibrium with the radiation at the given temperature ($h = 1$ for protons only, $h = 2.75$ for photons and e^+, e^- pairs when $T > m_e c^2$).

With muons and neutrinos included, $h \sim 9$. Formulae (3) apply to the initial stages, when $\rho_b \ll \rho$. We denote by n the number density of quarks relative to the number density of excess baryons $N = \rho_b/M$, where M is the baryon mass. The equilibrium number density when $T \ll m$ is

$$n_{\text{eq}} \approx \left(\frac{2}{N}\right)^{2/3} \frac{mT}{2\pi} e^{-(m-M/3)/T}. \tag{4}$$

This equation reminds one of the Saha equation, but for a system of three particles. The kinetic equation has the form

$$\frac{dn}{dt} \approx v\sigma_2 N(n_{\text{eq}}^2 - n^2), \tag{5}$$

where v is the mean velocity, σ_2 the cross section for collision of two quarks, antiquarks or diquarks, leading to a reduction of the number of such particles by unity. The factor N appears because the quark number density expressed in cm^{-3} is $Cq = nN$. It is clear that Cq changes both because of reactions and because of the general expansion, whereas n changes only because of reaction. The time for establishing equilibrium is $\tau \approx (v\sigma_2 N n_{\text{eq}})^{-1}$.

In order to determine the moment of "freezing," we must compare the time τ with the characteristic time for change of n_{eq} because of the expansion, described by formula (3). This characteristic time t_1 is gotten from the condition

$$t_1^{-1} = \frac{d \ln n_{\text{eq}}}{dt} \approx \frac{m}{T}\frac{1}{2t} = \frac{1}{2\theta t}, \tag{6}$$

where we use (4) and $T \sim t^{-1/2}$ from (3). The quantity $\theta = T/m$. (We neglect the change in the pre-exponential factor in (4).)

The whole period of expansion can be divided into two stages. In the first $t < t_1$ and $n \approx n_{\text{eq}}$. In the second stage $t > t_1$, $n > n_{\text{eq}}$ and one can neglect the creation of new quarks. At the time t_0 separating the two stages, one can approximately set $n \approx 2n_{\text{eq}}(t)$. Integration of the equation

$$\frac{dn}{dt} = -v\sigma_2 N n^2 \tag{7}$$

from $t = t_0$ gives for the relative number density after "quenching" the expression

$$\frac{1}{n(\infty)} = \int_{t_0}^{\infty} v\sigma_2 N\, dt \approx 2\sigma_2 v N_0 t_0. \tag{8}$$

The last result is obtained if we use the relation $N = N_0(t_0/t)^{3/2}$ (cf. (3)) and $\sigma_2 v = \text{const}$. Using t_0 from (3), we rewrite (8) in the form

$$n(\infty) = \sqrt{\frac{32\pi a}{3}} \frac{\sqrt{G}}{\sigma_2 v T_0 h_0^{1/2}} \frac{h_0 T_0^3}{N_0} = \sqrt{\frac{32\pi a}{3}} \frac{\sqrt{G}}{\sigma_2 v T_0} \left(\frac{h^2}{h_0}\right)^{1/2} \frac{T^3}{N}, \tag{9}$$

where T is the temperature of the radiation remaining at the present time (the expansion is assumed to be isentropic) while N is the average nucleon density at the present time. The value $h \approx 3$ refers to the time when only photons and the two kinds of neutrino pairs were left ν_e, $\overline{\nu}_e$, ν_μ, $\overline{\nu}_\mu$; $h_0 \approx 9$.

In units of the nucleon mass M we may take $\sigma_2 v \approx M^{-2}$, $T_0 \approx M$ (cf. below), $G = 0.6 \times 10^{-38}\, M^{-2}$. It then follows that

$$n(\infty) \approx 6 \cdot 10^{-19} \left(\frac{T^3}{M}\right). \tag{10}$$

Basically the low number density of quarks is a consequence of the smallness of the gravitational forces. To give a dimensionless characteristic of the gravitational interaction, we must write $GM^2/\hbar c$ (M is the nucleon mass), by analogy with $e^2/\hbar c$, the fine structure constant. It is qualitatively easy to understand why the weakness of the gravitational interaction results in a low number density: the rate of expansion in the initial stages must be chosen so that the kinetic energy of expansion overcame the gravitational attraction and enabled the material to go over from the value $\rho = \infty$ to the present value $\overline{\rho} = 10^{-30}$. The weakness of gravitation means a slow expansion and produces the conditions for the destruction of quarks.

On the other hand, the higher the temperature and entropy, the more different particles (photons, e^+, e^- pairs, etc.) occur per nucleon; in a more dilute system the quarks collide and die less frequently, and the number density of quarks per nucleon is higher.

The value of T^3/N taken per nucleon is now known with a very large uncertainty. If $T \approx 1$ K, while $N \approx 2 \times 10^{-7}$ cm^{-3}, which corresponds to the lower limit of the density including some galaxies (cf. [30] assuming the Hubble constant H = 100 km/s · Mpc), then $T^3/N \approx 10^9$. This corresponds to a hot model of the Universe. Direct measurements of the metagalactic background radio emission lead to the conclusion that $T < 3$ K.[5] But the value $T^3/N \approx 1$ also does not contradict the observations. Such

[5] The latest information, obtained by measuring radio noise at 7 cm, favors the value $T \sim 4\,K$ [47].

a small entropy corresponds to a model with initially cold matter consisting of free quarks and nucleons.[6] Thus, depending on the value of T^3/N, one can have a quark concentration after the primary expansion of 10^{-9}–10^{-18} also if at the beginning their concentration was close to unity. It should be emphasized that these numbers are rough estimates. In particular they may change drastically if we include inhomogeneity and anisotropy.

Let us now fix the values of t_0 and T_0, which are determined from the equation

$$(v\sigma_2 N n_{\text{eq}})^{-1} = 2\theta t \tag{11}$$

or

$$\theta e^{-1/\theta} \approx \sqrt{\frac{32\pi a}{3}\frac{G^{1/2}s^{1/3}}{v\sigma_2 m}\left(\frac{h}{h_0}\right)^{1/3}} \approx 5 \cdot 10^{-16}, \tag{12}$$

if $s \approx 10^9$. From this it follows that $\theta \approx 1/30$. If the quark mass is $m \approx 10$ GeV, the freezing temperature $T_0 \sim 300$ MeV, and the time of the freeze $t_0 \approx G^{-1/2}T_0^{-2} \approx 10^{-5}$ s. From (8) and (10) it follows that the value of T and consequently also of $n(\infty)$ are weakly dependent on the quark mass m. The point is that although the equilibrium number density n_{eq} depends exponentially on m and T, the reaction rate itself depends on n_{eq}. Thus the time t_0 turns out always to correspond to a definite n, which depends on m only as a power.

The whole process of freezing is completed at a temperature higher than 100 MeV, so that Coulomb barriers and Coulomb attraction of the quarks to nucleons play no part in the estimate of $v\sigma_2$, characterizing the cross section for the reaction between two quarks. Under these conditions nuclei do not exist.

4. Conservation of Quarks During the Evolution of the Galaxy

Earlier we estimated the burnout of quarks during the first microseconds of the Friedmann expansion. Now let us consider how the quark content must change during the process of further evolution of matter. This problem is of such great interest because an improvement of the upper limit for the quark content could give some limitations on the choice of cosmological models.

According to the present cosmological pictures, the primary gas developed condensations which, gradually fragmenting, gave the first galaxies. The gas in the galaxies changed into stars, of which the more massive ones went through their evolution rapidly, ejecting part of the gas, enriched in heavy

[6] It was already pointed out in [45] that in theories in which the baryons are regarded as composites, at sufficiently high density one cannot regard the gas as consisting of all the experimentally known particles ($p, N, \Lambda, \Sigma, \Xi$). One must instead speak of a gas of the "really" elementary particles, the p, N, Λ of the Sakata model, or the quarks, according to the present view.

elements, into interstellar space. In our Galaxy more than 98% of the gas has already been converted into stars. In the stars the process of burnout of the quarks in pair collisions has continued. But now the Coulomb interaction between quarks and nuclei and between the quarks themselves begins to be important.

For the Maxwell distribution the number of reactions per cm^3 per second is (cf., for example, [32])

$$C_1 C_2 F_{12} = C_{12} C_1 C_2 \left(\frac{Z_1 Z_2}{A_{12}} \right)^{1/3} T^{-2/3} \exp \left[-3 \left(\frac{\pi^2}{2} \alpha^2 Z_1^2 Z_2^2 \frac{m_{12}}{T} \right)^{1/3} \right], \tag{13}$$

where m_{12} is the reduced mass of the particles, A_{12} is the same quantity in fractions of the proton mass, $\alpha = 1/137$, $C_1 = C_2 = C(T)$ are the particle number densities, C_{12} is a constant for the particular reaction. If we take as the reaction parameter that for deuterium and express T in ergs, $C_{12} \approx 2 \times 10^{-20}$. From equations of the type of (7) and from (13) we find the number density of the remaining quarks

$$C(t) \approx \frac{2}{F_{qq}(t)}, \tag{14}$$

if, of course, the initial number density was higher.

For quarks with $Z = +2/3$ and $A \approx 5$ the number density after 10^9 years for $T = 10^6$ K will be $C_q \leq 10^{15}\ \mathrm{cm}^{-3}$, which amounts to 10^{-9} of the hydrogen concentration in the Sun's layers at this temperature. At higher temperatures the burning of quarks increases markedly. For $T = 10^7$ K, $C_q \lesssim 10^6\ \mathrm{cm}^{-3}$, i.e., 10^{-18} per gram in the corresponding layers.

For quarks with $Z = -1/3$, it follows from (13) that burnout occurs much faster. But these quarks can combine with nuclei to form a stable system. The charge of the quark-proton system is $+(2/3)e$ and the burnout of the quarks should now occur at approximately the same rate as for the case treated above. But at temperatures where annihilation is possible, the quarks will first be detached from the protons and bound to He or heavier nuclei. The burnout will no longer be important in this case because of the large values of Z and A. We make a quantitative estimate.

The binding energy of a quark to a nucleus is $Q = 2.76 Z_n^2 A$ keV, where Z_n is the charge number of the nucleus and A is the reduced mass of the system in units of the proton mass. According to the Saha formula, the relative number density of free and bound quarks is

$$\frac{C_q}{C_{qn}} = \frac{1}{C_n} \left(\frac{A m_n T}{2\pi} \right)^{3/2} e^{-Q/T}.$$

At $T = 10^6$ K the number density of quarks attached to protons is 10^9 times as great as that of the free quarks. But equilibrium is reached after 10^{-4}–10^{-5} s, so that even with a small fraction of free quarks, after $t \sim 10^5$ s they all go over to the heavier nuclei, in this case He. At higher temperatures the quarks will separate from the He but will attach to heavier elements. In all cases the time for going over to the heavier nuclei is much less than the burnout time.

At $T = 10^7$ K the fraction of quarks attaching to He is 10^9 times greater than that of the free quarks. Using the number densities of C, N, and O one can estimate that the time for all the quarks to go over to these nuclei is less than 10^7 s. Detachment from these nuclei requires a very high temperature. For example, for $T = 5 \times 10^7$ K, which exceeds the temperature in the interior of stars of the main sequence, the fraction of free quarks as compared with the quarks attached to O is 10^{-100}. Thus, transfer to heavier nuclei such as Fe practically does not occur; the fractions of Fe and O atoms having quarks are the same, but the Fe content is small compared to the O. If during explosion the temperature is raised above 10^9, the quarks go over to the Fe.

Summarizing, we may say that quarks with $Z = +2/3$ falling into a star are largely annihilated; the number density stays less than 10^{-18} per gram. Only those are left that stayed on the surface all the time. This is possible in stars having no convective zone, but since in the early and late stages almost all stars are convective, retention is improbable. Quarks with $Z = -1/3$, falling into a star, attach to the elements C, N, O and heavier ones. One should therefore look for spectra of atoms and molecules with C, N, O and heavier atoms, whose nuclei contain quarks. The atomic spectra should differ from the usual ones because of the change in the nuclear charge (the isotope shift is small) while the molecular spectra will be different because of the change in the vibration-rotation parameters of the system.

We now consider the quarks which do not remain in stars. They should either be free ($Z = +2/3$) or attached to hydrogen ($Z = -1/3$). We should mention that in the interstellar gas quarks and quark nuclei should attach to dust particles because of the presence of the charge. Thus their lines may also not be present in the spectrum of interstellar gas. Besides, the present methods of investigating the interstellar medium cannot in general give information about elements with low number density.

We now estimate the probability that during the process of formation of the Galaxy and the building up of stars no quarks entered the stars. In each process of star formation about half the condensing gas is converted into stars of low mass, which evolve slowly. The other half is converted into massive stars which go through their cycle rapidly and finally eject about half their mass into the interstellar space. The remainder collapses or forms a superdense star. Let us assume that 1/4 of the mass converted initially into stars is again ejected into the interstellar gas and mixes with the residue of the primary gas, after which star formation proceeds once more (cf., for example, the survey [33]). Such a gradual emission of gas is indicated by the gradual change in the chemical composition of stars and clusters with changes in their spatial and kinematic characteristics. Suppose that the fraction α of all the remaining gas is converted into stars in one cycle. This quantity is unknown; it may be of the order of 0.1–0.3, but its precise value is irrelevant. After n cycles, $(1 - \frac{3}{4}\alpha)^n$ of the gas remains, of which the fraction $a = [1 - \alpha/1 - (3\alpha/4)]^n$ did not pass through the star stage. At the present time 2% of the initial mass of the Galaxy is left in the form of gas. Thus $n\alpha \approx 5$ and $a \approx 0.25$–0.20 over the wide range of values of α from 0.1 to 0.3. Despite its naiveté, the computation shows that the gas in interstellar space should contain a considerable admixture of the original gas, and consequently the percentage content of quarks in the interstellar gas should be 10–20% of the initial value. As already pointed out, however, to detect them there is hardly possible.

The best conditions for experimental detection occur on planets like the Earth. The solar system apparently was formed during the process of contraction of a nebula whose central part became the Sun. The planets, in particular the inner ones, were formed mainly from dust [34, 35]. Thus the fraction of quarks in them should be reasonable. During the period of formation of the Earth, the temperature of most of the solar mass must have been low, since this was still during the stage of compression. The physical conditions during this period have not been carefully studied; it is difficult to judge the probability of mixing with deep layers and annihilation of quarks, if they have positive charge. The negatively charged quarks would have gone into He nuclei or heavier elements. It is more probable, however, that there was no significant burnout, since $T \sim 10^6$ deg was attained only at the very center of the condensing Sun, while the envelope was mainly convective. In addition the gas around the Earth was rarefied, and did not completely draw away the dust out of which the Earth was formed. Thus, if the quarks are negatively charged they should be sought in hydrogen or heavier elements, and their content there should be comparable to the initial value; if they are positive, their content can be either large or small, depending on the conditions of convection on the Sun during the period of formation of the Earth.

We should mention that the farther a planet is from the Sun, the higher the probability that quarks are retained in it. Meteorites that are the products of decay of comets are of great interest from this point of view. According to present notions, comets are retained on the periphery of the Solar system, go over under the influence of perturbations into orbits closer to the Sun, after which they gradually disintegrate [36]. Meteorites from such comets in all probability were not in the depths of the Sun during the period of formation of the solar system.

Since the annihilation of positively charged quarks occurs under the same conditions as the burning of deuterium, the deuterium content should serve as a check. However, its abundance on Earth is explained, as already mentioned, by the activity of the Sun during the period of condensation, when many of the cosmic rays were formed.

5. Possible Ways of Searching for New Particles

In searching for new particles one should consider the possibility of enriching or depleting samples by taking advantage of their physico-chemical properties.[7] Let us therefore consider some of the physico-chemical properties of the quarks and the hypothetical stable particles with integer charge, and discuss possible ways of searching for these particles.

Particles with integral charge $Z = +1(i^+)$ do not differ physico-chemically from the hydrogen isotopes. They are concentrated during the production of heavy water. Mass spectrometric investigation of heavy water samples gives a relative concentration $C < 10^{-10}$ based on hydrogen in the original water (before separation) [37, 38]. There is an important possibility of lowering this limit by several orders of magnitude. Optical spectroscopic methods for detecting such "hydrogen" under terrestrial conditions are not as good as the mass spectroscopic method. Optical searches extraterrestrially, using the isotopic shift are also very difficult, because this shift is usually very much smaller than the line width. The example of the search for deuterium in the Sun's spectrum shows that the sensitivity is no better than 10^{-3} of the content of the main isotope [39]. The molecular spectrum of hydrates depends strongly on the masses of the atoms, but it is observed only within a narrow interval of stellar spectral classes, limited from above by molecular dissociation and from below by the low brightness of stars.

The particles with integral charge $Z = -1(i^-)$, in particular, the antiparticles of the preceding ones, are created by energetic cosmic rays in the atmosphere, attach to nuclei in the atmosphere, jump from the protons to heavier nuclei, as in the case of muonic atoms. It is natural to look for them among the isotopes of carbon (C, i^-), nitrogen (N, i^-) and oxygen (O, i^-). Such atoms are formed as the result of reactions like $\mathrm{O} + p, i^- \to \mathrm{F}, i^- + \gamma$. The quark binding energy with such nuclei is of the order of several MeV.

The $q^{+2/3}$ or $q^{-1/3}$ quark (depending on which one is the lighter) and the corresponding antiquark are stable and can accumulate in the Earth. Atoms containing the quarks have an uncompensated electric charge and behave like ions. Solvation of the ions in polar solutions practically excludes the possibility of evaporation from water of the quarks or of molecules containing quarks. One can cite the example of the Li^+, Na^+, F^-, and Cl^- ions. All these ions have a noble gas structure, and the solubility in water is small. But the charge on the ions gives an energy of solvation of the order of

[7]In principle it is impossible, for example, to exclude the possibility of biological concentration of quarks by some organisms.

1.5 eV, which makes it impossible to drive them out of solution. For quarks the energy of solvation is lower in the ratio Z or Z^2, i.e., it is of order 0.5–1 eV for $Z = 2/3$ and 0.2–0.5 eV for $Z = -1/3$. At 100°C the corresponding value of $e^{-Q/T}$ is 10^{-14}–10^{-17} in the first case and 10^{-3}–10^{-7} in the second. The nonvolatility of the quarks must be considered in mass-spectroscopic searches for them; their number density in vapors is substantially less than in liquids. Upon distillation of the water the sample is purified of quarks. The experiments of Kohlrausch on the electrical conductivity of pure water [40] apparently give a relative number density of quarks less than 10^{-9}. Because of our earlier remarks, this estimate is even worse for ordinary water.

Ions containing quarks should be adsorbed from nonpolar materials (petroleums, oils) onto the surface of minerals containing petroleum, or onto filters or the glass and metallic walls of vessels containing oil. Loss of quarks during purification of oil should be considered in interpreting the Millikan experiment, which indicates a relative number density below 10^{-15}.

The technique of accurate determination of very small periodic forces, which has been successfully developed by V. B. Braginskiĭ [41], has been proposed by him for use in detecting single fractional charges in samples weighing up to 10^{-4} g, placed in a periodic electric field. This corresponds to a sensitivity of 10^{-19}.

The optical spectra of atoms containing quarks should be very characteristic. The L_α lines for $(q^{+2/3}e^-)$ and $(q^{-1/3}pe^-)$ atoms fall in the near ultraviolet $\lambda = 2750$ Å. The Hubble red shift for distant galaxies and quasars shifts this line to a convenient region for observation. Atoms of C, N, O, and Fe, which have a $q^{-1/3}$ quark attached to their nuclei, should have spectra altogether different from the usual spectra of these elements, since all the screening parameters of the electron shells are changed. Calculation of the spectra of such atoms represents a definite quantum mechanical problem. After the calculations have been made, one must make a serious search for these lines and also for the $\lambda = 2750$ Å line and molecular lines for unusual masses[8] in the spectra of various cosmic objects, entirely independently of all estimates of quark number densities, which at present cannot pretend to be reliable.

6. Conclusion

One of the main questions arising in connection with the possible existence of new stable particles, and quarks in particular, is why these particles have so far not been seen by us in nature. As the above arguments show, this is not at all surprising. In the process of the primary Friedmann expansion quarks were burned extensively and changed into nucleons; their number density becomes of order 10^{-9}–10^{-18} per nucleon, even if the quarks and nucleons had the same initial number densities. If the positively charged

[8]For analogous proposals for positronium and antiprotonium cf. [42, 43].

quarks are lightest, and consequently stable, their burnout continues in the interior of stars. The negatively charged quarks are "conserved," attaching themselves to nuclei. Thus the fact that quarks have so far not been seen can be regarded as an argument that the positively charged quarks are the stable ones. One should however remember that in the process of evolution a part of the matter does not go through the stage of being part of a star, and preserves the quark number density which was left after the initial stage of the Friedmann expansion.

In searching for thermalized quarks under terrestrial conditions one should take account of their physico-chemical peculiarities, such as solvation in water solutions and precipitation on the walls in nonpolar solvents. Under laboratory conditions the best methods for searching for quarks are to measure the elementary charges on macroscopic bodies and to use mass spectrometry. The latter method is especially effective in looking for new particles with integral charge. Optical spectroscopy can be used for quarks in extraterrestrial objects.

REFERENCES

1. *Gell-Mann M.*—Phys. Rev. Lett. **8**, 214 (1964).
2. *Zweig G.* CERN, Preprint (1964).
3. *Gursey F., Lee T. D.*—Phys. Rev. **B135**, 467 (1964).
4. *Franzini P., Lee J.*—Phys. Rev. Lett. **12**, 602 (1964).
5. *Gell-Mann M.*—Physics **1**, 63 (1964).
6. *Lee T. D.*—Nuovo Cim. **35**, 933 (1965).
7. *Okun L. B.*—Yader. Fiz. **1**, 297 (1965).
8. *Zeldovich Ya. B.*—Pisma v ZhETF **1**, 3 (1965).
9. *Okun L. B.*—ZhETF **11**, 1773 (1964).
10. *Adair R.* Coral Gables Conference on Symmetries Principles at High Energy. Univ. of Miami, W. H. Freeman, 36 (1964).
11. *Leipuner L. B., Chu W. T., Larsen R. S., Adair R. K.*—Phys. Rev. Lett. **12**, 423 (1964).
12. *Morrison D. R.*—Phys. Rev. Lett. **9**, 199 (1964).
13. *Bingham H. H.* et al.—Phys. Rev. Lett. **9**, 201 (1964).
14. *Hagopian V., Selove W., Ehrich R.* et al.—Phys. Rev. Lett. **13**, 280 (1964).
15. *Blum W., Brandt S., Cocconi V. T.* et al.—Phys. Rev. Lett. **13**, 353a (1964).
16. *Franzini P., Leontic B., Rahm D.* et al.—Phys. Rev. Lett. **14**, 196 (1965).
17. *Gilly L., Leontic B., Lundby A.* et al.—Proc. Rochester Conf., 808 (1960).
18. *Von Dar del G., Mermod R. M., Weber G., Winter K.*—Proc. Rochester Conf., 836 (1960).
19. *Cocconi V. T., Fazzini T., Fidecaro G.* et al.—Phys. Rev. Lett. **5**, 19 (1960).
20. *Baker W. F., Cool R. L., Jenkins E. W.*—Phys. Rev. Lett. **7**, 101 (1961).
21. *Bowen T., De Lise D. A., Kalbach R. M., Mortata L. B.*—Phys. Rev. Lett. **13**, 728 (1964).

22. *Sunyar A. W., Schwarzschield A. Z., Connors P. J.*—Phys. Rev. Lett. **136**, B1157 (1964).
23. *Jugaku J., Sargent W. L. W., Greenstein J. L.*—Astrophys. J. **134**, 783 (1961).
24. *Burbidge E. M., Burbidge G. R., Fowler W. A., Hoyle F.*—Rev. Mod. Phys. **29**, 548 (1957); *Layzer D.*—Phys. Rev. Lett. **15**, (1965); *Ginzburg B. L., Ozernoĭ L. M., Syrovatskiĭ S. I.*—Dokl. AN SSSR **154**, 557; *Ginzburg V. L, Ozernoĭ L. M., Syrovatskiĭ S. I.*—Quasistellar Sources and Gravitational Collapse. Chicago Univ., 937 (1965).
25. *Zeldovich Ya. B.*—UFN **80**, 357 (1963).
26. *McCrea W. H.*—Zs. Astrophys. **18**, 98 (1939).
27. *Zelmanov A. L.*—Tr. 6-go Soveshch. po voprosam kosmogonii [Reports on the 6-th Conference on Problems of Cosmogony]. Moscow: Izd-vo AN SSSR (1959).
28. *Gamow G.*—Phys. Rev. **70**, 572 (1946).
29. *Zeldovich Ya. B.*—Adv. Astron. and Astrophys. **3**, 242 (1965).
30. *Turtle A. J., Pugh J. E., Kenderdine S., Pauliny-Toth I. I. K.*—Month. Not. RAS **124**, 297 (1962).
31. *Allen C. W.* Astrophysical Quantities. University of London, Athlone Press, 284; *Oehm E.*—Bell System Tech. J. **40**, 1065 (1961); *De Grass J.*—Appl. Phys. **30**, 2013 (1959).
32. *Schwarzschild M.* Structure and Evolution of the Stars. Princeton (1958).
33. *Kaplan S. A., Pikelner S. B.* Mezhzvezdnaĭa sreda [Interstellar medium]. Moscow: Fizmatgiz (1963); see also *Kaplan S. A.* Interstellar Gas Dynamics. New York: Pergamon Press (1966).
34. *Shmidt O. Yu.* Chetyre lektsii o teorii proiskhozhdeniĭa Zemli [Four Lectures on the Theory of the Origin of the Earth]. Moscow (1950).
35. *Urey H. C.* The Planets. USA (1952).
36. *Oort J. H.*—Bull. Astron. Nederl. **11** 408, 91 (1960).
37. *Sherr R., Smith L. G., Bleakney W.*—Phys. Rev. **54**, 388 (1938).
38. *Kukavadze G. M., Memelova L. Ya., Suvorov L. Ya.*—ZhETF **49**, 689 (1965).
39. *Severnyi A. B.*—Astron. Zh. **34**, 328 (1957).
40. *Remy H.* Treatise on Inorganic Chemistry V.1. N.Y., Elsevier, 49 (1956).
41. *Braginskiĭ V. B.*—UFN **86**, 433 (1965); Pribory i tekhn. eksperimenta **3**, 130 (1964).
42. *Mohorovicic S.*—Astron. Nachr. **259**, 94 (1934).
43. *Vlasov N. A.*—Astron. Zh. **61**, 893 (1964).
44. *Hoyle F., Tuler A.*—Nature **224**, 1000 (1965).
45. *Zeldovich Ya. B.*—ZhETF **37**, 569 (1959).
46. *Zeldovich Ya. B.*—ZhETF **1**, 1 (1965).
47. Scientific American **213** 1, 44 (Editorial comment) (1965).
48. *Zeldovich Ya. B.*—UFN **86**, 303 (1965).

Commentary

This article is one of the approaches in the attack on quarks which Ya. B. organized in the mid-sixties. Of the same period is Ya. B.'s short review, unique in its simplicity and enthusiasm, "The Classification of Elementary Particles, and Quarks for Pedestrians,"[1] and his participation in the experimental search for free quarks.[2]

The upper limits achieved in the search for free fractional-charge particles completely excluded the possibility that the interactions of quarks are similar to the interactions of ordinary hadrons, and they led theoreticians to the idea of confinement. The above article is also one of the first in which Ya. B. established a new promising direction at the crossroads of cosmology, astrophysics and elementary particle physics. The role of this direction became widely recognized after it was realized that for many important questions of elementary particle physics the answers in principle could not be obtained in accelerators and required that investigators turn to the "hot laboratory of the early Universe."

[1] *Zeldovich Ya. B.*—UFN **86**, 303–310 (1965).

[2] *Braginskiĭ V. B., Zeldovich Ya. B., Martynov V. K., Migulin V. V.*—ZhETF **52**, 29–39 (1967).

26

Rest Mass of a Muonic Neutrino and Cosmology*

With S. S. Gershtein

Experimental estimates of the rest mass of the neutrino [1] have low accuracy and yield $m(\nu_e) < 200\,\text{eV}/c^2$ for the electron neutrino and $m(\nu_\mu) < 2.5 \times 10^6\,\text{eV}/c^2$ for the muon neutrino.

Cosmological considerations connected with the hot model of the Universe [2] make it possible to strengthen greatly the second inequality. Just as in the paper by Ya. B. Zeldovich and Ya. A. Smorodinskiĭ [3], let us consider the gravitational effect of the neutrinos on the dynamics of the expanding Universe. The age of the known astronomical objects is not smaller than 5×10^9 years, and Hubble's constant H is not smaller than 75 km/s-Mpc = $(13 \times 10^9\,\text{years})^{-1}$. It follows therefore that the density of all types of matter in the Universe is at the present time[1]

$$\rho < 2 \times 10^{-28}\ \text{g/cm}^3.$$

The space surrounding us is filled presently with equilibrium radiation of temperature 3°K [4]. It is proposed that this is "relic" radiation and is proof of the high temperature possessed by the plasma during the pre-stellar high-density period.

At a temperature of the order of 4 MeV for ν_e and of the order of 15 MeV for ν_μ, complete thermodynamic equilibrium existed between ν, γ, e^+, and e^-. The number of other particles in this equilibrium is small, except perhaps gravitons, which, however, have no effect on the arguments that follow. In thermodynamic equilibrium, the ratio of the number of fermions and

*Pisma v Zhurnal eksperimentalnoĭ i teoreticheskoĭ fiziki **4** (3), 174–177 (1966).

[1] We use the asymptotic formula

$$T = \pi/2H\sqrt{\rho/\rho_c}; \quad \rho_c = 3H^2/8\pi\sigma; \quad \rho = 3\pi/32\sigma T^2.$$

Other more complicated estimates based on an investigation of remote objects give a similar result:

$$q_0 = \rho/2\rho_c < 2.5; \quad H \leq 120\,\text{km/s}\cdot\text{Mpc},$$
$$\rho_c \leq 2.5 \times 19^{-29}\ \text{g/cm}^3, \quad \rho < 1.25 \times 10^{-28}\ \text{g/cm}^3.$$

antifermions with spin 1/2 to the number of photons is

$$[\nu_e] + [\overline{\nu}_e] = [\nu_\mu] + [\overline{\nu}_\mu] = [e^+] + [e^-] = 2\frac{\int (e^x + 1)^{-1} x^2\, dx}{\int (e^x - 1)^{-1} x^2\, dx}[\gamma] = 1.5[\gamma].$$

However, during the course of the cooling from $T > m_e c^2$ (for which these relations are written) to the present time, when $T \ll m_e c^2$, these relations change, since the annihilation of the e^+e^- increases the number of photons without changing the number of neutrinos per unit of co-moving volume [5]. At the present time we can expect

$$[e^+] + [e^-] = 0, \qquad [\nu_\mu] + \overline{\nu}_\mu] = [\nu_e] + [\overline{\nu}_e] = 0.5[\gamma].$$

At 3°K we have $[\gamma] = 550\ \mathrm{cm}^{-3}$, from which we obtain for the neutrino at the present time

$$[\nu_\mu] + [\overline{\nu}_\mu] = [\nu_e] + [\overline{\nu}_e] = 300\ \mathrm{cm}^{-3}.$$

Comparing with the density limit given above, we obtain

$$m_0(\nu_\mu) < 7 \times 10^{-31}\ \mathrm{g} = 400\ \mathrm{eV}/c^2$$

and the same for $m_0(\nu_e)$. Thus, we obtain no new information for the electron neutrino; for the muon neutrino, on the other hand, the cosmological considerations reduce the upper limit of the rest mass by three orders of magnitude.

In considering the question of the possible mass of the neutrino we have, naturally, used statistical formulas for the four-component ($m \neq 0$) particles. We know, however, that in accordance with the (V–A) theory, neutrinos having a definite polarization participate predominantly in weak interactions. Equilibrium for neutrinos of opposite polarization is established only at a higher temperature. This, incidentally, can change the limit of the mass by not more than a factor of 2.

A neutrino with non-zero rest mass can annihilate in accordance with the diagram

$$\nu_\mu + \overline{\nu}_\mu \underset{\text{weak}}{\longrightarrow} \mu^+ + \mu^- \underset{\text{el.-mag.}}{\longrightarrow} \gamma + \gamma$$

if $m(\nu_\mu) > m(e^\pm)$, and also either into 3γ or, in the second order of the weak interaction, into a $\nu_e + \overline{\nu}_e$ pair, if it is assumed the $m(\nu_\mu) > m(\nu_e)$. When $v < c$ the annihilation cross section behaves like $1/v$. Estimates show, however, that there is no time for noticeable annihilation to take place during the course of the cosmological expansion.

The momentum of interacting particles changes during the course of expansion like $1/R$, where R is the scale factor, independently of the presence and magnitude of the particle rest mass. At the present time the neutrino momentum should be of the same order (somewhat smaller) as the momentum of the relic photons, i.e., $p \approx 5 \times 10^{-4}\ \mathrm{eV}/c$.

If the neutrinos have a rest mass, then their velocity and the speed of sound in the neutrino gas are of the order of p/m, i.e., say 30 km/s at $m = 5\,\mathrm{eV}/c^2$ and 3 km/s at $m = 50\,\mathrm{eV}/c^2$. Strong gravitational perturbations should be produced in such a gas by the galaxies. It is possible that a more detailed analysis of these processes will allow us to lower the foregoing estimate of the upper limit of the neutrino mass.

This note is the result of the stimulating circumstances of the Summer School in Balaton-Vilagos, and we use this opportunity to express gratitude to the organizers of the school.

Received
June 4, 1966

REFERENCES

1. *Rosenfeld A. H., Barbaro Galtieri A., Barkas W. H.* et al.—Rev. Mod. Phys. **37**, 633 (1965).
2. *Gamow G.*—Phys. Rev. **70**, 572 (1946); **74**, 505 (1948); Rev. Mod. Phys. **21**, 367 (1949); *Gamow G.*—Vistas Astron. **2**, 1726 (1956); *Dicke R., Peebles P. J. E., Roll P. G., Wilkinson D. T.*—Astrophys. J. **142**, 414 (1965); *Zeldovich Ya. B.*—UFN **89**, 647 (1966).
3. *Zeldovich Ya. B., Smorodinskiĭ Ya. A.*—ZhETF **41**, 907 (1961).
4. *Penzias A. A., Wilson R. W.*—Astrophys. J. **142**, 419 (1965).
5. *Peebles P. J. E.*—Phys. Rev. Lett. **16**, 410 (1966).

Commentary

This article obtained for the first time an upper estimate of the mass of a neutrino on the basis of cosmological considerations. The paper had major influence on the further development of both cosmology and physics since, in essence, it had discovered a new, promising direction at the intersection of these two sciences. Abroad these ideas were taken up significantly later.[1,2] An earlier discussion of the relic neutrino sea under the assumption of lepton asymmetry and degeneracy was carried out at the beginning of the sixties,[3,4,5] before the discovery of the relic photons by Penzias and Wilson.

An especially large number of papers which develop the ideas of the above article were published after 1980, when there appeared information that the mass of the electron neutrino is possibly a quantity close to 30 eV.[6] In this case the bulk of the mass of the Universe is contained in the relic neutrinos, which should play

[1] *Cowsik R., McClelland J.*—Phys. Rev. Lett. **29**, 669–670 (1972).

[2] *Szalay A. S., Marx G.*—Astron. and Astrophys. **49**, 437–441 (1976).

[3] *Peres A.*—Progr. Theor. Phys. **24**, 149–154 (1960).

[4] *Pontekorvo B. M., Smorodinskiĭ Ya. A.*—ZhETF **41**, 239–243 (1961).

[5] *Smorodinskiĭ Ya. A., Zeldovich Ya. B.*—ZhETF **41**, 907 (1961).

[6] *Kozik V. S., Lĭubimov V. A., Novikov E. G.*, et al.—Yader. fiz. **32**, 301 (1980).

a key role in the formation of galactic clusters and their hidden mass; the first to point this out were A. Szalay and G. Marx.[7] These ideas are developed by Ya. B. and his coauthors in articles in the present book. Limitations on the total density of all forms of matter and energy which arise from information on the age of the Universe were applied by Ya. B. and his coauthors I. Yu. Kobzarev and L. B. Okun in another connection as well. The possibility is investigated of breaking the Universe up into domains (regions) which are distinguished by the sign of parity violation. An estimate is given of the specific density (per unit surface) of the walls between domains, and a conclusion is reached about the contradiction between this hypothesis and observations.[8]

Another important idea relating to neutrino cosmology was proposed by a student of Ya. B., V. F. Shvartsman.[9] He noticed that the rate of expansion of the Universe and, as a consequence, the abundance of primary helium in nature depends on the number of different types of neutrinos. Improvements on Shvartsman's estimate[10] led to the conclusion that, in addition to electron, muon and tau-neutrinos, no more than one extra type of neutrino can exist. A detailed discussion of the possible cosmological role of neutrinos, and also of hypothetical heavy neutral leptons, may be found in reviews published in the journal "Progress in Physical Sciences" (in Russian).[11,12]

[7] *Szalay A. S., Marx G.*—see footnote 2.
[8] *Voloshin M. B., Kobzarev I. Yu., Okun L. B.*—Yader. fiz. **20**, 12–29 (1974).
[9] *Shvartsman V. F.*—Pisma v ZhETF **9**, 315 (1969).
[10] *Steigman G., Schramm D., Gunn J.*—Phys. Lett. B **66**, 202–204 (1977).
[11] *Dolgov A. D., Zeldovich Ya. B.*—UFN **130**, 559–614 (1980).
[12] *Zeldovich Ya. B., Khlopov M. Yu.*—UFN **123**, 703–709 (1977).

27
The Cosmological Constant and Elementary Particles*

The hypothesis that the equations of general relativity contain the cosmological constant Λ of the order of $\Lambda \simeq +5 \cdot 10^{-56}\,\mathrm{cm}^{-2}$ has been recently advanced again [1–3]. A closed world is assumed, with a contemporary radius $R_1 \sim \Lambda^{-1/2}$, a Hubble constant $H_1 \sim c\Lambda^{-1/2}$, and a density $\rho_1 \sim \Lambda c^4/G$; the presence of Λ significantly slows down the expansion during the period corresponding to the redshift $z = 1.95$, about which the red shifts of the absorption lines in the quasar spectrum are grouped [4]. Corresponding to the given Λ is the concept of vacuum as a medium having a density $\rho_0 = \Lambda c^4/8\pi G = 2.5 \cdot 10^{-29}\,\mathrm{g/cm}^3$, an energy density $\varepsilon_0 = 2 \cdot 10^{-8}\,\mathrm{erg/cm}^3$, and a negative pressure (tension) $P_0 = -\varepsilon_0 = -2 \cdot 10^{-8}\,\mathrm{dyne/cm}^3$.

How is one to visualize a theory in which such properties of the vacuum are obtained from our notions regarding elementary particles? The starting point of such a theory are the formulas that give the required order of magnitude of ε_0, expressed in terms of the constants m, c, $\hbar$, and G, where m is the elementary-particle mass. Using the formulas of Eddington [5] and Dirac [6] for the quantities characterizing the contemporary Universe, and the connection between these quantities and Λ, we obtain

$$\Lambda \sim \frac{G^2 m^6}{\hbar^4}, \qquad \rho_0 \sim \frac{G m^6 c^2}{\hbar^4}, \qquad \varepsilon_0 \sim \frac{G m^6 c^4}{\hbar^4}. \tag{1}$$

We introduce the Compton wavelength of the elementary particle $\lambda = \hbar/mc$ and write

$$\varepsilon_0 \sim \frac{Gm^2}{\lambda}\frac{1}{\lambda^3}. \tag{2}$$

The latter formula corresponds to the assumption that the vacuum contains virtual pairs of particles with effective density $n \sim 1/\lambda^3$. It is assumed that the theory is such that the corresponding energy density is identically equal to zero. However, the energy of the gravitational interaction of these pairs (Gm^2/λ for one pair) does not vanish and yields precisely ε_0. In the relativistically invariant theory of vacuum, this ε_0 should correspond to $P_0 = -\varepsilon_0$.

Numerically, expression (2) with m equal to the proton mass yields a value 10^8 times larger than required. This may mean that (2) contains also

*Pisma v Zhurnal eksperimentalnoĭ i teoreticheskoĭ fiziki **6** (9), 883–884 (1967).

the weak-interaction constant. The dimensionless constant $g^1 \sim 10^{-5}$; in dimensional form ($g = 2 \cdot 10^{-49}$ erg/cm^3) it is assumed that[1]

$$\varepsilon_0 \simeq \frac{Ggm^8c^5}{\hbar^7} \simeq 10^{-5}\,\mathrm{erg/cm^3}. \tag{3}$$

Expressions (2) and (3) are related locally-measurable physical constants. They differ fundamentally from the Dirac-Eddington relations in that (2) and (3) presuppose neither variation of G nor the influence of the entire world (in the spirit of the Mach principle) on the local law (cf. [7]). The Dirac relations are obtained as approximately valid only for the present stage of the evolution of the world, soon after the cessation of the expansion. They are the consequences of the equations of general relativity with ε_0 and with the corresponding Λ.

It must be emphasized in conclusion that the final word with respect to the quantity Λ belongs to astronomic observations; it cannot be regarded as proved at present that $\Lambda \neq 0$.

I am grateful to N. S. Kardashev and I. S. Shklovskiĭ for suggesting the problem and for preprints of [2] and [3], and to A. L. Zelmanov, I. Yu. Kobzarev, and I. D. Novikov for valuable discussions.

[1]The interaction that violates time parity is probably even weaker.

Institute of Applied Mathematics
USSR Academy of Sciences. Moscow

Received
August 15, 1967

REFERENCES

1. *Petrosian V., Salpeter E., Szekeres P.*—Astrophys. J. **147**, 1222 (1967).
2. *Shklovskiĭ I. S.*—Astron. tsirkulyar **429** (1967)
3. *Kardashev N. S.*—Astron. tsirkulyar **430** (1967)
4. *Burbidge G.*—Astrophys. J. **147**, 851 (1967).
5. *Eddington A. S.*—Proc. Roy. Soc. London **133**, 605 (1931).
6. *Dirac P. A. M.*—Proc. Roy. Soc. London A **165**, 199 (1938).
7. *Zelmanov A. L.*—In: Fizicheskiĭ entsiklopedicheskiĭ slovar. Kosmologiĭa [Physics Encyclopedia. Cosmology]. Moscow: Sov. Entsiklopediĭa **2**, 490 (1962).

Commentary

In this note dimensional estimates of the ϵ_0-density of vacuum energy are proposed which in the units $\hbar = c = 1$ may be written in the form

$$\epsilon_0 \sim Gm^6 \quad \text{and} \quad \epsilon_0 \sim G_F G m^8$$

Here G is Newton's gravitational constant, G_F is Fermi's weak interaction constant, m is the characteristic mass of an elementary particle. For m equal to a proton mass, the first estimate of ϵ_0 proves to be 8 orders, and the second—3 to 4 orders larger than the upper limit allowed by data on the evolution of the Universe. Ya. B. returned several times to the question of the origin and possible cosmological role of the vacuum energy-momentum tensor; see his reviews.[1,2] He noted that Λ may be considered as the energy density and negative vacuum pressure also in the sense that for an expanding accompanying volume V the energy conservation law (first law of thermodynamics) is realized in the form $dE = d(\epsilon V) = -pdV$ for $p = -\epsilon$. Ya. B. particularly emphasized that the definition of a vacuum is a state with minimal energy, however, from this definition it does not at all follow that the minimum is equal to zero. The equality of the absolute value of Λ to zero (or its smallness) are known to us only from observations. A rigorous theoretical basis for this fact is still lacking.

In the last review[3] the idea of an inflationary universe was discussed; this idea was proposed by A. Guth and is being developed today in papers by a number of authors (S. Hawking, A. D. Linde, A. A. Starobinskiĭ).

[1] *Zeldovich Ya. B.*—UFN **95**, 209–230 (1968).
[2] *Zeldovich Ya. B.*—UFN **133**, 479–503 (1981).
[3] *Ibid.*

28
On the Concentration of Relic Magnetic Monopoles in the Universe*

With M. Yu. Khlopov

The modern concentration of relic 't Hooft-Polyakov monopoles is shown to amount to $\sim 10^{-19}$ cm^{-3}, being determined by the diffusion annihilation rate at $t < 10^{-5}$ s. To eliminate the contradiction with experimental upper limits some mechanism similar to quark confinement should forbid free monopole existence.

A. M. Polyakov [1] and G. 't Hooft [2] have demonstrated that in any non-abelian gauge theory containing electromagnetic interaction within a larger compact covering group, monopole [3, 4] type solutions exist with mass $M_0 \sim M_W/\alpha \sim 5 - 10$ TeV and magnetic charge $g = \hbar c/e$. In this note we evaluate the relic concentration of such monopoles in the Universe by analogy with calculations of the cosmological abundance of antibaryons [5, 6] and quarks [7].[1]

Consider first monopoles going out of thermodynamic equilibrium in the big-bang universe. The monopole (and antimonopole) concentration for monopoles, treated as particles with mass M_0, would be determined by thermodynamic equilibrium for sufficiently high temperatures (but still for $kT < M_0c^2$):

$$n = n_M = n_{\overline{M}} = \sqrt{\frac{2}{\pi^3}} \frac{M_0^{3/2}(kT)^{3/2}}{\hbar^2} \exp\left(-\frac{M_0c^2}{kT}\right). \tag{1}$$

However, in the theories with spontaneously broken gauge symmetry monopoles can be generated only after the phase transition to the asymmetric phase has occurred (see [8]), i.e., at $kT = kT_{\text{crit}} \sim M_W c^2 \ll M_0c^2$. If the monopole mass is determined by the vacuum expectation value of the Higgs field, monopoles will be in thermodynamic equilibrium, or even in excess, due to the small value of their initial mass. During the cosmological expansion

*Physics Letters **719B**, 239–243 (1978).

[1] An interesting paper [11] should be mentioned, where, based on experimental limits, cosmological constraints on the monopole mass and annihilation rate were obtained (see also [12]).

the temperature decreases and, simultaneously, the monopole mass grows as $\propto \sqrt{T_{\rm crit}^2 - T^2}$, and finally monopoles go out of thermodynamic equilibrium. As we will show, the cosmological monopole concentration does not depend substantially on the details of the monopole mass change. Thus, we take for an estimate $kT_{\rm crit} = M_W c^2$ and assume that the monopole mass is given by $M(T) = \sqrt{M_W^2 - (kT)^2/c^4}/\alpha$. To get the equilibrium concentration at $kT < M_W c^2(1 - \alpha^2/2)$ we must substitute $M(T)$ instead of M_0 in (1).

Define now the monopole annihilation rate. It is natural to think that the annihilation cross section is determined by Coulomb attraction of magnetic charges. At the temperature T monopole–antimonopole Coulomb attraction of magnetic charges is essential at distances $r \lesssim r_0 = g^2/kT$. If the mean free path of monopoles in a plasma $\lambda > r_0$, free monopole annihilation can be considered. In the opposite case ($\lambda < r_0$) annihilation is to be treated in the diffusion approximation. The monopole mean free path is determined by their scattering on charged particles $\lambda = (n_{\rm ch}\sigma)^{-1}$. The cross section of monopole multiple scattering at 90° is given according to [9] by $\sigma = 2 \cdot 10^{-33}\ {\rm cm}^2/\theta$, $\theta = kT/M_0 c^2$ and we get $\lambda < r_0$ at $t < 10^{-5}$ s and $\lambda > r_0$ at $t > 10^{-5}$ s for $M_0 = 10$ TeV and $g = \hbar c/e$.

To find the monopole annihilation rate in the diffusion approximation we consider diffusion of particles with magnetic charge $-g$ towards an absolutely absorbing sphere with radius $a \lesssim r_0$ and with magnetic charge $+g$. The diffusion equation is given by

$$\frac{\partial n(r,t)}{\partial t} = D\frac{1}{r^2}\frac{\partial}{\partial r}r^2\left[\frac{\partial n(r,t)}{\partial r} + \frac{g^2}{r^2 kT}n(r,t)\right], \tag{2}$$

where $D \approx (1/3)\lambda v$. For stationary ($\partial n(r,t)/\partial t = 0$) distribution of the diffusing particles with boundary conditions $n(\infty) = n_0$ and $n(a) = 0$, we have

$$n = 0 \quad \text{for} \quad r \le a,$$

$$n = n_0 \frac{1 - e^{r_0/r - r_0/a}}{1 - e^{-r_0/a}} \quad \text{for} \quad r \ge a. \tag{3}$$

Then the diffusion flux is given by

$$q = 4\pi r^2 D \frac{\partial}{\partial r} n(r) \approx 4\pi D r_0 n_M \qquad (n_M = n_0), \tag{4}$$

and the monopole diffusion rate in the diffusion approximation is equal to $n_M^2 4\pi D r_0$, and not to $n_M^2 \pi a^2 v$, as it is in the case of free monopoles. We recall that $r_0 = g^2/kT$.

With account of (4) the equation for the relative monopole concentration, $\nu = n_M/n_\gamma$, is given by [5–7].

$$\frac{d\nu}{dt} = -4\pi D r_0 n_\gamma (\nu^2 - \nu_{\rm eq}^2). \tag{5}$$

Solving (5) by iteration similar to [5–7] we obtain the relative monopole concentration ν_1 at time t_1 when monopoles go out of thermodynamic equilibrium. For n_{eq} given by (1) we have for constant monopole mass M_0,

$$\nu_1 = 2 \cdot 10^{-16}, \qquad t_1 = 2 \cdot 10^{-11}\ \text{s}, \qquad \theta_1 = \frac{kT_1}{M_0 c^2} = \frac{1}{42}. \tag{6}$$

and for the temperature dependent mass $M(T)$,

$$t_1' = 2 \cdot 10^{-10}\ \text{s}, \quad \theta_1' = \frac{kT_1'}{M_0 c^2} = \frac{1}{143}, \quad \frac{kT_1'}{M(T_1')c^2} = \frac{1}{41}. \tag{7}$$

We recall that $M_0 = \text{const} = M(T)|_{T \ll T_{\text{crit}}}$.

On the monopole's going out of thermodynamic equilibrium the monopole diffusion annihilation continues up to $t \sim 10^{-5}$ s, till $\lambda \lesssim r_0$. Equation (5) reduces to

$$\frac{d\nu}{dt} = -4\pi D r_0 n_\gamma \nu^2 = -A\theta^{1/2}\nu^2, \tag{8}$$

where $A = 2 \cdot 10^{28}\ \text{s}^{-1}$. Let us introduce the dimensionless parameter $\tau = t/t_1$. Then $\theta(t) = \theta_1 \tau^{-1/2}$ and we obtain as solution of (8)

$$\nu(t) = \nu_1 \frac{1}{1 + \frac{4}{3}A\theta_1^{1/2} t_1 (\tau^{3/4} - 1)\nu_1}. \tag{9}$$

While $\frac{4}{3}A\theta_1^{1/2} t_1 \nu_1 \sim 1/\theta_1 \gg 1$ as well as $\frac{4}{3}A{\theta_1'}^{1/2} t_1' \nu_1' \gg 1$ we have at arbitrary $\tau = t/t_1 > 1$ $(\tau' = t/t_1' > 1)$

$$\nu(t) = \frac{1}{\frac{4}{3}A\theta_1^{1/2} t_1 \tau^{3/4}} = \frac{1}{\frac{4}{3}A{\theta_1'}^{1/2} t_1' {\tau'}^{3/4}}. \tag{10}$$

Provided that $\theta_1^{1/2} \sim t_1^{-1/4}$ and $\tau^{3/4} \sim t_1^{-3/4}$, $\nu(t)$ is independent of t_1 (t_1') and consequently does not depend on the conditions of the monopole's going out of thermodynamic equilibrium. Note that due to the dependence on t the annihilation takes place all the time the diffusion approximation is valid. If for some reason, as seems rather improbable, monopoles had been produced in a nonequilibrium way only, (8) would have been valid even in this case. Such monopoles might have been produced due to fluctuations of the Higgs field vacuum expectation value [8, 10]. The corresponding relative concentration ν_1'' of produced monopoles depends essentially on the details of the considered gauge group. Still we may expect that $t_1'' \sim t_1'$, $\theta_1'' \sim \theta_1'$ and $\nu_1'' \sim \nu_1'$, so the solution (10) remains valid.

As we noted, after $t_2 = B^2 g^4/(G^{1/2}M^2)$, where B characterizes the cross section of monopole multiple scattering at 90° in a plasma, $\sigma = B/(TM)$ in units $\hbar = c = k = 1$, $\lambda > r_0$ and the diffusion approximation is not valid. We get

$$\nu_2 \equiv \nu(t_2) = \frac{G^{1/2}B}{g^2 M^{1/2}} \frac{1}{t_2^{3/4}} = \frac{G^{1/2}M}{B^{1/2}g^5} = 10^{-21}. \tag{11}$$

For $t > t_2$ the monopole annihilation cross section is given by $\sigma = \pi a^2$. We can define a as a maximal impact parameter for which the monopole motion turns out to be finite due to bremsstrahlung. While $g^2/hc \gg v/c$ for monopoles, their scattering is to be treated classically, and the radiation energy loss for monopole–antimonopole scattering is given by

$$\Delta E = kT \left(\frac{r_0}{\rho} \right)^5 \left(\frac{kT}{M_0 c^2} \right)^{3/2}.$$

So $\Delta E \gtrsim kT$ at $\rho \lesssim a$:

$$a = \rho_{\max} = \left(\frac{kT}{M_0 c^2} \right)^{3/10} r_0. \tag{12}$$

At $t > t_2$ the equation for the relative monopole concentration is given by

$$\frac{d\nu}{dt} = -\sigma v n_\gamma \nu^2, \tag{13}$$

or

$$\frac{d\nu}{d\tau} = -b\theta_2^{21/10} t_2 \nu^2 \tau^{-21/20},$$

where $b = 4 \cdot 10^{32}\ \mathrm{s}^{-1}$, $\theta_2 = 3 \cdot 10^{-8}$, $t_2 = 10^{-5}$ s; $\tau = t/t_2$.

For the concentration ν we have

$$\nu = \nu_2 \frac{1}{1 + 20\nu_2 b\theta_2^{21/10} t_2 (1 - \tau^{-1/20})}, \tag{14}$$

and

$$\nu_\infty = \nu_2 \frac{1}{1 + 20\nu_2 b\theta_2^{21/10} t_2} \approx \nu_2 \approx 10^{-21}.$$

Thus after $t_2 \sim 10^{-5}$ s monopole annihilation practically ceases. For the modern density of relic photons $n_\gamma = 400\mathrm{cm}^{-3}$ the modern monopole density amounts to

$$n_M = \nu_\infty n_\gamma = 4 \cdot 10^{-19}\ \mathrm{cm}^{-3}. \tag{15}$$

Further annihilation of monopoles could take place inside stars. It is essential, however, that the characteristic time of the monopole gravitational diffusion towards the protostellar center ($t_G = \sigma_{\mathrm{sc}} v_T/(Gm_p) \approx 10^{18} - 10^{20}$ s) exceeds greatly the time of collapse ($\sim 10^{12}$s). So, for the uniform distribution of monopoles in matter prior to the development of gravitational instability, monopole annihilation might have taken place only in that part of matter which undergoes transformations inside stars. It is difficult to give the exact value of the amount of matter f left in the interstellar space. Theoretical arguments give $f > 10\%$ and in any case f exceeds the mass of the interstellar gas in our galaxy (1%). So n_M cannot decrease more than by two orders of magnitude. Of course, different mechanisms leading to the lack of monopoles in the galaxy are not excluded (cf. small magnetic fields at the galaxy formation stage, etc.).

Nevertheless, with all reserves, the relic concentration obtained exceeds greatly the experimental upper limits on the amount of monopoles in the lunar and terrestrial matter and cosmic monopole fluxes $n_M < 10^{-30} - 10^{-38}$ cm^{-3} [13–17], as well as the estimates based on the energetics of the terrestrial and cosmic magnetic files [11, 18].

In summary, the hypothesis of free monopole existence seems to contradict the modern picture of the hot Friedman universe. The contradiction relates to the period which is not observed directly. So, if the Hagedorn hypothesis on the critical temperature had been valid, the contradiction would have been removed.

The discovery of free monopoles would be of the greatest importance not only for particle physics, but for cosmology as well, because it would cause us to abandon our main views on the big bang theory.

We are grateful to M. Y. Vysotskiĭ, S. S. Gershtein, A. D. Dolgov, A. G. Doroshkevich, A. D. Linde, I. Yu. Kobzarev, L. B. Okun, V. M. Chechetkin for helpful discussions.

Institute of Applied Mathematics
USSR Academy of Sciences. Moscow

Received
July 3, 1978

REFERENCES

1. *Polyakov A. M.*—Pisma v ZhETF **20**, 430 (1974).
2. *'t Hooft G.*—Nucl. Phys. B **79**, 276 (1974).
3. *Dirac P. A. M.*—Proc. Phys. Soc. **A 133**, 60 (1934); Phys. Rev. **74**, 817 (1948).
4. *Schwinger J.*—Phys. Rev. **144**, 1087 (1966).
5. *Zeldovich Ya. B.*—ZhETF **48**, 986 (1965); *Zeldovich Ya. B., Novikov I. D.* Stroenie i evoliutsiia Vselennoĭ [Structure and Evolution of the Universe]. Moscow: Nauka, 735 p. (1975).
6. *Chiu H. Y.*—Phys. Rev. Lett. **17**, 712 (1966).
7. *Zeldovich Ya. B., Okun L. B., Pikelner S. B.*—Phys. Lett. **17**, 1964 (1965).
8. *Kirzhnitz D. A., Linde A. D.*—Phys. Rev. Lett. **42B**, 471 (1972); Preprint IC/76/28.
9. *Amaldi E.* et al.—CERN Rep. **63**, 13 (1963).
10. *Kibble T. W. B.*—Preprint ICTP/75/5 (1976).
11. *Domogatskiĭ G. V., Zheleznykh I. M.*—Yader. Fiz. **10**, 1238 (1969).
12. *Adams P. J., Canuto V., Chiu H. Y.*—Phys. Lett. B **61**, 397 (1976).
13. *Eberhard P. H., Ross R. R., Alwares L. W.*—Phys. Rev. **D4**, 3260 (1971).
14. *Vant-Hull L.*—Phys. Rev. **173**, 1412 (1968).
15. *Kolm H. H., Villa F., Odian A.*—Phys. Rev. **184**, 1393 (1969).
16. *Fleischer R. L.* et al.—Phys. Rev. **184**, 1393 (1969).
17. *Schatten K. H.*—Phys. Rev. **D1**, 2245 (1970).
18. *Parker E. N.*—Astrophys. J. **160**, 383 (1970).

Commentary

In this article, written in 1978, the connection with Grand Unification models is not discussed directly. Nevertheless, it had a significant influence on the development of these models. The paper isolated the principle problem of cosmology based on Grand Unification models—the problem of relic magnetic monopoles. The main conclusion of the paper is formula (11) which describes the relic concentration of monopoles of arbitrary mass and magnetic charge, independently of the means of their formation, under the single condition of a sufficiently large initial concentration of monopoles. It remained to take only a small step—to substitute into this formula the mass 10^{16} GeV which is characteristic for monopoles predicted by the $SU(5)$ model and to verify that the initial concentration of monopoles formed in the phase shift from $SU(5)$ symmetric to $SU(3) \times SU(2) \times U(1)$ symmetric vacuum is sufficiently large and that formula (11) is valid. This in fact was done in the paper.[1] The subsequent flow of articles was devoted to analysis of the problem of "overproduction of monopoles" that arose. It became clear that the huge relic concentration of monopoles, apparently, cannot shrink by many orders in the course of their gravitational clustering.[2,3] Another possibility, noted already in the article here, is the confinement of monopoles, a mechanism for which has already been proposed.[4] But most fruitful turned out to be an approach[5] in which a delay in the phase transition led to a small initial concentration of monopoles, so that formula (11) ceased to be valid. The development of this approach combines the problem of relic magnetic monopoles with the problem[6,7,8] of homogeneity and isotropy of the universe, and with the problem of the formation of the initial perturbations—with the entire complex of problems which are being solved in the framework of inflationary models of the universe.

[1] *Preskill J. P.*—Phys. Rev. Lett. **43**, 1365–1368 (1979).
[2] *Goldman T., Kolb E. W., Toussaint D.*—Phys. Rev. D **23**, 867–875 (1981).
[3] *Polnarev A. G., Khlopov M. Yu.*—Astron. Zh. **58**, 706–716 (1981).
[4] *Linde A. D.*—Phys. Lett. B **96**, 293–296 (1980).
[5] *Guth A.*—Phys. Rev. D **23**, 347–356 (1981).
[6] *Ibid.*
[7] *Linde A. D.*—Phys. Lett. B **108**, 389 (1982).
[8] *Albrecht A., Steinhardt P. J.*—Phys. Rev. Lett. **48**, 1220–1223 (1982).

V

General Theory of Relativity and Astrophysics

29
Collapse of a Small Mass in the General Theory of Relativity*

A calculation of the equilibrium of a cold ideal Fermi gas in its own gravitational field made by Volkoff and Oppenheimer [1, 2] led to the following result: for a small number of neutrons ($N < 0.35\odot$) there is a single solution, for $0.35\odot < N < 0.750\odot$ there are two solutions, and for $N > 0.75\odot$ there is no solution at all (the symbol $\odot$ here means the number of neutrons in the sun).

It was assumed that the unique solution for $N < 0.35\odot$ is absolutely stable and that for such a value of N collapse is impossible. We shall show that this is not true.

By prescribing a sufficiently large density we can obtain for any given number N of particles a configuration with mass as close to zero as we please, and clearly less than the mass of the static solution. Such a configuration obviously cannot go over into the state of equilibrium (into the static solution), and consequently can only contract without limit.

Let us take an arbitrary spherically symmetrical distribution of motionless matter. We denote the particle density by n and the energy density by ε

*Zhurnal eksperimentalnoĭ i teoreticheskoĭ fiziki **42** (2), 641–642 (1962).

(ε includes the rest mass of the particles); n and ε are connected by the equation of state.

The metric is given by the expression (we everywhere set $c = 1$)

$$ds^2 = e^{\nu} dt^2 - e^{\lambda} dr^2 - r^2(\sin^2\theta d\varphi^2 + d\theta^2). \tag{1}$$

As is known from the equation for λ (cf. [3]) it follows that

$$e^{-\lambda(r)} = 1 - \frac{b}{r}\int_0^r \varepsilon(r) r^2\, dr, \tag{2}$$

where $b = 8\pi k$. The mass of the star is given by the expression

$$M = 4\pi \int_0^\infty \varepsilon(r) r^2\, dr, \tag{3}$$

and the number of particles by ($d\omega$ is an invariant volume element)

$$N = \int n\, d\omega = 4\pi \int_0^\infty n(r) e^{\lambda/2} r^2\, dr \tag{4}$$

Let us take the distribution of motionless matter given by the formulas

$$\varepsilon = \frac{a}{r^2}, \quad r < R; \qquad \varepsilon = 0, \quad r > R. \tag{5}$$

Then

$$e^{-\lambda} = 1 - ab, \quad r < R; \qquad e^{-\lambda} = 1 - \frac{abR}{r}, \quad r > R. \tag{6}$$

$$M = 4\pi aR, \qquad N = \frac{4\pi}{\sqrt{1-ab}} \int_0^R n r^2\, dr. \tag{7}$$

For an ultrarelativistic gas

$$\varepsilon = \hbar \left(\frac{3}{\pi^2}\right)^{1/3} n^{4/3}, \qquad n = \left(\frac{\pi^2}{3}\right)^{1/4} \left(\frac{\varepsilon}{\hbar}\right)^{3/4}. \tag{8}$$

Substituting (5) and (8) in (7), we get

$$N = \text{const} \cdot \frac{a^{3/4} R^{3/2}}{\sqrt{1-ab}},$$

$$R = \text{const} \cdot N^{2/3} a^{-1/2} (1-ab)^{1/3}, \qquad M = \text{const} \cdot N^{2/3} a^{1/2} (1-ab)^{1/3}. \tag{9}$$

It follows from this that $M \to 0$ for $a \to 1/b$, whatever the value of N.[1] This proves the assertion made above.

For a rough estimate of the energy barrier which separates the equilibrium solution with $M \leq Nm$ (m is the mass of the neutron) from the collapsing state, let us find the maximum M from (9). We get

$$M_{\max} \approx N^{2/3} \sqrt{\frac{\hbar}{k}}, \qquad \frac{M_{\max}}{Nm} \sim N^{-1/3} \frac{\sqrt{\hbar/k}}{m} \approx \left(\frac{N}{N_{\text{cr}}}\right)^{-1/3}, \tag{10}$$

where mN_{cr} is of the order of the maximum mass for which a solution exists, i.e., of the order of the mass of the sun. Consequently for systems consisting

[1] For small a one must not use the ultrarelativistic equation (8). For $a \to 0$, the mass $M \to Nm$.

of a small number of neutrons collapse may indeed be possible, but the height of the barrier is many times larger than the initial rest energy of the system. Since the barrier $\sim N^{2/3}$, its absolute value decreases (although the required density increases) when part of the body in question is compressed. All of the conclusions remain qualitatively unchanged when one takes interaction between the neutrons into account, and in particular even for the equation of state $\varepsilon \sim n^2$, which is the most rigid relation consistent with the theory of relativity [4].

In the use of the expression (1)–(4) it is not assumed that $n(r)$ and $\varepsilon(r)$ with zero velocity, $v = 0$, correspond to the static solution; the field equations give nonvanishing values of $\dot{\lambda}$, $\dot{\nu}$, $\dot{v}$, where the dot means differentiation with respect to time. Outside the body $(r > R)$ we have $\dot{\lambda} - 0$, so that the mass M measured from the external gravitational field remains unchanged during the process of evolution which ensues for a prescribed initial distribution which does not satisfy the conditions for equilibrium.

The distribution (8) used for the proof has singularities: $\varepsilon \to \infty$ for $r = 0$; ε has a discontinuous change from a/R^2 to 0 at $r = R$. It is easy to verify, however, that the result is not changed when one smoothes out these singularities, for example by replacing (5) by

$$\varepsilon = \frac{a}{\alpha^2 R^2} \qquad \text{for} \qquad r < \alpha R, \quad \alpha \ll 1;$$

$$\varepsilon = \frac{a}{r^2}\frac{R(1+\beta) - r}{2\beta R}, \qquad R(1-\beta) < r < R(1+\beta), \quad \beta \ll 1;$$

$$\varepsilon = \frac{a}{r^2}, \qquad \alpha R < r < R(1-\beta). \tag{11}$$

In the initial distribution (5) used in our argument, and also in the smoothed distribution (11) we have everywhere $e^{-\lambda} > 0$, $e^{\nu} > 0$, i.e., the metric is not singular and there are no difficulties of the sort associated with the Schwarzschild singularity ($e^{\lambda} \to \infty$, $e^{\nu} = 0$).

The author is grateful to N. A. Dmitriev, L. D. Landau, E. M. Lifshitz, and S. Kholin for valuable discussions.

REFERENCES

1. *Oppenheimer J. R., Volkoff G. M.*—Phys. Rev. **55**, 374 (1039).
2. *Landau L. D., Lifshitz E. M.* Statistical Physics. Reading, MA: Addison-Wesley (1969).
3. *Landau L. D., Lifshitz E. M.* The Classical Theory of Fields. New York: Pergamon (1971).
4. *Zeldovich Ya. B.*—ZhETF **41**, 1609 (1961).

Commentary

Research conducted in the forties through the sixties on the stability of cold stellar configurations with masses of the order of the sun's showed the existence of a critical mass of order $1.5\,M_\odot$ above which no stable configurations exist, and collapse is inevitable. Configurations with lower mass are stable. As is shown in the current article, in principle such masses may be forced to collapse if they are artificially compressed by external forces to a size of the order of their gravitational radius. However, for this it is first necessary to expend a huge amount of work. Therefore the state of collapse of a small mass is separated from a stable state by a gigantic energy barrier.

In fact, the energy barrier separating the stable state of a small mass from collapse may, in the classical theory, be arbitrarily small. The essence of the matter is that for transfer of a small mass to a state of collapse it is not necessary to increase its total mass-energy by the total work of external forces, but is sufficient to artificially form a huge fluctuation of the matter density at the center of the configuration. Here the mass of the fluctuation may be arbitrarily small and, consequently, the total work needed to create such a fluctuation is also small. The article above constructs the first concrete example of a configuration of a given number of nucleons with an arbitrarily small total gravitational mass, i.e., with a gravitational binding energy which is arbitrarily close in absolute value to the total rest mass-energy of the nucleons. This shows that, in principle, one may construct a machine using gravity forces which would extract from matter an energy almost equal to the rest energy $M_\odot c^2$, which is incomparably greater than the nuclear energy $0.01M_\odot^2$.

This article was also a first step towards the idea of the possibility of formation of primordial black holes with small masses in the early universe in the presence of sufficiently strong perturbations (see the article of Ya. B., I. D. Novikov, A. G. Polnarev and A. A. Starobinskiĭ[1] and references therein).

We note also that this short note is Ya. B.'s first paper on the general theory of relativity and simultaneously the last paper which Ya. B. was still able to discuss in detail with one of his teachers—Lev Davidovich Landau—a few days before the catastrophe which cut short Landau's scientific work.

[1] *Novikov I. D., Polnarev A. G., Starobinskiĭ A. A., Zeldovich Ya. B.*—Astron. Astrophys. **20**, 104–109 (1979).

29a
Gravitational Collapse of Nonsymmetric and Rotating Masses*

With A. G. Doroshkevich and I. D. Novikov

Collapse of nonsymmetric and rotating masses is considered. It is shown that the characteristic pattern of gravitational self-closing valid for the spherical case also holds in the general case. Moreover, collapse of a nonrotating body leads to a t^{-1} damping of quadrupole and higher field moments for an external observer. The field of a collapsing rotating body changes in a different manner. Metric changes related to a rotating local inertial system approach a nonvanishing constant. However, qualitatively the collapse picture remains the same as in the spherical case. Static nonspherical solutions of the Einstein equations are also investigated and in particular the properties of the $g_{00} = 0$ Schwarzschild surface in these solutions are analyzed.

1. Introduction

As is well known, stars with $M >\sim 1.6 M_\odot$ have an evolution such that they contract without limit. The theory of this phenomenon, for a simple model of a spherical body, has by now been clarified to a considerable extent (see the review [1], where references to original papers can be found). A characteristic feature of the process is the gravitational self-closing of a body, manifest in the fact that after contraction to a critical dimension, the gravitational field of the body does not let out either radiation or information. This critical dimension is determined by the gravitational radius $R_g = 2Gm/c^2$, where G is Newton's gravitational constant, c is the velocity of light, and m is the mass of the body.

In close relation with self-closing is the fact that, from the point of view of a remote observer, on approaching R_g the evolution slows down, and the observed picture approaches asymptotically (as $t \to \infty$) a certain limiting state, which, however, is not at all an equilibrium state. This apparent stoppage is a result of the slowing down of the time in the strong gravitational field, and for a contracting body the Doppler effect only intensifies this deceleration as seen by a remote observer. Thus, the apparent stoppage of

*Zhurnal eksperimentalnoĭ i teoreticheskoĭ fiziki **49**, 170–181 (1965).

the contraction is brought about by the same factors as the red shift of the emitted spectrum and self-closing.

This raises the question whether the picture is general, whether it has any special connection with the symmetry of the problem, and whether the reductions remain in force also in the general non-spherically symmetrical case. The statement that the picture remains qualitatively the same also in a nonspherically symmetrical collapse was advanced earlier [2] (see also [3] concerning the stability of the Schwarzschild solution). In this paper we present a proof of this far from obvious statement.

By way of a first attempt at finding the asymptotic nonspherical solution, it is natural to seek the stationary solutions by starting from the assumption that the collapse is seen by an external observer as a monotonic process and that as $t \to \infty$ all the $\partial/\partial t \to 0$. This assumption is proved in Sec. 3.

An analysis of the static solution outside the body shows that the deviation from the spherical solution, which is caused by a change in the source of the field, leads to the appearance of true singularities of space-time on the Schwarzschild surface $g_{00} = 0$. On the other hand, in the co-moving system of a contracting body with small initial deviations from sphericity in the density distribution, the instant when the surface of the body crosses the Schwarzschild surface is in no way specially distinguished, and is not accompanied by the appearance of true singularities either in the metric or in the density. A comparison of these results leads to the conclusion that the quadrupole and higher multipole moments of the external gravitational field attenuate during the relativistic stages of the collapse of an asymmetrical body.

Deviations in the stationary metric from sphericity, connected with the components g_0^α, i.e., with the rotation of the local inertial system relative to a far inertial system, and the fields induced in the source by the "rotational" motions, do not lead to singularities when $g_{00} = 0$. During the process of collapse these deviations do not vanish. We note that the "rotational motions" are not necessarily connected with the rotation of the body as a whole, and arise, for example, as a result of tangential velocities when an asymmetrical body is compressed. We present below a rigorous proof of the advanced considerations.

2. *Stationary Solutions*

In the spherically symmetrical case, the field is given by the known Schwarzschild solution and is static independently of the "spherically symmetrical" motion of the central mass which produces the field [4]. The solution contains a critical surface—the Schwarzschild sphere S_S, characterized by the condition $g_{00} \equiv 1 - R_g/R = 0$. Near this surface, the red shift of the radiation line, emitted by a source at rest and received by a remote observer, is given by the expression

$$\frac{\omega_{\text{obs}}}{\omega_{\text{rad}}} = \sqrt{g_{00}} \sim l/2R_g,$$

where l is a small distance from S_S,

$$dl = \sqrt{-g_{11}}\, dR \equiv \sqrt{\frac{R}{R-R_g}}\, dR, \quad l = 2\sqrt{R_g(R-R_g)}.$$

The observed frequency tends to zero as $l \to 0$. For a stationary external observer, a light ray and a trial particle can approach S_S only asymptotically, after an infinite time. For both the light ray and for the freely falling trial particle, this time is logarithmically infinite

$$t \sim \frac{R_g}{c} \ln \frac{R_g}{R-R_g}.$$

The four-dimensional space-time has no singularity on S_S, and in particular, when $R = R_g$ the curvature scalar $K = R_{\alpha\beta\gamma\delta}R^{\alpha\beta\gamma\delta}$, where $R_{\alpha\beta\gamma\delta}$ is the Riemann tensor, has a fully defined finite value $K = 12/R_g^4$. If the field source has dimensions smaller than S_S, then the Schwarzschild solution in the vacuum can be continued inside S_S into the so-called T-region [5, 6].

a) Static field with axial symmetry. Regge and Wheeler [3] considered the nonspherical problem in vacuum by the method of small perturbations superimposed on the Schwarzschild solution. From the solution of the equations for small perturbations, given in [3], we see that in the stationary case any perturbation that decreases at infinity increases without limit on approaching the Schwarzschild sphere of the unperturbed problem. It follows therefore that no matter how small the deviations from spherical symmetry at a finite distance from S_S, the method of small perturbations used by Regge and Wheeler [3] cannot give a correct answer up to S_S itself.

The static problem for some form of an axially symmetrical field of the quadrupole and higher multipoles was solved by Erez and Rosen [7] with the aid of Weyl's method [8]. The corresponding expression for the interval for the quadrupole field, with the error contained in [7] corrected by us, can be found in Appendix I. In this field, the surfaces of constant g_{00}, i.e., of constant gravitational potential, are singly-connected, closed, and embedded in one another so that they do not differ topologically from the spherically symmetrical case, where they were concentric spheres. However, as $g_{00} \to 0$ the metric of the surfaces $g_{00} = \text{const}$ differs radically from the metric of a sphere. In particular, for a positive quadrupole moment q (the body is elongated along the axis like a cucumber), the length of the equator tends to zero, and the length of the meridians to infinity as $g_{00} \to 0$. The area of the surface $g_{00} = \text{const}$ tends to infinity (but each surface with larger area lies completely inside the preceding surface with smaller area). The light and the freely-falling particle reach the surface $g_{00} = 0$ within a finite time of the external observer (see Appendix I). Finally, the invariant $K = R_{\alpha\beta\gamma\delta}R^{\alpha\beta\gamma\delta}$, which characterizes the total curvature of space-time, becomes infinite for $q \neq 0$ as $g_{00} \to 0$ like q^2/g_{00}.

These results are not limited to a quadrupole only, and are, as shown in Appendix II, general for any static axially-symmetrical solution.

b) External field of a rotating body. We now consider the deviations from spherical symmetry connected not with the change in the distribution of the masses in the field source, but with rotation. Kerr [9] obtained an exact solution of Einstein's equations in vacuum. This solution describes the field of a body of mass m with total angular momentum $M = amc$, where a is a constant with the dimension of length. For a body whose particles possess only rotational motion about a symmetry axis, the only nonvanishing-diagonal component of the metric, in a suitable coordinate system and in an external field, is g_{03}. This follows immediately from symmetry considerations and from the equivalence of the past and of the future. Kerr's solution contains non-removable off-diagonal components $g_{\mu\nu}$, in addition to g_{03}. Consequently, if this solution is realized as an external field of some stationary body, then the particles of the material of the body should execute not only rotational motion about the symmetry axis, but also some other motions (for example, such as rising at the poles and dropping at the equator), leading to non-equivalence of the past and the future. An analysis of Kerr's solution [15] leads to the following conclusions.

1) For arbitrarily small but non-vanishing a, the lengths of the "parallel" L on the surface $g_{00} = \text{const}$ [these lengths are proportional to $(-g_{33} + g_{03}^2/g_{00})^{1/2}$ at $\theta = \text{const}$ and $g_{00} = \text{const}$] tend to infinity as $g_{00} \to 0$. The asymptotic value of L is of the form

$$L = \frac{2\pi a \sin^2\theta}{\sqrt{g_{00}}}.$$

2) The precession of a gyroscope away from the body is determined by the known expression [4]:

$$\Omega^2 = c^2 a^2 R_g^2 R^{-6}(1 + 3\cos^2\theta).$$

Near S_S the precession in local time tends to infinity.

3) The scalar K, unlike the preceding type of deviations from spherical symmetry, does not have singularities on S_S and, in particular, we have on the equator, as in the Schwarzschild solution on S_S,

$$K = \frac{12}{R_g^4}, \qquad R_g = \frac{2Gm}{c^2}.$$

In this solution the field in vacuum can be continued inside S_S into the T-region. Kerr's solution has a space-time singularity (like Schwarzschild's solution) at $R = 0$.

4) A light ray traveling towards S_S in the direction of the pole, and light rays traveling in the plane of the "equator," reach S_S after a logarithmically

infinite time of the external observer. (The clocks are synchronized here against the trajectories of the rays.)

In Appendix III we give the field of a slowly rotating sphere with $a \ll R_g$. This solution is valid not only far away, where $R \gg R_g$, but also near S_S. In this solution of the equations of small perturbations superimposed on the Schwarzschild field, only the terms linear in a and the higher mechanical moments are retained in the corrections to the components $g_{\mu\nu}$, and the terms with a^2 and higher order have been discarded. Those of the effects of Kerr's solution on $g_{00} = 0$ which depend on the linear corrections to $g_{\mu\nu}$ are retained in this solution, too. In particular, we have here

$$K|_{g_{00}=0} = \frac{12}{R_g^4} < \infty,$$

and the rotation does not give terms of first order in a.

c) Schwarzschild sphere in an external quadrupole field There exist solutions of Einstein's equations in which there is a surface S_S which does not qualitatively differ at all from the Schwarzschild surface for the spherical case. In this case, however, the deviations from spherical symmetry should be brought about by the external field. For example, if we can consider a spherical mass in an external quadrupole field (which increases with increasing distance from the mass m) then the exact solution of Einstein's equations in vacuum is of the form given in Appendix IV. In this field the surface S_S is a Schwarzschild sphere deformed by the external fields, with all its properties.

3. Collapse of a Perturbed Spherical Dust Cloud

Let us consider the motion of the dust in a co-moving reference frame.[1] It is known (see [4]) that in spherically symmetrical motion in this reference frame the transition to S_S occurs within a finite time, and in this system S_S is no singularity whatever. The density of matter in this case is finite, and its order of magnitude is $\overline{\rho}_{\text{crit}} = 2 \times 10^{16} (M_\odot/M)^2$ g/cm^3. The spherically symmetrical motion of dust with small perturbations has likewise no singularities at this average density [10]. The invariant K here is finite. From a comparison with the invariant K of the stationary solution follows the conclusion, mentioned in the introduction, that the multipole moments of the external field attenuate during the course of the collapse.

It is shown in Appendix III that during the collapse of a rotating body, the "rotational-type" deviations from sphericity are conserved.

The foregoing considerations do not as yet exclude the possibility of the following situation.

[1]The results are valid also for a gas.

The body contracts, and after a finite proper time it passes through S_S with small perturbations, and then, already in the T-region, after being compressed to a high density and strongly deformed, it gives rise to strong perturbations of the metric of the surrounding space, making it possible for radiation to be emitted and for the body itself to expand again beyond S_S. It would seem that for a remote external observer, the question of the possibility of such a situation should not arise: after all, if the body crosses after a finite proper time S_S then this process stretches out for the external observer into an infinitely long one, and what happens afterwards is immaterial to the observer. Actually, however, this very conclusion, to which we are so accustomed, that the time of approach to S_S stretches to infinity, is obtained from the fact that the world line of the ray emitted from the surface of the body arbitrarily close to S_S, proceeds for an arbitrary long time (in the time of any system!) near the world line of the point S_S (see the figure). In our problem it is not at all obvious beforehand that the perturbations of the metric will not change after an infinitely long time the world line of the ray to such an extent that this ray and other rays, which have already been emitted after the surface of the body crossed S_S, could go to a remote observer. In other words, it must be proved that the going over to the asymptotic solution is a monotonic process and that any oscillations during the relativistic stage of collapse are already impossible for an external observer.

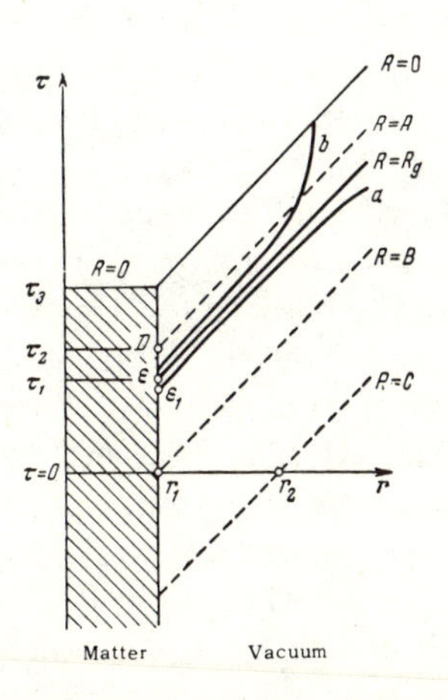

Dust sphere collapse in a freely falling reference frame (notation—Appendix VI); a, b are world lines of light rays. Ray a emitted from ε_1 near ε continues for a long time along $R = R_g$ (in a time measured in an arbitrary reference system).

We shall prove the following statement: assume that at the instant when the surface of the sphere crosses the S_S of the unperturbed problem the perturbations of the metric in the body, and the perturbations of the density and the velocity of the matter are small. Then, for an external observer, the picture of the contraction will be the same as in the case of an exactly spherical collapse—he sees the approach of the surface of the sphere to S_S as a process that stretches out to infinity, and the possibility of rays emitted by the surface of the body after crossing S_S is actually eliminated.

The proof (the details of which are given in Appendix V) consists in the following. We prove first that if in a co-moving freely-falling system of reference at some instant of proper time (close to the instant when the surface of the body crosses S_S) the perturbations in all of space are small and if the perturbations at space infinitely remain small in all the succeeding instants of time, then in all of space outside of S_S and (this is particularly important) also in the T-region of space-time near S_S, the perturbations will always remain small. Then, using the smallness of the perturbations of the metric inside S_S in the T-region, it is proved that a light ray can never leave this region and consequently an external observer will never find out what occurred after S_S was crossed, and the process of the approach of the surface of the body to S_S stretches out for him to infinity (see Appendix V).

This completes the proof of the statement. This result of our paper is important for a description of the picture of the collapse from the point of view of the external observer.

We note that this result cannot be obtained by the method of Regge and Wheeler [3], since they work in the Schwarzschild reference frame, which cannot be used when $g_{00} = 0$ or in the T-region.

In the proper time, the star can be compressed after crossing the S_S surface to tremendous densities, and the perturbations become colossal. But no matter what takes place there, this will never be manifest in the region of space-time to the right and below the dashed line $R = A$ in the figure, i.e., it will not be manifest in any way in the space outside S_S at any time t. This question is discussed in [6]. The conclusions of [16] contradict those of [10].

The conclusions obtained are obviously important primarily in attempts to attribute phenomena occurring in quasars (and also in supernovas) to the relativistic effects due to the collapse of large masses.

4. Collapse of an Asymmetrical Body From the Point of View of an External Observer

We have proved that a nonspherically symmetrical mass collapses for an external observer qualitatively in the same way as a spherical one. The change in the multipole moments during the course of contraction of the body should be accompanied by radiation of gravitational waves, but the energy carried by this radiation is small. The radiation of waves is a consequence of the change in the multipole moments, and cannot be regarded as the cause of their total damping. We note that in Newtonian theory, the moments also vary during the course of compression of the body, but for finite body dimensions they are finite. In Einstein's theory, a relativistic damping is superimposed on this change in the moments of the external field, due to the change in the dimensions of the contracting body.

Let us find the law governing the attenuation of q for an external observer during the course of the collapse. As shown in Appendix VI

$$q \sim \ln^{-1} \frac{R_g}{R - R_g},$$

but the approach to S_S proceeds like

$$t \sim \ln \frac{R_g}{R - R_g};$$

hence $q \sim t^{-1}$, i.e., the attenuation obeys a power law. The external observer "sees" (for example, with the aid of neutrino and antineutrino radiation) in the ultimate "cooled" state the finite nonsphericity of the distribution of the masses in the source of the field. However, this nonsphericity is not at all manifest in the external field.

The deviation of the limiting external field from the Schwarzschild field lies in the presence of components g_0^α which do not vanish during the process of contraction. These components cause quadratic deviations of other components of the metric from the Schwarzschild values. As was already noted in the introduction, the components g_0^α arise in the external field even in the absence of rotation of the body as a whole, for example as a result of tangential velocities arising when an asymmetrical body is compressed. When a sphere rotating like a rigid body collapses, the only component that differs from the Schwarzschild components is g_{03}, with $\partial g_{03}/\partial t = 0$.

Thus, g_{03} does not change in the external space during the course of the collapse of the body. For an external observer, the surface of the collapsing rotating sphere approaches asymptotically S_S after an infinite time. The sphere has time to execute only a finite number of revolutions. The external field in the terms linear in a remains constant all the time.[2]

Appendix I

We present the Erez and Rosen [7] solution of Einstein's equations for a static axially-symmetrical field in vacuum. The solution is presented after correction of the error that has crept into [7][3] which changes the final form of the formulas appreciably

$$ds^2 = e^{2\psi} dt^2 - m^2 e^{2\gamma - 2\psi}(\lambda^2 - \mu^2)\left(\frac{d\lambda^2}{\lambda^2 - 1} + \frac{d\mu^2}{1 - \mu^2}\right)$$
$$- m^2 e^{-2\psi}(\lambda^2 - 1)(1 - \mu^2) d\varphi^2,$$

$$\psi = \frac{1}{2}\left\{[1 + \frac{1}{4}q(3\lambda^2 - 1)(3\mu^2 - 1)] \ln \frac{\lambda - 1}{\lambda + 1} + \frac{3}{2} q\lambda(3\mu^2 - 1)\right\},$$

$$\gamma = \frac{1}{2}(1 + q + q^2) \ln \frac{\lambda^2 - 1}{\lambda^2 - \mu^2} - \frac{3}{2} q(1 - \mu^2)\left[\lambda \ln \frac{\lambda - 1}{\lambda + 1} + 2\right] + \frac{9}{4} q^2 (1 - \mu^2)$$
$$\times \left[(\lambda^2 + \mu^2 - 1 - 9\lambda^2\mu^2)\frac{1}{16}(\lambda^2 - 1) \ln^2 \frac{\lambda - 1}{\lambda + 1}\right.$$

[2] Of course, the theory of small perturbations gives only terms which are linear in a.
[3] The expression for γ given in [7] is in error.

$$+\frac{1}{4}(\lambda^2+7\mu^2-\frac{5}{3}-9\mu^2\lambda^2)\lambda\ln\frac{\lambda-1}{\lambda+1}$$
$$+\frac{1}{4}\lambda^2(1-9\mu^2)+\left(\mu^2-\frac{1}{3}\right)\Bigg]. \tag{I.1}$$

Here m is the mass of the body producing the field, q characterizes the quadrupole moment. The units used are chosen such that $c=1$ and $G=1$.

The scalar $K=R_{\alpha\beta\gamma\delta}R^{\alpha\beta\gamma\delta}$ for the metric (I.1) has for small q and for $\mu=0$ the following asymptotic form as $g_{00}\to 0$:

$$K=Bq^2g_{00}^{-1}+\frac{12}{R_g^4}, \qquad B=\text{const}.$$

We have written out the principal diverging term and the term that remains when $q=0$.

By virtue of the symmetry, the light rays at $\mu=0$ and $\mu^2=1$, which have initially a radial direction, will move all the time in this direction. Near $g_{00}=0$, the time of propagation of light from a certain point with $\lambda=\lambda_0$ to $g_{00}=0$ $(\lambda=1)$ will be

$$t=\text{const}\cdot(\lambda_0-1)^{q^2/8} \quad \text{for} \quad \mu=0,$$
$$t=\text{const}\cdot(\lambda_0-1)^{-q},\ q<0 \quad \text{for} \quad \mu^2=1.$$

This time is finite,[4] unlike in the case of the Schwarzschild field.

Appendix II

Weyl's equations [8] for an axially-symmetrical Einstein field in vacuum can be written in the form

$$\frac{1}{\rho}\frac{\partial}{\partial\rho}\rho\frac{\partial\psi}{\partial\rho}+\frac{\partial^2\psi}{\partial z^2}=0, \qquad \frac{\partial\gamma}{\partial\rho}=\rho\left[\left(\frac{\partial\psi}{\partial\rho}\right)^2-\left(\frac{\partial\psi}{\partial z}\right)^2\right],$$
$$\frac{\partial\gamma}{\partial z}=2\rho\frac{\partial\psi}{\partial\rho}\frac{\partial\psi}{\partial z}. \tag{II.1}$$

The coordinates ρ and z are connected with the coordinates λ and μ of Appendix I by the expressions

$$\rho=m\sqrt{(\lambda^2-1)(1-\mu^2)}, \qquad z=m\lambda\mu.$$

For sources of the type[5] $\sigma=\sigma(z)\delta(\rho)=0$, the solution of (II.1) is obviously the potential of a filament with linear density $\sigma=\sigma(z)$ in flat space. Near $g_{00}=0$, ψ and γ are written in the following manner:[6]

$$\psi=\sigma(z)\ln\rho, \qquad \gamma=\sigma^2(z)\ln\rho,$$

[4] An only exception is the case $q>0$, $\mu^2=1$.

[5] A source of only this type gives at finite distances from a singular surface small deviations from the spherical solution.

[6] An exception is the degenerate case of a point singularity [see [11], p. 269, formula (8.30)].

where $\sigma(z)$ is arbitrary. The expression for the metric is of the form

$$ds^2 = \rho^{2\sigma} dt^2 - \rho^{2\sigma(\sigma-1)}(d\rho^2 + dz^2) - \rho^{2(1-\sigma)} d\varphi^2.$$

The properties of this metric are analogous to those discussed in Appendix I. In particular, from the point with coordinates ρ_0, z_0, φ_0, moving along the line $z = z_0$ and $\varphi = \varphi_0$ with a velocity sufficiently close to that of light, we can reach $g_{00} = 0$ after a time

$$t = \frac{\rho_0^{[\sigma(z_0)-1]^2}}{[\sigma(z_0) - 1]^2}$$

measured with the clock of the external observer.

Appendix III

Let us consider the field of a rotating sphere in vacuum. The state of this sphere need not be static—the sphere can expand radially or contract. From symmetry considerations it is clear that in the case of weak rotation the only ones of the perturbations $h_{\mu\nu}$ of the components of the Schwarzschild solution in first approximation will be h_{03}, h_{13}, and h_{23} (the perturbations in the diagonal components are of second order of smallness). By means of a small coordinate transformation we can always cause one of these quantities to vanish: after a transformation $\varphi = \tilde{\varphi} + \xi$ the components h_{03}, h_{13}, and h_{23} receive increments

$$\Delta h_0^3 = \frac{\partial \xi}{\partial t}, \quad \Delta h_1^3 = \frac{\partial \xi}{\partial R}, \quad \Delta h_2^3 = \frac{\partial \xi}{\partial \theta}.$$

Let us cause h_{23} to vanish. We write out the non-trivial components

$$\delta R_{23} = -\frac{\partial}{\partial \theta}\left(\frac{\partial}{\partial t}\frac{g_{11}h_{03}}{\sin^2\theta} - \frac{\partial}{\partial R}\frac{g_{00}h_{13}}{\sin^2\theta}\right) = 0,$$

$$\delta R_{13} = -\frac{1}{R^2}\left(\sin\theta \frac{\partial}{\partial \theta}\sin^{-1}\theta \frac{\partial h_{13}}{\partial \theta} + 2h_{13}\right) - g_{11}\frac{\partial^2 h_{13}}{\partial t^2} + R^2 g_{11}\frac{\partial^2}{\partial R \partial t}\frac{h_{03}}{R^2} = 0,$$

$$-\delta R_{03} = -g_{00}\frac{\partial^2 h_{03}}{\partial R^2} - \frac{2}{R}h_{03}\frac{\partial g_{00}}{\partial R} - \frac{\sin\theta}{R^2}\frac{\partial}{\partial \theta}\sin^{-1}\theta \frac{\partial h_{03}}{\partial \theta} + g_{00}\frac{\partial}{\partial t}\left(\frac{\partial h_{13}}{\partial R} + \frac{2}{R}h_{13}\right) = 0. \quad \text{(III.1)}$$

To find the stationary solution we put

$$\frac{\partial h_{13}}{\partial t} = \frac{\partial h_{03}}{\partial t} = 0.$$

Then the solution of (III.1) takes the form

$$h_{13} = \psi(R)R^2 \sin^2\theta, \quad h_{03} = \frac{R_g}{R}\sum_n a_n f_n\left(\frac{R}{R_g}\right) P_n^1(\cos\theta)\sin\theta. \qquad \text{(III.2)}$$

Here $c = 1$, $G = 1$, $\psi(R)$ is arbitrary, $R_g = 2m$, $a_n = \text{const}$,

$$f_n(x) = x^3 u_n(x) \int \frac{dx}{x^4 u_n^2(x)},$$

$$u_n(x) = F(2+n; 1-n; 4; x),$$

F is the Gauss hypergeometric function (see [12]); P_n^1 is the first associated Legendre polynomial (see [12]). Asymptotically we have

$$f_n(x) \sim x^{1-n}, \quad x \gg 1.$$

Making now a small transformation $\tilde{\varphi} = \varphi - \psi(R)$, we obtain $h_{13} = 0$, and the only non-vanishing component is h_{03}, for which (III.2) holds true. This is exactly the field to which the field of a contracting rotating sphere can asymptotically come over as $t \to \infty$ ($R_{\text{sur}} \to R_g$).

The concrete form of the field in the vacuum is determined by the conditions for continuity of the internal solutions on the surface of the body. The continuity conditions which follow from the requirements that the field equations be satisfied on the boundary, necessitate that h_{03} be everywhere continuous. For a sphere rotating like a rigid body (but not necessarily stationarily—it can be deforming radially) this condition leads to $h_{03} \sim \sin^2\theta$ and $h_{13} \sim \sin^2\theta$ in vacuum. The first equation of (III.1) is then satisfied identically, while the solution of the two others, compatible with the boundary conditions, is transformed with the aid of a small coordinate transformation into

$$h_{03} = -\sin^2\theta \frac{2M}{R}, \qquad \text{(III.3)}$$

where $M = -am$ is the total momentum.

Thus, the external field of such a contracting sphere is constant (with respect to the terms linear in a). Expression (III.3) coincides in form with that given in [4] for a weak field. It is actually valid also in a strong field when $a \ll R_g$ (accurate to first order in a).

It is interesting to note that whereas the magnetic moment of a collapsing magnetic star attenuates [13], the field of the mechanical moment is conserved. This difference is explained in the following manner. The magnetic moment is connected with the current I, which tends to zero, for a Schwarzschild observer, as the rate of collapse approaches c and $R_{\text{sur}} \to R_g$. On the other hand, the mechanical moment remains unchanged, for although the velocity of rotation of the star v attenuates like I in the Schwarzschild system as $R_{\text{sur}} \to R_g$, the mass of the volume element for a local Schwarzschild observer increases with increasing rate of collapse. As the result, the moment $M \sim mvR$ remains unchanged.

Appendix IV

The solution of Einstein's equation in vacuum for a spherical mass m in an external quadrupole field (which increases with increasing distance from the mass m) is of the form (the notation is the same as in Appendix I):

$$\psi = \frac{1}{2}\ln\frac{\lambda-1}{\lambda+1} + \frac{1}{4}q(3\lambda^2-1)(3\mu^2-1),$$

$$\gamma = \frac{1}{2}\ln\frac{\lambda^2-1}{\lambda^2-\mu^2} - 3q\lambda(1-\mu^2) - \frac{9}{16}q^2(\lambda^2-1)(1-\mu^2)[9\mu^2\lambda^2-\lambda^2-\mu^2+1].$$

The surface $g_{00}=0$ is determined by the condition $\lambda=1$. The Gaussian curvature of this two-dimensional surface is

$$c_G = \frac{1}{4m^2}e^q[1+3q-12q\mu^2-9q^2\mu^2+9q^2\mu^4]$$

and is different for different μ, being everywhere finite. The constant external quadrupole field can be produced by remote masses which are secured to supports that prevent their displacement. Over a limited time interval, the same field can also be approximately produced by unfastened remote masses whose velocity of motion under the influence of the mutual gravitation is at first small, the field being almost static.

Appendix V

Let us consider the collapse of a spherical dust mass. We introduce in the dust a co-moving system. We continue this freely falling system beyond the boundary of the dust, using the known solution of Tolman (see [4]). For concreteness we shall assume that a point on the boundary of the dust falls with parabolic velocity, and that the density of matter inside the dust is uniform.[7] The metric inside the dust is the metric of Friedman's cosmological model (see [4]) with pressure equal to zero, while the metric outside the dust is the Lemaitre metric [14] with ds^2 in the form

$$ds^2 = d\tau^2 - [\tfrac{3}{2}(r-\tau+\tau_0)]^{-2/3}dr^2 - [\tfrac{3}{2}(r-\tau+\tau_0)]^{4/3}(d\theta^2+\sin^2\theta d\varphi^2). \quad (\text{V.1})$$

Here τ is the proper time, $\tau_0=$ const and depends on the origin of the time, r is the co-moving coordinate, $c=1$, $R_g=1$.

The space-time of this model is shown in the figure. The dashed lines are the lines $R=$ const, where $R=[\frac{3}{2}(r-\tau+\tau_0)]^{2/3}$ is the Schwarzschild coordinate.[8]

[7]If the collapse commenced far from R_g, then near R_g the velocity of the boundary is always close to parabolic. The extension of the proof to the case of motion of the boundary of the dust with elliptical or hyperbolic velocity and with a gradient of the dust density along the radius presents no difficulties.

[8]In the T-region (i.e., for $R<R_g$) R cannot be a space coordinate. See [6].

Assume that at the instant $\tau = 0$ (close to the instant τ_1 when the boundary of the dust crosses the Schwarzschild surface $R = R_g$), the perturbations of the density, of the velocity of matter and of the metric h_α^β are small for all $0 \leq r < \infty$. Further, assume that at an arbitrarily large $R =$ const the perturbations are always small (the latter is obvious). Then, first, h_α^β will always be small in the system in question when

$$R = [\frac{3}{2}(r - \tau + \tau_0)]^{2/3} > A,$$

i.e., to the right and below the dashed line $R = A$ in the figure; here A is some constant, $A < R_g$. Furthermore, the light ray leaving the dust after the instant τ_1 will never emerge outside the Schwarzschild surface $R = R_g$ (see the figure).

Let us prove the first statement. It is seen from (V.1) that in vacuum the components $g_{\alpha\beta}$ depend only on

$$R = [\frac{3}{2}(r - \tau + \tau_0)]^{2/3}.$$

Therefore, if we now consider as independent variables not r and τ but R and τ, then small perturbations of the metric in vacuum can be written in the form $h = \exp(i\omega\tau)f(R)$ (we shall henceforth omit the indices α and β). The function $f(R)$ depends on θ and φ, but this dependence is now immaterial and will not be considered.

The idea of the proof consists in the fact that from the smallness of the perturbations on the lines (see the figure) $D = r_1 - r_2$ and further along $R = C$, and from the form of h, it follows that h is small everywhere inside the region bounded by $R = A$, $R = C$, and $D = r_1 - r_2$.

We present a formal proof. The boundary of the dust crosses R_g at a finite density $\rho_c \approx 2 \cdot 10^{16}(M_\odot/M)^{-2}$. The solution of the equations of small perturbations inside the dust [13] shows that h increase without limit only when $\rho \to \infty$ and when $\rho = \rho_c$ it is finite. Thus, up to the instant τ_2 (which is still far from τ_3 when $\rho = \infty$), there will be $h < \varepsilon_1$ in the dust if $r < r_1$.

In a freely falling system in vacuum there are solutions which increase without limit on $R = R_g$. However, a correct formulation of the Cauchy problem excludes these solutions, and h is small near the surface of the sphere in vacuum up to $\tau = \tau_2$. We thus have in vacuum:

1) from the initial conditions: $h = f(R) < \varepsilon_2$ when $\tau = 0$, $r > r_1$,

2) owing to the smallness of the perturbations on the boundary of the dust: $h \leq \varepsilon_3$, when $0 < \tau \leq \tau_2$ when $r = r_1$.

It follows from 1) that $f(R) < \varepsilon_2$ when $R \geq B = [\frac{3}{2}(r_1 + \tau_0)]^{2/3}$ (see the figure).

It follows from 2) that $f(R) < \varepsilon_4$, where $\varepsilon_4 = \varepsilon_3/|\exp(i\omega\tau)|_{\max}$ when $0 \leq \tau \leq \tau_2$ and $A < R \leq B$ (see the figure).

We thus always have

$$f(R) < \varepsilon_5 \quad \text{for} \quad R > A, \quad \varepsilon_5 = \max(\varepsilon_2 \varepsilon_4). \tag{V.2}$$

Now, by condition, $h < \varepsilon_6$ at sufficiently large $R = \text{const} = C$ and for arbitrary $\tau > 0$:

$$h_{R=C} = e^{i\omega\tau} f(B) < \varepsilon_6, \quad \tau > 0.$$

Thus,

$$e^{i\omega\tau} < \frac{\varepsilon}{f(B)} = \varepsilon_6, \quad \tau > 0. \tag{V.3}$$

It follows from (V.2) and (V.3) that

$$h = e^{i\omega t} f(R) < \varepsilon_5 \varepsilon_7 = \varepsilon_8, \quad R > A,\ \tau > 0.$$

This proves the first statement.

We now prove the second statement. In an unperturbed metric (V.1), for any light ray (not necessarily traveling along the radius) in the T-region, when $R < R_g - F$, where F is an arbitrary constant smaller than R_g, the following inequalities hold true:[9]

$$\frac{d\tau}{dr} \geq \sqrt{\frac{-g_{00}}{g_{11}}} > 1 - N,$$

where $N = \text{const}$. This inequality denotes that the inclination of the ray is larger by a finite amount than the inclination of the line $R = R_g$ (see the figure). We have proved above that when $R > A$ the perturbations of the metric always remain small. It is clear that these perturbations change the value of $d\tau/dr$ of the ray little, and that the inequality

$$\frac{d\tau}{dr} > 1 - N$$

remains in force. Thus, the ray in the region $A < R < R_g$ never approaches $R = R_g$, and all the more, cannot cross it. Consequently, we have proved that in perturbed collapse the ray never emerges from the T-region.

Appendix VI

The axially symmetrical static quadrupole perturbations of the Schwarzschild metric as $g_{00} \to 0$ are written in the form ($c = 1$):

$$h_{00} \sim q \left(1 - \frac{R_g}{R}\right) \ln\left(1 - \frac{R_g}{R}\right),$$

$$h_{11} \sim q \left(1 - \frac{R_g}{R}\right)^{-1} \ln\left(1 - \frac{R_g}{R}\right),$$

$$h_{22} \sim h_{33} \sim q \ln\left(1 - \frac{R_g}{R}\right),$$

[9] We consider a ray for which $dr/d\tau > 0$.

where q is the quadrupole parameter of the perturbation.

In the collapse of a body with $q \neq 0$ in the co-moving system, all the quantities $h_{\mu\nu}$ are finite. Inasmuch as h_{22} and h_{33} are not transformed on going over from the co-moving system to the Schwarzschild system, it is obvious that as $R \to R_g$

$$q \sim \ln^{-1} \frac{R_g}{R_1 - R_g} \sim \frac{1}{t},$$

where R_1 is the position of the boundary of the collapsing body. Thus, in first order in q the perturbations in the diagonal terms vanish asymptotically.

However, the density perturbations in the collapsing body are accompanied, in the general case, by the appearance of terms h_{13}, h_{23}, and h_{12} in the synchronous reference frame [10]. This corresponds to the appearance of non-radial velocities, i.e., it is equivalent to some differential rotation with zero total momentum. Therefore in Schwarzschild coordinates there appear non-diagonal terms which depend on the time.

As was shown in Appendix III, h_0^α terms, describing nonspherically symmetrical motion of a central body, still remain asymptotically as $g_{00} \to 0$. Thus, when the body collapses with small deviations from spherical symmetry the external metric, in the limit as $g_{00} \to 0$, may differ in first order of perturbation theory from the Schwarzschild metric only in the terms $h_{0\alpha}$.

Received
December 16, 1964

REFERENCES

1. *Zeldovich Ya. B., Novikov I. D.*—UFN **84**, 377 (1964).
2. *Bergmann P. G.*—Phys. Rev. Lett. **12**, 139 (1964).
3. *Regge T., Wheeler J.*—Phys. Rev. **108**, 1063 (1957).
4. *Landau L. D., Lifshitz E. M.* The Classical Theory of Fields. New York: Pergamon (1971).
5. *Finkelstein D.*—Phys. Rev. **110**, 965 (1958).
6. *Novikov I. D.*—Soobshcheniĭa GAISH **132** (1964).
7. *Erez G., Rosen N.*—Bull. Res. Council Israel **F8**, 47 (1959).
8. *Weyl H.*—Ann. der Phys. **54**, 117 (1917); **59**, 185 (1919).
9. *Kerr R. P.*—Phys. Rev. Lett. **11**, 237 (1964).
10. *Lifshitz E. M., Khalatnikov I. M.*—UFN **80**, 391 (1963).
11. *Synge J. L.* Relativity, The General Theory. New York: Wiley (1960).
12. *Gradshtein I. S., Ryzhik I. M.* Tables of Integrals, Series, and Products. New York: Academic Press (1980).
13. *Ginzburg V. L.*—Dokl. AN SSSR **156**, 43 (1964).
14. *Lemaitre G.*—Ann. Soc. Sci., Bruxelles **A53**, 51 (1933).
15. *Kerr R. P.* Coll. Gravitational Collapse and Quasi-Stellar Sources, ed. by J. Robinson *et al.* Univ. of Chicago Press (1965).
16. *Penrose R.*—Phys. Rev. Lett. **14**, 57 (1965).

30
Analog of the Zeeman Effect in the Gravitational Field of a Rotating Star*

As is well known, in general relativity theory the gravitational potential is no longer a scalar (as in Newton's theory). The gravitational field of a rotating body differs from that of a body at rest, just as in electrodynamics a rotating charged body produces not only an electrostatic but also a magnetic field. Thirring and Lense have noted that an ideal gyroscope near a rotating body turns slowly relative to the inertial system at infinity, that is, it rotates relative to remote fixed stars. In order of magnitude, the rotation velocity Ω of a gyroscope on the surface of the body is equal to

$$\Omega \approx \omega \frac{R_g}{R},$$

where ω is the speed of rotation of the body and R is the radius, while R_g is the gravitational radius. At the Earth's pole Ω is equal to 0.1 angular second per year ($5 \cdot 10^{-7}$ radian/year= $1.6 \cdot 10^{-4}$ rad/s); observation of this effect would be an important verification of general relativity. The rotation of the plane of polarization of light was considered by Skrotskiĭ [1].

In this note we consider the effect of the variation of the gravitational field due to the rotation on the emission spectrum of atoms on the surface of the body and observed by a receiver far from the body.

The components of the gravitational field, in analogy with the magnetic field, produce in the spectrum changes similar to the Zeeman effect.

A line emitted by an atom with frequency ω_0 at the pole and received by a remote observer situated above the pole splits into two components with opposite circular polarization and with frequencies $\omega_0 + \Omega$ and $\omega_0 - \Omega$.

Unlike the classical magnetic Zeeman effect, the gravitational effect is universal, the splitting does not depend on the detailed properties of the system emitting the light, being the same for an atom and for a molecule and the same in the optical and in radio bands.

To prove this, let us consider a linear oscillator on the pole. We can imagine it to be secured to an ideal gyroscope[1] and to oscillate in a cen-

*Pisma v Zhurnal eksperimentalnoĭ i teoreticheskoĭ fiziki **1** (3), 40–45 (1965).

[1]The gyroscope axis lies in the horizontal plane and is perpendicular to the line drawn through the center of the body, the pole, and the observer, i.e., to the beam direction.

tral force field continuously in the plane in which the gyroscope axis lies. From the point of view of the observer on the pole, the oscillator emits a plane-polarized wave, which can be regarded as a superposition of two waves circularly polarized in opposite directions with equal frequency.

With respect to the remote observer, the gyroscope axis rotates with velocity Ω. Consequently, the plane of polarization rotates with the same velocity. Linearly polarized light with a rotating plane of polarization is obviously a superposition of two circularly polarized waves but with different frequencies $\omega_0 \pm \Omega$. We have thus proved that the light emitted by a charge oscillating in a central force field on a pole of a rotating body is received by a remote observer like a combination of rays with circular polarization, split in frequency. By virtue of the principle of correspondence between quantum theory and classical mechanics, it is obvious that this result remains valid for any atomic or molecular system. The effect can reach in principle an observable magnitude on the surface of a neutron star. In fact, for a mass of the order of that of the Sun the radius of the star is of the order of 10 km; the maximum speed ω of rotation of the star, corresponding to parabolic velocity at the equator, is of the order of $\sim 10^4$ s^{-1}. We then obtain (taking into account the density distribution) as much as $\Omega \sim 10^2$ s^{-1}. For a 21 cm radio line, $\omega_0 = 10^{10}$ s^{-1}, such a splitting $(10^{10} \pm 10^2)$ could be observed at contemporary accuracy. However, the actual observation is probably a hopeless task, since the surface of a neutron star is negligible and accordingly the radiation power in the long-wave band is negligible. There are other causes for broadening and shifting of the lines; the effect has opposite signs on the pole and at the equator.

Independently of the experiment, principal considerations of the existence of a gravitational Zeeman effect can be of interest from the point of view of deepening the analogy between the magnetic field and the corresponding terms in relativistic theory of gravitation. This analogy was independently pointed out by Smorodinskiĭ, who considered within the framework of general relativity a vector playing the role of a potential. The curl of this vector determines the local rotation of the inertial system.

An alternative description of the phenomenon consists in the fact that the photons which are left- and right-circularly polarized experience different red shifts in a gravitational field. Thus, we should have here a particular manifestation of the influence of the angular momentum of the particle (photon) on the motion of the particle in a gravitational field.

It is clear from the symmetry of the problem that this difference is due entirely to the rotation of the body that produces the gravitational field.

The change of Ω in the photon frequency ω_0 is independent of the latter and occurs essentially over a path of the order of 1/2 or 1/3 of the radius of the body. On Earth it amounts to approximately $2.5 \cdot 10^{-15}$ Hz over $(2-3) \cdot 10^8$ cm, i.e., 10^{-23} Hz/cm. This change can be compared with the red shifts of all the photons (right and left polarized) in the main static field of the Earth, measured by Pound and Rebka,

$$\frac{1}{\omega}\frac{d\omega}{dx} = \frac{g}{c^2} = 10^{-18}\,\mathrm{cm}^{-1}.$$

For photons with energy 14 keV and frequency $4 \cdot 10^{18}$ Hz, the change in frequency is 5 Hz/cm and the influence of the spin (circular polarization) of hard photons is immeasurably small. For a proton the influence of the direction of the spin on its weight, due to the Earth's rotation, is of the order of 10^{-28} of the weight of the proton.

Received
April 1, 1965

REFERENCES

1. *Skrotskiĭ G. V.*—Dokl. AN SSSR **114**, 73 (1957).

Commentary

The effect of rotation of a body on the gravitational field created by it, or (in the terminology of the general theory of relativity) on the metric of the surrounding body of space, has long been known. This effect bears the names of Lense and Thirring. To date the effect has not been observed experimentally. The most promising experiment seems to be that proposed by Braginskiĭ, Thorne and Polnarev.[1]

In the above note the most natural interpretation of the effect in a weak field is given.

It is shown that one may introduce the concept of gravitational magnetism and describe the effect as the interaction of the gravimagnetic moment of the body and the gravimagnetic moments of the atoms, particles, photons.

The interaction is universal and, like a magnetic field, leads to a Zeeman splitting of the levels.

[1] *Braginskiĭ V. B., Polnarev A. G., Thorne K. S.*—Phys. Rev. Lett. **53**, 863–866 (1984).

30a
The Hypothesis of Cores Retarded During Expansion and the Hot Cosmological Model*

With I. D. Novikov

The existence of bodies with dimensions less than $R_g = 2GM/c^2$ at the early stages of expansion of the cosmological model leads to a strong accretion of radiation by these bodies. If further calculations confirm that accretion is catastrophically high, the hypothesis on cores retarded during expansion [3, 4] will conflict with observational data.

Ambartsumyan has long held the view that stars and galaxies formed from hypothetical superdense D-objects, whose material never passed through a low-density state before conversion into ordinary celestial bodies [1, 2]. In 1964 one of the authors of this note advanced the hypothesis [3] that the D-objects might be individual pieces of matter, or cores, of an expanding Friedman universe, which for reasons that we shall not analyze here have been retarded in their expansion for an external observer, are located inside their Schwarzschild spheres of radius $R = R_g = 2GM/c^2$, and emerge from these spheres only after a long period has passed since the beginning of the general expansion. Subsequently, and independently, a similar hypothesis was advanced by Ne'eman [4].

This hypothesis was treated within the scope of a cosmological model in which the density of all forms of radiation was considered negligibly small.

The discovery of general cosmic radiation with $T \approx 3$ K would appear to confirm a hot cosmological model. This circumstance radically changes the situation regarding the retarded-cores hypothesis. If these bodies had existed at an early epoch in the expansion, when the mean radiation density ρ_r was large and much greater than the baryon density ρ_b, then accretion (gravitational capture) of radiation by the retarded cores might have been catastrophically strong.[1] Let us estimate this effect.

Stationary accretion of relativistic gas by a gravitating center of mass M may be regarded either as the motion of noninteracting particles (if the

*Astronomicheskiĭ Zhurnal **43** (4), 758–760 (1966).

[1] In a radiation-free model there is a solution in which a retarded core is surrounded by a region of lower density; the mean density does not change, and there is no accretion. Obviously, the reason for the difference in the hot-model case is the motion of photons, equivalent to a quantum-gas pressure $p = \varepsilon/3$, unlike the dust for which $p = 0$. Accretion has been considered in [5–7].

mean free path is greater than R_g) or as the motion of a continuous medium. Both approaches yield the same expression, in order of magnitude, for the mass variation of the body. The formula for non-interacting particles is

$$\frac{dM}{dt} = \frac{27}{4}\pi R_g^2 c\rho_r. \tag{1}$$

Substituting $\rho_r = 3/32\pi G t^2$, as is the case for the Friedman model, and integrating over the interval from $t = t_0$ to time $t = t_1$, when $\rho_r = \rho_b \simeq 10^{-17}$ g/cm^3, with $t_1 \approx 3 \cdot 10^{11}$ s (at subsequent times ρ_r will decline more rapidly, like $t^{-8/3}$, and its contribution will be small), we obtain

$$M_1 = \frac{M_0}{1 - \frac{81}{32}\frac{GM_0}{c^3}\left(\frac{1}{t_0} - \frac{1}{t_1}\right)}. \tag{2}$$

Equation (2) may yield infinite values of M_1, for

$$t_0 \leq \frac{\frac{81}{32}\frac{GM_0}{c^3}}{1 + \frac{81}{32}\frac{GM_0}{c^3 t_1}} \tag{3}$$

To a certain extent this singularity is a formal one since equation (2) is applicable only for a stationary flux, and it is clear that the mass of a core within the Schwarzschild sphere cannot increase because of accretion by more than the value of the mass that a perturbation wave can penetrate up to time t_1 at the velocity of light (approximately equal to the sound velocity), beginning at the time t_0 of collapse of the core, or at $t_0 = 0$ if the core has existed from the start of the expansion.

In the first place, then, M_1 should definitely be less than the mass of the radiations in the metagalaxy out to the optical horizon at time t_1;[2] that is, $M_1 \lesssim 10^{16} M_\odot$. This limit is very high; captured radiation will not participate in the cosmological expansion.[3]

Secondly, if the core has existed since the start of the expansion, then until the time when the optical horizon encompasses a mass of order M_0 of surrounding material the mass of the core will not grow appreciably; it is this time that should be used in equation (2) in place of t_0.

This epoch is determined by the relation

$$t_c = \frac{GM_0}{c^3}, \tag{4}$$

and corresponds to the mean density of the cosmological model, being equal to the characteristic density of gravitational self-closure for a body of mass M_0, namely $\rho_c = M_1 / (\frac{4}{3}\pi R_g^3)$.

Of course these considerations give only the order of magnitude of t_0. From formulas (3) and (4) it follows ($t_1 \gg t_0$), that t_c is of just the order of

[2]Or at the present time. The important thing is that the limit on M_1 is very high.

[3]Such a body is similar to a geon [8]: its mass is much greater than the rest mass of the baryons comprising it. The distinction from specific Wheeler models is that an essentially nonstationary collapsing formation is considered here.

the critical value t_0, giving a formal singularity, or more explicitly implying an accretion of radiation mass out to the current optical horizon.

Thus the answer depends essentially on the numerical coefficient in the exact nonstationary expression for dM/dt. If up to the time when the mean density equals the gravitational self-closure density $\rho_c = M_0 \; / \; (\frac{4}{3}\pi R_g^3)$ the expression for dM/dt differs from equation (1) by a factor significantly less than unity (more accurately, $< 32/81$), then the increase in the mass of the retarded cores due to accretion of radiation will be of the same order as the mass of the cores themselves;[4] otherwise the captured mass would tend to be catastrophically large.

The value of the factor should be determined by more detailed calculations, and it may depend on the particular assumptions made. One might show that the hypothesis of nuclei retarded in expansion and existing from the early stages of expansion is contrary to observation in certain respects.

Finally, we note that if gravitational collapse of any body occurs at a time when the mean radiation density is less than the gravitational self-closure density of the body, then accretion of radiation will already be unimportant.

Apart from the question of the accretion of radiation, the fraction α of all the material that suffered collapse at an early stage (when $\rho_r \gg \rho_b$) should be small. In fact, ρ_r/ρ_b will decrease during the expansion, while the radiation in the cores will not participate in the cosmological expansion but will have ρ_c/ρ_b = const, where ρ_c is the density of the collapsed masses averaged over all space. The observations now imply $\rho_c/\rho_b < 160$. One can thereby show that $\alpha < 10^{-6}(M/M_\odot)^{1/2}$, where M is the mass of the collapsed bodies. Similar considerations provide bounds on the initial spectrum of inhomogeneity fluctuations.

[4]In this event it would be interesting to study the observable effects associated with collisions between expanding cores and relativistic gas falling in from outside.

Received
March 14, 1966

REFERENCES

1. *Ambartsumyan V. A.* Scientific Works. Volume 2. Erevan (1960).
2. *Ambartsumyan V. A.* Rapport 13, Conseil de physique Solvay, Brussels (1964).
3. *Novikov I. D.*—Astron. Zh. **41**, 1075 (1964).
4. *Ne'eman Y.*—Astrophys. J. **141**, 1303 (1965).
5. *Salpeter E. E.*—Astrophys. J. **140**, 796 (1964).
6. *Zeldovich Ya. B.*—Dokl. AN SSSR **155**, 67 (1964).
7. *Zeldovich Ya. B., Novikov I. D.*—UFN **86**, 433 (1965).
8. *Wheeler J. A.*—Phys. Rev. **97**, 511 (1955).

30b
Physics of Relativistic Collapse*

With I. D. Novikov

1. Introduction

It is generally known that there exists an upper mass limit for an equilibrium spherical configuration of cold matter. Equilibrium for cold matter with $M \lesssim 1.2M_{\odot}$ is possible due to the pressure of degenerated electrons at the densities of $(10^5 - 10^{10})\,\mathrm{g/cm^3}$. It is white dwarfs which are well-known to be the stars corresponding to such an equilibrium configuration.

For $M \lesssim 2M_{\odot}$ equilibrium is ensured by the elasticity of the baryon gas, and the density has a value of the order of nuclear densities; these stars are named "baryon" or "neutron" stars.

There is no equilibrium configuration for cold matter with $M \gtrsim 2M_{\odot}$. In the course of cooling, the star in which nuclear reactions are over and which has retained its mass in excess of the critical one, loses its stability. The final stage of the stellar evolution is unlimited compression-collapse.[1] For an external observer the state tends asymptotically to the limit called "collapsed star."

What is the most mysterious thing—is why the latter two types of celestial bodies have not been discovered up to now.

The answer is to be searched for in the following directions:

1) It is possible that in the course of evolution the stars lose their extra mass through explosion or stellar wind. If this is the case, then no baryon or collapsing stars exist.

2) On the other hand, it may be that such objects exist, but they are not found because they are very difficult to observe. Now the next questions arise:

a) By what physical processes the very collapse would be followed and which are the observable properties of these processes?

*Supplemento Al Nuovo Cimento **IV** (4), 810–827 (1966).

[1]Review of Recent works see B. K. Harrison, K. S. Thorne, M. Wakano and J. A. Wheeler [1]; Ya. B. Zeldovich and I. D. Novikov [2–4]; monograph [3].

b) How can one detect a neutron or collapsed star a long time after the beginning of the catastrophic period?

While the first problem mainly concerns the nonrelativistic stage of stellar evolution, it is the relativistic theory of collapse which is essential for solving the two later problems.

Recent discoveries (quasars, explosions of galaxy cores, radio-galaxies) have suggested the source of energy for such processes to have a gravitational origin. The gravitational source is capable, in principle, to give ten to twenty times as much energy as a nuclear source does (up to 10% of the rest mass). The test of this suggestion is connected with the theory of collapse, too.

We shall first dwell on the kinematics of collapse and then on the observable properties of the neutron and collapsing stars.

2. Kinematics of Collapse

2.1 Spherically-symmetric collapse.—We recall some well-known facts. After the star loses its stability, the rate of its compression practically equals the hydrodynamical one (that is the rate of free fall) [2–5]. When in the course of compression the stellar radius R approaches the gravitational radius R_g, $(R-R_g) \sim R_g$, the relativistic effects will determine the picture of collapse. The theory of the phenomenon was set up long ago. The feature of the process is the gravitational self-locking which is expressed in that after compression up to R_g the gravitational field of the body fails to let out any radiation.

As to an external observer, the stellar radius R approaches R_g asymptotically for infinite time. Due to the effect of time dilation and Doppler effect all the processes in the star are getting paralyzed—as to this observer. The limiting state (which is in no sense equilibrium!) may be called the "hardened" state.

As to the observer at the collapsing star, its surface crosses the Schwarzschild sphere for a finite time and goes on contracting.

Only a finite number of photons leave the surface during the finite time; and since the motion toward the Schwarzschild sphere delays into infinity as to an external observer it is clear that the stellar radiation will decay asymptotically. The decay law is[2]

$$J = J_0 e^{-t/t_0}, \quad t_0 \approx \frac{R_g}{c}.$$

In the co-moving system of reference, continuously continued outside the surface of the collapsing star, neither the metric tensor, nor (what is especially significant) the curvature tensor have a singularity at the gravitational

[2]The exact value of t_0 has been estimated by M. A. Podurets [6]; it equals $t_0 = (3\sqrt{3}R_g)/4c$.

radius (the invariant $K = R_{iklm}R^{iklm} = 12/R_g^4$). This means that the moment of intersection with the Schwarzschild sphere is not distinguished as to the dynamics of the collapsing star itself, the mean density of matter at this moment,

$$\overline{\rho} = 2 \cdot 10^{16} \left(\frac{M}{M_\odot}\right)^2,$$

is quite modest for large masses.

The question arises as to whether the picture described is a general one, for it might be specially connected with the symmetry of the problem. Would the qualitative results remain valid as deviations from the exact spherical symmetry take place [7–11]? And, finally, what is the limiting picture (at $t \to \infty$) in the case of collapse of the asymmetric body? These questions are discussed in detail in [12]. We shall mainly give here the results and qualitative reasons without going into rigorous proofs.

2.2 Collapse of a body with deviations from spherical symmetry.—It is clear that a large deviation from the spherical symmetry, such as quick rotation, would not allow the body to contract as a whole up to R_g. So we shall consider small departures from the spherical symmetry.

Provided that at the beginning of contraction the perturbations of density and velocity of matter were small enough, then, while increasing with compression, they are still small by the moment when $R = R_g$, for small perturbations grow infinitely only if $\rho \to \infty$ [12, 14], and the moment of intersection with the Schwarzschild sphere is not distinguished dynamically.

Thus, in the system of the co-moving observer, the sphere's surface intersects the sphere $R = R_g$ when perturbations in matter and perturbations of the field itself are still small. The invariant K on the R_g differs but little from the undisturbed value $12/R_g^4$.

All the significant properties of the Schwarzschild surface are retained. In particular, no rays can pass through it to the external observer, i.e., self-locking takes place. What occurs *afterwards* in the co-moving system?

The compression goes on, perturbations grow up, but, because of the gravitational self-locking, this would not affect the region outside the Schwarzschild surface, a far observer would know nothing of that. It is obvious that due to the self-locking the limiting picture at $t \to \infty$ would be "hardened," as to the external observer, just like the case of the exact spherical symmetry. Thus in the limit all the $(\partial/\partial t) \to 0$, and the external field is stationary.

What is this external limiting field?

The first-order time-independent deviations from the spherical Schwarzschild field may be divided into 2 types:

a) "multipole" perturbations, which change only the spatial part of the metric tensor $g_{\alpha\beta}$ ($\alpha = 1, 2, 3$) and its temporal part g_{00}.

b) "rotational" perturbations, which are connected with the appearance of $g_{0\alpha}$-components in the field.

Regge and Wheeler [15] have pointed out that "multipole" time-independent perturbations have to increase infinitely when approaching the Schwarzschild surface.

It is shown in [12] that such perturbations lead inescapably to the appearance of real space-time singularities on the Schwarzschild surface $g_{00} = 0$, where the invariant $K = R_{iklm}R^{iklm} = \infty$.

But, as we have underlined already, as the sphere with perturbations contracts, the invariant K is finite on the gravitational radius R_g in the co-moving system of reference, and is close to $K_{undisturbed} = 12/R_g^4$, thus there are no singularities. The value of an invariant is independent of the reference system, consequently no multipole deviations are possible in the external field. Such deviations have to decay on the relativistic stage of collapse. As it was shown in [12], the multipole moments of the external field decay as t^{-1}.

The rotational type deviation changes in a quite different manner.

For the case of throughout-weak field (gravitational potential $\varphi \ll c^2$), the effect of rotation of the body on its field was studied long ago by Lense and Thirring [16].

Kerr has got and analyzed [10] an exact particular solution for the external field of a rotating body. It follows from the form of this solution [12] that the matter not only rotates, but circulates as well (roughly speaking, it moves up the poles and down near the equator).

To analyze the limiting field at collapse, it is necessary to obtain the general solution for the field of a rotating body.

In [12] the solution is given which contains the corrections to the Schwarzschild metrics for rotation of the central body. The solution applies as well to the case of a strong field (near R_g rotation is supposed to be weak as before: the angular momentum $I \ll McR_g$). To the first order values, the only distinction of the field under study from the Schwarzschild field is the appearance of a g_{03}-term which for the field of a rigidly rotating sphere assumes the form

$$g_{03} = -\frac{2I}{r}\sin\theta.$$

The curvature scalar of this field has no singularities on the Schwarzschild surface, contrary to the case of multipole deviations from the spherical symmetry. Here $K \approx 12/R_g^4$ (to the first order of the I/McR_g). There is no reason for the "rotational" disturbances to decay at collapse; as is shown in [12] g_{03} is constant during the collapse (to the first order of the I/McR_g), as it follows directly from the conservation of angular momentum (to first order of I/McR_g).[3]

So, the limiting field of the "hardened" rotating star contains g_0-components.

The question arises, how can one reconcile the overall rotational moment of the body, manifesting itself in the outside field, with the vanishing of the rotation velocity, as seen by the outside observer? One may remember that the radial velocity near R_g tends to c with a corresponding infinite increase of density as measured by a local observer; in any case the overall mass, as well as momentum, remain constant.

Note that $g_{0\alpha}$-components may appear in the external field of the compressing body even if it does not rotate as a whole. The reasons for this are the tangential components of the velocity of the contracting asymmetric body. These "vortex-like" perturbations also do not decay as t goes to infinity.

Thus, the collapse of the body with weak deviations from the spherical symmetry proceeds, as to the external observer, qualitatively in the same manner as the collapse of the exactly symmetric body. The main feature of the process is self-locking which leads to the state of the "hardened" star.

But what is the final fate of the collapsing star, as to the co-moving, not external, observer? What would happen to the star inside the Schwarzschild surface?

The metric in the space-time region inside the Schwarzschild surface is essentially time-independent, and the star cannot get to a stationary state.[4]

On the other hand, the star cannot expand once again, neither in the symmetric nor in the asymmetric manner, so as to get outside the Schwarzschild surface, into the region which is available for observations to the same external observer, for whom the collapse lasts for infinite time, without reaching any physical singularity.

It is argued by Penrose [11] that the compression results unavoidably in the actual physical space-time singularity.[5]

Directly opposite conclusions follow from Lifshitz, Sudakov and Khalatnikov paper [19]: the actual singularity cannot be reached because of the instability of the rising process of such a singularity, so that matter cannot compress up to infinite density. They stress that the most general solution which is applied to the general case contains no physical singularity.

[3]Two interesting effects concerning the field of a rotating star have been pointed out recently; the first is the gravitational analogue of the Zeeman effect [17]; the second is the loss of the angular momentum by the rotating star due to selectivity of trapping the particles flying nearby and distributed isotropically at infinity [18].

[4]The physical conditions inside R_g as depending upon the previous evolution are discussed further, in Sec. 4.

[5]His topological method gives no information about the details of the singularity: what is the law of density increase? Must every bit of matter be subjected to infinite compression? and so on. So there is a lot of work still to be done.

Thus, the question of the final state of the star, as by its proper time, is not clear.

It is possible that the development of the asymmetry would result in the strongest change of space-time geometry inside the Schwarzschild sphere or even in the change of topology. But no matter what has happened there, this would not affect the region outside the Schwarzschild surface, and the external observer would know nothing of that.

3. Physical Processes at Relativistic Collapse

3.1. Neutrino emission at the collapse of a cold star.—A star with the mass but little greater than the Chandrasekhar limit loses its stability and begins to collapse by the moment when it practically grows cold. In the course of hydrodynamical compression of such a star the degenerated electrons with Fermi energy higher than a certain limit induce inverse β-decay with atomic nuclei. "Neutronization" of the matter occurs.

If the matter compresses slowly enough, then each sort of the nuclei possesses its own critical density of matter, at which neutronization takes place. This density corresponds to the Fermi energy of the electrons being equal to the threshold of the neutronization reaction. This way, neutronization of the protons

$$\mathrm{p} + \mathrm{e}^- \to \mathrm{n} + \nu$$

begins at the density of $1.6 \cdot 10^7$ g/cm^3 (the reaction threshold is 1.25 MeV); neutronization of the iron

$$^{56}_{26}\mathrm{Fe} + \mathrm{e}^- \to {}^{56}_{25}\mathrm{Mn} + \nu$$

begins at 6.10^8 g/cm (reaction threshold is 3.3 MeV) (see Cameron [20], Salpeter [21]). The threshold corresponds to neutrino energy equal to zero.

The reaction of neutronization is connected with weak interaction and is relatively slow. At fast compression the neutronization will lag behind the equilibrium for a given density, that is the process will occur mainly at densities much greater than the threshold density and at higher Fermi energy of the electrons, the energy excess of the electrons being taken away by the neutrinos. Hence, it is a mechanism of formation of the highly energetic neutrinos.

Preliminary estimates [22] show that in the simplest case of the collapse of a hydrogen star neutronization occurs mainly at $\rho \approx 5 \cdot 10^9$ g/cm^3 (instead of $1.6 \cdot 10^7$ g/cm^3 in equilibrium) with the energy of neutrinos formed of 5–7 MeV.

In the case of a helium star collapse the neutrinos are emitted mainly at the density of 10^{12} g/cm^3 and have the energy of 30–40 MeV (the electron energy threshold $\approx$ 20 MeV). The number of neutrinos is one per two nucleons.

The rough estimation shows that the average cosmic flow of the neutrinos due to the stars which have already collapsed is of the order ≈ 0.01 of the flow of the high-energy sun neutrinos produced by the $^8\mathrm{B} \to {}^8\mathrm{Be} + \mathrm{e}^+ + \nu$ decay, with the upper boundary energy of 14 MeV.

Since the higher the neutrino energy, the greater the probability to detect them, the possibility is not excluded of the experimental detection of the neutrinos formed in the course of the collapse of stars. In this connection the projects of the experiments in which it would be possible to determine the energies and roughly the direction of the neutrinos are of special interest [22].

Note that at the collapse of a usual star neutronization occurs long before the gravitational radius is reached. As a matter of fact the neutronization of helium occurs at $\rho \approx 10^{12}$ g/cm^3, while R_g is reached at $\rho_g \approx 2 \cdot 10^{16}(M/M_\odot)^{-2}$. For this reason, at $M > 50M_\odot$ the neutrinos formed leave the star undergoing but a little red-shift.

3.2. Neutrino emission by high-temperature plasma at the collapse of massive stars.—If the mass of a star considerably exceeds the Chandrasekhar limit then the star loses stability and starts to collapse when its entropy S is still rather large.[6] In the course of the compression of such a star very high temperatures are achieved as $T \approx 10^9$°K and more. At such a temperature the electron-positron pairs exist in equilibrium. According to contemporary theory the process

$$\mathrm{e}^+ + \mathrm{e}^- \to \nu + \overline{\nu}$$

has to occur with a certain probability.

This reaction may be the powerful source of neutrino and antineutrino emission at the collapse of a hot star.

How much energy can the collapsing star emit by neutrinos?

Preliminary estimates [4, 7, 25–27] show that this amount is in no case large. The thing is that gravitational self-locking occurs for masses $M > 10^4 M_\odot$ earlier than the temperature rises high enough for the neutrino to take away any appreciable portion of the star mass. The approximate formula for the mass loss writes:

$$\frac{\Delta M}{M} \approx 8 \cdot 10^4 \left(\frac{M}{M_\odot}\right)^{-3/2}; \quad 10^4 < \left(\frac{M}{M_\odot}\right) < 10^6. \tag{1}$$

The temperature at the instant of self-locking is $T_9 \approx 4 \cdot 10^3 (M/M_\odot)^{-1/2}$. For greater masses neutrino losses are still less than that obtained from formula (1).

For $M \lesssim 2 \cdot 10^2 M_\odot$, the decrease of entropy of the compressing star due to the neutrino emission stops the further growth of the emission capacity in the course of the compression. The emitted energy turns out to be small

[6] The theory of critical state for cold matter is well known. In recent times Bisnovatyĭ-Kogan and Kagdan have extended it to the case of hot stars. For every mass there exists a definite critical point. For details see [24].

as well. The corresponding approximate formula writes:

$$\frac{\Delta M}{M} \approx 8 \cdot 10^{-17} S_0^{6/7},$$

where S_0 $(\text{deg}/10^9\,\text{erg} \times g)$ is the specific entropy of the stellar matter at the moment when collapse begins. S_0 in its turn depends on the mass of the star.

Below are the star mass losses by neutrino emission at collapse.

$M/M_\odot$	10	100	10^3	10^4	10^5	10^6	10^8
$\Delta M/M$	0.05	0.1	10^{-3}	$4 \cdot 10^{-4}$	10^{-5}	$3 \cdot 10^{-7}$	10^{-14}

The treatment given is actually but a rough approximation to the truth. At the same time the following conclusions are undoubtedly quite reliable:

1) Mass loss by neutrino emission at the collapse is always a small portion (not more than 5–10%) of the mass of the star, so that it cannot weaken considerably the gravity of the core. It cannot cause the explosion of the shell, contrary to Michel's suggestion [25, 28]: see addendum 1 at the end of this article.

2) At the collapse of the stars with a small mass $M \lesssim 3M_\odot$ neutrinos with an energy of 30–40 MeV are formed in the course of neutronization, the mass loss being about $\approx 1\%$ of the M.

3) At the collapse of stars with masses between $3M_\odot$ and $100M_\odot$ neutrinos and antineutrinos are formed with wide spectrum and mean energy of the order of 30–50 MeV; the mass loss is about $\approx 5\%$.

4) The total metagalaxy density of the neutrinos and antineutrinos formed at the collapse does not exceed $\approx 5\%$ of the averaged metagalaxy density of the collapsed stars.

Thus the registration of the high-energy neutrinos may turn out to be a method of detection of the spherically symmetric "soundless" collapse of the stars. See addendum 1 at the end of this article.

3.3. Accretion of the matter by neutron and collapsed stars.—The radiation of the collapsing star decays practically instantaneously because of the self-locking effect.

Thus the hardened stars are to be dark bodies interacting with the surrounding matter by means of the gravitation field solely. The star sucks the interstellar media in the sphere of its action.

This process used to be called accretion.

The main reason of one's interest in the process of accretion lies in the possibility of the emission of large energy amounts.

The matter falling onto the baryon star surface, where the potential

$$\varphi \approx (0.2\text{–}0.3)c^2,$$

gives up about 0.2–0.3 of its rest energy, that is many times as much as nuclear reactions could give.[7]

The rate of the hydrodynamical accretion by the star which is at rest to the gas is defined by the expression

$$\frac{dM}{dt} \approx 10^{15} \left(\frac{M}{M_{\odot}}\right)^2 \left(\frac{\rho_0}{10^{-24}}\,\mathrm{g\cdot cm^{-3}}\right) \left(\frac{a_0}{1}\,\mathrm{km/s}\right)^{-3},$$

ρ_0 being the density of the surrounding matter, a_0 is the sound velocity in it.

Note that ρ_0 may be much greater than the ordinary interstellar medium density because as a result of explosions and jets the compressing star may be surrounded by a condensed gas cloud: see addendum 2 at the end of this article.

On the sphere where the gravitational potential is of the order of a_0^2, the gas flow becomes hypersonic. Gas pressure is of no importance in the hypersonic flow; the gas movement is practically a free fall along the radius.

In the case where gas is accreted by the hardened star the accretion rate is determined by the same formula (1) (it does not depend on the stellar radius, because the "narrow" place of the flow is a sphere with potential equal to a_0^2). Gas moves to the Schwarzschild sphere in a supersonic flow along a radius, and, as seen by a motionless observer near R_g, the velocity of the flying-by gas is $V \to c$, when $r \to R_g$.

The density at the moment of intersection with the Schwarzschild sphere by the elementary gas volume is finite, as to the co-moving reference system, and is of the order of magnitude of [4]

$$\rho_m \approx \rho_0 \left(\frac{c}{a_0}\right)^3. \tag{1}$$

For $\rho_0 \approx 10^{-24}\,\mathrm{g/cm^3}$ and $a_0 = 1\,\mathrm{km/s}$, one finds $\rho_m \approx 3.10^{-8}\mathrm{g/cm^3}$.

For the distant observer, neither of the particles has crossed the Schwarzschild sphere, even for infinite time. Consequently, as to him, all the falling gas is accumulated in the space close to R_g. The capacity of the layer, adjacent to R_g, is thus infinite and the picture goes to the stationary one asymptotically.

It is of importance that the particles do not strike during their fall against any surface and, besides, do not intercollide while moving along the radii in a hypersonic flow. The kinetic energy of the translational movement is not transformed into the thermal energy or radiation. It is only the thermal gas energy, increasing in the course of compression, which can be irradiated. The compression of the matter is however rather moderate (see [1]) and this source of energy is small.

[7] As was shown by Bondi [29], the maximal possible gravitational red-shift for light, emitted by the surface of a static body is less than 0.6.

Thus the spherically symmetric gas accretion by the hardened star does not lead to any considerable energy production as well (and for the same reason) as the spherically symmetric collapse itself.

An accretion may proceed in an asymmetric way, however, as when the star moves through the medium (Salpeter [30]) or when discrete clots of matter fall toward the star (Zeldovich [31]). If the clots had different angular moments in respect to the star, then, colliding near the R_g, they will irradiate as much as about $0.2m_0c^2$ of the energy.[8] A part of this energy gets onto the Schwarzschild sphere and never returns; the other part, by getting over the gravitational field, goes into infinity. The calculations show that as much energy as $0.1mc^2$ may be irradiated, m being the mass of the falling matter. This value is seen to be of the same order as with accretion onto the neutron star.

The following circumstance will be noted as well. The accreted matter moves in the opposite direction to the light flow coming from those places of the star where the energy production results from the collisions.

At a certain light flow, which is to satisfy the condition

$$\frac{L}{M} = 3 \cdot 10^4 \frac{L_\odot}{M_\odot},$$

the gravitational force would be balanced by the light pressure; if the energy production at accretion exceeds this limit, then the accretion will be cut down, the energy production will decrease, that is, the mechanism is self-regulating [32]. It is not clear to what extent this regulation is stable; the possible causes of instability are pointed out in [32].

Finally, the situation is complicated by the magnetic fields "frozen" into the gas.

Thus, accretion of the matter makes it principally possible to observe cool baryon or hardened stars. The energy produced per gram of the falling matter at the accretion process is 10 to 20 times as large as with nuclear reactions. See addendum 2 at the end of this article.

3.4. Magnetic phenomena.—It is Ginsburg [33] who first called attention to the fact that the collapse of a star with a frozen-in magnetic field leads to a tremendous increase of this field on the surface of the star:

$$\frac{H}{H_0} \sim \left(\frac{R_0}{R}\right)^2.$$

In the late stages of the collapse when R goes to R_g, the magnetic moment of the star and the outside field however decay ($I_\mu \sim t^{-1}$ [33, 34]). The reason of it is the slowing down of all processes in the star, including the movement of charged particles (i.e., the currents), producing the magnetic field for a distant observer.

[8] In the case where the star moves with supersonic velocity with respect to the gas, the prolate surface of a shock wave is formed in the tail of the flow, close to the star, collisions of the plasma clots being replaced by shock waves, in which the energy is produced.

The change of the magnetic moment in the course of collapse leads to the external field radiation [35]. The amount of energy irradiated may be quite large (of the order of the field energy itself) and, for masses of $\sim 10^8 M_\odot$ with the original field of $\sim 10^{-6}G$ at $\rho = 10^{-29}$ g/cm^3, it is as large as up to 10^{56} erg.

An existence of the high-conductive interstellar medium, surrounding the star, is to lead to complicated hydromagnetic phenomena, the study of which is just beginning (see [4], the references are given there).

Some ideas, which are of interest insofar as the quasar problem is concerned, have been developed by Layzer [36]. He has considered the cloud in which the gravitation is balanced by the chaotic magnetic field, not by the gas or light pressure. See addendum 3 at the end of this article.

3.5. Discussion.—The observations of the physical phenomena, mentioned in the present report, may, in principle, allow us to find out the baryon and hardened stars which are predicted by the theory.

The question as to what portion of all the metagalaxy nucleons is contained in the hardened stars was raised in [37]. With arguments concerned with the age of the metagalaxy, one can conclude that the total density of all the forms of matter difficult to observe does not exceed 10^{-28} g/cm^3 [38]. Hoyle, Fowler and the Burbidges [7] (see also [39]) have resolutely raised the question of the possible existence of a large amount of hardened stars. If this were the case, one could try to find them out by their total effect on the dynamics of the stellar systems (spherical clusters, galaxies). But this is beyond the accuracy of present methods even at a 50 : 50 proportion of hardened and normal stars.

4. Lagging-Core Model

The situation contrary to collapse is permitted by GRT (General Relativity Theory).

At collapse the observer on the surface of the compressing star intersects the Schwarzschild sphere and reaches the central singularity after a finite time. Let us consider this phenomenon inverted in time. Now the star surface begins to expand from a point, intersects the Schwarzschild sphere after a finite proper time and expands further.

The distant observer at a definite moment begins to register the light, neutrinos and so on emitted by the expanding star. He sees the expansion beginning sharply from a point, and not asymptotically from a hardened state. The light observed is blue-shifted.

Thus, the picture, as seen by the external observer, is not merely the time-reversed picture of compression, but proceeds essentially otherwise. The reason for this, roughly speaking, is as follows. The phenomenon of the decay of the processes at collapse is explained by the action of both effect of time dilation in the strong field and Doppler effect. Both effects are such as to slow down the processes. At the expansion of the surface Doppler effect accelerates the processes in the star, as to the external observer, and overpowers the gravitational red-shift.

The existence of two types of motion, collapse and anticollapse, reveals an important property of a spherical gravitational field.

The physical continuation of space-time "inside" the Schwarzschild sphere (T-region)[9] is two-fold. For the case of a free space it was pointed out by Finkelstein [40], and for the case of matter-occupied regions in [41, 42]. With one continuation, any test particle or a light ray moves inward in the Schwarzschild sphere. With the other continuation, all the movements are directed outward. Coexistence of particles moving inward and outward is impossible in the T-region.

As shown in [41, 42], the choice between the two possible continuations of the Schwarzschild solution into the T-region is not arbitrary but is determined physically by the conditions in which this region has appeared. If it appears as a result of the sphere contracting up to a dimension less than the gravitational radius R_g, then all the movements in it will be inward. If one sets at the moment when the dimension is less than R_g the matter velocity to be outward, then any movement will be outward, and light rays will be let out of the Schwarzschild sphere and reach the external observer.

The compression in the T-region cannot change to expansion in the same outer space, even at infinite density.

One could try to give account of the quasar phenomena and of giant explosions with the help of the lagging-core model [43, 44].*

Consider the homogeneous isotropically expanding (Friedman) world. Expansion of the whole of matter begins in it at the same moment. Assume now that some regions (cores) lagged behind and did not expand for a certain period of the world time of the model. This time delay may be of arbitrary duration and is different for different cores.

When the expansion of these cores occurs, the core matter passes from under the gravitational radius and its energy transforms into the energy of cosmic rays and radiation, by interaction with the matter falling from outside. Non-simultaneous expansion of the lagged core matter is as well possible when separate shells of the core expand in sequence, that is, recurrent explosions and continuous matter emission are possible. The matter, falling from outside, has, possibly, been rejected by the core earlier with a velocity less than the parabolic one.

The mathematical model is constructed in [43] (see also [45]), which realizes the picture described above. The description of the expansion picture is given in [46]. Some conclusions of that work are repeated later in [47]. We

[9]The "T-region" term is explained by that in this region rigid reference systems are impossible; the field ($g_{\mu\nu}$) there is essentially time-dependent.

*Now this possibility seems improbable; see paper 34a below.—*Editor's Note*

should like to emphasize that the "anticollapse" hypothesis is based on the conventional physical laws solely and in a sense develops Ambarsumian's ideas on the possibility of the existence of massive D-bodies over a long period of time terminating with subsequent explosions [48].

On the other hand, this hypothesis has nothing to do with Hoyle's conception [7, 49] of continual creation of matter (baryon charge increasing at expansion). The gravitational action of the cores, lagged with expansion, is unalterable for all the time. We should like to stress that the contents of this hypothesis is the supposition of lagging of a part of the Friedman world and the construction of the mathematical model of such a situation rather than the constatation of the possibility of changing the sign of t, that is, of changing compression into expansion. The latter is well known and has been stressed, e.g., in the works of Kalitzin [50].

The hypothesis under discussion could be applied to the cosmology as a possible way of evolution of giant systems, such as the metagalaxy (if such systems actually exist, of course). As to the application of this hypothesis in order to explain quasar's phenomena or explosions of galaxy cores, it would be noted that though it is in a position to explain the giant energy production and contradicts neither observational data, nor the laws of physics, still it is based on a quite unusual initial condition (superdense state). The quasar's puzzle is more likely, in the authors' opinion, to be solved without making use of such unusual assumptions as is expansion delay.

We should like to stress once more that the decision will depend on the observational data. At last, it is to be mentioned that the discovery of the background cosmic radiation with $T \approx 3°$K, which proves the hot cosmological model, changes the situation as to the hypothesis considered. In early stages of the expansion of the metagalaxy, radiation density being much greater than the baryon density, a strong accretion (gravitational capture) of the radiation by the lagging cores is possible.[10]

5. *Equilibrium and Collapse of a System of Point-Masses Interacting by Gravitation Alone*

5.1. Equilibrium.—Consider a spherically-symmetric equilibrium system of particles of mass m, which are moving in common self-conformed gravitational field, collisions being absent. Such a system models a spherical star-cluster or an elliptical galaxy $E0$.

For the system of given mass under study the upper density limit exists corresponding to mean velocity of the order of $0.5c$, beyond which an equilibrium solution is impossible [4, 51].

In fact, there exists a minimum radius for a test particle circular orbit

[10]See *Zeldovich Ya. B. and Novikov S. D.*—Astron. Zh. **43** 4 (1966); Ann. Rev. Astr. Astrophys. **5** (1967).

in GRT, the mass of the central body being given. For a stable orbit this minimum radius is $R_{\min} = 3R_g$; the kinetic energy of a particle movement on this orbit is $0.16c^2$. Consequently, at system dimensions less than $\approx 3R_g$, peripheral particles cannot move along finite trajectories, so that there is no equilibrium solution. As is well known (Chandrasekhar [52]), the Maxwell energy distribution is established by collisions.

In a system (almost) without collisions the thermal conductivity is (almost) infinite, so that temperature space distribution accords to the thermodynamical equilibrium, in GRT: $T = T_\infty \sqrt{g_{00}}$ instead of $T = \text{const}$ as in the Newton case.

The equilibrium problem has, however, no solution for the case of strict Maxwell distribution, due to the fact that the particles are present with energy large enough to let them go into infinity.

For this reason one looks for the quasi-equilibrium with modified distribution function, i.e., the Maxwellian one, which is cut off at the energy of, say, $mc^2 - \frac{1}{2}T_\infty$.

The results of integration of the equilibrium equations suggest the following conclusions.

1) In the critical state the mass defect is $\Delta m/m = 0.037$ and temperature: $T_\infty \approx 0.22mc^2$.

2) The mass and the central density ρ_0 are related in the critical state by

$$(\rho_0)_{\text{crit}} \approx 10^{16} \left(\frac{M_\odot}{M}\right)^2 .$$

5.2. Evolution and collapse.—The system of point masses, being in quasi-equilibrium, would evolve slowly owing to the factors neglected when solving the equilibrium problem. It is well known that in collisions the particles have a small change of acquiring enough energy to be evaporated. The characteristic evaporation time is

$$\tau_\varepsilon \approx A \left(\frac{v^2}{Gm}\right)^2 v^{-1} n^{-1},$$

v being the average velocity of the particles, n the number density of the particles, A a numerical factor $A \approx e^6/10\pi \approx 10$. At high velocities (upward from $0.01c$) there are new powerful factors of evolution: 1) the gravitational radiation (W_r) and 2) the two-body capture in head-on collisions (W_c). Their ratio to the evaporation (W_e) is given by

$$W_c : W_r : W_e = 16\left(\frac{v}{c}\right)^2 : \left(\frac{2}{45}\right)\left(\frac{v}{c}\right)^5 : \frac{1}{40} .$$

The capture equals the evaporation at $v/c \approx 0.04$, the gravitational radiation is never of great significance.

After the system is brought to the critical state by the slow evolution, the collapse begins.

As was mentioned above, the cause for the catastrophical compression is the absence of stable finite orbits for $R > 3R_g$. Derangement and fall of some particles results in increasing the mass inside the orbits of the other particles, which will derange them in turn, and so on. The avalanche-like catastrophical compression will occur with the characteristic time of order of the time of revolution of peripheral particles.

It is to be stressed that the idealized problem considered above is far from being an adequate description of the process of evolution of stellar systems. The stars cannot be idealized as point masses for near collisions.

On the late stages, when v is comparable with c, the main factor of evolution are direct collisions of stars, which are to result in powerful explosive processes, gas jets, and so on. The direction of evolution does not change, however.

So, evolution of the stellar systems has, like evolution of the stars, to finish by relativistic collapse.

Addenda

1) S. A. Colgate and R. H. White: Astrophys. J. **143** 3 (1966), suggest that during collapse the inner portion stops, a receding shock wave is formed which emits neutrinos with $E \sim 80$ MeV. These neutrinos are absorbed by the outer part, heat it and cause the explosion. It is not clear to what extent the result is due to assumptions about the equation of state at high density, combined with Newton's law of gravitation.

2) The baryon or hardened star can be a component of a double star. In this case it can be discovered by the perturbation of motion of the other, normal star (Guseinov, Zeldovich, Astron. Zh. **43**, 313 (1966)). In a double system the accretion on the collapsed star is felt by the stellar wind from the normal star. It shall lead to strong X-ray and γ-ray emission. In the case of accretion the falling gas is heated by a stationary shock wave and radiates like an optically thin hot layer (bremsstrahlung) instead of the black-body radiation. The last investigations of Scorpius-XR1 seem to be in accord with this picture.

3) This line of investigation (see end of Sec. 3.5) is developed by Ozernoĭ Z. M., Astron. Zh. **43** 2 (1966); Bardeen J. M. and Anand S. P. S., Ap. J. **144**, 959 (1965); Bisnovatyĭ-Kogan G. S., Novikov I. D. and Zeldovich Ya. B., Astron. Zh. (in press). For the underlying theory of supermassive stars see Hoyle F. and Fowler W., Nature **197**, 533 (1963); Fowler W., Rev. Mod. Phys. **36**, 545, 1104 (1964); Zeldovich Ya. B. and Novikov I. D. [2].

Received
September 4, 1964

REFERENCES

1. *Harrison B. K., Thorne K. S., Wakano M., Wheeler J. A* Gravitational Theory and Gravitational Collapse. Chicago (1965).
2. *Zeldovich Ya. B., Novikov I. D.*—UFN **84** 3 (1964); **86** 3 (1965).
3. Quasi-Stellar Sources and Gravitational Collapse. Chicago (1965).
4. *Zeldovich Ya. B., Novikov I. D.* Relativistic Astrophysics, a monograph. Moscow: Nauka (in press).
5. *Podurets M. A.*—Dokl. AN SSSR **154**, 300 (1964).
6. *Podurets M. A.*—Astron. Zh. **41** 6 (1964).
7. *Hoyle F., Fowler W., Burbidge E. M.*—Astrophys. J. **139**, 909 (1964).
8. *Eden G., Rosen N.*—Bull. Res. Council Israel **F8**, 47 (1959).
9. *Bergmenn P. G.*—Phys. Rev. Lett. **12**, 139 (1964).
10. *Kerr R. P.*—Phys. Rev. Lett. **11**, 522 (1963); (see [3], p. 99).
11. *Penrose R.*—Phys. Rev. Lett. **14**, 57 (1965).
12. *Doroshkevich A. G., Zeldovich Ya. B., Novikov I. D.*—Sov. Phys. JETP **36** 1 (7) (1965).
13. *Lifshitz E. M.*—Sov. Phys. JETP **16**, 587 (1946).
14. *Lifshitz E. M., Khalatnikov I. M.*—UFN **30** 3 (1963).
15. *Regge T., Wheeler J. A.*—Phys. Rev. **108**, 1063 (1957).
16. *Thirring H.*—Phys. Zeits. **19**, 33 (1918); **22**, 29 (1921).
17. *Zeldovich Ya. B.*—Sov. Phys. JETP, Letters to the Editor **1** 3 (1965).
18. *Doroshkevich A. G.*—Astron. Zh. **42** 6 (1965).
19. *Lifshitz E. M., Sudakov V. V., Khalatnikov I. M.*—Sov. Phys. JETP **40**, 1817 (1961); Phys. Rev. Lett. **6**, 311 (1961).
20. *Cameron D. G. W.*—Astrophys. J. **130**, 916 (1959).
21. *Salpeter E. E.*—Astrophys. J. **134**, 669 (1961).
22. *Zeldovich Ya. B., Guseinov O. H.*—Dokl. AN SSSR **162** 4 (1965); Sov. Phys. JETP Letters **1** 4 (1965).
23. *Reines F., Woods R. M.*—Phys. Rev. Lett. **14**, 20 (1965).
24. *Bisnovatyĭ-Kogan G. S., Kagdan Ya. M.*—Astron. Zh. **43** 4 (1966).
25. *Zeldovich Ya. B.*—Astron. Tsirkulyar **250**, 1-Y11 (1963).
26. *Zeldovich Ya. B., Podurets M. A.*—Dokl. AN SSSR **156**, 57 (1964) (Sov. Phys. Doklady **9**, 373 (1964)).
27. *Misner C. W.*—Phys. Rev. **137**, B 1360 (1965).
28. *Michel F. C.*—Phys. Rev. **B 133**, 329 (1964).
29. *Bondi H.*—Proc. Roy. Soc. **A 282**, 303 (1964).
30. *Salpeter E. E.*—Astrophys. J. **140** (2), 796 (1964).
31. *Zeldovich Ya. B.*—Dokl. AN SSSR **155**, 67 (1964) (Sov. Phys. Doklady **9**, 195 (1964)).
32. *Zeldovich Ya. B., Novikov I. D.*—Dokl. AN SSSR **158**, 811 (1964) (Sov. Phys. Doklady **8**, 834 (1965)).
33. *Ginsburg V. L.*—Dokl. AN SSSR **156**, 43 (1964); (see [3]).
34. *Ginsburg V. L., Ozernoĭ L. M.*—Sov. Phys. JETP **47** 9 (1964).
35. *Novikov I. D.*—Astron. Tsirkulyar USSR **290** (1964).
36. *Layzer D.*—Astrophys. J. **141**, 837 (1965).
37. *Zeldovich Ya. B., Smorodinskiĭ Ya. A.*—Sov. Phys. JETP **41**, 90 (1961).

38. *Sandage A. R.*—Astrophys. J. **133**, 355 (1961).
39. *Novikov I. D., Ozernoĭ L. M.*—Preprint, Lebedev Phys. Inst. **A-17** (1964).
40. *Finkelstein I. D.*—Phys. Rev. **110**, 4 (1958).
41. *Novikov I. D.*—Vestnik Moscow Univ. **111** 6 (1962).
42. *Novikov I. D.*—Vestnik Moscow Univ. **111** 5 (1962); Communications Shternberg Astron. Inst. USSR **132** (1964); Astron. J. USSR **40**, 772 (1963).
43. *Novikov I. D.*—Astron. Zh. **41** 6, 1075 (1964).
44. *Ne'eman Y.*—Astrophys. J. **141**, 1303 (1965).
45. *Novikov I. D.*—Sov. Phys. JETP, Letters to the editor **3** 5, 223 (1966).
46. *Novikov I. D., Ozernoĭ L. M.*—Dokl. AN SSSR **150**, 1019 (1963).
47. *Faulkner J., Hoyle F., Narlikar J. V.*—Astrophys. J. **140**, 1100 (1964).
48. *Ambartsumyan V. A.* Scientific Works **2**, Erevan (1960).
49. *Hoyle F., Narlikar S.*—Proc. Roy. Soc. **274**, 4 (1963).
50. *Kalitzin N. S.*—International Conference on Relativistic Theories of Gravitation **2**. London (1965).
51. *Zeldovich Ya. B.*—Sov. Phys. JETP **42** (6), 1667 (1962).
52. *Chandrasekhar S.* Principles of Stellar Dynamics, Univ. of Chicago Press (1942); Stochastic Problems in Physics and Astronomy; Rev. Mod. Phys. **15**, 1 (1943).

30c

Neutrinos and Gravitons in the Anisotropic Model of the Universe*

With A. G. Doroshkevich and I. D. Novikov

Anisotropic cosmological solutions are widely discussed at the present time [1] in connection with the chemical composition of prestellar matter [2] and the primordial magnetic field [3]. These solutions approach asymptotically Friedman's isotropic solution and in this respect they do not contradict the observations.

The picture of anisotropic expansion is radically altered when weakly-interacting particles (neutrinos and gravitons) are taken into account in the composition of prestellar matter.

The decrease in the anisotropy may be oscillatory. Nor can we exclude an appreciable increase in the present-day average energy of the relict neutrinos and gravitons compared with the isotropic model. This energy may exceed the present-day energy of relict electromagnetic photons, corresponding to a temperature 3°K, and may even make the observation of neutrinos possible.

The reason for the singular behavior of weakly-interacting particles (we shall henceforth refer, for brevity, to neutrinos only) lies in the fact that in the absence of interaction the freely flying particles in an anisotropically expanding universe transform their momentum in such a way that the momentum components vary differently in the directions of the different axes. The equilibrium spherically-symmetrical momentum distribution of the particles changes into preferred motion along one axis (but with equal components in both directions).

As a result, the neutrino energy density decreases not as rapidly as in the anisotropic model filled with matter having isotropic pressure (we shall call the latter the Pascal model). The slower decrease in the energy density gives rise to an earlier gravitational influence (compared with the Pascal model) of the energy density on the expansion law itself (as is well known [4], in the limit as $\rho \to \infty$ the presence of matter has no effect on the deformation law).

On the other hand, if the anisotropy is accompanied by compression along one of the axes during the early stage, some of the neutrinos and antineutrinos acquire such an energy that the probability of their irreversible trans-

*Pisma v Zhurnal eksperimentalnoĭ i teoreticheskoĭ fiziki **5** (4), 119–121 (1967).

formation into electrons and positrons becomes noticeable. This leads to an increase in the entropy of the medium. (The formulas are given at the end of this note.) Owing to such "heating," the energy density of the photons and of the pairs decreases more slowly than in the Pascal model, and has the same order of magnitude as the neutrino energy density.[1]

Starting with some instant τ, the energy density already exerts an effect on the expansion. If by that instant $\varepsilon_\nu \gg \varepsilon_{\gamma,e^\pm}$, then the cosmological model is described by a solution in which the energy density is determined only by neutrinos moving along one axis. Such a solution is presented at the end of this note. The main property of the solution is rapid expansion in the direction of neutrino motion and slow expansion in the transverse directions. As a result, the neutrino energy is rapidly decreased along the axis because of the red shift, and the principal role in the energy density is assumed by the photons and pairs. The solution rapidly becomes isotropic. The isotropization time is of the order of τ. However, the neutrino momentum distribution remains strongly anisotropic. Disregarding the influence of other weakly-interacting particles, the energy density of the neutrinos and of the γ photons in the anisotropic model should at the present time be of the same order of magnitude, but the neutrino flux is strongly anisotropic and the average neutrino energy is much higher than the γ-quantum energy.

In the special case of very small α (see the formulas at the end of the article), the "heating" by the anisotropic neutrinos is insignificant, the approach to the isotropic solution may be oscillatory, the present-day distribution of the neutrino momenta is isotropic, and the neutrino energy is of the order of the γ-quantum energy.

We did not take into account other weakly-interacting particles (such as gravitons), which may interact with one another and with the remaining particles in a manner different than the neutrinos. These particles may give rise to strong and prolonged oscillations of the anisotropy, survival of a large number of high-energy particles to the present time, and other phenomena. Instability of the directional neutrino flux and the flux of other free particles, leading to scattering, is also possible.

A complete analysis of all possible variants (in particular, with inclusion of the magnetic field) is quite difficult and has not yet been carried out completely. At the same time, it is clear even now that unless account is taken of the role of weakly-interacting particles, it is impossible to consider the theory of the anisotropic universe, and in particular the question of the chemical composition of prestellar matter.

In conclusion, we present a few formulas.

Near the singularity, the solution does not depend on the presence of matter, and takes for the homogeneous plane anisotropic model the form

[1] Provided the exponent of the time in the dependence of the distance on the time is not especially small on the axis along which the compression takes place (see the end of this note). This limitation is implied throughout.

[5]:

$$ds^2 = c^2dt^2 - t^{2p_1}dx_1^2 - t^{2p_2}dx_2^2 - t^{2p_3}dx_3^2. \tag{1}$$

We put $\alpha = -\min(p_1, p_2, p_3)$, $0 < \alpha < 1/3$ [4]. The energy density in the Pascal stage (prior to the separation of the neutrino) changes like $\varepsilon = kt^{-4/3}$. In Friedman's isotropic model, the neutrinos are separated at the instant $t \equiv \tau' \simeq 10^{-1}$ s [6]. The instant of neutrino separation in model 1 is expressed in terms of the constant $\theta = (\kappa k)^{-3/2}$, namely $t_0 \simeq \theta(\tau'/\theta)^{9/4}$ (κ is Einstein's gravitational constant). The separation is followed by a non-Pascal stage. During this stage the entropy growth is given by

$$\frac{S}{S_0} = \left\{1 + \frac{4\alpha}{1-\alpha}\left[\left(\frac{t}{t_0}\right)^{(1-\alpha)/3} - 1\right]\right\}^{3/4}. \tag{2}$$

For $t \gg t_0$ the ratio of the neutrino energy density to that of the γ photons and pairs is $\varepsilon_{\gamma e^\pm}/\varepsilon_\nu \simeq [4\alpha/(1-\alpha)] + (t_0/t)^{(1-\alpha)/3}$. The instant τ, when the influence of the energy density on the solution begins and the entropy growth ends is expressed in terms of the constant θ:

$$\tau \sim \theta\left(\frac{\tau'}{\theta}\right)^{(9/4)[(1-\alpha)/(3-\alpha)]} \tag{3}$$

when $\theta > \tau'$ and we have simply $\tau = \theta$ and no entropy growth when $\theta < \tau'$. When the instant τ is reached (for small values of α, $\alpha > (t_0/t)^{(1-\alpha)/3}$), the density ratio is $(n_\nu/n_{\gamma e^\pm}) \simeq (\tau'/\theta)^{(3/8)[(3+5\alpha)(3-\alpha)]}$, and the ratio of the average particle energies is

$$\frac{E_\nu}{E_\gamma} \simeq \frac{n_\gamma}{n_\nu} \simeq \left(\frac{\tau'}{\theta}\right)^{(-3/8)[(3+5\alpha)/(3-\alpha)]}.$$

For $t > \tau$ we have

$$S(t) \simeq S(\tau); \quad \frac{n_\nu}{n_\gamma} \simeq \left.\frac{n_\nu}{n_\gamma}\right|_{t=\tau}; \quad \frac{E_\nu}{E_\gamma} \simeq \left.\frac{E_\nu}{E_\gamma}\right|_{t=\tau}$$

The cosmological solution itself, with a neutrino moving along the x_3 axis, and with $T_3^3 = -T_0^0 = \varepsilon$ and $T_1^1 = T_2^2 = 0$, takes the form

$$ds^2 = c^2dt^2 - a_1^2dx_1^2 - a_2^2dx_2^2 - a_3^2dx_3^2;$$

$$a_1 = a_{10}y^{1/2+p}; \quad a_2 = a_{20}y^{1/2-p}; \quad a_3 = a_{30}y^{p^2-1/4}e^y; \qquad -\frac{1}{2} \le p \le \frac{1}{2};$$

$$t = \tau_0 \int y^{p^2-1/4}e^y dy; \qquad \kappa\varepsilon = \tau_0^{-2}y^{-1/2}e^{-2y}; \qquad 0 < y < \infty.$$

The instant τ corresponds to $y = 1/4 - p^2$. When $t \gg \tau$, the neutrino energy decreases like $\varepsilon \approx t^{-2}$.

Received
November 23, 1966

REFERENCES

1. *Thorne K. S.*—Preprint CST, California (1966).
2. *Hawking S. W., Tayler R. J.*—Nature **209**, 1278 (1966).
3. *Zeldovich Ya. B.*—ZhETF **48**, 986 (1965), Sov. Phys. JETP **21**, 656 (1965).
4. *Lifshitz E. M., Khalatnikov I. M.*—UFN **80**, 391 (1963), Sov. Phys. Uspekhi **6**, 495 (1964).
5. *Schücking E., Heckmann O.*—11 Conseil de Physique, Solvay, Bruxelles (1958).
6. *Zeldovich Ya. B.*—UFN **89**, 647 (1966), Sov. Phys. Uspekhi **9** (1967).

30d
Physical Limitations on the Topology of the Universe*

With I. D. Novikov

In general relativity theory, rejection of the simple assumption of a flat Euclidean space raises naturally the question that the topology of space (three-dimensional as well as four-dimensional space-time) can differ from the simple topology of flat space for an open world or the topology of a sphere for a closed world.

Shortly following Einstein's first paper [1] on the cosmological problem, in which he constructed a static cosmological model with a closed spherical three-dimensional space, Klein [2] indicated that a three-dimensional space with the same metric can also be elliptic,[1] i.e., it can have on the whole other properties than a spherical space.

The question of the connectivity of the space as a whole and of its topology is continuously mentioned in the literature (see [3]).

The nonstationary nature of the Universe and the probable existence of a singularity in the past obviously limits the region accessible to observation and hinders a direct observational investigation of the topology, say by observing the same remote object in opposite directions. Interest attaches therefore to those limitations that can be imposed on the topology by starting from considerations that do not depend on astronomic observations.

One such limitation is the requirement that causality be satisfied. This requirement is incompatible with manifolds containing closed timelike world lines (see [4, 5]).

The purpose of the present note is to emphasize that recent discoveries in the physics of elementary particles, which make possible absolute definitions of "right" and "left," show that a real physical three-dimensional space cannot be non-orientable (this cannot be refuted *a priori*). It is known that non-orientable three-dimensional spaces are contained, for example, among the 18 possible spaces of constant zero curvature (among both the open and the closed ones)[2] [6].

*Pisma v Zhurnal eksperimentalnoĭ i teoreticheskoĭ fiziki **6** (7), 772–774 (1967).

[1] A space is called elliptic if diametrically opposite points of a three-dimensional sphere are identified in it.

[2] There are no non-orientable spaces among the three-dimensional spaces of positive constant curvature.

Suveges [6] emphasizes that in a non-orientable space inversion is a continuous formation (and not discrete, as in an orientable space).

In this note we draw some physical conclusions from this remark by Suveges.

In a non-orientable space there exists a contour such that circulation over it transforms a right-hand system into a left-hand one, i.e., circulation over such a contour is equivalent to the operation of space reflection (P). This fact was noted also by Schild [11] in a discussion of a paper by Zumino at the Pisa Conference of 1964.

In 1956, discovery of parity nonconservation in weak interactions made it possible to define uniquely the concepts of a "right" and "left" system, and therefore a real physical space cannot be non-orientable.[3]

This statement cannot be made if it is assumed that a circuit over a non-oriented contour transforms a particle into an antiparticle, i.e., the circuit corresponds to combined parity (CP), since, according to Landau's hypothesis, it is precisely such a combined inversion which conserves the right-left symmetry of empty space.

Thus, in a non-orientable space, if CP-invariance holds, the question of whether two remote particles are identical or are a particle and antiparticle depends on the path along which they are brought to the same point.

The discovery of CP-invariance in K_2^0-meson decay allows us to predict an absolute difference between matter and antimatter (see [8]) and by the same token excludes any possible existence of non-orientable continuous space. It is obvious that in such a space the properties of particles should change jumpwise on going over a non-orientable contour, which is impossible.[4]

We emphasize that the foregoing conclusions do not depend on the non-stationary nature of the co-moving space in cosmology, nor do they depend on whether we are able to go around the world, in a cosmological model that expands from a singularity, by moving not faster than light along a closed contour that gives space reflection. Indeed, we can consider a chain of particles situated along this contour, compare the properties of particles that lie alongside each other, and arrive at the same contradiction.

[3]The fact that parity nonconservation should be of significance to astrophysics was emphasized by Pontecorvo [7]. Shapiro [9] related parity nonconservation with non-orientability of space in the small ($l \sim 10^{-17}$ cm); the importance of the topological properties of space to physics was emphasized by Wheeler a number of times [10]. In this note we speak of non-orientability on a cosmic scale.

[4]If the mirror-particle hypothesis turns out to be valid (see [8]), then symmetry between right and left will be restored. In this case a non-orientable space becomes possible under the condition that a circuit over a non-orientable contour corresponds to a CPA inversion (A-transformation into a mirror particle).

Received
June 25, 1967

REFERENCES

1. *Einstein A.*—Sitzgsber, preuss. Akad. Wiss. **142** (1917).
2. *Klein F.*—Ges. Abh. **1**; Gotinger Nachr. **394** (1918).
3. *Heckman O., Schucking E.*—In: Sternsysteme, Handb. d. Physik **53**, Springer (1959).
4. *Gödel K.*—Rev. Mod. Phys. **21**, 447 (1949).
5. *Synge J. L.* Relativity, The General Theory. New York: Wiley (1960).
6. *Suveges M.*—Acta Phys. Hung. **20**, 273 (1966).
7. *Pontecorvo B. M.*—Vopr. kosmogonii **9**, 132 (1963).
8. *Okun L. B.*—UFN **89**, 603 (1966).
9. *Shapiro I. S.*—UFN **61**, 313 (1957).
10. *Wheeler J. A.* Neutrinos, Gravitation, and Geometry. Bologna: Tipografia Comp. (1960).
11. *Schild A.*—Suppl. to Nuovo Cimento **4**, 412 (1966).

31

Particle Production in Cosmology*

According to rough semiquantitative estimates, the production of elementary particles near a singularity in an anisotropic cosmological model is capable of resulting in an energy density sufficient for isotropization of the expansion. If the aforementioned estimates are correct then, when account is taken of particle production, the power-law asymptotic form of a singularity of the Kasner type turns out to be internally contradictory as applied to cosmology, and all that can remain is the degenerate case of a fictitious singularity with exponents 1, 0, 0 or an isotropic Friedman singularity. Particle production may turn out to be of importance for the explanation of the presently observed ratio of the total number of particles (mainly photons) to the number of baryons.

The question of particle production was considered in the cosmological problem as applied to an isotropic (Friedman) solution in a paper by Parker [1].

In the present paper we consider a singularity of the Kasner type (see [2])

$$ds^2 = c^2dt^2 - t^{2p_1}d\xi^2 - t^{2p_2}d\eta^2 - t^{2p_3}d\zeta^2. \tag{1a}$$

The role of (1a) as a prototype of the most general solution with a singularity and for the description of the initial state of the evolution of the Universe, was considered in [3–5].

From the point of view of a local Newtonian observer in a space with coordinates $x = t^{p_1}\xi$, $y = t^{p_2}\eta$, $z = t^{p_3}\zeta$ we have a gravitational potential (see [6, 7])

$$\Phi = -\frac{p_1(p_1-1)x^2 + p_2(p_2-1)y^2 + p_3(p_3-1)z^2}{2t^2}, \quad \Delta\Phi = 0. \tag{1b}$$

Let us assume that $p_1 < p_2 < p_3$ so that $p_1 < 0$ and the potential along the x axis has a maximum at the origin, $\Phi = -kx^2$, $k > 0$. We can assume that such a potential is favorable for the production of pairs moving in opposite directions along the x axis with an ever increasing velocity. In analogy with the production of charged pairs in a static electric field (cf. e.g., [8]) we find the half-width of the barrier r from the condition

$$\Phi(r) = \Phi(0) - c^2, \qquad r = \frac{c}{\sqrt{k}}. \tag{2}$$

If the width is smaller than the Compton wavelength of the particle, then we can expect the production probability to be independent of the rest mass

*Pisma v Zhurnal eksperimentalnoĭ i teoreticheskoĭ fiziki **12** (11), 443–445 (1970).

of the particle, and to be given by an expression that follows from dimensionality considerations:

$$\frac{dn}{dt}\left[\frac{1}{\text{cm}^3\,\text{s}}\right] = k^2 c^{-3} = \frac{p_1^2(p_1-1)^2}{t^4 c^3}. \tag{3}$$

An analysis based on the method of [1] leads to a similar result. It is possible to state the problem rigorously only if a static metric is specified at $t = \pm\infty$ and the singularity is replaced by continuous and finite expressions for the metric coefficients in the region $-t_0 < t < t_0$.

A static metric is necessary at $t = -\infty$ for an unambiguous definition of the vacuum and at $t = +\infty$ for an unambiguous determination of the number of produced particles. The elimination of the singularities in the region $|t| < t_0$ is necessary in order to obtain a single-valued solution.

As shown in [1] and [9], particle production is determined by the ratio $b(t = +\infty)/b(t = -\infty)$ of the quantity $b(t)$ satisfying an equation of the type

$$\ddot{b} + \omega^2(t)b = 0, \tag{4}$$

where ω is the frequency of the given proper mode of the field describing the particles.

It seems paradoxical that the particles are produced separated by a distance $2r$ exceeding ct (see formula (2)).

This paradox is the consequence of a classical description in the language of particles having definite coordinates, phenomena that are essentially linked inseparably with the wave properties of matter, namely, the passage under a potential barrier.

An analogous paradox arises also in the production of charged particles by an electrostatic field E; the classical trajectories are given by the expression

$$x_\pm = x_0 \pm \sqrt{\frac{(mc^2)^2}{(eE)^2} + c^2(t-t_0)^2}.$$

The sum of the momenta p_+ and p_-, where $p = mc\beta/\sqrt{1-\beta^2}$, $\beta = (dx/dt)/c$ is $p_+(t) + p_-(t) = 0$. This property, just like the equation for the trajectories, is Lorentz-invariant. In any system, however, the interval between the world points of electron and positron production is space-like. Although the production of the two particles of the pairs is interrelated, since e^+ or e^- cannot be produced separately, a causal connection is impossible in the classical description. Let us turn now to the gravitation problem.

In the case of a spatially homogeneous problem, the spatial dependence of the wave field is given by the plane wave $\exp(k_1\xi + k_2\eta + k_3\zeta)$, so that the wave vector and the frequency (at $m = 0$) are given by the expression

$$\frac{\omega^2}{c^2} = |\mathbf{k}|^2 = k_1^2 t^{-2p_1} + k_2^2 t^{-2p_2} + k_3^2 t^{-2p_3}. \tag{5}$$

An equation of the type (4) solves the problem of the behavior of the classical wave field in the specified time-dependent metric.

The number of photons is an adiabatic invariant [10] of the classical field. Violation of adiabatic invariance in the classical problem, $|b_+|/|b_-| > 1$, in accord with the correspondence principle, describes the production of pairs of particles in vacuum in quantum field theory.

Approximate solutions of (4) are (we set $\omega(t=\infty) = \omega(t=-\infty)$ and $n = (b_+^2 - b_-^2)/b_-^2$

$$n \cong \int_{-\infty}^{+\infty} \frac{d\ln\omega}{dt} e^{2i\int \omega dt} dt \equiv \frac{1}{2i}\int_{-\infty}^{+\infty} \omega^{-2}\frac{d\omega}{dt} e^{2i\int \omega dt} dt$$
$$\cong \frac{1}{i}\int_{-\infty}^{+\infty} \frac{1}{\sqrt{\omega}} \frac{d^2\sqrt{\omega}}{dt^2} e^{2i\int \omega dt} dt. \qquad (6)$$

The answer depends mainly on the vicinity of the singularity. It can be roughly assumed that the production takes place near t_0 and that particles with $\omega(t_0) < t_0^{-1}$ are produced with a probability of the order of 1. The density of such particles is in this case

$$n\left[\frac{1}{\text{cm}^3}\right] = k^3 = \frac{\omega^3}{c^3} = \frac{1}{c^3 t_0^3}. \qquad (7)$$

This expression coincides by the order of magnitude with that obtained from (3) by substituting $n(t_0) \cong t_0(dn/dt)|_{t=t_0}$. The difference lies only in a factor that depends on p_1 and vanishes when $p_1 = 0$. This difference is not surprising. The metric with p_1 has only a fictitious singularity, but in the formulation of the problem (with a changeover from contraction to expansion near the singularity) such a transition introduces a nonzero curvature. The average particle energy at the instant of production, corresponding to (7), is of the order of $\hbar\omega \sim \hbar/t_0$, so that the energy density is $\varepsilon_0 \sim \hbar/c^3 t_0^4$.

Let us stop to discuss the place occupied by spontaneous production of particles in a gravitational field in general relativity theory (GRT).

The classical equations

$$R_{ik} - \frac{1}{2} g_{ik} R = \kappa T_{ik} \qquad (8)$$

are not compatible with the production of particles, since they lead to the identity $T_{ijk}^k \equiv 0$. Let the initial state be vacuum, and let T_{ik} and its derivatives be equal to zero on the hypersurface $t = \text{const}$ or $t = -\infty$. It then follows from $T_{ijk}^k = 0$ that the vacuum is always conserved.

Consequently, particle production is of necessity connected with corrections to the GRT, or more accurately with quantum corrections, since $\varepsilon \sim \hbar$.

The quantum corrections to the GRT equations were considered in several papers, on the basis of the classical equations [11, 12], and most recently in [13], as well as within the framework of a new approach [14] to the derivation of the GRT equations. The corrections considered there were "real" corrections of the order of the square of the curvature and of higher order, i.e., the nonlinear change in the elasticity of the vacuum.

Particle production is an "imaginary" correction, having the meaning of "viscosity" of the vacuum.

Let us turn to the cosmological consequences of particle production. The period of intense production at t_0 is followed by expansion with a decrease of the energy density, following the law

$$\varepsilon = \left(\frac{t}{t_0}\right)^{-\alpha}, \qquad \frac{4}{3} \geq \alpha \geq 1 - |p_1|, \tag{9}$$

where the exponent α depends on the assumption made concerning the interaction between the particles [15].

Let us find now the instant t_1 when this energy density is sufficient to exert a gravitational action on the metric and to transform the Kasner solution into a Friedman solution. From the condition

$$\varepsilon(t_1) \approx \frac{c^2}{Gt^2} \tag{10}$$

we obtain

$$t_1 = (c^5 G^{-1} \hbar^{-1} t_0^{4-\alpha})^{1/(2-a)} = (t_0^{4-\alpha} t_p^{-2})^{1/(2-a)}, \tag{11}$$

where $t_p = 10^{-43}$ s. Consequently, if we start the calculation with $t_0 \sim t_p$, then the transition to the Friedman solution occurs practically at the same instant and the Kasner solution terminates here in suicide. This leaves open the question of the situation in the case of the exponents 1, 0, 0. Apparently it is necessary to make a deeper analysis of the singularities. Within the framework of the Friedman model, according to Parker [1], no massless particles are produced if $p = \epsilon/3$ and $a(t) \sim \sqrt{t}$. Let us consider an initial state as a cold world with an extremely rigid [16] equation of state (m is the baryon rest mass):

$$p = \varepsilon = n^2 m^4 c^5 \hbar^{-3}, \qquad a(t) \sim t^{1/3}.$$

The spontaneous production of (pairs of) particles in such a metric leads to a growth of the entropy. Stipulating $t_0 = t_p$, we obtain in order of magnitude, the dimensionless entropy

$$s \cong \sqrt{\frac{\hbar c}{GM^2}} \sim 10^{18}$$

in place of the observed $s \sim \gamma/n_b \sim 10^8$, see [6, 17]. It is not excluded that some modification of the equation of state yields the correct value of s.

I take the opportunity to thank V. A. Belinskiĭ, V. L. Ginzburg, B. Ya. Zeldovich, D. A. Krizhnits, N.B. Narozhny, A. I. Nikishov, I.D. Novikov, A. M. Perelomov, and V. S. Popov for numerous discussions.**

**The commentary follows article 35

Institute of Applied Mathematics
USSR Academy of Science. Moscow

Received
September 23, 1970

REFERENCES

1. *Parker L.*—Phys. Rev. 183, 1057 (1969).
2. *Landau L. D., Lifshitz E. M.* The Classical Theory of Fields. New York: Pergamon (1971).
3. *Belinskiĭ V. A., Khalatnikov I. M.*—ZhETF **56**, 1700 (1969).
4. *Misner C. W.*—Phys. Rev. Lett. **22**, 1071 (1969).
5. *Khalatnikov I. M., Lifshitz E. M.*—Phys. Rev. Lett. **24**, 76 (1970).
6. *Zeldovich Ya. B., Novikov I. D.* Relativistic astrophysics. Univ. of Chicago Press (1971).
7. *Doroshkevich A. G., Zeldovich Ya. B., Novikov I. D.*—ZhETF **60** 1 (1971).
8. *Narozhnyi N. B., Nikishov A. K.*—Yader. Fiz. **11**, 1084 (1970).
9. *Perelomov A. M., Popov V. S.*—ZhETF **56**, 1375 (1969).
10. *Zeldovich Ya. B.*—Dokl. AN SSSR **163**, 1359 (1965).
11. *De Witt B. S.*—Phys. Rev. **162**, 1254 (1967).
12. *Feynman R. L.*—In: Report at Warsaw Conference on Relativity. Warsaw (1962).
13. *Ginzburg V. L., Kirzhnits D. A., Lyubushin A. A.*—ZhETF **60** 1 (1971).
14. *Sakharov A. D.*—Dokl. AN SSSR **117**, 1 (1967).
15. *Doroshkevich A. G., Zeldovich Ya. B., Novikov I. D.*—ZhETF **53**, 644 (1967).
16. *Zeldovich Ya. B.*—ZhETF **41**, 1609 (1961).
17. *Zeldovich Ya. B.*—UFN **89**, 617 (1966).

32

On the Possibility of the Creation of Particles by a Classical Gravitational Field*

With L. P. Pitaevskiĭ

In a recent article, Hawking [1] pointed out that "if T^{ab} satisfies a physically reasonable condition ... a space time ... which is empty at one time must be empty at all times." Knowing about annihilation of pairs of particles into gravitons and vice versa, he traces the discrepancy "to the difficulty of defining a local energy-momentum operator for the matter fields in a curved space-time" and "to the fact that we have quantized the matter fields but not the metric" (citations from [1], p. 301, 302 and 305). More precisely, he asserts that if at an initial moment of time the energy density T_0^0 is zero in some region (as well as other components of T_k^i), then it will remain zero there henceforth. The proof is based on the inequality

$$T_0^0 \geq |T_\alpha^\beta| \tag{1}$$

for the components of the energy-momentum tensor.

Our purpose in this note is to point out that the inequality (1), while valid for the part of T_k^i corresponding to particles already in existence, is in general violated by the part of T_k^i corresponding to the polarization which the classical gravitational field causes in the quantized particle vacuum (i.e. the electron-positron vacuum). It is precisely the latter part of T_k^i which describes the creation of particles.

As a simple example, we discuss the creation of particles by a weak gravitational field in a linear approximation. Let

$$g_{ik} = \overset{0}{g}_{ik} + h_{ik}. \tag{2}$$

The weak perturbation h_{ik} gives rise, as a result of polarization of the vacuum, to a contribution to the energy-momentum tensor

$$T_{ik}^{(1)} \sim h_{ik}.$$

This contribution can be calculated by the methods of quantum field theory. In the linear approximation in h_{ik}, it is described by diagrams (considering the right hand side of the equations we do not need to add a second wavy line (graviton propagator) to the left of the diagram):

*Communications in mathematical physics **23**, 185–188 (1971).

$$T_{ik}^{(1)} = \text{[diagram: loop with wavy line]} \qquad (3)$$

where the wavy line stands for the field h_{ik}, the smooth lines are the propagators for the created particles, and the dots are the vertices for the interaction of these particles with the field. We shall not write out here the corresponding analytical expressions which are different for different sorts of particles. We emphasize only the fact that they are non-zero.

Inasmuch as it is a question of principle, we work with small perturbations of Minkowskian metric,

$$g_{00} = 1, \quad g_{11} = g_{22} = g_{33} = -1.$$

For a weak field, we may expand h_{ik} in plane waves and consider each Fourier component separately; thus we set

$$h_{ik} \sim e^{-ikx}. \qquad (4)$$

Then the relationship between $T_{ik}^{(1)}$ and h_{ik} is described by the graph (3):

$$T_{ik}^{(1)} = a_{iklm} h^{lm}. \qquad (5)$$

(The field components which can create particles are, of course, those with $k^2 = \omega^2 - \mathbf{k}^2 > 0$. For instance we must have $k^2 > 4m^2$ for the creation of an electron-positron pair.)

In the approximation linear in the field, $T_{ik}^{(1)}$ must satisfy an equation of continuity in the simple form

$$\frac{\partial T_i^{k(1)}}{\partial x^k} = 0. \qquad (6)$$

In this connection we note some symmetry properties of a_{iklm}. First of all we must have

$$a_{iklm} = a_{ikml} = a_{kiml} = a_{mlki}.$$

The first two equations are obvious from the definition; the third corresponding to the Onsager symmetry principle follows directly from the diagram (3). Equation (6) gives

$$k^i a_{iklm} = 0 \qquad (7a)$$

and then also

$$k^l a_{iklm} = 0. \qquad (7b)$$

The last equations express the peculiar gauge invariance of the theory. In fact, for an infinitesimal coordinate transformation (ξ is an arbitrary infinitesimal vector)

$$h_{ik} \to h_{ik} + ik_i \xi_k + ik_k \xi_i,$$

the tensor $T_{ik}^{(1)}$, in an approximation linear in h and ξ, should be invariant, and this is ensured by (7b). Consequently, a_{iklm} can be expressed in terms of two scalar functions

$$\begin{aligned} a_{iklm} = a_1(k^2)[(g_{ik}^{(0)}k^2 - k_i k_k)(g_{lm}^{(0)}k^2 - k_l k_m)] \\ + a_2(k^2)[(g_{il}^{(0)}k^2 - k_i k_l)(g_{km}^{(0)}k^2 - k_k k_m) \\ + (g_{kl}^{(0)}k^2 - k_k k_l)(g_{im}^{(0)}k^2 - k_i k_m)]. \end{aligned} \tag{8}$$

For a plane wave with $k^2 > 0$ it is always possible to choose a system of coordinates in which the field is homogeneous ($\mathbf{k} = 0$, $k^2 = \omega^2$). Then it follows from (6) that

$$-i\omega T_0^{0(1)} = 0, \qquad T_0^0 = 0.$$

This in any case contradicts (1) and resolves Hawking's paradox. In the next order of approximation in powers of h we have

$$\overline{\frac{\partial T_0^{0(1)}}{\partial x^0}} = \overline{\frac{1}{2}\frac{\partial h_{\alpha\beta}}{\partial x^0} T^{\alpha\beta(1)}} \tag{9}$$

where the bar over T denotes an average.

In complex form

$$\frac{\partial T_0^0}{\partial x^0} = \frac{\omega h_{\alpha\beta} h^{\gamma\delta} \operatorname{Im}(a_{\gamma\delta}^{\alpha\beta})}{4}. \tag{10}$$

where Im means imaginary part.

The right-hand side of (10) is positive, and as one can easily show, is exactly the energy of the particles created by the field in unit time. We note also that the direct calculation of $\operatorname{Im} a$ from diagram (3) leads to a convergent expression, and needs no renormalization.

From (10) it can be seen that in the chosen system of coordinates $T_0^0 \sim h^2$, while $T_\alpha^\beta \sim h$. This is fully analogous to the usual quantum-mechanical situation, when the change of conserved quantity, say the energy of an atom, is of order E^2 in an electric field E, while, for example, the dipole moment of the atom contains terms which are linear in the field. We note also the obvious analogy between formula (10) and the familiar expression for the dissipation of the energy of an electromagnetic field in a medium whose dielectric constant has an imaginary part.

In conclusion we emphasize that in a free gravitational field for plane waves, of course, $k^2 = 0$ so that our discussion is concerned with the creation of particles by a gravitational field due to matter. $T_{ik}^{(1)}$ is then only a part of the full energy-momentum tensor—that part which is due to the field. This is, however, connected only with the linear approximation and we believe does not affect our main assertion. Already in the second order matter can be created by a free gravitational field, e.g., by two plane transverse gravitational waves, propagating in different directions. A phenomenological description of this process can be developed in a similar fashion. The components which arise once again will satisfy the inequality (1). It should here be emphasized that there can hardly be any doubt about the possibility of creation of particles by a classical gravitational field. The problem of defining the local energy momentum operator in a slightly curved space has the same kind of difficulty as in electrodynamics; as is well known, these difficulties can be overcome by a renormalization technique, leaving finite expressions with all needed properties. Indeed, the quantum process of conversion of gravitons into particles is possible—as mentioned already in [1]. But bosons in large number constitute a classical field, and the conversion process may of course take place in the presence of large numbers of gravitons.

The creation of particles in a non-stationary metric may be important for cosmology; see [2] and [3]. In this connection our discussion of the possibility of calculating the probability of creation in the approximation of an unquantized gravitational field (i.e., of an unquantized metric) has a central significance. A general discussion of pair-creation in external gravitational and electric fields can be found in papers of one of the authors [4]. The importance of pair creation for anisotropic metric evolution is stressed in [5].

Institute of Applied Mathematics
Institute of Physical Problems
USSR Academy of Science. Moscow

Received
March 18, 1971

REFERENCES

1. *Hawking S.*—Communs. Math. Phys. **18**, 301 (1970).
2. *Parker L.*—Phys. Rev. **183**, 1057 (1969).
3. *Zeldovich Ya. B.*—Pisma v ZhETF **12**, 443 (1970).
4. *Zeldovich Ya. B.* In: "Magic without Magic", Ed. Klauder. N.Y., 277–288 (1972).
5. *Zeldovich Ya. B., Starobinskiĭ A. A.*—ZhETF **61** 2161–2175 (1971).

Commentary

This paper by L. P. Pitaevskiĭ and Ya. B. was of quite fundamental value. When it was written papers by L. Parker[1] and A. A. Grib[2] had already become known in which the creation of particles by a gravitational field was studied. However, not long before the above paper an article, quoted here, by S. Hawking appeared with a proof of the impossibility of particle creation. This proof used the same assumption as the commonly accepted proof of the inevitability of singularity, namely, the assertion of energy dominance of particles and fields. In the present

[1] *Parker L.*—Phys. Rev. **183**, 1057–1068 (1969).
[2] *Grib A. A., Mamaev S. G.*—Yader. Fiz. **10**, 1276–1281 (1969).

work L. P. Pitaevskiĭ and Ya. B. prove that polarization of a vacuum can violate the condition of energy dominance. In this way a difficult paradox was eliminated and the way opened for further study of particle creation. Ya. B.. and A. A. Starobinskiĭ used this in subsequent papers of the cycle,[3,4] as did Hawking himself, who developed a theory of particle creation by the gravitational field of a black hole.

It was later realized that violation of energy dominance by polarization of a vacuum made it possible to construct self-consistent cosmological solutions without singularity. An example of such a solution is given in a paper by A. A. Starobinskiĭ;[5] references to previous works may also be found there.

The ideas of a connection between polarization of a vacuum and particle creation, both for gravitational and electromagnetic fields, are presented in a very graphic, detailed and lively manner in Ya. B.'s remarkable review article.[6]

[3] *Zeldovich Ya. B., Starobinskiĭ A. A.*—ZhETF **61**, 2161–2175 (1971).
[4] *Zeldovich Ya. B., Starobinskiĭ A. A.*—Pisma v ZhETF **26**, 373–377 (1977).
[5] *Starobinskiĭ A. A.*—Phys. Lett. B **91**, 99–102 (1980).
[6] *Zeldovich Ya. B.*—In: Magic without Magic. San-Francisco, p 277–288 (1972).

33
Particle Production and Vacuum Polarization in an Anisotropic Gravitational Field*

With A. A. Starobinskiĭ

Particle production and vacuum polarization of a scalar field with arbitrary mass are considered in a strong external gravitational field with a homogeneous spatially-flat nonstationary metric. The finite renormalized average values of the energy-momentum tensor are found in the general anisotropic case. It is established that vacuum polarization and production of particles of a massless field* are absent only in the isotropic case. The behavior near the singularity is investigated; in the case of isotropic collapse the exact asymptotic forms are determined. It is shown that the average values found for anisotropic collapse increase like t^{-4}, and the contraction law may be substantially altered when the reciprocal effect of the scalar field on the metric is taken into account. The polarization operator is calculated to first order in h for the case of a weak gravitational field.

Statements made prior to 1960 [1–3] about the possibility of particle production in empty space are of interest as an indication of an important area of investigation, but they do not contain precise physico-mathematical treatments of the problem.

The structure of the field corrections to the theory of gravitation [4–7] was investigated in a number of articles during the following ten-year period, right up to attempts to derive a complete theory of gravitation from a consideration of vacuum polarization [8]. The question of the cosmological constant has also been studied from this point of view [9,10].

In recent years particle production in strong gravitational fields, in particular near the cosmological singularity, has been considered in a number of articles [11–16]. Interest in this problem has been "warmed-up" by the theory of a "hot" Universe containing, as is well known, a large preponderance of neutral particles over charged particles. It is also important to consider particle production during the process of relativistic collapse.

As is well known, the most general solution of the problem of collapse near the singularity turns out to be locally anisotropic [17]. Cosmological

*Zhurnal eksperimentalnoĭ i teoreticheskoĭ fiziki **61** (6), 2161–2175 (1971).

*The statement about polarization is incorrect. See the end of the commentary following Art. 35.–*Editor's Note*

solutions are also known in which the expansion is anisotropic at first, near the singularity, and only later does the transition to isotropic expansion, which is observed at the present time, occur (the simplest of these solutions is the Heckmann-Schüking solution). Interest in such models has recently increased [18–20].

The object of this article is primarily the investigation of the production of scalar particles in strong anisotropic gravitational fields, specifically near a singularity.

In the presence of a strong gravitational field, it is natural to treat it in the classical (not the quantum) approximation. On the contrary, the particles which are produced (photons, e^+e^- pairs, $\nu\overline{\nu}$ pairs, $N\overline{N}$ pairs, etc.) necessarily must be described within the spirit of the theory of quantized fields. The same also refers to gravitons; this means that it is advantageous to represent the metric as a superposition of slowly and rapidly varying quantities (compare with the investigation of gravitational waves in the article by Isaacson [21]) and it is necessary to quantize only the latter quantities.

Vacuum polarization and particle production must be investigated concurrently; and what is more, so long as the metric does not become four-dimensionally flat, it is impossible to separate in the formulas the particles which are actually produced from the virtual particles responsible for the polarization. Paradoxes arise [22] if this fact is not taken into consideration; however, these paradoxes can be resolved in analogy to the situation in electrodynamics [23].

It is found that the energy density and the other components of the energy-momentum tensor of the produced particles increase rapidly upon approach to an anisotropic singularity. At a characteristic time $t_p = \sqrt{G\hbar/c^5}$, allowance for particle production may substantially alter the asymptotic form of the collapse; however, such complete calculation with allowance for the back reaction of the produced particles on the metric has yet been made.

It is necessary to emphasize that the isotropic case (uniform contraction or expansion in all directions), which was treated in [11–14], is degenerate. Upon approach to an isotropic singularity, particle production ceases (see Sec. 3) and its reaction on the metric is small; however, in the anisotropic case the situation is just the opposite. The reasons for this difference are related to conformal invariance of the fields.

The important concept of conformal invariance [24], that is, invariance with reference to a conformal transformation of the metric $ds^2 \rightarrow v^2(x^i)ds^2$, exists in the theory of wave fields (classical or quantum fields—it doesn't make any difference). When v depends on the coordinates, a change occurs in the properties of space, such as the transformation of flat space into curved space (but not into every curved space).

It is obvious that the particle's rest mass violates the conformal invariance of its wave equation, since the length $\hbar/mc$ associated with the rest mass

m does not transform in proportion to the scaling factor v. Only the fields of particles with $m = 0$ can be conformally invariant. For such particles a remarkable result is obtained: they are not produced (which is shown in the articles by Bronnikov and Tagirov [12] and by Parker [13]) and they do not give any contribution to the vacuum polarization in conformally-flat space-time. For particles with $m \neq 0$ the contribution at large momenta is proportional to the fourth power of the mass, which leads to the removal of the divergences from the polarization and to the smallness of the pair production.

The most important class of Friedman's cosmological solutions pertains to conformally-flat metrics. However, the more general singular solutions and, in particular, the simplest of these, whose three scaling factors a, b, c along the three spatial axes depend on the time in different ways (see the metric (1) given below), are not conformally-flat. In such a metric one can expect strong particle production, independent of the mass (that is, particle production also occurs even for $m = 0$), increasing like t^{-4} in the case of power-law dependent a, b, and c [16]. The form of the answer, $\varepsilon \sim \hbar c^{-3} t^{-4}$, follows from dimensional considerations: the action of the gravitational field on the particles and on the other fields does not contain the gravitational constant G; the energy density is constructed in a unique way out of $\hbar$, c, and t. One can formulate the result in the same way as the well-known properties of the viscosity of a gas, namely, the absence of pair production at $m = 0$ in the isotropic case means that the so-called second viscosity, associated with the divergence of the velocity and the change of volume, is equal to zero for a vacuum of relativistic particles. However, the first viscosity of the vacuum, associated with shear, does not vanish—the relations are the same as for a gas. We have carried out specific calculations for the metric (1), in which three-dimensional space (the section $t = \text{const}$) is homogeneous and three-dimensionally flat. The quantum theory can be treated in the Hamiltonian formalism owing to the distinguished role of the time. The problem reduces to solving the classical wave equation.[1]

The proposed method is general covariant. In fact, the wave equation and the energy-momentum tensor of the scalar field are covariant. However, the choice of the quantum state, with respect to which the averaging is carried out, is not unique. As such a state we take the vacuum of the scalar field, which can be properly defined only in flat space-time (ST). However, if an arbitrary ST adjoins such an ST, then by solving the wave equation (and also the "n-equation" necessary for renormalization) one can uniquely determine the energy-momentum tensor T_{ij} of the considered wave field resulting

[1]This connection between the production of particles in the quantum theory and the consideration of the classical equation is quite natural, since for Bose particles the classical theory is the exact asymptotic limit of the quantum theory for a large number of particles. Amplification of the classical wave due to nonadiabaticity and parametric resonance corresponds to the creation of particles—the quanta of the field.

from the vacuum. In curved ST it is not clear how to separate the contribution made to T_{ij} by previously created particles from the contribution due to vacuum polarization; however, perhaps this question pertains more to philology than to physics.

The considered form of the metric (1) contains both the conformally-flat case as well as the opposite case, which is not conformally reducible to the flat case. Further, in the mentioned case divergences of the polarization part of the energy-momentum tensor take place in full amount, and therefore a procedure of renormalization and elimination of the infinities is required. For this purpose a method is developed which is a modification of the Pauli-Villars method. Each individual "wave" with a given m and a wave vector $\mathbf{k}$ is associated with an n-wave with similarly increased nm and $n\mathbf{k}$ and an amplitude diminished by $\sqrt{n}$ times. The energy-momentum tensor $T_{ij}^{(n)}(\mathbf{k})$ of the n-wave and the necessary number of its derivatives with respect to n are subtracted from the $T_{ij}(\mathbf{k})$ of the physical particles under consideration in such a way that $\int d^3k T_{ij}^{\mathrm{Reg}}(\mathbf{k})$ converges, and then the limiting transition $n \to \infty$ is made. The increase of not only m, but also of $\mathbf{k}$, substantially facilitates the procedure. Just like the T_{ij}, the quantities $T_{ij}^{(n)}$ are constructed from the solutions of the wave equation; therefore each of the quantities satisfies the conservation law $T^{ij}_{;j} = 0$. Therefore the renormalized quantities T_{ij}^{Reg} identically satisfy the conservation law. In comparison with the Pauli-Villars method, the principal advantage consists in the fact that in the new method one can trace the renormalized contribution of each individual wave with a given $\mathbf{k}$. The subtracted quantities correspond to an n-particle, whose trajectory is the classical limit (ray) of the considered wave with given values of m and $\mathbf{k}$, which is not the case in the Pauli-Villars method. Real production of n-particles obviously does not occur in the limit $n \to \infty$.

In order to study the general properties of particle production, the case of small changes of the metric is instructive:

$$a^2(t) = 1 + h_1(t), \quad b^2(t) = 1 + h_2(t), \quad c^2(t) = 1 + h_3(t);$$
$$h_\alpha(\pm\infty) = 0, \quad |h_\alpha| \ll 1.$$

In this case, by having specified the vacuum at $t = -\infty$, one can uniquely determine the number of particles and all of the properties at $t = +\infty$, that is, after completion of the process. In this connection the answer for $t = +\infty$ turns out to be convergent for continuous h_α. The number of produced particles is connected with the imaginary part of the vacuum polarization. This imaginary part is finite and does not need renormalization. However, the real part for $t \neq \pm\infty$ turns out to be divergent, and a finite answer is obtained only after renormalization.

The consideration of small perturbations of the metric also enables us to classify quantities according to powers of the small parameter h. It is found that the spatial components $T_{\mu\nu}$ of the energy-momentum tensor are proportional to the first power of h, whereas the energy density $\varepsilon = T_{00}$ is proportional to h^2. The energy conservation law, which has schematically the form $\dot{T}_{00} \sim \dot{h}T_{\mu\nu}$, is therefore satisfied. Only for $t \to +\infty$ do the spatial components $T_{\mu\nu}$ of first order in h vanish, leaving $T_{\mu\nu}(+\infty) \sim h^2$. Thus, the condition of "energy dominance" (the relation $|T_{\mu\nu}| < T_{00}$) is violated in the process of particle production; only after completion of the process do we deal with a definite number of produced real particles and the indicated condition is satisfied [22, 23].

Above, the production of particles was treated in a metric which was specified beforehand and without taking account of the reaction of the produced particles on the metric. By virtue of the equations of the general theory of relativity (GTR), the given metric corresponds to a quite definite energy-momentum tensor $T_{ik(\mathrm{ext})}$ of the "external" matter (in particular, it is possible that $T_{ik(\mathrm{ext})} = 0$). In this sense it is not completely correct to call the process under study particle production in vacuum; we are actually investigating particle production in the presence of "external" matter. However, only the gravitational interaction of the "external" matter with the produced particles is taken into consideration here.

The particular case of the Kasner metric with power-law dependencies of a, b, and c on t, and with the exponents p_1, p_2, p_3 satisfying the well-known relations, corresponds to the absence of "external" matter. Incidentally, if we gradually make the transition from the Minkowski metric to the Kasner metric, then external matter is certainly necessary. The natural way of transition to the Kasner metric gives the Heckmann-Schücking solution with $a(t) \sim t^{p_1}(t+t_1)^{2/3-p_1}$ and so forth. In this case as $t \to 0$ the "external" matter does not disappear, its density increases like t^{-1}, and its role is small in comparison with the major terms ($\sim t^{-2}$) in the equations of GTR. From the methodological point of view the formulation of the problem of particle production in a given metric is not altered thereby. The method of renormalization, verified with a weak field as an example, turns out to be applicable also to the problem of production in an asymptotically Kasner metric. We mention the case $p_1 = 1$, $p_2 = p_3 = 0$, whose importance has been insisted on by V. A. Belinskiĭ. Here space-time is four-dimensionally flat and is easily transformed to the Minkowski form. However, the metric with the transition region $a(t) \to \mathrm{const}$ for $t < t_0$ and $a(t) \sim |t|$ for $t > t_0$, $b = c = \mathrm{const}$, is essentially non-flat for $t \sim t_0$; "external" matter exists in it. As a consequence of the nonlocal nature of the theory, the influence of the transition at $t = t_0$ also appears for $t > t_0$. Therefore, the nonvanishing result for $p_1 = 1$ and $p_2 = p_3 = 0$ in such a formulation of the problem does not violate any general principles.

The present article does not claim to be an exhaustive investigation of the problem. The immediate fundamental problem, which is in principle solvable (although it is also difficult) within the framework of the concepts developed here, is the systematic investigation of the collapse to a singularity with a self-consistent calculation of the reaction of the produced particles and of the vacuum polarization. As indicated above, such an effect is small in the isotropic case; however, it may become large in the case of anisotropic collapse. It is very probable that the reaction of particle production on the metric leads to isotropization of the contraction, the last stage of the approach to the singularity is switched over to the tracks of the quasi-isotropic solution described by Lifshitz and Khalatnikov [17]. On the other hand, it is improbable that the back reaction might lead to the replacement of contraction by expansion. For this to happen it would be necessary for the energy density to be negative at some stage.

Finally, let us mention problems which are natural components of the general problem, but which require new ideas: 1) a general-covariant formulation of the theory, in particular a formalism without a distinguished role for the time; 2) allowance for the direct (electromagnetic, strong, etc.) nongravitational interaction between the particles; 3) the most difficult and important problem—the cosmological problem of emergence from the singularity, and the formulation of the initial conditions in the singular state. It is possible that this problem is inseparable from the general problem of the quantization of the metric, where the separation into high- and low-frequency dependencies on the coordinates is no longer satisfactory because of the contraction of the horizon. Here one can also express the tempting hypothesis that, just as in the case of collapse, the reaction of vacuum polarization on the gravitational field leads to the transition of an anisotropic expansion of the Kasner type into quasi-isotropic expansion at $t \sim t_p$ (see [16] for the appropriate estimates).

1. Quantization of the Field and the Mean Value of the Energy-Momentum Tensor

Let us consider a real scalar field $\varphi(x^i)$ in a homogeneous cosmological model with a spatially-flat nonstationary metric:

$$ds^2 = dt^2 - a^2(t)dx_1^2 - b^2(t)dx_2^2 - c^2(t)dx_3^2, \tag{1}$$

where a, b, and c are certain given non-negative functions of the time. We take the Lagrangian density of this field in the form

$$L = \frac{1}{2}[g^{ik}\varphi_{,i}\varphi_{,k} - (m^2 - \frac{R}{6})\varphi^2], \tag{2}$$

where $R = g^{ik}R_{ik}$ is the scalar curvature (we use a system of units in which $\hbar = c = 1$; the notation for products and the choice of signs in the definitions of g_{ik}, R_{iklm} and R_{ik} coincide with those adopted in [25]). The corresponding field equation is

$$\varphi_{;i}^{;i} + (m^2 - \frac{R}{6})\varphi = 0. \tag{3}$$

It is conformally invariant for $m = 0$ [24], that is, under the conformal transformation of the metric

$$(ds = v(x^i)d\tilde{s}, \quad g_{jk} \to \tilde{g}_{jk} = v^{-2}(x^i)g_{jk})$$

and under the corresponding transformation of the field function

$$(\varphi \to \tilde{\varphi} = v(x^i)\varphi)$$

equation (3) preserves its form. Tagirov and Chernikov [11] advance arguments in favor of the necessity of choosing the scalar field Lagrangian in the form (2).

The energy-momentum tensor of this field [11] is of the form

$$T_{ik} = \varphi_{,i}\varphi_{,k} - g_{ik}L + \frac{1}{6}(R_{ik} + g_{ik}\Box - \nabla_i\nabla_k)\varphi^2, \tag{4}$$

where $\Box = g^{ik}\nabla_i\nabla_k$; ∇ denotes the operator of covariant differentiation. This tensor possesses the following properties:

$$T_{ik} = T_{ki}, \quad T \equiv T_i^i = m^2\varphi^2, \quad T^{ik}_{;k} = 0. \tag{5}$$

We note that in the case of flat space-time T_{ik} differs from the usually used energy-momentum tensor of a scalar field by the quantity $(1/6)(g_{ik}\Box - \nabla_i\nabla_k)\varphi^2$, which has the form of a divergence.

The metric is assumed to be given and is not quantized. We carry out quantization of the scalar field by the standard method, using the Hamiltonian formalism (for the case of the isotropic Friedman model, this was done by Bronnikov and Tagirov [12] and also by Parker [13]), namely: we introduce the canonical equal-time commutation relations

$$\begin{aligned} &[\varphi(\mathbf{x},t), \varphi(\mathbf{x}',t)] = [\pi(\mathbf{x},t), \pi(\mathbf{x}',t)] = 0, \\ &[\varphi(\mathbf{x},t), \pi(\mathbf{x}',t)] = i\delta^{(3)}(\mathbf{x}-\mathbf{x}'), \end{aligned} \tag{6}$$

where the generalized momentum of the field is given by $\pi = \partial(\sqrt{-g}L)/\partial\dot{\varphi} = V\dot{\varphi}$, $V \equiv \sqrt{-g} = abc$.

Relations (6) are satisfied if we represent the function $\varphi(x^i)$ in the form

$$\varphi(x^i) = \frac{1}{(2\pi)^{3/2}} \int d^3k[\hat{A}_k\varphi_k(t)e^{-i\mathbf{kx}} + \hat{A}_k^+\varphi_k^*(t)e^{i\mathbf{kx}}]. \tag{7}$$

Here $\mathbf{x} = (x^1, x^2, x^3)$ and $\mathbf{k} = (k_1, k_2, k_3)$ is the constant wave vector; the function $\varphi_k(t)$ satisfies the equation

$$\ddot{\varphi}_k + \frac{\dot{V}}{V}\dot{\varphi}_k + \left(\omega_k^2(t) - \frac{R}{6}\right)\varphi_k = 0, \tag{8}$$

where

$$\omega_k^2(t) = m^2 + \frac{k_1^2}{a^2} + \frac{k_2^2}{b^2} + \frac{k_3^2}{c^2},$$

and the condition

$$\dot{\varphi}\varphi^* - \dot{\varphi}^*\varphi = \frac{i}{V} \tag{9}$$

(a dot indicates differentiation with respect to t), and the A_k are some constant operators with the commutation relations:

$$[\hat{A}_k \hat{A}_{k'}] = [\hat{A}_k^+ \hat{A}_{k'}^+] = 0, \quad [\hat{A}_k \hat{A}_{k'}^+] = \delta^{(3)}(\mathbf{k} - \mathbf{k}'). \tag{10}$$

Further, we shall everywhere assume that, as $t \to -\infty$, a, b, and $c \to$ const and space-time becomes flat, and also the initial condition for $\varphi_k(t)$ is

$$\varphi_k(t)\big|_{t\to\infty} = \frac{e^{-i\omega_{k0}t}}{\sqrt{2\omega_{k0}V_0}},$$

where $\omega_{k0} = \omega_k(-\infty)$ and $V_0 = V(-\infty)$. Only in this case do the operators $\hat{A}_k$ and $\hat{A}_k^+$ coincide with the operators for the annihilation and creation of quanta of the free field at $t = -\infty$ and one can introduce the constant Heisenberg state vector $|0\rangle$ (satisfying the condition $A_k|0\rangle = 0$ for all $\mathbf{k}$), which is the properly defined vacuum of the scalar field for $t = -\infty$. A complete system of in-states for $t = -\infty$ is introduced in similar fashion.

Because of the nonstationary nature of the metric (1), in the general case it is impossible to identify the operators $\hat{A}_k$ and $\hat{A}_k^+$ with the creation and annihilation operators for $t > -\infty$, and the state $|0\rangle$ is not the vacuum state for $t > -\infty$. The question as to how to define the creation and annihilation operators and the "vacuum" state for all values of t was investigated by Bronnikov and Tagirov [12] in the case of the isotropic Friedman model, and they showed that the answer is not unique (in this connection, see also the articles by Grib and Mamaev [14]).

For the problems considered in the present article, this ambiguity is unimportant, since we are only interested in the average value $\langle 0|T_i^k(t)|0\rangle$. In a space with the metric (1), the average values of the diagonal components of the energy-momentum tensor are different from zero:

$$\mathcal{E} \equiv \langle 0|T_0^0|0\rangle = \frac{1}{(2\pi)^3}\int d^3k\Big\{\frac{1}{2}(|\dot{\varphi}_k|^2 + \omega_k^2|\varphi_k|^2) + \frac{1}{6}(R_0^0 - \frac{1}{2}R)|\varphi_k|^2 + \frac{1}{6}\frac{\dot{V}}{V}\frac{d}{dt}|\varphi_k|^2\Big\}, \tag{11}$$

$$\mathcal{P}_1 \equiv -\langle 0|T_1^1|0\rangle = \frac{1}{(2\pi)^3}\int d^3k\Big\{\frac{k_1^2}{a^2}|\varphi_k|^2 + \frac{1}{2}(|\dot{\varphi}_k|^2 - \omega_k^2|\varphi_k|^2) - \frac{1}{6}(R_1^1 - \frac{1}{2}R)|\varphi_k|^2 - \frac{1}{6V}\frac{d}{dt}(V\frac{d}{dt}|\varphi_k|^2) + \frac{1}{6}\frac{\dot{a}}{a}\frac{d}{dt}|\varphi_k|^2\Big\}$$

and so forth.

It is clear that in such a formulation the quantum problem reduces to a consideration of the classical wave equation (3) for $\varphi(x^i)$ or equation (8) for $\varphi_k(t)$.

We immediately note that the vacuum expectation values (11) include the energy density and the pressure of the zero-point oscillations of the vacuum and necessarily diverge. Therefore, in what follows we investigate the renormalized quantities

$$\varepsilon = \mathcal{E} - \mathcal{E}_0, \quad p_\alpha = \mathcal{P}_\alpha - \mathcal{P}_{\alpha 0} \qquad (\alpha = 1, 2, 3),$$

where

$$\mathcal{E}_0 = \frac{1}{(2\pi)^3 V} \int d^3k \frac{\omega_k}{2}, \quad \mathcal{P}_{10} = \frac{1}{(2\pi)^3 V} \int d^3k \frac{k_1^2}{2a^2\omega_k}$$

and so forth; $\mathcal{E}_0$ and $\mathcal{P}_{\alpha 0}$ do not depend on the rate of change of the metric, and they diverge like k^4.

The renormalized energy density ε and pressure p_α satisfy the following conservation law:

$$\frac{1}{V}\frac{d}{dt}(V\varepsilon) = -\left(p_1\frac{\dot{a}}{a} + p_2\frac{\dot{b}}{b} + p_3\frac{\dot{c}}{c}\right). \tag{12}$$

The question of their convergence will be investigated in the following section.

Because of the spatial homogeneity of the metric (1), the total momentum of the scalar field, $G_\alpha = \int d^3x\sqrt{-g}T^0_\alpha$, is conserved. In the case of a complex scalar field $\varphi(x^i)$ (whose quantization is carried out in analogous fashion) the total charge $Q = \int d^3x\sqrt{-g}J^0$ is also conserved, that is, the particles are created in pairs with opposite charges and momenta.

2. The General Anisotropic Case

Let us make the following change of the variable t and of the field function φ:

$$\eta = \int \frac{dt}{V^{1/3}(t)}$$

$$\varphi(\eta, \mathbf{x}) = \frac{\chi(\eta, \mathbf{x})}{V^{1/3}(\eta)}, \qquad \varphi_k(\eta) = \frac{\chi_k(\eta)}{V^{1/3}(\eta)}. \tag{13}$$

We introduce the notation $v(\eta) = V^{1/3}$, $g_1(\eta) = a/v$, and $g_2(\eta) = b/v$, then equation (8) reduces to the form

$$\chi''_k + [\Omega_k^2 + Q(\eta)]\chi_k = 0, \tag{14}$$

where

$$\Omega_k^2 = v^2\omega_k^2 = m^2v^2 + \frac{k_1^2}{g_1^2} + \frac{k_2^2}{g_2^2} + g_1^2g_2^2k_3^2,$$

$$Q(\eta) = \frac{\left(\frac{a'}{a} - \frac{b'}{b}\right)^2 + \left(\frac{a'}{a} - \frac{c'}{c}\right)^2 + \left(\frac{b'}{b} - \frac{c'}{c}\right)^2}{18}$$

$$= \frac{\left(\frac{g_1'}{g_1}\right)^2 + \left(\frac{g_2'}{g_2}\right)^2 + \frac{g_1'g_x'}{g_1g_2}}{3},$$

and a prime denotes differentiation with respect to η. It is clear that $Q(\eta) = 0$ only in the isotropic case.

We have obtained the equation of a classical oscillator with a variable frequency.

Let us consider the special case of a metric whose evolution is such that $a, b, c|_{\eta=-\infty} = a, b, c|_{\eta=+\infty} = 1$. Let us take a single mode $\mathbf{k}$. As $\eta \to -\infty$ let the function $\chi(\eta)$ corresponding to this mode have the form $\chi(\eta) = e^{-i\Omega_0 \eta}$, where $\Omega_0 = \sqrt{m^2 + \mathbf{k}^2}$. Then, as $\eta \to +\infty$ this same function χ has the asymptotic form

$$\chi = \alpha e^{-i\Omega_0 \eta} + \beta e^{i\Omega_0 \eta},$$

where $|\alpha|^2 - |\beta|^2 = 1$, $\beta \neq 0$ in the general case. Thus, amplification of the wave occurs—its energy increases by $1 + 2|\beta|^2$ times. The same thing also refers to the second elementary wave: if $\chi = e^{i\Omega_0 \eta}$ as $\eta \to -\infty$, then for $\eta \to +\infty$ one has

$$\chi = \alpha^* e^{i\Omega_0 \eta} + \beta^* e^{-i\Omega_0 \eta}.$$

The energy of this wave also increases by $1+2|\beta|^2$ times. An arbitrary linear combination of both waves with different signs of the frequency for $\eta \to -\infty$ obviously can be both intensified and weakened.

The wave equation is invariant with regard to the replacement of η by $-\eta$; one can easily construct the initial combination

$$\chi = C_1 e^{-i\Omega_0 \eta} + C_2 e^{i\Omega_0 \eta},$$

for $\eta \to -\infty$, which gives a decrease of the energy on emergence, that is, for $\eta \to +\infty$. However, if for $\eta \to -\infty$ the ratio of the moduli of C_1 and C_2 is fixed and the answer is averaged over the relative phase $(\operatorname{Arg}(C_1/C_2))$, then one again obtains an increase of the total energy by that same factor of $1 + 2|\beta|^2$ times.

From the quantum point of view, the energy increase associated with this process implies the creation of new quanta of the field. In the classical theory the increase of energy is proportional to its initial magnitude by virtue of the linearity of the field equations. The quantum theory of Bose particles is equivalent to a classical theory with a nonvanishing energy $(\hbar\Omega_0/2)$ of the state without any particles, and therefore gives a non-zero result for the production of particles from this state. Since the vacuum is a state with an undetermined phase, the energy always only increases.

It is obvious that β is a measure of the nonadiabaticity of the process. The departure from adiabaticity decreases with increase wave vector $\mathbf{k}$ and frequency Ω_k; therefore $|\beta|^2$ turns out to be a rapidly decreasing function of $|\mathbf{k}|$. For the calculations it is convenient to write the corrections to the adiabatic approximation in explicit form; therefore, in order to solve equation (14) we shall use the method of Lagrange.

Thus, we shall seek the solution of equation (14) in the form

$$\chi_k(\eta) = \frac{\alpha_k(\eta)e_-}{\sqrt{2\Omega_k}} + \frac{\beta_k(\eta)e_+}{\sqrt{2\Omega_k}},$$

$$e_\pm = e^{\pm i \int \Omega_k \, d\eta}, \tag{15}$$

with the additional condition

$$\chi_k'(\eta) = -i\Omega_k \left(\frac{\alpha_k(\eta)e_-}{\sqrt{2\Omega_k}} - \frac{\beta_k(\eta)e_+}{\sqrt{2\Omega_k}} \right) \tag{16}$$

(that is, the derivative of χ_k is taken as if α_k, β_k, and Ω_k did not depend on the time). From (9), (15), and (16) it follows that

$$|\alpha_k(\eta)|^2 - |\beta_k(\eta)|^2 \equiv 1. \tag{17}$$

As a result, instead of the single second-order differential equation (14) for χ_k, a system of two linear first-order equations is obtained for α_k and β_k:

$$\alpha_k' = \left(\frac{\Omega_k'}{\Omega_k} - \frac{iQ}{\Omega_k} \right) \frac{e_+^2 \beta_k}{2} - \frac{iQ\alpha_k}{2\Omega_k},$$

$$\beta_k' = \left(\frac{\Omega_k'}{\Omega_k} + \frac{iQ}{\Omega_k} \right) \frac{e_-^2 \alpha_k}{2} + \frac{iQ\beta_k}{2\Omega_k} \tag{18}$$

with the initial conditions: for $\eta = -\infty$ ($t = -\infty$) the quantities $\alpha_k = 1$, $\beta_k = 0$.

Sometimes it is convenient to change from the two complex variables α_k and β_k, which are related by the condition (17), to three real variables:

$$s_k = |\beta_k|^2, \quad u_k = \alpha_k \beta_k^* e_-^2 + \alpha_k^* \beta_k e_+^2, \quad \tau_k = i(\alpha_k \beta_k^* e_-^2 - \alpha_k^* \beta_k e_+^2). \tag{19}$$

For these variables one can obtain a system of three linear equations:

$$\frac{ds_k}{d\eta} = \frac{1}{2}\frac{\Omega_k'}{\Omega_k} u_k + \frac{1}{2}\frac{Q}{\Omega_k}\tau_k,$$

$$\frac{du_k}{d\eta} = \frac{\Omega_k'}{\Omega_k}(1 + 2s_k) - \left(\frac{Q}{\Omega_k} + 2\Omega_k \right) \tau_k, \tag{20}$$

$$\frac{d\tau_k}{d\eta} = \frac{Q}{\Omega_k}(1 + 2s_k) + \left(\frac{Q}{\Omega_k} + 2\Omega_k \right) u_k$$

with the initial conditions: $s_k = u_k = \tau_k = 0$ for $\eta = -\infty$. From the asymptotic form of the solution of the system (20) (or the system (18) which is equivalent to it), which is given in Appendix II, it is seen that as $\Omega_k \to \infty$

$$s_k \sim \Omega_k^{-2}, \quad u_k \sim \Omega_k^{-2}, \quad \tau_k \sim \Omega_k^{-1},$$

so that ε and p_α diverge like k^2 at the upper limit [also see formula (22)].

Therefore, an additional renormalization is required in the anisotropic case. We shall use the method discussed in the introduction to this article. Let $T_{ij}(\mathbf{k}, m)$ be the energy-momentum tensor for the single mode $\mathbf{k}$, and let $T_{ij}^{(n)} = n^{-1}T_{ij}(n\mathbf{k}, nm)$ be the energy-momentum tensor for the n-wave. Then the renormalized energy-momentum tensor for the given mode will be given by

$$T_{ij}^{\mathrm{Reg}}(\mathbf{k}, m) = \lim_{n\to\infty}\left[T_{ij}(\mathbf{k}) - T_{ij}^{(n)}(\mathbf{k}) - \frac{\partial}{\partial(n^{-2})}T_{ij}^{(n)}(\mathbf{k}) - \frac{1}{2}\frac{\partial^2}{\partial(n^{-2})^2}T_{ij}^{(n)}(\mathbf{k})\right]. \tag{21}$$

By expanding $T_{ij}(\mathbf{k})$ in inverse powers of Ω_k as $\Omega_k \to \infty$, one can easily show that for large values of Ω_k we have $T_{ij}^{\mathrm{Reg}}(\mathbf{k}) \sim \Omega_k^{-5}$; so the total energy-momentum tensor of the field, $T_{ij}^{\mathrm{Reg}} = \int d^3k T_{ij}^{\mathrm{Reg}}(\mathbf{k})$, converges. In addition, for $|\mathbf{k}| \to 0$ and $m \neq 0$, $|T_{ij}^{\mathrm{Reg}}(\mathbf{k})| < \infty$. We note that the first subtraction corresponds to discarding the energy-momentum tensor of the zero-point oscillations of the vacuum, which was done at the end of Sec. 1. It is clear from the method of regularization that the tensor T_{ij}^{Reg} satisfies the conservation law: $T_{\mathrm{Reg};j}^{ij} = 0$.

The present method of renormalization is not applicable when $m = 0$, owing to the appearance of an infra-red logarithmic divergence (the analogous phenomenon is observed in quantum electrodynamics). Really, this divergence is fictitious, and evidently it can be eliminated by the same method as in electrodynamics.

With the aid of this procedure, one can determine the finite renormalized average values of the field's energy-momentum tensor (for $m \neq 0$):

$$\varepsilon_{\mathrm{Reg}} = \frac{1}{(2\pi)^3 v^4}\int d^3k\left\{\Omega_k(s_k - s_k^{(2)} - s_k^{(4)}) - \frac{1}{2}\frac{Q}{\Omega_k}\left[s_k - s_k^{(2)} + \frac{1}{2}(u_k - u_k^{(2)})\right]\right\},$$

$$\begin{aligned} p_{1\mathrm{Reg}} = \frac{1}{(2\pi)^3 v^4}\int d^3k\bigg\{&\frac{k_1^2}{g_1^2\Omega_k}(s_k - s_k^{(2)} - s_k^{(4)}) \\ &+ \frac{1}{6\Omega_k}\left(3\frac{k_1^2}{g_1^2} - \Omega_k^2\right)(u_k - u_k^{(2)} - u_k^{(4)}) + \frac{1}{6\Omega_k}\left[\left(\frac{g_1'}{g_1}\right)' - Q\right] \\ &\times\left[s_k - s_k^{(2)} + \frac{1}{2}(u_k - u_k^{(2)})\right] - \frac{g_1'}{6g_1}(\tau_k - \tau_k^{(1)} - \tau_k^{(3)})\bigg\} \text{ etc.,} \end{aligned} \tag{22}$$

where $s_k^{(2)}$, $s_k^{(4)}$, $u_k^{(2)}$, $u_k^{(4)}$, $\tau_k^{(1)}$, $\tau_k^{(3)}$, represent the corresponding terms of the asymptotic expansion of s_k, u_k, and τ_k in powers of Ω_k^{-1} for $\Omega_k \to \infty$ (see Appendix II); the superscripts indicate the order of decrease with respect to Ω_k^{-1}.

The case of a weak gravitational field (a small difference between the metric (1) and the Minkowski metric) is considered in Appendix I.

The quasi-Euclidean Friedman model is an interesting and important special case; in this model one can assume $a = b = c = v$, and then the metric (1) is isotropic and conformally-flat:

$$ds^2 = v^2(\eta)(d\eta^2 - dx_1^2 - dx_2^2 - dx_3^2).$$

As a consequence of this, $Q(\eta) \equiv 0$ and all the formulas in Sec. 2 simplify considerably.

In the first place, in the isotropic case for $k \equiv |\mathbf{k}| \to \infty$ we have $s_k \sim k^{-6}$, $u_k \sim k^{-4}$, $\tau_k \sim k^{-3}$ so that the quantities ε and p, introduced at the end of Sec. 1, turn out to be finite (that is, the first subtraction is already sufficient for regularization). They have the following form:

$$\varepsilon = \frac{1}{(2\pi)^3 v^4} \int d^3k \Omega_k s_k,$$

$$p = \frac{1}{(2\pi)^3 v^4} \int \frac{d^3k}{3\Omega_k} \left(k^2 s_k - \frac{m^2 v^2}{2} u_k \right), \tag{23}$$

where $\Omega_k = \sqrt{k^2 + m^2 v^2}$. From the expression for ε it follows that one can interpret s_k as the average number of pairs (real and virtual) created in the mode $\mathbf{k}$.

Furthermore, for particles with $m = 0$, one will have $\Omega_k' = 0$ and $\beta_k \equiv 0$, so that $\varepsilon = p = 0$. Thus, in the isotropic case with an arbitrary dependence of $v(\eta)$ not only the production of real massless particles does not occur, which is shown in articles [12, 13], but also vacuum polarization is completely absent. We emphasize that in the general anisotropic case $\beta_k \neq 0$ for $m = 0$, which is clear from the system (18).

3. Behavior of the Energy Density and Pressure of a Scalar Field as the Singularity is Approached

The investigation of the behavior of the field near a singularity is of the greatest physical interest. Let us consider the case of collapse, when we have flat space-time at $t = -\infty$ and a singularity at $t = 0$. Let the quantities a, b, and c be power-law functions of t as $t \to 0$.

In the anisotropic Kasner case the following region gives the basic contribution to the integrals in formula (22): $\omega_k \sim 1/t$, where $s_k \sim 1$, and for $t \to 0$ and $|mt| \ll 1$ we have $\varepsilon \sim p_\alpha \sim t^{-4}$ (possibly with logarithmic accuracy). The exact coefficients associated with ε and p_α have not yet been determined. Since the components of the Riemann tensor increase like t^{-2} as $t \to 0$, but the energy density of the "external" matter increases like $t^{-4/3}$ (for an ultrarelativistic gas), it follows that as the characteristic time $t \sim t_p = \sqrt{G}$ is approached the reciprocal effect of the created particles on the metric becomes large and it may substantially alter (of course, when the self-consistent problem is considered) the subsequent course of the contraction.

In the degenerate case of the isotropic contraction of space, the following case is of physical interest: $a(t) \sim t^q$ for $t \to 0$; $0 < q < 1$. Let us choose η such that $\eta = 0$ when $t = 0$; then $a(\eta) \sim \eta^{q_1}$, where $q_1 = q/(1-q)$. From equation (II.2) (see Appendix II) it is seen that if $ma(\eta) \ll k \equiv |\mathbf{k}| \ll 1/\eta$

(and for a sufficiently small value of η, any arbitrary k falls in this region), then s_k, u_k, and $\tau_k \to$ const; therefore pair production stops as $\eta \to 0$.

The reason for this is very simple: $\omega_k \approx k/a$ as $\eta \to 0$ (that is, one can neglect the mass) but particles with $m = 0$ are generally not created during isotropic collapse. Upon further contraction the created particles behave as a relativistic gas with the equation of state $p = (1/3)\varepsilon$, so that for $t \to 0$ and $|mt| \ll 1$ we find

$$\varepsilon = 3p = \frac{1}{2\pi^2 a^4} \int_0^\infty dk k^3 n(k) \sim m^4 |mt|^{-4q},$$

where $n(k) = \lim_{\eta\to\infty} s_k$. One can show that $n(k) \sim k^{-1}$ as $k \to 0$, and $n(k) \sim k^{-4/(1-q)}$ as $k \to \infty$; therefore the written integral converges. The corresponding expression for the asymptotic density of particles (also including vacuum polarization) is given by

$$n(t) = \frac{1}{2\pi^2 a^3} \int_0^\infty dk k^2 n(k) \sim m^3 |mt|^{-3q}.$$

If $a(t) \sim \sqrt{|t|}$ $(q = 1/2)$ for $|t| < t_0 \gg 1/m$ (the external matter is an ultra-relativistic gas), then the ratio of the energy of the created particles to the energy of the external matter will be $\varepsilon/\mathcal{E}_{\text{ext}} \sim Gm^2 \ll 1$ (we are considering "ordinary" elementary particles with masses from 10^{-24} to 10^{-27} g).

We note that in the isotropic case one can set up the opposite "cosmological" problem, specifying the "vacuum" state at $t = 0$ with the aid of the condition: $s, u, \tau|_{t=0} = 0$ (or $C_1 = 1/(2\sqrt{k})$, $C_2 = \sqrt{k}/2$; see equation (II.2)). In the anisotropic case s, u and $\tau \to \infty$ as $t \to 0$ and it is impossible to impose such an initial condition.

The authors express their deep gratitude to L. P. Pitaevskiĭ for constant attention to this research and for valuable suggestions. We also take this opportunity to thank the participants in the Joint Astrophysics Seminar, the Seminar of the Quantum Theory Department of the Physics Department of Moscow State University, and the Seminar of the All-union Research Institute of Physico-technical and Radio Measurements for their interest and criticism.

Appendix I. Weak Gravitational Field

Let us consider the case of a weak external field (that is, there is a small difference between the metric (1) and the Minkowski metric) and let us calculate ε and p_α to first order in $h(t)$ using perturbation theory. The following anisotropic case is of the greatest interest:

$$a^2(t) = 1 + h_1(t), \quad b^2(t) = 1 + h_2(t), \quad c^2(t) = 1 + h_3(t),$$

where

$$|h_\alpha(t)| \ll 1, \quad h_\alpha(t)|_{t=\pm\infty} = 0, \quad \sum_{\alpha=1}^{3} h_\alpha = 0.$$

to first order in h_α, in the Fourier representation we have

$$\varepsilon_{\text{Reg}} = 0, \quad \sum_{\alpha=1}^{3} p_{\alpha\text{Reg}} = 0, \quad p_{\alpha\text{Reg}}(q) = A(q)h_\alpha(q), \tag{I.1}$$

where

$$h(t) = \frac{1}{2\pi}\int_{-\infty}^{\infty} h(q)e^{iqt}\,dq, \quad p(t) = \frac{1}{2\pi}\int_{-\infty}^{\infty} p(q)e^{iqt}\,dq.$$

The imaginary part of $A(q)$ is finite and does not require renormalization:

$$\operatorname{Im} A(q) = -\frac{1}{1920\pi}(q^2 - 4m^2)^2\sqrt{1 - \frac{4m^2}{q^2}}\,\theta(q^2 - 4m^2)E(q), \tag{I.2}$$

where

$$\theta(x) = \begin{cases} 1, & x > 0 \\ 0, & x < 0 \end{cases}, \quad E(x) = \begin{cases} 1, & x > 0 \\ -1, & x < 0 \end{cases}.$$

The quantity $\operatorname{Im} A(q)$ determines the energy at $t = +\infty$ in the second-order approximation (that is, the energy of real created particles):

$$\varepsilon(+\infty) = -\frac{1}{2}\int_{-\infty}^{\infty} dt \sum_{\alpha=1}^{3} p_\alpha \dot{h}_\alpha = -\frac{1}{4\pi}\int_{-\infty}^{\infty} dq\, q \operatorname{Im} A(q)\sum_{\alpha=1}^{3}|h_\alpha(q)|^2 \geq 0. \tag{I.3}$$

The real part of the polarization operator, which has been regularized with the aid of three subtractions, has the form

$$\operatorname{Re} A(q) = \frac{1}{960(2\pi)^2}\frac{q^6}{m^2}\left[\frac{1}{\gamma^3} - \frac{7}{3\gamma^2} + \frac{23}{15\gamma} - \frac{(1-\gamma)^{5/2}}{\gamma^{7/2}}\operatorname{arctg}\sqrt{\frac{\gamma}{1-\gamma}}\right] \tag{I.4}$$

for $0 \leq \gamma \leq 1$; $\gamma = q^2/4m^2$;

$$\operatorname{Re} A(q) = \frac{1}{960(2\pi)^2}\frac{q^6}{m^2}\left[\frac{1}{\gamma^3} - \frac{7}{3\gamma^2} + \frac{23}{15\gamma} - \frac{(\gamma-1)^{5/2}}{\gamma^{7/2}}\ln(\sqrt{\gamma} - \sqrt{\gamma - 1})\right]$$

for $\gamma > 1$.

The asymptotic form of $\operatorname{Re} A(q)$ is given by:

$$1)\quad \operatorname{Re} A(q) = \frac{1}{960(2\pi)^2}\frac{1}{7}\frac{q^6}{m^2} \quad \text{for} \quad \gamma \ll 1;$$

$$2)\quad \operatorname{Re} A(q) = -\frac{1}{960(2\pi)^2}\cdot 2q^4 \ln\frac{q^2}{m^2} \quad \text{for} \quad \gamma \gg 1. \tag{I.5}$$

Also the following expressions in the coordinate representation are instructive:

$$\varepsilon(+\infty) = -\frac{1}{2}\int_{-\infty}^{\infty} dt \sum_{\alpha=1}^{3} \dot{h}_\alpha(t) \int_{-\infty}^{t} h_\alpha^{\mathrm{VI}}(\tau)\mathcal{K}(t-\tau)\, d\tau, \tag{I.6}$$

where $\mathcal{K}(t-\tau) \sim (t-\tau)\ln(t-\tau)$ for particles with $m = 0$. The quantity $\int_{-\infty}^{t} \cdots$ plays the role of $p'_\alpha = -T'^{\alpha}_{\alpha}$, a component of the stress tensor. However, this is only a part of p_α; it is obvious that the addition of terms of the form $h_\alpha^{(2n)}(t)$ to p'_α does not change the value of $\varepsilon(+\infty)$. It is precisely such terms, which do not give any contribution to $\varepsilon(+\infty)$, which are affected by renormalization.

We note that the p_α found are proportional to h_α, provided $\sum_\alpha h_\alpha = 0$, just as for the viscous stresses associated with the deformation of an incompressible liquid. However, viscosity would give $p_\alpha \sim \dot{h}_\alpha$; vacuum polarization (and even just its imaginary part) has a more complicated nonlocal dependence on the form of h as a function of t.

Appendix II. The Asymptotic Behavior of the Solution of Equations (20)

Let us consider the case of large momenta ($\Omega_k \to \infty$) and let us expand the solution of the system of equations (20) in an asymptotic series in powers of Ω_k^{-1}. This expansion is valid in the quasiclassical region: $|\Omega'_k| \ll \Omega_k^2$. In this region $|s_k|$, $|u_k|$, $|\tau_k| \ll 1$. It is found that $\tau = \tau^{(1)} + \tau^{(3)} + \ldots$, $s = s^{(2)} + s^{(4)} + \ldots$, $u = u^{(2)} + u^{(4)} + \ldots$ (here and in what follows, the subscript k is omitted), where

$$\tau^{(1)} = \frac{1}{2}\frac{\Omega'}{\Omega^2}, \quad u^{(2)} = \frac{1}{2\Omega}\left[\frac{d\tau^{(1)}}{d\eta} - \frac{Q(\eta)}{\Omega}\right],$$

$$s^{(2)} = \frac{1}{2}\int\left[\frac{\Omega'}{\Omega}u^{(2)} + \frac{\Omega(\eta)}{\Omega}\tau^{(1)}\right]d\eta = \frac{1}{16}\frac{\Omega'^2}{\Omega^4},$$

$$\tau^{(3)} = -\frac{1}{2\Omega}\left[\frac{du^{(2)}}{d\eta} - 2\frac{\Omega'}{\Omega}s^{(2)} + \frac{Q(\eta)}{\Omega}\tau^{(1)}\right],$$

$$u^{(4)} = \frac{1}{2\Omega}\left[\frac{d\tau^{(3)}}{d\eta} - \frac{Q(\eta)}{\Omega}(2s^{(2)} + u^{(2)})\right],$$

$$s^{(4)} = \frac{1}{2}\int\left[\frac{\Omega'}{\Omega}u^{(4)} + \frac{Q(\eta)}{\Omega}\tau^{(3)}\right]d\eta \tag{II.1}$$

and so forth. The superscript inside the brackets indicates the order of decrease in powers of Ω_k^{-1} for $\Omega_k \to \infty$.

Near the singularity ($\eta = 0, t = 0$) one can obtain another important asymptotic form of the system (18), (20). Let

$$\lim_{t\to 0} \frac{a, b, c}{t} = \infty,$$

then for sufficiently small η one has

$$\left| \int_0^\eta \Omega_k \, d\eta \right| \ll 1.$$

for arbitrary k. In this essentially non-quasiclassical region we have

$$\alpha = C_1 \left(\sqrt{\Omega} - \frac{i}{\sqrt{\Omega}} \int Q(\eta)\, d\eta \right) + \frac{C_2}{\sqrt{\Omega}},$$

$$\beta = C_1 \left(\sqrt{\Omega} + \frac{i}{\sqrt{\Omega}} \int Q(\eta)\, d\eta \right) - \frac{C_2}{\sqrt{\Omega}},$$

$$s = |C_1|^2 \left[\Omega + \frac{1}{\Omega} \left(\int Q\, d\eta \right)^2 \right] + \frac{|C_2|^2}{\Omega} + \frac{i}{\Omega}(C_1^* C_2 - C_1 C_2^*) \int Q\, d\eta - \frac{1}{2},$$

$$u = 2|C_1|^2 \left[\Omega - \frac{1}{\Omega} \left(\int Q d\eta \right)^2 \right] - \frac{2|C_2|^2}{\Omega} - \frac{2i}{\Omega}(C_1^* C_2 - C_1 C_2^*) \int Q\, d\eta,$$

$$\tau = 4|C_1|^2 \int Q\, d\eta + 2i(C_1^* C_2 - C_1 C_2^*), \tag{II.2}$$

where C_1 and C_2 are complex numbers (depending on $\mathbf{k}$), satisfying the condition: $C_1^* C_2 + C_1 C_2^* = 1/2$. In the case of power-law dependencies for the a, b, and c we have $Q(\eta) \sim \eta^{-2}$ and $s \approx -2u \sim 1/\Omega\eta^2$; $\tau \sim \eta^{-1}$ as $\eta \to 0$; in the isotropic case s, u, and $\tau \to$ const as $\eta \to 0$.

By joining the two asymptotic expressions in the region

$$\left| \int_0^n \Omega_k \, d\eta \right| \sim 1,$$

one can derive the law $\varepsilon \sim p_\alpha \sim t^{-4}$, which was cited in Sec. 3.**

**The commentary follows article 35.

Institute of Applied Mathematics
USSR Academy of Science. Moscow

Received
June 25, 1971

REFERENCES

1. *Hoyle F.*—Month. Not. RAS **108**, 372 (1948); **109**, 365 (1949).
2. *Ivanenko D. D., Sokolov A. A.* Kvantovaîa teoriîa polya [Quantum Field Theory], Part 2. Moscow: Gostekhizdat, 780 p. (1952).
3. *Stanyukovich K. P.* Gravitatsionnoe pole i elementarnye chastitsy [The Gravitational Field and Elementary Particles]. Moscow: Nauka, 311 p. (1965).

4. *Wheeler J. A.* Neutrinos, Gravitation, and Geometry. Bologna: Tipografia Comp. (1960).
5. *Feynman R. P.*—Acta Phys. Pol. **24**, 697 (1963).
6. *De Witt Br. S.*—Phys. Rev. **160**, 1113 (1967); **162**, 1195, 1239 (1967).
7. *Ginzburg V. L., Kirzhnits D. A., Lyubushin A. A.*—ZhETF **60**, 451 (1971).
8. *Sakharov A. D.*—Dokl. AN SSSR **177**, 70 (1967).
9. *Zeldovich Ya. B.*—Pisma v ZhETF **6**, 883 (1967); UFN **95**, 209 (1968).
10. *Sakurai J.*—In: Elementarnye chastitsy i kompensiruĭushchie polya [Elementary Particles and Compensating Fields]. Moscow: Mir, 42 (1964).
11. *Tagirov E. A., Chernikov N. A.*—Preprint OIYaI P2-3777. Dubna (1968).
12. *Bronnikov K. A., Tagirov E. A.*—Preprint OIYaI P2-4151. Dubna (1968).
13. *Parker L.*—Phys. Rev. Lett. **21**, 562 (1968); Phys. Ref. **183**, 1057 (1969); Phys. Rev. **3**, 346 (1971).
14. *Grib A. A., Mamaev S. G.*—Yader. Fiz. **10**, 1276 (1969); XV Int. Conf. on High-Energy Phys. Kiev Vol. 2, 809 (1970).
15. *Sexl R. U., Urbantke N. K.*—Phys. Rev. **179**, 1247 (1969).
16. *Zeldovich Ya. B.*—Pisma v ZhETF **12**, 443 (1970).
17. *Lifshitz E. M., Khalatnikov I. M.*—UFN **80**, 391 (1963); *Belinskiĭ V. A., Lifshitz E. M., Khalatnikov I. M.*—UFN **102**, 463 (1970).
18. *Misner C. W.*—Phys. Rev. Lett. **22**, 1071 (1969); Phys. Rev. **186**, 1328 (1969).
19. *Doroshkevich A. G., Novikov I. D.*—Astron. Zh. **47**, 948 (1970).
20. *Doroshkevich A. G., Zeldovich Ya. B., Novikov I. D.*—ZhETF **60**, 3 (1971).
21. *Isaacson R. A.*—Phys. Rev. **166**, 1263 (1968).
22. *Hawking S.*—Comm. Math. Phys. **18**, 301 (1970).
23. *Zeldovich Ya. B., Pitaevskiĭ L. P.*—Comm. Math. Phys. **21**, 185 (1971).
24. *Penrose R.*—In: Relativity, groups and topology, Ed. by C. and B. de Witt. N.Y., London, 565 (1964).
25. *Landau L. D., Lifshitz E. M.* The Classical Theory of Fields. New York: Pergamon (1971).

34

Generation of Waves by a Rotating Body*

An axially-symmetrical body rotating inside a resonator cavity is capable of amplifying certain oscillation modes inside the resonator, transferring the rotation energy to these oscillations.

The frequency of the amplified oscillations is not an integer multiple of the angular velocity of the body, and the instantaneous state of the resonator does not depend on the time, so that the phenomenon in question differs from the parametric resonance.

In scattering of a plane wave incident on the rotating body, it is advisable to expand the wave into spherical (or cylindrical) waves with different values of momentum projection on the rotation axis. In the scattering, the waves with (sufficiently large) momentum parallel to the rotation vector become amplified, and all others become attenuated. In the presence of an external reflector with small losses (resonator), the amplification following single scattering may turn into generation. The linear velocity on the surface of the rotating body obviously is smaller than the speed of light, $v = \beta c$, $\beta < 1$. The amplified waves have an angular dependence $\exp(in\varphi)$, where $n > \beta^{-1}$. It follows therefore that the radius of the body is smaller than n wavelengths by at least a factor of β; this means that the body is inside the zone in which the wave amplitude decreases more rapidly than $(r/\lambda)^n$. Therefore at small β the gain is exponentially small, like $\beta^{\beta^{-1}}$ or even weaker.

The foregoing refers to a body made of a material that absorbs waves when at rest; the conditions for amplification and generation are obtained after transforming the equations to the moving system. A similar situation can apparently arise also when considering a rotating body in the state of gravitational relativistic collapse.

The metric near such a body is described by the well-known Kerr solution. The gravitational capture of the particles and the waves by the so-called trapped surface replaces absorption; the trapped surface ("the horizon of events") is located inside the surface $g_{00} = 0$. Finally, in a quantum analysis of the wave field one should expect spontaneous radiation of energy and momentum by the rotating body. The effect, however, is negligibly small, less than $\hbar\omega^4/c^3$ for power and $\hbar\omega^3/c^3$ for the decelerating moment of the force (for a rest mass $m = 0$, in addition, we have omitted the dimensionless function β). The rotating body is regarded classically, and none of the foregoing applies to particles with quantized angular momentum.

*Pisma v Zhurnal eksperimentalnoĭ i teoreticheskoĭ fiziki **14** (8), 270–272 (1971).

Further generalization to the case of fermions, including charged ones, is also possible. The rotating body produces spontaneous pair production in the case when the body can absorb one of the particles, while the other (anti)particle goes off to infinity and carries away energy and angular momentum. All that is necessary is that the momentum carried away be sufficient to draw energy away from the body, which requires a definite value of the impact parameter b; the region between the surface of the body and a cylinder of radius b is the barrier. Finally, there is a possible variant in which particle absorption is replaced by scattering of the particle by the material of the body. Obviously, rotation alone will not lead to pair production without interaction with the body.

To prove all the statements, let us consider the simplest case of a scalar field ψ. In vacuum ψ satisfies the equation $\Box\psi - m^2\psi = 0$. In an absorbing medium in a coordinate system where the medium is at rest we have $\Box\psi + \alpha(\partial\psi/\partial t) - m^2\psi = 0$, where α characterizes the damping.

In a system where the medium moves along the x axis the Lorentz transformation yields

$$\alpha\frac{\partial\varphi}{\partial t} \to \alpha\gamma\left(\frac{\partial\psi}{\partial t} - \beta\frac{\partial\psi}{\partial x}\right); \quad \gamma = \frac{1}{\sqrt{1-\beta^2}}.$$

Let us consider the cylindrical case $\psi = f(r)\exp(i\omega t + in\varphi)$.

Let $x = r\varphi$ be reckoned along the circle over which the rotation takes place, $\beta = r\Omega/c$, where Ω is the angular velocity of the body.

In this case the additional term in the wave equation inside the rotating body (under its surface) is equal to

$$\psi\alpha\gamma\left(i\omega + \frac{i\beta nc}{r}\right) = \psi i\alpha\gamma(\omega + n\Omega).$$

Consequently, the additional term reverses sign at $n\Omega < -\omega$, i.e., at $n < 0$ and $|n| > \omega/\Omega$. The medium operates effectively as an amplifier and not as an absorber with respect to waves with such values of n.

For an n-pole, the boundary of the wave zone corresponds in order of magnitude to $r_\omega = |n|\lambda\!\!\!^{-} = |n|c/\omega$; the radius of the body is $r = v/\Omega = \beta c/\Omega$. From the inequality needed for amplification we get $r/r_\omega < \beta$, the body lies deep inside r_ω. The condition for wave amplification coincides with the following simple energy criterion: the photon energy is $E = \hbar\omega$, the angular momentum of the photon in the n-pole state is $\mu = n\hbar$ (we neglect the spin ± 1), thus $\mu\Omega > E$ means that the decrease in the body's rotational energy is larger than the energy of the emitted photon. The constant $\hbar$ does not enter into the solution, its previous use being a tribute to the modern method of expression in an era when "quantum mechanics helps understand classical mechanics."

I take the opportunity to mention a stimulating discussion with Misner, Thorne, and Wheeler of the problem of extracting energy from a rotating collapsing body, and useful discussions with G. A. Askar'yan, G. A. Grinberg, B. Ya. Zeldovich, P. L. Kapitza, and I. I. Sobel'man. I am grateful to all of them.

Institute of Applied Mathematics
USSR Academy of Science. Moscow

Received
July 9, 1971

Commentary

An interesting example of how a single idea may be fruitfully used in such seemingly distant areas as applied classical electrodynamics and the quantum theory of black holes is given by Ya. B.'s paper "Generation of Waves by a Rotating Body." In hydrodynamics we know of the effect of amplification of a wave in reflection off the interface between two media which is moving at a velocity greater than the phase velocity of sound (super-reflection). Direct generalization of this effect to the case of electromagnetic waves in a vacuum is impossible since matter cannot move at a superluminal velocity. However, Ya. B. noted that "superluminal *rotation*" of a body is possible: if in a vacuum a monochromatic electromagnetic wave with temporal and angular dependence $\sin(m\varphi - \omega t)$, where m is an integer, falls onto a rotating body, then the angular rotation velocity of the body Ω may certainly be greater than the angular phase velocity of rotation of the wave ω/m even though the linear velocity of rotation Ωr is obviously less than the speed of light. In the paper it is shown that under the condition $\omega < m\Omega$ electromagnetic waves are amplified during reflection from a rotating conducting cylinder.* The presence of non-zero conductivity of the cylinder which leads to dissipation of energy inside of it and to increase of the entropy is essential. The energy necessary for the process of amplification is borrowed from the rotational mechanical energy of the cylinder which is thus transformed partly into energy of the electromagnetic wave and partly into heat.

This problem was studied in greater detail in subsequent papers by Ya. B.[1] and by Ya. B. and A. A. Starobinskiĭ[2] which consider the cases of two different polarizations of the electromagnetic field, as well as of low and high conductivity.

This same idea was applied by Ya. B. to black holes, which led to prediction of the effect of amplification of waves (including electromagnetic and gravitational waves) which are scattered off a rotating black hole. Here one may also introduce the quantity Ω having the sense of the angular velocity of rotation of the black hole, although a black hole has no real solid surface. The criterion of existence of the amplification effect then again takes the form $\omega < m\Omega$. This assertion was rigorously proved (and the magnitude of the effect quantitatively calculated) in later papers by A. A. Starobinskiĭ[3] and by A. A. Starobinskiĭ and S. M. Churilov.[4]

*Note Ya. B.'s unfortunate choice of a negative n.

[1] *Zeldovich Ya. B.*—ZhETF **62**, 2076–2082 (1972).

[2] *Zeldovich Ya. B., Starobinskiĭ A. A.*—In: Voprosy Matem. Fiziki [Problems of Math. Physics]. Leningrad: Nauka, 35–48 (1976).

[3] *Starobinskiĭ A. A.*—ZhETF **64**, 48–57 (1973).

[4] *Starobinskiĭ A. A., Churilov S. M.*—ZhETF **65**, 3–11 (1973).

Here the condition of entropy increase is replaced by the condition of growth of the surface area of the event horizon of the black hole.

Finally, developing the analogy between the amplification of a classical wave and quantum particle creation, Ya. B. predicted the effect of quantum particle creation (photons, gravitons, electrons, positrons, and others) by a rotating black hole. This effect goes together with another effect of particle creation by a black hole discovered three years later by S. Hawking (the latter effect has a different nature and exists in the case of a non-rotating black hole as well).

34a
Quantum Effects in White Holes*

With I. D. Novikov, A. A. Starobinskiĭ

The influence of quantum effects (particle production near the singularity) on the external properties of a white hole is considered. It is shown that spontaneous particle production radically changes the properties of a white hole. In the big-bang model of the Universe, a white hole separates from the surrounding expanding matter of the Universe in the epoch $t \approx r_g/c$, where r_g is the gravitational radius measured by an external observer after the separation of the white hole. Excluding extremely artificial choices of initial conditions, a white hole either blows up practically without delay at the epoch $t \lesssim r_g/c$ or never explodes at all. An appeal to the energy of explosions of white holes for an explanation of the energy production in astrophysical phenomena seems extremely unlikely.

Introduction

A white hole is a hypothetical object expanding from a singularity to the outside of its Schwarzschild horizon. One of the authors [1] and later also Ne'eman [2] have shown that the existence of such bodies (within the framework of classical general relativity) is possible in an expanding Universe. According to this hypothesis white holes are nuclei of matter of the Friedman cosmological model which have been retarded in their cosmological expansion (owing to local inhomogeneities of the initial conditions). The delay (retardation) of the expansion in the time of an external observer can be prescribed arbitrarily. The magnitude of the delay and also the composition and the kinetic energy of the emitted matter are given as initial conditions at the singularity (or arbitrarily close to the singularity). The delayed core is surrounded by a cavity of lower density (a vacuole) or, in the limiting case, by vacuum.

It is tempting (but not necessary!) to compare the hypothesis of white holes with a series of not yet completely explained astrophysical phenomena, in particular quasars and explosions in galactic nuclei. In the present paper an attempt is made to take into account quantum phenomena which must occur in a white hole. The theory predicts spontaneous particle production inside the white hole near the singularity of space-time. The produced

*Zhurnal eksperimentalnoĭ i teoreticheskoĭ fiziki **66**, 1897–1910 (1974).

particles have an extremely strong influence on the metric of space-time and interact with the expanding matter exterior to the white hole, as well as with the matter of the retarded nucleus, and thus radically change the whole phenomenon. As will be shown below, it seems very unlikely that one can appeal to white hole explosions in order to explain energy production in astrophysical phenomena.

Let us consider the situation in more detail. The general theory of relativity introduces the concept of a non-Euclidean space-time continuum. After giving up the idea of Euclidean space it is not hard to make the next step and introduce a change in the topology of space. Thus have appeared the ideas of Eddington and others on "channels" (wormholes) connecting "our" space with "other" spaces and of the possibility of "injection" of matter and energy through such a channel. We first stress that the presence of such channels does not allow one to change the mass of isolated structures by injection of energy from another space. There is a rigorous theorem asserting that the total mass of any object measured by a far away observer by means of the gravitational field can change only on account of an influx of mass (energy) through a remote sphere of *our* space, a sphere which surrounds the object in question.

The idea of white holes differs from Eddington's idea. A single space with a simply connected topology is considered. The evolution of the Universe as a whole is taken into account. The expansion starts from a singularity. Within the white hole, in the vacuum of the vacuole ($R_1 < R < R_2$) the singularity is of the Schwarzschild type in the simplest case and outside the vacuole ($R > R_2$) the singularity is of the Friedman type (cf. Fig. 1).

The singularity is everywhere spacelike—both outside and inside the white hole—i.e., there exists a frame in which this singularity is simultaneous. However, on account of the inhomogeneity of the singularity there appears a retardation (delay) (according to the clock of an external observer) of the expansion of part of the matter and this part expands later for an external observer. Depending on the magnitude of the length of the Schwarzschild singularity (the quantity $R_2 - R_1$) the retardation between the delayed matter and the external matter can be arbitrarily large for an external observer.

In the case of matter with vanishing pressure $p = 0$ an exact solution has been constructed in [1] for the case when in an isotropic homogeneous Universe some spherical mass M is replaced by a white hole of the same mass. In this case the unperturbed cosmological solution survives in the external region.

In the case when the retarded matter has a nonvanishing pressure p, in particular $p = \varepsilon/3$, mathematical difficulties arise in the consideration of the interior region of the vacuole, difficulties which are not problems of principle and do not call into question the existence of solutions of the type of a "white hole." However, if $p \neq 0$ in the whole surrounding Universe, the external solution gets perturbed. Indeed, there is a discontinuity of the pressure in the solution at the boundary between the unperturbed Universe and the vacuum Schwarzschild solution. The discontinuity in pressure leads to a flow of plasma from the exterior into the interior region. The white hole leads to an accretion of plasma for any ratio of the proper mass of the hole to the mass torn off from the cosmological solution. So far there are only rough estimates of the effect, and there is no complete picture of the phenomenon. It remains to find out whether, taking accretion into account, one may use the hypothesis of white holes for the explanation of astrophysical phenomena.

In addition to the accretion problem there exists another side of the problem of the possible existence of white holes, related, first to the fact that in a white hole an extremely strong anisotropic expansion of space exists in the vacuole, near the Schwarzschild singularity, in distinction from the isotropic expansion outside the white hole in the isotropic Universe, and second, related to the fact that with the "white hole hypothesis" a remote observer will "see" the Schwarzschild singularity throughout all the delay time. In a white hole the so-called T_+-region is situated underneath the gravitational radius r_g, where all particles move only from the singularity towards r_g and can get out from under r_g to an external observer (in this connection, cf. [3] and the review in [4]).

The first circumstance—anisotropic expansion near the singularity—leads, as shown in [5, 6], to a strong spontaneous particle production. The second circumstance—the fact that this occurs in an expanding T_+ region—gives rise to the possibility that the produced particles are ejected beyond r_g to an external observer, and consequently can lead to a spontaneous decrease of the mass of the white hole. Indeed, as will be shown in detail below, the produced particles very strongly modify the metric of space-time below r_g. As a result, in the real case of a big-bang Universe, the produced particles turn out to be "locked-in" in the white hole first, during the very early stages of the expansion of the Universe for $t \ll r_g/c$, on account of the pressure of the surrounding gas, and later, for t of the order of r_g/c and $t > r_g/c$, owing to gravitational self-closure: the gravitation of the produced particles does not allow them to go out beyond r_g. Owing to the outflow of particles a white hole can lose only an insignificant fraction of its mass.

It is very important that all the changes related to the produced particles lead to impeding the explosion of the delayed core if the retardation is larger than r_g/c. Thus, the main conclusion is that in a hot (big-bang) Universe a white hole consists of a mass of particles produced near a Schwarzschild singularity under which is "buried" a delayed core. The matter consisting of these particles expands and for a remote observer the outer boundary of the matter approaches r_g from within asymptotically.

For times $t \gg r_g/c$ any emission of radiation or outflow of matter from the white hole decays exponentially very fast (in spite of the fact that r_g is

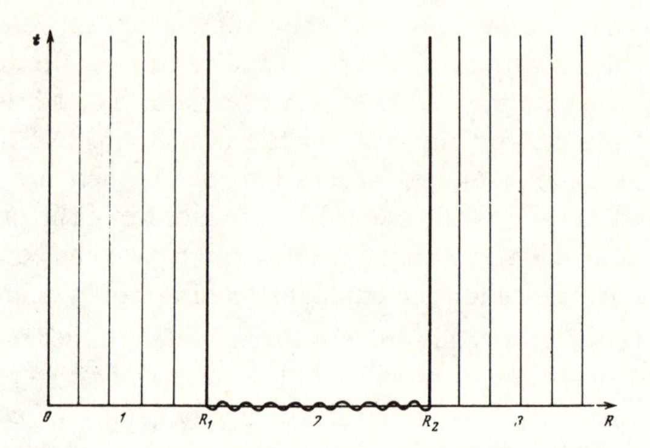

Fig. 1. The structure of space-time near a white hole singularity without particle production: 1–the singularity of the delayed (retarded) core, 2–the Schwarzschild singularity, 3–the Friedmann singularity. The thin vertical lines indicate regions filled by dustlike matter.

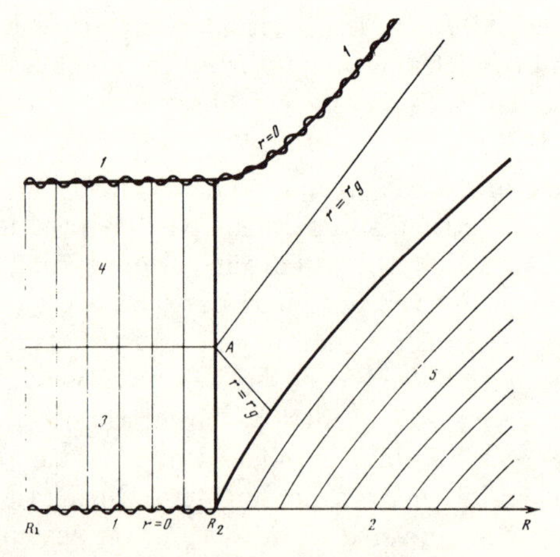

Fig. 2. The complete space-time for a white hole in the Friedmann model ($p = 0$), including the phase of compression of the dustlike matter of the white hole: 1–Schwarzschild singularity, 2–Friedmann singularity, 3–phase of expansion of the white hole matter, 4–phase of its compression, 5–expanding matter of the Friedmann Universe.

approached in the process of expansion!). Thus, although the white hole can exist arbitrarily long, for an external observer it acquires the characteristics of a black hole: there occurs a freezing of processes and the boundary of matter tends to r_g. However, a qualitative distinction is the fact that such an object appears as a result of a (quantum) explosion from the interior and the singularity to r_g, and not as a result of gravitational collapse towards r_g of initially rarefied matter.

In the following sections we analyze in detail the quantum effects in white holes which have been enumerated above.

1. Calculation of Particle Production

We first calculate the production of particles near the singularity in the vacuole. The vacuole which surrounds the delayed core is considered to be empty near the singularity. Near the Schwarzschild singularity there occurs particle production. The production of particles is a nonlocal effect. Since the Schwarzschild singularity is spacelike and homogeneous, the center of mass of each volume element of produced particles must be at rest in a reference frame in which the space is homogeneous and the singularity is simultaneous. Such a frame is the T system of Novikov [3] with the metric (we set $c = 1$)

$$ds^2 = dt^2 - e^{\lambda} dR^2 - r^2(d\theta^2 + \sin^2\theta d\varphi^2), \tag{1}$$

where λ and r depend only on t.

As the singularity is approached we have $r \to 0$. We choose $t = 0$ at the singularity. Asymptotically, as $t \to 0$, the metric (1) corresponds to the Kasner axially symmetric solution:

$$r \sim t^{2/3}, \quad e^{\lambda/2} \sim t^{-1/3}. \tag{2}$$

Particle production in such a metric has been calculated in [6]. The process occurs near the singularity $t = 0$. In order to obtain a finite expression for the energy density of the produced particles it is necessary to switch on the production at some $t_0 > 0$. It is likely that t_0 must be of the order of the Planck time $t_{\mathrm{Pl}} = (G\hbar)^{1/2}$, $t_0 \sim t_{\mathrm{Pl}}$. For the time being we leave the magnitude of t_0 arbitrary. The production process dies out very rapidly as t increases: $d\varepsilon/dt \propto t^{-5}$. Therefore after $t \approx (3\text{–}5)t_0$ one may consider the particle production completed. At a time of order t_0 the energy density of the produced particles is [5, 6]

$$\varepsilon \approx \frac{\hbar}{t_0^4} = \frac{t_{\mathrm{Pl}}^2}{G t_0^4}. \tag{3}$$

After the production (occurring practically at the time t_0) there occurs a decrease of the energy density ε at subsequent times owing to the expansion of the system (1). The energy density decreases adiabatically and for a further detailed calculation one must know the equation of state of the produced particles.

2. A Model Solution for $p = 0$

We start from the simplest assumption that the pressure vanishes identically ($p = 0$) in the delayed core, in the matter created near the singularity, and in the external matter of the expanding Universe. These assumptions are blatantly false, but, as we shall see, they allow one to construct an exact solution on which one can clearly follow the important features surviving also in the more complicated case of nonvanishing pressure.

Thus, we shall assume that everywhere $p = 0$. Fig. 1 represents space-time near the singularity of the white hole surrounded by the cold $p = 0$ matter of the expanding Universe, furthermore without taking into account the particles produced near the singularity, i.e., according to Novikov's original 1964 hypothesis [1]. According to the particle-production hypothesis [5, 6], particles are produced near the Schwarzschild singularity (where the deformation of space-time is highly anisotropic) and are not produced either near the singularity in the delayed core, nor near the Friedman singularity (where the deformation of space-time is isotropic).[1] The process of matter production near the Schwarzschild singularity has been discussed in the preceding section. Let us follow through the further expansion of the produced matter. We first recall that at $p = 0$ the external matter of the Universe to the right of the Lagrangian coordinate R_2 does not at all affect the internal solution [4]. The influence of the boundary R_1 can be essential, since, depending on the initial conditions, this boundary may have a velocity relative to the produced matter. Moreover, the motion of this boundary and its influence on the whole solution determines the explosion of the retarded nucleus. However, we shall take this influence into account later, and for the moment we consider the Schwarzschild singularity sufficiently extended in the Lagrangian coordinate R and shall consider created matter which has not yet been reached by the influence of the boundary R_1.

In order to describe the expansion of the produced particles it is necessary to solve the Einstein equation for the metric (1) (cf. Appendix I, Eqn. (I.1)) with the initial conditions

$$\frac{r}{r_g} = e^{-\lambda} = \left(\frac{3t}{2r_g}\right)^{2/3} \quad \text{for} \quad t = t_0 \ll r_g, \tag{4}$$

where $r_g = \text{const.}$[2] This solution is written in the parametric form:

$$r = \frac{1}{2} r_g (1 - \cos\xi), \quad t = \frac{1}{2} r_g (\xi - \sin\xi),$$
$$e^{\lambda/2} = \cot\left(\frac{\xi}{2}\right) + \alpha\left[1 - \frac{1}{2}\cot\left(\frac{\xi}{2}\right)\right], \tag{5}$$
$$8\pi G\varepsilon = \alpha r^{-2} e^{-\lambda/2}, \quad 0 \le \xi \le 2\pi.$$

The constant α is determined from the condition (3):

$$\alpha \approx r_g t_{\text{Pl}}^{-1} \left(\frac{t_{\text{Pl}}}{t_0}\right)^3 \sim \frac{r_g}{t_{\text{Pl}}} \gg 1. \tag{6}$$

[1]Strictly speaking this refers only to zero mass particles; however, the production of particles with $m \neq 0$ near the Friedman singularity does not affect the subsequent expansion, if $Gm^2/\hbar \ll 1$, and can therefore be neglected.

[2]The function e^λ is defined up to a scale transformation; the coefficient of e^λ in Eq. (4) has been selected from convenience considerations; r_g is the initial gravitational radius of the white hole.

The solution (5) describes the expansion of the whole produced matter from the instant $t = t_0$ to the size of the gravitational radius r_g and its subsequent compression to the singularity. The whole solution, including the region to the right of the Lagrangian coordinate R_2, is schematically represented in Fig. 2. For an external observer the boundary of the expanding matter approaches the gravitational radius only after an infinite time, asymptotically. The phase of matter compression ($\pi \leq \xi \leq 2\pi$) is not observed by an external observer at all.

Thus, in the indicated approximation ($p = 0$) there is no flux of produced matter beyond the gravitational radius. Although the singularity produces particles via quantum effects, the matter never gets out from under r_g.

We stress two important properties of the solution (5) which persist also for solutions with other equations of state of matter. The first property: at the maximum of expansion of r, when $dr/dt = 0$, the matter energy density ε is finite, independent of the initial conditions and equals

$$\varepsilon = \frac{1}{8\pi G r_{\text{max}}^2} = \frac{1}{8\pi G r_g^2}. \tag{7}$$

The second property is: the path traversed by a signal with the velocity of light along the radial coordinate R from the instant of beginning of expansion to the instant t_1, when $dr/dt = 0$, is finite and is of the order of r_g, independent of the initial conditions.

The second property leads directly to an important result. If a portion of the Schwarzschild singularity is sufficiently long in the Lagrangian coordinate R, i.e., if in the frame (5)

$$R_2 - R_1 > R_{\text{hor}} = \int_0^{t_1} dt\, e^{-\lambda/2} = \frac{r_g}{\alpha} 2 \int_0^{\pi/2} \frac{\sin^2 z}{1 - z \cot z} dz \approx 7\frac{r_g}{\alpha} \tag{8}$$

for $\alpha \gg 1$, then the signal emitted at $t = 0$ at the left boundary R_1 is manifestly unable to reach the right boundary R_2 over the whole time of expansion of this boundary (up to the time when R_2 reaches the point A in Fig. 2; then $\xi = \pi$) no matter what velocity $v \leq 1$ it had. This means that a signal can never get out from under r_g to an external observer.

Thus, if $R_2 - R_1 > R_{\text{hor}}$, then no matter what conditions are imposed on the expanding nucleus, by "matching" it at the initial instant t_0 to the boundary R_1, this nucleus can never expand beyond r_g to an external observer, i.e., for such an observer the white hole does not explode.

It is further clear that in order that a signal from R_1 should reach an external observer with a large delay time τ with respect to the clock of the external observer started at the moment of beginning of expansion of the external matter of the Universe, it is necessary that the length $R_2 - R_1$ be only slightly smaller than the critical length R_{cr} corresponding to an infinite retardation time (the reaching of the boundary R_2 at A by a signal from

R_1).[3] The approximate formula has the form

$$R_{cr} - (R_2 - R_1) \approx R_{cr} e^{-\tau/2r_g}, \tag{9}$$

where the quantity R_{cr} depends on the speed of the signal. If the speed of the signal tends to the speed of light, $R_{cr} \to R_{hor}$.

Thus, the enumerated properties make extremely unlikely a prolonged delay of the expansion of the matter of a white hole with subsequent explosion, since this would require an artificial readjustment of the initial data. The white hole either explodes practically instantaneously, during a time $t \lesssim r_g$ (for $R_2 - R_1 < R_{cr}$),[4] or it never explodes (for $R_2 - R_1 > R_{cr}$).

3. Massless Particles in a White Hole

Let us consider a more realistic equation of state for particles produced near the Schwarzschild singularity. We assume that the produced particles do not interact with each other and move with the speed of light along the radial coordinate. Then $T_0^0 = -T_1^1 = \varepsilon$, and all other $T_i^k = 0$. In order not to consider accretion problems we shall assume, as before, that the white hole is placed in a cold Universe, i.e., that outside the white hole, at $R > R_2$, the pressure of matter vanishes. As in the case of dust, we shall not take into account the influence of the boundary R_1 for the time being (we consider first that it is sufficiently remote).

The problem reduces to the calculation of the motion of free massless particles produced at $t = t_0$ with a density (3) on the segment $R_1 < R < R_2$. One can solve this problem exactly and the solution is given in Appendices I and II. At the time t_0 the produced particles form two noninteracting fluxes of equal density moving against each other with the speed of light. In the region of produced particles to which the decompression wave has not arrived yet the solution describes an expansion followed by a contraction of the system. These particles never get out from under r_g. The particles flying to the right which have been reached by the decompression wave before r reaches its maximal value in the expansion are emitted from under r_g towards an external observer. The flow of these particles carries off mass. Thus, in this model the white hole inevitably quickly decreases its mass due to quantum effects. The calculation carried out in Appendices I and II shows that the ratio of the initial mass M_0 of the white hole (determined by the conditions near the singularity) to the final mass M_1 after emission, is

$$\frac{M_1}{M_0} = \frac{r_{max}}{r_g} = \left(\frac{\ln\gamma}{\gamma}\right)^{1/2}, \quad \gamma \approx \left(\frac{r_g}{t_{Pl}}\right)^{4/3} \left(\frac{t_{Pl}}{t_0}\right)^{10/3} \gg 1. \tag{10}$$

[3] The delay (retardation) is determined by an observer situated at a constant distance from the white hole.

[4] But not specially $R_{cr} - (R_2 - R_1) \ll R_{cr}$, when the equality (9) holds.

We note that M_0 is determined by the properties of the "vacuum" solution near the Schwarzschild singularity (the magnitude of r at the time t_0) and M_1 is determined by r at the time of maximum expansion of the homogeneous metric ($2GM_1 = r_{\rm max}$). As we shall see in this model problem, the mass decreases quite substantially.

If the white hole explodes after a long delay, the energy released will be much smaller than the energy "introduced" into the singularity, which is carried away by the produced particles. One may not, however, apply this conclusion directly to astrophysics, since in this model the white hole is placed in a cold Universe. As we shall see in the following section, placing the white hole in a "hot Universe" is quite essential for the interpretation.

We finally note that a second property of the solution of the preceding section holds for the solution under consideration, namely: the path traversed along the radial coordinate in the matter of the produced particles is finite. It follows immediately (as in the preceding section) that if the left boundary is sufficiently remote

$$R_2 - R_1 > R_{\rm hor} = \frac{1}{2} r_g \gamma^{-1} \ln \gamma, \tag{11}$$

the white hole will never explode. The condition (9) is necessary for a long delay of the explosion.

We consider now another assumption on the equation of state of the produced particles. We assume that owing to the interaction of the particles with each other the produced particles exhibit Pascal pressure $p = \varepsilon/3$, i.e., $T_1^1 = T_2^2 = T_3^3 = -T_0^0/3$. In the rest we maintain the conditions of the problem the same as in the preceding case. A simple exact solution for all space-time does not exist in this case, since from the boundary R_2 matter flows to the right under the influence of hydrodynamic forces, and a decompression wave travels along the matter to the left with a speed $3^{-1/2} \leq v < 1$.

However, an exact solution can be found for the matter which has not yet been reached by the decompression wave. This solution is listed in Appendix I (cf. Eq. (I.4)). Qualitatively it coincides with (5). The energy carried away by the outgoing matter can be estimated in the following matter. We shall assume that from the surface which was initially attained by the decompression wave matter is emitted with the speed of light and with a density equal to the density of particles moving along the radius from R_1 to R_2 at that moment. Under these assumptions

$$\frac{M_1}{M_0} = \frac{r_{\rm max}}{r_g} = 2\beta^{-1/2}, \quad \beta \approx \left(\frac{r_g}{t_{\rm Pl}}\right)^{2/3} \left(\frac{t_{\rm Pl}}{t_0}\right)^{5/3} \gg 1. \tag{12}$$

In this case the mass loss is also quite significant.

Other properties of this solution are analogous to the case of massless free particles which has been discussed above.

4. A White Hole in the Hot Universe

Finally, let us consider a white hole situated in a hot Universe. We shall assume that the equation of state in the matter outside the white hole is $p = \varepsilon/3$. In addition we shall consider, as before, the apparently most realistic case when the particle production instant is $t_0 \approx t_{\mathrm{Pl}}$. In this case the most acceptable approximation for the equation of state of the produced particles will be the isotropic pressure $p = \varepsilon/3$. The expansion of the matter of produced particles (cf. Appendix I) becomes isotropic practically just after t_0 and remains isotropic up to a time slightly smaller than t_1, when r attains the maximum. The expansion of produced particles during the period $t_0 < t < t_1$ occurs isotropically according to a law coinciding with the expansion law of the external matter of the hot Universe.

Thus, in the interval $t_0 < t < t_1$ the presence of the boundary R_2 has practically no influence on the solution, since the conditions on both sides of the boundary R_2 are practically the same.

Created matter expands up to the scale of the gravitational radius $r_{\max}$ which is much smaller than r_g, the one "put into" the singularity (determined from r at the time t_0 by means of the initial condition (4)). Thus, the mass of the white hole decreases in the course of expansion. But this does not signify an outflow of mass from the white hole (no mass flows through the boundary R_2 in the given case, up to $t \sim t_1$ the boundary R_2 is unimportant). The decrease of mass is caused by the work done by pressure forces at the surface $R = R_2$ during the expansion. In the same manner the mass of any given co-moving volume of the Friedman hot Universe decreases in the course of expansion. Indeed, the mass of the co-moving volume of the hot model varies according to the law (cf. [7]):

$$M = \frac{M_0 a_0}{a}, \tag{13}$$

where $a(t)$ is a scale factor. If one uses for the white hole the relations

$$r_{\max}^2 = r_g r_{\mathrm{Pl}}, \quad r_{\mathrm{Pl}} = r_g^{1/3} t_{\mathrm{Pl}}^{2/3}, \tag{14}$$

one obtains

$$\frac{r_{\max}}{r_{\mathrm{Pl}}} = \left(\frac{r_g}{t_{\mathrm{Pl}}}\right)^{1/3}.$$

Substituting into (13) $r_{\max}/r_{\mathrm{Pl}}$ in place of a/a_0 we obtain (12). Thus, in the hot Universe the mass of the white hole does not flow out of the hole throughout the period of expansion, but decreases only on account of the work done by the pressure forces in accordance with the decrease of the mass of any co-moving volumes during expansion of the hot Universe.

This can be shown another way. Indeed, the second equation of the system (I.1) of Einstein equations for the metric (1) has the integral (taking into account the initial condition (4))

$$r\left(\frac{dr}{dt}\right)^2 = r_g - r + 8\pi G \int_{r(t_0)}^{t} r^2 T_1^1 \, dr. \tag{15}$$

It follows that

$$r_{\max} = r_g + 8\pi G \int_{r(t_0)}^{r_{\max}} r^2 T_1^1 \, dr \tag{16}$$

(We note that $r_{\max} = r_g$ for $T_1^1 = 0$). If the pressure in the hot Universe is equal to the pressure of the produced particles in the white hole then the integral in the right-hand side of Eq. (16) equals exactly the work done by the pressure forces on the white hole in its expansion. $T_1^1 \leq 0$, so that $r_{\max} \leq r_g$.

Only at a later stage, when $t \sim t_1$ the expansion of matter to the left of the boundary R_2 begins to differ markedly from the expansion of the external matter. This is the process of separation of the white hole. In order to determine exactly the outflow or influx of mass, hydrodynamic calculations become necessary here. However, it seems that this process does not modify substantially the final mass of the white hole (cf. in this connection [4]). All conclusions of the preceding sections on the necessity of extremely specific initial conditions for a long retardation of the explosion of the white hole (Eq. (9)) remain valid also for a white hole in a hot Universe.

The question whether a white hole could explode after a long delay with the production of energy of the order $E = Mc^2$ requires separate consideration.

5. Conclusions

Spontaneous particle production in the vicinity of the singularity in a white hole is of principle importance. In the big-bang model of the Universe the white hole "separates" from the surrounding space at the epoch $t \sim t_1 \approx r_{\max}/c$. The white hole either explodes before t_1 or (excluding degenerate initial conditions given by Eq. (9)) it never explodes.

In light of what was said, it seems unlikely that one can explain in terms of explosions of white holes astrophysical phenomena of the type of quasar explosions or explosions of galactic nuclei. In addition it seems that the existence in nature of so-called "bare" singularities [8] is impossible, since they will immediately wrap themselves in a "fur" of produced particles.

Appendix I

The Einstein equations for the metric (1) have the form

$$8\pi G T_0^0 = 2\frac{\dot{a}}{a}\frac{\dot{r}}{r} + \left(\frac{\dot{r}}{r}\right)^2 + \frac{1}{r^2}, \tag{I.1}$$

$$8\pi G T_1^1 = 2\frac{\ddot{r}}{r} + \left(\frac{\dot{r}}{r}\right)^2 + \frac{1}{r^2},$$

$$8\pi G T_2^2 = 8\pi G T_3^3 = \frac{\ddot{a}}{a} + \frac{\ddot{r}}{r} + \frac{\dot{a}}{a}\frac{\dot{r}}{r},$$

where $a = e^{\lambda/2}$ and the dot denotes differentiation with respect to t.

For dustlike matter ($T_1^1 = T_2^2 = T_3^3 = 0$) an exact solution of the system (I.1) is given by Eq. (5); then $r = r_{\max} = r_g$ for $t = t_1 = r_g/2$ ($\zeta = \pi$). If $\alpha > \pi^{-1}$, and in the particle production problem $\alpha \gg 1$, then for $t = \pi r_g$ ($\zeta = 2\pi$), as well as for $t = 0$ ($\zeta = 0$) the metric has a singularity of the Kasner type with exponent $(-1/3, 2/3, 2/3)$.

The exact solution of the system (I.1) for matter with an equation of state $p = \varepsilon/3$ ($T_1^1 = T_2^2 = T_3^3 = -T_0^0/3$) is of the form[5]

$$dt = \frac{3}{2}\left(\frac{r}{a}\right)^{1/3} d\zeta, \qquad z = a^{2/3} r^{1/3}, \tag{I.2}$$

$$|B - z^2| = C|A - \zeta|, \qquad 8\pi G\varepsilon = B(ar^2)^{-4/3},$$

$$r = r_0 + 3C^{-2}(Bz - \frac{1}{3}z^3),$$

where A, B, C are arbitrary constants ($B, C > 0$). If $C = 0$ the last equation of the system (I.2) must be replaced by

$$r = \frac{3}{2}B^{-1/2}(A\zeta - \frac{1}{2}\zeta^2).$$

Taking into account the initial conditions (4) we select the arbitrary constants in the following manner:

$$A = r_g^{2/3}, \quad B = \beta r_g^{2/3}, \quad C = |\beta - 1|, \quad r_0 = -r_g\frac{3\beta - 1}{(\beta - 1)^2}, \tag{I.3}$$

where $\beta > 0$. Then (I.2) can be rewritten in the form (for $\beta \neq 1$)

$$a^{2/3} r^{1/3} = [r_g^{2/3} + (\beta - 1)\zeta]^{1/2},$$

$$r = (\beta-1)^{-2}\{[r_g^{2/3}+(\beta-1)\zeta]^{1/2}[-(\beta-1)\zeta+r_g^{2/3}(3\beta-1)]-r_g(3\beta-1)\}, \tag{I.4}$$

$$dt = \frac{3}{2}\left(\frac{r}{a}\right)^{1/3} d\zeta, \quad 8\pi G\varepsilon = \beta r_g^{2/3}(ar^2)^{-4/3}.$$

[5] In a slightly different form this solution was first found by Shikin [9].

From the equation (3) it follows that in the case of expansion of matter consisting of the produced particles

$$\beta \approx \left(\frac{r_g}{t_{\rm Pl}}\right)^{2/3} \left(\frac{t_{\rm Pl}}{t_0}\right)^{8/3} \sim \left(\frac{r_g}{t_{\rm Pl}}\right)^{2/3} \gg 1, \qquad \text{(I.5)}$$

therefore we shall consider in the sequel just this case.

For $t \ll r_g\beta^{-1/2}$ the spatial curvature of the metric (1) is unimportant and by the substitution $\zeta = r_g^{2/3}\beta^{-1}(\cosh^4\lambda - 1)$ the solution (I.4) transforms into an exact solution for the spatially flat model of Bianchi type I, filled with matter with the equation of state $p = \varepsilon/3$, which was found in the paper of Doroshkevich [10]. Then, for $t \ll r_g\beta^{-3/2}$ the expansion is Kasner-like with exponents $(-1/3, 2/3, 2/3)$, and for $r_g\beta^{-3/2} \ll t \ll r_g\beta^{-1/2}$ it is Friedman-like ($a \sim r \sim t^{1/2}$). For $t \sim r_g\beta^{-1/2}$ the spatial curvature of the metric (1) becomes essential.

For $t \gg r_g\beta^{-3/2}$ and $\beta \gg 1$ the solution (I.4) can be written in the form

$$e^\lambda \equiv a^2 = \frac{\beta^2\zeta}{3r_g^{2/3} - \zeta}, \qquad r = \left(\frac{\zeta}{\beta}\right)^{1/2} (3r_g^{2/3} - \zeta), \qquad \text{(I.6)}$$

$$t = 3\left(\frac{3}{\beta}\right)^{1/2} r_g \left[1 - \left(1 - \frac{\zeta}{3r_g^{2/3}}\right)^{3/2}\right].$$

At the instant of maximal expansion

$$\zeta = r_g^{2/3}, \quad r = r_{\rm max} = 2r_g\beta^{-1/2}, \quad t = t_1 = r_g\beta^{-1/2}(3^{3/2} - 2^{3/2}).$$

For $\zeta = 3r_g^{2/3}$, $t = 3^{3/2}r_g\beta^{-1/2}$ the metric has a singularity of the Kasner type with exponents $(-1/3, 2/3, 2/3)$. The size of the horizon along the R axis at the instant of maximal expansion is

$$R_{\rm hor} = \int_0^{t_1} \frac{dt}{a(t)} = 8r_g\beta^{-3/2}. \qquad \text{(I.7)}$$

We now construct a solution of the system (I.1) for the case when the produced particles are free and move only along the radius, i.e., $T_2^2 = T_3^3 = 0$ and $T_1^1 = -T_0^0 = -\varepsilon$. Then $8\pi G\varepsilon = \gamma a^{-2}r^{-2}$, where

$$\gamma = \left(\frac{r_g}{t_{\rm Pl}}\right)^{4/3} \left(\frac{t_{\rm Pl}}{t_0}\right)^{10/3} \sim \left(\frac{r_g}{t_{\rm Pl}}\right)^{4/3} \gg 1. \qquad \text{(I.8)}$$

For $t \ll r_g\gamma^{-1/2}$ one may neglect the spatial curvature, i.e., the term r^{-2} in the system (I.1), and then the solution corresponding to the initial conditions (4) is of the form

$$r = (2r_g\eta)^{1/2}, \quad a = \left(\frac{r_g}{2\eta}\right)^{1/4} \exp\left(\frac{\gamma\eta}{2r_g}\right), \quad dt = a(\eta)\, d\eta. \qquad \text{(I.9)}$$

The solution (I.9) coincides with a solution describing a flux of free particles in a Bianchi type I metric, solution of which was found in [11]. For $t \ll r_g\gamma^{-3/4}$ the solution is Kasner-like with exponents $(-1/2, 2/3, 2/3)$ and for $r_g\gamma^{-3/4} \ll t \ll r_g\gamma^{-1/2}$ it is of the form (up to terms of the form $\ln\ln t$ and $\ln\ln\gamma$)

$$a = \frac{\gamma}{2r_g}t, \qquad r = \frac{2r_g}{\gamma^{1/2}}\ln^{1/2}\left(\frac{t}{r_g}\gamma^{3/4}\right). \tag{I.10}$$

In order to investigate the subsequent evolution of the system it is convenient to take the first equation and the sum of the first and second equations of the system (I.1) and reduce them to the form

$$\frac{dr^2}{d\zeta} = \frac{2}{a^2}(r_g - \zeta), \qquad \frac{da^2}{d\zeta} = \frac{\gamma - a^2}{r_g - \zeta} - \frac{r_g - \zeta}{r^2}, \tag{I.11}$$

where

$$\zeta = \int_0^t a\,dt.$$

Then in the region $r_g\gamma^{-3/4} \ll t \ll r_g(\gamma^{-1}\ln\gamma)^{1/2}$ one can neglect the term $(r_g - \zeta)/r^2$ and the solution is of the form

$$a = \frac{\gamma}{2r_g}t, \qquad r^2 = \frac{4r_g^2}{\gamma}\ln\left(\frac{t}{r_g}\gamma^{3/4}\right) - \frac{t^2}{2}. \tag{I.12}$$

At the instant of maximal expansion

$$\zeta = r_g, \quad r = r_{\max} = r_g\left(\frac{\ln\gamma}{\gamma}\right)^{1/2}, \quad t = t_1 = \frac{2r_g}{\gamma^{1/2}}.$$

In the region $r_g\gamma^{1/2} \ll t \ll \frac{1}{2}\pi r_g(\gamma^{-1}\ln\gamma)^{1/2}$ we have $a \gg \gamma^{1/2}$, $\zeta \ll r_g$, therefore in the system (I.11) one can omit all terms containing γ and r_g. Then the solution is of the form

$$r^2 = (r_g\ln\gamma - \zeta)^2(\gamma\ln\gamma)^{-1}, \quad a^2 = \zeta(r_g\ln\gamma - \zeta)^{-1}\gamma\ln\gamma, \quad dt = a^{-1}\,d\zeta. \tag{I.13}$$

For $t = \frac{1}{2}\pi r_g(\gamma^{-1}\ln\gamma)^{1/2}$ $(\zeta = r_g\ln\gamma)$ the metric has a singularity of the Kasner type with exponents $(-1/3, 2/3, 2/3)$.

The metric (I.12) joins smoothly to the metrics (I.9), (I.10) and (I.13) respectively from the left and from the right (up to terms of the form $\ln\ln(tt_1^{-1})$. Thus, a complete solution of the system (I.1) for the case of a flux of free particles has been constructed.

Appendix II

Let us construct a solution which describes a centrally-symmetric outflow of matter from a white hole with the speed of light into the vacuum (or into the cold Universe filled with matter at zero pressure). This solution has the form

$$ds^2 = 2\,dr\,du - \left(\frac{f(u)}{r} - 1\right) du^2 - r^2(d\theta^2 + \sin\theta\, d\varphi^2), \tag{II.1}$$

where $f(u)$ is some function which can be determined from the equation

$$R^r_u = -\frac{1}{r^2}\frac{df}{du} = 8\pi G T^r_u. \tag{II.2}$$

The other Einstein equations for the metric (II.1) are identically satisfied for any choice of the function $f(u)$ if all the components of the energy-momentum tensor except T^r_u vanish. The metric (II.1) is a generalization of the solution of Vaidya [12] to the case when there exist both R and T regions. For $r > f(u)$ the coordinate r has a space-like character and for $r < f(u)$ it has a time-like character. The coordinate u is isotropic (light-like) and has the meaning of a retarded coordinate. For $f(u) = r_g = \text{const}$ and $r > r_g$ the metric (II.1) reduces to the Schwarzschild metric by means of the transformation

$$u = t - r - r_g \ln(r - r_g). \tag{II.3}$$

Thus, the function $f(u)$ has the meaning of a variable gravitational radius and shows how the mass enclosed in a certain volume changes with time due to the outflow of the matter. It follows from (II.2) that if $T^r_u \geq 0$, $f(u)$ is a nonincreasing function.

Let the light-like geodesic $u = 0$ pass through the point $t = t_0$, $R = R_2$. Since the production of particles takes place in the region $R < R_2$ one must assume that $f(u) = r_g$ for $u \leq 0$. It follows from the conservation law $T^k_{i;k} = 0$ that

$$T^r_u = r^{-2} g(u),$$

hence

$$f(u) = r_g - 8\pi G \int_0^u g(u)\, du. \tag{II.4}$$

In order to determine the function $f(u)$ and $g(u)$ it is necessary to match the metric (II.1) with the homogeneous metric (1) (the solution (I.9)–(I.13)). Since in the case under consideration the matter consists of free massless particles, the decompression wave moves to the left with the speed of light according to the metric (1). Therefore the matching must be done along a light-like geodesic, the equation of which in the metric (II.1) has the form

$$2dr = [f(u)/r - 1]\, du, \qquad r = r(t_0) \quad \text{for} \quad u = 0, \tag{II.5}$$

and in the metric (1):

$$R = R_2 - \int_{t_0}^{t} dt \exp[-\tfrac{1}{2}\lambda(t)]. \tag{II.6}$$

The continuity condition for the energy flux of the produced particles through the moving boundary (II.5), (II.6) yields

$$g(u(t))\,du/dt = r^2\varepsilon(t)\,dr/dt, \tag{II.7}$$

where $\varepsilon(t)$ is the energy density of the matter consisting of the particles created in the homogeneous metric (1), $r(t)$ is determined from the solution (I.9)–(I.13) and the function $u(r(t))$ from the condition (II.5). Hence

$$f(u) = r_g - 8\pi G \int_{t_0}^{t(u)} r^2 \frac{dr}{dt}\varepsilon\,dt = r_g - 8\pi G \int_{r(t_0)}^{r(u)} r^2\varepsilon\,dr, \tag{II.8}$$

where $t(u)$ is determined from the equation

$$du = 2r\left(r_g - r - 8\pi G \int_{r(t_0)}^{r} r^2\varepsilon\,dr\right)^{-1} dr = 2\left(\frac{dr}{dt}\right)^{-1} dt \tag{II.9}$$

(in the derivation of which use has been made of the relation (15)).

It follows from (II.8) and (II.9) that for $t \to t_1$, when $dr/dt \sim t - t_1 \to 0$, $u \to \infty$ and

$$2GM_1 \equiv f(\infty) = r_{\max}.$$

A similar relation between M_1 and $r_{\max}$ can be obtained also in the case when the matter consisting of the particles produced in the homogeneous metric (1) has the equation of state $p = \varepsilon/3$, assuming that 1/6 of all the particles moves along the R axis to the right with the speed of light.

Institute of Applied Mathematics
L. D. Landau Institute for Theoretical Physics
USSR Academy of Science. Moscow

Received
January 21, 1974

REFERENCES

1. *Novikov I. D.*—Astron. Zh. **41**, 1075 (1964).
2. *Ne'eman Y.*—Astrophys. J. **141**, 1303 (1965).
3. *Novikov I. D.*—Soobshcheniĭa GAISh **132**, 3, 43 (1964).
4. *Zeldovich Ya. B., Novikov I. D.* Teoriĭa tyagoteniĭa i evolyutsiĭa zvezd [Theory of Gravitation and Stellar Evolution]. Nauka (1971).
5. *Zeldovich Ya. B.*—Pisma v ZhETF **12**, 443 (1970).
6. *Zeldovich Ya. B., Starobinskiĭ A. A.*—ZhETF **61**, 2161 (1971).
7. *Zeldovich Ya. B., Novikov I. D.* Relativistic Astrophysics. Univ. of Chicago Press (1971).
8. *Hawking S. W., Ellis G. F. R.* The Large Scale Structure of Space-Time, Cambridge Univ. Press (1973).
9. *Shikin I. S.*—Dokl. AN SSSR **176**, 1048 (1967).
10. *Doroshkevich A. G.*—Astrofizika **1**, 255 (1965).
11. *Doroshkevich A. G., Zeldovich Ya. B., Novikov I. D.*—ZhETF **53**, 644 (1967).
12. *Vaidya P. C.*—Nature **171**, 260 (1953).

35

On the Rate of Particle Production in Gravitational Fields*

With A. A. Starobinskiĭ

The local rate of production of massless particles is calculated in second-order perturbation theory in a weakly anisotropic homogeneous cosmological model, and also in a weak inhomogeneous gravitational field, in which the condition $q^i q_i \geq 0$ is satisfied for the wave vectors with nonzero Fourier components. The rate turns out to be proportional to the local values of the invariants of the curvature tensor.

It is well known that particle-antiparticle pairs, including those with zero rest mass, can be produced by an alternating gravitational field. Different aspects of these effects were considered in a large number of papers. In particular, we have previously [1] calculated the mean value of the energy-momentum tensor (EMT) of created scalar particles in the case when the external classical gravitational field is described by a homogeneous anisotropic Bianchi type-I metric. This mean value turns out to be a complicated non-local causal functional of the space-time metric. It includes a contribution from the real produced particles and the polarization of the vacuum of the quantum field, and in the general case the splitting of the unified mean value of the EMT into an EMT of the real particles and an EMT that describes the polarization of the vacuum is not general-covariant and unique. The reason is the ambiguity in the definition of the concept of a particle as a field quantum (in contrast to the operator of the EMT of the field) in Riemannian space-time.

If, however, the space-time is flat at $t = \pm\infty$, then we can uniquely define the in- and out-vacuums and calculate the number of real particles produced from the in-vacuum in all of space during the entire time interval $-\infty < t < \infty$. In this paper we point out that in some cases it is possible to obtain the local rate of production of massless particles per unit volume and per unit time, and this rate depends only on the local values of the invariants of the Riemann tensor.

As the first example we consider a weakly anisotropic homogeneous cosmological Bianchi type model, and calculate the local rate of production of

*Pisma v Zhurnal eksperimentalnoĭ i teoreticheskoĭ fiziki **26** (5), 373–377 (1977).

massless conformally covariant particles $(-g)^{-1/2}[d((-g)^{1/2}n)/dt]$ to second order in the anisotropy (i.e., to first order in the small ratio $C_{iklm}C^{iklm}/R_{iklm}R^{iklm}$, where C_{iklm} is the Weyl conformal tensor). The space-time metric takes the form[1]

$$ds^2 = dt^2 - a^2(t)[(1+h_1(t))dx_1^2 + (1+h_2(t))dx_2^2 + (1+h_3(t))dx_3^2]. \quad (1)$$

where $|h_\alpha| \ll 1$, $\sum_{\alpha=1}^{3} h_\alpha = 0$. Introducing, in accordance with [1], the variable $\eta = \int dt/a(t)$ and, in the case of production of scalar particles, the field function $\chi = \varphi a$, we obtain for the space-like Fourier component (to first order in h_α) of χ:

$$\chi_k'' + k^2\chi_k = V_k(\eta)\chi_k, \quad (2)$$

where

$$V_k(\eta) = \sum_{\alpha=1}^{3} h_\alpha(\eta)k_\alpha^2, \quad k^2 = \sum_{\alpha=1}^{3} k_\alpha^2;$$

$\mathbf{k} = \{k_\alpha\}$ is a constant wave vector. If the quantum field was in a vacuum state at $t = -\infty$ $(\eta = -\infty)$, then $\chi_k = \exp(-ik\eta)$ as $\eta \to -\infty$. Then (2) can be represented in the form of an integral equation:

$$\chi_k(\eta) = \frac{1}{k}\int_{-\infty}^{\eta} \sin k(\eta-\eta_1)V_k(\eta_1)\chi_k(\eta_1)\,d\eta_1 + e^{-ik\eta}. \quad (3)$$

As $t \to +\infty$ $(\eta \to +\infty)$, when $h_\alpha \to 0$, we have $\chi_k(\eta) = \alpha_k \exp(-ik\eta) + \beta_k \exp(ik\eta)$, where $|\alpha_k|^2 - |\beta_k|^2 = 1$. Then

$$a_k = 1 + \frac{i}{2k}\int_{-\infty}^{\infty} e^{ik\eta}V_k(\eta)\chi_k(\eta)\,d\eta, \quad (4)$$

$$\beta_k = -\frac{i}{2k}\int_{-\infty}^{\infty} e^{-ik\eta}V_k(\eta)\chi_k(\eta)\,d\eta. \quad (5)$$

The total density of the real produced particles is connected with β_k at $\eta \to +\infty$ by the formula:

$$n = (2\pi)^{-3}a^{-3}\int d^3k|\beta_k|^2. \quad (6)$$

Solving (3) by iteration and substituting $\chi_k(\eta)$ in (5), we find that to second order in h_α we have as $\eta \to +\infty$:

$$n = \frac{1}{1920\pi a^3}\int_{-\infty}^{\infty} d\eta \sum_{a=1}^{3}(h_a'')^2 = \frac{1}{960\pi a^3}\int_{-\infty}^{\infty} d\eta\, a^4 C_{iklm}C^{iklm} \quad (7)$$

(the prime denotes differentiation with respect to η). Therefore the local rate of production of massless scalar particles in the metric (1) is:[2]

$$\frac{1}{\sqrt{-g}}\frac{d}{dt}(\sqrt{-g}\,n) = \frac{1}{960\pi}C_{iklm}C^{iklm} \geq 0 \quad (8)$$

[1] We choose a system of units in which the speed of light is equal to unity.

[2] Also, the result (8) can be easily obtained from the formulas given in the Appendix 1 of [1].

It can be shown by the same method that an analogous formula holds also for the production of photons and neutrinos. The only difference is that the numerical coefficient in the right-hand side of (8) is $1/320\pi$ for neutrinos and $1/80\pi$ for photons.

We consider now the production of gravitons in the Friedman isotropic model ($h_\alpha = 0$), when furthermore the ratio $R^2/R_{iklm}R^{iklm}$ is small (R is the scalar curvature). Using the equation for gravitational waves [2], which reduces to the form:

$$\chi_k'' + (k^2 - \frac{a''}{a})\chi_k = 0,$$

and the methods described above, we arrive at the following result:

$$\frac{1}{\sqrt{-g}}\frac{d}{dt}(\sqrt{-g}n_g) = \frac{R^2}{288\pi} \geq 0.$$

At $R = 0$ and $C_{iklm} = 0$, no gravitons are produced—this result was noted by Grishchuck [3].

We note, however, that in the next higher orders in the small parameter

$$\frac{C_{iklm}C^{iklm}}{R^{iklm}R^{iklm}}$$

(or $R^2/R_{iklm}R^{iklm}$) the density of the produced particles $n(\eta = +\infty)$ is not expressed in terms of a single integral of local quantities with respect to η, and therefore the local formulas (8) and (9) cannot be generalized to include higher orders of perturbation theory.

We proceed to the case of a weak inhomogeneous gravitational field. Let the space-time metric be of the form $g_{ik} = \eta_{ik} + h_{ik}$, where η_{ik} is the Minkowski metric and $|h_{ik}| \ll 1$. Then the total probability of pair production, to second order in h, is

$$W = \langle \mathbf{0}|S^{(1)+} \cdot S^{(2)}|\mathbf{0}\rangle = -2Re[\langle \mathbf{0}|S^{(2)}|\mathbf{0}\rangle], \qquad (10)$$

where $S^{(1)}$ and $S^{(2)}$ are the corresponding terms of the S-matrix expansion in powers of h. Calculating S by perturbation theory, in analogy with the procedure used in [4], we obtain:

$$W = \frac{\alpha}{8\pi}\frac{1}{(2\pi)^4}\int d^4q[R_{ik}(q)(R^{ik}(q))^* - \frac{1}{3}|R(q)|^2]\theta(q^0)\theta(q^iq_i), \qquad (11)$$

where the coefficient α is equal to $1/60$ in the case of massless conformal scalar particles, $1/20$ in the case of neutrinos, and $1/5$ in the case of photons (for details of the calculations see [5]). On account of the factor $\theta(q^iq_i)$, the integral in (11) cannot be transformed in the general case into a single integral with respect to d^4x. If, however, the condition $q^iq_i \geq 0$ is satisfied at $R_{ik}(q) \neq 0$, then the θ function of q^iq_i in (11) can be omitted, and then (taking into account the known fact that the invariants $C_{iklm}C^{iklm}$ and $2(R_{ik}R^{ik} - \frac{1}{3}R^2)$ differ by a total derivative) W reduces to the form

$W = \int w(x^i) d^4x$, where:

$$w(x^i) = \frac{\alpha}{32\pi} C_{iklm} C^{iklm}, \tag{12}$$

which again leads to (8) in the homogeneous case.

At the present time, during the Friedman stage of the evolution of the Universe, no neutrinos or photons are produced, and the rate of production of gravitons, as follows from (9), is vanishingly small, $\sim 10^{-106}\ \mathrm{cm}^{-3}\,\mathrm{sec}^{-1}$. Near the singularity we have $(-g)^{-1/2}[d((-g)^{1/2}n)/dt] \sim t^{-4}$, which yields $(d\varepsilon/dt) \sim t^{-5}$, in agreement with the results of [1]. Formula (8) is valid also in the case of particles with mass, provided that $\hbar^{-4}m^4 \ll |R_{iklm}R^{iklm}|$.

Institute of Applied Mathematics
USSR Academy of Sciences
L. D. Landau Institute of Theoretical Physics
USSR Academy of Sciences. Moscow

Received
August 15, 1977

REFERENCES

1. *Zeldovich Ya. B., Starobinskiĭ A. A.*—ZhETF **61**, 2161 (1971).
2. *Lifshitz E. M.*—ZhETF **16**, 587 (1946).
3. *Grishchuk L. P.*—ZhETF **67**, 825 (1974).
4. *Sexl R. U., Urbantke H. K.*—Phys. Rev. **179**, 1247 (1969).
5. *Starobinskiĭ A. A.* Ph.D. Dissertation, Inst. Theor. Phys. USSR Acad. Sci., Chernogolovka (1975).

Commentary

We will comment on a series of papers devoted to the study of quantum-gravitational processes of particle creation and vacuum polarization in strong gravitational fields near the cosmological singularity. The effect of particle creation was known in isotropic models of the universe, however, there it does not lead to significant cosmological consequences. In the paper "Particle Creation in Cosmology" Ya. B. showed that the effect of particle creation becomes uniquely important in a situation where at the beginning matter was nonexistent (or insignificant), i.e., in anisotropic vacuum solutions such as Kasner's. There this effect leads to rapid isotropization of the solution.

At the time this paper was written it was not yet clear whether the effect of particle creation in a vacuum is consistent with the generalized conservation law $T^k_{i;k} = 0$. The answer to this question was found in a paper by Ya. B. and L. P. Pitaevskiĭ, and proved to be positive: the particle creation effect is indeed consistent with the condition $T^k_{i;k} = 0$, but at the same time the condition of "energy dominance" must be violated for components of the matter energy-momentum tensor, in particular, the pressure proves to be greater than the energy.

This is possible due to the fact that the complete matter energy-momentum tensor T_i^k includes both the energy-momentum tensor of real particles and the vacuum polarization, and in the general case it is impossible to separate these two contributions. Therefore, in the calculation of the full mean value $\langle T_i^k \rangle$ particle creation and vacuum polarization must be studied together. A program for calculating $\langle T_i^k \rangle$ which includes the correct regularization of this quantity was realized for anisotropic, homogeneous models of the Universe in a paper by Ya. B. and A. A. Starobinskiĭ, "Particle Production and Vacuum Polarization in an Anisotropic Gravitational Field," which served as the basis for numerous subsequent papers. There exists, however, a situation in which one may nevertheless separately calculate the contribution of real particles which are created. For them, in a paper by Ya. B. and A. A. Starobinskiĭ, "On the Rate of Particle Creation in Gravitational Fields," an elegant formula is obtained which expresses the rate of creation of real particles from a vacuum in terms of geometric invariants which are constructed from Weyl and Ricci tensors.

We note that the regularization applied using n-waves (see **33**) differs from the method of Pauli-Villars. The massless physical field is associated with an also massless regularization field. In this way the important property of massless fields—their conformal invariance—is preserved. In a conformally-flat space-time creation of real particles does not occur, but nevertheless local polarization of the vacuum occurs.

The opposite assertion, which appears several times in the paper, is incorrect. Polarization in this case is called a "conformal anomaly."[1] However, it is remarkable that (as was understood later) the "n-wave regularization" proposed in paper 33 correctly reproduces the value of the conformal anomaly for massless particles if the limit $m \to 0$ is taken after (and not before) the calculation of $\langle T_{ik} \rangle_{\text{Reg}}$ using the general expression (22) of this paper.

[1] *Davis P. C. W., Fulling S. A., Christensen S. M., Bunch T.S.*—Annal. of Physics **109**, 108–120 (1977).

36

A Hypothesis Unifying the Structure and Entropy of the Universe*

A hypothesis about the averaged initial state and its perturbations is put forward, describing the entropy of the hot Universe (due to damping of short waves) and its structure (clusters of galaxies due to long wave perturbations).

A hypothesis is put forward, assuming that initially, near the cosmological singularity, the Universe was filled with cold baryons. The averaged evolution was described by the uniform isotropic expansion, according to Friedman solution and the equation of state of cold baryons.

Superimposed on this averaged picture are initial fluctuations of baryon density and corresponding fluctuations of the metric. One unique value (approximately 10^{-4}) of non-dimensional amplitude of metric fluctuations, scale-independent, describes two different, most important properties of the Universe—its structure and its entropy. The density fluctuations are in inverse proportion to the square of the scale at a given moment of time. The fluctuations of small scale of the order of the mean distance between two neighboring baryons first increase, soon they are transformed into acoustic waves, i.e., phonons, propagating in the baryonic fluid.

The damping of short acoustic waves is accompanied by their transformation into various modes of excitation and relaxation to thermodynamic equilibrium with high entropy per baryon. This line of reasoning with one parameter adjusted (the initial metric perturbation 10^{-4}) leads to the ratio of photons to baryons characteristic for the hot Universe.

The relaxation into thermodynamic equilibrium occurs early, at $t \ll \hbar/mc^2$. Therefore the well-known scenario of hot Universe evolution is conserved, including the hadronic era with plenty of antibaryons, the nucleogenesis leading to 25–30 percent He^4, the radiation dominated era giving blackbody 2.7 K radiation.

Particularly the hadronic era with the ratio of baryons to antibaryons $B : \bar{B} = 1 + 10^{-8}$ seemed to be the most unnatural in the standard scenario of the hot Universe. Why is the ratio not 1? Why is the departure from unity (10^{-8}) everywhere positive?

In our hypothesis initially $B : \bar{B} = 1 : 0$ everywhere. Thermal excitations add pairs $B = \bar{B}$, whose number is 10^8 times greater than the number

*Monthly Notices of the Royal Astronomical Society **160**, 1p–3p (1972).

of initial baryons. Therefore the puzzling $1 + 10^{-8}$ ratio is explained by harmless (although arbitrary) density and metric fluctuations.

The second line of reasoning concerns long wave fluctuations of metric and density.

The characteristic time when $\lambda = ct$ for these perturbations occurs very late in the radiation dominated era or even in the era of neutral gas. The theory originated by Lifshitz [1] has been worked out in detail in the last years. The spectrum proposed is analogous to that given by Harrison [2].

The amplitude of metric fluctuation of the order of 10^{-4}–10^{-5} leads to formation of clusters of galaxies with characteristic mass $\sim 10^{13}\,M_\odot$ not long ago, at the redshift between $Z \sim 2$–20.

Smaller scale perturbations are damped [3] and on a bigger scale the perturbations are not great enough to build isolated clusters of mass much greater than $10^{13}\,M_\odot$. The hypothesis with $b \sim 10^{-4}$ is in accord with the low limit of fluctuations of relict 2.7 K blackbody radiation [4].

What seems to be most fascinating is the fact that one unique flat spectrum of metric perturbation with one arbitrary constant leads to the calculation of the 'content' of the Universe (its entropy, primeval chemical composition, photon-baryon ratio, etc.) and to the calculation of the 'structure' of the Universe (the mass and density of clusters, the degree of uniformity on greater scale).

By interpolation small initial perturbations in the intermediate region $10^{-24}g < M < 10^{13}M_\odot$ are predicted. No regions with antimatter preponderance are expected. The unsolved questions are reduced to: (1) Why was the averaged Universe cold and Friedmanian, and (2) Why was the spectrum of metric perturbations flat and of the order of 10^{-4}.

The rough calculation of entropy is shown below. Units with $\hbar = c = 1$ are used so that the proton mass has the same dimensions as inverse length, cm^{-1}, the Newtonian gravitation constant $G = \mathrm{cm}^2$. The dimensionless parameter $Gm^2/\hbar c = g = 10^{-38}$, all other numbers $(2, 3, \pi, e)$ are omitted. Metric perturbations[1] are designated b. The index zero corresponds to the Planckian moment $t_0 = G^{1/2} (= 10^{-43}\,\mathrm{s})$. The equation of state of cold baryons is taken in the form $p = \mathcal{E} = n^2 m^{-2}$ valid at $n > m^3$ [5], where n is the baryon density; nm and $n^{4/3}$ contributions in p, $\mathcal{E}$, important at low n are omitted. The averaged solution is $\rho = p = \mathcal{E} = G^{-1} \cdot t^{-2}$, $n = mt^{-1}G^{-1/2}$. The zero values are $(t = t_0 = G^{1/2})$

$$\rho_0 = G^{-2} (= 10^{93}\,\mathrm{g\,cm}^{-3}), \quad n_0 = mG^{-1}.$$

The shortest waves considered have $\lambda_{\mathrm{min}} = n^{-1/3}$, therefore $\lambda_{0,\mathrm{min}} = G^{1/3}m^{-1/3}$, their transformation into acoustic waves occurs at t_1 such that $t_1 = \lambda_{1,\mathrm{min}} = n_1^{-1/3}$; $\lambda_{1,\mathrm{min}} = t_1 = G^{1/4}m^{-1/2} = t_0 g^{-1/4} \gg t_0$. To metric fluctuations b

[1]It is not the Fourier amplitude b_κ but the $b = (|b_\kappa|^2\kappa^3)^{1/2}$ for $\kappa = \lambda^{-1}$, the same is true for $\delta\rho$.

correspond density fluctuations

$$\delta\rho = bG^{-1}\lambda^{-2}.$$

The acoustic energy at $t = t_1$ is

$$\mathcal{E}_{ac} = \rho^{-1}(\delta\rho)^2 = b^2 G^{-3/2} m.$$

After thermodynamic relaxation the thermal energy, temperature and entropy S per unit volume are $\mathcal{E}_{th} = T^4 = b^2 G^{-3/2} m$; $S = T^3 = G^{-9/8} m^{3/4} b^{3/2}$. Dividing by baryon density n_1, we obtain the entropy per baryon

$$S_b = b^{3/2} G^{-3/8} m^{-3/4} = b^{3/2} g^{-3/8} = 10^{14} b^{3/2}$$

so that $b = 10^{-4}$ gives $S_b = 10^8$ approximately the observed value. In symbols to obtain the empirical law [6] $S_b = g^{-1/4}$, we need $b = g^{1/12}$. Only excitations connected with baryon movement are considered, because other types of excitations (gravitational waves, electromagnetic, mesons) are conformally invariant (exactly, or approximately, neglecting rest mass) and therefore do not increase in conformal-flat averaged evolution [7–10]. Density fluctuations increase due to gravitational instability, although the corresponding metric fluctuations remain constant in the time interval $t_0 < t < t_1$ [1, 11].

Recently Rees [12] proposed a theory of entropy increase due to damping of strong perturbations by shock waves in the radiation dominated era.

The resemblance of our approaches is obvious: both of us are asking for perturbations of the Friedman solution, which would explain entropy (due to short waves) and structure (due to long waves). But Rees's proposal needs great perturbations. The chemical composition of primeval matter ($H : He^4$) and the blackbody spectrum [13] are in accord with unperturbed (or weakly-perturbed) Friedman Universe. There is a danger that this accord will be destroyed in Rees's picture. On the other hand the defect of my hypothesis consists in the use of the unproven equation of state of baryons up to the limit of modern theory.

No *a priori* preference can be given to small or big perturbation theories—the analyses of observations is the unique approach to the problem.

My gratitude is due to A. A. Starobinskiĭ for valuable discussions and help.

Institute of Applied Mathematics
USSR Academy of Sciences

Received
September 4, 1972

REFERENCES

1. *Lifshitz E. M.*—ZhETF **16**, 587 (1946).
2. *Harrison E. R.*—Phys. Rev. D **1**, 2726 (1970).
3. *Silk J.*—Astrophys. J. **151**, 569 (1968).

4. *Zeldovich Ya. B., Sunyaev R. A.*—Astrophys. Space Sci. **7**, 201 (1970).
5. *Zeldovich Ya. B.*—ZhETF **41**, 1609 (1961).
6. *Zeldovich Ya. B., Novikov I. D.*—Astron. Zh. **43**, 758 (1966).
7. *Parker L.*—Phys. Rev. Lett. 21, 562 (1968).
8. *Parker L.*—Phys. Rev. 183, 1057 (1969).
9. *Parker L.*—Phys. Rev. Lett. 28, 705 (1972).
10. *Zeldovich Ya. B., Starobinskiĭ A. A.*—ZhETF **61**, 2161 (1971).
11. *Zeldovich Ya. B., Novikov I. D.* Relativistic Astrophysics. Univ. of Chicago Press (1971).
12. *Rees M.*—Phys. Rev. Lett. **28**, 1669 (1972).
13. *Sunyaev R. A., Zeldovich Ya. B.*—Astrophys. Space Sci. **9**, 368 (1970).

Commentary

In the above article, following Harrison (see Ref. [2]), a specific initial perturbation spectrum is considered. In Lifshitz's terminology (Ref. [1]), the subject is scalar perturbations.

It was then noted that such a spectrum is self-similar and fractal. In simple physical language this means that perturbations of the metric are identical at any scale.

Ya. B. and I. D. Novikov[1] put forward a proposal of equality of the amplitude (in the mean) of scalar and tensor perturbations. They showed that the finite vector (vortex) perturbations are incompatible with the Friedman asymptotic form of the singularity.

It later turned out that the Harrison–Zeldovich spectrum is a result of the inflationary model of the early universe. However, the absolute value of the inhomogeneity of the metric has still not been obtained form the theory and must be determined from observations of relic radiation and the structure of the universe—see Sections VIII and IX of the present book.

[1] *Zeldovich Ya. B., Novikov I. D.*—Astron. Zh. **46**, 775–778 (1969).

36a
Cosmological Fluctuations Produced Near a Singularity*

The perturbations of a uniform Friedmanian universe, leading to galaxy formation, are explained by the strings formed during the symmetry loss of vacuum of a complex Higgs field with mass characteristic of grand unification. Difficulties are pointed out inherent to phase transition, particle decay and black hole evaporation as sources of growing perturbations.

Observational cosmology presents at least three puzzles. Why is the Universe (1) nearly Friedmanian—with metric perturbations less than 1 percent on all scales yet investigated; (2) why is it nearly flat: even a curvature of the order of the inverse horizon today corresponds to average curvature of the order of 10^{-30} of the inverse horizon near the singularity; (3) why is the Universe not exactly uniform, having initial perturbations of the order of $1 - 0.1$ percent on the scale of clusters of galaxies (and perhaps on other scales too).

We do not ask the fourth question—why does the Universe exist, but assuming that (1) and (2) are granted somehow, we concentrate on the third question on the origin of initial perturbations.

We assume the exact uniformity of the metric and composition at some moment, when the time is several times larger than Planckian, 10^{-43} s, so that a classical metric is meaningful. The metric involves Einstein's equations of motion and density of matter. At extremely high density full local equilibrium is established, therefore non-uniform initial composition is incompatible with the assumed uniformity of the total density [1]. The degree of density uniformity is even better than the uniformity of the metrics, $\delta\rho/\rho = (ct/\lambda)^2 \delta g/g$ for $\lambda \gg ct$ and is a quasi-isotropic (i.e., locally Friedmanian) solution.

The fluctuations are less than those in equilibrium with an external isothermal bath; there is not enough time to establish normal thermodynamical fluctuations characteristic of a stationary system. To be exact, if such a bath existed and a stationary non-expanding Universe could be brought in contact with the bath, the equilibrium would be unstable and no finite thermodynamical perturbations would establish on scales larger than the Jeans length. The onset of fluctuations by random dissipative forces would give

*Monthly Notices of the Royal Astronomical Society **192**, 663–667 (1980).

an initial amplitude to the growing modes of gravitational instability. The overall picture does not contradict general relativity, if a large cosmological constant is allowed.

The build-up of fluctuations due to viscosity during expansion is too slow in a uniform plasma to be the source of perturbations needed for galaxy formation from an initial state with uniform metric in the evolutionary Friedmanian universe. The discrepancy is most flagrant for large scales in space.

A new hope is connected with modern broken symmetry field theory. As pointed out by Kirzhnitz and Linde [2] the general feature of these theories is the symmetry restoration at ultra-high temperature. During cooling, phase transition must occur to the cold broken symmetry situation. The phase transition is likely to induce non-uniformity much larger than ideal gas fluctuations.

In this paper I will try to use topological excitations, particularly strings, as the immediate cause of perturbations.

Strings with mass per unit length of the order of $\sigma = m^2$ if $\hbar = c = 1$ or $\sigma = m^2 c/\hbar$ g cm^{-1} are predicted to be formed after cooling in the simplest version of one scalar complex field interacting with one vector field [3]. They are similar to Abrikosov magnetic strings in type II superconductors, being consequences of the Ginzburg-Landau theory of superconductivity.

Kibble [4] points out that the linear mass density in theories of Weinberg-Salam type is too small to be of cosmological importance. With $m =$ 100 GeV, $\sigma = 10^{-6}$ g cm^{-1}, assuming string density of the order[1] of $1/c^2t^2$ one obtains mass density $\rho_s = 10^{-27}t^{-2}$ g cm^{-3} which is negligible compared with the average density of the Universe $(5-8)\cdot 10^5 t^{-2}$ g cm^{-3}.

But the now fashionable GUT—grand unification theories—involve particle masses which are sizable fractions of the Planckian mass $M = G^{-1/2} =$ 10^{-5} g. If the characteristic mass is written as $m = \varepsilon M$, ε say of the order of 10^{-2}, then $\sigma = \varepsilon^2/G$ and $\rho_s = \varepsilon^2/Gt^2$ which should be compared with the critical density $\rho_c = 1/6\pi G t^2$ or $\rho = 3/32\pi G t^2$ (depending on $p = 0$ or $p = \rho c^2/3$).

One obtains $\rho_s/\rho_c = 25\varepsilon^2 \sim 2.5\cdot 10^{-3}$ for $\varepsilon = 10^{-2}$. This is approximately the amplitude of perturbations needed. Of course, the Universe with strings is rather strange.

Strings frozen in plasma would change their density like a^{-2}, i.e., like t^{-1} in radiation-dominated or $t^{-4/3}$ in a matter-dominated universe (a—the radius of the Universe, $d\ln a/dt = H$ where H is the Hubble constant).

In this approximation the strings would soon be dominant. The tension along a string is equal to its energy per unit length. Therefore the equation of state of a chaotic ensemble of strings is $p = -\rho c^2/3$. Expansion is connected with work done on pulling strings.

[1]The order of magnitude estimate $c^{-2}t^{-2}$ for surface density comes from the belief that strings are moving with velocity of the order of c and that they intercommute and annihilate when they meet one another.

The order of magnitude estimate given above with ρ_s proportional to t^{-2} and to ρ implies more complicated behavior of strings.

They are not simply pulled, but they actively diminish their length due to annihilation of closed loops, rectification of curved parts of strings and recommuting of intersecting strings. Due to all these processes they are moving (which increases the average $\bar{p}$ over $-\rho c^2/3$) and they are giving energy to the surrounding plasma. At $\rho_s/\rho \sim 10^{-3}$ the input of energy seems to be admissible from the point of view of spectral distortion of cosmic microwave radiation [5].

The most difficult question concerns the local effects of these exotic strings on the microwave background. The mechanics of strings is not clear. Their gravitational potential is of the order[2] of $\varphi \cong c^2\varepsilon^2 \ln(t/t_{\rm pl})$, with $t_{\rm pl} = (G\hbar/c^5)^{1/2} = 10^{-43}$ s. A closed loop at rest is characterized by the average potential $\bar{\varphi} = c^2 l\varepsilon^2/r$; with $l = 2\pi r$ this is $\bar{\varphi} = 2\pi c^2\varepsilon^2$.

When the loop is shrinking slowly due to friction by surrounding plasma, φ remains the same. But if the loop is shrinking without friction, its effective mass is conserved, the decrease of length is compensated by the kinetic energy of transverse motion (the motion along the string is unobservable, the string is Lorentz-invariant to this motion; formally this is connected with the equality of tension and linear density). For shrinking loops $\sigma' = \sigma/\sqrt{1-\beta^2}$, $\sigma' l' = \sigma l_0$, $m = \text{const}$. If a circular loops shrinks remaining a circle, $l' = 2\pi r$ and finally the loop transforms in a black hole when $r = \varepsilon^2 r_0 \ll r_0$. But it is a degenerate case. An irregular loop will oscillate and it is difficult to estimate the probability of a black hole formation in the general case, especially with account of plasma friction.

It must be stressed that the underlying physical theory, i.e., the broken symmetry and grand unification concept are not yet confirmed. Moreover, possible variants of the theory give diverging predictions about the existence of strings [6]. Therefore all said above is no more than a tentative hypothesis. Short comments should be added on previous work aiming at explanation of perturbations.

In a very naive cold universe picture [7], solid hydrogen breakup was considered. The physical idea was fantastic, but an important methodical point remains: local events give rise to a quadratic spectrum on a large scale $\sqrt{(\delta\rho/\rho)^2_k} \sim k^2$, $\delta M/M \sim M^{-7/6}$.

This is a general law, connected with conservation of mass and momentum. It was discussed in a review article by Zeldovich [8].

Kirzhnitz and Linde [2] first predicted a second-order transition, which goes smoothly and does not induce large scale perturbations as a consequence of broken vacuum symmetry. But in a later work they pointed out the

[2]The gravitational potential on the surface of a string of density σ, radius r and mean distance from other strings R is of order $|\varphi| = G\sigma \ln(R/r)$. Inserting $\sigma = \varepsilon^2/G$, $r = r_{\rm pl}/\varepsilon = ct_{\rm pl}/\varepsilon$, $R = ct$ and neglecting ε in the logarithmic term we obtain this estimate.

possibility of a first-order transition with supersaturation and nucleation [9, 10].

The discrete broken symmetry [11] leads to formation of walls between domains with opposite signs of CP violation. These domains were shown to be too heavy to exist in Nature and this is an argument against the underlying CP-violation explanation [4, 12].

Press [13] published an important work. The task of obtaining perturbation from first principles is formulated. But the immediate cause of perturbations is thought to be the energy density depending on the phase distribution after the establishment of broken symmetry. In the usual theories the phase gradient is totally compensated by the gauge field. No phase gradient terms occur unless one works without the gauge field, admitting the phase to be a long-range Goldstone scalar field. No such field is known in Nature. Instead of a distributed phase gradient, strings are formed when the field topology is appropriate.

Hogan [14] assumes the existence of heavy particles with abnormally long decay time. Perhaps the small primordial black holes (PBH) can play the same role. Hogan stresses the statistical character of the radioactive decay. A distribution of PBH mass would give dispersion of their evaporation time.

But there is a very important point about theories of this type, which is shared by theories of random nucleation in first-order phase transitions.

Common to all these theories is the variation in the equation of state of the matter filling the Universe. Between initial and final ultra-relativistic gas $p = \rho c^2/3 - \pi$, one has a period plagued with heavy particles, PBH or nucleation. It could be expressed as $p = \rho c^2/3 - \pi$, with π being the deviation $\pi = \pi(\rho, x)$ such that $\pi = 0$ at $\rho > \rho_i$ and $\rho < \rho_f$: it exists in a limited density region. In contrast to the main part $p = \rho c^2/3$, the contribution described by π is fluctuating, it is randomly distributed in space. It is thought that the π on x dependence will generate the perturbations. We are interested in perturbations on a scale much larger than the momentary value of the horizon ($\lambda \gg ct$) when the random equation of state perturbation is present, $\pi \neq 0$.

For perturbations on such a scale one can neglect the interaction of the domains under consideration. Every single domain follows the flat Friedman evolution pattern with the given local average equation of state.

Of course $\rho(t)$ depends on the $\pi(t)$ and variations of $\pi(x,t)$ are leading to definite non-uniform, x-dependent variations of the density, $\rho = \bar{\rho}(t) + \delta\rho(x,t)$.

Order of magnitude estimates give density perturbations at the moment when pressure fluctuations end

$$\left|\frac{\delta\rho}{\rho}\right|_{\rho<\rho_f, t=\text{const}>t_f} \cong \frac{\pi(\rho_i - \rho_f)}{c^2(\rho_i + \rho_f)^2}.$$

They are proportional to the fluctuations of pressure (due for example to random fluctuations of the moment of particle decay or new phase formation).

It is tempting to use these density perturbations as seeds of subsequent galaxy formation, etc. The pressure fluctuations are statistically independent in regions causally not connected ($\lambda > ct$ at the fluctuation epoch). Therefore one would expect white noise, i.e., flat Fourier spectrum of $\delta\rho/\rho$ in this mechanism, which is even more than that needed for the large-scale structure of the Universe.

But, in fact, all these expectations are erroneous! The point is, that in this case, the density perturbations are compensated by local Hubble constant perturbations. To demonstrate this behavior, we integrate the Friedman equations with the result that the solution after the end of the equation of state perturbation is $\rho = 3/32\pi G(t+\tau(x))^2$ and fluctuations of equations of state are changing the density due to their influence on the constant $\tau(x)$, i.e.,

$$\frac{\tau(x)}{t_f} \cong \frac{\pi(\rho_i - \rho_f)}{2c^2(\rho_i + \rho_f)^2}.$$

But the shift of a solution by a time τ variable in space, $\tau = \tau(x)$, produces a decreasing perturbation mode.[3] This point has already been clarified in an early review article of the author [8]. This is seen immediately by decomposition of the formula for ρ, giving

$$\frac{\delta\rho}{\rho} = \frac{-2\tau(x)}{t+\tau}$$

decreasing like t^{-1} at $t \gg t_f > \tau$. This is not to say that the π-fluctuations are totally vanishing.

In the second approximation involving interaction of the adjacent regions, the growing modes of perturbations is excited. But its amplitude is smaller compared with the total amplitude in the ratio $(ct_f/\lambda_f)^2 \sim (\kappa_f ct_f)^2$. Instead of the flat spectrum, corresponding to white noise, the growing perturbations are characterized by the amplitude proportional to κ^2. This actually kills the hope of explaining the structure of the Universe on a galactic scale by early random processes. The temperature $\sim$ 100 GeV corresponds to $z \sim 3 \cdot 10^{14}$, $t \sim 10^{-10}$ s. A perturbation now on scale of 1 Mpc $= 3 \cdot 10^{24}$ cm had λ of the order of 10^{10} cm when ct was 3 cm. Therefore, the dimensionless $(ct/\lambda)^2$ factor is equal to 10^{-19}.

In a strongly exaggerated case assume $\delta\rho/\rho \sim 1$ on the causal scale $\sim$ 1 cm. This would give in white noise approximation $\delta\rho/\rho \sim (ct/\lambda)^{3/2} \sim 10^{-28}$. The amplification of growing perturbations in radiation dominated regime is $t \sim z^2 \sim 10^{29}$. This seems to be enough to give $\delta\rho/\rho \sim 1$ or 10 now, at $z = 0$ with the white noise amplitude—but the extra 10^{-19} factor

[3] This approach to perturbations was initiated by Barenblatt and Zeldovich [15].

in the amplitude of the growing mode makes the perturbations hopelessly small.

The fluctuations of the equation of state are generating entropy perturbations which actually have a white noise flat spectrum but they do not grow until decoupling at the modest $z \sim 10^3$.

This analysis confirms that what one needs are just filament tensions in order to move matter. So we return to the first part of the paper.

My thanks are due to A. D. Linde, L. B. Okun and A. S. Schwartz for consultations, to M. Khlopov and R. A. Sunyaev for discussions and help. The preprints of Press and Hogan stimulated the work.

Institute of Applied Mathematics
USSR Academy of Sciences

Received
September 4, 1972

REFERENCES

1. *Ellis J., Gaillard M., Nanopoulos D.*—Phys. Lett **B80**, 360 (1979).
2. *Kirzhnitz D., Linde A.*—ZhETF **67**, 1263 (1974).
3. *Nielsen N., Olesen P.*—Nucl. Phys. **B61**, 45 (1973).
4. *Kibble T. W. B.*—J. Phys. **A9**, 1387 (1976).
5. *Sunyaev R., Zeldovich Ya. B.*—Astrophys. Space Sci. **7**, 20 (1970).
6. *Polyakov A. M.*—ZhETF **68**, 1975 (1975).
7. *Zeldovich Ya. B.*—ZhETF **43**, 1037 (1962).
8. *Zeldovich Ya. B.*—Adv. Astr. Astrophys. **3**, 242–379, esp. 346, ed. Kopal, Academic Press (1965).
9. *Kirzhnitz D., Linde A.*—Ann. Phys. N.Y. **101**, 195 (1976).
10. *Linde A.*—Rept. Progr. Phys. **42**, 389 (1979).
11. *Lee T. D.*—Phys. Rev. **D8**, 1226 (1973).
12. *Zeldovich Ya. B., Kobzarev J, Okun L.*—ZhETF **67**, 3 (1974).
13. *Press W. Spontaneous Production of the Zeldovich Spectrum of Cosmological Fluctuations*, preprint, Harvard University, Cambridge, USA (1979).
14. *Hogan C. G. Particle Physics and Cosmic Homogeneity*, preprint, Institute of Astronomy, Cambridge (1979).
15. *Barenblatt G., Zeldovich Ya. B.*—Dokl. AN SSSR **130**, 115 (1959).

37

Complete Cosmological Theories*

With L. P. Grishchuk

1. Introduction

Modern cosmology successfully describes the main features of the Universe but uses for that some specific initial data which do not yet have a reasonable explanation. There is no theory capable not only of describing the observable world but giving also an answer why and for what reasons the Universe has these or other properties. The most important unsolved issue is the nature of the cosmological singularity whose existence is predicted by the classical (not quantum) relativistic theory of gravity. For a thorough analysis of the singularity one needs a quantum theory of gravity which is yet to be constructed. Because of the lack of such a theory one has to start doing cosmology from classical stages by introducing different initial conditions and comparing their consequences with real observations.

In this talk an attempt is suggested to give a rational explanation of the origin of those initial conditions which lead to the so-called standard cosmological model. Cosmological theories which pretend to describe the Universe from the "very beginning" (including the quantum-gravitational stage) and up to the present time we shall call complete. It is clear that there is not yet any finished theory of such a kind, but the very fact that there are some reasonable considerations and arguments (presented below) which permit one to make an outline of such a theory we consider as some progress. The scenario suggested below contains many separate ingredients known already in the literature; however the full picture, as far as we know, has not yet been presented. Being conscious of the vague nature of some of the suggested conclusions, we consider this talk as a challenge to the highbrow experts on quantum gravity gathered here, as an attempt to call their attention to the problems which appear to be important for cosmology.

*Preprint of the Institute for Space Research, AS USSR Pr-176, Moscow, 1982; an abbreviated version of the article was published in the book *Kvantovaĭa teoriĭa gravitatsii* [Quantum Theory of Gravitation], Moscow: AS USSR Nuclear Research Institute, 39–47 (1982) ; see also *Quantum Structure of Space and Time*, eds. M. J. Duff and C. J. Isham, Cambridge Univ. Press, Cambridge (1982), p. 409.

2. Present State of the Universe

It is useful to recall some properties of the observed world which seem to reflect conditions in the very early Universe. These properties can be used as landmarks in the process of extrapolating backward in time. On the other hand, they should represent the outcome of a complete cosmological theory.

It is believed that the most important properties of the observable world are the following [1–4]:

1. Large scale homogeneity and isotropy. The distribution of galaxies in phase space is such that they do not show any noticeable inhomogeneity or anisotropy after averaging over scales of the order of 100 Mpc. The most convincing manifestation of the large scale homogeneity and isotropy is the absence of angular variations in the microwave background radiation, $\Delta T/T \lesssim 10^{-4}$. The uniformity of the radiation emitted by the primordial plasma and coming to us from different directions is quite mysterious. According to the standard cosmological model elements of the primordial plasma sufficiently separated could not have been causally connected in the past even if one extrapolates their histories up to the cosmological singularity. More formally, one of the elements of the primordial plasma lies outside the particle horizon of the other element. This is known as the horizon problem [5].

2. Closeness of the mean density to the critical density. A variety of astronomical data indicate that the mean matter density, ρ_m, is less than the critical one, $\rho_c = 3/8\pi G H_0^2$, and $\rho_m \approx (10^{-1} - 3 \cdot 10^{-2})\rho_c$. Other data (for instance the possible existence of massive neutrinos) do not exclude the possibility that the total density ρ_t may in fact be of the same order of magnitude or even higher than ρ_c. In any case, the ratio $\Omega = \rho_t/\rho_c$ seems to be close to unity. In terms of Newtonian mechanics one can say that the kinetic energy of expansion is almost equal, at the present time, to the gravitational potential energy. Such a balance between kinetic and potential energy is quite mysterious since it implies that, say, in the epoch of the nucleosynthesis the equality of kinetic and potential energy was satisfied with an incredible relative accuracy, of the order of 10^{-15}. In terms of the relativistic equations this means that the spatial curvature term was dynamically negligible during all the past history of the Universe [6]. This fact is known as the flatness problem, though it would be more appropriate to call it the problem of the closeness of Ω to unity.

3. Existence of structure in the form of galaxies and their clusters. The Universe is obviously inhomogeneous at scales characteristic of galaxies and their clusters. According to contemporary views this structure was formed as a result of the growth of small initial perturbations. Near cosmological singularity, in the limit $t \to 0$, the initial perturbations of the metric tensor do not vanish and approach a constant value [7]. In order to produce the

actual structure, the initial perturbations had to have a specific spectrum and a specific amplitude. In particular, some arguments in favor of the fractal type spectrum [9] were given in [8]. Thus, a complete cosmological theory should contain an explanation of the overall properties of the Universe as well as origin of its structure.

4. Baryon asymmetry and specific entropy. The observable matter consists of baryons, and there is no indication in the Universe of any large amount of antibaryons. The ratio of the number density of photons, n_γ, to the number density of baryons, n_b, is $n_\gamma/n_b \approx 10^9$. The large value of the specific entropy $s = 4n_\gamma/n_b$ motivates the use of the words "hot Universe." Baryon asymmetry and the actual value of the specific entropy will probably find a satisfactory explanation within the scope of grand unified theories, provided the application of the standard cosmological model is justified up to the energies of the order of 10^{17} GeV. Generation of baryon asymmetry in an expanding Universe may become possible because of the CP-violation and the baryon number nonconservation. This problem is surveyed by Dolgov and Zeldovich [10].

3. Outline of a Complete Scenario

Let us present briefly a complete scenario, leaving the details and references for the other parts of the paper.

We assume that in the initial state there was nothing except zero (vacuum) fluctuations of all physical fields including gravitational. Since the notions of space and time are classical, in the initial state there were no particles, no space, no time. Speaking about time, one can say, loosely, that there was a time when there was no time.

It is assumed that as a result of a fluctuation there appeared a classical 3-dimensional closed geometry. The finiteness of the 3-volume is a necessary condition for such a process. Since there were no real particles yet, the dynamical evolution of this geometry was governed by vacuum polarization effects caused by the external gravitational field. For the first time there appears a notion of classical space-time. It is natural to expect that all the characteristic parameters of the world were Planckian, i.e., classical space-time comes into being at the limit of applicability of classical general relativity.

During some interval of its evolution the Universe existed as a space-time very close to that of De-Sitter. It is well known that De-Sitter space-time is as symmetric as Minkowski space-time; it admits a 10-parameter group of motions. The line element of De-Sitter space-time is

$$ds^2 = c^2dt^2 - a^2(t)[dr^2 + \sin^2 r(d\theta^2 + \sin^2\theta d\varphi^2)] \tag{1}$$

where $a(t) = r_0 \operatorname{ch}(ct/r_0)$, $r_0 = \text{const}$. In such a space-time the vacuum

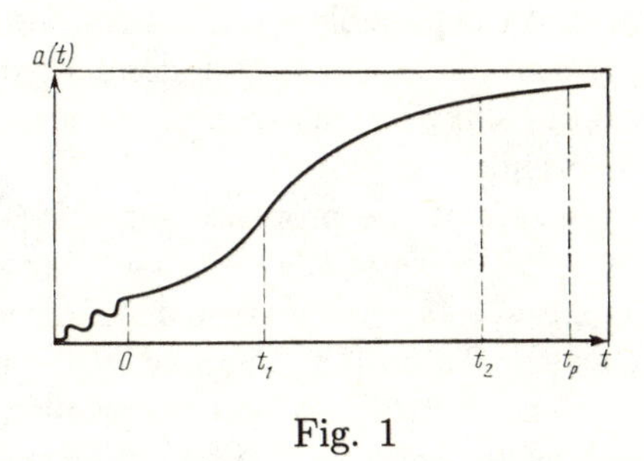

Fig. 1

polarization effects produce the expectation value $\langle 0|T_{\mu\nu}|0\rangle$ which is proportional to the De-Sitter metric tensor. Thus, De-Sitter space-time emerges as a self-consistent solution of the Einstein equations.

According to our scenario the moment of appearance of the Universe corresponds to $t = 0$ (see Fig. 1). It is reasonable to suppose that the world appeared with small deviations, $h_{\mu\nu}$, from the metric tensor (1). Similarly to the scale factor $a(t)$, perturbations $h_{\mu\nu}$ have also induced the vacuum polarization of all physical fields and were governed by it. In a closed space one can represent $h_{\mu\nu}$ as a discrete set of mode functions. Among them there is the lowest mode which corresponds to the perturbation of the scale factor: $a(t) \to a(t) + \delta a(t)$. It does not violate homogeneity and isotropy of the background space-time. In fact, it diminishes the size of the group of motions from 10 up to 6 parameters. The other modes introduce small inhomogeneities and anisotropies. As is usual in problems of this kind one can expect that after a certain time the lowest mode will be the most excited one. It may ensure the change of the scale factor everywhere in the space simultaneously while the higher frequency modes may provide the initial perturbations necessary for the formation of the large-scale structure in the Universe.

We assume that by the time $t = t_1$, during a short transition period, the expansion law of the Universe will change from De-Sitter to Friedman regime, $a(t) \sim t^{1/2}$, and the Universe will get filled with various particles having dominant equation of state $p_r = \varepsilon_r/3$. Such a drastic change could happen for the following reasons. The lowest mode, after a sufficient increase, ensures the transition of the scale factor everywhere in the space simultaneously. Apart from this, the perturbations $h_{\mu\nu}$ make possible the creation of different particles by the external gravitational field. As is known, particle creation is a quadratic effect in $h_{\mu\nu}$, while the vacuum polarization is already present in the linear approximation. Created particles fill the Universe with matter. One can expect that, after $t = t_1$, the Universe will be described by the standard radiation-dominated cosmological model, which contains small perturbations of matter, radiation and gravitational field.

All these processes occur long before the time when the baryon number asymmetry is generated and the possible phase transitions take place. Thus, early nucleosynthesis, transition to the matter dominated stage, growth of density perturbations, and galaxy formation could proceed according to the usual ideas about these events.

What are the achievements of the proposed scenario? First, it answers the question about the initial state of the Universe. According to this scenario, the Universe originates, in a certain sense, from nothing (the first idea of this kind was expressed by Tryon [12]). Second, it includes a sufficiently long De-Sitter stage, which resolves the horizon problem and the $\Omega \approx 1$ problem similar to the way it occurs in a different scenario considered by Guth [6]. Third, it provides a natural explanation of the origin of the density perturbations with spectrum and amplitude necessary for the formation of the observed large scale structure. These perturbations can grow from the initial fluctuations $h_{\mu\nu}$ appearing in the De-Sitter stage. Moreover, it appears to be sufficient to take $h_{\mu\nu}$ with the minimal possible amplitude, i.e., at the level of quantum fluctuations.

4. Spontaneous Birth of the Closed World

A widespread point of view is that the cosmological singularity can be avoided by quantum effects in an external gravitational field. This becomes possible since the vacuum polarization effects violate the energy dominance condition so important for the singularity theorems [13]. Singularity-free solutions of this kind do exist but they are unstable; they cannot be extrapolated back in time up to $t = -\infty$. We take another point of view. We assert that the cosmological singularity must be replaced by an essentially quantum-gravitational process which can be called spontaneous birth of the Universe.

A quantitative formulation of this process can be, probably, done by using Feynman's path integral approach. For quantum gravity it was developed by Wheeler [14], Hawking [15] and others.

The states of the gravitational field can be characterized by 3-dimensional geometries—the initial one, $g_1^{(3)}$, and the final one $g_2^{(3)}$. The probability amplitude for passing from the initial state to the final state is given by the path integral

$$\left\langle g_2^{(3)} \mid g_1^{(3)} \right\rangle = \int d[g^{(3)}] \exp[iI(g^{(3)})]$$

For simplicity we restrict ourselves to Friedman-Robertson-Walker metrics

$$ds^2 = c^2 dt^2 - a^2(t)\left[dr^2 + f^2(r)\left(d\theta^2 + \sin^2\theta d\varphi^2\right)\right].$$

(Such a restriction brings to mind Penrose's [16] hypothesis on the vanishing of the Weyl tensor at the initial singularity.) As the initial state we accept a 3-geometry with zero volume, i.e., $a_i = 0$. As the final state we accept a 3-geometry with finite volume, of the order of the cube of Planckian length, i.e., $a_f \approx ct_{\rm pl}$. (In the line element (1) this state corresponds to $t = 0$, $r_0 = ct_{\rm pl}$.) Compactness of the 3-space is a necessary condition for such a transition. Indeed, the probability amplitude A can be expressed as

$$A \sim \int d[g^{(3)}] \exp[-\int L\, dvdt]$$

A nonvanishing value for A can only be obtained if the integration over 3-volume gives a finite quantity. Thus, only a closed world could, probably, appear as the result of such a quantum-mechanical jump [17]. (It is worth mentioning that after introducing the notion of probability in the context of the "creation of the Universe" one can hardly avoid speculations on the validity of the anthropic principle which selects one of them. For recent discussion see [18].)

An additional argument in favor of such a process is the fact that the birth of a closed world preserves the quantum numbers of the vacuum: zero value for the total energy, total electric charge, etc. [12]. According to present views the baryon charge is not strictly a conserved quantity. Therefore, the creation of the Universe with zero total baryon charge does not prevent the generation of baryon asymmetry during subsequent evolution. Transition to the final state characterized by the presence of classical gravitational field can be probably viewed as a kind of spontaneous breakdown of symmetry in the spirit of modern quantum field theory. However, in our case, the field which acquires a nonzero classical value is not a field given in a space-time but the space-time (gravitational field) itself.

In some papers the birth of an expanding Universe is regarded as the result of quantum-mechanical tunneling from an initial configuration which is considered as the ground state of the system. Specifically, the question has been studied whether quantum decay of Minkowski space-time or De-Sitter space-time with constant cosmological Λ-term is possible [19–21]. In our opinion these suggestions have the following disadvantages. First, they do not avoid the question of the origin of the initial classical space-time. Second and more important, within the scope of standard general relativity, Minkowski and De-Sitter space-times inside the event horizon possess minimal energy and, therefore, are stable classically as well as quantum-mechanically [22, 23]. Nevertheless, in the context of more complicated theories (like Kaluza-Klein theory, for example) the ground state ($M^4 \times S^1$) may happen to be unstable with respect to quantum mechanical tunneling [24, 25].

5. *Intermediate De-Sitter Stage*

According to the scenario the created world was governed by the vacuum polarization of all physical fields. It means that its dynamical evolution was described by a self-consistent solution of the Einstein equations:

$$R_{\mu\nu} - \frac{1}{2} g_{\mu\nu} R = \frac{8\pi G}{c^4} \langle 0 | T_{\mu\nu} | 0 \rangle \tag{2}$$

The evaluation of the right-hand side of these equations and the search for self-consistent solutions, in particular, De-Sitter solutions, were undertaken in the works [26–30, 11]. A. A. Starobinskiĭ [11] considered self-consistent solution as a means to avoid the cosmological singularity. He has also pointed out the possibility of transition of such a solution into the Friedman stage (see also [31]). In our scenario, the De-Sitter stage has not only an end but also a beginning. It is an intermediate stage which joins the act of the quantum birth of the Universe with the radiation.

In the papers mentioned above the expectation value $\langle 0 | T_{\mu\nu} | 0 \rangle$ was determined by the conformal anomaly of the full energy-momentum tensor. At the present time it is not clear if the conformal anomaly is an inevitable feature of quantum field theory in curved space-time (see [32, 33]). However, the appearance of a De-Sitter stage for one reason or another seems justifiable anyway since the necessary source term in (2) can be provided by massive or massless nonconformal particles.

An intermediate De-Sitter stage is also a likely feature of another scenario appealing to the Higg's fields and the phase transitions in the early Universe [6,34-37]. However, De-Sitter stage of that origin could only occur much later in the history of the Universe and we expect that it will hardly change the consequences of the previous De-Sitter stage governed by the vacuum polarization effects.

Since the De-Sitter solution may be so important for cosmology it is worthwhile to recall certain properties of it.

It is well known [37] that the De-Sitter solution can be represented as a hyperboloid with arbitrary radius r_0, embedded into 5-dimensional pseudo-Euclidean space-time. The frame of reference (1) covers the whole hyperboloid, i.e., the whole De-Sitter space-time. Spatial sections $t =$ const are 3-dimensional spheres. The section $t = 0$ is different from the others in that it has the minimal volume. However the points of this section are by no means distinguished from other points of the space-time. Because of the high symmetry of De-Sitter space-time it can be covered by an infinite number of other frames of reference where the metric will again have the form (1) but the spatial section of minimal volume will be different. In fact such a section can include an arbitrary point of the space-time and can have an arbitrarily oriented time-like normal vector in it.

One or another choice of time (i.e., choice of spatial sections) is very important since, by assumption, at certain moments changes will occur in the expansion law, and one particular moment of time, $t = 0$, marks the act of spontaneous creation of the world. Which spacelike sections should be chosen? According to views put forward here, the created world was closed.

Therefore, the privileged spatial sections are closed spaces corresponding to $t = \text{const}$ in the presentation (1) of De-Sitter space-time. The 3-sphere $t = 0$ is physically distinguished by the very event of creation. It seems natural to assume that the transition to the Friedman stage occurs at a hypersurface $t = t_1$ which is "parallel" to the hypersurface $t = 0$.

If the self-consistent De-Sitter solution was governed by the conformal anomaly then it follows from the Einstein equations (2) that $r_0 = c/H$, where H^{-1} is proportional to t_{pl} and depends on contributions of all fields. It is important to note that if the duration of the De-Sitter stage was 70–100 characteristic periods H^{-1} then, as will be shown below in detail, the horizon and $\Omega \approx 1$ problems could be completely eliminated.

6. Transition to the Friedman Stage

Ignoring, at the present time, the possible causes by which the De-Sitter stage is replaced by a Friedman one, let us consider the joining of these solutions at $t = t_1$ (see Fig. 1). During the De-Sitter stage vacuum polarization effects give an effective energy density ε_v and pressure p_v where $\varepsilon_v = 3c^2H^2/8\pi G$ and $p_v = -\varepsilon_v$. During the Friedman radiation-dominated stage the equation of state is $p_r = \frac{1}{3}\varepsilon_r$ and the energy density behaves as $\varepsilon_r \sim a^{-4}$. After transition, at $t = t_2$, to matter-dominated stage the density of matter goes like $\rho_m \sim a^{-3}$. Let us denote the present values of density and scale factor by ρ_p, a_p.

The $\Omega \approx 1$ problem is solved as follows: Ω_p satisfies the relation

$$\Omega_p - 1 = \frac{k}{E-k}, \qquad E \equiv \frac{8\pi G \rho_p a_p^2}{3c^2},$$

where $k = +1$ for a closed world. In order to have $\Omega_p \approx 1$ it is necessary that E be of the order of unity. By using the junction conditions one obtains $\rho_p a_p^2 = \rho_2 a_2^3/a_p = \rho_1 a_1^4/a_2 a_p$. At $t = t_1$ one has $a_1 = r_0 chHt_1 \approx \frac{c}{2H}e^{Ht_1}$ and $\rho_1 = 3H^2/8\pi G$. Assuming $a_p \approx 10^{29}$ cm, $c/H \approx 10^{-32}$ cm, $a_2 \approx 10^{-4}a_p$ one concludes from the equation $E/2 \gtrsim 1$ that $e^{Ht_1} \gtrsim 10^30$ which means that there should be $Ht_1 \gtrsim 70$. It follows that Ω_p will be of the order of unity if $Ht_1 \gtrsim 70$.

The horizon problem is eliminated by the fact that the particle horizon increases practically up to the future event horizon already during the time interval from $t = 0$ to $t = t_1$ (i.e., during the De-Sitter stage). Indeed, by the time $t = t_1$, a light ray emitted at $t = 0$ covers the distance $l \approx a_1(\frac{\pi}{2} - 2e^{-Ht_1})$. Therefore, all the particles now accessible for observations could have been in causal contact long ago. It should be mentioned that the usefulness of the De-Sitter solution for eliminating the horizon and $\Omega \approx 1$ problems was first considered by Guth [6] though in his case De-Sitter solution was realized in a different cosmological epoch and for different reasons.

7. Small Perturbations and Galaxy Formation

Similarly to the background De-Sitter solution small variations to it also satisfy equations (2). It is convenient to introduce the variable η by the relation $d\eta = cdt/a(t)$. Then, $a(\eta) = r_0/\cos\eta$. If the function $f(\eta)$ represents a perturbation with the proper wavenumber n then the typical equation for it is

$$f'' + f\left[n^2 - \frac{2+C^2}{\cos^2\eta}\right] = 0$$

where C^2 is a constant depending on the actual parameters of the conformal anomaly. Equations of this form occur often in the theory of amplification of classical waves and quantum particle creation in external gravitational fields. On the basis of past experience it can be expected that perturbations with low eigenvalues will guarantee the departure of the scale factor from that of the De-Sitter scale factor, while the (density) perturbations with large n (which correspond to the size of the present day clusters of galaxies) will increase up to the necessary amplitude. This important issue is not yet worked out though some encouraging suggestions have been proposed in recent papers [38–40].

8. Localized Creation of an Open World

The instability of the background De-Sitter solution with respect to the lowest mode, $a(t) \to a(t)+\delta a(t)$, implies that the transition to the radiation-dominated stage will occur simultaneously in the whole space. In this case the spatial sections corresponding to the co-moving frame of reference during the Friedman stage will be closed. However, it cannnot be excluded that an instability could be related to the fact that in the polarized vacuum there is $\varepsilon_v > 0$, $p_v > 0$, and zero entropy, $s = 0$. It can happen that for this medium it will be energetically more favorable to form an expanding bubble [42] inside of which there will be $\varepsilon_r > 0$, $p_r > 0$, $s > 0$ after certain time. The walls of the bubble expand almost with the velocity of light and "cut out" a region of the background space-time restricted by a light cone. It is natural to think that, inside this region, the transition to the radiation-dominated stage will take place at some moment of time $\tau = \tau_1 = \text{const}$, where τ is measured along the world-lines which emanate from the event $\tau = 0$ where the localized perturbation first originated. But the spatial sections $\tau = \text{const}$ are hyperbolic, open. In terms of τ—time a piece of De-Sitter space-time inside the light cone can be described by

$$ds^2 = c^2 d\tau^2 - r_0^2 \text{sh}^2 c\tau/r_0 [d\chi^2 + \text{sh}^2\chi(d\theta^2 + \sin^2\theta d\varphi^2)].$$

By assumption, the perturbation has originated at the point $\tau = 0$, $\chi = 0$. If the transition happens at $\tau = \tau_1$ then the spatial sections with uniform density, pressure and temperature will be open. It means that in this case we would find ourselves living in an open universe. Again, in order to have $\Omega_p \approx 1$, i.e., in order that the Universe not be "too open," it is necessary to have a transition to the radiation-dominated stage not earlier than at $\tau = \tau_1$, where $e^{H\tau_1} \approx 10^{30}$.

Does a possibility of such a localized perturbation mean that the idea of the spontaneous birth of the world is superfluous? Can one maintain the conception of a complete De-Sitter space-time which exists from $t = -\infty$ and which is a background for many localized perturbations transforming later in many open "universes"? Such an idea is considered by Gott [21]. It seems to us that the idea of the spontaneous birth of a closed world is necessary, i.e., one cannot regard the De-Sitter stage as existing from $t = -\infty$. This conclusion is connected with the observation that the origin of localized perturbations should be characterized by a finite probability per unit of time per unit of volume. Because of the full symmetry of De-Sitter space-time this probability should be the same at all world points. But in this case the total probability is infinite since the integration should include all previous moments of time, up to $t = -\infty$. One can avoid this difficulty only by assuming that the classical stage of the complete De-Sitter solution starts not from $t = -\infty$ but from some finite $t = 0$.

9. *Gravitons and Gravitinos in the Early Universe*

Apparently the only way to check the hypothesis on physical conditions in the very early Universe will be the search for a primordial gravitational-wave background. Gravitational waves and gravitons (in contrast to other known massless fields and particles) have the remarkable ability to be amplified and to be created by conformally flat gravitational field [43]. As a consequence of this process a nonthermal background of gravitons should exist now. Predictions on their spectrum and intensity depend on specific properties of the gravitational field in the very early Universe [43,44]. These conclusions about gravitons follow from standard general relativity; however, they also hold in supergravity theories [45]. Contemporary experimental power is not sufficient for detection of such a gravitational wave background; however the situation can, hopefully, improve in the near future.

We are indebted to B. DeWitt, S. Hawking, A. Polyakov, V. G. Mukhanov, G. V. Chibisov, A. D. Linde, J. Hartle and especially A. A. Starobinskiĭ for valuable discussions.

REFERENCES

1. *Zeldovich Ya. B.*—Adv. Astron. and Astrophys. **3**, 241 (1965).
2. *Zeldovich Ya. B., Novikov I. D.* Stroenie i evolyutsiia Vselennoĭ [Structure and evolution of the Universe]. Moscow: Nauka, 735 pp. (1975).
3. *Weinberg S.* Gravitation and Cosmology. New York: Wiley (1972).
4. *Peebles P.* The Large-Scale Structure of the Universe. Princeton Univ. Press (1980).
5. *Misner C. W., Thorne K. S., Wheeler J. A.* Gravitation. San-Francisco: Freeman and Co. (1973).
6. *Guth A.*—Phys. Rev. **23D**, 347 (1981).
7. *Lifshitz E. M.*—ZHETF **16**, 587 (1946).
8. *Zeldovich Ya. B.*—Month. Not. RAS **160**, 1 (1972).
9. *Mandelbrot B. B.* Fractals: Form, Chance and Dimension. San-Francisco: Freeman and Co. (1977).
10. *Dolgov A. D., Zeldovich Ya. B.*—Rev. Mod. Phys. **53**, 3 (1981).
11. *Starobinskiĭ A. A.*—Phys. Lett. **9113**, 99 (1980).
12. *Tryon E. D.*—Nature **246**, 396 (1973).
13. *Zeldovich Ya. B., Pitaevskiĭ L. P.*—Comm. Math. Phys. **23**, 185 (1971).
14. *Wheeler J. A.* Geometrodynamics. New York: Acad. Press (1962).
15. *Hawking S. W.*—Nucl. Phys. **B138**, 349 (1978).
16. *Penrose R.*—In: General Relativity/ Ed. S. W. Hawking, W. Israel. Cambridge Univ. Press (1979).
17. *Zeldovich Ya. B*—UFN **133**, 479 (1981).
18. *Rees M. I.*—Quart. L. Roy. Astron. Soc. **22**, 109 (1981).
19. *Brout R., Englert F., Gunring E.*—Ann. Phys. **115**, 78 (1978).
20. *Atkatz D., Pagels H.*—Prepr. Rep. Numb. RU81/B/2 (1981).
21. *Gott J. R.*—Prepr. Princeton (1981).
22. *Witten E.*—Comm. Math. Phys. **80**, 381 (1981).
23. *Abbot L. F., Deser S.*—Prepr. TH 3136—CERN (1981).
24. *Witten E.*—Prepr. Princeton (1981).
25. *Lapedes A., Mottola E.*—Prepr. (1981).
26. *Dawker J. S., Critchley R.*—Phys. Rev. **D 13**, 3224 (1976).
27. *Davies P. C. W.*—Phys. Lett. **68B**, 402 (1977).
28. *Davies P. C. W., Fulling S. A., Christensen S. M. et al.*—Ann. Phys. **109**, 108 (1977).
29. *Fischetti M. V., Hartle J. B., Hu B. L.*—Phys. Rev. **D20**, 1757 (1979).
30. *Mamaev S. G., Mostapenko V. M.*—ZhETF **78**, 15 (1980).
31. *Gurovich V. Ts., Starobinskiĭ A. A.*—ZhETF **77**, 1699 (1979).
32. *DeWitt B. S.*—In: General Relativity/Ed. S. W. Hawking, W. Israel. Cambridge Univ. Press (1979).
33. *Christensen S. M., Duff M. J., Gibbons G. W. et al.*—Phys. Rev. Lett. **45**, 161 (1980).

34. *Linde A. D.*—Prepr., Lebedev Inst. N 229, Moscow (1981).
35. *Linde A. D.*—Prepr., Lebedev Inst. N 265, Moscow (1981).
36. *Hawking S. W., Moss I. G.*—Prepr. Cambridge, DAMTI (1981).
37. *Hawking S. W., Ellis G.* The Large Scale Structure of Space-Time. Cambridge Univ. Press (1973).
38. *Mukhanov V. F., Chibisov G. V.*—Pisma to ZhETF **33**, 549 (1981).
39. *Kompaneets D. A., Lukash V. N., Novikov I. D.*—Prepr. Space Research Inst., AS USSR Pr-652 (1981).
40. *Starobinskiĭ A. A.*—Pisma to ZhETF **34**, 460 (1981).
41. *Gurovich V. Ts.*—Pisma to Astron. Zh. **8**, 532 (1982).
42. *Colemen C.*—Phys. Rev. **D15**, 2929 (1977).
43. *Grishchuk L. P.*—ZhETF **67**, 825 (1974).
44. *Starobinskiĭ A. A.*—Pisma to ZhETF **30**. 719 (1979).
45. *Grishchuk L. P., Popova A. D.*—ZhETF **77**, 1665 (1979).

Commentary

Besides the papers noted in the text, several additional articles should be mentioned which contain statements on the possibility of "gemmation" of the closed universe from flat space-time[1] and early statements (before the well-known work by Guth)[2] on the cosmological significance of a "vacuum-like" equation of state.[3,4]

The subsequent development of ideas associated with a quantum origin of the Universe are related to the construction of the "wave" function of the Universe[5] and to the analysis of the role of classical scalar fields.[6,7]

[1] *Fomin P. I.*—Dokl. AN USSR **A 9**, 831 (1975).
[2] *Guth A.*—Phys. Rev. **230**, 347–356 (1981).
[3] *Gurevich L. E.*—Astrophys. Space Sci. **38**, 67–78 (1975).
[4] *Gliner E. B., Dymnikova I. G.*—Pisma v Astron. Zh. **1**, 7–9 (1975).
[5] *Hartle J. B., Hawking S. W.*—Phys. Rev. **28D**, 2960–2975 (1983).
[6] *Linde A. D.*—ZhETF **87**, 369–374 (1984).
[7] *Linde A. D.*—Rep. Progr. Phys. **47**, 925 (1984); UFN **144**, 177–214 (1984).

37a
Inflationary Stages in Cosmological Models with a Scalar Field*

With V. A. Belinskiĭ, L. P. Grishchuk, I. M. Khalatnikov

Homogeneous isotropic cosmological models with a massive scalar field are studied. It is shown that inflationary stages of evolution are characteristic of most solutions in these models.

We provide an analysis of all solutions of gravity equations for a homogeneous isotropic universe with a scalar field. Scalar fields of different nature are used as a mechanism ensuring the inflationary (de Sitter) stage in the evolution of the early universe (A survey of the up-to-date state of the theory of the inflationary universe and references to preceding works is given in [1]). The role of the inflationary stage in the solution of cosmological problems is clearly described in [2].

Here we deal with the simplest case of a massive (with mass m) field φ in the universe with scale factor $a(t)$. For the energy density and pressure of the field we have

$$\varepsilon = \frac{1}{2}\dot{\varphi}^2 + \frac{1}{2}m^2\varphi^2, \quad p = \frac{1}{2}\dot{\varphi}^2 - \frac{1}{2}m^2\varphi^2. \tag{1}$$

It is clear from these formulas that at $\dot{\varphi}^2 \ll m^2\varphi^2$ the effective equation of state is $p = -\varepsilon$, which points to the possibility of the emergence of the quasi-de Sitter stage, i.e., the stage at which $a(t)$ grows according to a law close to the exponential law. In the opposite case at $\dot{\varphi}^2 \gg m^2\varphi^2$ and, in particular, at $m = 0$ we have the equation of state for stiff matter $p = \varepsilon$. Cosmological solutions for this case are studied in [3]. This equation has been derived in [4] for a different case. Finally in the regime of oscillating φ the averaged p turns to zero. In other words, in different regimes the homogeneous field φ possesses different equations of state. This paradoxical situation has been studied in [5] on the basis of the model where gravitation of the field φ is ignored, the Hubble parameter $H = \dot{a}/a$ is assumed to be constant and $|H| \gg m$. At $H =$ const the equation for φ (see eq. (2) below) is solved exactly. In [5] it is shown that at expansion, i.e., at $H > 0$, all solutions for the field φ (except one) behave so that the equation of state tends to

*Physics Letters **155B**, 232–236 (1985); see also Zhurnal eksperimentalnoĭ i teoreticheskoĭ fiziki **89**, 346-360 (1985).

$p = -\varepsilon$. At contraction, *vice versa*, all solutions (except one) behave so that the equation of state tends to $p = \varepsilon$.

It has already been mentioned in [1] that if there is a field φ in the early universe with values exceeding $m_P = G^{-1/2}$, where m_P is the Planck mass and if, besides, the initial value of $\dot{\varphi}$ is sufficiently low, then the quasi-de Sitter expansion stage is realized. However, problems concerning the degree of generality of solutions which possess an inflationary stage, the emergence of these stages at large initial values of $\dot{\varphi}$, quantitative characteristics of "favorable" and "unfavorable" cases, modifications in the set of possible solutions due to nonzero spatial curvature, and finally the problem of the probability of the different solutions have not been investigated. To settle down part of these problems we employ the methods of the qualitative theory of dynamical systems.

We shall consider solutions of the classical equations for $\varphi(t)$ and $a(t)$ at all t, up to singularities, which are typical of the cosmological models, although in the closed universe there are also some solutions without singularities. The values of the fields at which $\varepsilon \approx m_P^4$ (which we call the quantum boundary) can be regarded as the boundary of applicability for classical solutions. The initial data for the classical stages of the evolution should be given on this boundary. These data can be represented as a consequence of the solution corresponding to the quantum gravity problem in the region $\varepsilon \gtrsim m_P^4$. Here we shall stick to the ideas of [6] according to which the cosmological singularity should be somehow replaced by "an act of spontaneous creation of the universe."[1]

Leaving for a while problems of quantum mechanical analysis and the discussion of the probability of these or those initial data on the quantum boundary, it is important to find out the fate of the solutions characterized by arbitrary initial values of φ and $\dot{\varphi}$ in the classical regime. Of particular interest are closed models since for them the concept of "creation" is more sensible. The results of this work testify to the fact that the inflationary stage is a fairly general property of solutions under study and thus the concept of quantum "creation" of the universe with a subsequent inflationary stage does not require that any particularly special conditions should be imposed.

The equations of gravity and the scalar field are:

$$\ddot{\varphi} + 3H\dot{\varphi} + m^2\varphi = 0, \tag{2}$$

$$\dot{H} + H^2 = \frac{m^2}{9s^2}(m^2\varphi^2 - 2\dot{\varphi}^2), \tag{3}$$

$$H^2 + ka^{-2} = \frac{m^2}{9s^2}(m^2\varphi^2 + \dot{\varphi}^2),$$

[1]Some articles have already been devoted to the problem of quantum creation of the universe (see [7] and references therein).

where $s = (12\pi)^{-1/2} m m_{\mathrm{P}}$ and $k = 0, +1, -1$.These equations form a three-dimensional system in the phase space φ, $\dot{\varphi}$, H. It is, however, more convenient to use these variables alongside with the dimensionless quantities x, y, z, η: $\varphi = 3sm^{-1}x$, $\dot{\varphi} = 3sy$, $H = mz$, $t = m^{-1}\eta$. The surface of the cone $x^2 + y^2 - z^2 = 0$ corresponding to the case $k = 0$ separates the regions containing trajectories of the open model and the closed model. Trajectories of the flat model lie on the cone.

Trajectories can cross the plane $H = 0$ only at $k = 1$, which corresponds to the moments of extrema of the scale factor ($a = 0$). For models with $k = -1, 0$ expansion and contraction cannot alternate in classical solutions and they should be considered separately. All trajectories describing contraction at $k = -1, 0$ can be obtained from trajectories describing expansion by means of symmetry transformations which the original system of equations possesses

$$\text{(a)} \qquad \eta \to -\eta, \quad z \to -z, \quad x \to -x,$$

$$\text{(b)} \qquad \eta \to -\eta, \quad z \to -z, \quad y \to -y, \tag{4}$$

At $k = 0$ the system is simplified. We shall study only expanding models, $H > 0$. The trajectories lying on the cone are mapped onto the x,y-plane. In the finite region of variation of x, y there is only one singular point—the origin $(x, y) = (0, 0)$. It is a stable focus and the asymptotics of the solution in its vicinity $\eta \to +\infty$ is

$$x = \frac{2}{3\eta}\sin(\eta - \eta_0), \quad y = \frac{2}{3\eta}\cos(\eta - \eta_0),$$

$$z = \frac{2}{3\eta}, \quad \eta_0 = \text{const}, \tag{5}$$

i.e., describes the final stages of expansion without bounds. All the remaining singular points lie at infinity and to find out their properties it is possible to employ the standard method of constructing the "Poincaré sphere."

The results are illustrated in Fig. 1. Fig. 1a shows the behavior of trajectories in the x,y-plane and Fig. 1b shows the mapping of this phase diagram onto a circle with unit radius. On the boundary of this circle corresponding to infinity, $x^2 + y^2 = \infty$, the system has 4 singular points: 2 repulsive knots K_1 and K_2 and two saddle points S_1 and S_2. In the center of the circle there is the stable focus F with the asymptotics (5). All trajectories go out from infinitely remote singular points and then are woven around F. Near K_1 we have

$$\varphi = \frac{s}{m}\ln\frac{t}{t_0}, \quad H = \frac{1}{3t}, \quad t \to +0, \tag{6}$$

where $t_0 = \text{const}$.

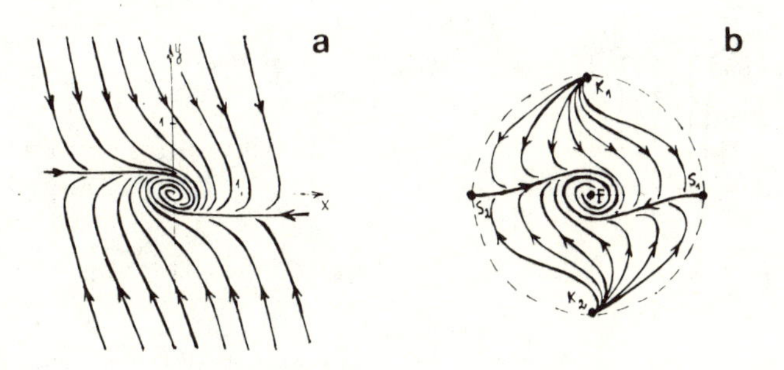

Fig. 1

Near S_1 the asymptotics of the solution corresponding to the outgoing separatrix S_1F is

$$\varphi = -st, \quad \dot{\varphi} = -s, \quad H = -\frac{1}{3}m^2 t, \quad t \to -\infty.$$

In the $\varphi,\dot{\varphi}$-plane the separatrix S_1F has a horizontal asymptote to which S_1F tends at $\varphi \to \infty$.

The asymptotics near K_2 and S_2 as well as the properties of the separatrix S_2F are similar and follow from the symmetry properties.

Since the equation of state $p = -\varepsilon$ is realized in the vicinity of S_1F at sufficiently large values of φ (specified below), it is in this region that the emergence of inflationary stages should be expected. Under $\dot{\varphi}^2 \ll m^2\varphi^2$ (equivalent to $|\dot{H}| \ll H^2$) we approximately have

$$H = \frac{m^2}{3s}\varphi. \tag{7}$$

From (2) and (7) it is possible to obtain the ratio of $a(t)$ at a certain initial moment t_i to $a(t)$ at a certain finite moment t_f. Since

$$\frac{a(t_f)}{a(t_i)} = \exp\left(\int_{t_i}^{t_f} H(t)\, dt\right)$$

then

$$\frac{a(t_f)}{a(t_i)} = \left|\frac{\dot{\varphi}(t_i) + s}{\dot{\varphi}(t_f) + s}\right|^{1/3};$$

hence it is clear that a considerable growth of $a(t)$ is only possible for those trajectories which approach the separatrix S_1F closely enough at moment t_f.

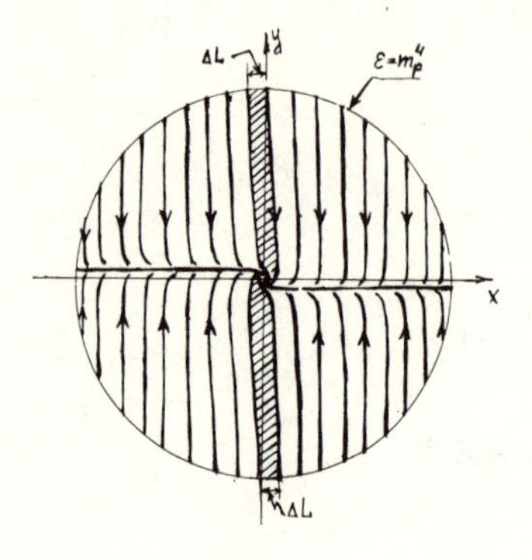

Fig. 2

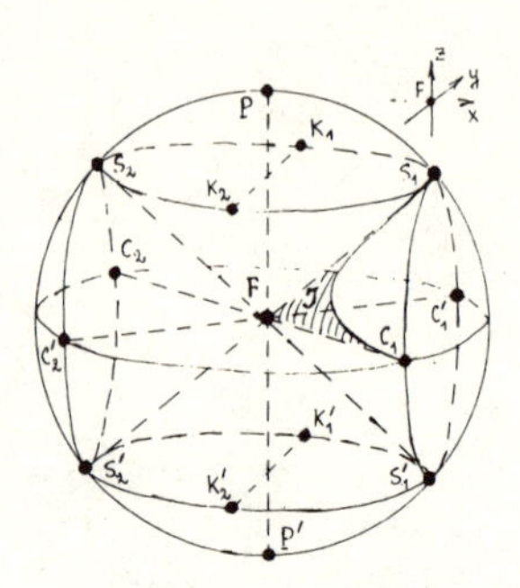

Fig. 3

Now let us confine ourselves to the study of parts of trajectories inside the quantum boundary. It is known that at the inflationary stage $a(t_f)/a(t_i)$ should be of the order of 10^{30}. The ratio of this order can be obtained only on those parts of trajectories which start at $|\dot{\varphi}+s| \sim m m_{\mathrm{P}}$ and terminate at $|\dot{\varphi}+s| \sim 10^{-90} m m_{\mathrm{P}}$. As for parts of trajectories where $|\dot{\varphi}|$ changes from values $\sim m|\varphi| \sim m_{\mathrm{P}}^2$ up to values $|\dot{\varphi}+s| \sim m m_{\mathrm{P}}$, $a(t_f)/a(t_i)$ does not exceed the value $(m_{\mathrm{P}}/m)^{1/3}$ on these parts. Since additional cosmological constraints require that $m/m_{\mathrm{P}} \sim 10^{-5} - 10^{-6}$ then the growth of $a(t)$ by $(m_{\mathrm{P}}/m)^{1/3}$ times is actually negligibly small in comparison with the growth necessary. It is also possible to show that on the parts of trajectories where $\dot{\varphi}^2 \gg m^2\varphi^2$, the growth of $a(t)$ is also very small and it is determined by the same value $(m_{\mathrm{P}}/m)^{1/3}$.

The ratio $a(t_f)/a(t_i)$ can also be expressed in terms of the initial and finite values of φ:

$$\frac{a(t_f)}{a(t_i)} = \exp[2\pi m_{\mathrm{P}}^{-2}(\varphi_i^2 - \varphi_f^2)]$$

If we take $\varphi_f \sim m_{\mathrm{P}}$, then it suffices to have $\varphi_i \sim (3-4)m_{\mathrm{P}}$ for the inflationary stage to be realized (see [1]).

The problem to what extent the inflationary stage is a property characteristic of these solutions is most simply solved by means of a precisely drawn phase diagram in the region inside the quantum boundary. The phase diagram at large scales (up to the quantum boundary) is represented by Fig. 2. The trajectories are practically vertical beyond the central region with radius $\sim m_{\mathrm{P}}$ and turn sharply near the separatrices and go along them as far as the central region where they are woven around F.

Favorable trajectories (i.e., possessing the required inflationary stage) are those for which $\varphi_i > (3-4)m_{\mathrm{P}}$. These trajectories start everywhere on the circumference of the quantum boundary except for two of its sections near the y-axis (see Fig. 2). The number of unfavorable trajectories, i.e., starting from points at these sections is small since the length of each of these sections $\Delta L \sim (6-8)m_{\mathrm{P}}$ whereas the length of the whole quantum circumference is $L \sim 2\pi m_{\mathrm{P}}^2 m^{-1}$. Thus the ratio $\Delta L/L$ determining the measure of unfavorable trajectories is

$$\frac{\Delta L}{L} \approx \frac{m}{m_{\mathrm{P}}} \ll 1.$$

This conclusion testifies the inflationary regimes in models with $k = 0$ to be very general. Note that these results are also qualitatively valid for the variant of the scalar field with a potential $\sim \lambda\varphi^4$.

Let us discuss models with $k \neq 0$. In the finite region of variation of x, y, z the system has only one equilibrium state—the origin $(x, y, z) = (0, 0, 0)$. The remaining singular points lie at infinity. The phase space can be made compact and can be represented as a sphere of unit radius with the center at the origin. There are 14 singular points on the surface of the sphere (see Fig. 3). But only 4 of them are essentially different: one point of each group, P, K, S and C. The properties of all remaining points are obtained by an appropriate combination of symmetry transformations (4).

A two-dimensional bunch of trajectories (a two-dimensional separatrix) goes out from the saddle point P inside the phase space. Near P we have

$$\varphi = \varphi_0 - \frac{1}{8}m^2\varphi_0 t^2 \qquad (\varphi_0 = \text{const}),$$

$$H = t^{-1}, \qquad a \sim t, \qquad t \to +0.$$

Among these solutions there is a purely vacuum solution whose trajectory is the axis PF. From the saddle points S_1 and S_2 only one trajectory goes out inside the sphere, i.e., the separatrices of the flat model $\mathrm{S}_1\mathrm{F}$ and $\mathrm{S}_2\mathrm{F}$. The points K_1 and K_2 eject trajectories both into the phase space of the flat model and into the space of the open and closed models. These knots correspond to initial cosmological singularities and the asymptotics near them is universal (for K_1 see (7)).

Four additional singular points of saddle character are located at the equator of the sphere. The asymptotics of the solutions near C_1 and C_2 is

$$\varphi = \text{const} \cdot \exp(-mt/\sqrt{2}), \quad H = \frac{m}{\sqrt{2}},$$

$$a \sim \exp(mt/\sqrt{2}), \quad t \to \infty.$$

Now the quantum boundary is the surface of a cylinder, $x^2+y^2 = 8\pi m_{\mathrm{P}}^2/3m^2$ ($\varepsilon = m_{\mathrm{P}}^4$). For the non-flat models the surface of the boundary is two-dimensional since in comparison with the case $k = 0$ it is necessary to specify another independent parameter, i.e., the initial value of H or a.

It is evident that the trajectories of the expanding models with $k = -1$, starting at the surface of the cylinder but close to its intersection with the cone, will have qualitatively the same properties as the trajectories of the flat model. Most of them approach and then go by along the separatrices S_1F and S_2F and are subjected to prolonged inflation. Part of the remaining trajectories will approach the cone in the regions of the separatrices S_1F and S_2F inside the classical region and will experience an inflationary regime. Others will reach the region of oscillations near the focus F without experiencing inflation.

Thus the inflationary stage is an inevitable intermediate stage of most solutions of the open model, although it is difficult to give a numerical estimate of their measure.

In the case of the closed model $k = 1$ trajectories may cross the plane $H = 0\,(z = 0)$. There appear points of regular maxima $a(t)$ ($a_{\max}$) or minima $a(t)$ ($a_{\min}$). These points are strictly separated. It is possible to show that any trajectory intersecting the plane $z = 0$ in the region $x^2 > 2y^2$ has $a_{\min}$ there. The trajectories intersecting the plane $z = 0$ at points $x^2 < 2y^2$ have $a_{\max}$ at these points. On any section of the trajectory inside the cone with the equation $x^2 - 2y^2 - z^2 = 0$ whose surfaces are positioned horizontally H grows with time. Outside the cone H falls whereas on the cone itself $\dot{H} = 0$.

To find out the degree of generality of inflationary stages in the closed model it is important to find the points of the cone $\dot{H} = 0$ where trajectories can go out. Determining a curve at the surface of the cone where $\ddot{H} = 0$ it is possible to show that in the expansion phase $H > 0$ and at $\varphi > 0$ an exit from the cone is only through a narrow region on its surface denoted by J (shaded in Fig. 3). At large values of $|\varphi|$ which we are mainly interested in, the trajectories going out from the cone enter the vicinity of the separatrices S_1F and S_2F and the respective solutions experience a prolonged inflationary regime. Other trajectories approach S_1F and S_2F without entering inside the cone $\dot{H} = 0$. We assume that almost all solutions possess the required inflationary stage.

Alongside with the solutions studied there are numerous trajectories along which it is possible to pass directly (avoiding the region near the separatrices S_1F and S_2F) from the initial singularity K to the collapse K$'$, either through one expansion maximum or by experiencing a finite number of oscillations of the scale factor between $a_{\max}$ and $a_{\min}$.

A closed chain of trajectories on the boundary of the phase space of the system (e.g., $S_1FS_2'S_2FS_1'S_1$) implies the possibility of the existence of periodic solutions represented by closed trajectories in the vicinity of this chain. The realization of these solutions requires a special choice of initial data since on the section FS_2', for example, going towards the saddle points S_2' a trajectory should go along the unstable solution FS_2' for a long time. It is possible to show that the trajectories of the solutions which are t-symmetric with respect to $a_{\max}$ and $a_{\min}$ cross the plane $z = 0$ at the points $x = 0$ and $y = 0$ correspondingly. The existence of solutions which are t-symmetric with respect to both $a_{\max}$ and $a_{\min}$ is also possible. This has been previously mentioned in [8]. The possible existence of infinitely oscillating nonperiodic solutions has been discussed in [9]. As has been asserted in [8], periodic solutions represent a discrete set. Under such conditions the existence of trajectories of which one end starts or terminates in the singularities K, K′ and the other end is woven around one of the periodic solutions also becomes probable. Besides, the existence of trajectories performing a finite number of rotations around the periodic solutions and passing successively from one to another is not excluded. The existence of classes of trajectories with the above-described properties in a dynamical system usually leads to a stochastic character of the regime of its behavior.

Thus we come to the conclusion that the inflationary stage is a fairly general property of the system considered, whereas a perpetually bouncing universe (even without additional known physical objections) corresponds to a degenerate case. The investigation of more realistic cases, i.e., inhomogeneous initial conditions and creation of particles, amplifies the validity of the proof concerning the low probability of the creation of the bouncing universe.

After we had completed this work, we became aware of [10] where similar problems are dealt with.

Institute for Physical Problems
USSR Academy of Sciences

Received
January 21, 1985

REFERENCES

1. *Linde A. D.*—Rep. Prog. Phys. **47**, 925 (1984).
2. *Guth A.*—Phys. Rev **D23**, 347 (1981).
3. *Belinskiĭ V. A., Khalatnikov I. M.*—ZhETF **63**, 1121 (1972).
4. *Zeldovich Ya. B.*—ZhETF **41**, 1609 (1961).
5. *Zeldovich Ya. B.*—Soviet Scientific Review **7**, ed. R. Sunyaev. Harwood (1988).
6. *Grishchuk L. P., Zeldovich Ya. B.*—In: Quantum Structure of Space and Time, ed. M. Duff and C. Isham, London: Cambridge Univ. Press (1982).
7. *Linde A. D.*—ZhETF **87**, 369 (1984).
8. *Hawking S. W.*—Prepr. DAMTP, July (1983).
9. *Page D.*—Prepr., Physics Dept., Pennsylvania State Univ. (1984).
10. *Kofman L. A., Linde A. D., Starobinskiĭ A. A.*—Phys. Lett., to be published.

VI

Neutron Stars and Black Holes

Accretion

38

The Fate of a Star and the Release of Gravitational Energy Under Accretion*

It is well known that cold matter cannot resist the compressive action of gravitational attraction if its mass is much greater than the mass of the Sun. This result was obtained by Oppenheimer and Volkoff [1] in 1938 in a study of the degenerate neutron gas in the Einstein theory of gravitation (the "general theory of relativity") and is to be found in textbooks [2]. This result is not affected qualitatively by any assumptions concerning the interactions of elementary particles in the presence of high matter density.

The general theory of relativity makes a radical modification of the picture of the dynamics of condensation. In the classical theory an infinite density is attained after a finite time; one may suppose that an expansion results from this, or a shock wave arises, traveling outward from the center and ejecting a portion of the matter.

As was shown by Oppenheimer and Snyder [3], the general theory of relativity leads to the conclusion that an infinite density is indeed reached after a finite proper time (as measured by an observer moving with any particle of the star). However, one must take the change in the time scale into account if signals are propagated between the star and an external observer located

*Doklady Akademii Nauk SSSR **155** (1), 67–69 (1964).

outside the star's gravitational field. The red shift of lines emitted from the surface of a star is a special case of this change in the time variation. It turns out that for an external observer, the external surface of a star only attains the so-called gravitational (Schwarzschild) radius $r_g = 2GM/c^2$ asymptotically at $t \to \infty$. For every particle (for example, the central one) it is possible to determine the moment at which the particle must emit a signal which will arrive at the external observer as $t \to \infty$. One may speak in this sense of the gravitational self-collapse of a star [4].

At the moment of emission of the signal, the density for each particle is less than the characteristic value

$$\rho_g = \frac{3M}{4\pi r_g^3} = \frac{3c^6}{32\pi M^2 G^3} = 1.8 \cdot 10^{16} \left(\frac{M}{M_\odot}\right)^{-2} \frac{\text{g}}{\text{cm}^3}.$$

As a result, the attainment of an infinite density during the condensation is not observable, and the acquisition of information is accomplished long before this moment. Hence, the question of what happens after $\rho = \infty$ has even less meaning.

All of these conclusions remain valid when pressure is taken into account [5], even for the case of hot matter. One should regard the entropy of the matter as constant in studying the problem of the existence and stability of mechanical equilibrium in the presence of pressure and gravitational attraction. Equilibrium corresponds to a minimum of the total energy for a given entropy (and also, of course, for fixed number of conserved particles—the baryons). The entropy may be considered as constant also during a rapid collapse.

For each value of entropy S there is a number of equilibrium configurations of a hot gas (stars) of different masses M; such configurations exist, however, only for masses less than the critical mass. The value of the critical mass M_c increases with increasing S; $M_c = M_c(S)$. For $S = 0$, $M_c = M_c(0) \cong M_\odot$. Stars with $M > M_\odot$ may be found in a state of mechanical equilibrium to the extent that they are burning and to the extent that, the densities being equal, the pressure of the hot matter exceeds that of the cold matter [6].

As a result, the final stage in the evolution of every nonrotating star whose mass is much greater than that of the Sun consists of an uncontrollable condensation, i.e., a collapse.*

The brightness of a hot star drops off exponentially quite rapidly during gravitational self-collapse, with a time constant of the order r_g/c, i.e., 10^{-4} s for $M \sim 10M_\odot$. As a result, collapsed stars must be dark bodies which interact with the surrounding medium only via their gravitational field. Since the star's radiation, and therefore its loss in mass, during the collapse is small [4, 7], the gravitational field of a collapsing star at large distances is not different from the field of the same star prior to condensation. The question has been previously raised [8] as to what portion of all nucleons in the Universe exist

*See commentary.

at a given moment in dark condensed stars. From considerations relating to the problem of the growth of the Universe, one obtains only an inequality for the overall density $\bar{\rho} < 2 \cdot 10^{-28}$ g/cm^3, whereas the density for normal, visible stars is $\rho \cong (0.3 - 1) \cdot 10^{-30}$ g/cm^3. Hoyle, Fowler, and Burbidge [7] have decisively raised the question of whether a large number of dark stars is present. According to their estimate the mass of such stars is several times larger than the mass of the luminescent stars, which puts the mean density close to the upper limit [8]. The interest in the catastrophic condensation of stars arose in connection with the discovery of optically intense distant radio sources [9, 10], and Hoyle's hypothesis [11] that these sources are superstars with masses of the order of $10^8\, M_\odot$. Until recently, the source of the energy of the powerful radio galaxies was not understood; ideas such as the annihilation of matter and antimatter, collisions between galaxies, and the simultaneous explosion of many supernova turned out to be untenable. How can the condensation of superstars lead to the emission of necessarily gigantic quantities of energy? According to Hoyle *et al.* [7] the condensation is accompanied by fluctuations, whose density reaches 10^{30} g/cm^3; ultrarelativistic particles are emitted when the density reaches its maximum. It is impossible to agree with this point of view, merely because the break at $\rho_m = 10^{30}$ g/cm^3 depends on a hypothetical κ-field with very strange properties [7, 12]. For an external observer the growth in the density as a result of gravitational self-collapse stays asymptotically at a quite modest value of the order of 2-200 g/cm^3 (for $M = 10^8 - 10^7\, M_\odot$).

An alternative mechanism of energy emission is examined, in the present note, which is associated with an infall of the external mass in the gravitational field of a collapsing star.

As a freely falling particle approaches the gravitational radius its velocity approaches that of light.

The relative velocity is of the order c in the collision of two particles at a distance of the order (no less!) of the gravitational radius. The energy radiated by relativistic particles during a collision can therefore be αmc^2; the energy carried out to infinity is less because of the red shift and the geometrical factor: a part of the radiation and the particles falls to the star. The net amount of energy which escapes is $\alpha\beta mc^2$, where m is the mass of the particle, $\alpha < 1$, $\beta < 1$. This product attains a maximum which is of the order 0.1 mc^2 for a collision at $r = 1.5\, r_g$. It is essential that the colliding particles have different angular momenta relative to the star; the numbers given refer to the case $\mathbf{M}_1 = -\mathbf{M}_2$. If the collisions are infrequent, the energy emission is proportional to the collision cross section for a given velocity distribution of the particles far away from the star. If the cross section is large, making the mean free path small compared to the dimensions of the star, the motion of the particles is governed by collisions, making it necessary to use a hydrodynamic description of the motion. At the same time it appears that for spherically symmetric motion the gravitational energy is transformed principally into kinetic energy of radial motion; the amount of energy which can be radiated amounts to a negligibly small fraction of the rest mass of the falling matter.

However, if a star is embedded in a directed supersonic mass flow at a great distance from the star, the picture of the motion is radically changed. A stationary shock wave develops on the side of the star opposite the direction of the incident matter flow. Near the star the velocity change at the front of the wave is of the order of c and an appreciable portion of rest mass of the matter being compressed by the wave is radiated. According to Bernoulli's theorem, not even a small part of the matter can be ejected to infinity with a velocity greater than the initial velocity far from the star. But when the cloud of matter arrives at the star, surrounds it and strikes the rear side, the motion ceases to be stationary and the cumulative ejection of part of the matter with a velocity of the order of c becomes possible.

In conclusion we note that the "particles" of which we have been speaking are not necessarily atoms and molecules, but may be localized plasma concentrations with a frozen magnetic field; the production of relativistic electrons by collisions then becomes especially probable. When examining the flow of matter from a distance one should evidently not visualize it as interstar gas and dust with $\bar{\rho} = 10^{-25}$ g/cm^3, which would give a small output.

After the collapse of one member of a closely bound pair of stars, one may regard the matter of the other member as the falling material. This can be that part of the shell of the collapsing star itself which was ejected before the instant of gravitational self-collapse; in addition to the matter which acquires a hyperbolic velocity, part of the ejected matter can be stored in distant, but closed orbits.

The idea of infall in a powerful gravitational field as a source of the radiated energy of radiosources was advanced in its most general form by I. S. Shklovskiĭ [13].

We take this opportunity to express our sincere gratitude to I. D. Novikov and I. S. Shklovskiĭ for numerous discussions.

REFERENCES

1. *Oppenheimer J., Volkoff G.*—Phys. Rev. **55**, 374 (1938).
2. *Landau L. D., Lifshitz E. M.* Statistical Physics. Reading, MA: Addison-Wesley (1969).
3. *Oppenheimer J., Snyder H.*—Phys. Rev. **56**, 455 (1939).
4. *Zeldovich Ya. B.*—Astr. Tsirkulyar **250**, July 1 (1963).
5. *Podurets M. A.*—Dokl AN SSSR **154** 2 (1964).
6. *Zeldovich Ya. B.*—Vopr. Kosmogonii **9**, 80 (1963).
7. *Hoyle F., Fowler W., Burbidge G., Burbidge M.*—Relativistic Astrophysics, preprint (1963).

8. *Zeldovich Ya. B., Smorodinskiĭ Ya. A.*—ZhETF **41**, 907 (1961).
9. *Matthews T., Sandage A.*—Publ. Astron. Soc. Pacific **74**, 406 (1962).
10. *Schmidt M.*—Astrophys. J. **136**, 684 (1962); Nature **197**, 1040 (1963).
11. *Hoyle F., Fowler W.*—Nature **197**, 533 (1963).
12. *Hoyle F., Narlikar J. V.*—Proc. Roy. Soc. **273** 1352, 1 (1963).
13. *Shklovskiĭ I. S.*—Astron. Zh. **39**, 591 (1962).

Commentary

In the problem of accretion onto a white dwarf or neutron star the release of energy is, in the final analysis, related to the falling of matter onto the surface. In the case of a black hole there is no such mechanism; an individual particle falls, carrying with it all the energy.

It was necessary to bring into consideration the collision of at least two particles. This paper even proposes a realistic variant: in a binary system the ordinary star provides a flow of gas around the black hole, with the formation of a radiating shock wave in the downstream flow.

In this article, for the first time the idea of also detecting black holes by luminescence of matter moving in their strong gravitational field was proposed. This idea decisively changed the attitude of observers to the problem of black holes. Earlier, black holes had been considered as non-observable consequences of an unproven extrapolation of an unreliable theory. After the paper by Ya. B. and O. Kh. Guseinov (**42**) on the celestial mechanics of a binary system with a black hole and particularly after the paper above, the search and study of black holes in our Galaxy become the most important task of stellar astronomy. Within the framework of this task a concrete mechanism was indicated for the radiation of energy in the falling of a gas which makes a black hole observable. It works in the case of a supersonic flow of gas around a black hole. The radiation is related to the formation of a radiation shock wave behind the black hole. It was noted that this mechanism may work in binary star systems where one of the components is a black hole.

The paper played a great role in establishing the theory of accretion and the theory of compact X-ray sources. Today the energy-release mechanism proposed in the above article is considered to be one of the three most important variants of accretion flows in the neighborhood of relativistic stars together with disk or spherically symmetric accretion (see, e.g., the reviews).[1,2]

At the same time, and independently, the idea was published by E. Salpeter.[3]

We note an inaccuracy at the beginning of the article: a star with a large mass may not be transformed into a black hole if in the course of evolution (slow and explosive) it loses a sufficient portion of its mass. On the other hand, a neutron star may be transformed into a black hole by accumulating mass in the course of accretion. The problem of the overall number of black holes in our and other galaxies has still not been solved.

[1] *Lightman A.P., Shapiro S. L., Rees M. J.*—In: Phys. and Astrophys. of Neutron Stars and Black Holes, 786–827 (1978).

[2] *Sunyaev R. A.*—In: Phys. and Astrophys. of Neutron Stars and Black Holes, 697–763 (1978).

[3] *Salpeter E. E.*—Astrophys. J. **140**, 796–800 (1964).

38a
An Estimate of the Mass of a Superstar*

With I. D. Novikov

Analysis of the observational data on superstars like 3C 273 shows [1] that the optical continuum spectrum of these objects is emitted by a central object with a size of the order of $2 \cdot 10^{16}$ cm, while the emission lines arise in an outer envelope with a scale of a few parsecs or more. The radiation is probably produced in the innermost central region of the central object, which we shall call the nucleus. The plasma surrounding the nucleus we shall conventionally call the atmosphere. In order to estimate the mass of the superstar, we shall consider the forces acting on the plasma atmosphere. We shall assume that the force of gravity is balanced by the force of radiation pressure. This assumption, some basis for which will be given below, leads, as we shall show, to a mass for a superstar of the order of $10^8\, M_\odot$.

The initial quantity is the total luminosity of the object. According to [2], for 3C 273 the flux of energy at the surface of the Earth is $q_e = 3.5 \cdot 10^{25}\,\mathrm{erg\,cm^{-2}\,s^{-1}}$*. The redshift $\lambda_1/\lambda_2 = 1.158$, with $H = 100\mathrm{km\,s^{-1}\,Mpc^{-1}}$ corresponds to a distance $R = 5 \cdot 10^8$ pc, which gives the total luminosity of the object[1]

$$Q = 4\pi R^2 q_e = 2 \cdot 10^{46}\ \mathrm{erg\,s^{-1}}. \quad (1)$$

For other objects the total luminosity has the same order of magnitude.

At a distance r from the center the flux is $q = Q/4\pi r^2$. Compton scattering of photons by electrons occurs in the plasma. Due to this process, the time-averaged force acting on a single electron is

$$\bar{F}_1 = \frac{\sigma q}{c}, \quad (2)$$

where the cross-section $\sigma = 6.7 \cdot 10^{-25}\ \mathrm{cm}^2$.

We make the following obvious remarks: this expression does not depend on the frequency of photons as long as $\hbar\omega \ll mc^2$, and does not depend on the angular distribution of photons. Indeed, one can obtain the expression (2) by considering one electron, on which photons from a central point source

*Doklady Akademii Nauk SSSR **158**, 811–814 (1964).

*This value is incorrect. It should be $6.7 \cdot 10^{-10}$ erg/s · cm²—*Editors' note.*

[1]Cosmological corrections are in this case not large, but they are substantial for more distant objects.

are falling. Let us denote the density of photons by n_ω. The energy flux $q = n_\omega c\hbar\omega$; the number of scatterings per unit time is $\nu = n_\omega c\sigma$; in each collision the photon gives up its momentum $p = \hbar\omega/c$; the average momentum of a scattered photon being zero, the force $F_1 = p\nu$; whence (2) follows.

We now consider the opposite case of diffusion of radiation in a medium whose optical depth exceeds unity. In this case the flux

$$q = -D\frac{d\varepsilon}{dr},$$

where ε is the energy density of radiation, and the diffusion coefficient

$$D = \frac{1}{3}lc = \frac{1}{3}\frac{c}{n_e\sigma},$$

where n_e is the electron density.

The force acting on an electron is

$$F_1 = -\frac{1}{n_e}\frac{dP}{dr},$$

Substituting $P = \varepsilon/3$, for almost isotropic radiation, we again obtain expression (2). In highly ionized plasma, Compton scattering appears as a fundamental process, causing a radiation drag.[2] Due to the force preserving electroneutrality, a mass μ/A, $A = 6 \cdot 10^{23}$ accompanies each electron.

Equating the force of gravity and the force of radiation drag on each electron,

$$\frac{GM\mu}{Ar^2} = \frac{\sigma q}{c} = \frac{\sigma Q}{4\pi c r^2},$$

we obtain the mass of the body: $M = 3 \cdot 10^{41}$ g $= 10^8 M_\odot$ (for $\mu = 1$).*

We have essentially used the usual equation of hydrostatic equilibrium for a star in which radiation pressure exceeds gas pressure. As is clear from the foregoing, in order for this to be applicable it is not necessary in reality to have a body with a large optical depth and a nearly isotropic radiation field, for which Pascal's law holds. Furthermore, it is not absolutely necessary for the star as a whole to be in a state of hydrostatic equilibrium. In the approximation that gas pressure may be neglected, the plasma is in a state of neutral equilibrium for *any* radial distribution $n = n(r)$. The condition of equilibrium does not determine the density of matter in the neighborhood of the center; the condition depends only upon the directly observed energy flux, irrespective of its mechanism of generation.

The well-known changes in the flux from 3C 273 indicate that an optical depth of order 1 is reached at a comparatively short distance from the center. For subsequent estimates, we shall take a radius of 1 light-week—i.e., $r = 2 \cdot 10^{16}$ cm. At this distance $q = 4 \cdot 10^{12}$ erg cm^{-2} s^{-1}, which corresponds to $T_{\rm eff} = 1.6 \cdot 10^4$ K $= 1.5$ eV.

[2] We call this force radiation drag, to emphasize that it is caused by the flow of energy relative to matter.

* *Editor's note.* Zeldovich appears to have rediscovered the "Eddington Limit" here.

We shall estimate the density of matter at r_0 from the condition that a large increment of distance gives an optical depth of order unity:

$$n_e \sigma \Delta r = 1 \tag{3}$$

Substituting $\Delta r = r_0$, we find $n_e = 10^8\,\text{cm}^{-3}$.

We shall imagine that the density of matter changes considerably over a distance of order r. Then the gradient of gas pressure is of order

$$\frac{dP_m}{dr} \approx \frac{P_m}{r} = n_e kT \frac{1 + 1/\mu}{r}.$$

The force due to one electron is

$$F_g = kT \frac{1 + 1/\mu}{r}.$$

Substituting $r = 2 \cdot 10^{16}$ cm and $T_{\text{eff}} = 1.6 \cdot 10^4$ K, we arrive at the conclusion that this force is a million times less than that of radiation pressure. For smaller radii it grows as $r^{-5/4}$ for $\rho = \text{const}$, or as $r^{-7/4}$ for $\rho \sim r^{-2}$; i.e., more slowly than gravity and radiation drag. We will show that in such conditions one can neglect all other processes of interaction of matter with light (except for Compton scattering).

We will consider a hydrogen plasma. Compton scattering will play the determining role, provided

$$\frac{\sigma n_e}{\bar{\sigma}_{ph} n_1} > 1, \tag{4}$$

where $\bar{\sigma}_{ph}$ is the effective cross-section for photoionization of a hydrogen atom in the given conditions; n_1 is the density of neutral atoms. Since the color temperature [3, 4] (see also [5]) corresponds to the effective temperature at the distance r_0, one can use Saha's formula to estimate n_e/n_1.[3]

By definition, the effective cross-section $\bar{\sigma}_{ph}$ is

$$\bar{\sigma}_{ph} = \int_0^\infty q_\nu \sigma_{ph\,\nu}\, d\nu \Bigg/ \int_0^\infty q_\nu\, d\nu.$$

For $kT \ll I$ (I being the ionization potential of hydrogen),

$$\bar{\sigma}_{ph} \approx 10^{-18} (I/kT)^3 e^{-I/kT};$$

for $kT \gg I$,

$$\bar{\sigma}_{ph} \approx 10^{-18} (I/kT)^3.$$

The result is that the condition (4) becomes

$$n_e < 10^{12} \Theta^{9/2}\,\text{cm}^{-3}, \tag{5}$$

where Θ is the temperature in electron volts. The inclusion of trace of heavy elements and free-free transitions slightly reduces the numerical coefficient

[3] According to the estimate of I. S. Shklovskiĭ [6] and the data of [5], the spectrum has a nonthermal character; this changes the estimates of n_e/n_1 and $\bar{\sigma}_{ph}$. However the final inequality (5) is fulfilled by a large margin, and no reasonable variation in n_e/n_1 and $\bar{\sigma}_{ph}$ changes the conclusions.

on the right-hand side of the inequality (5). We see that for n_e found from the condition (3), inequality (5) is fulfilled by a huge margin.

We will now consider the regions with $r < r_0$. The following estimates are of an illustrative character and have been made with certain assumptions. The inequality turns out to be fulfilled by a large margin, so the final conclusion is not in doubt. Here the optical depth is large and the temperature distribution is determined by the equation of radiative diffusion

$$\frac{d}{dr}\left(\frac{1}{4}acT^4\right) = \frac{3}{4}\frac{\sigma\mu}{A}\rho\frac{Q}{4\pi r^2},$$

where a is Stefan's constant.

In a strictly stationary state with constant velocity the mass flux is constant and $\rho \sim r^{-2}$. We shall assume such a density distribution. Then from the equation of diffusion it follows that for $r < r_0$ one gets, instead of (5), the expression

$$n_e < 10^{12}\Theta_0^{9/2}(r_0/r)^{27/8}\,\mathrm{cm}^{-3},$$

where Θ_0 is the temperature at r_0.

Since the left-hand side of the inequality is proportional to r^{-2}, this condition is even better fulfilled for all $r < r_0$. Consequently everywhere for $r < 2 \cdot 10^{16}$ cm Compton scattering is the dominant process of interaction of radiation with matter.

We shall now turn to the regions with radius greater than a light week. Here the degree of ionization is determined by the formula with a factor for the dilution of radiation $W \approx r_0^2/4r^2$:

$$\frac{n_e}{n_1} = \left(\frac{n_e}{n_1}\right)_0 \frac{(n_e)_0 W}{n_e} e^{-\tau}, \tag{6}$$

where $(n_e/n_1)_0$ is the degree of ionization at r_0 and τ is the optical depth. For the effective cross-section $\bar{\sigma}_{ph}$ we will take $\bar{\sigma}_{ph} = \bar{\sigma}_{ph}(r_0) \approx 10^{-19}\,\mathrm{cm}^2$. If $\rho \sim r^{-2}$ for $r > r_0$, then the right-hand side of (6) is constant (τ is small) and condition (4) is fulfilled—i.e., here too Compton scattering dominates. If, however, the distribution of plasma is such that beyond some radius r the condition (4) is not satisfied, then the radiation drag becomes much greater than gravity, and the matter is driven away outside. For example, let $n_e = \mathrm{const} \approx 10^5\,\mathrm{cm}^{-3}$, as this follows from Shklovskii's estimate for the outer envelope [1].[4] Then $n_e/n_1 \approx (2.5 \cdot 10^{44}/r^2)e^{-\tau}$. Substituting this value into (4), we find that at $r \approx 10^{19}$ cm, photoionization begins to dominate and the matter is driven out.

One may suppose that the source of the radiated energy is accretion, i.e., the infall of matter, in the plasma state, into the immediate neighborhood

[4]According to [5], $n_e \approx 10^7\,\mathrm{cm}^{-3}$. In this case the outer envelope cannot be uniform and it appears to be a thin hollow sphere with $R = 10^{19}$ cm [6]. Our estimates are concerned with an atmosphere, i.e., with a deeper region, not with this envelope; we emphasize their illustrative nature.

of a massive collapsed nucleus [7]. We recall that for $M = 10^8 M_\odot$, the gravitational (Schwarzschild) radius is $3 \cdot 10^{13}$ cm, i.e., it occupies only a small fraction of the conjectured outer radius of the plasma. Assuming that in accretion it is possible to radiate away up to 10% of the rest energy of the infalling material, we obtain the expenditure of matter needed to pay for the energy loss:

$$\frac{dm}{dt} = \frac{Q}{c^2}\frac{1}{0.1} = 2 \cdot 10^{26}\,\mathrm{gm\,s^{-1}} = 3\mathrm{M}_\odot\,\mathrm{yr}^{-1}.$$

If the plasma is hanging, supported by radiation drag, then it is conceivable that reduction in the flux of radiation will cause a collapse of the plasma, i.e., an increase in accretion. This will once again increase the radiation flux, which reduces the accretion: the system possesses a mechanism of self-regulation. This mechanism can result in oscillations. For a rough estimate of their period, one should take the orbital period of a particle in the field of a mass $10^8\mathrm{M}_\odot$ at a distance r_0, which gives $t \approx 3\,\mathrm{yr}$. It would be interesting to equate this period with the period $\sim 3\,\mathrm{yr}$ observed in 3C 273 and 3C 48 [8, 9]. Finally, we will note that the plasma has an azimuthal instability which does not depend on the gravitational attraction of its separate parts.

Let us imagine that $n_e = n_e(r) f(\theta, \phi)$ depends on the angle. Then the radiation flux also can be proved to depend on angle, $q = (Q/4\pi r^2)\psi(\theta, \phi)$, ψ being larger where f is smaller. The equilibrium will be broken; the plasma will be driven out where it is more rarefied, and will sink where it is denser. Gas pressure and magnetic forces will try to smooth out the plasma.

The principal results are the following:

1. The assumption that radiation pressure balances the force of gravity in the atmosphere of a superstar leads to an estimate for the mass of the nucleus of a superstar of $\approx 10^8\,\mathrm{M}_\odot$.
2. In the outer parts of the atmosphere, where the pressure of radiation on matter is principally determined by photoionization processes and not by electron scattering, the radiation forces abruptly increase, the equilibrium is broken, and the outer parts of the atmosphere are blown off.
3. A possible mechanism for the generation of the superstar's energy is the accretion of the matter of the atmosphere in the immediate vicinity of a collapsed nucleus. The amount of accreted material necessary to provide the observed luminosity is $\approx 3\,\mathrm{M}_\odot\,\mathrm{yr}^{-1}$. This mechanism is self-regulating.

Received
July 16, 1964

REFERENCES

1. *Shklovskiĭ I. S.*—Astron. Zh. **41**, 176 (1964).
2. *Oke J.*—Nature **197**, 1040 (1963).
3. *Arkhipova V. P., Kostyakova Ye. V., Sharov A. S.*—Astr. Tsirkulyar **251** (1963).
4. *Sandage A.*—Ap. J. 139, 416 (1964).
5. *Dibai E. A., Pronik V. I.*—Astr. Tsirkulyar **286** (1964).
6. *Shklovskiĭ I. S.*—Astron. Zh. **41**, 801 (1964).
7. *Zeldovich Ya. B.*—Dokl. AN SSSR **155**, 67 (1964).
8. *Sharov, Yu. N. Efremov*—Inform. Bull. Var. Stars **23**, Commission 27 of IAU (1963); *Sharov A. S., Efremov Yu. N.*—Astron. Zh. **40**, 950 (1963).
9. *Smith H. J., Hoffleit D.*—Nature **198**, 650 (1963).

39
Nuclear Reactions in a Super-Dense Gas*

It is shown that nuclear reactions occurring in cold hydrogen at densities of 10^4–10^6 g/cm^3 via barrier penetration proceed with a probability which is quite noticeable on an astrophysical scale. This fact puts a limit on the possible compression of cold hydrogen, since a celestial body cannot last more than 10^8 years at a density of $0.7 \cdot 10^5$g/cm^3. Such a density is reached in cold hydrogen under the action of gravitation for a mass close to that of the Sun.

It is known [1–3] that the thermonuclear reactions $p + p = \mathrm{D} + e^+ + \nu$, $p + \mathrm{D} = \mathrm{He}^3 + \gamma$ occur in stars; in addition, at high temperature we have the reaction $\mathrm{He}^3 + \mathrm{He}^3 = \mathrm{He}^4 + p + p$, and at high density the reactions $\mathrm{He}^3 + e^- = \mathrm{T} + \nu$, $\mathrm{T} + p = \mathrm{He}^4 + \gamma$.

Schatzman [4] first noted that at high densities and low temperatures deviations from the usual expressions for the rate of thermonuclear reactions should begin to appear, and gave general expressions for the case of a degenerate nuclear gas.

Let us consider reactions occurring in hydrogen compressed to a density of 10^4–10^6 g/cm^3, at a temperature below 10^6 degrees. Under these conditions hydrogen is a solid, i.e., that the protons are located at the lattice points of a crystal; thermal motion and thermonuclear reactions can be neglected.

It is well known that the Coulomb repulsion of the nuclei, which hinders nuclear reaction, can be overcome by virtue of the quantum-mechanical phenomenon of barrier penetration (tunnel effect). The probability of tunneling is exponentially small, and the exponent depends more strongly on the width than on the height of the barrier.

The separation at which nuclear reaction begins is of order $3 \cdot 10^{-13}$; at this distance the Coulomb energy of two protons is 0.5 Mev. At a hydrogen density of 10^4–10^6 g/cm^3, the distance between protons is $6.5 \cdot 10^{-10}$–$1.35 \cdot 10^{-10}$ cm (1000 times as great as nuclear distances and only 10–50 times smaller than the distance between the nuclei in a hydrogen molecule). At this distance, the Coulomb energy of two protons is 400–2000 ev.

Thus a compression to such densities hardly lowers the height of the barrier; however, the wide low energy region, which is most difficult to overcome,

*Zhurnal eksperimentalnoĭ i teoreticheskoĭ fiziki **33** (4), 991–993 (1957).

is gotten rid of through the internal pressure, and the rate of the nuclear reaction $p+p=\mathrm{D}+e^{+}+\tilde{\nu}$ turns out to be entirely perceptible on an astronomical scale. To calculate the reaction rate at zero temperature and a given density, we find the equilibrium distance r_0 between nuclei at the densest packing. Energy as a function of the distance between nuclei r we approximate by the expression $e^2/r+e^2/(2r_0-r)$. The initial energy is $2e^2/r_0+E_0$, where $E_0=\hbar\omega/2=e\hbar\sqrt{Mr_0^3}$ is the zero point vibration energy, and M is the proton mass.

The exponential factor for barrier penetration is

$$B=\exp\{-\varphi(4e/\hbar)\sqrt{2M_r r_0}\},$$

where M_r is the reduced mass, equal to $M/2$, and φ is determined by numerical integration. For small ε,

$$\varphi=0.70-0.17\varepsilon\ln(1/\varepsilon),\quad \varepsilon=E_0r_0/e^2.$$

We determine the probability $\psi^2(0)$, for finding two protons at the same point in the absence of nuclear interaction, approximately, by assuming that we have a diatomic molecule [5] with equilibrium separation r_0 and zero point oscillation energy E_0

$$\psi^2(0)=BE_0M_r^2e^2/\hbar^4.$$

Finally we determine the reaction rate by using Salpeter's [2] calculations: the reaction probability is $p(\mathrm{s}^{-1})=w\psi^2(0)$. The constant w is related to the reaction cross section σ (for the case where the reaction occurs in the laboratory, using a beam of particles), by the formula (for singly-charged ions),

$$\sigma=\frac{w}{v}\frac{2\pi e^2}{hv}\exp\left(-\frac{2\pi e^2}{hv}\right)=w\frac{\pi Me^2}{h}\frac{B}{E}=\frac{S}{E}B.$$

where v is the velocity, E the energy, and S is given in [2], formula (7). Comparing formulas (9) and (39) of [2], we find $S=4\cdot10^{-19}$ barn · ev, so that $w=5\cdot10^{-40}\,\mathrm{cm}^3/\mathrm{s}$. After the reaction of formation of deuteron, the formation of He^3 should follow extremely rapidly. At the densities we are considering, the maximum energy of electrons E_e (in the degenerate Fermi gas) is greater than the tritium decay energy; therefore the reaction $\mathrm{He}^3+e^-=\mathrm{T}+\nu$ also goes rapidly [2], and after it the reaction $p+\mathrm{T}=\mathrm{He}^4+\gamma$. Thus each initial reaction of two protons soon leads to the liberation of ~ 25 Mev of energy. A summary of the results for two densities is given in the table.

ρ, g/cm^3	r_0, cm	E_0, ev	$-\log B$	p, s^{-1}	H, $\frac{\mathrm{erg}}{\mathrm{g\cdot s}}$	E_e, ev	P, $\frac{\mathrm{dyne}}{\mathrm{cm}^2}$	T, ev
$5\cdot10^5$	$1.65\cdot10^{-10}$	125	8.5	$4\cdot10^{-17}$	1000	$2.6\cdot10^5$	$5\cdot10^{22}$	400
$0.7\cdot10^5$	$3.1\cdot10^{-10}$	45	11.6	10^{-20}	0.25	$0.7\cdot10^5$	$0.2\cdot10^{22}$	150

Here H is the rate of energy liberation, P is the pressure of the degenerate electron gas, T is that temperature at which the same reaction rate would be reached through the thermonuclear mechanism [2].

The considerations presented above impose a limit on the possible compression of cold hydrogen. In fact, even at a density of $0.7 \cdot 10^5$ (the lower line in the table), with zero initial temperature we will reach a temperature of 150 v after $6 \cdot 10^7$ years, and the speed of the reaction will roughly double, so that a celestial body consisting of cold hydrogen at a density of $0.7 \cdot 10^5$ cannot last more than 10^8 years.

From the well known formula of Emden (cf. [6]), the density of cold hydrogen reaches $0.7 \cdot 10^5$ g/cm^3 at a pressure of $2 \cdot 10^{21}$ dyne/cm^2 under the action of gravitation, at the center of a body whose mass is $1.5 \cdot 10^{33}$ g, i.e., 0.75 times the mass of the Sun.

In 1938, Landau [7] pointed out that at high density a "neutron condensation" occurs. For hydrogen, the limiting energy of the reaction $e^- + p = n + \tilde{\nu}$ is 0.75 Mev. The corresponding density is 10^7 g/cm^3.

The tunneling reaction sets in much earlier. Consequently, the neutron condensation begins only after the conversion of the hydrogen to helium (or to heavier nuclei), and correspondingly at an even greater density than that considered in [6].

The reactions $p + \mathrm{D}$, $p + \mathrm{T}$, $\mathrm{D} + \mathrm{D}$, $\mathrm{D} + \mathrm{T}$ can also occur via the cold mechanism and require lower pressures. However, this process will never be of practical importance: the required pressures are so high that, under terrestrial conditions, they can be realized only in a non-stationary state, in an extremely small volume and for extremely short times. For equal expenditure of energy or equal pressure, the tunnel reaction is far inferior to the thermonuclear reaction in heated matter.

I take this opportunity to express my thanks to D. A. Frank-Kamenetskiĭ and A. I. Lebedinskiĭ for interest in the work and for valuable comments.

Received
April 29, 1957

REFERENCES

1. *Bethe H. A., Critchfield C. L.*—Phys. Rev. **54**, 248 (1938).
2. *Salpeter E. E.*—Phys. Rev. **88**, 547 (1952).
3. *Frank-Kamenetskiĭ D. A.*—Dokl. AN SSSR **104**, 30 (1955); UFN **58**, 415 (1956).
4. *Schatzman E.*—J. de Phys. et le Radium **8**, 46 (1948).
5. *Zeldovich Ya. B., Sakharov A. D.*—ZhETF **32**, 947 (1957).
6. *Landau L. D., Lifshitz E. M.* Statistical Physics. Reading, MA: Addison-Wesley (1969).
7. *Landau L. D.*—Nature **141**, 333 (1938).

Commentary

In the course of research on processes of stellar collapse accompanying the transformation of matter into neutrons, it was natural to pose the question of hydrogen neutronization.

However, as is shown in the note above, hydrogen sooner enters into reactions leading to the formation of helium, even in a cold state due to sub-barrier effects.

In the paper there is reference to the work by Schatzman, who posed the problem of reactions in a degenerate nuclear gas. However, later Ya. B. found an article by Wildhack, "The Transformation of Protons into Deuterium as an Energy Source in Dense Stars" [Phys. Rev. **57**, 81–86 (1940)]. In this article picnonuclear (i.e., at high density) reactions and their astrophysical consequences are investigated in detail.

Wildhack's article is first mentioned in the monograph of Ya. B. and I. D. Novikov, "Theory of Gravitation and the Evolution of Stars" [Moscow: Nauka, 484 p. (1972)].

We note also the possibility mentioned at the end of the paper of carrying out reactions of hydrogen isotopes at lower pressures. This set of ideas overlaps with papers on muon catalysis of nuclear reactions of hydrogen isotopes.

40

Neutronization of He^{4}*

With O. Kh. Guseinov

As is well known, at high densities, when the electron energy becomes sufficient for the inverse β process, the neutronization reaction begins in matter [1].

The first step in the investigation of the kinetics of this process was made by Frank-Kamenetskiĭ [2]. During the time of the collapse of a star, owing to the neutronization of matter, high-energy neutrinos will be emitted and these may be experimentally detectable. In an earlier note [3] we considered the process of collapse with neutronization of cold hydrogen. Estimates for other elements were extremely crude. An estimate under the assumption of an annual collapse of ten stars in our galaxy with masses 2–3 times the Sun's mass yielded a high-energy neutrino flux (10–30 MeV) amounting to several percent of the solar flux (the neutrino from Be^8 decay, with maximum energy 14 MeV). The neutrino energy was underestimated in the cited note. Let us obtain a more accurate expression for the energy of the neutrinos produced during the course of neutronization of helium.

The production of high-energy neutrinos upon collapse of a cold star is connected with the process

$$e^- + \mathrm{He}^4 = \mathrm{T} + n + \nu \tag{1}$$

The threshold energy of this process is $Q = 22.1$ MeV $= 43.4\,mc^2$. This reaction is followed by the "easier" reaction $e^- + \mathrm{T} = 3n + \nu$. The course of the reaction (1) is made complicated by the fact that the H^4 nucleus does not exist and that the neutronization is accompanied by emission of a neutron. Nor does the H^4 apparently exist like a virtual state [4]. It is therefore natural to assume in a first approximation that the matrix element depends neither on the neutrino energy nor on the neutron energy, nor on the angle between them, and the probability of the reaction is therefore assumed proportional to the phase volume.

For a given density ρ_e of electrons with energy E, the total kinetic energy of the products of the reaction (1) is $E - Q$; it is made up of the neutrino energy E_ν and the n+T kinetic energy relative to the center of inertia of this system, $E_1 = E - Q - E_\nu$. The energy motion of the T + n center of mass (of the order of 1 MeV) is neglected. In view of the fact that T and n are

*Pisma v Zhurnal eksperimentalnoĭ i teoreticheskoĭ fiziki **1** (1), 11–17 (1965).

nonrelativistic particles, their phase volume is proportional to $(E_1)^{1/2}dE_1$. Thus, at a given electron energy E and for a given electron density ρ_e, the differential probability of a process with production of a neutrino in the energy interval from E_ν to $E_\nu + dE_\nu$ is of the form

$$dW = \rho_e K E_\nu^2 \sqrt{E - Q - E_\nu}\, dE_\nu. \tag{2}$$

From this we obtain for the total probability

$$W = \rho_e K \int_0^{E-Q} E_\nu^2 \sqrt{E - Q - E_\nu}\, dE_\nu = \rho_e B (E - Q)^{7/2}$$
$$= \rho_e B (mc^2)^{7/2} (E' - Q')^{7/2}, \tag{3}$$

where K and B are constants, E' and Q' the dimensionless energies expressed in units of mc^2, and $m = m_e$.

To determine the constant B we use the analogy between the process of interest to us and the reaction

$$\mu^- + \mathrm{He}^4 = T + n + \nu, \tag{4}$$

which was investigated experimentally [5].[1]

The experimentally obtained probability of this reaction, $W_\mu = 370 \pm 50\,\mathrm{s}^{-1}$, pertains to muons in the $1s$ state in the nuclear field. Setting up an expression analogous to (3) for the probability W_μ of a process with a μ meson,

$$W_\mu = B|\psi(0)|^2 (m_e c^2)^{7/2} (E' - Q')^{7/2}, \tag{5}$$

we obtain B. According to the universal weak interaction hypothesis, B should be the same for the electronic and muonic processes.

We now go from the reaction with an electron of specified energy E to the case of neutronization by a degenerate relativistic electron gas:

$$W_F = \frac{B}{\pi^2}\left(\frac{\hbar}{mc}\right)^{-3} (mc^2)^{7/2} \int_{Q'}^{E'_F} (E' - Q')^{7/2} E'^2\, dE',$$
$$E_f = mc^2 \left(\frac{\rho}{\mu_e \cdot 10^6}\right)^{1/3}. \tag{6}$$

Substituting the expression for B obtained from the experimental data on the muon reaction, we obtain

$$W_F = W_\mu \left[\left(\frac{m_e}{m_\mu}\right)^3 \left(\frac{e^2}{\hbar c}\right)^3 \frac{(m_{\mathrm{He}} + m_\mu)^3}{m_{\mathrm{He}}^3 \pi z_{\mathrm{He}}^3}\right] (E'_\mu - Q')^{-7/2} \int_{Q'}^{E'_F} (E' - Q')^{7/2} dE'$$
$$= W_\mu \cdot 9.5 (y - 1)^{9/2} (0.154 y^2 + 0.056 y + 0.012), \tag{7}$$
$$y = \frac{E_F}{Q} = \left(\frac{\rho}{\mu_e \cdot 1.7 \cdot 10^{11}}\right)^{1/3}.$$

[1]The average number of neutrons in one event of the reaction $\mu^- + \mathrm{He}^4$ is approximately 1.2, from which it follows that the reactions $\mu^- + \mathrm{He}^4 = \mathrm{D} + 2n + \nu$ and $\mu^- + \mathrm{He}^4 = p + 3n + \nu$ constitute less than half of all the cases.

It is assumed that the next act following the process $e^- + \mathrm{He}^4 = \mathrm{T} + n + \nu$, namely $e^- + \mathrm{T} = 3n + \nu$, occurs practically instantaneously, in accordance with the fact that the nucleus T is much weaker and more "friable" than He^4.

Using the relations for free fall

$$\rho = \frac{1}{6\pi G(t_0 - t)^2} = \frac{8 \cdot 10^5}{(t_0 - t)^2}; \qquad dt = 4.5 \cdot 10^2 \rho^{-3/2} d\rho,$$

we obtain an approximate equation for the neutronization kinetics (x is the fraction of the non-decaying He^4)

$$\frac{dx}{d\rho} = -3 \cdot 10^5 \frac{x}{\rho^{3/2} Q'^2} \left(\frac{\rho x}{2 \cdot 10^6}\right)^{2/3} \left[\frac{1}{Q'} \left(\frac{\rho x}{2 \cdot 10^6}\right)^{1/3} - 1\right]^{9/2}.$$

Integration of this equation from $\rho = 0$, $x = 1$ yields $\rho = \rho_t = 1.7 \cdot 10^{11}$, $x = 1$ (threshold); $\rho = 7.5\rho_t$, $x = 0.86$; $\rho = 15\rho_t$, $x = 0.5$; $\rho = 60\rho_t$, $x = 0.16$. Neutronization of the bulk of the He^4 mass in the free-fall regime occurs at a Fermi energy double the threshold value, i.e., 45 MeV. This means that neutrinos with energies up to 35 MeV are produced in the process $e^- + \mathrm{T} = 3n + \nu$. Their registration probability is 10–20 times greater than that of the threshold neutrinos from B^8 decay expected to be observed in the spectrum of the Sun; these neutrinos can be distinguished from the solar neutrinos, if the detector registers the neutrino energy and, albeit roughly, their direction [6].

The emission of a neutrino with energy up to 35 MeV from a collapsing star occurs at a density on the order of 10^{12}–10^{13} g/cm^3. This density must be compared with the critical value [7, 9]

$$\rho_g = 1.8 \cdot 10^{16} (M/M_\odot)^{-2}$$

at which gravitational self-closure* takes place. It is clear that at $M < 50M_\odot$, i.e., for the overwhelming majority of stars, the high-energy neutrinos produced by neutronization have time to leave the star without becoming noticeably weakened by the gravitational field.

At the same time, the density $\rho \sim 3 \cdot 10^{12}$ g/cm^3 is still appreciably smaller than nuclear density, so that the analysis made above, without account of nuclear interaction, is perfectly justified. Concerning emission of thermal neutrinos following collapse see [7] and [10].

Received
April 6, 1957

*I.e., when an object enters its own gravitational horizon.

REFERENCES

1. *Landau L. D., Lifshitz E. M.* Statistical Physics. Reading, MA: Addison-Wesley (1969).
2. *Frank-Kamenetskiĭ D. A.*—ZhETF **42**, 875 (1962).
3. *Zeldovich Ya. B., Guseinov O. Kh.*—Dokl. AN SSSR **162** 4 (1965).
4. *Baz A. I., Goldanskiĭ V. I., Zeldovich Ya. B.*—UFN **85**, 415 (1965).
5. *Foldy L., Walecka.*—In: Congr. Intern. Phys. Nucl. P. **II**, 1168 (1964).
6. *Reines F., Woods R. M. Jr.*—Phys. Rev. Lett. **14**, 20 (1965).
7. *Zeldovich Ya. B.*—Astron. Tsikulyar **250**, I–VII (1963).
8. *Hoyle F., Fowler W.*—Relativistic Astrophysics, Preprint (1963).
9. *Zeldovich Ya. B., Novikov I. D.*—UFN **84**, 377 (1964).
10. *Zeldovich Ya. B., Novikov I. D.*—UFN **86**, 447 (1965).

Commentary

The formation of neutron stars should be accompanied by the transformation of protons (both free and bound in nuclei) into neutrons with electron capture and neutrino emission.

In the present note for the first time, even though in a rough fashion, the dynamics of this process and the spectrum of the emitted neutrinos are investigated. It is shown that the neutrinos acquire a large energy, up to 20–30 MeV, and the reasons for this are graphically explained. A proposal is put forward for experimentally discovering the neutrino burst in the collapse of a star in the vicinity of the Sun.* Later, the calculations of neutrino and antineutrino emission in the collapse of cold and hot stars, taking into account dispersion and other factors, were refined many times. The reader may acquaint himself with the present state of the problem in the review by V. S. Imshennik and D. K. Nadezhin.[1]

[1] *Imshennik V. S., Nadezhin D. K.*—In: Astrophys. Space Phys. Rev., Harwood Acad. Publ. GmbH **2**, 73–161 (1983); Itogi Nauki i Tekhniki [Summaries of Science and Technology]. Moscow: VINITI **21**, 63–129 (1982).

**Editor's note.* This prediction was confirmed in 1987 when neutrinos were detected from a supernova in the nearby Magellenic Clouds.

41

Neutrino Luminosity of a Star in Gravitational Collapse in the General Theory of Relativity*

With M. A. Podurets

The theory of gravitational collapse, whose principles were developed in 1937–8 [1, 2] has recently again attracted attention in relation to the question of a number of collapsed stars in the Universe [3–6], the dynamics of collapse, and the energy losses, in particular by neutrino emission during collapse [4, 6–8].

Oppenheimer and Snyder [2] found that the surface of the star intersects the gravitational radius r_0 in a finite proper time τ, but an external observer sees the surface as approaching r_0 asymptotically, $r \to r_0$ as $t \to \infty$ (the time being the same as the coordinate time t for a distant observer). The luminosity of the surface tends asymptotically to zero in spite of the finite surface temperature. The star is gravitationally self-closed.

Here we consider the law of gravitational self-closure for neutrino radiation, which is generated mainly at the center of the star. In this case there is no Doppler effect (governed by the motion of the surface) nor any change of solid angle.

Rough allowance (as to order of magnitude) has been made [4, 7] for gravitational self-closure; the characteristic density $\rho_g = M/v_g$, with $v_g = \frac{4}{3}\pi r_0^3$, has been deduced, the integration of the energy loss by an element of mass being terminated when the density reaches ρ_g.

From the solution for the collapse it is simple to find the paths of light or neutrino rays ($ds = 0$) propagating radially. The time τ_1 when the last ray escapes from the center, passes through the surface when $r = r_0$ and reaches an external observer when $t = \infty$, does not coincide with the moment when ρ_g is reached.

The radiation intensity transforms in accordance with the law

$$I(t) = Q(\tau)\left(\frac{d\tau}{dt}\right)^2, \qquad (1)$$

in which τ is the proper time at the origin of the ray (center of the star), t is the coordinate time for a remote observer at rest when this ray reaches

*Doklady Akademii nauk SSSR **156** (1), 57–60 (1964).

him, $Q(\tau)$ is the power of the radiation emitted by the matter at the center of the star in energy terms, $dE = Qd\tau$, and $I(t)$ is the power reckoned from the energy passing through the remote sphere containing the observer,

$$dE = 4\pi r^2 i(t)dt = I(t)dt.$$

We note that formula (1) contains the square of $d\tau/dt$, which is less than 1 and tends to zero for $t \to \infty$, so

$$\int_{-\infty}^{\tau_1} Q(\tau)\, d\tau < \int_{-\infty}^{+\infty} I(t)\, dt.$$

In calculating the number of the particles conserved along the ray (neutrinos), we have

$$n(t) = \nu(\tau)\frac{d\tau}{dt}, \qquad N = \int_{-\infty}^{\tau_1} \nu(\tau)\, d\tau = \int_{-\infty}^{\infty} n(t)\, dt,$$

in which ν is the number of neutrinos emitted in unit proper time τ, n is the number of neutrinos passing through the remote sphere in unit coordinate time, and N is the total number of particles (invariant).

In calculating the energy we have to allow for the redshift, i.e., the change of energy of each individual neutrino during motion from the potential well towards the remote observer. The energy varies in the same way as the frequency (in accordance with $\varepsilon = \hbar\omega$), which gives one more power in $d\tau/dt$.

Then to calculate the luminosity from (1) we need to know the path of radial light rays within the sphere and outside it; within the sphere it is more convenient to use not the Schwarzschild coordinate system (t, τ), but the comoving one (τ, R). With this in view we multiply and divide the right-hand side in (1) by $(dt_1/d\tau_1)^2$, in which dt_1 is the change in time between two rays along the boundary of the star and $d\tau_1$ is the same interval of proper time. Then

$$I(t) = Q(\tau)\left(\frac{d\tau}{d\tau_1}\right)^2 \left(\frac{d\tau_1}{dt_1}\right)^2 \left(\frac{dt_1}{dt}\right)^2. \tag{2}$$

The factor $d\tau_1/dt_1$ is deduced from the law of transformation of time along the boundary and is

$$\frac{d\tau_1}{dt_1} = \frac{r_1 - r_0}{r_1} \frac{1}{\sqrt{\dot{r}_1^2 + 1 - r_0/r_1}}, \tag{3}$$

in which r_1 is the radius of the boundary, the dot denoting differentiation with respect to the proper time τ_1.

We can easily find dt_1/dt from the equation for the paths of light rays *in vacuo*, $r + r_0 \ln(r/r_0 - 1) - t = \text{const}$ (here and henceforth $c = 1$):

$$\frac{dt_1}{dt} = \frac{\sqrt{\dot{r}_1^2 + 1 - r_0/r_1}}{\sqrt{\dot{r}_1^2 + 1 - r_0/r} - \dot{r}_1}. \tag{4}$$

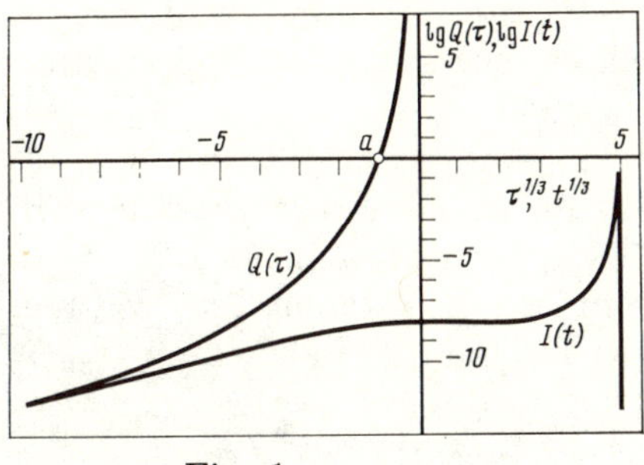

Fig. 1

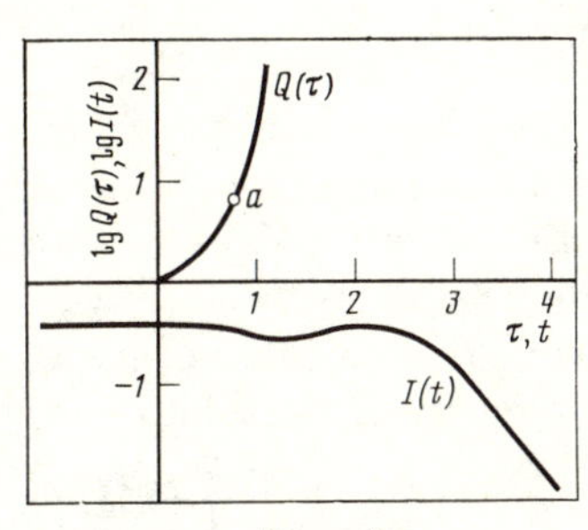

Fig. 2

Combining (2) and (4), we have

$$I(t) = Q(\tau)\left(\frac{d\tau}{d\tau_1}\right)^2\left(1-\frac{r_0}{r}\right)^2\frac{1}{[\sqrt{\dot{r}_1^2+1-r_0/r}-\dot{r}_1]^2}. \tag{5}$$

All the quantities appearing here are functions of τ_1, so t also should be expressed in terms of τ_1. The relation is

$$t = r + r_0\ln\left(\frac{r}{r_0}-1\right) - r_1 - r_0\ln\left(\frac{r}{r_0}-1\right) + t_1. \tag{6}$$

Then for $r_1 \to r_0$,

$$I(t) \propto \left(\frac{r_1}{r_0}-1\right)^2, \quad t \propto -2r_0\ln\frac{1}{r/r_0-1} \tag{7}$$

[the last is readily found from (3)]. The asymptote for the luminosity of the collapsing star is then of the form

$$I(t) \propto e^{-t/r_0}. \tag{8}$$

We see that the process of collapse appears to an external observer to take an indefinitely long time, but the characteristic time of the decay of luminosity is very small, equal to the gravitational radius divided by the velocity of light. We have $r_0 \approx 3\cdot 10^5$ cm for the Sun, and the time for the luminosity to decay by a factor e is 10^{-5} s.

We now consider some numerical examples.

The rate of loss of energy is taken as $Q \propto T_9\rho$, which is true for $T_9 > 5$ (T is temperature and ρ is density of the particles). If the ratio of the specific heats is $\gamma = 4/3$, then $Q \propto \rho^2$.

Two cases may be distinguished:

1. Collapse of dust ($p = 0$). The motion is described by Tolman's [10] solution, which we take in its simplest form:

$$r = \left(\frac{3}{2}\right)^{2/3} r_0^{1/3}R\tau^{2/3}, \quad e^{\omega/2} = \left(\frac{3}{2}\right)^{2/3} r_0^{1/3}\tau^{2/3}, \quad \rho = \frac{1}{6\pi\tau^2}. \tag{9}$$

Here $k = 1$, $R = 1$ at the boundary of the star. The interval is $ds^2 = d\tau^2 - e^{\omega} dR^2 - r^2(d\theta^2 + \sin^2\theta d\varphi^2)$. Collapse occurs at $\tau = 0$, and gravitational self-closure at $\tau = 2r_0/3$.

2. Collapse of a star consisting of a cold Fermi gas, initially in a state of rest; the initial density distribution differs slightly from that in a star with the maximum possible mass $M_{\max} \simeq 0.73 M_{\odot}$, so the total mass is $0.98 M_{\odot}$. Numerical calculations have been presented for such a star [6]. For dust, the simplicity of the solution enables us to put (5) and (6) in finite form

$$I(t) = q\frac{[1 + (2/3)^{1/3} r_0^{1/3} \tau_1^{-1/3}]^2}{\tau_1^{4/3}} \left[\tau_1^{1/3} - (3/2)^{2/3} r_0^{1/3} \,/\, 3\right]^2,$$

$$t = r + r_0 \ln\left(\frac{r}{r_0} - 1\right) + \tau_1 - \left(\frac{3}{2}\right)^{2/3} r_0^{1/3} \tau_1^{2/3}$$
$$+ 3\left(\frac{4}{9}\right)^{1/3} r_0^{2/3} \tau_1^{1/3} - r_0 \ln\left[1 + \left(\frac{3}{2}\right)^{1/3} r_0^{-1/3} \tau_1^{1/3}\right]^2.$$

Here q is a factor of proportionality, while the arbitrary constant in $t(\tau_1)$ is taken to be such that $\tau_1 = t_1$ for $\tau_1 \to -\infty$.

Figure 1 shows the radiation power $Q(\tau)$ and the luminosity of the star $I(t)$ as functions of the proper and coordinate times respectively; the abscissas are $\tau^{1/3}$ and $t^{1/3}$, the ordinates being $\log Q(\tau)$ and $\log I(t)$. The unit of length (and time) is $\frac{9}{4} r_0$; the unit for Q and I is $q/(\frac{9}{4} r_0)^4$. Point a on $Q(\tau)$ denotes the point above which $Q(\tau)$ is inaccessible to the external observer; point a corresponds to $\tau_1 = -1$, gravitational closure occurring at $\tau_2 = -8/27$. In these units

$$\int_{-\infty}^{-8/27} Q\, d\tau = 12.81; \quad \int_{-\infty}^{-1} Q\, d\tau = \frac{1}{3}; \quad \int_{-\infty}^{+\infty} I\, dt = 0.0882.$$

Figure 2 shows analogous curves for a collapsing star with $M = 0.98 M_{\odot}$ consisting of a cold Fermi gas; it is assumed that the star is at rest for $\tau < 0$. The unit of luminosity is the strength of a source in the star at rest.

Note. The table below gives the loss of mass by stars as a result of neutrino emission, as well as T and ρ for the moment of gravitational self-closure. This table is from [6]. The last line has been added to give the losses reduced by the factor $\int_{-\infty}^{\infty} I\, dt \,/\, \int_{-\infty}^{\tau_2} Q\, d\tau$.

$M/M_{\odot}$	10^2	10^4	10^5	10^6	10^8
T_9	360	36	11.3	3.6	0.5
ρ, g/cm^3	$1.8 \cdot 10^{12}$	$1.8 \cdot 10^8$	$1.8 \cdot 10^6$	$1.8 \cdot 10^4$	1.8
$\Delta M/M$	55	0.055	$1.7 \cdot 10^{-3}$	$4.5 \cdot 10^{-5}$	10^{-12}
$(\Delta M/M)_1$	0.38	$3.8 \cdot 10^{-4}$	$1.17 \cdot 10^{-5}$	$3.1 \cdot 10^{-7}$	$7 \cdot 10^{-15}$

Received
February 11, 1964

REFERENCES

1. *Oppenheimer J. R., Volkoff G.*—Phys. Rev. **55**, 374 (1939).
2. *Oppenheimer J. R., Snyder H.*—Phys. Rev. **56**, 455 (1939).
3. *Zeldovich Ya. B., Smorodinskiĭ Ya. A.*—ZhETF **41**, 907 (1961).
4. *Hoyle F., Fowler W., Burbidge E., Burbidge G.*—Relativistic Astrophysics, Preprint (1963).
5. *Wheeler J.*—Solvay Conference, Bruxelles (1958).
6. *Novikov I. D., Ozernoĭ L. M.*—Preprint P. N. Lebedev Fiz. Inst. AS USSR, Moscow (1964).
7. *Zeldovich Ya. B.*—Astron. Tsirkulyar. **250** (1963).
8. *Podurets M. A.*—Dokl. AN SSSR **154** 2 (1964).
9. *Zeldovich Ya. B.*—Dokl. AN SSSR **154** 3 (1964).
10. *Tolman R.*—Proc. Nat. Acad. Sci. US Phys. Sci. **20**, 3 (1934).

Commentary

This article relates to the early period of understanding the observational consequences of the general theory of relativity applied to the last catastrophic stages of the evolution of stars.

The most important result consists in the definition of what the authors loosely called the "line of the last gasp."

The specifics of the metric of relativistic collapse consists in the fact that the final compression is infinite, and a singularity appears. However, the law of photon propagation (if we ignore their scattering) or of neutrinos is such that information comes only from matter at a particular stage, with the final compression still far from the singularity.

The explanation for this circumstance, which today seems trivial, was of much significance in the establishment of the theory of stellar collapse.

For a systematic treatment of the problem of collapse, see the monograph by Ya. B. and I. D. Novikov.[1]

[1] *Zeldovich Ya. B., Novikov I. D.* Relativistic Astrophysics. Univ. of Chicago Press (1971).

42

The Evolution of a System of Gravitationally Interacting Point Masses*

With M. A. Podurets

The evolution of a system of heavy point particles is studied. The system evolves slowly through a sequence of quasiequilibrium states, the properties of which are determined by the equations of equilibrium in general relativity (GR). It is shown that the evolution inevitably ends in a collapse.

Effects leading to evolution are discussed: evaporation of particles and, specific to GR, emission of gravitational radiation and capture. The applicability of the results to astronomical objects is discussed.

Let us consider a system composed of point masses, i.e., particles that interact only via the force of universal gravitation.

In the Newtonian approximation, the problem has been exhaustively treated in a number of papers, among which we should note the paper by V. A. Ambartsumyan [1], the insufficiently well-known paper by L. E. Gurevich [2], and the recent numerical calculations of [3].

Briefly, the situation is the following: true statistical equilibrium is impossible because the statistical integral diverges both when a particle moves out to infinity, as well as when two particles approach one another indefinitely. However, the evaporation of particles proceeds slowly and we can consider near equilibrium states of the system with given energy E (negative), number of particles N, and angular momentum J (in the following, we consider the case $J = 0$). In this state, each particle basically moves in the common averaged gravitational field of all the other particles. Occasionally particles collide and exchange energy, which leads to the establishment of an energy distribution which differs little from Maxwellian for $\varepsilon < 0$ (ε is the particle energy). More rarely, particles acquire an energy $\varepsilon > 0$ in a collision and evaporate. As the evaporation proceeds, E and N change and the system evolves, moving through a sequence of almost equilibrium states; the law of evolution is of the form

$$N \sim (t_0 - t)^{2/7}, \quad \sqrt{\bar{u}^2} \sim (t_0 - t)^{-1/7}, \quad t_0 \sim \frac{GN^2 m}{u_0^3}.$$

*Astronomicheskiĭ zhurnal **42** (5), 963–973 (1965).

The formation of close particle pairs requires triple collisions[1] and occurs rarely by comparison with evaporation. The almost equilibrium system is stable relative to small perturbations in the density distribution. In the present note we discuss the changes that arise when the velocities of the particles are comparable with that of light. In doing this, we must replace the Newtonian theory of gravitation by the general theory of relativity (GR). It is clear that such a situation will inevitably arise during the last stages of evolution of the system even if during the initial stage $v/c \ll 1$.

The principal results are the following:

1. Evolution is governed not only by evaporation, but also by the specific occurrence of close nonstationary pairs in binary collisions with perihelion of the order of the Schwarzschild gravitational radius of a particle.

2. The sequence of almost-equilibrium states terminates with a critical state with a definite value of $\bar{v}/c \sim 1$ at which stage a large fraction of the system contracts (collapses) during a time of the order of one revolution in orbit.

1. Equilibrium Configurations of a System of Point Masses

Let us proceed to the derivation of these results starting from the second. For discussion of the almost-equilibrium states, we will initially neglect collisions and consider the motion of particles in a self-consistent centrally-symmetric gravitational field.

The static field in GR is characterized by the two functions $\nu(r)$ and $\lambda(r)$ in the expression for the interval

$$dS^2 = c^2 e^\nu dt^2 - e^\lambda dr^2 - r^2(d\theta^2 + \sin^2\theta\, d\varphi^2). \tag{1.1}$$

The ensemble of particles moving in this field will be described statistically by means of the distribution function $N(x^i, p_k)$ describing the distribution of particles in phase space. The meaning of the invariant distribution function is given by the equality [4]

$$dn\, dS = mN\, dx\, dp,$$

where $dx\, dp$ is the invariant element of volume of the eight-dimensional phase space, dn is the number of world lines intersecting dx in the direction $p + dp$, dS their length inside the volume element dx, and m the rest mass of the particles.

In a static field, the static solution of the kinetic equation can only depend on the integrals of the motion: particle energy—cp_0 (including the gravitational potential energy) and the angular momentum. Assuming that the distribution is independent of the direction of the velocity, we will restrict ourselves to the simplest case $N = N(p_0)$.

[1] The third particle carries away energy.

We can now write the energy-momentum tensor for the particles

$$T_{ik} = \int p_i p_k N \frac{dp}{\sqrt{-g}}. \tag{1.2}$$

The integration is best carried out in the spherical system of coordinates of momentum space:

$$p_0 = -\sqrt{-g_{00}}\sqrt{B^2 + m'^2c^2},$$
$$p_1 = \sqrt{g_{11}} B\cos\theta,$$
$$p_2 = \sqrt{g_{22}} B\sin\theta\cos\Phi,$$
$$p_3 = \sqrt{g_{33}} B\sin\theta\sin\Phi,$$
$$0 < B < \infty, \quad 0 < m' < \infty, \quad 0 < \theta < \pi, \quad 0 < \Phi < 2\pi.$$

Of course, we assume that all of the particles have the same mass, i.e., that $N \sim \delta(m'^2 - m^2)$. After this, the integration with respect to the variables m', θ, Φ can be carried out, leaving only one quadrature in B.

Finally, we write down the mixed components $T_i^{\ k}$ (the common dimensionless factor is left out as it can be included in N),

$$T_0^0 = -\frac{m^2}{c^2} e^{-\frac{3}{2}\nu} \int_{mc^2 e^{\nu/2}}^{\infty} N(\eta)\eta^2 \sqrt{e^{-\nu}\eta^2 - m^2c^4}\, d\eta, \tag{1.3}$$

$$T_1^1 = T_2^2 = T_3^3 = \frac{1}{3}\frac{m^2}{c^2} e^{-\frac{1}{2}\nu} \int_{mc^2 e^{\nu/2}}^{\infty} N(\eta)(e^{-\nu}\eta^2 - m^2c^4)^{3/2} d\eta, \tag{1.4}$$

$$T_{01} = T_{02} = T_{03} = 0. \tag{1.5}$$

Here, $\eta = -cp_0$.

It can be seen that the system of particles moving in a gravitational field with a distribution independent of the angular momentum behaves as a gas with energy density $\varepsilon = -T_0^0$ and pressure $p = T_1^1$ obeying Pascal's law.

The special feature of this gas is the fact that the expressions for ε and p explicitly contain the magnitude of the gravitational field $\nu(r)$. It should also be noted that, as was to be expected, the energy-momentum tensor (1.3)–(1.5) satisfies the equilibrium equation $T_{i;k}^k = 0$ identically. Analogously to the tensor T_{ik}, we can introduce the vector current

$$j_k = \int p_k N \frac{dp}{\sqrt{-g}}. \tag{1.6}$$

In our case, $j_1 = j_2 = j_3 = 0$, while the component j_0 depends on the number density n of the particles

$$ncu_0 = j_0.$$

Calculations yield

$$n = \frac{m^2}{c^2} e^{-\nu} \int_{mc^2 e^{\nu/2}}^{\infty} N(\eta)\eta\sqrt{e^{-\nu}\eta^2 - m^2c^4}\, d\eta. \tag{1.7}$$

The whole system is characterized by two important parameters, the mass M and the rest mass M_0,

$$M = \frac{4\pi}{c^2} \int \varepsilon r^2 \, dr, \tag{1.8}$$

$$M_0 = 4\pi m \int n e^{\lambda/2} r^2 \, dr. \tag{1.9}$$

The difference $(M - M_0)c^2 = -\Delta M c^2$ corresponds to the negative Newtonian energy and makes it impossible for the system to fly apart.

Let us adopt the Maxwellian distribution function

$$N = A e^{-\eta/T}, \tag{1.10}$$

where T is the temperature in energy units.

The temperature is constant because there are no collisions, i.e., the thermal conductivity is infinitely high.[2]

With this distribution function, the problem of finding the equilibrium configurations has no solution because the density does not become zero as $r \to \infty$. This difficulty reflects the fact that the Maxwellian distribution always contains particles with $\eta > mc^2$ which can escape to infinity.

Therefore, we should search for equilibrium states, using a modified distribution function $N(\eta)$ such that when

$$\eta > \eta_0 < mc^2, \quad N(\eta) = 0.$$

It appears reasonable to assume that

$$\eta_0 = mc^2 - \frac{T}{2}. \tag{1.11}$$

It should be noted that for one simple form of the function, the problem has been solved in the literature. Indeed, when Oppenheimer and Volkoff (OV) [5] consider the ideal degenerate Fermi gas at zero temperature, this corresponds to

$$N(\eta) = K = \text{const} \quad \text{at} \quad \eta \le \eta_0 < mc^2, \qquad N(\eta) = 0 \quad \text{at} \quad \eta > \eta_0.$$

In the solution found for the Fermi gas we can theoretically exclude collisions between particles and consider the star (which the above authors had in mind) as an aggregate of particles each of which moves along an orbit in the general gravitational field independently of the others [13].

What is the difference between the OV problem and the problem of the system of particles being considered in the present paper? In the case of the Fermi gas the solution is exact when the collisions between particles are taken into account: no pair of particles with $\eta_1 < \eta_0$, $\eta_2 < \eta_0$ can give rise to a pair with other values η_3, η_4 after the collision since one or both of the states

[2]The magnitude of T which can be called "the temperature measured by an infinitely-remote observer" is constant throughout the system. The temperature at a given point measured by a local observer located at this point is given by $T_{\text{ex}} = Te^{-\nu/2}$ and in equilibrium varies with position as must be the case in GR.

3 and 4 lie below η_0 and hence are occupied, so that according to the Pauli principle the transition $1, 2 \to 3, 4$ is forbidden. In our problem of classical nonquantized particles there is no Pauli principle and collisions will disrupt the form of $N(\eta)$, bringing $N(\eta)$ closer to the Maxwellian distribution and giving rise to evolution of the system.

Another difference is the fact that in the OV problem the constant K has a definite value expressed in terms of Planck's constant $\hbar$, $K = 2/(2\pi\hbar)^3$. On the other hand, when classical particles are being considered, the value of K is arbitrary.

The OV solution can be easily generalized to the case of arbitrary K, since the equations have a group of similarity transformations: if K is changed, $KR^2 = const$ being kept fixed, where R is a characteristic dimension, then the solution remains valid.

Let us now proceed direction to the search for equilibrium solutions for a system of point particles. The problem can be reduced to the integration of Einstein's equations with the energy-momentum tensor derived above. The equations are of the form

$$e^{-\lambda}\left(\frac{1}{r}\frac{d\nu}{dr} + \frac{1}{r^2}\right) - \frac{1}{r^2} = \frac{8\pi G}{c^4}p,$$

$$e^{-\lambda}\left(\frac{1}{r}\frac{d\lambda}{dr} - \frac{1}{r^2}\right) + \frac{1}{r^2} = \frac{8\pi G}{c^4}\varepsilon,$$

where

$$p = \frac{1}{3}\frac{m^2}{c^2}e^{-\nu/2}\int_{mc^2e^{\nu/2}}^{mc^2-\frac{1}{2}T} Ae^{-\eta/T}(e^{-\nu}\eta^2 - m^2c^4)^{3/2}\,d\eta,$$

$$\varepsilon = \frac{m^2}{c^2}e^{-\frac{3}{2}\nu}\int_{mc^2e^{\nu/2}}^{mc^2-\frac{1}{2}T} Ae^{-\eta/T}\eta^2\sqrt{e^{-\nu}\eta^2 - m^2c^4}\,d\eta,$$

$$n = \frac{m^2}{c^2}e^{-\nu}\int_{mc^2e^{\nu/2}}^{mc^2-\frac{1}{2}T} Ae^{-\eta/T}\eta\sqrt{e^{-\nu}\eta^2 - m^2c^4}\,d\eta.$$

Here, condition (1.11) has already been used. The calculation is performed as follows. We adopt a value for the temperature T and a constant A; we start the integration at the center, assuming as is usual that $\lambda(0) = 0$ and $\nu(0) = \nu_0$ which is determined from the conditions at the free boundary; we carry out the integration until p and ε become zero (free boundary), i.e., up to the point $r = R$ where the condition $e^{\nu/2} = 1 - \frac{1}{2}T/mc^2$ is satisfied. The choice of the required value of ν_0 is made so as to match the solution on the boundary with the Schwarzschild solution in vacuum, $\nu(R) + \lambda(R) = 0$.

It is convenient to introduce as parameters the dimensionless temperature $\tilde{T} = T/mc^2$ and the rest mass density at the center $\rho_0 = n_0 m$ instead of A, and also to introduce the dimensionless variable $\tilde{r} = r/r_0$, where

$\tilde{T}$	$\tilde{n}_0$	$\tilde{R}\tilde{n}_0^{1/2}$	$\tilde{M}\tilde{n}_0^{1/2}$	$\Delta\tilde{M}\tilde{n}_0^{1/2}$
0.01	$1.035 \cdot 10^{-46}$	0.0939	$4.682 \cdot 10^{-4}$	$1.050 \cdot 10^{-6}$
0.05	$7.494 \cdot 10^{-11}$	0.2162	$5.208 \cdot 10^{-3}$	$5.821 \cdot 10^{-5}$
0.1	$5.820 \cdot 10^{-6}$	0.3113	$1.518 \cdot 10^{-2}$	$3.205 \cdot 10^{-4}$
0.2	$4.538 \cdot 10^{-3}$	0.4818	$4.578 \cdot 10^{-2}$	$1.645 \cdot 10^{-3}$
0.22	$9.571 \cdot 10^{-3}$	0.5147	$5.340 \cdot 10^{-2}$	$1.984 \cdot 10^{-3}$
0.25	$3.070 \cdot 10^{-2}$	0.6015	$7.050 \cdot 10^{-2}$	$2.519 \cdot 10^{-3}$
0.27	$7.836 \cdot 10^{-2}$	0.7091	$8.928 \cdot 10^{-2}$	$2.524 \cdot 10^{-3}$
0.27	0.1840	0.9210	0.1151	$1.257 \cdot 10^{-3}$
0.25	0.2416	1.141	0.1337	$-4.61 \cdot 10^{-4}$
0.22	0.2570	1.485	0.1544	$-2.16 \cdot 10^{-3}$
0.15	$9.517 \cdot 10^{-2}$	2.911	0.2095	$-1.57 \cdot 10^{-3}$
0.1	$1.033 \cdot 10^{-2}$	6.552	0.3185	$5.45 \cdot 10^{-3}$
0.1	$2.711 \cdot 10^{-2}$	12.29	0.5985	$1.01 \cdot 10^{-2}$
0.125	0.3281	14.97	0.9062	$4.52 \cdot 10^{-3}$
0.135	0.6368	15.03	0.9807	$7.88 \cdot 10^{-4}$

$\tilde{r}_0 = (\tilde{n}^0/G\rho_0)^{1/2}c$; in this case the dimensionless particle number density at the center

$$\tilde{n}_0 = e^{-\nu_0} \int_{e^{\nu_0/2}}^{1-\tilde{T}/2} e^{-x/\tilde{T}} x \sqrt{e^{-\nu_0} x^2 - 1}\, dx$$

depends only on the dimensionless temperature $\tilde{T}$. Having made such a choice of units, we find that the problem contains only one parameter $\tilde{T}$ (neither m nor ρ_0 appear), while $\nu_0 = \nu_0(\tilde{T})$.

Thus, having used only similarity properties, we have obtained important relationships for the dependence of the main quantities—the system radius R, mass M, and mass defect ΔM—on the density ρ_0 at the center,

$$R = \tilde{R}\tilde{n}_0^{1/2} G^{-1/2} c \rho_0^{-1/2},$$

$$M = \tilde{M}\tilde{n}_0^{1/2} G^{-3/2} c^3 \rho_0^{-1/2}, \tag{1.13}$$

$$\Delta M = \Delta\tilde{M}\tilde{n}_0^{1/2} G^{-3/2} c^3 \rho_0^{-1/2}.$$

In these formulas, the dimensionless quantities marked by a tilde depend only on the dimensionless temperature $\tilde{T}$, while their numerical values are obtained by the integration of the equilibrium equations (1.12) written in dimensionless form.

The results of the numerical integration are given in the table.

The main result required for the understanding of the evolution of the system consists in the absence of static solutions for sufficiently high temperatures $\tilde{T} > \tilde{T}_{\text{crit}}$, where $\tilde{T}_{\text{crit}} = 0.273$. A review of other properties of the solution is given in the appendix.

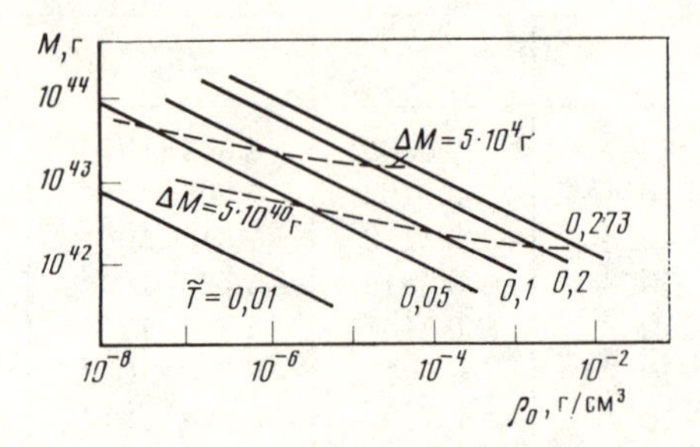

Fig. 1. Evolution in the (M, ρ_0)-plane. Solid curves are the equilibria for different $\tilde{T}$; dotted lines give the evolution left to right, ending at $\tilde{T}_{\text{crit}} = 0.273$.

2. Direction and Rate of Evolution

The evaporation of particles leads to an increase in the density and temperature, so that the critical value of the temperature is inevitably reached, after which a rapid collapse ensues. That $\tilde{T}_{\text{crit}}$ is inevitably reached is easily seen from a consideration of evolution in the (M, M_0) plane. Each temperature $\tilde{T}$ corresponds to a straight line $M(M_0)$ and these lines fill the sector from $M = M_0$ for $\tilde{T} = 0$ to $M = 0.964M_0$ for $\tilde{T} = \tilde{T}_{\text{crit}}$. For the evaporation of "parabolic" particles (those reaching the point at infinity with zero velocity), $dM/dM_0 = 1$ (or $\Delta M = \text{const}$) and $\tilde{T}_{\text{crit}}$ is always attained provided that initial $\tilde{T} > 0$. In fact, the evaporating particles are "hyperbolic," $dM/dM_0 > 1$ and the evolution takes place even more rapidly.

Figure 1 shows the curves $\Delta M = \text{const}$ describing evolution from a rarefied and cold state.

Typically new processes which are encountered in the relativistic problem are gravitational radiation and the capture of particles in binary collisions.

We begin by considering radiation depending on the motion of particles in the general (self-consistent) gravitational field. Since in this case $M_0 =$ const, this process is represented in the (M, M_0) plane by a vertical descent. As is known, the intensity of gravitational radiation is given by

$$w = \frac{1}{45}\frac{G}{c^5}(\dddot{Q})^2. \tag{2.1}$$

Let us substitute in this expression the value of the quadrupole moment Q associated with the motion of one particle, $Q = ma^2$, where a is the average distance of the particle from the center. We will replace the time derivatives by a multiplication by the frequency which in order of magnitude is $\omega = v/a$ (this is exact for a circular orbit); thus we have

$$W = \frac{1}{45}\frac{Gm^2v^6}{a^2c^5}. \tag{2.2}$$

We now have to express the average radius a in terms of the particle velocity and the number of particles. In order of magnitude, we have

$$\frac{GM}{a} = \frac{GNm}{a} = v^2.$$

Eliminating a, we find that

$$W = \frac{1}{45}\frac{v^{10}}{GN^2c^5}. \tag{2.3}$$

Finally, in order to obtain the rate of particle energy loss (in our units, this will be an inverse time) we divide W by the kinetic energy of the particle,

$$\frac{1}{\tau} = \frac{2W}{mv^2} = \frac{2}{45}\frac{v^8}{GN^2mc^5}. \tag{2.4}$$

Let us compare this quantity with the frequency of the principal motion,

$$\omega = \frac{v}{a} = \frac{v^3}{GNm},$$

$$\frac{1}{\omega\tau} = \frac{2}{45}\frac{1}{N}\left(\frac{v}{c}\right)^5. \tag{2.5}$$

Thus, because the radiation emitted by the individual particles is not coherent, the energy loss by gravitational radiation is a slow process even at relativistic velocities $v/c \sim 1$; it contains the factor $1/N$ and, consequently, takes place over a period of not less than N revolutions. The relativistic collision of two particles of equal masses has not yet been considered exactly. As a first approximation, one considers the motion of a test (light) particle in the Schwarzschild field of a second (heavy) particle. S. A. Kaplan [6] has classified the possible orbits. A review of the problem, together with the discussion of gravitational radiation has recently been made by I. D. Novikov and one of the authors [7].

In the collision of two particles with an angular momentum of the order of $mcr_g = 2Gm^2/c$, a spiral orbit occurs instead of a Newtonian hyperbolic orbit and the two particles fall toward one another. From the point of view of an external observer, the two particles approach one another asymptotically to a distance of the order of the gravitational radius. This process can be called coalescence. The energy balance of the system of particles as a whole will now contain only the kinetic energy of motion of the center of mass of the two particles, but not the energy of their relative motion. Consequently, as far as the system is concerned, the energy of relative motion of the particles $\sim mv^2/2$ is lost; the cross section of this process is governed by the impact parameter b

$$mvb = 2mcr_g = \frac{4Gm^2}{c}, \quad b = \frac{4Gm}{vc}, \quad \sigma_c = \pi b^2 = \frac{16\pi G^2m^2}{v^2c^2}. \tag{2.6}$$

The inverse of the time for the energy to change as the result of coalescence is given by

$$\frac{1}{\tau} = n\sigma_c v, \tag{2.7}$$

where n is the number density of particles.

Expressing all quantities in terms of v, N, m, we obtain

$$n = \frac{N}{a^3} = \frac{v^6}{G^3N^2m^3}, \quad \frac{1}{\tau_c} = \frac{16\pi v^5}{GN^2mc^2}. \tag{2.8}$$

The energy loss due to gravitational radiation occurring as the result of binary collisions decreases rapidly with increasing impact parameter.

This loss in one collision does not exceed the kinetic energy of relative motion, as otherwise the collision is classified as a coalescence.

With gravitational radiation taken into account, the probability of coalescence occurring increases. However, a significant increase in the probability occurs only at such values of v/c when the total effect is small; therefore, in practice we can restrict ourselves to the expression for the energy loss due to coalescence as the result of gravitational radiation emission by a pair of particles,

$$\frac{dE_c}{dt} = \frac{-mv^2N}{2\tau_c}. \tag{2.9}$$

The evolutionary processes characteristic of GR must be compared with the principal classical evolutionary process, namely, evaporation. The inverse characteristic time of evaporation is of the order of

$$\frac{1}{\tau_e} = n\sigma_e ve^{-u/\theta} = n\alpha\pi G^2m^2e^{-6}v^{-5},$$

where the factor α in effect takes distant collisions into account, $\alpha \sim 1$ (the argument of the exponent has been taken from Ambartsumyan's paper [1]).

A comparison of the above expressions leads to the following result: the rates of evolution on account of evaporation (W_e), coalescence (W_c), and radiation (W_r) are in the ratio

$$W_e : W_c : W_r = \frac{1}{40} : 16\frac{v^2}{c^2} : \frac{2}{45}\frac{v^5}{c^5}$$

As was to be expected, the effects due to GR are small when $v/c \ll 1$. However, because of the numerical factors, coalescence becomes as important as classical evaporation when v/c is as low as 0.04. Gravitational radiation, on the other hand, is never important.

The increase in the rate of evolution as the result of coalescence has the result that in a statistical ensemble of systems whose evolution by hypothesis began at different times, the number of systems with $v/c > 0.04$ will decrease.

3. Catastrophic Contraction

After the system has reached the critical state as the result of slow evolution, catastrophic contraction begins. The key to the understanding of contraction is the examination of the properties of the trajectories of test particles in the Schwarzschild field of a point mass. It should be recalled that for an angular momentum less than the value

$$J = 2mr_g c = \frac{4GmM}{c}$$

there are not finite trajectories in GR. In this case, the last critical circular trajectory has the radius $3r_g$.

Let us consider a test particle moving in a circular trajectory. Let us suppose that during its motion the mass of particles present inside the sphere whose equator is the given orbit gradually increases. In such a process, the angular momentum of the test particle remains constant, while its orbit gradually contracts. However, after the ratio of the angular momentum and the mass inside the sphere exceeds the critical value, the trajectory of the test particle becomes a spiral. It should be noted that for such a change from a circular orbit it is not necessary for the mass inside the orbit to be a point mass or to be concentrated inside its own gravitational radius, since the radius of the critical orbit is three times the gravitational radius.

In the problem of the motion of particles in a self-consistent field, the collapse of the orbits of some particles will lead to an increase of the field acting on the other particles, whose orbits will collapse in turn, etc. Thus, there will be an avalanche-type catastrophic contraction of the system once it has reached the critical state.

Near the critical state, the contraction develops exponentially as e^{τ/τ_k}, with a characteristic time τ_k of the same order of magnitude as the time taken for a particle to complete one revolution. Consequently, the catastrophic collapse takes place much more rapidly than evolution, since the characteristic time of evolution is greater by at least a factor of N, where N is the number of particles.

The most important factor in the development of catastrophic collapse is the presence of quasi-elliptic orbits, i.e., orbits in which the maximum and minimum distances of the particle from the center are not equal.

In fact, let us imagine a system in which all particles move in circular orbits; the planes of the orbits are randomly distributed so that the model has a spherically-symmetrical density distribution. Suppose that as the result of collisions individual particles fall towards the center and that the mass inside each sphere increases. As a result of this, orbits with $r < r_1$ become unstable [6] and the corresponding particles start moving in spirals towards the center: however, the mass acting on particles with $r > r_1$ will not increase and there will be no avalanche-type increase in the perturbation.

If, however, a particle moving between $r'_{\text{min}} \leq r' \leq r'_{\text{max}}$ has moved to a spiral orbit, this change will affect all particles for which $r''_{\text{min}} < r'_{\text{max}}$; these particles in their turn will affect the next layer of particles with $r'''_{\text{min}} < r''_{\text{max}}$, etc.

Elliptic orbits physically link the layers intersected by these orbits.

It should be noted that the isotropic distribution of particles in phase space, $N = N(\eta)$, takes care of the presence of elliptic orbits. It was pointed out above that this distribution leads to Pascal's law—isotropic pressure. The component T_1^1 responsible for the transfer of momentum along the radius depends on the radial components of the momentum of the particles. These components are absent in the case of purely circular motion. This means that T_1^1 depends on the elliptic trajectories present. It is obvious that the transfer of momentum through the surface elements of the sphere perpendicular to r is a necessary condition for the occurrence of catastrophic contraction when the inner parts of the system are unable to support the weight of the outer ones and contract under the action of the external pressure.

Detailed calculation of the dynamics of catastrophic contraction, i.e., the collapse of a system of particles is very complicated. The problem is more difficult than the nonstationary hydrodynamical problem, since in the problem concerning particles the state of matter at each point is determined not by several quantities (density, pressure, temperature, velocity), but by a function of the momentum vector.

In the almost equilibrium state, a good approximation would be to consider the function isotropic, i.e., depending only on the magnitude and not on the direction of the momentum vector or, in other words, depending only on the energy.

In the catastrophic-collapse problem, the situation is decidedly non-isotropic. Interesting qualitative features can be expected to make their appearance. There is a rapid contraction of the major portion of the mass.

On the other hand, particles which in the initial almost-equilibrium state were moving near the outer boundary of the system in orbits close to circular ones will not be drawn into this contraction. After the contraction, these particles will create a cloud surrounding the main mass. The main mass will rapidly contract to its gravitational radius; the escape of particles and the emission of gravitational radiation will rapidly and asymptotically tend to zero [8, 9]. The particles in the cloud, following the laws of slow evolution, will only gradually fall into the collapsed mass.

It is very desirable to perform calculations for such a model.

4. Applicability of the Model to Stellar Systems

The solution of the strictly-formulated problem of the evolution of a system of point masses interacting only gravitationally does not give a complete answer to the real question of the evolution of stellar systems.

The analysis carried out above is to a large extent of methodological interest and the conclusions arrived at should be applied to astronomical

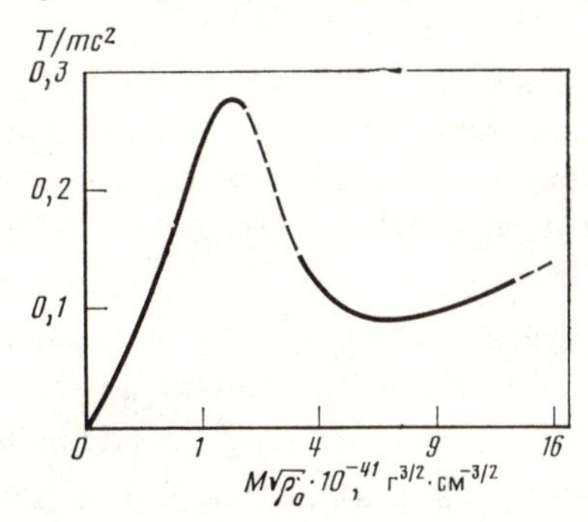

Fig. 2. Equilibrium solutions. Solid lines: $\Delta M > 0$; dashed: $\Delta M < 0$.

objects with a great deal of caution.

The average density of the system in a critical state corresponds to a mean distance between stars of the order of

$$r = 3r_g N^{2/3},$$

where r_g is the gravitational radius of a star. Consequently, for large N the distance between stars may be sufficiently large, although in view of the fact that the larger the value of N, the slower the rate of evolution, there is much less interest in the case of large N.

On the other hand, whatever the value of N, collisions between stars leading to evolution of the system occur when stars approach one another to distances of the order of Gm/v^2. In the relativistic region with $v/c \sim 1$, the approach distance becomes comparable with the gravitational radius.

This means that even if the particles represent white dwarfs or neutron stars, their dimensions are such that they cannot be considered as point masses. In real collisions there must be strong interactions between the envelopes of the stars, ejection of gas streamers, and many other phenomena, which are beyond the scope of the present paper. The situation is somewhat improved if we take into account the well-known fact that distant encounters play a big role in the case of Coulomb or Newtonian interactions.

The main conclusion of the paper—the absence of static solutions with "temperatures" above a certain limit and the change from evolution to a catastrophic collapse will, it seems, remain in force even when all of the complicating factors are taken into account.

Appendix

It can be seen from the table that the solutions are multivalued at sufficiently high temperatures. The multiple solutions are given in the lower part of the table. It is found possible to represent the whole set of results in the form of a single function. Let us notice the fact that the quantity $\tilde{M}\tilde{n}_0^{1/2}$ (or $M\rho_0^{1/2}$) varies monotonically and this makes it possible for us to use it as the independent variable. The variation of T/mc^2 with $M\rho_0^{1/2}$ is shown in Fig. 2. Using these variables, we discover a remarkable analogy with the static solution for stars, if in the corresponding graph we plot the dependence of M on ρ_0 at constant entropy throughout the whole star [5, 10, 11].

In accordance with the above analogy, for example, the quantity $\Delta M\rho_0^{1/2}$ reaches its maximum value at $\tilde{T} = \tilde{T}_{\rm crit}$ and then decreases, becoming zero on the descending branch of the curve $\tilde{T}(M\rho_0^{1/2})$, and finally becoming negative.

As the argument increases, the function $\tilde{T}$ does not decrease monotonically, but oscillates with diminishing amplitude. These oscillations are analogous to the oscillations of the function $M(\rho_0)$ for stars first discovered by N. A. Dmitriev and S. A. Kholin [12] in Oppenheimer and Volkoff's solutions.

There is a difference here, however, in that the mass defect does not remain everywhere negative; it is positive, for example, on the second ascending branch of the curve $\tilde{T}(M\rho_0^{1/2})$. Sections of the curve where $\Delta M < 0$ have been indicated by the dashed line.

It can be shown that the mathematical nature of the oscillations is the same in the two cases. N. A. Dmitriev and S. A. Kholin have shown that for large ρ_0, where the equation of state for the gas is $p = \alpha\epsilon$, the system of equilibrium equations can be reduced to one first-order equation. This equation has one singular point, the focus, and the coiling of the integral curve around it gives rise to the oscillations of $M(\rho_0)$. It is not difficult to see that in the case of our "gas," the limiting equation of state for large $\tilde{n}$ (or large $e^{-\nu}$) can be written in the form

$$p = \frac{1}{3}e^{-2\nu}\varphi(\tilde{T}), \quad \tilde{\varepsilon} = e^{-2\nu}\varphi(\tilde{T}), \quad \tilde{n} = e^{-3/2}\psi(\tilde{T}),$$

i.e., our "gas" behaves as an ultrarelativistic gas[3] for which $\gamma = 4/3$. Using the equation of state, we can derive from the equilibrium equations exactly the same first-order equation as was investigated in [12].

As in the case of stars, it seems that we can assert that the solutions corresponding to the first ascending branch of the curve $\tilde{T}(M\rho_0^{1/2})$ are stable. The point $\tilde{T} = \tilde{T}_{\rm crit}$ corresponds to neutral equilibrium and solutions for larger $M\rho_0^{1/2}$ are unstable, each successive extremum of the curve $\tilde{T}(M\rho_0^{1/2})$ being associated with the appearance of a new form of instability [12].

[3]In the other limiting case $\tilde{n} \to 0$, we find that $\gamma = 5/3$.

Received
January 13, 1965

REFERENCES

1. *Ambartsumyan V. A.*—Uchenye Zapiski LGU **22**, 19 (1938).
2. *Gurevich L. É.*—Voprosy Kosmogonii **11**, 151 (1954).
3. *Michie R. W., Bodenheimer P. H.*—Month. Not. RAS **125**, 127 (1963); **126**, 278 (1963).
4. *Tauber G. E., Weinberg J. W.*—Phys. Rev. **122**, 4 (1961).
5. *Oppenheimer J. R., Volkoff G.*—Phys. Rev. **55**, 374 (1939).
6. *Kaplan S. A.*—ZhETF **19**, 851 (1949).
7. *Zeldovich Ya. B., Novikov I. D.*—Dokl. AN SSSR **155**, 1033 (1963).
8. *Zeldovich Ya. B., Podurets M. A.*—Dokl. AN SSSR **156**, 57 (1964).
9. *Podurets M. A.*—Astron. Zh. **41**, 1090 (1964).
10. *Zeldovich Ya. B.*—Voprosy Kosmogonii **9**, 157 (1963).
11. *Ambartsumyan V. A., Saakyan G. S.*—Voprosy Kosmogonii **9**, 91 (1963).
12. *Dmitriev N. A., Kholin S. A.*—Voprosy Kosmogonii **9**, 254 (1963).
13. *Zeldovich Ya. B.*—ZhETF **42**, 1641 (1962); *Zeldovich Ya. B., Novikov I. D.*—UFN **84**, 377 (1964).

Commentary

Systems of point masses in their own self-consistent gravitational field (in a self-consistent metric) have been investigated for a long time, from the first years of the general theory of relativity. The present article stands out in its consideration of the evolution of such a system, without artificial assumptions about circular orbits. The authors' conclusion about relativistic collapse in the late stages of the evolution is indisputable and remains in force today.

This conclusion is of great importance for astrophysics as a whole since the nuclei of quasars (and, in certain cases, the nuclei of galaxies) are black holes with a large mass. There is every reason to believe that such black holes formed not by the collapse of a single super-massive gas superstar, but precisely by the collapse of a cluster of individual stars. The above work considers collapse due to evaporation of stars which take away more than average energy and due to the gravitational radiation of stars in collision. Later works found that for a cluster consisting of ordinary stars (not neutron stars and not black holes with masses of order 3–30 $M_\odot$), the collision of shells of stars plays a major role. Finally, we must regretfully note that the idea of explaining the phenomenon of quasars by accretion on a supermassive black hole did not arise in Ya. B.'s group, even though all the individual ingredients of this idea were discussed from various points of view.

In Ya. B.'s group there followed the construction of mathematical models of star clusters with a large gravitational red-shift (see works by Ya. B. and M. A. Podurets[1] and G. S. Bisnovatyĭ-Kogan[2]).

[1] *Zeldovich Ya. B., Podurets M. A.*—Astrofizika **5**, 223–234 (1968).
[2] *Bisnovatyĭ-Kogan G. S., Zeldovich Ya. B.*—Astrofizika **5**, 223–234 (1969).

43

Collapsed Stars in Binary Systems*

With O. Kh. Guseinov

A method for detecting collapsed stars which are members of spectroscopic binary systems is proposed. Several pairs are selected from among the spectroscopic binary systems with invisible companions, in which one may suppose that their components are collapsed stars.

Collapsed stars (CS), whose existence is predicted by the general theory of relativity, can be detected only through their gravitational field (see review and references to original literature in [1]). A comparison of the total mass of a cluster to the mass of its visible stars for the purpose of determining the masses of CS is very inconclusive in view of the unreliability of the mass determinations. One would like to discover an individual CS. In principle this is possible if a CS happens to belong to a binary system having an ordinary star (OS) for the other component.

A survey of the literature allowed us to select several pairs among the spectroscopic binaries with invisible companions in which one can suppose that the companion is a CS. We list the data for these pairs (Table 1). Reference data are given in the first columns—the designation of the star and its coordinates. Then the observational data follow—the apparent magnitude m, and the spectral type of the observable star, the period of the binary p, the half-amplitude of the periodic change in velocity K_1, and $a_1 \sin i$, the projection of the semimajor axis of the orbit onto the plane of the sky, which is calculated by means of celestial mechanics, and $MF = \mathcal{M}_2^3 \sin^3 i/(\mathcal{M}_1 + \mathcal{M}_2)^2$ in $\mathcal{M}_\odot$, the subscript 1 referring to OS, 2—to CS. The remaining quantities were determined very unreliably: We can find the mass of the OS, $\mathcal{M}_1$, corresponding to its spectral type and luminosity class. Assuming an inclination $i = 90°$, then $\sin^3 i = 1$, and we find the mass of the CS, $\mathcal{M}_2$. Next to it is given $\mathcal{M}_2'$, calculated for the average value $\sin^3 i = 2/3$. Finally the absolute parallax π is given and the literature is cited. We note that assuming $i = 90°$ for given MF and $\mathcal{M}_1$ gives a minimum value for the mass of the second star. Among the pairs listed in the table, the invisible companion has a mass $\mathcal{M}_2$ larger than the mass of the observable ordinary star $\mathcal{M}_1$. One could advance the hypothesis that the invisible components of the indicated spectroscopic binaries are CS. It is also possible that the invisible component of the last star in the table is a neutron star.

*Astronomicheskiĭ zhurnal **43** (2), 313–315 (1966).

However, such a conclusion is by no means categorical. According to a remark by I. D. Novikov, the invisible component of a spectroscopic binary can produce a spectrum without noticeable spectral lines on the photograph and be "invisible" only in this sense. L. I. Snezhko mentions that among binaries there often occur stars which have passed through the giant phase, ejected a hydrogen envelope, and are burning helium. In this case the luminosity of the OS would be far greater than on the main sequence,[1] and the secondary—invisible—star could be located on the main sequence and nevertheless be invisible by contrast with the brighter (although less massive) primary.

No.	Star	α_{1950}	δ_{1950}	m_1	Sp	p, Dn	K_1, km/sec	$a_i \sin i$, 10^6 km	MF	$\mathcal{M}_1$	$\mathcal{M}_2$	$\mathcal{M}_2''$	π''	m_2	Ref.
1	HD 187399	$19^h46^m,6$	$+29° 17'$	7.7	B9	27.97	104.5	37.6	2.72	4.4	7.1	9.6			[3]
2	+40°1196	05 03 ,7	+40 53	8.1	B3	3710	31.5	1528	10.4	10	22	29			[4]
3	HD 30353	04 45 ,2	+43 12	7.7	cAS_p	359.7	51.3	244	4.5	12.6	15.1	19			[5]
4	HD 193928	20 17 ,8	+36 36	9.7	WN6	21.64	130	—	4.94	10	14.2	18			[6]
5	π Cep A	23 06 ,0	+75 07	4.56	gG1	556.2	23.02	169	0.623	3	2.76	3.4	0.11	6.1	[7]
6	ξ Pav	18 18 ,6	−61 31	4.25	gM1	2214	17.92	526	1.188	7	5.9	7.2	015	3	[8]
7	α Her B	17 12 ,4	+14 27	5.39	F8	51.58	36.12	25.62	0.258	1.4	1.22	1.48	006	10.2	[9]

Absorption lines of hydrogen are observed in the spectrum of the first star—HD 187399, and they are shifted in the direction of negative velocity. The author ascribes these lines to a scattered cloud of hydrogen surrounding the orbit of the double star. In the case of the third star—HD 30353—the authors remark on the star's lack of hydrogen and the unusually large value of the mass function. It is known that the motion of star +40°1196 as determined from lines of hydrogen, calcium, and iron is greatly different.

The main purpose of the present paper is to draw the attention of observers to the objects indicated above (and analogous ones). Along with a careful study of the spectra, it would be very important to determine the parallax of these stars or at least its lower limit. Knowing the distance to a star, it would be possible to determine its absolute magnitude and ascertain whether the secondary star is invisible for a trivial reason—the anomalously large luminosity of the primary. In the last cases [5, 6, 7], where the parallax is known, we found m_2—the apparent magnitude which a star of mass $\mathcal{M}_2$ would have if it were located on the main sequence. This magnitude is determined very roughly, since in addition to

[1]Such a situation is mentioned rather often among binaries [10].

possible errors in the determination of the mass $\mathcal{M}_2$, there are also possible uncertainties in the parallaxes, for example, due to incorrect allowance for proper motion of the OS.

It is possible to try to detect the proper motion of a star (of the order of 0".01—at the limit of observational accuracy), and also to check with the aid of the Michelson interferometer whether or not the star is a visual double. One could search for specific phenomena associated with the gas motion in the field of a CS [1]. The proof of the actual existence of even a single collapsed star would have great fundamental significance for the entire theory of stellar evolution.[2] Even a negative result, i.e., the explanation of all the pairs without resorting to CS, would be of definite interest: from statistical comparisons one would be able to conclude that the evolution OS stars of large mass does not lead to the condition of a collapsed star, evidently, due to the regularity of the mass loss in certain stages of evolution;[3] also the onset of the collapse can be delayed because of the rotation of the stars. This is already indicated by the large number of visual and eclipsing pairs which we find in comparison to the suspected CS among the stars with masses greater than $2M_\odot$. In [13] the authors remark that the presence of white dwarfs in young clusters (and in particular in binary systems) points to the possibility of the ejection of matter, occurring in a time of the order of 10^4–10^5 years.

In conclusion it is necessary to emphasize again the great fundamental importance of the detection of collapsed stars. By virtue of this every case in which there is a possibility of demonstrating the existence of such an unusual object should become the subject of a careful and comprehensive critical investigation.

We express our gratitude to P. G. Kulikovskiĭ, I. D. Novikov and L. I. Snezhko for help and discussion.

[2]The possibility of the detection of collapse from the emission of energetic neutrinos is examined in [11].

[3]Cameron [12], for example, discussed such a possibility.

Received
October 18, 1965

REFERENCES

1. *Zeldovich Ya. B., Novikov I. D.*—UFN **83** (3), 377 (1964); **86** (3), 477 (1985).
2. *Becvar A.*—Atlas coeli, II (1959); *Moore J. H., Neubauer F. I.*—Lick Observ. Bull **20**, 521 (1948).
3. *Merrill P. W.*—Astrophys. J. **110**, 59 (1949).
4. *Melustey S. W.*—Astrophys. J. Suppl. **4**, 35 (1959); *Schmidt-Naler T.*—Veröff Univ. Sternwarte Boon **70** (1964); *Merrill W.*—Astrophys. J. **79**, 343 (1934).
5. *Heard J. E., Boshko O.*—Astrophys. J. **121**, 192 (1955); *Heard J. E.*—Publs. D. Dunlap Observ. Univ. Toronto **2** 9, 269 (1962).

6. *Hilther W. A.*—Astrophys. J. **101**, 356 (1945).
7. *Harper W. E.*—Victoria Publ. **3** 8, 189 (1924).
8. *Bateson F. M., Jones A. F., Philpott D. A.*—Circ. NZAS **107** (1960).
9. *Bouigne R.*—Ton. Ann. **25**, 69 (1957).
10. *Sahade I.*—In: Modeles d'etoiles et evolution stellaire. Liege **16**, 76 (1960).
11. *Zeldovich Ya. B., Guseinov O. Kh.*—Dokl. AN SSSR **162** 4, 791 (1965).
12. *Cameron A. G.*—In: Modeles d'etoiles et evolution stellaire. Liege **16**, 76 (1960).
13. *Auer L. H., Wolf N. I.*—Astrophys. J. **142**, 182 (1965).

Commentary

This work provoked in its time an enormous response and helped to establish a large number of observational programs in a number of observatories. Following the ideas of this article, V. Trimble and K. Thorne[1] compiled an extensive list of "black hole candidates."

It should nevertheless be noted that, due to the successes of X-ray astronomy, the search for black holes took a different route.

However, in order to distinguish a black hole from a neutron star, even today the criterion of a mass exceeding a certain limit, about 3 solar masses, is used.

Together with the mention of the role of accretion on neutron stars (see **44**), the present work stimulated the theory and observation of the final products of evolution.

[1] *Trimble V. L., Thorne K. S.*—Astrophys. J. **156**, 1013 (1969).

44
X-Ray Emission Accompanying the Accretion of Gas by a Neutron Star*

With N. I. Shakura

A discussion is given of the accretion of gas by a neutron star and the spectrum of the radiation emitted in this process. The entire treatment applies to the spherically symmetric case. A systematic and internally consistent method is developed for calculating the physical parameters of the atmosphere of a neutron star and the electromagnetic radiation spectrum. Calculations are performed under two assumptions regarding the mean free path of the incident protons: 1) that the protons are decelerated by pair collisions in the atmosphere, and 2) that collective plasma oscillations would serve to reduce the mean free path. The computed spectrum differs considerably from a Planck curve, particularly in the second case.

Introduction

Neutron stars were born at the tip of the theoretician's pen over 30 years ago, but a convincing identification of such a star has yet to be made. The same is also true of stars in a state of relativistic collapse.

Calculations indicate that neutron stars would develop as a result of catastrophic contraction to a state of intensive oscillations and with a high temperature. The initial state at the time when stability is lost (prior to the catastrophe) has been discussed by Bisnovatyĭ-Kogan [1], and calculations of the catastrophic stage itself have been performed by Arnett [2]. However, the heat would rapidly be dissipated by neutrino radiation [3], and the oscillations would also readily decay because of gravitational radiation [3].

Accretion of gas from the surrounding space could ensure a very prolonged maintenance of X-ray emission by a neutron star.

The lifetime of a neutron star, if measured as the period during which its mass increases, say, from 1 to $1.5\,M_{\odot}$, would be $3 \cdot 10^8$ yr for a luminosity $L = 10^{37}$ ergs/s. The energy released per unit mass of infalling material,

*Astronomicheskiĭ zhurnal **46** (2), 226–236 (1969).

equal to the gravitational potential at the surface, would be roughly (0.1–0.2)c^2, a value 10–20 times the energy released in nuclear reactions. These considerations were advanced in 1964 and at that time a distinctive, self-regulating accretion mechanism was described: the pressure of the radiant flux would restrict the accretion rate to a value of the order of $1.5 \cdot 10^{-8} M_\odot$ per year, corresponding to a "critical" luminosity $L_c \approx 6 \cdot 10^4 L_\odot$ for neutron stars with $M \approx M_\odot$.

It has been pointed out [5–8] that in addition to interstellar gas, some of the gas ejected at the time when a neutron star was formed may be subject to accretion, or gas flowing away from the secondary component if the neutron star belongs to a binary system.

Shklovskiĭ [9] has analyzed the radiation of Sco X-1, representing it as the bremsstrahlung of optically thin gas layers at different temperatures. He claims to have proved the gas-accretion mechanism in a binary star directly from observational data, and thereby independently of prior theoretical hypotheses.

Cameron and Mok [10] have considered accretion by white dwarfs and have pointed out that soft X-rays might be emitted in this case as well. One can show without difficulty that the lifetime should then be substantially shorter, of order $3 \cdot 10^5$ yr for $L \approx 10^{37}$ erg/s, since the gravitational potential is smaller at the surface of the white dwarf than of the neutron star. Possibly nuclear reaction bursts and the eruption of gas [11] would lengthen this period.

Melrose and Cameron [12] have recently considered the generation of fast particles and plasma oscillations when gas is accreted by a neutron star in conjunction with the instability for $L \approx L_c$. This mechanism has been applied to explain the compact radio source in the Crab Nebula.

To conclude this survey we may mention the distinctive accretion by a collapsing ("frozen") star [7]; in this instance spherically symmetric accretion in general would not, in the hydrodynamic approximation, be accompanied by the release of energy.

We would also point out that it is meaningful to consider accretion only for stars that are in the closing stage of their evolution, since in the case of stars whose energy source is nuclear burning a phenomenon opposite to accretion occurs—the "stellar wind."

The great diversity of accretion effects will be evident from this brief survey. Yet until recently no accurate solutions have been obtained for even the simplest and most idealized problems.

In this paper we shall consider the accretion of gas by a neutron star and the radiation arising in this process. The entire treatment will refer to the simplest idealized case of spherically symmetric gas motion. Adopting a law for the deceleration of the incoming particles, we shall find the distribution of temperature, density, and pressure in the layer where particle deceleration occurs. We shall then find the spectrum of the emergent radiation. The calculation will be performed in two approximations: 1) for particle deceleration by collisions and 2) with allowance for plasma instability, serving to reduce the mean free path.

We shall not take into account the influence of the magnetic field, either that frozen into the incoming gas or the field of the star. We assume that $L < L_c$, so that we may neglect the influence of radiation on the incident flux and the inverse action of the rarefied incoming gas on the emission spectrum. A preliminary report of the results was presented at the 13th General Assembly of the International Astronomical Union [13].

1. *Equations for the Energy Balance, Temperature Distribution, and Density*

All the numerical computations are very approximate in character and have been made for a neutron star with $M \approx M_\odot$, $R \approx 10^6$ cm, and a luminosity $L_1 = 1.3 \cdot 10^{37}$ erg/s for $L_2 = 1.3 \cdot 10^{36}$ erg/s, equal to 0.1 or 0.01 of the critical luminosity respectively. A stream of ionized hydrogen impinges on the surface of the neutron star; according to the free-fall law the velocity $v = (2GM/R)^{1/2}$ of the stream reaches $\approx 0.5c$. The mass flux $dM/dt = 4\pi\rho v R^2 = 10^{17}$ g/s or 10^{16} g/s, corresponding to a density in the incident stream of $5 \cdot 10^{-7}$ g/cm^3 or $5 \cdot 10^{-8}$ g/cm^3 at the surface.

As it decelerates in the atmosphere of the neutron star the stream loses its kinetic energy, which is transformed into radiation. Extreme limits on the temperature can easily be given. Under laboratory conditions, when a flow encounters an obstacle a shock wave is formed, kinetic energy is transformed into thermal energy, and radiation is slowly released. In this approximation we find a temperature $T = 10^{12}$K along a Hugoniot adiabat. We can obtain another estimate from the energy balance, assuming that the neutron star radiates like a blackbody:

$$L = \frac{GM}{R}\frac{dM}{dt} = 4\pi R^2 b T_{\text{eff}}^4, \quad b = 5.75 \cdot 10^{-5}\,\text{erg}/(\text{cm}^2 \cdot \text{deg}^4).$$

We thereby have $T_{\text{eff}} = 1.15 \cdot 10^7$K for $L = 0.1L_c$ or $T_{\text{eff}} = 6.5 \cdot 10^6$K for $L = 0.01L_c$.

The enormous disparity between these two estimates shows how important a task the detailed study of the phenomenon is. The essential aspect of the problem is that the protons in the incident stream are not decelerated instantaneously. They transfer their energy to a comparatively extended layer of the atmosphere. Let us consider the thermal balance of this layer, or more accurately, of the electrons located in this layer. They receive their energy from the incoming protons (either directly or through protons in the layer which receive their energy from the incoming protons); the electrons release their energy by bremsstrahlung or through the inverse Compton effect.

We introduce the quantity $y = \int_{x_0}^{\infty} \rho(x)\,dx$; y has the meaning of the amount of material lying above a given point. This integral contains the density $\rho(x)$ in the material in the neutron-star atmosphere, which is essentially at rest.

Without making the mean free path of the incident particles more specific at this point, we shall denote it by y_0.

Then in the first approximation the energy released per gram of atmospheric material is $W = Q/y_0$ for $y < y_0$ and $W = 0$ for $y > y_0$, where $Q = L/4\pi R^2$ is the energy flux per unit surface area of the star. The thermal balance of the electrons will be determined by the arriving energy W and the expenditure to bremsstrahlung,

$$W' = 5 \cdot 10^{20} \sqrt{T_e}\rho \tag{1.1}$$

as well as the energy lost to Comptonization, that is, the change in the energy of the photons because of scattering by Maxwellian electrons [14, 15]:

$$W'' = \frac{4\varepsilon c \sigma_c k T_e}{m_p m_e c^2}; \tag{1.2}$$

here ε is the radiant energy density and $\sigma_c = 6.65 \cdot 10^{-25}$ cm^2.

These expressions do not take the inverse processes into consideration. Evidently in an equilibrium radiation field with $T_{ph} = T_e$, $W' = W'' = 0$. We shall allow for the inverse processes by introducing effective radiation temperatures T' and T''; the energy-balance equation will then take the form

$$\frac{Q}{y_0} = 5 \cdot 10^{20} \sqrt{T_e}\rho \left(1 - \frac{T'}{T_e}\right) + 6.5\varepsilon T_e \left(1 - \frac{T''}{T_e}\right). \tag{1.3}$$

The quantities T' and T'' depend on the spectrum of the radiation, but below, in the first approximation, we shall adopt $T' = T'' = T_{ph} = (\varepsilon/a)^{1/4}$; $a = 7.8 \cdot 10^{-15}$ erg/cm$^3 \cdot$ deg^4.

The radiant energy density is determined by the diffusion equation:

$$q = Q(y - y_0)/y_0 = -c\,d\varepsilon/3\sigma dy \qquad y < y_0, \tag{1.4}$$

with $\sigma = 0.38$ cm^2/g the opacity of fully ionized plasma and q the radiant energy flux. For $y > y_0$, $q = 0$ and $\varepsilon =$ const. For fully ionized hydrogen plasma this equation is valid independently of the emission spectrum.

If we take $\varepsilon = \sqrt{3Q}/c$ at the boundary of the atmosphere, where $y = 0$, we have

$$\begin{aligned} \varepsilon &= \frac{Q}{c}\left\{\sqrt{3} + 3\sigma y_0 \left[\frac{y}{y_0} - \frac{1}{2}\left(\frac{y}{y_0}\right)^2\right]\right\}, \qquad 0 < y < y_0, \\ \varepsilon &= \frac{Q}{c}\left\{\sqrt{3} + \frac{3}{2}\sigma y_0\right\}, \qquad y > y_0. \end{aligned} \tag{1.5}$$

In the interior, where $y \gg y_0$, complete thermodynamic equilibrium will be established, with a temperature which we can find from the formula $\varepsilon = aT^4$. Using Eqs. (1.5), we find that for $y \gg y_0$, $T \approx L^{1/4} y_0^{1/4}$. At the surface, where $\rho \to 0$ and $W' \to 0$, the temperature of the electrons will be determined by Comptonization and will not depend on the luminosity, since $\varepsilon \sim Q \sim L$.

We can find the density distribution of the material trivially by using the equation of hydrostatic equilibrium:

$$P = \frac{2\rho kT}{m_p} = \left(\frac{GM}{R^2} + \frac{\rho_0 v^2}{y_0}\right) y, \qquad 0 < y < y_0;$$

$$P = \frac{2\rho kT}{m_p} = \frac{GMy}{R^2} + \rho_0 v^2, \qquad y > y_0. \tag{1.6}$$

The first term in the sum represents the weight of the atmospheric material; the second, the momentum imparted by the incident particles to the layer y_0.

If we now adopt a flux $Q_1 = 10^{24}$ erg/cm^2·s or $Q_2 = 10^{23}$ erg/cm^2·s at the surface and a mean free path y_0, we can obtain the temperature and density distribution in the atmosphere of the neutron star. The determination of the value of y_0 remains the fundamental difficulty in the problem.

1a. Deceleration of protons by collisions with individual particles. The kinetic energy of the incident protons can vary within wide limits, depending on the gravitational potential of the neutron star. For a mass $M \sim M_\odot$, $E_p \sim 100$–300 MeV. The mean free path of such protons in fully ionized plasma will be determined by Coulomb collisions, and we obtain $y_0 \sim 5$–30 g/cm^2. For our numerical computations we shall adopt $y_0 \sim 20$ g/cm^2. Since in the interior $T \sim y_0^{1/4}$, a different value for y_0 will not materially alter the physical conditions for $y \gg y_0$, while at the surface the electron temperature will not depend on y_0. To be sure, for small y_0 the contribution of Comptonization to the energy exchange will increase, and this will affect the emergent spectrum, but we shall take the influence of Comptonization into account in another limiting case (Sec. 1b). For $y_0 \sim 20$ g/cm^2 at depths where $y \gg y_0$ we have $T_1 = 1.5 \cdot 10^7$K or $T_2 = 8.6 \cdot 10^6$K for the luminosities adopted. The electron temperature at the surface is $\approx 10^8$K in both cases. We shall obtain the temperature distribution in the y_0 layer by solving Eqs. (1.3), (1.5), and (1.6) numerically. The results of the computations are displayed in Fig. 1; here we give also the effective temperatures obtained from the general energy estimates.

We introduce the quantity

$$\eta = \left(\int_0^{y_0} W''\, dy\right) \Big/ Q. \tag{1.7}$$

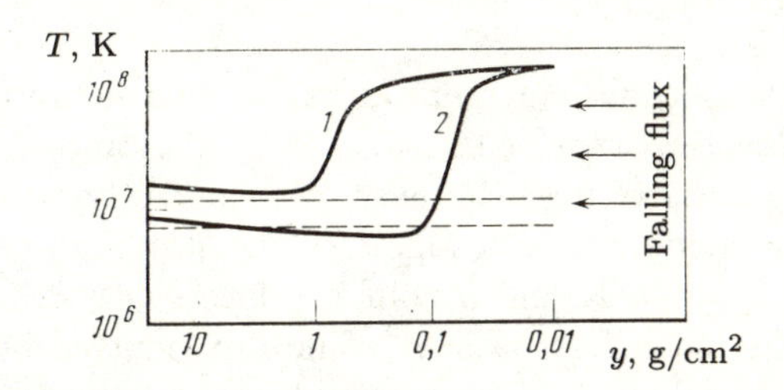

Fig. 1. Electron temperature distribution in the atmosphere of a neutron star for luminosities (1) $L_1 = 1.3 \cdot 10^{37}$ erg/s and (2) $L_2 = 1.3 \cdot 10^{36}$ erg/s. Dashed lines denote effective temperatures derived from general energy balance; y_0 is taken to be 20 g/cm^2.

which represents the fraction of energy released by electrons to Comptonization. For $y_0 \sim 20$ g/cm^2, $\eta < 0.05$ if $L_1 = 0.1 L_c$ or $\eta < 0.01$ if $L_2 = 0.01 L_c$, so that the contribution of Comptonization to the overall heat exchange is not large; accordingly, Comptonization will have a small effect on the emergent radiation.

1b. The influence of plasma oscillations. It is well established that when a beam of charged particles passes through a plasma, oscillations will appear in the plasma, primarily at the Langmuir frequency $\nu_{\rm pl} = (\pi e^2 n/4m_e)^{1/2}$. The plasma oscillations will interact efficiently with the ions in the beam, decelerating the beam.

The reduction in the mean free path due to plasma instabilities will lead to a rise in the electron temperature. We may regard the rise in temperature as restricted by the circumstance that for $v_{T_e} \geq v = (2GM/R)^{1/2}$ the rate of generation of plasma oscillations will be subject to Landau decay [16]. We shall assume below that a regime of plasma oscillations and a value of y_0 are established such that $v_{T_e} = v = (2GM/R)^{1/2}$. The electron temperature determined by this condition will be $\theta \sim 10^9$K.

The mean free path for deceleration will be found from the condition that the heating and losses to radiation and Comptonization occur at a given temperature θ. Using Eqs. (1.3), (1.5) and (1.6), and setting $\theta = 10^9$K throughout the y_0 layer, we obtain $y_0 \sim 2$ g/cm^2.

We can determine the temperature of the material in the interior, for $y \gg y_0$, in the same manner as previously: $T = (\varepsilon/a)^{1/4}$. This yields $T_1 = 1.1 \cdot 10^7$K or $T_2 = 6 \cdot 10^6$K for the different luminosities.

In this case, then, the temperature in the surface layer increases by almost an order of magnitude; the fraction of Comptonization rises ($\eta = 0.96$ for L_1 or $\eta \sim 0.7$ for L_2), and hard X-ray photons of energy 50–100 keV appear (bremsstrahlung from the hot layer). The emergent spectrum will definitely be non-Planckian.

2. Spectrum of the Radiation

In a real atmosphere the emergent radiation flux is determined by solving the integro-differential transfer equation with allowance for the change in frequency through scattering. However, to a perfectly satisfactory approximation we may divide the atmosphere of the star into a relatively "cool" half-space with $T \sim 10^7$K and a thin hot layer with $\theta \sim 10^8$K or 10^9K, depending on the mean free path y_0. The flux emitted by the isothermal half-space with the relatively low radiation temperature will be Comptonized by the hot electrons in the thin layer; the radiation of the hot layer itself will be added to this flux.

It is readily seen that for a large part of the X-ray spectrum the coefficient $\sigma = 0.38$ cm^2/g for scattering by free electrons in the atmosphere of a neutron star will be many times larger than the true absorption coefficient, which is here determined by free-free transitions:

$$\kappa_v = 1.14 \cdot 10^{56} \frac{1 - e^{-h\nu/kT}}{T^{1/2}\nu^3} \rho \, \text{cm}^2/\text{g}. \tag{2.1}$$

If scattering is included, the radiation flux from the isothermal homogeneous half-space is

$$F_\nu \approx \pi B_\nu(T) \sqrt{\kappa_\nu/(\sigma + \kappa_\nu)} \tag{2.2}$$

and the Wien region, where $h\nu > kT$ (in this region the inequality $\sigma/\kappa_\nu \gg 1$ will be satisfied with large margin), $F_\nu \sim \nu^{3/2} \exp(-h\nu/kT)$. For comparison, the radiant flux from a blackbody ($\sigma = 0$) will have $F_\nu \sim \nu^3 \exp(-h\nu/kT)$. However, the atmosphere of the neutron star is not homogeneous; its density increases very rapidly with depth. This circumstance somewhat mitigates the effects of scattering, and $F_\nu \sim \nu^2 \exp(-h\nu/kT)$. A calculation of the spectrum in the Eddington approximation, for $\sigma > \kappa_\nu$, is given in Appendix 1. For $\sigma < \kappa_\nu$ the half-space radiates like a blackbody. In our case this condition is satisfied in the frequency range $h\nu < kT$.

It is very difficult to arrive at an exact solution to the problem of the spectrum with allowance for the change in frequency because of scattering. The problem is formulated as an integral equation for a function of the coordinates (x) and the wave vector of a photon; in the plane case the function takes the form $F(x, p, \mu)$, where p is the modulus of the wave vector and $\mu = p_x/p = \cos(\mathbf{p}, x)$.

If there is no change in frequency the function $F_p(x, \mu)$ is found independently for each p, and very efficient methods are known.

We present below an approximate method in which we use the solution to the problem of scattering with a change of frequency in the nonstationary but spatially homogeneous case, that is, for $F(p, t)$.

The exact solution includes a determination of the time τ that the photons remain in the high-temperature zone; in the geometry adopted here we may speak of the distribution of photons with respect to the quantity τ. We shall

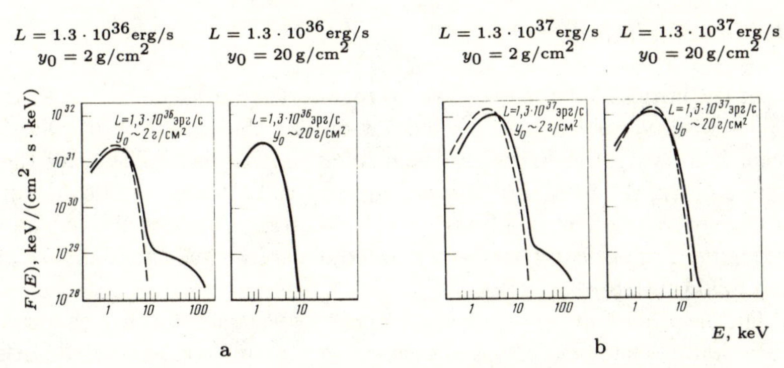

Fig. 2. Neutron star energy spectra computed for different values of the parameters L and y_0. Dashed lines are the spectra for black bodies with equal overall luminosity.

replace this distribution below by a single mean value $\bar{\tau}$, and shall use the solution $F(p, \bar{\tau})$ to the homogeneous problem. We shall find the value of $\bar{\tau}$ from energy considerations. It was shown above that even if we do not know the spectrum we can find the energy that the electrons release by the mechanism of Comptonization.

The problem of the Comptonization of radiation in a homogeneous medium for scattering by electrons whose mean energy is many times larger than the mean energy of the photons has an analytic solution [17]. The change in the spectrum depends on a unique quantity $t = \bar{\tau}(kT_e/m_e c^2)$, or in the general case of a variable temperature $t = \frac{k}{m_e c^2} \int_{t_1}^{t_2} T\, dt$.

We can also express in terms of this quantity the change in the total energy radiation per unit volume: $F_2(t_2) = e^{4t} F_1(t_1)$.

In the problem of the radiation of an atmosphere we identify the quantity F_2 with the total radiant flux emerging outward from the hot layer, and the quantity F_1 with the radiant flux generated in the cool half-space and penetrating into the hot layer. We thereby find the effective value of t, which we use to transform the radiation spectrum of the cool layer; after the Comptonization process one finally obtains for the emergent spectrum

$$F_2(\nu) = \int_0^\infty K(\nu, \nu', t) F_1(\nu')\, d\nu', \tag{2.3}$$

where

$$K = \left(\frac{\nu}{\nu'}\right)^3 \frac{1}{\nu'} \exp\left\{-\frac{\left[\ln\frac{\nu}{\nu'} + 3t\right]^2}{4t}\right\}.$$

The mathematical details are all given in Appendix 2. Qualitatively, the influence of the Comptonization on the hot electrons consists in redistributing the photons: the number of soft photons decreases with increasing number of hard photons.

However, in the region of still harder photons the bremsstrahlung of the hot layer plays the leading role, as is evident from Fig. 2.

3. Plasma Radiation

In the optical, infrared, and radio regions the radiation of heated ionized gas is weak simply because it cannot exceed the equilibrium radiation in accordance with the Rayleigh-Jeans equation: $h\nu \ll kT$. Despite the high temperature, the total flux of low-frequency radiation is small as the surface of the neutron star is small; that a neutron star may be visible in the X-ray region but not optically has been pointed out some time ago (Ambartsumyan and Saakyan [18]).

On the other hand, evidently it is in the low-frequency range that coherent radiation mechanisms play a decisive role. In this connection the question of plasma instabilities should be considered in the phenomenon as pictured here. It is well recognized that such instabilities will appear for practically any deviations from statistical equilibrium. In the spherical problem there is a compelling factor in the incident stream that serves to induce a departure from a Maxwellian distribution: in free fall a plasma volume element will become elongated along the radius and will contract in the perpendicular direction. In the collisionless problem the distribution in momentum space will become anisotropic (this remark is due to Sagdeev, who predicted it with reference to the "solar wind").

In our case, however, collisions of electrons with photons, i.e., the Compton effect, will not only maintain the electron temperature in the incident stream at a constant level, of order 1 keV, but will moreover ensure an isotropic Maxwellian distribution for the electrons. We may assume that no strong plasma oscillations are present in the incident stream.

Under the influence of the incident stream, strong plasma oscillations would probably be generated in the atmosphere, materially reducing the mean free path of the protons (see Sec. 1b). These oscillations should be transformed into electromagnetic waves with considerable efficiency. In the zone where the protons are decelerated the density changes from $5 \cdot 10^{-7}$ to $3 \cdot 10^{-3}$ g/cm^3, corresponding to $y_0 \sim 2$ g/cm^2. The plasma frequency changes from $4.4 \cdot 10^{12}$ to $3.4 \cdot 10^{14}$ Hz; λ, from $7 \cdot 10^5$ to 8000 Å.

The optical depth in the incident stream will be greater than unity for frequencies $\nu \lesssim 10^{14}$ Hz, so that the shortest-wave region will be of interest; rough estimates indicate that some of the optical radiation may perhaps be of plasma origin, with this portion being confined to a negligible solid angle.

It remains unclear whether the plasma oscillations in the atmosphere will generate fast electrons, and what the subsequent fate of these electrons would be. We may presume that the oscillations serving to reduce the mean free path of the protons would simultaneously reduce the path of the electrons as well, impeding the escape of fast electrons toward the stream and reducing the electron heat conductivity.

In this investigation we shall assume that the luminosity is significantly lower than the critical value, so that the influence of the outgoing photon flux on the incident particle stream will be small. Thus we shall not consider here the question of the possible instability of the stream for $L \approx L_c$ as formulated previously [6]. Cameron and Melrose [12] regard this instability as the probable cause of the generation of particles and radio emission in the compact Crab Nebula source.

4. Interaction Between the Emergent Radiation and the Incident Stream

Let us estimate the optical depth τ in the incident stream relative to Compton scattering. (For photons of energy $E \geq 1$ keV the absorption is negligible.)

Using the free-fall law $v \sim r^{-1/2}$ and the conservation law $\rho v r^2 = \text{const}$ for the stream, we obtain the density distribution $\rho \sim r^{-3/2}$ in the incident stream.

We now have for τ

$$\tau = \sigma_c \int_R n(R) \left(\frac{R}{r}\right)^{3/2} dr = 2\sigma_c n(R) R. \tag{4.1}$$

Thus $\tau = 0.4$ for $L = 1.3 \cdot 10^{37}$ erg/s, and is determined basically by the layers that are located in the immediate vicinity of the stellar surface. Clearly the spectral energy distribution will remain practically unchanged.

However, this is fully adequate for the Compton effect to keep the temperature in the incident stream to $T \sim T_{ph} \sim 10^7$K out to distances $r \sim 10^{10}$ cm.

The temperature T_{ph} will be determined not by the integrated radiation density at a given point [see Eq. (1.3)], which is greatly weakened by dilution, but by the spectral composition of the incoming radiation. Let us now set up the energy-balance equation. For 1 g of incident material we have

$$R^* \frac{dT}{dt} = 6.5\varepsilon(T_{ph} - T) - 5 \cdot 10^{20}\sqrt{T}\rho + R^* \frac{T}{\rho} u \frac{d\rho}{dr}, \tag{4.2}$$

$$R^* = 8.3 \cdot 10^7 \text{ erg/(mole} \cdot \text{deg)}.$$

Here in addition to Compton scattering we have included bremsstrahlung and the change in energy due to isothermal compression of the material during the fall. We may rewrite Eq. (4.2) in the following way:

$$\frac{dT}{dt} = \frac{10^{19}}{r^2} \frac{L}{L_c}(T_{ph} - T) - 10^{12}\sqrt{T}\rho + \frac{Tu}{r}. \tag{4.3}$$

By direct substitution one can readily show that out to distances $r \sim 10^{10}$ cm the term for the Compton effect will be decisive. Supersonic flow will not be destroyed in the process. We can estimate the bremsstrahlung

$L' = 10^{-27}T^{1/2}ME$ of this layer:

$$ME = 4\pi \int_R n^2(R) \left(\frac{R}{r}\right)^3 r^2\, dr = 4\pi n^2(R) R^3 \ln(r/R) \qquad (4.4)$$

for $r \sim 10^{10}$ cm, $ME \sim 10^{54}$ cm^{-5} and $L' \sim 3 \cdot 10^{30}$ erg/s, amounting to $\approx 10^{-7}$ of the total luminosity of the neutron star.

For $r > 10^{10}$ cm the energy balance and radiation of the rarefied incident gas will depend appreciably on the assumptions regarding the distant regions. Are we concerned with accretion from the interstellar gas, with gas flowing out of the secondary component of a binary star, or with gas that has previously been ejected in a supernova explosion? We may mention the possibility that emission lines of H, He II, N III, and C III may appear in this region, as observed in the spectrum of the optical object identified with Sco X-1. We hope to undertake a detailed analysis of these assumptions in forthcoming papers.

Discussion of the Results

The X-ray spectra obtained above depend on two parameters, the luminosity L (or the mass flux dM/dt associated with it) and the distance y_0 for deceleration of the stream. The general form of the spectrum is such that by assigning parameters we can obtain a satisfactory agreement with experiment (for example, for Sco X-1, $L \sim 10^{37}$ erg/s and $y_0 \sim 2$ g/cm^2 at a distance ~ 320 pc [solid line] in Fig. 3). A summary of the data has been given in Morrison's survey [19].

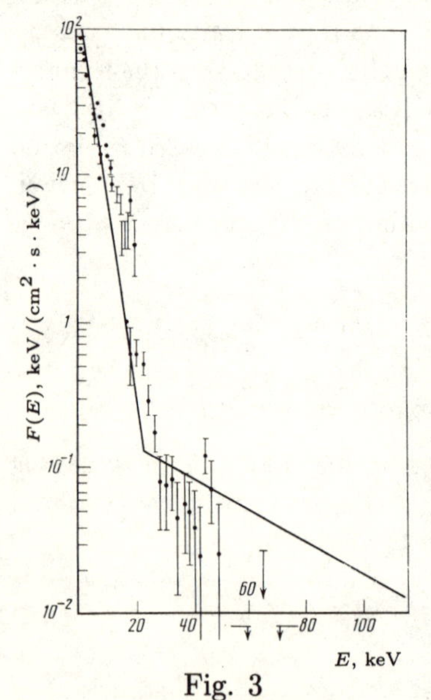

Fig. 3

However, an agreement of this kind cannot be considered a proof that the model is correct, because of the idealized formulation of the problem and the simplifying assumptions. We may recall the most important aspects: spherical symmetry, the absence of a magnetic field, the shortening of the path by plasma oscillations, and the neglect of electron heat conductivity because of the same plasma oscillations. It is these simplifications, not the rough character of the mathematical calculations, that limit the reliability and accuracy of the conclusions.

The main result of this investigation is that for an idealized model it has been possible to find a systematic, internally consistent solution in which the temperature distribution and the distribution and spectrum of the electromagnetic radiation are taken into account. A realistic theory of accretion, including also the remote region responsible for the amount of mass flow, would utilize the results described above, although such a theory might yield different predictions regarding the spectrum.

Only at the end of such a route might one be able to ascertain whether the observed point X-ray sources are in fact neutron stars, and whether neutron stars are entitled to pass from the realm of hypothetical objects into the class of reliably identified kinds of stars.

Far more detailed study ought to be accorded accretion by white dwarfs and by collapsed stars, as well as nonthermal mechanisms for accelerating electrons which might lead to the emission of X-rays.

We would like in conclusion to point out the possibility that γ photons might be emitted in an accretion process, in principle enabling neutron stars to be distinguished from other point X-ray sources. As first indicated in previous papers [7], if a neutron star has the maximum possible potential the free-fall energy will be sufficient for π mesons to be generated, and their decay can yield γ photons of energy 20–60 MeV. It would also be possible for γ photons of energy ~ 2 MeV to be emitted through the reaction of proton capture by a neutron, which in turn would be knocked out by an incident proton of a He^4 nucleus.

Calculations show [20] that the maximum possible ratio of the energy radiated in the form of γ photons (20–60 MeV) to the X-ray luminosity is $\sim 10^{-4}$. For Sco X-1 the flux of γ photons at the Earth would be nearly two orders of magnitude below the sensitivity of currently available equipment.

Appendix 1. Radiation of an Isothermal Half-Space for $\sigma \gg \kappa_\nu$

We shall use the transfer equation

$$\mu \frac{dI_\nu}{dy} = (\kappa_\nu + \sigma) I_\nu - \sigma J_\nu - \kappa_\nu B_\nu. \tag{A.1.1}$$

Here $J_\nu = \int_{4\pi} I_\nu \, d\Omega / 4\pi$ is the mean intensity at a given point and B_ν is the Planck function.

The transfer equation may evidently be used in this form if the scattering is isotropic and occurs without change of frequency. Moreover, operating in the standard manner for a calculation in the Eddington approximation, we introduce $H_\nu = (1/4\pi) \int I_\nu \mu \, d\Omega$ and $K_\nu = (1/4\pi) \int I_\nu \mu^2 \, d\Omega$, obtaining from Eq. (A.1.1)

$$\frac{dH_\nu}{dy} = \kappa_\nu (J_\nu - B_\nu),$$

$$\frac{dK_\nu}{dy} = (\sigma + \kappa_\nu) H_\nu \approx \sigma H_\nu. \tag{A.1.2}$$

We shall assume that the relation $K_\nu \approx J_\nu/3$ is satisfied for the whole atmosphere; the second equation may then be written as $dJ_\nu/dy = 3\sigma H_\nu$, and from the system of two first-order equations we obtain the equation

$$\frac{d^2 J_\nu}{dy^2} = 3\sigma\kappa_\nu (J_\nu - B_\nu). \tag{A.1.3}$$

An isothermal atmosphere in a gravity field is distributed according to the barometric law

$$\rho = \rho_1 + \frac{y}{h} \tag{A.1.4}$$

where $h = kT/\mu_e m_p g$ is the scale height of the homogeneous atmosphere and ρ_1 is the density at the boundary of the isothermal half-space. If we introduce in Eq. (A.1.3) the density ρ in place of y as the variable, we have

$$\frac{d^2 J_\nu}{d\rho^2} = 3\sigma\kappa_\nu h^2 (J_\nu - B_\nu). \tag{A.1.5}$$

We may rewrite Eq. (A.1.5) as follows:

$$\frac{d^2 J_\nu}{d\rho^2} = \frac{3\sigma\kappa_\nu h^2}{\rho}\rho(J_\nu - B_\nu) \tag{A.1.6}$$

setting $k_\nu = 3\sigma\kappa_\nu h^2/\rho$. Evidently k_ν will be constant for a given frequency and will not depend on the density.

Introducing now the variable $\xi = k_\nu^{1/3}\rho$, we obtain the Airy equation

$$\frac{d^2 J_\nu}{d\xi^2} = \xi(J_\nu - B_\nu), \tag{A.1.7}$$

whose solution may be written in terms of the Airy function:

$$J_\nu = B_\nu + CA_i(\xi) + DB_i(\xi). \tag{A.1.8}$$

$D = 0$ because of the condition that the solution be bounded as $y \to \infty$.

Using the condition $H_\nu(0) = J_\nu(0)/2$, we can determine the emergent flux of radiation

$$F_\nu = \pi B_\nu 4k_\nu^{1/3} \Bigg/ \left[2k_\nu^{1/3} - 3\sigma h \frac{A_i(k_\nu^{1/3}\rho_1)}{A_i'(k_\nu^{1/3}\rho_1)}\right]. \tag{A.1.9}$$

As $\rho_i \to 0$, $A_i/A_i' \approx 1.4$; for large frequencies $2k_\nu^{1/3}$ will be small, and $F_\nu \approx (3\kappa\nu/\rho\sigma^2 h)^{1/3}\pi B_\nu$.

Appendix 2. Change in Frequency Through Comptonization

Following Kompaneets [14] and Weymann [15], we may write the interaction of photons with electrons for $kT \gg h\nu$ as a kind of diffusion equation in frequency space, for a population number $n = I_\nu c^2/2h\nu^3$:

$$\frac{\partial n(x,t)}{\partial t} = \frac{1}{x^2}\frac{\partial}{\partial x}x^4\frac{\partial n}{\partial x}, \quad x = \frac{h\nu}{kT_e} \tag{A.2.1}$$

The total number of photons will be $N = a\int_0^\infty nx^2\,dx$. Multiplying both members of Eq. (A.2.1) by x^2 and integrating with respect to x, we obtain $dN/dt = 0$, so that the total number of photons will be conserved in the scattering process.

The expression for the total energy is $E = a\int_0^\infty nx^3\,dx$. Multiplying both sides of the equation by x^3 and integrating with respect to x, we obtain $E_2 = E_1 e^{4t}$, the change in the total through Comptonization. Knowing $\Delta E = E_2 - E_1$, we can determine the effective value of t from the total energy balance (see Sec. 2). We then proceed in the manner described elsewhere [17].

In order to obtain an analytic expression for the change in energy at each frequency of the spectrum, we introduce in place of x the variable $y = \ln x$, and then transform to the variable $z = 3t + y$.

After this replacement Eq. (1.2.1) reduces to the heat conductivity equation

$$\frac{\partial n}{\partial t} = \frac{\partial^2 n}{\partial z^2}. \tag{A.2.2}$$

The solution of Eqs. (A.2.1, A.2.2) has the form

$$n(x,t) = \frac{1}{\sqrt{4\pi t}}\int_{-\infty}^{\infty} n(y)\exp\left\{-\frac{(\ln x + 3t - y)^2}{4t}\right\}dy. \tag{A.2.3}$$

One finally obtains for the intensity

$$I_\nu = \frac{1}{\sqrt{4\pi t}}\int_0^\infty I(\nu')\left(\frac{\nu}{\nu'}\right)^3 \exp\left\{-\frac{(\ln \nu/\nu' + 3t)^2}{4t}\right\}\frac{d\nu'}{\nu'}. \tag{A.2.4}$$

REFERENCES

1. *Bisnovatyĭ-Kogan G. S.*—Astron. Zh. **43**, 89 (1966).
2. *Arnett W. D.*—Canad. J. Phys. **45**, 1621 (1967); **44**, 2553 (1966).
3. *Chiu H. Y.*—Ann. Phys. **26**, 364 (1964).
4. *Chao Wai-yin*—Astrophys. J. **147**, 664 (1967).
5. *Zeldovich Ya. B.*—Dokl. AN USSR **155**, 67 (1964).
6. *Zeldovich Ya. B., Novikov I. D.*—Dokl. AN USSR **158**, 811 (1964).
7. *Zeldovich Ya. B., Novikov I. D.*—UFN **86**, 447 (1965); Relativistic Astrophysics. Univ. of Chicago Press (1971).

8. *Novikov I. D., Zeldovich Ya. B.*—Suppl. Nuovo Cimento **40**, 810 (1966).
9. *Shklovskiĭ I. S.*—Astron. Zh. **44**, 930 (1967); Astrophys. J. Lett. **148**, L1 (1967).
10. *Cameron A. G. W., Mok M.*—Nature **215**, 464 (1967).
11. *Saslaw W. C.*—Month. Not. RAS **138**, 337 (1968).
12. *Melrose D. B., Cameron A. G. W.*—Prepr. (1968).
13. *Zeldovich Ya. B.*—Report to 13th Gen. Assembly, IAU, Prague (1967).
14. *Kompaneets A. S.*—ZhETF **31**, 876 (1956).
15. *Weyman R.*—Phys. Fluids **8**, 2112 (1965).
16. *Vedenov A. A., Velikhov E. P., Sagdeev R. Z.*—UFN **73**, 701 (1961).
17. *Zeldovich Ya. B., Sunyaev R. A.*—UFN **6**, 123 (1970).
18. *Ambartsumyan V. A., Saakyan G. S.*—Voprosy kosmogonii **9**, 91 (1963); Astron. zh. **37**, 193 (1960).
19. *Morrison P.*—Ann. Rev. Astron. Astrophys. **5**, 325 (1967).
20. *Shvartsman V. F.*—Astrofizika (in press).

Commentary

This article is today still one of the most widely quoted in astrophysics literature on the theory of accretion. It was written more than two years before the launching of the satellite "Uhuru," which brought about a revolution in observational X-ray astronomy. This present article prepared astrophysicists for many observational facts which were later investigated in detail by the instruments of the satellites "Uhuru," "Ariel," "ANS," "SAS-III" and "HEAO-I."

1. The problem of proton braking in the atmosphere of a neutron star acquired special importance in connection with the discovery of X-ray pulsars. Dozens of theoretical papers are devoted to the detailed study of this question which was first posed in the present article. Of particular interest is the study of the possibility of formation of collisionless shock wave near the surface of a neutron star, as well as the analysis of the effect of the strong magnetic field of neutron stars on the motion of particles in plasma.

2. The article above was the first to solve the problem of the spectrum of radiation emitted from the hot isothermic atmosphere of a star with dominating Thomson scattering in an opaque medium. The solution found by the authors became widely known and now, after the discovery of X-ray bursters, it is a very important object for astrophysical applications.

3. This article was the first to point out the very important role of the Compton effect in the formation of radiation spectra of X-ray sources. The analytical solution presented in the article (II.2.3) was published by Ya. B. and R. A. Sunyaev not in *Uspekhi fizicheskikh nauk*, but in the journal *Astrophysics and Space Science* (**49**).

45

Neutron Stars and "Black Holes"*

The orbiting of satellites outside of the atmosphere has created a foundation for the development of X-ray astronomy. Discrete X-ray sources (including X-ray pulsars) with luminosities 1000 times that of the Sun and even higher have recently been discovered within the limits of our Galaxy. In a number of cases, these sources form (together with ordinary stars that emit light in the optical band) binary systems or, more precisely, close pairs. It has become possible as a result to determine the mass of the X-ray star. The aggregate of data indicates that the X-ray source is in some cases a neutron star and in others a star in a state of relativistic collapse, i.e., a cooled star or, to use the current term, a "black hole."

Let us examine the observational material with the source Hercules X-1 as our example. This source has a period of 1.7 days (about 40 hours), of which about 8 hours are spent in eclipse, with the ordinary star between the source and the observer on the earth. The period of the velocity variation of the ordinary star coincides with the period of the eclipses, and this confirms duplicity.

The X-radiation has a short period of 1.24 sec, i.e., a period in the pulsar range. The 1.24-second period indicates that the source is a rotating neutron star (its mass according to orbital observations is suitable, $\sim 0.8\,M_\odot$, where $M_\odot$ is the Sun's mass, $2 \cdot 10^{33}$ g), with a magnetic field that controls the directivity of the X-radiation.

This radiation is linked to the incidence and impingement of gas flowing across from the ordinary star onto the surface of the neutron star.

Simple energy considerations demonstrate that it is the accretion (and not the rotation) that is the energy source feeding the X-ray emission. The 1.24-sec period corresponds to a comparatively slow rotation of the neutron star; there would be enough rotational energy for only a few years.

This is the fundamental difference between Hercules X and the pulsar in the Crab Nebula (PC), which also emits X-rays. The period of PC is 0.03 sec, with the result that its kinetic energy of rotation is 1000–2000 times greater, and, in addition, PC is in fact young (only 1000 years). In this case, the energy balance between radiation and rotational-energy loss converges.

The X-ray source in the Hercules X-1 system illuminates the ordinary star;

*Paper at the Anniversary Scientific Session of the Division of General Physics and Astronomy, USSR Academy of Sciences, 22 November 1972—Uspekhi fizicheskikh nauk **110** (2), 441–443 (1973).

at the surface of the latter, the energy of the X-radiation (which exceeds the intrinsic luminosity of the star) is transformed to optical radiation. As a result, the illuminated part is several times brighter than the dark part. Over the revolution cycle of the binary system, we see the illuminated and dark parts by turns, and this explains the brightness curve of the system and the changes in its light. Heating of the illuminated part causes a gas outflow that intensifies the impingement of gas onto the neutron star. There is another 36-day period associated with the periodic rotation of the pulsar's magnetic axis; this period is not of great fundamental importance and is not discussed here.

There are no eclipses in the binary system Cygnus X-1. Analysis of the Doppler shift of the lines, the brightness curves, and the spectral characteristics of the optical star with consideration of the laws of mechanics leads to the conclusion that the mass of the X-ray source is on the order of $10\,M_\odot$. According to theory, a compact body of this mass must be in a state of relativistic collapse. Falling of gas onto the "black hole" is accompanied by X-radiation when the incident gas carries rotational momentum. This situation occurs in a binary system. In first approximation, the gas particles revolve around the "black hole" (on circular Keplerian orbits under the influence of the attraction to the "black hole"). The progressive decrease in the radius of the orbit takes place comparatively slowly as the gas yields its rotational momentum in interaction with the gas on neighboring orbits (due to friction). Typically, the gas forms a disk around the "black hole." The friction in this disk is accompanied by release of energy and X-radiation. Only in the immediate vicinity [of the "black hole"], where the rotational momentum drops below the critical value, are the circular orbits replaced by a short spiral, followed by accretion of gas particles onto the surface of the "black hole"—at the so-called gravitational radius. But this most dramatic part of the process is virtually unaccompanied by radiation of energy.

The total radiated energy is 6 to 20% of the rest mass of the incident gas, which exceeds the energy release that could be obtained from nuclear reactions.

The X-radiation has quasiperiodic fluctuations with a period of less than a second. These fluctuations can be explained by the presence of bright points on the surface of the disk and by the Doppler effect from the rapid rotation of the disk. Their investigation, which is in its initial stage, can, in principle, yield valuable information on the gravitational field of the "black hole." This is because the familiar Schwarzschild solution is valid only for a nonrotating star. A stationary, but not static field appears on the collapse of a rotating star (the Kerr solution). The existence of a gravitational analog of the magnetic field is predicted in the relativistic theory of gravitation.[1]

[1]A "black hole" has no true magnetic field; it essentially sucks in the magnetic field, just as the gravitational field does with light and neutrinos in the case of a "black hole." Therefore, in particular, a "black hole" does not exhibit pulsar phenomena.

The gravimagnetic field influences the orbits of the gas particles, their period of revolution, and the energy released. In principle, these new properties of the gravitational field could be detected in the study of the fluctuations.

An X-ray source with a power 100,000 times that of the Sun has been discovered in the Magellanic Cloud. It may be assumed that here again the source is a "black hole" in a binary system with an ordinary star.

Thus, X-ray astronomy has presented us with data of enormous interest and importance. The significance of X-ray astronomy arises out of the fact that compact relativistic objects with energy release in a small region of space inevitably accelerate particles to high energies, develop high temperatures, and emit waves of high frequency. X-ray astronomy has opened a passable path to the investigation of such objects as the "black holes," where fundamentally new situations arise and the structure of the four-dimensional complex of space-time is radically altered.

This situation must be regarded as fundamentally new, although the equations from which it arises are the long-established equations of general relativity theory. Let us briefly examine the history of the problem. Before the war, Zwicky, Landau, Oppenheimer, and their colleagues had established the possible existence of neutron stars and "black holes". The theory of these objects was improved by a number of authors, including Ambartsumyan and Saakyan, Cameron, Bethe, and others. It was established that neutron stars cool down within a few years, and that "black holes" vanish from sight within less than a thousandth of a second.

Our group, the Theoretical Astrophysics Division of the Institute of Applied Mathematics, USSR Academy of Science, advanced the idea that it might be possible to detect an ultrastrong gravitational field around relativistic objects. The problem of investigating the external matter and its accretion as separate from the emission of the object itself was stated; note was taken of the advantageous position of double stars in this respect. The properties of the objects were evaluated in a number of papers, of which we note a few, most of them by Soviet scientists. Migdal: The Superfluidity of the Nuclear Matter of which Neutron Stars (Pulsars) consist; Ginzburg and Ozernoĭ: Drawing-in of the Magnetic Field by the Black Hole; Kerr (USA): Solution of the GTR Equations for Rotating Bodies, Including "Black Holes", with Angular Momentum; Doroshkevich, Zeldovich and Novikov: The Stability of this Solution. Guseinov was concerned with the search for binary systems with relativistic stars. The concept of accretion was refined. Shklovskiĭ treated an X-ray source in Scorpio as a neutron star in a binary system. Burbidge, Prendergast, and Lynden-Bell began the treatment of disk accretion (although for other objects—white dwarfs, galactic nuclei, and quasars).

The discovery of the pulsars came as a surprise to the theoreticians. Working from it as a premise, they were able to recognize neutron stars in the pulsars and use the stability of their periods for precise conclusions as to the structure of these stars. General confidence in the theory of the late stages in evolution was enhanced.

The theoreticians played the next round: two years prior to the corresponding observations, Schvartsman predicted that a pulsar in a binary system would periodically radiate X-ray pulses due to accretion.

Much work was done within a short time after the observations of binary (including eclipsing-binary) X-ray sources—in Hercules, in Cygnus, and elsewhere. Sunyaev and Basko examined the vaporization of a gas under the action of a flux of X-rays. Sunyaev and Shakura considered the spectrum of the disk. Sunyaev suggested quasiperiodic fluctuations as offering a way to investigate the "black hole."

Novikov and the American physicist Thorne developed a consistent relativistic theory of disk accretion. Lyutyĭ, Cherepashchuk, and Kurochkin, specialists on variable and binary stars at the Sternberg Institute (the State Astronomical Institute of Moscow State University), became involved and, together with members of our group, interpreted the optical observations of the "ordinary companions" of relativistic stars. A new and extremely rapidly developing branch of astronomical science, to which Soviet astrophysicists have rendered meritorious service, has emerged.

General information may be found in *The Theory of Gravitation and the Evolution of the Stars*, by Ya. B. Zeldovich and I. D. Novikov (Nauka, 1972). This book is significantly stronger in the area that is the subject of this paper than the book by the same authors, *Relativistic Astrophysics* (Nauka, 1967). On the other hand, recent results are being set forth in a steadily rising stream of communications in *Astrophysical Journal Letters*, *Astrophysical Journal*, *Astronomy and Astrophysics*, *Astrophysics and Space Science*, *Astronomicheskiĭ Tsirkulyar*, *Astronomicheskiĭ Zhurnal*, *Astrofizika*, and other journals and to an even greater degree in preprints of the work of Soviet and foreign scientists.

VII

Interaction of Matter and Radiation in the Universe

46
The "Hot" Model of the Universe*

The year 1965 has brought a most important discovery in astronomy. Measurements at short and ultra-short wavelengths ($\lambda = 7$, 3, and 0.25 cm) have demonstrated the presence of an isotropic radiation, i.e., independent of the observation direction, corresponding to a temperature near 3°K. At the time this article was written, the following were available: a brief communication published in July 1965 on measurements at $\lambda = 7.3$ cm, a note on measurements at $\lambda = 3$ cm (March 1966), and a report of unpublished measurements by a different group at $\lambda = 0.25$ cm, also by radio methods.[1]

A study of the optical absorption spectrum of the cyanide radical (CN) in galactic interstellar gas has also confirmed the presence of radiation of 0.254 cm wavelength, the intensity of which corresponds to 3°K. It is curious that the results of measurements pertaining to CN were known already in the forties, but then they remained an unexplained paradox. In the investigated wavelength region, the observed radiation is $10^2 - 10^5$ times larger than expected from the known sources such as stars and radiogalaxies.

The possible existence of such radiation was predicted by Friedman's theory of the expanding universe. Since 1948, within the framework of this

*Uspekhi fizicheskikh nauk **89** 4, 647–653 (1966). Due to limited space in this volume, the basic text of the review is given without the appendices. For a translation of the full article, see Soviet Physics Uspekhi **9** (4), 602–617 (1967).

[1]Measurements at 20 cm were also made very recently.

theory, a hot model has been under consideration, in which it is assumed that matter in the prestellar state is characterized by a large entropy. The entropy concept has in this case a simple and intuitive meaning: when thermal equilibrium obtains, and the entropy is large, there are many photons for each atom, and it turns out that the entropy is directly proportional to the number of photons. The already-mentioned measurements lead to a value of approximately 10^9 photons/atom. The entropy, meaning also the number of photons, is conserved during the expansion. However, the photon energy decreases upon expansion, in accordance with the fact that the wavelength increases in the same proportion in which all distances between any specified pair of particles or pair of galaxies increase during the course of expansion.

Thus, extrapolation to the past leads to the notion of a plasma in which the number of photons per unit volume exceeds in giant fashion the number of atoms or, more accurately, the number of nucleons and electrons in the same volume. The temperature of this plasma changes during the expansion. Let us assume that there was a moment when the density was infinite, and let us reckon the time from that moment, approximately 10^{10} years ago. From the equations of mechanics and from the presently known relation between the number of photons and the number of atoms per unit volume, we can find the temperature and write the corresponding time dependence. At the instant $t = 1$ s, the temperature was approximately 1 MeV, i.e., 10^{10} degrees; in addition to photons in equilibrium, there were almost as many pairs of electrons and positrons; the nuclei could not exist then, and the number of protons and neutrons was practically equal; collisions with electrons and positrons led to transformations of protons to neutrons and vice versa.

With continued expansion, the positrons disappeared. A part of the neutrons decayed, and the remainder joined the protons and yielded finally a composition of $\sim$ 70% hydrogen and 30% helium by weight (i.e., approximately 10% of helium by the number of atoms). There should also remain from that period neutrinos and antineutrinos in amounts approximately equal to the number of photons, and with the same average energy, corresponding at the present time to several degrees, i.e., approximately 10^{-3} eV. It was assumed a few years ago that the content of helium in matter that has not been subjected to processing in the stars is much smaller than 30%, and this seemed a major difficulty for the hot model.

Also formulated a few years ago was a hypothetical "cold" model, which gives 100% hydrogen and total absence of radio emission of the type recently discovered. It is obvious that in the light of the latest data the cold-model hypothesis should be discarded, although the question of the primary content of helium still remains unclear from the observational point of view and calls for further research.[2]

[2]According to private communications, old stars with low helium content have been observed very recently. The reliability of these communications is uncertain.

Even more paradoxical conclusions follow from the hot model for earlier times and higher temperatures. It follows from the theory that there was a period (when $T \geq 10^{13}$ deg), when there were many nucleons and antinucleons, and from this point of view the present-day nucleons are the result of a small excess of nucleons over antinucleons at this early stage. If heavier particles—quarks—exist, then they should be present in noticeable amounts in equilibrium, and as an estimate up to 10^{-9}–10^{-10} quarks/hydrogen atom have survived to the present time; this is more than the average concentration of gold (10^{-12}–10^{-13}) and radium (10^{-18}, likewise, per hydrogen atom). It must be emphasized that the conclusion concerning nuclear reactions at $t = 1$ s and later are practically independent of the assumptions concerning the earlier stage, concerning the nature of quarks and antinucleons, and whether some density higher than 10^6 g/cm^3 and a temperature higher than 10^{10} deg was attained. Even under these conditions, all the processes are rapid; no matter what the initial composition we specify, equilibrium is established at that time practically instantaneously; the system will forget the initial composition, and further development of events will not depend on the assumptions concerning what had occurred at $t < 1$ s. We recall that the theory of the hot universe is essentially an extrapolation, to past states, of the Universe surrounding us, whose properties are being investigated at the present time. Like any other extrapolation, the closer it is to us, i.e., the less remote the state considered in the past, the more reliable it is.

After the period of nuclear reactions (1 s $< t <$ 100 s), the second characteristic landmark is $t = 3 \cdot 10^6$ years, corresponding to a temperature of 3000 – 4000°K. While the hydrogen and helium are more or less uniformly mixed with the radiation, prior to that time they were ionized. The radiation density exceeds the density of ordinary matter. The radiation pressure is high; the "matter," i.e., the electrons and ions (nuclei), interacts strongly with the radiation. This interaction with the radiation prevents formation of stars and galaxies. Gravitational forces that gather matter into the region where the density of matter already exceeds the average density ("gravitational instability") are unable to overcome the radiation pressure, which increases upon compression of the matter. Only after $t = 3 \cdot 10^6$ years does recombination of the electrons and protons into hydrogen atoms occur, and presumably formation of stars and galaxies begins. This leads to a concept whereby the first generation of the stars constitutes a small fraction of the total mass of matter; perhaps the term "protostars" is more appropriate here, since the conditions for their formation differ radically from the conditions under which stars are formed at the present time; accordingly, their properties can differ, too.

It is thus assumed that the protostars can rapidly release an energy sufficient to heat the remaining, larger portion of matter. A considerable fraction of this matter remains in the form of hot ionized plasma and does not condense into stars. Thus, the concept of the hot model of the Universe is related to an independent hypothesis, that of the existence (at the present time!) of hot ionized intergalactic gas of density ($\sim 10^{-29}$ g/cm^3) tens of times larger than the density of matter in the stars, averaged over the entire space. This hypothesis is supported by the recent discovery of unusually bright objects—quasars. The spectrum of one of the most remote of the presently known quasars, 3C-9, has peculiarities that lead to the conclusion that the intergalactic medium contains $6 \cdot 10^{-11}$ atoms of neutral hydrogen per cm^3 (density 10^{-34} g/cm^3).[3] This hydrogen absorbs the Lyman α line with wavelength $\lambda = 1216\mathring{A}$. The 3C-9 source is so far that the red shift increases all the wavelengths by a factor 3.012! Because of this, the far ultraviolet region of the spectrum, with wavelength $\lambda = 1216\mathring{A}$ (which is not transmitted by the atmosphere), is perceived by us as $\lambda = 3662\mathring{A}$ and can be observed by terrestrial telescopes. The closer to us the considered layer of intergalactic gas, the smaller the red shifts of the absorption line; on the whole, we obtain an absorption band extending for the terrestrial observer from $\lambda = 3.012 \cdot 1216 = 3662\mathring{A}$ toward the shorter wavelengths. This density of the neutral hydrogen, 10^{-34} g/cm^3, pertains to the vicinity of 3C-9, i.e., not only to a spatially remote object, but also to the remote past. It is assumed that during that period the density of the ionized gas was $(2-5) \cdot 10^{-28}$ g/cm^3, i.e., the neutral gas amounted to less than 10^{-6} of the entire gas, corresponding to gas temperatures of the order of 10^6 deg. The gas was also quite rarefied, transparent, and far from being in equilibrium with the overall cosmic radiation, whose temperature at that instant was of the order of 10°K. In the period under consideration ($t \sim 2 \cdot 10^9$ years), the gas was heated only by the energy of such objects as stars and galaxies.

A distinction must be made between two related hypotheses, which essentially are different in scope:

1) *The hot model of the Universe*—temperatures from 10^{10}°K at $t = 1$ s to 3°K now. Isotropic radio emission at centimeter and millimeter wavelengths is the consequence of the hot model. The radio emission is the direct successor of powerful radiation in the compressed hot plasma, whose density during the earlier stage exceeded by many times the density of matter; it is quite immaterial in this case whether we are dealing with the same individual photons; in the state of total thermal equilibrium, some photons are absorbed, others are emitted, still others are scattered, but all these processes are balanced in such a way that neither the total energy of the medium nor the spectrum of the photons changes. This is the situation in

[3]More recent measurements of a different source have not confirmed this conclusion, but this does not exclude the presence of ionized hydrogen (see below).
Editor's note. This footnote remains correct in the sense that no smoothly distributed neutral hydrogen has been found. But we now know that intergalactic space contains innumerable "clouds" of partially neutral gas.

the case of high density. At low density, absorption and emission of photons becomes insignificant, their scattering becomes insignificant, and it is possible to follow the fate of each individual photon. In any case, the presently observed radio photons are either the same photons or the descendants of the photons, which, by assumption, had an energy on the order of 1 MeV at $t = 1$ s and whose energy decreased by virtue of the red shift during the course of expansion, i.e., essentially by virtue of the Doppler effect. These photons are currently referred to as "relic radiation." It is important that neither stars nor radiogalaxies nor the hot interstellar gas can give anything approaching the properties of relic radiation: the energy density of the relic radiation is too high, and its spectrum is not similar to either the spectra of the stars or to the spectrum of the radio sources.[4] This indeed proves the cosmological, relic origin of the radio emission discussed.

2) *Hot intergalactic plasma*—This concept certainly pertains to a period later than $t = 3 \cdot 10^6$ years; the plasma is heated to $10^5 - 10^6$ degrees and is no longer in equilibrium with the relic radiation. This means that such a plasma needs heat sources.

The question of interaction between a hot plasma and the relic radiation is quite interesting. Energy exchange between them is at present small, but it could have been larger earlier. Energy estimates show that the plasma could not produce the radiation which we regard as relic (with its density and spectral composition). On the other hand, the same energy balance calculations show that the plasma could not have been heated as early as shortly after $t = 3 \cdot 10^6$ years. The plasma electrons scatter radiation photons because of the Compton effect. It is curious that an optical thickness of the order of unity is reached in a distance corresponding to a redshift of the order of 6–8. The isotropic relic radiation does not change its properties during this scattering; the energy of the electrons and of the photons are such that the scattering is not accompanied by any noticeable exchange of energy and does not alter the spectrum. However, observation of remote discrete sources—quasars with large red shift—is made difficult by the electron scattering. At the same time, since the scattering cross section is independent of the frequency, it is very difficult to prove with assurance the presence of Compton scattering and to determine the concentration of the free electrons.

The most important would be observation of helium in the composition of the intergalactic gas and measurement of its temperature by determining the emission of soft X-ray photons. At the present time measurements yield only the inequality $T < 2 \cdot 10^6 {}^\circ$K for the temperature of the intergalactic gas.

A very difficult problem is the observation of relic neutrinos and antineu-

[4]According to a most recently discussed hypothesis, practically all the nuclear energy was released in the stars during the earlier stage, and the light of the stars was absorbed by dust particles which transformed it to thermal radiation. For many reasons, however, this hypothesis must be regarded as unlikely.

trinos. To do this, it is necessary to improve the existing accuracy by a factor 10^6!

At the same time, we must emphasize the tremendous significance of such an experiment. Observation of neutrinos in the required amount and with the expected spectrum would confirm the hot-model notions concerning the very early stage at $t < 1$ s, a density larger than 10^6 g/cm^3, and a temperature higher than 1 million volts. Indeed, measurement of relic neutrinos would be the "experiment of the century"!

The hot model poses problems of tremendous importance and difficulty to the theoreticians. These include primarily:

a) The question whether it is possible to construct (using quantum concepts) a theory of transition from compression at $t < 0$ to expansion at $t > 0$.

b) The specific entropy of matter in the hot model (characterized by the number 10^9 photons/atom) was regarded above as an initially specified characteristic of this model, as one of those initial conditions that must be stipulated for the integration of the equations. Can we ask: Why is the entropy precisely this, not larger or smaller? Only by understanding the origin of the large entropy can we explain satisfactorily how an almost charge-symmetrical state is obtained with a gigantic number of antinucleons and nucleons, but at the same time with a definite small excess of nucleons. In fact, if there were only nucleons when $t < 0$, and these somehow produced a large entropy, then at low density this entropy appeared in the radiation. In such a case the subsequent production of $N + \overline{N}$ pairs upon contraction, and the presence of $N + \overline{N}$ pairs (with conservation of the excess of nucleons N) at the start of expansion at $t > 0$ are consequences of known laws.

c) It is curious that the hot model leads to a reformulation of the problem of existence of superdense bodies within their own gravitational radius. These bodies must gather radiation from their surroundings and with this they increase their mass catastrophically. The hypothesis that superdense bodies have existed initially should in some manner take account of the interaction between these bodies and the radiation surrounding them.

d) We note some concrete problems pertaining to the state of large density: What is the situation with vortex motion and magnetic fields in such a state? What is the degree of homogeneity and fluctuation in this state?

e) The theory of formation and evolution of "ordinary" objects (stars, galaxies, clusters, quasars) should also be developed with allowance for relic radiation.

f) We recall, finally, the most vital question which still remains unanswered: Should we imagine the evolution of the Universe as 1) a single expansion from a special singular state, or 2) a single contraction from $t = -\infty$, $\rho = 0$ via $t = 0$, $\rho = \infty$ (or 10^{93} g/cm^3) and subsequent expansion, continuing to the present time, or else 3) an infinite sequence of cycles of contraction and expansion? It is impossible to choose from among these hypotheses on the basis of philosophical considerations. Each agrees with the laws of physics known to us; the laws of physics are not connections between sensations experienced by a thinking person, but objectively existing recognizable laws of the external world. In full accordance with this, the answer to our question, i.e., the choice between the different hypotheses, must be based on an objective natural-scientific study of the Universe surrounding us, both by observation and by development of a physical theory that accounts for the observational data.

We can advance theoretical evidence (connected with the increase of entropy) against an infinite repetition of such cycles. It is curious that by virtue of the laws of the general theory of relativity, the situation is not at all similar to the ordinary concept of "thermal death." In fact, during the extent of each cycle, the Universe can be inhomogeneous, consist of stars, and bear very little resemblance to the constant average density and temperature picture corresponding to the maximum entropy in a system without gravitation. The repetition of these cycles requires that the density at the present time be larger than a definite critical value $\rho_c \sim (1-2)\cdot 10^{-29}$ g/cm^3. With this, the Universe should be geometrically closed.[5] The entropy increases in each cycle, and consequently also the energy per atom in its co-moving coordinate frame; the period and the swing of the cycle, i.e., the maximum radius at the instant of transition from expansion to contraction, also increase.

However, the last word belongs to observational astronomy, namely, measurement of the overall density of matter (including the most difficultly observed forms—from neutrino and gravitational waves to collapsing stars and galaxies) will determine the properties and fate of the Universe. As already mentioned, expansion gives way to contraction if $\overline{\rho} > \rho_c$. If $\overline{\rho} < \rho_c$, we are left only with the choice between a single expansion or a single contraction that has given way to expansion.

The value of ρ_c itself depends on the Hubble constant, $U = Hr$, viz., $\rho_c = 3H^2/8\pi G$, where $H \sim 75 - 100$ km/s Mpc $\cong 0.3 \cdot 10^{-17}$ s^{-1}, and $G = 6.7\cdot 10^{-8}$ cm^3/g$\cdot$s is the gravitational constant. Therefore ρ_c is known with sufficient accuracy. The main contribution to $\overline{\rho}$ is probably made by the intergalactic plasma, and therefore its study is the most important practical problem.

Let us turn from hypotheses, which have been described in increasing order of boldness (meaning also with decreasing reliability), to the actual observational aspect of the situation.

[5]Closure at $\rho > \rho_c$ is deduced as a consequence of the general relativity equations relating the curvature of space with the density of matter. Additional hypotheses are assumed here, that there is no so-called cosmological term in the equations, that the distribution of matter is homogeneous, and that the expansion is isotropic (equivalence of all directions). These three hypotheses cannot be regarded as rigorously proved, but at any rate they do not contradict the observations, and have recently gained support in connection with measurements of the relic radiation.

Is there not a discrepancy between several observations at three or four different wavelengths and the grandiose nature of the conclusions? What consequences can be deduced directly from the experimental data? These two questions are not asked together accidentally, it is understandable that the tremendous resonance and the significance of the experiments referred to depend on their comparison with the entire system of views of evolutional cosmology, with the theory of A. A. Friedman, and with more information on the interaction between matter and radiation and on nuclear reactions. We must dwell on this, because quite recently the following point of view was advanced: Friedman's theory, i.e., the idea of general cosmological evolution, is an approximation. It was useful at a definite stage and led to the discovery of the red shift, but for the last 30–40 years this theory has not given any new results; it is not fruitful and further development should follow the line of replacing it with new concepts, and so on.

However, the development of cosmology in recent, and even, more accurately, in the last year, has not confirmed this forecast, meaning that it likewise has not confirmed the estimate given above for the past. New facts (the relic radiation and the chemical composition) can be understood and fitted within the framework of the applicability of Friedman's theory to an ever more remote past, and this theory is justified for states which are ever closer to singular. This theory is fruitful, it offers guidelines for observations, and without it no fundamental significance would be attached to the singularities of the CN spectrum.

Turning to direct consequences of observations that are not connected with the theory, we must emphasize the isotropy of the radiation and the agreement with the form of the equilibrium spectrum (Planck curve) within the limits of the investigated part of the spectrum.* The isotropy of the radiation, irrespective of any definite theory and of any assumptions concerning the remote past, is evidence of the identity of the conditions in different directions away from us. The weak interaction of the radiation in question (7–0.25 cm) with dust, neutral atoms, or plasma leads to conclusions pertaining to much larger distances than could be deduced from statistics of remote discrete objects. The presently measured photons have experienced scattering at a distance corresponding (on the average) to a red shift of the order of $z = 6$ or 8, i.e., to $t \sim t_0/15$, where t_0 is the present-day age. This means that the expansion has been going on isotropically at least since then.

The second consequence is that the Earth (Sun, Galaxy) does not have a large velocity relative to the radiation field; a velocity of 10,000 km/s would give an anisotropy of 10% at a wavelength 7 or 3 cm and 20% at 0.25 cm. Such an anisotropy could presumably be detected. The Planck spectrum of the relic radiation (which must be verified further to $\lambda \sim 0.05$ cm) can yield indirect confirmation of the existence of a period when light and plasma

**Editor's note.* The entire spectrum was observed by the U.S. "COBE" satellite in 1990 and confirmed to be perfectly Planckian to within one part in 1000.

were in equilibrium, a period pertaining to an earlier stage. The properties of the relic radiation then allow a deeper look into the Universe, a check and direct confirmation for the principles of contemporary cosmological views—isotropy and homogeneity of practically the entire part of the Universe that can be presently observed.

The foregoing review was aimed at presenting a general description of the status of cosmology after the sharp turn which has occurred quite recently. This review is aimed primarily at the non-specialist, i.e., the physicist, but not the astronomer. To facilitate the perception of the entire picture, we did not interrupt the exposition with technical details, formulas, priority questions, or references to the original literature.

However, a reader interested in this problem or some of its individual parts needs such material. It is all included in the appendices which follow the main text of the review and were purposely written in a dry manner, so as to compensate for the non-rigorous journalistic style of the review.

Commentary

This article was written almost immediately after the discovery of relic radiation. It presents almost all the ideas which were later developed both observationally and theoretically. And today it is an outstanding introduction to the articles of this and the next section of this book.

In this article, apparently for the first time, the problem was clearly posed of the amazing composition of the Universe in the early stages, when the concentrations of nucleons and antinucleons were almost identical; however, a small excess of nucleons ensured after the annihilation the presently observed existence of nucleons without antinucleons.

We note the clear rejection of the "cold" model (see the next article of this section).

Because of space restrictions in this volume, only the basic text of the review is given, without the appendices, which are full of figures, formulas and references to the original works.

Other review papers by Ya. B.[1,2] played a major role in the development of cosmology both in the USSR and abroad; these papers, together with the above paper, served as a basis for writing the sections devoted to cosmology in later works.[3,4]

[1] *Zeldovich Ya. B.*—UFN **80**, 357–390 (1963).

[2] *Zeldovich Ya. B.*—In: Advances in Astron. Astrophys. **3**, 241–379 (1965).

[3] *Zeldovich Ya. B., Novikov I. D.* Relativistic Astrophysics. Univ. of Chicago Press (1971).

[4] *Zeldovich Ya. B., Novikov I. D.* Stroenie i evoliutsiia Vselennoi [Structure and Evolution of the Universe]. Moscow: Nauka, 352 p. (1975).

47
Prestellar State of Matter*

In accordance with the presently observed expansion of the nebulae, it is deemed probable that in the earlier stages of the evolution of the universe there existed a homogeneous isotropic Friedman nonstationary solution with the density of matter decreasing from an infinite value at the initial instant.

Let us consider the state of matter in the initial stage of evolution. We assume that the matter consists of a mixture of protons, electrons, and neutrinos in equal amounts, and that the entropy is low. Then at high density (on the order of nuclear density) and at zero temperature the neutrinos and the electrons form a degenerate relativistic Fermi gas, and owing to the helicity (two-component nature) of the neutrino their Fermi energy E_{F_ν} exceeds the Fermi energy of the electrons; at a density $2.5 \cdot 10^{38}$ particles/cm^3 we have $E_{F_\nu} = 500$ MeV and $E_{F_\nu} = 400$ MeV. The process $e^- + p = n + \nu$, which leads to the formation of neutrons at high density of matter in the stars [1], turns out to be forbidden here, since the neutrino states that are energetically obtainable in this process are occupied. In the uniform model (closed or open) the neutrinos do not depart anywhere. Upon expansion such a substance turns into pure cold hydrogen.

According to modern views, it is deemed most probable that it is precisely pure hydrogen which is the initial substance from which the stars were originally formed, and that the heavier elements were the result of nuclear processes in stars.

Earlier Gamow, Alpher, and Herman [2, 3] suggested that in the initial state matter consists of neutrons or of approximately equal amounts of neutrons and protons (see also [4]) and is at such a high temperature, that the radiation density is gigantically in excess of the nucleon density. This point of view, on the basis of which the authors attempted even to reconstitute the presently observed abundance of the elements, leads to insurmountable contradictions. In the prestellar stage they obtain a large amount of helium (about 10–20%) and deuterium (about 0.5%). The radiation density (energy/c^2) remains approximately equal to the nucleon density after matter has expanded to the modern average nucleon density 10^{-30} g/cm^3. These deductions are incompatible with the observations; the notion of matter consisting of protons, electrons, and neutrinos is the only one possible.[1] At a density many times larger than nuclear, when the Fermi energy of the protons becomes comparable with their rest mass, various processes of neutron

*Zhurnal eksperimentalnoĭ i teoreticheskoĭ fiziki **43** (5), 1561–1562 (1962).

[1]This deduction is motivated in detail in an article by the author, submitted to the collection "Voprosy Kosmogonii" [Problems of Cosmogony].

and hyperon productions become possible in principle (see [5]) ($p \to \Sigma^+$, $p \to n + \pi^+$, $p + e^- = n + \nu$). However, the expansion is slow (see [4]) compared with the relaxation time of these processes,[2] and therefore only p, e^-, and ν remain by the time nuclear density is reached.

For the theory of the decay of homogeneous matter into individual clusters corresponding to the galaxies, an important factor may be the inhomogeneities of the density, arising during the course of expansion in the density interval 0.5–2 g/cm^3, where metallic hydrogen is transformed into molecular hydrogen [6], and where a phase transition from the solid body to the gas at density < 0.07 g/cm^3 is also possible. This question needs further investigation.

[2]We note that the process inverse to neutron formation proceeds not as the decay of a free neutron, but under the influence of a Fermi gas of neutrinos, and therefore is practically terminated within the expansion time.

REFERENCES

1. *Landau L. D.*—Dokl. AN SSSR **17**, 301 (1937).
2. *Gamow G.*—Rev. Mod. Phys. **21**, 367 (1949).
3. *Alpher R., Herman R.*—Rev. Mod. Phys. **22**, 153 (1950); Ann. Rev. Nucl. Sci. **2**, 1 (1953).
4. *Hayaschi C.*—Progr. Theor. Phys. **5**, 224 (1950).
5. *Ambartsumyan V. A., Saakyan G. S.*—Astron. Zh. **37**, 193 (1960).
6. *Abrikosov A. A.*—Astron. Zh. **31**, 112 (1954).

Commentary

Ya. B. began actively to work on cosmology even before the discovery of relic radiation. At first he took an uncritical approach to individual indications in the literature regarding the low content of primordial helium-4 and the low temperature of the background radiation.

Rough calculations of the cosmological nucleosynthesis lead Ya. B. to the erroneous conclusion that the hot model of the universe contradicted observations.[1] The problem of finding another cosmological model arose, one which preserved Friedman's theory of an expanding universe. A variant of such a model is given in the present article.

The article represents certain historical interest. After the discovery of relic radiation, Ya. B. became an ardent proponent of the hot model of the universe (see his review article **46**).

[1]*Zeldovich, Ya. B.*—Atomn. Energ. **14**, 92–99 (1983).

48

Recombination of Hydrogen in the Hot Model of the Universe*

With V. G. Kurt and R. A. Sunyaev

The considerable emission of energetic photons during recombination of hydrogen in an expanding universe leads to a slowing down of the recombination and to a distortion of the relic radiation spectrum in the Wien region. The energy exchange between electrons and radiation in the Compton effect maintains the temperature of matter equal to that of the radiation up to a time corresponding to a red shift of $z \sim 150$, and this leads, in particular, to a change in the time dependence of the Jeans wavelength of the gravitational instability of a homogeneous medium in an expanding universe.

In the hot model of the Universe [1, 2] it is assumed that at an early stage of the expansion the fully ionized plasma is in equilibrium with radiation. Cooling as a result of expansion leads to recombination. Since in the course of the expansion there is conservation of specific entropy $T^3/n = 3 \cdot 10^6 \deg^3 \mathrm{cm}^3$ [2], it can be easily verified[1] that according to the Saha equilibrium formula a 50% degree of ionization is attained at a temperature $T \sim 4100°$K and a density $n = p + H \approx 2.6 \cdot 10^4 \,\mathrm{cm}^{-3}$ (the letters p, H, and e denote the densities of the corresponding particles).

The density of photons in the Universe is much greater than the density of ions, electrons, and atoms ($n_\gamma/n \sim 10^8$). At first sight it seems to be impossible to have any sort of an inverse reaction of matter on the blackbody radiation. As will be shown later, the distortion of the spectrum indeed turns out to be small, but owing to considerably more complicated causes.

The point is that the temperature for the effective recombination is considerably lower than the ionization potential: $kT \approx I/40 \approx h\nu_\alpha/30$ (ν_α is the frequency of the L_α-photon), and therefore the density a of the energetic photons with $\nu > nu_\alpha$ constitutes a small part of the total radiation density and is approximately a factor of 200 lower than the density of protons and of atoms:

$$a = \frac{8\pi}{(hc)^3} \int_{E_0}^{\infty} E^2 e^{-E/kT} \, dE = \frac{8\pi E_0^2 kT}{(hc)^3} e^{-E_0/kT}.$$

*Zhurnal eksperimentalnoĭ i teoreticheskoĭ fiziki **55** (1), 278–286 (1968).

[1] Calculations are carried out on the basis of the following assumptions regarding the present-day situation: $T = 3\,\mathrm{K}$, $n = 10^{-5}\,\mathrm{cm}^{-3}$.

For $T \sim 4100°$K the value of a is about $10^2\,\text{cm}^{-3}$. According to the Saha formula for $e = p = H$ (50-percent ionization) we have

$$e = \frac{n}{2} = \frac{(2\pi m_e kT)^{3/2}}{h^3} e^{-I/kT}$$

and the ratio of the number of electrons to the number of photons of energy above the ionization threshold I is equal to $[\pi(mc^2)^3 kT/2^3 I^4]^{1/2}$. This ratio does not depend on the absolute value of the hydrogen density: the smaller the value of n, the lower is the temperature for 50-percent ionization, and the smaller is the factor $\exp(-I/kT)$ which enters into the expression for the number of quanta of energy above threshold. The number of photons with $\nu > \nu_\alpha$ is larger than the number of photons with $h\nu > I$; since $h\nu_\alpha = 3I/4$, we get $a \sim n^{3/4}$ and $a/n \sim n^{-1/4}$, i.e., depends only weakly on n.

If each recombination act were accompanied by the emission of an energetic photon, then the density of photons would grow rapidly in comparison with the Planck spectrum. At the present time the temperature of the equilibrium radiation is equal to 3°K [2], so that all the wavelengths have increased since the time of the recombination by approximately a factor of 1400, and one might expect anomalies in the relic spectrum for wavelengths $\lambda \sim 10^{-2}$ cm. However, the assumption concerning the emission of an energetic photon in the case of each recombination is incorrect in the present situation; the inverse absorption of L_α-photons and of harder photons takes place. As will be shown later a noticeable retardation of the recombination of electrons and protons occurs, since the super-equilibrium density of photons with $\nu > \nu_\alpha$ leads to an above-equilibrium density of excited hydrogen atoms which are easily ionized by soft photons ($I_2 = I/4 = 3.4\,\text{eV}$). A decisive role is played by the two-photon transition $H^* \to H + \gamma_1 + \gamma_2$, the astrophysical role of which was considered in [3, 4]. Below we give an approximate theory of the recombination and of the spectrum arising as a result.

1. Dynamics of Recombination

Under dynamical equilibrium, the principal processes are the photoionization process $H + \gamma \to p + e$ and the inverse process of recombination with emission of a photon. Regardless of whether the recombination proceeds directly to the ground state or by means of a cascade via excited states $p+e \to H^* + \gamma_1$, $H^* \to H + \gamma_2$, a single photon is emitted, of energy equal to or greater than $h\nu_\alpha$. The same photon is absorbed in the case of ionization.[2] The processes indicated above leave constant the sum $\Sigma = e + a$, where e is the electron density and a is the density of photons with $\nu > \nu_\alpha$.

We take the cosmological expansion into account; in this case the quantity

$$\Sigma' = V(e + a'),$$

[2] Along with the direct ionization $h\nu + H = p + e$, we also consider the stepwise ionization: $h\nu + H \to H^*$, $H^* + \text{soft photon} = p + e$.

is conserved, where V is the instantaneous physical volume corresponding to a constant co-moving volume, $Vn = \text{const}$, and a' is the number of photons of frequency exceeding the frequency

$$\nu' = \nu_\alpha \frac{1+z_0}{1+z'} = \nu_\alpha \left(\frac{n'}{n_0}\right)^{1/3},$$

which varies in accordance with the red shift in the course of expansion. It is well known that the expansion transforms the equilibrium spectrum with temperature T_0 into an equilibrium spectrum with $T' = T_0(n'/n_0)^{1/3}$. We choose the instant of time T_0 which corresponds to a 50-percent ionization with $e/a_0 \sim 100$, and the instant T' such that $e/n \ll 1$ when recombination has practically been finished. In this case it follows from $\Sigma' = \text{const}$ that

$$a' = a_0 \left(1 + \frac{e}{a_0}\right) \left(\frac{n'}{n_0}\right)^{1/3}.$$

The equilibrium spectrum corresponds to a density $a' = a_0(n'/n_0)^{1/3}$, and consequently in the approximation assumed by us one would obtain at the later stages a spectrum with a' 100 times larger than the value for the equilibrium distribution. At the present time $\nu'_\alpha \approx 10^{-3}\nu_\alpha$, i.e., this maximum possible violation of equilibrium pertains to radiation of wavelength of the order of 10^{-2} cm. The expansion, leaving Σ' unchanged, alters Σ in accordance with the fact that a) all the densities are decreased and b) a fraction of the photons "falls below the threshold ν_α" as a result of the red shift (their energy becomes less than the energy of a L_α-photon). It is easy to construct the equation for Σ under this assumption

$$\frac{d\Sigma}{dt} = -3\mathcal{H}\Sigma - \nu_\alpha \mathcal{H} f(\nu_\alpha),$$

since $d\nu/dt = -\mathcal{H}\nu$, where f is the function describing the distribution of the photons with respect to frequencies, $\mathcal{H}$ is the instantaneous value of the Hubble constant, $\mathcal{H} = -d\ln n^{1/3}/dt$. Stipulating the Wien form of the spectrum

$$f \sim e^{-h\nu/kT}, \quad a = f(\nu_\alpha)\frac{kT}{h},$$

we obtain

$$\frac{d\Sigma}{dt} = -3\mathcal{H}\Sigma - \mathcal{H}h\nu_\alpha \frac{a}{kT}.$$

Knowing the value of Σ, it is easy to reconstruct the whole situation. Indeed, the fast processes guarantee dynamic equilibrium between the electron density and the density of different excited states of the hydrogen atoms; transitions between them occur as a result of the absorption and emission of low-energy photons, the number of such photons is great and they, doubtlessly, are in an equilibrium corresponding to the overall temperature of the radiation. This means that

$$H^* = Kep = Ke^2, \tag{1}$$

where $K = K(I^*, T)$ is evaluated by means of the Saha formula for the binding energy of the excited state. However, such a relation does not hold between e and H; the density e is greater than would correspond to the Saha formula. Consequently, $f(\nu_\alpha)$ likewise does not correspond to equilibrium with a common temperature T.

Owing to the large cross section for the absorption of L_α-photons, the equilibrium $e + p \rightleftharpoons H^* + \gamma_1$, $H^* \rightleftharpoons H + \gamma_\alpha$ is established at which we have

$$f(\nu_\alpha) = \text{const}\frac{H^*_{2P}}{H} = \text{const}\frac{ep}{H}e^{I_{2P}/kT},$$

$$a = kT\frac{f(\nu_\alpha)}{h} = \text{const}\cdot T\frac{ep}{H}e^{I_{2P}/kT}$$

(for details cf. the Appendix). Here H^*_{2P} is the density of hydrogen atoms in the $2P$-state.

The relation connecting a with e and p together with the trivial conditions $e = p$ and $e + p + H = n$, enables us to express in an elementary fashion in terms of Σ, n, and T all the quantities e, p, f, and a of interest to us, as well as the hydrogen atom density in all the excited states.[3] In particular, this pertains also to H^*_{2S}—the density of hydrogen atoms in the $2S$-state. Taking the two-photon radiation into account, the equation for Σ assumes the form

$$\frac{d\Sigma}{dt} = -3\mathcal{H}\Sigma - \mathcal{H}h\nu_\alpha\frac{a}{kT} - w(H^*_{2S} - He^{-h\nu_\alpha/kT}), \tag{2}$$

where w is the known transition probability, $w = 8\,\text{sec}^{-1}$, while the last term takes into account the possibility of the two-photon excitation of an atom from the ground state to H^*_{2S}. Strictly speaking, one should add to w the contributions of all other two-photon processes $3S \to 1S$, $3D \to 1S$, ..., and also of the transitions from the continuum, leaving aside, however, only those processes in which the energy of each of the two photons is less than the energy of the L_α-photon.

In the problem of recombination during a cosmological expansion, n and T are known functions of the time. In the approximate theory the problem reduces to the integration of one differential equation with several algebraic relations between the quantities entering into it.[4]

Thus, we obtain all the quantities of interest to us as functions of the time. In order to obtain the radiation spectrum at the present time we note that at each instant the spectral density has no discontinuity at a frequency ν_α;

[3]These densities are small and may be left out of account in the sums used to evaluate Σ and n.

[4]At each instant of time, for appropriate values of T and n, there is a definite equilibrium value of Σ_e which would be realized in the course of time if the expansion were stopped, i.e., at $\mathcal{H} = 0$. It is evident that at $\Sigma = \Sigma'_e$ the solution of all equations for e, p H^*_{2S}, etc. will identically give equilibrium values, and for the present time one would obtain exactly the Planck expression, or more exactly its limiting expression—the Wien formula $f(\nu) = \text{const}\cdot\nu^2\cdot\exp(-h\nu/kT)$, since stimulated processes have not been taken into account. Since $h\nu/kT \gtrsim 30$, the difference is negligibly small.

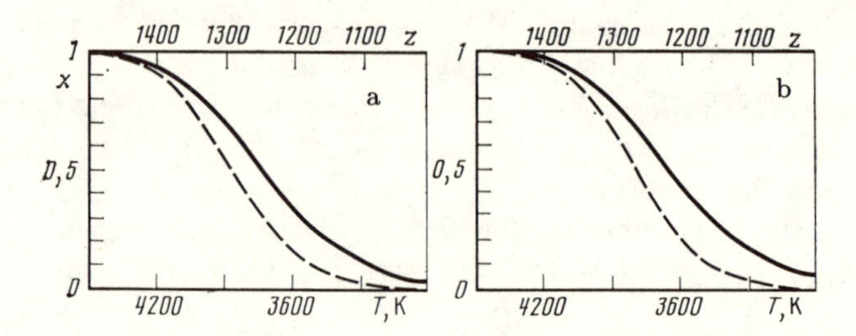

Fig. 1. Deviation of the degree of ionization from equilibrium $x = p/(p + H)$. For comparison recombination curves are shown for different values of the present density of matter in the Universe: a) $\Omega = 1$, $\rho = 2 \cdot 10^{-29}$ g/cm^3; b) $\Omega = 0.05$, $\rho = 10^{-30}$ g/cm^3.

on the other hand, at later times the photons of frequency $\nu < \nu_\alpha$ undergo only a red shift in the approximation under consideration. Therefore to each value of $f(\nu_\alpha)$ at the present moment t' there corresponds at the present time $f(\nu = \nu_\alpha/(1+z')) = (1+z')^{-2} f_\alpha(t')$. Here z' is the red shift corresponding to t', and the power of $(1 + z')$ corresponds to the fact that f is given as the density of photons per unit volume (physical, not co-moving!) per unit frequency.

2. Numerical Calculations

As shown in the Appendix, the rate of passage of L_α-photons below the threshold is given by

$$w_\alpha H^*_{2P} = \frac{H^*_{2P} 8\pi \mathcal{H}}{3H\lambda^2_\alpha} [1/\mathrm{cm}^3 \cdot \mathrm{sec}]$$

In the case of weak ionization (when $H \sim n$)

$$w_\alpha = \frac{8\pi\mathcal{H}}{3\lambda^3_\alpha H} \approx \frac{8\pi\mathcal{H}}{3\lambda^3_\alpha n} = \frac{8\pi\mathcal{H}_0}{3\lambda^3_\alpha n_0 (1+z)^{3/2}}$$
$$= \frac{1.55 \cdot 10^3}{(1+z)^{3/2}} [1/\mathrm{sec}].$$

(Here we have taken into account that $\mathcal{H} = \mathcal{H}_0 (1+z)^{3/2}$ and $n = n_0(1+z)^3$, where $\mathcal{H}_0$ and n_0 are the present values of Hubble's constant and of the density). It is evident that in the region of interest to us $z \sim 10^3$, $w_\alpha \ll w$, and the rate of recombination is determined by the two-photon decays of the $2S$ level.

Neglecting the passage of L_α-photons below the threshold and taking into account the fact that $a < e$, we simplify Eq. (2):

$$\frac{dp}{dt} = -3\mathcal{H}p - w[H_{2S} - He^{-h\nu_\alpha/kT}]. \tag{3}$$

We introduce dimensionless variables: the red shift z and the degree of ionization $x = e/(e + H) = e/n$; substituting $H_{2S} = K(I^*_2, T)e^2$ from (1), we obtain

$$\frac{dx}{dz} = \Omega^{1/2} w n_c z^{1/2} e^{I/4kT_0 z} \left[x^2 - (2\pi m_e k T_0)^{3/2} (1-x) \frac{\exp(-I/kT_0 z)}{\Omega h^3 n_c z^{3/2}} \right] \quad (4)$$

where $n_c = \rho_c / m_p = 3\mathcal{H}_0^2 / 8\pi G m_p = 10^{-5}\,\mathrm{cm}^{-3}$ is the present critical density, $\Omega = n_0 / n_c$; and it has been taken into account that

$$\frac{dt}{dz} = \frac{\mathcal{H}_0^{-1}}{(1+z)^2 (1+\Omega z)^{1/2}} \approx \frac{\mathcal{H}_0^{-1}}{z^{5/2} \Omega^{1/2}}.$$

The expression in brackets in (4) is evidently equal to zero in the case of thermodynamic equilibrium, since it represents the Saha formula.

The results of numerical integration are shown in Fig. 1, which shows the equilibrium degree of ionization and the degree of ionization obtained upon integrating (4).

3. The Asymptotic Behavior of Recombination

When the temperature has dropped significantly, the energy of the equilibrium photons is insufficient to maintain equilibrium between excited levels of the hydrogen atom and the free electrons, and each recombination event leads to the emission of an energetic photon. The rate of photoionization and upwards diffusion along the energy axis (which decreases exponentially as the temperature falls) becomes smaller than the rate of two photon decay of the $2S$ level starting with $z \sim 870$ ($T \sim 2500°$K). At $z \sim 700$, the rate of the passage of L_α-photons below the threshold and the rate of photoionization of the $2P$ level become equal, i.e., at $z \sim 700$ each recombination event[5] leads to a decrease in the degree of ionization. Utilizing the results of integration of (4) up to $z \sim 870$, we obtain the asymptotic behavior of the recombination from the equation

$$\frac{dp}{dt} = -\alpha(t) p^2$$

or

$$\frac{dx}{dz} = \alpha(z) x^2 \Omega^{1/2} z^{1/2} \mathcal{H}_0^{-1} n_c. \quad (5)$$

In the temperature range where the Compton effect for the photons of the relic radiation incident on electrons maintains the electron temperature equal to the radiation temperature (cf. Sec. 5), $T \approx T_0(1+z)$ and the recombination coefficient [5]

$$\alpha = \alpha_{2P} + \alpha_{2S} = \frac{\alpha_0}{\sqrt{T}} \approx \frac{2.5 \cdot 10^{-11}}{\sqrt{T}}\,\mathrm{cm}^3/\mathrm{sec}$$

[5]As a result of cascade transitions, the recombining electron must fall into a $2S$ level or a $2P$ level.

is proportional to $z^{-1/2}$, i.e.,

$$\frac{dx}{dz} = \frac{\alpha_0 \Omega^{1/2} \mathcal{H}_0^{-1} n_c x^2}{\sqrt{T_0}} = Ax^2,$$

whence

$$x(z) = \frac{1}{C - Az}.$$

For $\Omega = 1$ numerical calculation gives for $z \sim 870$ the value $x \sim 2 \cdot 10^{-3}$. Taking into account the fact that up to $z \sim 700$ the $2P$ level is in equilibrium with the radiation and that $\alpha_{2S} \sim \alpha/3$ [5], we obtain for $z \sim 700$ the value $x \sim 3.3 \cdot 10^{-4}$. For smaller values of z we have $x(z) = [3 \cdot 10^3 + 43(700 - z)]^{-1}$. However, at $z \sim 150$ the temperature of matter is no longer the same as the temperature of radiation (cf. Sec. 5), and beginning from that instant, $\alpha(z) \propto z^{-1}$ and $x(z = 0) = 2.5 \cdot 10^{-5}$.[6]

The ratio of the probabilities of the processes $H + p \to H_2^+ + h\nu$ and $H + e^- \to H^- + h\nu$ to the probability of radiative recombination in the temperature range of interest to us is much smaller than the ratio $p/H = e/H = x$ [5]. Therefore these processes can be neglected. The weakness of the processes $H + p \to H_2^+ + h\nu$ and $H + H \to H_2 + h\nu$ indicates the negligible ratio of the density of H_2 to that of H in the intergalactic medium.

4. Distortion of the Relict Radiation Spectrum

Emission of energetic photons in two-photon decays of the $2S$ level must distort the shape of the spectrum of the observed relic radiation $F(\nu_0)$ in the Wien region. If j_ν is the spectral brightness per unit volume and $\nu = \nu_0(1 + z)$, then according to [7] the distortion of the spectrum is

$$\Delta F(\nu_0) = \frac{c}{\mathcal{H}_0} \int_0^{z_{\max}} \frac{j_\nu(z) dz}{(1+z)^5 (1 + \Omega z)^{1/2}}.$$

In the case of two-photon emission

$$j_\nu(z) = h\nu A \left(\frac{\nu}{\nu_\alpha}\right) \frac{N_{2S}(z)}{4\pi},$$

where $A(\nu/\nu_\alpha)$ is the probability of emission of a photon of frequency ν per unit frequency interval and is tabulated in [4], with

$$w = \int_0^{\nu_\alpha} A\left(\frac{\nu}{\nu_\alpha}\right) d\nu.$$

Therefore we have

$$\Delta F(\nu_0) = \frac{ch\nu_0}{4\pi \mathcal{H}_0} \int_0^{\nu_\alpha/\nu - 1} \frac{N_{2S}(z) A(\nu_0(1+z)/\nu_\alpha)}{(1+z)^4 (1+\Omega z)^{1/2}} dz.$$

[6]The residual degree of ionization in the hot model of the Universe has been obtained earlier by Ozernoĭ under the assumption of complete thermodynamic equilibrium, subsequent quenching, and recombination at $T \propto (1+z)^2$; he obtained $x(z=0) = 10^{-4}$ [6]. It is evident that our discussion should increase $x(z=0)$ considerably.

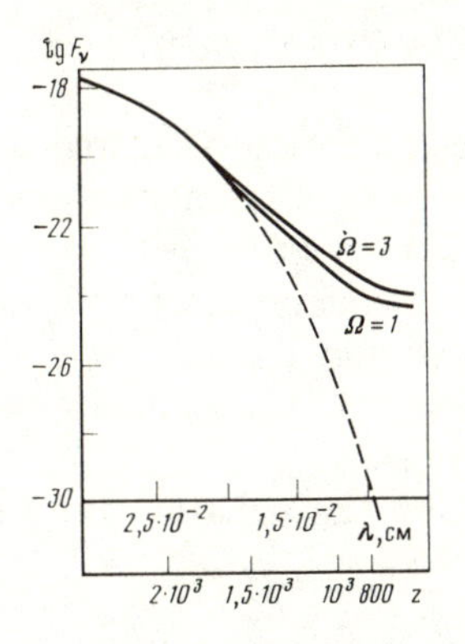

Fig. 2. Distortion of the spectrum of relic radiation in the recombination of hydrogen (F_ν, erg/cm^2·s·Hz·stereorad). The dashed line is Planck's curve.

Starting with $z \sim 700$, each recombination event leads to emission of an energetic photon, and therefore the asymptotic behavior of $F(\nu_0)$ has the form [7] for $\nu_\alpha/\nu_0 = z_\alpha \ll 700$:

$$F(\nu_0) = \frac{ch}{4\pi\mathcal{H}_0}\Omega^{3/2}z^{3/2} \propto (z_\alpha)x^2(z_\alpha) \approx 6 \cdot 10^{-29}\Omega^{3/2}\frac{\nu_\alpha}{\nu},$$

and for $\nu_\alpha/\nu \sim 700$

$$F(\nu_0) = 6 \cdot 10^{-28}\frac{\nu_\alpha}{\nu}.$$

The calculated shape of the spectrum is given in Fig. 2. Instead of the expected excess by a factor of 100–200 compared to the equilibrium spectral density in a narrow region, we have an entirely different picture.

For $\lambda < 1.6 \cdot 10^{-2}$ cm, the radiation due to the recombinations always is considerably greater than the equilibrium radiation, but is small in absolute value. Apparently in this domain even such weak factors as radiation by dust and by galaxies in the infrared range would turn out to be more essential. Moreover it is necessary to take into account the distortion of the spectrum as a result of the inverse Compton effect of the radiation involving hot electrons for $z \sim 100$ [8], if by that time the secondary heating of the gas has already taken place, and also the bremsstrahlung pertaining to the hot intergalactic gas.

It should be emphasized once more that the obtained considerable intensification of the spectrum compared to the equilibrium value refers to the region of the spectrum in which the total number of photons and their energy constitute a very small part of the total equilibrium radiation, therefore the change in the intensity referred to above corresponds to a negligible change in the effective temperature. However, as can be seen from Fig. 2, a measurement of the flux for $\lambda < 2 \cdot 10^{-2}$ cm enables us to establish an upper limit for the density of matter in the Universe.

5. Energy Exchange Between Electrons and Radiation

A slowing down of the rate of recombination compared to the Saha formulation leads to the fact that the density of the non-recombined electrons (which is practically independent of the total density of matter) remains sufficient, up to $z \sim 150$, to maintain the temperature of the matter equal to the temperature of the radiation. The energy is transferred from the photons to the electrons by the Compton effect; collisions equalize the temperature of the electrons and the hydrogen atoms.

The rate of energy exchange between the radiation and the electrons [9] is given by

$$L = 4\sigma_0 k (T_\gamma - T_e)\frac{\mathcal{E}_\gamma}{m_e c},$$

where σ_0 is the Thomson cross section, $\mathcal{E}_\gamma \sim (1+z)^4$ is the radiation energy density, T_e and T_γ are the temperatures of the electrons and of the radiation. If there were no Compton exchange of energy between electrons in the radiation, then the electron temperature after recombination would fall adiabatically with $\gamma = 5/3$, while the radiation temperature would fall adiabatically with $\gamma = 4/3$, i.e., $T_e \sim \rho^{2/3} \sim (1+z)^2$, while $T_\gamma \sim \rho^{1/3} \sim (1+z)$. From this it follows that

$$\frac{d(T_\gamma - T_e)}{dz} = \frac{T_\gamma}{1+z} - \frac{2T_e}{1+z}.$$

The limiting value $z_{\min}$ up to which $T_\gamma \sim T_e$, could be obtained by equating $d(T - T_e)/dz$ and

$$L(z) = L(t)\frac{dt}{dz} = \frac{4\sigma_0 k T(z)(1+z)^2}{m_e c \mathcal{H}_0 (1+\Omega z)^{1/2}}.$$

However, there are many more neutral hydrogen atoms than electrons and collisions with the gas-kinetic cross section entirely suffice to transfer energy to them from the electrons, and therefore in order to maintain the temperature of the matter it is necessary to transfer energy from the radiation to the electrons at a rate which is bigger by a factor $1/x$, i.e.,

$$\frac{kT(z)}{1+z} \approx \frac{4\sigma_0 k T(z) x (1+z)^2}{m_e c \mathcal{H}_0 (1+\Omega z)^{1/2}},$$

and from this it follows that $\left[4\sigma_0 \Omega^{-1/2} x(\Omega, z) z^{3/2} \,/\, mc\mathcal{H}_0\right] = 1$. Consequently $z_{\min}$ depends only weakly on Ω and is equal to 150–200.

Adiabatic variation of the temperature of matter with $\gamma = 4/3$ heating for such a long period must have an effect on the Jeans wavelength of gravitational instability in a homogeneous expanding universe. The Jeans mass remains constant after recombination, while in the absence of interaction between the relic radiation and plasma it falls off in accordance with a power law [10].

The authors are grateful to L. Domozhilova and A. G. Doroshkevich for their aid in carrying out the numerical calculations.

Appendix

We obtain the rate of passage of photons below the threshold accompanying an expansion of the Universe. The equation for the spectral density of photons in the line L_α has the following form:

$$\frac{df_\nu}{dt} = A_\nu \varphi(\nu) H^*_{2P} - B_\nu \varphi(\nu) f_\nu H,$$

where A_ν and B_ν are the Einstein coefficients, and $\varphi(\nu)$ is the line profile. Since for an expanding universe $\nu = \nu_0 (t_0/t)^{2/3}$ and

$$\frac{\partial \nu}{\partial t} = -\frac{2\nu_0 t^{2/3}}{3t_0^{5/3}} \approx -\frac{2\nu}{3t} = -\mathcal{H}\nu,$$

it follows that

$$\frac{df_\nu}{dt} = \frac{\partial f_\nu}{\partial t} - \frac{\partial f_\nu}{\partial \nu}\mathcal{H}\nu.$$

Solving the characteristic equation,

$$\frac{df_\nu}{B_\nu \varphi(\nu) H (f^* - f_\nu)} = -\frac{d\nu}{\mathcal{H}\nu},$$

where

$$f^* = \frac{H^*_{2P} A_\nu}{H B_\nu} = \frac{H^*_{2P} 8\pi h \nu^3}{3Hc^3},$$

we obtain
$$f_\nu = f^* \left(1 - e^{-(n-p)B_{\nu_\alpha}/\mathcal{H}\nu_\alpha} \int_{+\infty}^{\nu} \varphi(\nu)\, d\nu\right),$$

whence it can be seen that outside the profile $f_\nu \approx f^*$.[7] Now one can find the number of photons passing below the threshold. It is given by

$$-f^* \frac{d\nu/dt}{h\nu} = \frac{\mathcal{H} f^*}{h} = \frac{H^*_{2P} 8\pi \nu^3 \mathcal{H}}{3Hc^3}\ \mathrm{cm}^{-3} \cdot \mathrm{sec}^{-1}.$$

[7] One need not take the induced radiation into account, since $H_{2P} \ll H$ as a result of the low temperature.

Institute of Applied Mathematics
USSR Academy of Sciences

Received
December 27, 1967

REFERENCES

1. *Gamow G.*—Phys. Rev. **70**, 572 (1946).
2. *Zeldovich Ya. B.*—UFN **89**, 647 (1966).
3. *Kipper A. Ya.*—Astron. Zh. **27**, 321 (1950).
4. *Spitzer L., Greenstein L.*—Astrophys. J. **114**, 407 (1965).
5. Atomic and Molecular Processes, edited by D. R. Bates, Academic Press (1962).
6. *Ozernoĭ L. M.*—Dissertation, GAISh (1966).
7. *Kurt V. G., Sunyaev R. A.*—Kosmichesk. Issled. **5**, 573 (1967).

8. *Weymann R.*—Astrophys. J. **145**, 560 (1966).
9. *Weymann R.*—Phys. Fluids **8**, 2212 (1965).
10. *Doroshkevich A. G., Zeldovich Ya. B., Novikov I. D.*—Astron. Zh. **44**, 295 (1967).

Commentary

The history of this article is as follows: in the Spring of 1967 V. G. Kurt, who at that time was studying extra-atmospheric observations of planets and the interplanetary medium in the L_α line of hydrogen ($\lambda_0 = 1216\,\text{Å}$), set Ya. B. and R. A. Sunyaev a problem—where did the L_α photons which formed in the recombination of hydrogen in the Universe go and shouldn't a narrow bright line with wavelength $\lambda_0(1 + z_r) \sim 170$ mkm be observed in the spectrum of relic radiation? We recall that even in the pioneering work of G. Gamow (Ref. 1 in the references above) it was noted that the hydrogen recombination in the Universe should occur in agreement with Saha's formula of ionization equilibrium. Substitution of the parameters of the Universe (the temperature of the relic radiation and the average density of matter) into this formula showed that the recombination occurred at $T_r \sim 4000^\circ K$, i.e., at a red-shift $z_r \sim 1500$.

Kurt's question attracted the attention of his future coauthors. Estimates showed that the drift of photons in the L_α line to the far wing of the line due to the cosmological red-shift and expansion of the line due to multiple resonance scattering occurs relatively slowly. The high density of photons in the line leads to high occupation of excited levels and slows the rate of recombination. It unexpectedly turned out that the most effective mechanism is not the radiation of photons in the L_α line and their drift into the red end, but a two-photon decay of the $2S$-level. In the paper an equation is derived which describes the course of recombination; its numerical solution is given and distortions of the relic radiation spectrum in recombination are calculated.

In 1967 in the USA the Texas symposium on relativistic astrophysics took place; I. S. Shklovskiĭ and I. D. Novikov participated in the work there. They told of this work, its main idea and result. The well-known American astrophysicist Peebles repeated all the calculations and published them,[1] noting in the introduction that he had borrowed the ideas and main result from remarks by I. S. Shklovskiĭ at the Texas symposium and the review by I. D. Novikov and Ya. B.[2]

The present paper proved quite important for cosmology. Its results have entered into many books (see, e.g., the work by Peebles[3] and by Ya. B. and Novikov[4]).

The remainder concentration of free electrons found in the paper (those which did not manage to recombine in cosmological time) turned out to be sufficient to support a temperature of electrons and ions equal to the temperature of radiation up to $z \sim 400$. Later D. A. Varshalovich, V. K. Khersonskiĭ and R. A. Sunyaev[5]

[1] *Peebles P. J. E.*—Astrophys. J. **153**, 1–11 (1969).

[2] *Novikov I. D., Zeldovich Ya. B.*—Ann. Rev. Astr. Astrophys. **5**, 627 (1967).

[3] *Peebles J.*—Physical Cosmology. Princeton Univ. Press (1971).

[4] *Zeldovich Ya. B., Novikov I. D.*—Stroenie i evolĭutsiĭa Vselennoĭ [Structure and Evolution of the Universe]. Moscow: Nauka, 736 p. (1975).

[5] *Varshalovich, D. A., Khersonskiĭ V. K., Sunyaev R. A.*—Astrofizika **17**, 487–493 (1981).

showed that a small concentration of heavy elements (C, O, N, etc.) leads to a similar effect. We note, however, that in the cosmological situation at large z a noticeable abundance of heavy elements is improbable.

The non-instantaneous character of the recombination leads to intensified damping of mean-scale perturbations because of the effects of radiative viscosity and heat conduction.[6,7] The scale of the surviving perturbations grows. And, of most significance, the delay of recombination leads to smoothing of small-scale angular fluctuations of the relic radiation (this was first noted by Ya. B. and R. A. Sunyaev in another article in this book (**66**), and is taken into account in all subsequent calculations). Later, Longair and Sunyaev[8] and Ya. B. and Sunyaev (the above-mentioned article **66**) found the asymptotic analytic solution describing the course of recombination at $1000 < z < 1500$, when the concentration of electrons is less than that of neutral atoms). The course of recombination of helium was calculated.[9] V. K. Dubrovich[10] noted that in the relic radiation spectrum weak spectral details should exist which are related to transitions between excited levels in a hydrogen atom. Yu. E. Lĭubarskiĭ and R. A. Sunyaev[11] indicated that in the presence of strong initial distortions of the spectrum of relic radiation such details may prove to be observable. In addition they calculated the course of recombination in various cosmological models and, in particular, taking into account the hidden mass concentrated in weakly-interacting particles. B. Jones and R. Wyse[12] published detailed calculations of recombination.

Comparison of observational data on interstellar deuterium with the results of calculations of nucleosynthesis in the hot model of the Universe give evidence of the low present mean density of baryons in the Universe, $n \sim 10^{-7}\,\mathrm{cm}^{-3}$. This value is significantly smaller than those taken in the present article, $n \sim 10^{-5} - 3 \cdot 10^{-7}\,\mathrm{cm}^{-3}$. In the papers by Lĭubarskiĭ and Sunyaev[11] and Jones and Wyse[12] the calculations are carried out for $n \sim 10^{-7}\,\mathrm{cm}^{-3}$.

[6] *Silk J.*—Astrophys. J. **151**, 459–471 (1968).
[7] *Chibisov G. V.*—Astron. Zh. **49**, 74–84 (1972).
[8] *Longair M. S., Sunyaev R. A.*—Nature **223** N5207, 719–721 (1969).
[9] *Sato H., Matsuda T., Takeda H.*—Suppl. to Prog. Theoret. Phys. **N49**, 11–82 (1971).
[10] *Dubrovich V. K.*—Pisma v Astron. Zh. **1**, 3–4 (1975).
[11] *Lĭubarskiĭ Yu. E., Sunyaev R. A.*—Astron. and Astrophys. **123**, 171–182 (1983).
[12] *Jones B. J. T., Wyse R. F. G.*—Month. Not. RAS **205**, 983-1007 (1983).

49

The Interaction of Matter and Radiation in a Hot-Model Universe*

With R. A. Sunyaev

In this paper we continue the investigation initiated by Weymann as to the reason why the spectrum of the residual radiation deviates from a Planck curve. We shall consider the distortions of the spectrum resulting from radiation during the recombination of a primeval plasma. Analytical expressions are obtained for the deviation from an equilibrium spectrum due to Compton scatter of hot electrons. On the basis of the observational data it is concluded that a period of neutral hydrogen in the evolution of the Universe is unavoidable. It is shown that any injection of energy at $t > 10^{10}$ s (red shift $z < 10^5$) leads to deviation from an equilibrium spectrum.

1. Introduction

More than three years have already passed since the discovery of residual radiation. In this time dozens of experimental papers have already appeared, direct measurements have been carried out, and upper limits to the temperature of the residual radiation have been obtained in the wavelength region between 50 cm to $5.6 \cdot 10^{-2}$ cm. The combined results of this work are shown in Fig. 1. The temperature of the radiation, according to the present results at 3.2, 1.58 and 0.86 cm, is evaluated as $2.68^{+0.09}_{-0.14}$K [1, 2].

Values of the excitation temperatures of the rotational levels of the CN, CH and CH^+ molecules, which are of twenty years standing, made it possible to put an upper limit on the temperature of the radiation in the microwave region [3] and to evaluate it at $\lambda = 0.263$ cm [3, 4]. This evaluation, together with new measurements in the millimeter range ($\lambda = 3.3$ mm) have established that the observed spectrum of residual radiation is fundamentally different from the spectrum of black-body radiation. Attempts to explain the observed background in the centimeter range by the integrated radiation of sources having an anomalous discontinuity in the spectrum in the 50–30 cm range are doomed to failure (see, for example, [5]). It seems also that attempts to explain the origin of residual radiation by the interaction of the optical radiation of hypothetical primeval stars with dust are impossible

*Astrophysics and space science **4**, 301–316 (1969).

to uphold. In a previous paper [6] it was shown that such an explanation meets with significant difficulties.

The background radio emission has already given cosmologists significant information about the isotropy of the expansion of the universe [7] and about the early stages of its evolution, supporting the hot universe model [8].

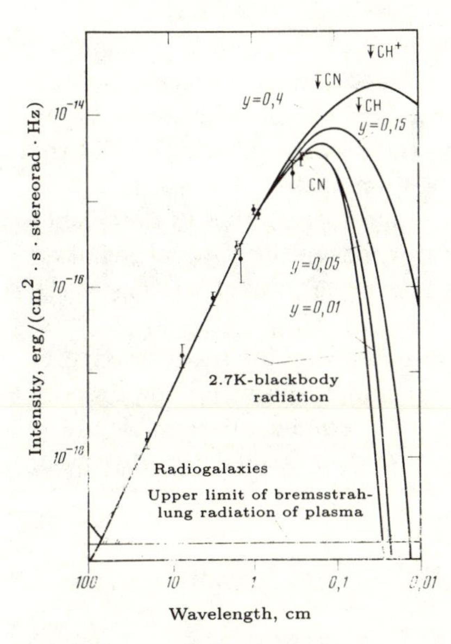

Fig. 1. Possible distortions of relic radiation under the Compton effect with hot electrons; existing experimental data are superposed (see references in [3, 21, 26].

An examination of the physical processes which lead to the formation of the spectrum of residual radiation shows that practically any energy produced during the expansion of the Universe up to the time $t \sim 10^{10}$ s will be transmuted in full, the plasma and the radiation will at all times be in thermodynamic equilibrium, the spectrum of the radiation will be Planckian. In the interval $10^{10} < t < 3 \cdot 10^{17}$ s the interaction of radiation and electrons can no longer transform a non-Planckian spectrum to a Planckian one. From this it follows that the residual radiation presently observed was produced before the time $t \sim 10^{10}$ s, corresponding to a red shift of $z \sim 10^5$.

In the following we shall accept the existence of full equilibrium for $z > 10^5$. This equilibrium is preserved up to $z \sim 1500$. The point is that, in agreement with the hot Universe model, in the early stages of expansion the full ionized plasma existed in equilibrium with the radiation. Later cooling of the plasma must have led to the recombination of hydrogen accompanied by the distortion of the residual radiation spectrum in its 'blue' region (at $h\nu/kT_r > 30$, that is at $\lambda < 0.02$ cm for a present-day observer). We shall introduce the results of the calculations carried out in [9].

In the process of further expansion of a strictly homogeneous model, the hydrogen would necessarily remain neutral. However, we know that in fact matter combines to form stars, galaxies and clusters of galaxies—in other words, it is known that a significant inhomogeneity plays a part. It seems that such inhomogeneity leads to the ionization of the gas possibly even at a much earlier stage than that in which present-day structure was formed.

Observations in our neighborhood up to the region corresponding to a red shift $z \sim 2$ have not disclosed significant densities of neutral hydrogen. At the same time, they do not rule out the existence of a hot ionized metagalactic gas.[1] This supports the hypothesis regarding the ionization of the gas. The existence of the metagalactic gas is compatible with the hot Universe model if we postulate a second heating and ionization of the gas, which no longer exists in equilibrium with the residual radiation afterwards.

A very serious qualitative question now arises: Is it possible to maintain that there existed a stage of neutral hydrogen between the period of equilibrium ionized gas at $z \sim 1500$ and the non-equilibrium ionized gas postulated at the present time? At what moment, on the average (at what z) did the second ionization take place? Weymann [13] has noted that the radiation of the gas after ionization together with Compton scattering of the quanta from electrons should lead to significant distortions of the residual radiation spectrum and should convey information about the time of heating of the gas.

We have been able to find an analytical expression for the distortion of the spectrum as a result of Compton effect and to prove on the basis of existing measurements that a neutral hydrogen period is unavoidable in the evolution of the Universe. A part of the results presented here has been previously published in [9, 14].

2. The Distortion of the Spectrum During the Recombination of Hydrogen

In the hot Universe model it is proposed that in the early stages of expansion, the fully ionized plasma is found in equilibrium with radiation. Cooling during expansion leads to recombination. In accordance with Saha's equation, a state of ionization of 50%: $e = p = H$, is achieved at $T_r \sim 4000°$K at a period corresponding to a red shift $z \sim 1500$.

The density of quanta n_γ in the Universe is much larger than the density of ions, electrons and atoms

$$\frac{n_\gamma}{p+H} = \frac{0.244}{\Omega n_{\rm cr}(kT_r/hc)^3} = 4.15 \cdot 10^7 \Omega^{-1},$$

where $\Omega = \rho/\rho_{\rm crit} = n/n_{\rm crit}$.

However, the temperature at which the recombination effectively proceeds is significantly less than the temperature of ionization and excitation of the first level in a hydrogen atom

$$kT_r \simeq \frac{I}{40} \simeq \frac{h\nu_\alpha}{30}$$

[1] Moreover, the most recent results from observations of the soft X-ray background are interpreted as revealing a metagalactic gas with a temperature in the range 10^5–10^6 K and with a density of the order of the critical density $\sigma_{\rm crit} = 3H_0^2/8\pi G = 2 \cdot 10^{-29}$ g/cm^3 [10, 11]. It is true, as is pointed out by one of the authors (R.S.), that the existence of very extended bridges between galaxies and the distribution of neutral hydrogen on the periphery of galaxies casts doubt upon such an interpretation [12].

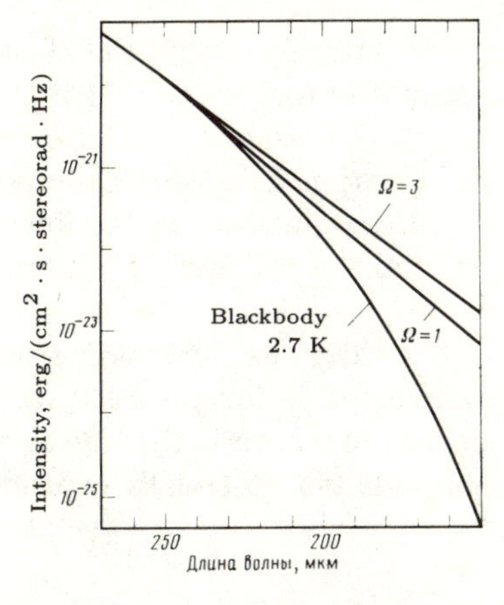

Fig. 2. Distortion of the spectrum in the recombination of hydrogen.

where ν_α is the frequency of a $L\alpha$ quantum. Therefore, in the period of recombination, the density of energy quanta with $\nu > \nu_\alpha$ comprises a small part of the general quantum density, and represents a density less than that of ions and atoms by a factor of about 200.

V. G. Kurt has noted that if every recombination were accompanied by the emission of an energetic quantum, then the density of these quanta would rise sharply in comparison with that under the Planck spectrum. Since all wavelengths have increased by a factor of 1500 from the time of recombination, it would be legitimate to expect an anomaly in the residual spectrum for wavelengths $\lambda \sim 10^{-2}$ cm.

This question has been investigated in detail in [9].[2] The hypothesis of production of an energetic quantum at each recombination is not accurate in the given situation, since there is absorption of $L\alpha$ and photons of higher energy. As a result of this there occurs a marked slowing of the recombination of electrons and protons, since a super-equilibrium concentration of photons with $\nu > \nu_\alpha$, engenders a super equilibrium concentration of excited hydrogen atoms, which are easily ionized by soft photons ($I_2 = I/4 = 3.4$eV).

A decrease in the concentration of protons and electrons existing in thermodynamic equilibrium with excited states occurs principally by the two-quantum decay of the metastable $2S$ level of the hydrogen atom. The production of energetic quanta in the two-quantum decay of the $2S$ level should distort the appearance of the spectrum of the observed residual radiation in the microwave range. Zeldovich [9] carried out a calculation of the distortion of the spectrum for the case $T_r = 3$K. In Fig. 2 curves showing the distortion of the spectrum for $T_r = 2.7$K are presented. A marked deviation $\Delta J/J \sim 30\%$ may already be expected at $\lambda \sim 230\mu$. For $\lambda < 230\mu$ the recombination radiation always markedly outweighs the equilibrium radiation. Clearly the radiation of radio-galaxies and stars in the infra-red is markedly less.

[2]The results of this work were discussed in [15, 16]. The results of Peebles [17] are in good agreement with ours.

Estimates of the thermal radiation of the metagalactic plasma and of the distortion due to Compton effect from thermal electrons show that if the heating of the gas occurred at $z < 20/\Omega$, then in the region $\lambda \sim 240\mu$ these would be considerably weaker than the recombination radiation. Only the radiation of dust in galaxies (existing theoretical evaluations of the intensity of this radiation are extremely uncertain) and the radiation of infra-red objects may prove to be stronger.

Measurements in the wavelength region $\lambda \sim 200\mu$ are very interesting since, first of all, they can give information about the period $z \sim 1000$, and, secondly, they enable the evaluation of the density of matter in the Universe. As is seen in Fig. 2, the flux of recombination radiation essentially depends on the density of matter.

3. Bremsstrahlung Radiation

It is known that if ionization is determined by electron impacts, then up to a temperature of $T_e \sim 10^4$K a hydrogen plasma will remain practically neutral, while at greater temperatures, the degree of ionization quickly increases. This fact, together with the observed low brightness temperature of the thermal radiation of the gas (see below) testifies to its low optical depth up to the period corresponding to a red shift $z \sim 10^4$. For this reason, in the following evaluation of the thermal radiation of the metagalactic plasma we shall not take into account self-absorption, that is, we shall use the formula for the radiation of a thin sheet.

The volume emissivity coefficient for free-free transitions [18] is

$$E_{ff}(\nu) = 5.44 \cdot 10^{-39} g T_e^{-1/2} e^{-h\nu/kT_e} n_e^2 \ \mathrm{erg/(cm^3 \cdot s \cdot ster \cdot Hz)}, \qquad (1)$$

where the Gaunt factor for $h\nu \gg kT_e$ is $g = 1$ and for $h\nu \ll kT_e$ is $g = \sqrt{3}/\pi[\ln(4kT/h\nu) - 0.577]$ according to [19]. Our basic interest will be in the case $h\nu \ll \kappa T_e$ when the spectrum of radio-emission of the metagalactic plasma is flat, facilitating its identification

$$E_{ff}(\nu) = 5.44 \cdot 10^{-39} g T_e^{-1/2} n_e^2 \ \mathrm{erg/(cm^3 \cdot s \cdot ster \cdot Hz)}, \qquad (2)$$

Since the spectrum of residual radiation has a maximum in the wavelength range $\lambda \sim 0.1$cm, it follows that the traces of thermal radio emission of the plasma should be sought in the long wavelength $\lambda \sim 50$ cm and short wavelength $\lambda \sim 0.03$ cm regions.

In the wavelength region $\lambda \sim 50-100$ cm the contribution of the combined radio emission of discrete sources, galaxies, radiogalaxies and quasars to the radio background of the Universe is comparable to the contribution due to residual radiation. According to Bridle [20] the brightness temperature of the extragalactic component of the radio background at 170 cm amounts to 30 ± 7K. Its spectral index is 0.7 ($T_b \sim 3$K at $\lambda = 75$ cm).

The contribution of radiation from our Galaxy to the radio-background strongly depends on direction and is comparable to the contribution due to residual radiation up to $\lambda \sim 40-60\,\text{cm}$. The measurements by Howell and Shakeshaft [21] at the wavelengths $\lambda = 75\,\text{cm}$ and $\lambda = 50\,\text{cm}$ and for $\lambda = 15$, 20 and 30 cm by Pelyushenko *et al.* [22] made it possible to set an upper limit to the intensity of thermal radio emission of the metagalactic plasma

$$J_\nu < 10^{-19}\ \text{erg}/(\text{cm}^2 \cdot \text{s} \cdot \text{ster} \cdot \text{Hz})\ (T_b(50\ \text{cm}) < 1\ \text{K})$$

4. The Distortion of the Spectrum of Residual Radiation as a Result of Compton Scattering from Thermal Electrons

This phenomenon has been examined in detail by Weymann [13] who calculated numerically several variants of the distorted spectrum. We were able to obtain an analytical expression for the distortion of the spectrum due to Compton effect of thermal electrons (the derivation is presented in the appendix). In the derivation we used formulae obtained earlier by Kompaneets [23] and by Weymann [24].

For the frequency derivative of the occupation number

$$n(\nu, T) = \left(\frac{8\pi h\nu^3}{c^3}\right)^{-1} \frac{dE}{d\nu}$$

in the non-relativistic approximation $kT_e \ll m_e c^2$ and $kT_r \ll m_e c^2$ (in this case the change of frequency of a quantum in each scattering process is due to the thermal velocity of the electron, that is, as a result of the Doppler effect) both Kompaneets and Weymann give the formula

$$\frac{\partial n}{\partial t} = \frac{kT_e}{m_e c^2} \frac{n_e \sigma_e}{c} (x^1)^{-2} \frac{\partial}{\partial x^1} (x^1)^4 \left(\frac{\partial n}{\partial x^1} + n + n^2\right), \tag{3}$$

where T_e and T_r are the electron and radiation temperatures, n_e is the electron density, σ is the Thomson cross-section, and $x^1 = h\nu/\kappa T_e$. In the case $T_e \gg T_r$ this equation simplifies to the form

$$\frac{\partial n}{\partial t} = \frac{kT_e}{m_e c^2} \frac{n_e \sigma_e}{c} x^{-2} \frac{\partial}{\partial x} x^4 \frac{\partial n}{\partial x}. \tag{4}$$

It is useful to change to the variable $x^1 = h\nu/\kappa T_r$ since during the expansion of the Universe the frequency and temperature of the radiation change according to the same rule $\nu \sim (1+z)$ and $T_r \sim (1+z)$. Now it is possible to introduce, following Kompaneets, the dimensionless variable

$$y = -\int_{t_0}^{t} \frac{kT_e}{m_e c^2} n_e \sigma_0 c\, dt = \int_0^{\tau} \frac{kT_e}{m_e c^2}\, d\tau, \tag{5}$$

where $\tau = \int \sigma_0 n_e\, dl$ is the optical depth due to Thompson scattering. Going from t and n_e to the red shift z and the dimensionless density $\Omega = \rho/\rho_{\text{crit}}$

we have

$$\frac{\partial n}{\partial y} = x^{-2}\frac{\partial}{\partial x}x^4\frac{\partial n}{\partial x}, \tag{4'}$$

$$y = \Omega n_{\rm crit}\sigma_0 c H_0^{-1}\int_0^{z_{\rm max}} \frac{kT_e(z)}{m_e c^2}\frac{1+z}{\sqrt{1+\Omega z}}dz \tag{5'}$$

where H_0 is the present value of Hubble's constant; it is taken that

$$\frac{dt}{dz} = \frac{cH_0^{-1}}{(1+z)^2\sqrt{1+\Omega z}}.$$

Equation (4′) is easily solved for small values of the parameter y when the deviations from a Planck curve are small, by inserting in the right-hand side the unperturbed Planck function $n_0(x) = 1/(e^x - 1)$. Then, as shown in the Appendix, the relative distortion of the spectrum is expressed by the equation

$$\frac{\Delta n}{n_0} = \frac{\Delta J}{J_0} = xy\frac{e^x}{e^x - 1}\left\{\frac{x}{\tanh(x/2)} - 4\right\}. \tag{6}$$

The physical meaning of the degree of dependence of the distortion upon frequency is clear; conservation of the overall number of quanta is accompanied by an increase in the mean energy of a photon; at the same time the flux in the Rayleigh-Jeans domain decreases ($x < 3.83$) and increases in the microwave domain. But we know neither the original nor the distorted flux. We determine the temperature of the residual radiation by measurements in various parts of the spectrum, with greatest accuracy in the Rayleigh-Jeans domain. It is useful to transfer our attention to an expression into which enters only measured values of temperature. Taking into account the decrease of temperature in the Rayleigh-Jeans region, we obtain (see Appendix)

$$\frac{T(x) - T_{\rm RJ}}{T_{\rm RJ}} = y\left\{\frac{x}{\tanh(x/2)} - 2\right\};$$

$$\frac{J(x) - J(x, T_{\rm RJ})}{J(x, T_{\rm RJ})} = \frac{xye^x}{e^x - 1}\left\{\frac{x}{\tanh(x/2)} - 2\right\}. \tag{7}$$

The amount of distortion, as expected, increases with increasing frequency. In the microwave domain for large x, where even $y \ll 1$ may lead to marked deviations from the Planck curve, as is shown in the Appendix

$$\frac{J(x)}{J(x, T_{\rm RJ})} = \exp\left(x + \frac{(\eta_0 - \eta_0^2/2)}{2y}\right)\cdot\frac{1}{\sqrt{1-\eta_0}}, \tag{8}$$

where $-\eta_0 e^{-\eta_0} = 2xye^y$. This formula is accurate for $y < 1$. Finally, for a given y, the following relation holds true for the energy density of the radiation

$$E = \frac{4\pi}{c}\int_0^\infty J_\nu\,dx = \sigma T_{\rm RJ}^4 e^{12y}.$$

Part of the exponential index, $8y$, is connected with the fact that the observed temperature in the Rayleigh-Jeans domain is lower than the original temperature $T_{RJ} = T_0 e^{-2y}$; and the second part, $4y$, owes its existence to the increase in the mean photon energy in the Compton process $E = E_0 e^{4y} = \sigma T_0^4 e^{4y}$. The lowest values of y from present observations are obtained from the Equations (7) and (8) using the given measurements of the excitation temperature of the levels of the molecule CN and the direct measurements of the flux at the wavelength $\lambda = 0.33$ cm. The latest measurements by Bortolot *et al.* [25] gave $T_r = 2.83 \pm 0.15$K at $\lambda = 0.263$ cm and from the Princeton measurements it follows that, at $\lambda = 0.33\,cm$, $T_r = 2.46 \pm 0.44$K. These results agree within measurement errors with the measurements in the Rayleigh-Jeans region $T_{RJ} = 2.68^{+0.09}_{-0.14}$K; therefore, it cannot be ruled out that $y = 0$. However, within these errors the results do not exclude values of $y < 0.15$ either, which can lead to a marked distortion of the spectrum in the microwave domain and to an enormous ($e^{12y} \sim 6$) increase in the energy density relation

$$0.25 \text{ eV/cm}^3 \leq E < 1.5 \text{ eV/cm}^3.$$

In this way, the lack of exact measurements of the residual radiation in the region of the maximum leads to a large indeterminacy in the most important quantity E. At the present time, the calculation of the X-ray emission and thermal balance of the metagalactic gas, both linked to the inverse Compton effect, are based upon the value of $E = 0.25$ eV/cm^3.

5. *The Inevitability of a Period of Neutral Hydrogen in the Evolution of the Universe*

It has already been mentioned in the second section that at $z \sim 1500$ an agreement with the hot Universe model required that a recombination of the hydrogen should take place. In further expansion the hydrogen was to remain neutral. However, observations in the 21-cm line and measurements of $L\alpha$ absorption in the spectra of distant quasars indicate the absence of neutral metagalactic hydrogen.

Let us make a rough estimate of the $z_{\rm max}$ at which the second heating took place. Previously, we showed that the heating of the gas could not have taken place earlier than $z \sim 300$ [14]. In fact, the existence of an experimental limit to the radio-emission of the gas (see Section 3)

$$\int \frac{E_{ff}\, dl}{(1+z)^3} < J_\nu, \tag{10}$$

and the limit to the energy losses of the gas by the most efficient inverse Compton-effect of thermal electrons on the residual radiation

$$\int L^- \, dt < w \tag{11}$$

enable us to set boundaries to the period of existence of the non-equilibrium ionization of the gas. In Equation (11) in agreement with [24]

$$L = 4\sigma_0 k(T_e - T_r)\frac{\sigma T_r^4}{m_e m_p c} = 3 \cdot 10^{-12}(1+z)^4 T_e(z) \text{ erg/(g} \cdot \text{s)}, \qquad (12)$$

and E_{ff} in (10) is replaced by (2) in which the Gaunt factor is taken as $g = 10$, since the measurements are carried out in decimeter region with a gas temperature of the order 10^5–10^6K. The inequalities (10) and (11) give us a system of two functionals

$$\Omega^2 \int_0^{z_{\max}} \frac{(1+z)T_e^{-1/2}(z)}{\sqrt{1+\Omega z}}\, dz < 1.8 \cdot 10^{19} J_\nu, \qquad (10')$$

$$\int_0^{z_{\max}} \frac{(1+z)^2 T_e(z)}{\sqrt{1+\Omega z}}\, dz < 10^{-6} w. \qquad (11')$$

The bremsstrahlung radiation of an ionized plasma is strong for a low electron temperature, and Compton losses are significant at high temperatures. The system functionals (10′) and (11′) enable us to find the extremal function

$$T_e(z) = 1.4 \cdot 10^{-18} \left(\frac{w}{J_\nu}\right)^{2/3} \Omega^{4/3} z^{-2/3}$$

and the highest possible value of the moment of heating $z_{\max} = 1.1 \cdot 10^5 (J_\nu^4 w^2/\Omega^5)$. In the case of any other dependence $T_e(z)$ for the same values J_ν and w one of the inequalities (10) or (11) leads to lower values of z for the heating. Sunyaev [14], assuming only nuclear sources of energy and an original chemical composition corresponding to a hot universe model, pointed out the inevitability of a period of neutral hydrogen in the evolution of the Universe ($z_{\max} < 300$). Recently, however, the interesting work of Ozernoi and Chernin [27] appeared in which attention is brought to the possibility of heating of the gas by residual turbulence and to the accompanying production of energy, markedly exceeding both the nuclear energy and the rest mass energy (mc^2). In such a theory the criteria (11) disappear. Compton effect of electrons heated by turbulence (or by other sources of unlimited—or more accurately—unknown potential) on the residual radiation should distort its spectrum. The presence of an upper limit to the distortion and to the 'effective scattering thickness' (Section 4)

$$y = \sigma_0 \Omega n_{\text{crit}} c H_0^{-1} \frac{k}{m_e c^2} \int_0^{z_{\max}} \frac{1+z}{\sqrt{1+\Omega z}} T_e(z)\, dz < 0.15, \qquad (13)$$

together with the limit to bremsstrahlung radiation of the gas leads to a second system of functionals:

$$\Omega^2 \int_0^{z_{\max}} \frac{1+z}{\sqrt{1+\Omega z}} T_e^{-1/2}(z)\, dz < 1.8 \cdot 10^{19} J_\nu, \qquad (10')$$

$$\Omega \int_0^{z_{\max}} \frac{1+z}{\sqrt{1+\Omega z}} T_e(z)\, dz < 9 \cdot 10^{10} y. \qquad (13')$$

The extremal of this system turns out to be the function

$$T = 2.9 \cdot 10^{-6} \left(\frac{\Omega y}{J_\nu}\right)^{2/3}$$

(that is, a constant temperature, independent of z) and

$$z_{\max} = 1.3 \cdot 10^{11} J_\nu^{4/9} y^{2/9} \Omega^{-7/9}.$$

Using the existing limits $J_\nu < 10^{-19}$ erg/(cm$^2 \cdot$ s $\cdot$ ster $\cdot$ Hz) and $y < 0.15$, we obtain $z_{\max} < 300\Omega^{-7/9}$ while $T_e = 3.8 \cdot 10^6 \Omega^{2/3}$K. In this way, if the mean density of matter in the Universe exceeds $\sigma = 0.12\sigma_{\text{crit}} \simeq 2 \cdot 10^{-30}$ g/cm^3 then from the observations follows the inevitability of a period of neutral hydrogen in the Universe. During this loss of energy, the coincident outlay of energy to heat the matter has an upper bound $w < 2.5 \cdot 10^{18} \Omega^{-16/9}$ erg/g.

6. The Formation of the Spectrum of Residual Radiation

Any production of energy during the expansion of the Universe leading to the raising of the temperature of the plasma also distorts the spectrum of the residual radiation. However, there exists a moment t_1 such that the interaction of the mass and radiation in further expansion erases all distortion and makes the spectrum of the radiation closely Planckian. However, heating of the electrons after this moment leads to deviations from the Planck curve. As Weymann and Kompaneets have shown, Compton scattering of radiation from Maxwellian electrons with $T_e > T_r$ leads to the establishment of a Bose-Einstein distribution with a smaller number of photons than in the Planck distribution

$$n(x) = (e^{\alpha+x} - 1)^{-1}, \qquad \alpha \geq 0, \qquad (14)$$

in which the time taken to approach a quasi-equilibrium distribution [14] is characterized by the quantity $y' = 4y \cong 1$, where y is given by formula (5). Under the conditions of an expanding universe with a density of matter $\sigma = \Omega\sigma_{\text{crit}}$ and with an electron temperature $T_e \simeq T_r(1+z)$, y' takes a value of the order unity for $z \sim 10^4\Omega^{-1/5}$.

We notice that for $z_0 = 4 \cdot 10^4\Omega$ the energy densities of matter and radiation become equal, and for $z > z_0$ the rate of expansion changes, it proceeds more rapidly, that is, as

$$\frac{dt}{dz} \sim z^{-3} \quad \text{instead of} \quad z^{-2.5}.$$

Due to this fact the case $\Omega \ll 1$ results in an increase in the value of z for which $y' \sim 1$ is achieved.

The spectrum of the residual radiation observed at the present time is closely Planckian. In order to obtain it from a Bose-Einstein distribution, a creation of photons is unavoidable, since the Compton effect leaves their number unchanged. We neglect the double Compton effect. At sufficiently low frequencies the optical depth due to bremsstrahlung processes is always large and such photons exist in thermodynamic equilibrium with electrons. Following Kompaneets, it is possible to find the frequency limit x_0, at which the Compton process acts upon a photon more quickly than it (the photon) can be absorbed as,

$$\frac{y'}{\tau'} \sim 1, \tag{15}$$

where $\tau' = \int \kappa(\nu) n_e^2 \, dl$ is the optical thickness due to bremsstrahlung processes. The absorption coefficient [18]

$$\kappa(\nu) = 3.8 \cdot 10^8 \frac{g(\nu)}{T_e^{1/2} \nu^3} (1 - e^{-h\nu/kT_e}).$$

The relation (15) is satisfied for

$$x_0 \left(\frac{\ln 4}{x_0} \right)^{-1/2} \approx 80 z^{-3/4} \Omega^{+1/2}, \quad z > 10^3. \tag{15'}$$

Knowing the overall number of photons in the residual radiation $N = 0.244(2\pi\kappa T_r hc)^3$ it is possible to find the red shift corresponding to the moment t_1 which we are seeking by solving the equation

$$\frac{4\pi\Omega n_{\rm crit}^2}{h} \int_0^{z_1} \frac{1}{N(z)} \int_{z_0}^{\infty} \frac{E_{ff}(z)}{x} \, dx \, \frac{dt}{dz} \, dz \sim 1;$$

$$\Omega^2 z_1^{1/2} \ln^2 \left(\frac{z_1 \Omega^{-2/3}}{500} \right) \approx 10^4, \tag{16}$$

where x_0 is determined by the relation (15) and $E_{ff} = \kappa(\nu)(B_\nu - J_\nu)$ by formula (1). From (16) $z_1 = 10^5 \Omega^{-4}$, and $t_1 \simeq 4 \cdot 10^9 \Omega^{+8}$ s. Thus, practically any production of energy during the process of expansion of the Universe up to the moment $t_1 \sim 4 \cdot 10^9$ s will be fully transformed; the plasma and radiation will at all times be in thermodynamic equilibrium, and the spectrum of the radiation will be Planckian. In the interval $10^{10} < t_1 < 3 \cdot 10^{17}$ s the interaction of the radiation with the electrons cannot any longer change a non-Planckian spectrum to a Planckian one.

From this it follows that the residual radiation presently observed was formed up to the moment $t \sim 10^{10}$ s, corresponding to a red shift $z \sim 10^5$. For $\Omega \ll 1$ the moment is $t_1 \ll 10^{10}$ s, however it is clear that the intensive production of electron-positron pairs for $t < 300$ s rarely increases the optical depth due either to bremsstrahlung or Compton processes.

Appendix

1. The formula for the distortion of the spectrum by Compton scattering of thermal electrons (for small distortions). If we introduce the variables

$$y = \int_0^{z_{\max}} \frac{\kappa T_e(z)}{m_e c^2} \frac{d\tau}{dz} dz, \qquad \text{(see Section 4)}$$

(τ is the optical thickness due to Compton effect) and $x = h\nu/kT_r$ (clearly independent of the magnitude of the red shift at which the scattering occurred) it is then easy to obtain, for $T_e \gg T_r$ from the expressions written down by Weymann and Kompaneets, the following equation for the occupation index

$$n(x,t) = \frac{c^3}{8\pi h\nu^3} \frac{dE}{d\nu}$$

($dE/d\nu$ is the spectral density of the radiation energy), and

$$\frac{\partial n}{\partial y} = x^{-2} \frac{\partial}{\partial x} x^4 \frac{\partial n}{\partial x}. \qquad \text{(I)}$$

Taking it that the deviations from the Planck function are small, we substitute in the right-hand side of (I) $n_0 = (e^x - 1)^{-1}$. Then

$$\frac{\partial n}{\partial y} = \frac{e^x}{(e^x-1)^2} x \left\{ \frac{x}{\tanh(x/2)} - 4 \right\}$$

and

$$\frac{\Delta n}{n_0} = \frac{\Delta J}{J_0} = \frac{e^x}{e^x - 1} xy \left\{ \frac{x}{\tanh(x/2)} - 4 \right\} \qquad \text{(II)}$$

The distortion in the Rayleigh-Jeans region of the spectrum is found by letting $x \to 0$ to be

$$\frac{\Delta J}{J} \xrightarrow[x\to 0]{} -2y. \qquad \text{(III)}$$

From experiments, the temperature in the Rayleigh-Jeans region of the spectrum is known, therefore it is better to present (II) in the form

$$\frac{\Delta n}{n(x, T_{\mathrm{RJ}})} = \frac{\Delta J}{J(x, T_{\mathrm{RJ}})} = \frac{xye^x}{e^x - 1} \left\{ \frac{x}{\tanh(x/2)} - 2 \right\}. \qquad \text{(IV)}$$

Since

$$\frac{\Delta J}{J} = \frac{d \ln J}{d \ln T} \cdot \frac{\Delta T}{T},$$

then

$$\frac{\Delta T}{T_{\mathrm{RJ}}} = y \left\{ \frac{x}{\tanh(x/2)} - 2 \right\}. \qquad \text{(V)}$$

2. The formula for the distortion of the spectrum through the Compton effect of thermal electrons (the asymptote for large x). With the substitution $\xi = \ln x$ Equation (I) becomes

$$\frac{\partial n}{\partial y} = 3\frac{\partial n}{\partial \xi} + \frac{\partial^2 n}{\partial \xi^2};$$

and going over to the variables $z = 3y+\xi$, y we can reduce the latter equation to the equation of heat conduction

$$\frac{\partial n}{\partial y} = \frac{\partial^2 n}{\partial z^2}. \tag{VI}$$

The solution of Equation (I), (VI) with the initial condition $n(x, y = 0)$ takes the form of a Planck function, and has the form

$$n(x,y) = \frac{1}{\sqrt{4\pi y}} \int_{-\infty}^{+\infty} n(t) \exp[-(\ln x + 3y - t)^2/4y]\, dt \tag{VI$'$}$$

It is easy to find the form of the renormalization factor necessary for the calculations of the distortion in the Rayleigh-Jeans region. For Compton effect in the Rayleigh-Jean region $n(x) = 1/x$, $n(t) = e^{-t}$, it is necessary to set $n(x') = 1/x'$,

$$\begin{aligned} n(x,y) &= \frac{1}{\sqrt{4\pi y}} \int_{-\infty}^{+\infty} n(t) \exp[-(\ln x + 3y - t)^2/4y]\, dt \\ &= \int_{-\infty}^{+\infty} \frac{1}{\sqrt{4\pi y}} \exp[-y - (\ln x + 3y - t)^2/4y]\, dt \\ &= \frac{1}{\sqrt{4\pi y}} e^{-2y-\ln x} \int_{-\infty}^{+\infty} \exp[-(t - \ln x - y)^2/4y]\, dt = x^{-1}e^{-2y}, \end{aligned}$$

from which an obvious renormalization is (it is interesting to compare with (III))

$$x' = xe^{2y}. \tag{VII}$$

Now

$$n(x', y) = \frac{1}{\sqrt{4\pi y}} \int_{-\infty}^{+\infty} n(t) \exp[-(\ln x' - t + y)^2/4y]\, dt,$$

and the substitution $\eta = t - \ln x' - y$ gives

$$n(x', y) = \frac{1}{\sqrt{4\pi y}} \int_{-\infty}^{+\infty} \frac{1}{\exp(xe^{\eta+y}) - 1} e^{-\eta^2/4y}\, d\eta. \tag{VIII}$$

For $x \gg 1$ it is possible to neglect the quantity unity in the denominator. The function $f = xe^{\eta+y} + \eta^2/4y$ has a minimum at

$$-\eta_0 e^{-\eta_0} = 2xye^y. \tag{IX}$$

Expanding it in a Taylor series around η_0

$$f = f(\eta_0) + \frac{1}{2} f''(\eta_0)(\eta - \eta_0)^2,$$

we obtain

$$n(x', y) = \frac{1}{\sqrt{4\pi y}} \cdot \sqrt{\frac{2\pi}{f''(\eta_0)}} e^{-f(\eta_0)}.$$

Finally

$$\frac{J(x',y)}{J(x')} = \frac{1}{\sqrt{1-\eta_0}} \exp\left(\frac{\eta_0 - \eta_0^2/2}{2y} + x'\right), \tag{X}$$

where the parameter $\eta_0(x,y)$ is given by (IX). For $x \gtrsim 1$ the error is decreased by the introduction of the multiplier

$$(\exp\{xe^{\eta_0+y}\} - 1)^{-1}.$$

For $xy < \frac{1}{2}$ the exponent in f may be expanded in the series $e^{\eta+y} = 1+\eta+y+\frac{1}{2}(\eta+y)^2+\ldots$. Then for $x > 1$ we obtain

$$\frac{J(x',y)}{J(x')} = \frac{1}{\sqrt{1+2xy}} \exp[xy(x-1)/(1+2xy)]. \tag{XI}$$

The asymptotes of (X) and (XI) coincide with the exact solution (IV) for $y \to 0$.

3. The change in total energy of the residual radiation due to Compton effect of thermal electrons.

Again we revert to Equation (I)

$$\frac{\partial n}{\partial y} = x^{-2}\frac{\partial}{\partial x}x^4\frac{\partial n}{\partial x}.$$

Since the expression for the total energy of the radiation takes the form

$$E = DT_0^4 \int nx^3\,dx,$$

then multiplying both sides of (I) by x^3 and integrating with respect to x we find, following Weymann and Kompaneets,

$$\frac{1}{DT_0^4}\frac{\partial E}{\partial y} = \frac{\partial}{\partial y}\int_0^\infty x^3 n\,dx = \int_0^\infty x\frac{\partial}{\partial x}x^4\frac{\partial n}{\partial x}\,dx;$$

but

$$\int_0^\infty x\frac{\partial}{\partial x}x^4\frac{\partial n}{\partial x}\,dx = x^5\frac{\partial n}{\partial x}\Big|_0^\infty - \int_0^\infty x^4\frac{\partial n}{\partial x}\,dx$$
$$= -x^4 n\Big|_0^\infty + 4\int_0^\infty x^3 n\,dx = 4\frac{E}{DT_0^4}.$$

Finally, $E = E_0 e^{4y}$, where

$$E_0 = DT_0^4\int_0^\infty \frac{x^3}{e^x - 1}\,dx$$

is the energy of Planckian radiation having the same temperature in the Rayleigh-Jeans region. Taking the renormalization $x' = xe^{2y}$, we have

$$E = \sigma T_{\mathrm{RJ}}^4 e^{12y}. \tag{XII}$$

Institute of Applied Mathematics
USSR Academy of Sciences. Moscow

Received
December 30, 1968

REFERENCES

1. *Wilkinson D.*—Phys. Rev. Lett. **19**, 1195 (1967).
2. *Stokes R. A., Partridge R. B., Wilkinson D. T.*—Phys. Rev. Lett. **19**, 1199 (1967).
3. *Thaddeus P., Clauser J. E.*—Phys. Rev. Lett. **16**, 819 (1966).
4. *Field G. B., Hitchcock J. L.*—Phys. Rev. Lett. **16**, 817 (1966).
5. *Pariiskiĭ Yu. N.*—Astron. Zh. **45**, 279 (1968).
6. *Zeldovich Ya. B., Novikov I. D.*—Astron. Zh. **44**, 663 (1967).
7. *Partridge R. B., Wilkinston D. T.*—Nature **215**, 70 (1967).
8. *Dicke R. H., Peebles P. J. E., Roll P. G., Wilkinson D. T.*—Astrophys. J. **142**, 414 (1965).
9. *Zeldovich Ya. B., Kurt V. G., Sunyaev R. A.*—ZhETF **55**, 278 (1968).
10. *Henry R. C., Fritz G., Meekins J. E.* et al.—Astrophys. J. **153**, L11 (1968).
11. *Vainshtein D. A., Sunyaev R. A.*—Kosmich. Issled. **6**, 635 (1968).
12. *Sunyaev R. A.*—Astrophys. Lett. **3**, 33 (1969).
13. *Weymann R.*—Astrophys. J. **145**, 560 (1966).
14. *Sunyaev R. A.*—Dokl. AN SSSR **179**, 45 (1966).
15. *Novikov I. D., Zeldovich Ya. B.*—Ann. Rev. Astron. Astrophys. **5**, 627 (1967).
16. *Shkovskiĭ I. S.*—Proceedings of IV Texas Conference on Relativistic Astrophysics (1967).
17. *Peebles P. J. E.*—Astrophys. J. **153**, 1 (1968).
18. *Allen C. W.*—Astrophysical Formulae. London: Athlone Press (1963).
19. *Karzas W. J., Latter R.*—Astrophys. J. Suppl. **6**, 167 (1961).
20. *Bridle A. N.*—Month. Not. RAS **136**, 219 (1967).
21. *Howell T. F., Shakeshaft J. R.*—Nature **216**, 753 (1967).
22. *Pelyushenko S. A., Stankevich K. S.*—Astron. Zh. **46**, 283 (1969).
23. *Kompaneets A. S.*—ZhETF **31**, 876 (1956).
24. *Weymann R.*—Phys. Fluids **8**, 2112 (1965).
25. *Bortolot V., Clauser J. F., Thaddeus P.*—Phys. Rev. Lett. **22**, 307 (1969).
26. *Boynton R. E., Stokes R. A., Wilkinson D. T.*—Phys. Rev. Lett. **21**, 462 (1968).
27. *Ozernoĭ L. M., Chernin A. D.*—Astron. Zh. **44**, 1131 (1967).

Commentary

Plasma of temperature from 10^8 to 4000°K which fills the Universe is characterized by complete ionization of hydrogen and deuterium. Helium, present in a quantity of only about 7% (by the number of nuclei), captures electrons even earlier. Neither hydrogen nor helium have low-lying levels which are collisionally excited. In addition, even for complete ionization, the concentrations of nuclei and electrons are negligible (10^9 times smaller) compared to the concentration of photons.

Thus, in a wide range of levels we may neglect processes occurring in the collisions of electrons and nuclei compared to those in collisions of photons with electrons. The primary role is played by Compton scattering of photons on electrons.

In general form the problem of the evolution of the spectrum and establishment of thermodynamic equilibrium in such a situation was first posed by A. S. Kompaneets in a paper written in 1947 and published in 1956 (cited in the above article).

The present article applies Kompaneets' equation to the cosmological situation.

The chief result of the paper is the analytical solution (given in the Appendix and denoted there by an asterisk) of Kompaneets' equation (neglecting recoil and induced effects). This solution describes the time evolution of the radiation spectrum as a result of Compton interaction with hot $kT_e \gg h\bar{\nu}$ Maxwellian nonrelativistic electrons. It made it possible to find distortions of the relic radiation spectrum which might occur if certain energy release processes were superimposed on the Friedman expansion of the plasma. Thus it was possible to obtain important restrictions on energy release in the early Universe using data on the relic radiation spectrum. The paper became widely known (see review articles on the present state of the problem by R. A. Sunyaev, Ya. B.,[1] and by L. Danese *et al.*[2]

Before this paper, calculations of distortion of the spectrum were carried out by numerical methods (see Ref. 13 by Weymann cited in the present article).

The solution (7) given in the article served as a basis for calculations of the effect of lowering the brightness of the relic radiation in directions toward galaxy clusters with hot intergalactic gas (the article in the present volume by Ya. B. and Sunyaev).

The Zeldovich–Sunyaev solution found wide application in papers on the theory of X-ray sources. It served as the basis for calculations of Comptonization of radiation in spherically-symmetric accretion on neutron stars (**44**), and is used in calculations of the spectra of X-ray radiation of accretion discs[3,4] and in estimates of the characteristic times and variation of their radiation.[4,5]

A second important result is its conclusion of the necessity of a period of neutral hydrogen in the Universe. While accounting for contributions of weakly-interacting particles to the mean density of matter in the Universe may modify the estimates given in the article, the method and primary result are of great interest.

The third important result is the posing of the problem of the formation of the observed black-body spectrum of relic radiation. This problem was solved by the authors in the next article of this volume.

[1] *Sunyaev R. A., Zeldovich Ya. B.*—Ann. Rev. Astron. Astrophys. **18**, 537–560 (1980)

[2] *Danese L., De Zotti G.*—Rev. Nuovo Cim. **7**, 277 (1977).

[3] *Sunyaev R. A., Titarchuk L. G.*—Astron. Astrophys. **86**, 121–138 (1980).

[4] *Shakura N. I., Sunyaev R. A.*—Astron. Astrophys. **24**, 337–355 (1973).

[5] *Pozdnyakov L. A., Sobol I. M., Sunyaev R. A.*—In: Astrophys. and Space Phys. Rev. Soviet Sci. Rev. **2**, Harwood Acad. Publ. GmbH, 189–331 (1983).

50

The Interaction of Matter and Radiation in a Hot-Model Universe II. Distortion of the Relic Radiation Spectrum*

With R. A. Sunyaev

Heating [1] of the primeval plasma prior to the epoch of recombination results in distortions in the Rayleigh-Jeans region of the microwave relic radiation spectrum ($\lambda \sim 1 - 60\,\text{cm}$, or more exactly $\lambda = 2.5\Omega^{-7/8}\,\text{cm}$). The present observational data allow limits to be set to such energy injection from which follow upper limits to (a) the amount of antimatter in the Universe; (b) the parameters of primeval turbulence; and (c) the adiabatic fluctuation spectrum for small masses ($M < 10^{11}\, M_\odot$).

If the heating takes place prior to the epoch $t = 10^{10}\Omega^{12/5}\,\text{s}$ (and in particular at the annihilation of electron-positron pairs at $T \sim 10^8 - 10^{10}\text{K}$, $t < 300\,\text{s}$), no observable distortions are expected in the relic radiation spectrum. Here $\Omega = \sigma/\sigma_{\text{crit}}$ is the dimensionless average density of matter in the Universe.

In the first part of this series [1] distortions of the relic spectrum excited by the discharge of energy at the stage of expansion after recombination of hydrogen in the Universe were discussed. Below we consider the period before recombination with a fully ionized plasma. The discharge of energy at this stage leads to deviations in the Rayleigh-Jeans region of the Planck spectrum. The point is that during this period bremsstrahlung processes are not able to maintain complete dynamic equilibrium between matter and radiation. Meanwhile the transfer of energy between electrons and photons due to scattering occurs sufficiently quickly for a Bose-Einstein distribution to be established with density of photons which is smaller than in the Planck distribution corresponding to the same energy density of radiation [2, 3]

$$n(x) = (e^{x+\mu} - 1)^{-1}; \qquad \mu \geq 0. \tag{1}$$

here and in what follows the notation is the same as in [1]; n is the occupation number.

At this stage Compton scattering maintains the equilibrium between the temperatures of the plasma and radiation $T_e = T_r = T$ and $x = h\nu/kT$.

*Astrophysics and space science **7**, 20–30 (1970).

The difference from complete equilibrium is connected with the chemical potential of photons $m = -\mu kT$. However, (1) is not applicable everywhere. In the region of small frequencies $x < x_0 \geq 1$, where the optical depth due to bremsstrahlung processes is large, the Rayleigh-Jeans distribution is maintained. The value x_0 will be given below and it will be shown that the region where the largest possible deviations from the Planck curve occur is $\lambda \sim 1 - 60$ cm. The absence of significant distortions of the relic spectrum in this region establishes an upper limit to the energy release when the temperature of the primeval plasma changes from 10^5–10^7K (depending on the matter density in the Universe) to $T \sim 3 \cdot 10^4$K, i.e., long before the moment of recombination of the primeval plasma. Thus it is possible to place bounds in particular on the energy dissipation of turbulent motions and sound waves connected with adiabatic density perturbations and also on the release of energy by the annihilation of antimatter.

1. *Small Deviations from the Planck Distribution of Photons*

We consider the situation when a slow change of temperature occurs due to a flow of heat. The kinetic equation for photons with $kT_e \ll m_e c^2$ including the variation of electron temperature (but for the time being without taking account of Friedman expansion) has the form

$$\frac{\partial n(x,t)}{\partial x} = a\frac{1}{x^2}\frac{\partial}{\partial x}x^4\left[\frac{\partial n}{\partial x} + n^2 + n\right] + \frac{ke^{-x}}{x^3}[1 - n(e^x - 1)] - x\frac{\partial n}{\partial x}\frac{\partial \ln T_e}{\partial t}, \quad (2)$$

where $x = h\nu/kT_e$. The first term describes the change of frequency due to Compton scattering for which $a = \sigma_0 c n_e kT_e/m_e c^2$ [2, 3] and the second—bremsstrahlung together with the corresponding reverse and induced processes where

$$k = \frac{8\pi}{3}\frac{e^6 h^2 g(x) n_e^2}{m_e(6\pi m_e kT_e)^{1/2}(kT_e)^3} = 1.25 \cdot 10^{-12} g(x)\frac{n_e^2}{T_e^{3.5}} \quad (3)$$

and $g(x)$ is the Gaunt factor. Finally the third term is connected with the fact that the temperature T_e enters into a determination of the variable x. In fact, if no processes occurred among the photons, then $\partial n(\nu,t)/\partial t|_\nu \equiv 0$ and for $n(x)$ we have

$$\left.\frac{\partial n}{\partial t}\right|_x = \left.\frac{\partial n}{\partial t}\right|_\nu + \left.\frac{\partial n}{\partial x}\frac{\partial x}{\partial t}\right|_\nu = \left.\frac{\partial n}{\partial t}\right|_{\text{phys}} - \frac{\partial n}{\partial x}\frac{x}{T}\frac{dT}{dt}, \quad (4)$$

where $\left.\partial n/\partial t\right|_{\text{phys}}$ corresponds to the first two terms in Equation (2).

The general properties of Equation (2) are obvious: the first term is equal to zero not only for a Planck distribution $n(x) = (e^x - 1)^{-1}$, but also for $n(x) = (e^{x+\mu} - 1)^{-1}$, i.e., for a Bose-Einstein equilibrium distribution with a given number of photons. The reason is that the Compton effect does not change the number of photons, although it redistributes the photons in frequency. The second term reaches zero only in true equilibrium $n = (e^x - 1)^{-1}$. The third term describes the perturbing influence of the energy supply in the case when this energy is given primarily to the electrons.

To the equations for photons we should add the equation for the electron temperature which depends on the rate of heat supply Q. The zeroth approximation in which an equilibrium of photons, equality of electron and radiation temperatures and a small heat capacity of electrons is assumed, gives

$$\frac{d}{dt}\sigma T^4 = Q \tag{5}$$

Since Compton scattering is proportional to the first power of the electron density, but bremsstrahlung processes—to the second power, $k \ll \alpha$ in the case of a rarefied plasma (for $n_e < 2.7 \cdot 10^{12} T_e^{4.5}$ [3]. In this case it may be assumed that the next approximation is of the form $n = (e^{x+\mu} - 1)^{-1}$ and it is possible to reduce the problem to finding $\mu(t)$. Taking such a distribution and assuming that μ is small, we write equations for n and the radiation density[1] $\mathcal{E}$ expanded in series in

$$n = (e^{x+\mu} - 1)^{-1} = (e^x - 1)^{-1} - \mu \frac{e^x}{(e^x - 1)^2}, \tag{6}$$

$$\begin{aligned} \mathcal{E} &= \sigma T^4 G \int_0^\infty (e^{x+\mu} - 1)^{-1} x^3\, dx = \sigma t^4 \left(1 - \mu G \int \frac{e^x}{(e^x - 1)^2} x^3\, dx\right) \\ &= \sigma T^4 (1 - 1.1\mu) \approx \sigma T^4 (1 - \mu). \end{aligned} \tag{7}$$

We neglect the energy density included in the electrons and nuclei. Thus the equation of energy balance (5) gives one relation between dT/dt and $d\mu/dt$

$$4\frac{d\ln T}{dt} - \frac{d\mu}{dt} = \frac{Q}{\sigma T^4}, \tag{8}$$

A second relation may be found from considering the balance of the total number of photons per unit volume

$$\begin{aligned} N &= HT^3 F \int_0^\infty (e^{x+\mu} - 1)^{-1} x^2\, dx = HT^3 \left(1 - \mu F \int \frac{e^x}{(e^x - 1)^2} x^2\, dx\right) \\ &= HT^3 (1 - 1.35\mu) \approx HT^3 (1 - \mu). \end{aligned} \tag{9}$$

In the limiting case $at \gg 1$ and $k = 0$ an exact form of the spectrum $n = (e^{x+\mu} - 1)^{-1}$ is ensured and at that time the number of photons is strictly constant

$$\frac{1}{N}\frac{dN}{dt} = \frac{3 d\ln T}{dt} - \frac{d\mu}{dt} = 0, \tag{10}$$

[1]Here and below F and G are dimensionless numbers

$$G^{-1} = \int_0^\infty (e^x - 1)^{-1} x^3\, dx = \frac{\pi^4}{15}; \quad F^{-1} = \int_0^\infty (e^x - 1)^{-1} x^2\, dx = 2.404.$$

and, correspondingly, in this limiting case

$$\frac{d\ln T}{dt} = \frac{1}{3}\frac{d\mu}{dt} = \frac{Q}{\sigma T^4}. \tag{11}$$

The problem consists in working out the equation for μ including bremsstrahlung which limits the growth of μ, but in the limit, after the cessation of energy release and the passage of sufficient time, which leads to $\mu = 0$, i.e., to complete equilibrium. It is natural to express the derivative dN/dt by using the equation

$$N = HT^3F\int n(x,t)x^2\,dx,$$

$$\frac{dN}{dt} = HT^3F\left[\int\frac{dn}{dt}x^3\,dx + \frac{3d\ln T}{dt}\int nx^2\,dx\right]. \tag{12}$$

In this case the first and third terms of Equation (2) do not contribute so that

$$\frac{dN}{dt} = HT^3Fk\int\frac{e^{-x}}{x}[1-n(e^x-1)]\,dx. \tag{13}$$

Now it would seem natural to substitute into this equation $n = (e^{x+\mu}-1)^{-1}$ and expand it in a series in small μ, taking only the first two terms of the expansion (see Equation (6)). However, in this case the integral on the right-hand side of Equation (13) blows up. Therefore, a more precise procedure is required in this case. The physical reason is clear: for quite small frequencies (small x) due to the factor $1/x^3$ the rate of establishment of equilibrium grows and therefore the expression for n with constant μ is no longer applicable for all x, for arbitrarily small k.

Let us take $n = (e^{x+\mu}-1)^{-1}$ with μ dependent on x and, in order to find this dependence[2] in the region of small x, consider in a given region both bremsstrahlung processes and the Compton effect, but neglect the nonstationarity, i.e., the term $\partial n/\partial t$. As previously, we consider μ to be small and obtain

$$\begin{aligned} 0 &= \frac{a}{x^2}\frac{d}{dx}x^4\left[\frac{dn}{dx}+n^2+n\right] + \frac{ke^{-x}}{x^3}[1-n(e^x-1)] \\ &= \frac{a}{x^2}\frac{d}{dx}x^4\frac{e^x\,d\mu/dx}{(e^x-1)^2} + \frac{ke^{-x}}{x^3}\frac{\mu e^x}{e^x-1}. \end{aligned}$$

Since x is small, we substitute $e^x \to 1$ and obtain

$$-\frac{a}{x^2}\frac{d}{dx}x^2\frac{d\mu}{dx} + \frac{k}{x^4}\mu = 0. \tag{14}$$

The solution of this equation is elementary,

$$\mu(x) = \mu_0 e^{\pm\frac{1}{x}\sqrt{k/a}}. \tag{15}$$

[2]Above—see Equation (1)—μ was considered to be independent of x. We will show that for $x > x_0$, actually $\mu(x) \to \text{const}$.

We discard the solution with the plus in the exponential while the second solution, presented previously by Kompaneets (1956) [2], has the necessary properties, $\mu(x) = 0$ for $x \to 0$ and $\mu = \text{const} = \mu_0$ for $x > \sqrt{k/a}$. Thus we have found a natural limiting frequency, below which the solution with $\mu = \text{const}(x)$ is incorrect. The value of μ, independent of x, which was introduced previously, is exactly μ_0 of the equation, and $\sqrt{k/a} = x_0$ is the limit of the region where μ is constant.

The deviation of $n(x,t)$ from the expression with $\mu = \text{const}(x) = \mu_0$ does not significantly change the expressions for N and $\mathcal{E}$. Thus the left hand sides of the differential equations for dT/dt and $d\mu/dt$ are not changed. The sole purpose of introducing a more precise form for μ in the region of small x is to obtain a correct expression for the rate of free-free production of photons. We find

$$\frac{1}{N}\frac{dN}{dt} = k\int_0^\infty \mu(x)\frac{1}{x^3(e^x-1)}x^2\,dx = k\mu_0\int_0^\infty e^{-1/x\sqrt{k/a}}\frac{dx}{x(e^x-1)}$$
$$\approx k\mu_0\sqrt{\frac{a}{k}} = \mu_0\sqrt{ak}. \qquad (16)$$

Hence we obtain the system

$$4\frac{d\ln T}{dt} - \frac{d\mu}{dt} = \frac{Q}{\sigma T^4},$$
$$3\frac{d\ln T}{dt} - \frac{d\mu}{dt} = \mu\sqrt{ak}. \qquad (17)$$

The exact solution of this system has the form

$$\mu(t) = 3\int_0^t \frac{Q}{\sigma T^4}(\tau)\exp\left\{-4\int_t \sqrt{a(p)k(p)}\,dp\right\}d\tau. \qquad (18)$$

Under quasi-stationary conditions ($\mu = \text{const}$)

$$\mu = \frac{3}{4\sqrt{ak}}\frac{Q}{\sigma T^4}. \qquad (19)$$

From the system (17) it also follows that, in the absence of heating,

$$\mu = \mu(t_0)e^{-4\sqrt{ak}(t-t_0)}$$

or, more precisely, if the difference from unity of the coefficient with μ in (7) and (9) are taken into account

$$\mu = \mu(t_0)e^{-2\sqrt{ak}(t-t_0)}. \qquad (20)$$

Making use of (18), it is easy to find the maximum deviation of the spectrum from a Planck distribution in the region of long wavelengths which at the critical wavelength $x_0 = \sqrt{k/a}$ should be

$$\frac{\Delta n}{n} = \frac{\mu e^{x_0}}{(e^{x_0}-1)^2} : \frac{1}{e^{x_0}-1} \simeq \frac{\mu_0 e^{-1}}{x_0} = \mu\sqrt{a/k}e^{-1} = \frac{3}{4k}\frac{Q}{\sigma T^4}e^{-1}. \qquad (21)$$

The maximum deviation turns out in general to be independent of the rate of the Compton change of frequency a if the condition $a \gg k$ is satisfied. Only the value of the frequency (wavelength) at which the maximum deviation occurs depends on the rate a.

2. Distortion of the Spectrum of Relic Radiation Due to the Release of Energy at Different Stages of Expansion of the Universe

We present rough approximations for Friedman's cosmological model with the present radiation temperature $T = 2.7\text{K}$ and an average energy density $n_e = 10^{-5}\Omega\,\text{cm}^{-3}$. In place of time we introduce the variable z (red shift). In the region $z > 4 \cdot 10^4\Omega$, where the energy density of radiation exceeds the rest mass of matter $\rho_\gamma > \rho_m$, but for $kT \ll m_e c^2$, we have

$$T = 2.7z\text{K}; \quad \mathcal{E} = 4 \cdot 10^{-13} z^4\,\text{erg/cm}^3,$$

$$n_e = 10^{-5}\Omega z^3\,\text{cm}^{-3}; \quad \frac{kT}{m_e c^2} = 4.55 \cdot 10^{-10} z,$$

$$\rho_\gamma = 4.5 \cdot 10^{-34} z^4 = \frac{4.5 \cdot 10^5}{t^2}\,\text{g/cm}^3;$$

from which we find the 'hydrodynamic' time from the beginning of the expansion

$$t = 3 \cdot 10^{19} z^{-2}\,\text{s} \quad (\text{if } 1 \le z < 4 \cdot 10^4\Omega, \text{ then } t = 2 \cdot 10^{17}\Omega^{-1/2} z^{-1.5}).$$

We also introduce values characterizing the interaction of matter and radiation:

$$\tau_c = \frac{1}{\sigma_0 n_e c} = \frac{5 \cdot 10^{18}}{\Omega z^3}\,\text{s}; \quad a = \sigma_0 n_e c \frac{kT}{m_e c^2} = 9.1 \cdot 10^{-24}\Omega z^4\,\text{s}^{-1},$$

and, in agreement with (3),

$$k = 3.9 \cdot 10^{-24} g(x)\Omega^2 z^{5/2}\,\text{s}^{-1}.$$

As shown by Kompaneets [2] and Weymann [3,4], the characteristic time for the establishment of a Bose-Einstein distribution is determined by the condition[3] $4y = 4\int a(t)\,dt \sim 1$ from which we find that this time $t_a \sim 3 \cdot 10^{11}\Omega\,\text{s}$ corresponds to a red shift

$$z_a \approx 10^4\Omega^{-1/2}. \tag{22}$$

The observed effects will be formed in this time period. The inclusion of the Friedman expansion of the Universe in Equation (2) does not change the

[3]More precisely, the characteristic time of establishment of a Bose-Einstein distribution depends logarithmically on frequency; hence in our case we must take into account the factor $\ln 4/x_0$ which changes our results insignificantly in the following examples because y rapidly increases with z.

conclusions of point 6; this is connected with the fact that in the absence of physical processes, redistributing the photons in energy relative to each other, for any group of photons $x = h\nu/kT$ and n are not changed in the course of the expansion. It is easy to obtain the mathematical formulation of this assertion from Equation (3) in the work of Weymann (1966) [4].

We introduce a dimensionless quantity for the release of energy

$$q = \frac{Qt}{\sigma T^4}, \tag{23}$$

and through it express the maximum distortion of the Planck spectra (21) at $x = x_0$

$$\frac{\Delta n}{n} = \frac{\Delta T}{T} = \frac{3}{4}\frac{qe^{-1}}{kT} = \frac{2500q}{\Omega^2\sqrt{z}}. \tag{24}$$

Substituting into (24) z_a from (22), we find the expected maximum distortion from the Planck spectrum

$$\left(\frac{\Delta n}{n}\right)_a = \left(\frac{\Delta T}{T}\right)_a = 25q\Omega^{-2.25}. \tag{25}$$

It is attained at

$$x_a = \sqrt{\frac{k}{a}} \approx 200\Omega^{1/2} z_a^{-3/4} = 0.2\Omega^{7/8},$$

$$\lambda_a = \frac{1}{x_a} : \frac{hc}{kT} = 2.5\Omega^{-7/8} \text{ cm}. \tag{26}$$

Since

$$\frac{1}{45} < \Omega < 3, \quad 60 \text{ cm} > \lambda_a > 1 \text{ cm},$$

λ_a lies in a well investigated region.[4] The existing experimental data (see the references in [1]) indicate that $\Delta n/n = \Delta T/T < 10\%$ in the region $1 < \lambda < 21$ cm ($0.1 < \Omega < 3$), which corresponds to

$$q = 4 \cdot 10^{-3}\Omega^{1.75}. \tag{27}$$

At longer wavelengths $21 < \lambda < 60$ cm ($0.1 > \Omega > 1/45$), $\Delta T/T < 30\%$ which gives an upper limit on the energy release

$$q < 10^{-2}\Omega^{1.75}. \tag{28}$$

The limit (27) corresponds to

$$\mu_0 = \mu_0(z_0) < 5 \cdot 10^{-2}\Omega^{7/8}. \tag{29}$$

If the energy release occurs earlier than $t_b = 10^{10}\Omega^{12/5}$ s ($z_b = 5.4 \cdot 10^4\Omega^{-6/5}$), then the distortions are removed according to the rule (20)

$$\mu = \mu(z) \exp\left\{-\left(\frac{z}{z_b}\right)^{1.25}\right\}. \tag{30}$$

[4]Taking into account the Gaunt factor in (3), we obtain 30 cm > λ > 1 cm for the same region of Ω.

As soon as the above condition for stationarity is fulfilled, then the change of spectra is conditioned not by the instantaneous value, but the average value of Q, and it is necessary to understand the value $Qt/\sigma T^4$ in (23) as $\int [Q(t)/(\sigma T^4(t)]\, dt$, connected with the time when z changes from $z_b \simeq 5.4 \cdot 10^4 \Omega^{-6/5}$ to $z_a \sim 10^4 \Omega^{-1.2}$. The effects, connected with the release of energy at earlier stages, may be calculated with formula (30). The physical result of this section is that the energy release in the period $z \sim 10^4$ may not exceed $\frac{1}{250}$ of the energy density of radiation in this period.

3. Limits on the Physical Conditions in the Initial Plasma

Unfortunately, the rapid disappearance of deviations of the spectra of relic radiation for $z > 10^7$ does not allow a detection of the effects connected with the powerful release of energy by annihilation of electrons and positrons in the period $z \sim 10^9$. In principle, the degree to which effects are diminished may provide information about the average density of matter in the Universe. However, even in the case of a universe filled only by radiation the observed effects are negligibly small, $\mu \sim 10^{-8}$: in the early stage of annihilation due to the high density of electrons and positrons are quickly eliminated, and in the late stage there is an insignificant release of energy. The release of energy due to the burning of 30% of the initial matter into helium in the course of nuclear reactions at $z \sim 10^8 - 10^9$ is insignificant in comparison to the total energy of radiation in that period and therefore it also does not lead to significant deviations ($\mu \ll 10^{-8}$). We introduce a series of examples for the use of estimates on the energy release in the Universe obtained above.

A. Initial Turbulence

Ozernoĭ and Chernin [5] extensively developed the idea of Weizsaecker [6] about the presence of initial turbulence in the Universe. Following Landau and Lifshitz [7][5] we estimate the release of energy due to the dissipation of turbulent motions

$$Q \sim \frac{(\Delta u)^3}{c} \frac{\text{erg}}{\text{g} \cdot \text{s}}, \qquad (31)$$

or

$$q \sim \left(\frac{\Delta u}{c}\right)^3 \frac{ct}{l},$$

where l is the maximum scale of turbulence and Δu is the turbulent velocity of this scale. Comparing (31) with (27) and (28), we see that at the moment

[5]At the stage before recombination of hydrogen with $z \gg 10^3$ the speed of sound $a_{\text{sound}} = c/\sqrt{3}$ and the initial turbulence should be subsonic; otherwise, the release of energy would be too great at the stage after recombination of hydrogen (see [1]).

$z_a \sim 10^4 \Omega^{-1/2}$,

$$\frac{\Delta u}{c} < 0.15 \Omega^{7/12} \left(\frac{l_a}{ct_a} \right)^{1/3}. \tag{32}$$

B. Antimatter in the Universe

Harrison [8] considered a charge-symmetric model of the Universe with initial fluctuations of the baryon charge from which a spatial division of matter and antimatter follows after the annihilation of baryons in the 'hot' model of the Universe, i.e., at $T < 10^{12}$K. Later, but before the moment of recombination, the friction between matter and radiation impedes the motion of matter and therefore annihilation is impeded; regions with matter and antimatter (future objects and antiobjects) exist until $z \sim 10^3$. According to the remarks of Bardeen (see the review of Field [9]) after the recombination of hydrogen the regions with one sign of the baryon charge would quickly become isolated and would form objects with a mean density of order $10^{-20}\,\mathrm{g \cdot cm^{-3}}$ which considerably exceeds the observed density of matter in galaxies and clusters of galaxies ($M > 10^8 M_\odot$). It will be shown below that the presence of such regions with small masses contradicts the bounds obtained on energy release. The point is that in spite of the strong connection between matter and radiation, the matter will diffuse into separate regions with different baryon charges, annihilate, and the release of energy will quickly be picked up by radiation leading to a distortion of its spectrum.

The coefficient of ambipolar diffusion of a plasma with temperature T in the field of radiation with an energy density $\mathcal{E}$ is equal to

$$D = \frac{2kT_e}{\sigma_0 \mathcal{E}} = \frac{2kT_0}{\sigma_0 \mathcal{E}_0 z^3}.$$

The size of the zone r from which plasma can reach the location of division of matter and antimatter in hydrodynamic time is determined from the condition [7]

$$r^2 = 6Dt.$$

The characteristic size of the region with mass M equals $R = \sqrt[3]{3M/4\pi\rho}$. The annihilating zone of size r in which energy is released is

$$q \sim \frac{3r}{R} \frac{\rho_m}{\rho_\gamma}. \tag{33}$$

Using the limits (27) and (28), we find that observations of relic radiation contradict the initial division of the Universe into regions of baryons and antibaryons if their mass is smaller than $3 \cdot 10^9 \Omega^{5/2} M_\odot$. This assertion and the remarks of Bardeen create great difficulties for the hypothesis of Harrison [8].

In the general picture of entropy perturbations [10, 11], the presence of fluctuations of the density of baryons is hypothesized which in the interesting ranges $M \sim 10^5 - 10^6 M_\odot$ does not exceed the limit $\delta/\delta\rho \sim 0.01$ since ρ is of one sign everywhere and there is no antimatter. Here it is possible to pose the question of extrapolating $\delta\rho/\rho$ to small volumes; this question will be considered separately [12].

C. Adiabatic Density Perturbations

Before the moment of recombination the Jeans mass exceeded $10^{16} M_\odot$ and sound waves represented adiabatic density perturbations of small scales. Silk (1968) [13] noted that the friction between matter and radiation at the stage before recombination leads to dissipation of the energy of sound waves and to smoothing out of the density perturbations with mass scales less than 10^{10}–$10^{12} M_\odot$. From this time the scale grows according to the law $M \sim t^3$ with a simultaneous dissipation of energy. In the interval between $z_a \sim 10^4 \Omega^{-1/2}$ and $z_v = 5.4 \cdot 10^4 \Omega^{-6/5}$ oscillations connected with masses $M \sim 5 \cdot 10^5 \Omega^{4.9} - 10^9 \Omega^{7/4}$ are damped. The limits on energy release (27) and (28) give an upper limit on the amplitude of adiabatic density perturbations in this mass region. The energy of sound waves is

$$E \sim \rho u^2, \tag{34}$$

where

$$u = \frac{\Delta\rho}{\rho} a_s \approx \frac{1}{\sqrt{3}} \frac{\Delta\rho}{\rho} c$$

is the velocity of matter. Now it is easy to estimate the energy release

$$q = \frac{1}{3} \left(\frac{\Delta\rho}{\rho} \right)^2 . \tag{35}$$

In agreement with (27) we obtain

$$\left. \frac{\Delta\rho}{\rho} \right|_{M > 5 \cdot 10^5 \Omega^{4.9}} < 10^{-1} \Omega^{7/8}. \tag{36}$$

Taking the amplitude of the perturbations in the mass region of $10^{11} M_\odot$ at the moment of hydrogen recombination, it is easy to obtain from (36) limits on the spectrum of adiabatic fluctuations (for details, see Zeldovich and Sunyaev [12]).

Institute of Applied Mathematics
USSR Academy of Sciences. Moscow

Received
November 6, 1969

REFERENCES

1. *Zeldovich Ya. B., Sunyaev R. A.*—Astrophys. Space Sci. **4**, 301 (1969).
2. *Kompaneets A. S.*—ZhETF **31**, 876 (1956).
3. *Weymann R.*—Phys. Fluids **8**, 2112 (1965).
4. *Weymann R.*—Astrophys. J. **145**, 560 (1966).
5. *Ozernoĭ L. M., Chernin A. D.*—Astron. Zh. **44**, 1131 (1967).
6. *Weizsaecker C. F.*—Astrophys. J. **114**, 165 (1951).
7. *Landau L. D., Lifshitz E. M.* Mechanics of Continuous Media. New York: Pergamon (1984).
8. *Harrison E. R.*—Phys. Rev. Lett. **18**, 1011 (1967).
9. *Field G. B.* Stars and Stellar Systems, Vol. IX, ed. by A. and M. Sandage (1968).
10. *Doroshkevich A. G., Zeldovich Ya. B., Novikov I. D.*—Astron. Zh. **44**, 295 (1967).
11. *Peebles P. J. E., Dicke R. H.*—Astrophys. J. **154**, 898 (1968).
12. *Zeldovich Ya. B., Sunyaev R. A.*—Astrophys. Space Sci. **9**, 368 (1970).
13. *Silk J.*—Astrophys. J. **151**, 459 (1968).

Commentary

This paper was the first to point out the existence of specific distortions of the relic radiation spectrum related to energy release in the stage preceding recombination. It solves the problem of the formation of the black-body spectrum with simultaneous effects of Comptonization and bremsstrahlung processes.

Subsequent development of the idea expressed in this paper may be found in works by A. F. Illarionov and R. A. Sunyaev.[1,2] It turned out that in a universe with low baryon density, the double Compton-effect may play a greater role than bremsstrahlung processes.[3,4]

Let us turn to the spatial distribution of temperature.

Small-scale perturbations in the plasma which fills the Universe decay in radiation-dominated plasma. Such perturbations do not yield fluctuations in the presently observed relic microwave background radiation.

The significance of this paper is in the fact that study of the spectrum of relic radiation makes it possible to bound from above the amplitude of adiabatic density perturbations, as well as the amount of antimatter and other energy sources.

The limit on energy release in the early Universe proved useful in calculations of the mass spectrum of the first low-mass black holes[5] and in the analysis of properties of unstable, long-lived elementary particles in cosmology (see the review by A. D. Dolgov and Ya. B. Zeldovich).[6]

[1] *Illarionov A. F., Sunyaev R. A.*—Astron. Zh. **51**, 698–711 (1974).
[2] *Illarionov A. F., Sunyaev R. A.*—Astron. Zh. **51**, 1162–1176 (1974).
[3] *Sunyaev R. A., Zeldovich Ya. B.*—Ann. Rev. Astron. Astrophys. **18**, 537–560 (1980).
[4] *Danese L., De Zotti G.*—Riv. Nuovo Cim. **7**, 277 (1977).
[5] *Carr B. J.*—Astrophys. J. **201**, 1–19 (1975).
[6] *Dolgov A. D., Zeldovich Ya. B.*—UFN **130**, 559–614 (1980).

In the standard picture there is no energy release, nor are there black holes or unstable particles. Because of this, the result—an exactly Planckian spectrum—looks trivial. However, in historical perspective, and taking into account possible future unexpected developments, this is not at all so. Strong acoustic waves could also have been an energy source.

Direct observations yield a smallness of $\Delta T/T$ only at scales exceeding 1 Mps (on today's distance scale), which corresponds to an angle of several arc minutes. The limitation is determined not by the resolution of the instruments, but by the smoothing and damping of small-scale perturbations in the course of hydrogen recombination in the Universe.

Any judgments about any kind of smaller-scale inhomogeneity would have remained unreliable extrapolations if it were not for the present paper.

As examples of new questions we note variants with evaporation of black holes and with unstable particles; for the appropriate estimates we may again unsheathe the still-valid arguments of papers I and II.

Another side of the matter is the transfer of the methods and ideas to galaxy clusters, X-ray sources, etc., where the methods developed in these two articles prove quite useful. We note also that the search for distortions in the relic radiation spectrum continues. It is an important part of the research program planned for the satellite "COBE."

VIII

Formation of the Large-Scale Structure of the Universe

51

The Theory of the Large-Scale Structure of the Universe*

Introduction

The Godfather of psychoanalysis Professor Sigmund Freud taught us that the behavior of adults depends on their early childhood experiences. In the same spirit, the problem of cosmological analysis is to derive the observed present day situation and structure of the Universe from certain plausible assumptions about its early behavior. Perhaps the most important single statement about the large scale structure is that there is no structure at all on the largest scale—1000 Mpc and more. On this scale the Universe is rather uniform, structureless and isotropically expanding—exactly as in the simplified pictures of Einstein, Friedman, Humason, Hubble, Robertson, Walker. On the other hand there is a lot of structure on the scale of 100 or 50 Mpc and less. There are clusters and superclusters of galaxies.

Much work has been done on the classification of these bodies into "richness classes" and attempts have been made to deduce from observations a "mass function" giving the distribution of matter among clumps of various sizes and masses. There is a firmly established division between regions with enhanced, higher-than-average density of stars and radio-sources and regions

*Large-Scale Structure of the Universe. Moscow: Mir, 452–464 (1981) (Proceedings of the Symposium of the International Astronomical Union No. 79, Tallin, 1978). English edition: Large Scale Structure of the Universe. Dordrecht: Reidel, 409–421 (1979).

with density lower than average. In recent years correlation functions have been used to characterize the relation between density enhancements and the linear scales of the distribution of galaxies in space.

A systematic effort of measuring red shifts (optical and 21 cm radio) of thousands of galaxies has resulted in confirmation of Hubble's law. Surprisingly it is approximately valid for smaller distances than those characteristic for the density inhomogeneities. The redshift measurements have opened the way for disentangling the three-dimensional structure of the Universe, as opposed to the two-dimensional projection of astronomical objects on the celestial sphere.

The present symposium has really opened up a new direction in the search for geometrical patterns governing the distribution of luminous matter in space. We hear about ribbons or filaments along which clusters of galaxies are aligned; the model of a honeycomb was presented with walls containing most of matter; the presence of large empty spaces (holes—not black holes of course) was emphasized. Cosmological theory must be aware of this information and try to use it in order to discriminate between various proposed schemes. Let us briefly characterize those schemes which seem to us most promising at the present time. There are two extreme assumptions which can be made about the initial density perturbations. The first concerns an ideal unperturbed metric connected by General Relativity with ideal homogeneity of the overall density in the early radiation dominated Universe. The perturbations consist of an inhomogeneous distribution of "matter"—of baryon excess—on a background of homogeneous radiation. Therefore the ratio γ/B (photons per baryon) varies from place to place. But the specific entropy of matter is proportional to this ratio and therefore those perturbations are called "entropy perturbations."

The second type of perturbation consists of common motion of photons and baryons. These perturbations conserve entropy and therefore they are called "adiabatic."

A departure from the main line of this report is permissible in the introduction. The actual value of the ratio γ/B which is of order 10^8 or 10^9 is most important for cosmology. The closed or open geometry of the Universe as a whole depends on this number.

Is it possible that in due time this number will be calculated by elementary particle theories, taking into account the lack of exact symmetry between particles and antiparticles as indicated by laboratory experiments (the so-called CP-violation, 1964) and also baryon non-conservation predicted by some theories? In this case it is conceivable that the γ/B ratio is constant everywhere, just because the physical constants are everywhere the same. But this argument is not very strong. It is equally possible that γ/B depends on the interrelation of external physical constants and the local properties of the space-time metric; in this case the γ/B ratio does not have to be a constant.

Let us return to conventional cosmology. At the present moment we do not see any better policy than to make plausible assumptions about the size, amplitude and character of the initial perturbations, to develop logically and mathematically all the consequences of these assumptions and to compare them with observations.

In this report we shall investigate adiabatic perturbations—the second type, according to the classification given above. This investigation has been carried out approximately the last ten years by our group, which includes Doroshkevich, Sunyaev, Novikov, Shandarin, Sigov, Kotok and others. We use important theoretical results obtained by Lifshitz, Bonnor, Silk, Peebles, Yu and others.

We consider several phases in the development of the perturbations:

1) acoustic oscillations of the radiation-dominated plasma and their attenuation before and during recombination;

2) the growth of small perturbations in the neutral gas;

3) the non-linear growth of perturbations leading to the formation of compressed gas layers—pancakes;

4) the further fate of pancakes, the interaction of pancakes, their decay into galaxies and protoclusters of galaxies.

The first two points are investigated using the "merry old" linear theory of perturbations. In 3) and 4) an approximate nonlinear theory is widely used and also numerical simulation. The statistical side of the problem is considered. Radio astronomical predictions are made. The main result is most encouraging: the adiabatic perturbation spectrum possesses a definite cut-off wavelength as already pointed out by Silk. We now see that this critical wavelength is also reflected in the cell structure of the Universe.

1. The Theory of Perturbation Growth

A plausible featureless initial spectrum of density fluctuations in the radiation-dominated plasma is assumed. Due to photon viscosity and damping during recombination the final Fourier spectrum of growing perturbations in the neutral gas is given by

$$\overline{\left(\frac{\delta\rho}{\rho}\right)^2_k} = b_k^2 = a_k^2 k^n e^{-kR_c}.$$

The critical length R_c depends on 1) the radiation density during recombination taking account of the specific effects of $L\alpha$ reabsorption through the $2s \to 1s + 2\gamma$ metastable hydrogen decay, 2) the Compton cross-section for scattering of photons by electrons, 3) the matter density or γ/B ratio. The best calculations give the characteristic length (multiplied by $(1 + z_{\rm rec})$ in order to account for the expansion from recombination to the present epoch)

$$R_c = 8\,\mathrm{Mpc} \quad \text{for} \quad \Omega = 1, \qquad R_c = 40\,\mathrm{Mpc} \quad \text{for} \quad \Omega = 0.1.$$

The wavelength λ_c is determined by $\lambda_c = 2\pi/k_c$, $k_c R_c = 1$ so that $\lambda_c = 2\pi R_c$.

The index n and the average value of a_k^2 are adjusted to fit the observed picture. But independent of this adjustment, due to the exponential damping factor e^{-kR_c} we are sure that the surviving fluctuations are very smooth. It is immediately clear that in the adiabatic theory early formation of stars or globular clusters or even of galaxies is impossible. The large-scale density enhancements must grow, and only thereafter is their fragmentation into smaller objects possible.

The second qualitative feature is that gas motion occurs under the influence of gravitation alone. The pressure forces, which depend on gradients, are negligible on the large scales involved. In this case the growth of perturbations is especially simple: they grow in amplitude due to gravitational instability and increase in linear dimensions, conserving their form. The density perturbations and the peculiar velocity (the excess over the Hubble velocity) are given by

$$\frac{\delta\rho}{\rho} = f\left(\frac{r}{R}\right)\varphi(t), \qquad \mathbf{u} = \mathbf{U}\left(\frac{r}{R}\right)\psi(t),$$

where $\varphi(t) = t^{2/3} \sim (1+z)^{-1}$ and $\psi(t) = t^{1/3} \sim (1+z)^{-1/2}$ for $\Omega = 1$ with $R = \text{const} \cdot t^{2/3} \sim (1+z)^{-1}$.

Obviously the density field and velocity field are connected by the continuity equation

$$\frac{\partial}{\partial t}\left(\frac{\delta\rho}{\rho}\right) \sim f \sim \operatorname{div}\mathbf{u} \sim \operatorname{div}\mathbf{U},$$

and by the equation of motion in which the perturbation of the potential by the perturbed density is included.

It is important to realize that already in the linear theory the extra compression in places with positive and growing $\delta\rho/\rho$ is anisotropic: the three components of the divergence of the peculiar velocity are not equal

$$\frac{\partial u_x}{\partial x} \neq \frac{\partial u_y}{\partial y} \neq \frac{\partial u_z}{\partial z}.$$

There is also shear, $\partial u_x/\partial y \neq 0$, etc.—but of course no vorticity $\partial u_x/\partial y - \partial u_y/\partial x = 0$ because the motion is due to potential (gravitational) forces. The anisotropy of the deformation due to peculiar velocities is easily understood by tidal action. The nearby density distribution distorts the motion at the point under consideration.

A natural way to build an approximate theory, exact in the linear region and also good enough in non-linear situations, is to use the Lagrangian formulation. The position of every particle in space (i.e., its Eulerian coordinates) $\mathbf{r}$ is given as a function of time t and the initial position (i.e., the Lagrangian coordinate) ξ.

The solution with growing perturbations only is written

$$\mathbf{r} = a(t)[\xi + b(t)\psi(\xi)].$$

The first term $a\xi$ describes the Hubble expansion $\dot{a}/a = H$, the second term $ab\psi(\xi)$ describes the displacement of every particle from its legitimate unperturbed position. $b(t)$ is a growing function, $b(t) \sim t^{2/3}$. The perturbation due to gravitation $\psi(\xi)$ is of potential type.

Analytical and numerical studies confirm that this is a good approximation—less than 20–30% errors occur in highly nonlinear situations; the proofs are in our original papers.

Given the formula for $\mathbf{r}(\xi, t)$, it is easy to write down the velocity of every particle

$$\mathbf{u} = \left(\frac{\partial \mathbf{r}}{\partial t}\right)_{\xi}, \quad \mathbf{u}_{\text{pec}} = \mathbf{u} - H\mathbf{r} = a\dot{b}\psi(\xi),$$

and also the density of matter

$$\rho = \rho(\xi, t) \sim \left(\frac{\partial \mathbf{r}}{\partial \xi}\right)^{-1},$$

where $\partial \mathbf{r}/\partial \xi$ is the Jacobian, i.e., the determinant of the partial derivatives.

Using $\psi = \text{grad}_{\xi}\varphi$ and choosing coordinate axes which diagonalize the deformations, using the notation

$$\frac{\partial^2 \varphi}{\partial \xi_1^2} = -\alpha, \quad \frac{\partial^2 \varphi}{\partial \xi_2^2} = -\beta, \quad \frac{\partial^2 \varphi}{\partial \xi_3^2} = -\gamma, \quad \alpha > \beta > \gamma$$

we obtain

$$\rho = \bar{\rho}\frac{1}{(1 - b\alpha)(1 - b\beta)(1 - b\gamma)}.$$

With φ determined by the initial small perturbation field we find the particles where α has local maxima $\alpha_{m1}, \alpha_{m2}, \ldots$. The condition $1 - b(t_i)\alpha_{mi} = 0$ determines the moment when infinite density is obtained for the i-th particle.

From the density formula we see that this infinity is due to the intersection of trajectories of adjacent particles lying on the ξ_1 coordinate axis. At the moment t_i the perturbation along the other two axes ξ_2, ξ_3 is finite.

The approximate theory predicts the formation of thin dense gas clouds. They grow due to fresh gas falling onto their flat boundary and being compressed and heated by shock waves. They also spread sideways due to new intersections of trajectories of adjacent particles.

The picture outlined above was already known at the time of the Krakow IAU Symposium No. 63, "Confrontation of cosmological theories with observational data." Qualitatively they are described in the report by Doroshkevich *et al.*; formulae and detailed analyses were given in our original papers,

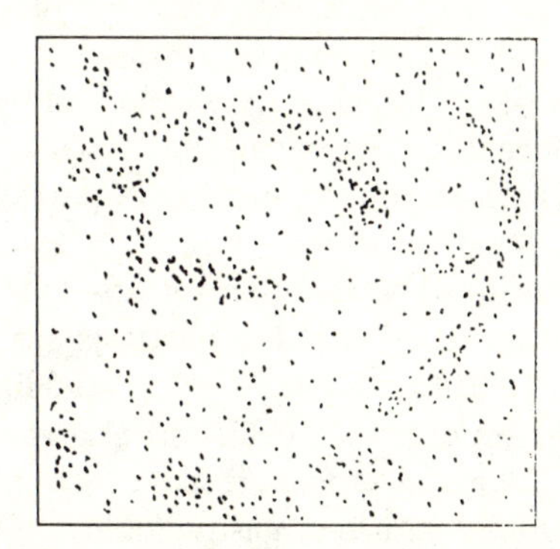

Fig. 1

and also in the book by Zeldovich and Novikov "Structure and Evolution of the Universe," published in Russian in 1975 and prepared for publication in English by Chicago University Press. These are mentioned in this report for the sake of completeness and to make it possible to read this report without using references.

Now we turn to the results obtained after the Krakow Symposium, partly published in *Astronomicheskiĭ Zhurnal* and partly in preprints of the Institute of Applied Mathematics. These results are most important in connection with optical and radio astronomical observations.

2. Late Phases of the Development of Perturbations and Cell Structure

Numerical calculations were necessary to follow the density perturbations' evolution in the later stages, when more than half the matter was brought together into the dense phase. During the lecture in Tallinn a movie was shown made by an electronic computer display. Here, in written form, only selected pictures can be shown.

There is one movie (corresponding to Fig. 1) calculated using the approximate nonlinear theory for two-dimensional perturbations. The initial spectrum has a sharp cut-off on the short wave—and also long wave end; it is flat (on the average) within the excited interval. The individual Fourier coefficients in this interval are taken at random according to a normal Gaussian distribution.

Comparing these calculations with others it must be stressed that the potential, the velocity and density contributions are calculated for a continuous medium, not for a finite number of discrete point masses. The calculations are not exact and the initial conditions somewhat artificial (two dimensions, flat spectrum). But these departures from the ideal calculation are not of the sort which arise when a finite number of discrete masses is considered.[1]

[1]It is $\sqrt{N}$ in two dimensions and $\sqrt[3]{N}$ in three dimensions which are important in incorporating shortwave perturbations involuntarily in N-body calculations just due to

The pictures in the movie contain a finite number of points. But these points are test particles for visualizing the motion and density distribution. The potential used in the calculation corresponds to a continuum or, in other words, to a calculation with an infinite number of particles with inertial and gravitational mass.

The calculation is continued to the moment after the first intersection of trajectories occurs. It is assumed that the particles are non-interacting and one layer can penetrate through another. The sticking together of particles, their physical, non-gravitational collisions and the formation of shock waves are not included. Therefore the pancakes in this picture are somewhat thicker than would be found in a real gasdynamical calculation. Still they are rather thin, distinctly different from the spherical or irregular clumps predicted in a simplified approach.

As time goes on, the pancakes spread laterally and they intersect. In Figure 1 a typical net structure is seen: matter is mostly concentrated in thin filaments, the inner regions are empty and divide up the network.

It seems plausible that a three-dimensional calculation will lead to a cell or honeycomb structure with matter concentrated in the walls surrounding large disconnected empty spaces. The intersection of walls could give enhanced density along lines.

It is not yet clear if such types of structure, or at least its remnants are discernible in the observational data of Joeveer, Einasto, Tago and Seldner, Siebers, Groth and Peebles. It is well known that the human eye has the property of finding lines and other patterns in random assemblies of points. One example is the Sciapparelli channels on Mars, but even more striking are the constellations—figures of humans and beasts found by the ancients in the distribution of stars on the sky. Therefore one must be very cautious in interpreting the observations. One must find some mathematical algorithm to distinguish between superclusters as quasispherical clumps and the honeycomb structure. It is possible that correlation functions (two points, three points, etc.) are not the best method for this particular task.

There are obvious difficulties: 1) the walls must fragment into separate galaxies and clusters of galaxies. The turbulence inside the pancakes and the gravitational interaction of the fragments must partly wash out the structure. 2) In investigations of the three-dimensional structure we use redshift as a measure of distance. Because of peculiar velocities, this procedure is not exact. These points need further investigation. But with all these uncertainties, one point must be stressed: the occurrence of cells in theoretical calculations is not an artifact due to the use of an approximate theory.

Calculations of another type were carried out and used to make the second part of the movie. The motion of $(128)^2 \approx 16\,000$ points in two dimensions was calculated numerically. The potential for every distribution of points was calculated using Poisson's equation $\Delta\varphi = 4\pi G\rho$ with some smoothing

the discrete character of the mass distribution.

and interpolation on the smallest scale. Periodicity on the largest scale was assumed: points intersecting from inside the wall of the square reappeared on the opposite wall. The periodicity condition was also used in the potential calculation.[2]

Again a flat spectrum of perturbations with cut-offs from both sides was used in formulating the initial conditions. The results of numerical simulations are practically indistinguishable from the results of the approximate theory. The characteristic pattern with thin walls and disconnected empty spaces depends on the cut-off of the short waves—this is our firm conclusion. It is confirmed by the fact that the average linear dimension of the empty spaces is approximately equal to the cut-off wavelength, $2\pi/k_{\max}$.

Therefore this pattern is characteristic of the adiabatic theory with the exponential cut-off at short wavelengths, which results from the matter-radiation interaction.

For entropy perturbations, there is no cut-off except the Jeans' mass for neutral hydrogen corresponding to masses of the order of globular clusters. Therefore probably no net or cell-structure will occur in this case. Corresponding numerical computations with sufficient accuracy are still lacking. Perhaps the two-dimensional case will be easier to handle and still be meaningful.

We are optimistic about the prospects for discriminating between entropy and adiabatic perturbations by means of investigations of the large-scale structure of the Universe.

A three-dimensional calculation was done for adiabatic perturbations with an exponentially cut-off spectrum, but it is the visualization of the results which is the bottleneck in this case. This report is written at the moment when this work is still in progress.

3. Statistical Properties of the Birth of Pancakes and Counts of QSO

The birth of an individual pancake occurs at the moment of intersection of trajectories, i.e., mathematically speaking at the moment when the smallest denominator in the density expression $[1 - b(t)\alpha]$ vanishes so that $\rho \to \infty$. Therefore we must find the local maxima of $\alpha = -\partial^2\varphi/\partial\xi_1^2$. Due to the statistical character of the problem the answer is also given in statistical terms. The function $P(\alpha_m)$ gives the density of local maxima of given amplitude

$$dN\,(\mathrm{cm}^{-3}) = P(\alpha_m)d\alpha_m.$$

In the case of a cut-off spectrum P is proportional to R_c^{-3}. The dependence on α_m is universal, given the normal Gaussian law of density perturbations. But $P(\alpha_m)$ is not a simple Gaussian function, because in calculations

[2]The force proportional to r^{-1} and potential proportional to $\log r$ are characteristic of two-dimensional gravitation; to be exact we are working with infinite bars, not points.

of α_m we are performing the nonlinear operation of diagonalization of the $\partial^2\varphi/\partial\xi_i\partial\xi_k$ matrix and we are choosing the maxima. Doroshkevich has obtained

$$P(\alpha_m) \sim \alpha^5 e^{-n\alpha^2}, \qquad (\alpha > 1/\sqrt{n}).$$

Using the connection between the amplitude of the maximum α_m and the moment t_m we can obtain the birth function $F(t)$ or $f(z)$ giving the number of pancakes born per unit co-moving volume per unit time or unit of z.

The newborn pancakes have small vorticity and low temperature. The formation of compact objects and brightest galaxies is easiest just at the birthplace of the pancake and even before the growth of the ends of the pancake.

Therefore it is plausible to identify the birthrate of pancakes with the birthrate of the brightest known compact objects—quasars. If the life time of quasars is short and independent of their absolute age, then at every epoch the concentration of quasars is proportional to their birthrate. The high power α^5 before the Gaussian factor leads the $(1+z)^7$ dependence of the density of quasars as an intermediate asymptote in the case of a flat Universe. At high z the power law is cut off by the Gaussian exponent. By the choice of a single constant (corresponding to the amplitude of the perturbations) it is possible to obtain a good fit of the pancake birth rate curve to Longair's results on radio source counts and Schmidt's data on quasar evolution.

Still, the similarity between the radio source and quasar evolution and the birthrate of pancakes should not be overestimated. The power laws involved refer to different regions of z. The birth of cold pancakes occurs from some high z (of the order of 10 or 20) up to $z \sim$4–3. It is well known that for $z < 4$ the gas is totally ionized; therefore even if pancakes are formed, their physical properties are totally different as compared with genuine pancakes formed from cold initial gas. On the other hand, the observed counts of radio sources and quasars refer to the range $0 < z < 4$; at $z > 4$ instead of evolution there is a cut-off or stagnation. This question needs further investigation.

Another statistical test concerns the two-point correlation function. The adiabatic pancake theory does not contradict the most interesting part of Peebles' correlation function $\delta \sim \xi^{-1.7}$ in the region near $\delta \sim 1$. We refer to original papers for quantitative confirmation.

The general outlook seems to be that the adiabatic theory does not contradict the observations.

4. The Crucial Tests and Further Problems

Still the absence of contradiction is not positive proof. In order to distinguish between the entropy and adiabatic theories one needs direct observation. The observation of very early globular clusters and galaxies at $z > 30$ to $z \sim 100$ or 200 would be strong evidence in favor of the entropy perturbation theory with further clumping of the initial small mass objects into clusters of galaxies. If hot gas clouds of primordial composition (H + He) are found, identifiable with pancakes, this would be a strong argument for the adiabatic theory. Fully ionized very hot gas could be detected by its x-ray emission and by distortions of the Planckian background radiation spectrum (cooling in the Rayleigh-Jeans region). The medium-hot hydrogen gives redshifted 21 cm radiation.

In any case, the controversy with the observed limits on $\Delta T/T$ of the relic radiation fluctuations must be solved—but this is needed for *all* variants. Entropy perturbations predict $\Delta T/T$ only 2 or 3 times less than adiabatic perturbations. The study of those perturbations which are directly connected with the structure of the Universe is the most rewarding part of the problem.

Extrapolating from Krakow through Tallinn to the next symposium somewhere in the early eighties one can be pretty sure that the question of the formation of galaxies and clusters will be solved in the next few years.

What remains is the wider question of the overall spectrum of perturbations including the smallest scale damped in the very early radiation dominated or hadronic era and of the longest perturbations, whose amplitude remains small even now. Is the power law spectrum without any characteristic length valid? New, indirect observational tests are needed. Still the major theoretical questions remain unsolved: what is the fundamental theory of the initial perturbations? And what is the ultimate reason for the homogeneous and isotropic expansion from the singularity which is the background for the perturbations?

REFERENCES

1. *Doroshkevich A. G., Sunyaev R. A., Zeldovich Ya. B.* Confrontation of Cosmological Theories with Observational Data/Ed. M. S. Longair. Dordrecht, Boston (1974).
2. *Zeldovich Ya. B., Novikov I. D.* Stroenie i evolyutsiĭa Vselennoĭ [Structure and Evolution of the Universe]. Moscow: Nauka (1975).
3. *Zeldovich Ya. B.*—Astron. and Astrophys. **5**, 84 (1970).
4. *Doroshkevich A. G.*—Astrofizika **6**, 581 (1970).
5. *Sunyaev R. A., Zeldovich Ya. B.*—Astron. and Astrophys. **20**, 189 (1972).
6. *Doroshkevich A. G., Ryabenkiĭ V. S., Shandarin S. F.*—Astrofizika **9**, 257 (1973).
7. *Doroshkevich A. G., Shandarin S. F.*—Astrofizika **9**, 549 (1973).
8. *Doroshkevich A. G.*—Astrophys. Lett. **14**, 11 (1973).
9. *Doroshkevich A. G., Shandarin S. F.*—Sov. Astron. USSR **18**, 24 (1974).
10. *Shandarin S. F.*—Astron. Zh. **51**, 667 (1974).

11. *Doroshkevich A. G., Shandarin S. F.*—Sov. Astron. USSR **19**, 4 (1975).
12. *Sunyaev R. A., Zeldovich Ya. B.*—Month. Not. RAS **171**, 375 (1975).
13. *Doroshkevich A. G., Shandarin S. F.*—Month. Not. RAS **175**, 15P (1976).
14. *Doroshkevich A. G., Zeldovich Ya. B., Sunyaev R. A.* Obrazovanie i evolyutsiĭa galaktik i zvezd [Formation and Evolution of Galaxies and Stars]/ed. S. B. Pikelner. Moscow: Nauka (1976).
15. *Doroshkevich A. G., Shandarin S. F.*—Preprint IPM AN USSR **3**. Moscow (1976).
16. *Doroshkevich A. G., Shandarin S. F.*—Month. Not. RAS **179**, 95 (1977).
17. *Doroshkevich A. G., Shandarin S. F.*—Month. Not. RAS **182**, 27 (1978).
18. *Doroshkevich A. G., Saar E. M., Shandarin S. F.*—Preprint IPM AN SSSR **72**. Moscow (1977).
19. *Doroshkevich A. G., Shandarin S. F.*—Preprint IPM AN SSSR **73**. Moscow (1977)
20. *Doroshkevich A. G., Shandarin S. F.*—Astron. Zh. **54**, 734 (1977).
21. *Doroshkevich A. G., Shandarin S. F.*—Preprint IPM AN SSSR **84**. Moscow (1977).
22. *Doroshkevich A. G., Seer E. M., Shandarin S. F.*—In: Krupnomasshtabnaĭa struktura Vselennoĭ [Large Scale Structure of the Universe]. Moscow: Mir (1981).
23. *Seldner M., Siebers R., Groth E. J., Peebles P. J. E.*—Astron. J. **82**, 249 (1977).
24. *Joeveer M., Einasto J., Tago E.*—Preprint A-1, Struve Astrophysical Observatory, Tartu (1977).

Commentary

In this review paper the state of the problem toward the end of the seventies is presented in an accessible and sufficiently complete manner. The report is a good introduction to the theory of the formation of structure. The editors felt it wise to preface this general report to the series of particular papers in which individual problems are first developed. For numerical calculations, see the article by A. G. Doroshkevich *et al.*[1]

Directions of future work are noted here with great scientific foresight, having indeed become central in the following years.

In addition, a certain limitedness of the report should be noted. Ya. B. works from the principle of the most economical construction of the theory, introducing only those laws and particles whose existence had been reliably established by the end of 1970.

The development of particle physics has led today to an enormous variety of propositions: heavy neutrinos; axions; supersymmetric particles, stable and unstable; a non-zero cosmological constant; etc. Astronomical considerations on their part insistently demand some form of "hidden mass."

The beginnings of this direction are reflected in papers published in this volume (articles **59–61**). The end of this path is still shrouded in the mists of the future ...

We emphasize here that many of the most important results, such as the concept of "pancakes" and a cellular structure with voids, remain valid in many variations which go beyond the bounds of the picture presented in this report.

[1] *Doroshkevich A. G., Kotok E. V., Novikov I. D.*, et al.—Month. Not. RAS **192**, 321 (1980).

52

The Energy of Random Motions in the Expanding Universe*

With N. A. Dmitriev

A differential equation is set up which connects the kinetic energy of random motions with a suitably defined change of the gravitational energy due to inhomogeneities of the density. An inequality is obtained which gives an upper limit on the energy of random motions developed owing to gravitational instability.

1. Statement of the Problem

The observed Universe differs from the isotropic and homogeneous cosmological model developed by A. A. Friedman in two respects: 1) the actual density distribution is not uniform, $\rho(r,t) \neq \bar{\rho}(t)$, and 2) the actual velocity distribution of the matter differs from Hubble's law: $\mathbf{u} \neq H\mathbf{r}$. The velocity of random (or peculiar) motion is defined as the difference $\mathbf{v} = \mathbf{u} - H\mathbf{r}$.

It is well known that in the Friedman model the momentum of an individual particle which has the random velocity $\mathbf{v}$ decreases with time[1] in inverse proportion to the increase of the radius R of the Universe: $|\mathbf{p}| = \text{const}/R$. This theorem is sometimes applied to the random motions of galaxies [1]. Present random velocities are of the order of 200 km/s. It is assumed that at the time t_0 of the formation of the galaxies the mean density of matter in the Universe was equal to the present mean intragalactic density, $\bar{\rho}(t_0) \approx 10^{-25}$ g/cm^3, whereas at the present time $\bar{\rho}(t_1) \approx 10^{-30}$ g/cm^3. Consequently,

$$R(t_0) = R(t_1)\left[\frac{\bar{\rho}(t_1)}{\bar{\rho}(t_0)}\right]^{1/3} = 0.02R(t_1).$$

The conclusion is drawn from this that at the time of formation of the galaxies the random velocity was $200/0.02 = 10\,000$ km/s. Such a random velocity is clearly too large; this estimate is regarded by Hoyle as a difficulty in the application of the Friedman theory to the actually observed Universe.

An obvious error in the argument is that the change of the random velocity is treated as if each individual particle moves in the gravitational field of

*Zhurnal eksperimentalnoĭ i teoreticheskoĭ fiziki **45** (4), 1150–1155 (1963).

[1]The red shift of the spectra of distant objects can be regarded as a special case in which the theorem is applied to photons.

matter uniformly distributed in space. Actually the nonuniform distribution of density has a decided effect on the law according to which the velocity changes.

As a simple example we can consider the revolution of the Earth around the sun; obviously the expansion of the Universe does not diminish this velocity. The same applies to any stationary system of bodies in which the mean distances remain constant. Another and more instructive example is that of small perturbations of the density and velocity imposed on the homogeneous isotropic model. In this case, as E. M. Lifshitz has shown [2], there are two solutions: an increasing disturbance,

$$\frac{\delta\rho}{\bar{\rho}} \sim t^{2/3}, \quad \mathbf{v} \sim t^{1/3},$$

and a decreasing disturbance,

$$\frac{\delta\rho}{\bar{\rho}} \sim t^{-1}, \quad \mathbf{v} \sim t^{-4/3}.$$

Consequently, in the expanding Universe decreasing velocities of random motion are not the only possibility.

For a definite relation between the phase of the random velocity and that of the density perturbation the velocity increases during the expansion. An obvious cause of this is the gravitational instability of homogeneous matter expanding strictly according to Hubble's law. A deviation from homogeneity is accompanied by a decrease of the gravitational energy of the random motions.

The problem of the present paper is to obtain a general relation between the kinetic energy of the random motions and a quantity which is to characterize the deviation of the actual density distribution from uniformity, without confining ourselves to small perturbations. Such a relation can be of interest in connection with the hypothesis that at an early stage of evolution the Universe exactly satisfied the equations of the Friedman model. Suppose that then some small perturbations (thermodynamic fluctuations, or phenomena at phase transitions [3],[2] or some other causes) produced small deviations from uniformity and caused small random velocities. We assume that after this all processes developed through the action of gravitational processes alone. Then there must be a definite relation between the present energy of random motions and the inhomogeneities of the density distribution which are now observed. Unfortunately, the present state of the observational data scarcely makes it possible to come to definite conclusions. Moreover, besides the gravitational forces, the release of thermonuclear energy in stars may also have some effect. Therefore, making a sober estimate of the significance of our results, we must suppose that they are mainly of a methodological and negative character: the only thing shown clearly is that

[2]We note that according to subsequent calculations we have made the phase transitions have less effect than was assumed in [3].

the observed random velocities cannot be regarded as a refutation of the applicability of the Friedman model in the past.

2. The Result, and Limiting Cases

Our main result can be written in the following form:

$$\frac{1}{R}\frac{d}{dt}[R(T+F)] = -T\frac{1}{R}\frac{dR}{dt}. \tag{1}$$

Here R is the radius of the Universe and T is the kinetic energy of the random motions:

$$T = \int \frac{1}{2}\rho v^2\, dV.$$

The quantity F is given by the expression

$$F = -\frac{\kappa}{2}\int\int \frac{(\rho_1-\bar\rho)(\rho_2-\bar\rho)}{r_{12}}\, dV_1 dV_2,$$

$$\rho_1 = \rho(r_1,t), \quad \rho_2 = \rho(r_2,t), \quad \bar\rho = \bar\rho(t),$$

$$r_{12} = |\mathbf{r}_1 - \mathbf{r}_2|, \quad dV_1 = d^3\mathbf{r}_1, \quad dV_2 = d^3\mathbf{r}_2,$$

where κ is the Newtonian gravitational constant.

This result refers to a state which is homogeneous and isotropic, but only on the average, on a sufficiently large scale L. Therefore when the integration is over a volume $V \gg L^3$ the quantities T and F are both proportional to V. For the quantity T this is immediately obvious; for the quantity F we must note that at distances $r_{12} > L$ there is no correlation between $\rho_1 - \bar\rho$ and $\rho_2 - \bar\rho$. Fixing the point $\mathbf{r}_1$, we note that for $r_{12} > L$ the quantity $\rho_2 - \bar\rho$ is of varying sign and is zero over a large volume, so that there is no contribution to the integral from $r_{12} > L$. Consequently, F is in order of magnitude

$$F \approx \kappa \int (\rho_1 - \bar\rho)^2 L^2 dV_1,$$

and it is seen that F is also proportional to the volume of integration V.

In considering the change of T and F with time, we suppose that the volume of integration always contains the same quantity of matter, $M = \bar\rho V$, so that it follows that $V = \text{const}\cdot R^3$, and the volume V expands along with the general Hubble expansion. In this sense we can speak of the specific values (per unit mass) $T_1 = T/M$ and $F_1 = F/M$, which differ from T and F only by a constant factor.

The quantities T_1 and F_1 are "local" quantities on scales larger than L. The quantity R is conceptually connected with ideas about the curvature of space, which are characteristic features of the general theory of relativity. Actually, however, our theorem is local in nature, and we can eliminate R. To do so we note that $d\ln R/dt = H$, where H, the Hubble constant, is a quantity definable locally in terms of the distribution of the mean velocity near the volume considered. Thus we can rewrite the theorem in terms of local quantities:

$$\frac{d(T_1 + F_1)}{dt} + H(T_1 + F_1) = -HT_1.$$

Let us consider various limiting cases.

1) Uniform density, $\rho = \bar{\rho}$, $F = 0$. We find

$$\frac{dT_1}{dt} = -2HT_1 = -2T_1 \frac{d \ln R}{dt}.$$

From this we have at once

$$T_1 = \frac{\text{const}}{R^2},$$

corresponding to the fact that in this case the random velocity falls off as $1/R$.

2) The matter has divided into clumps of constant volume which occupy a small part of space and do not expand with the general expansion (only the distance between clumps increases). In this case $F_1 \approx \text{const} \approx -m/r$, where m is the mass of a clump and r is its radius;

$$\frac{dT_1}{dt} = -2HT_1 - HF_1.$$

T_1 tends asymptotically to the limit

$$T_1 = -\frac{1}{2}F_1.$$

This result is the virial theorem for the individual clumps.

3) The theorem holds in the theory of small perturbations. The formulas given in Section 1 refer to a flat world, i.e., to the case $R \sim t^{2/3}$, $H = 2/3t$, where t is measured from the instant of infinite density. In the increasing disturbances

$$v \sim t^{1/3}, \qquad T \sim t^{2/3},$$

$$\frac{\delta\rho}{\bar{\rho}} \sim t^{2/3}, \qquad \delta\rho \sim t^{2/3}t^{-2} = t^{-4/3},$$

$$L \sim t^{2/3}, \qquad F \sim (\delta\rho)^2 L^2 V \sim t^{-8/3}t^{4/3}t^2 = t^{2/3}.$$

Substituting in (1), we get

$$T_1 = -\frac{2}{3}F_1.$$

In a similar way we have for the decreasing disturbances

$$T_1 \sim F_1 \sim t^{-8/3}, \qquad T_1 = -3/2F_1.$$

4) When perturbations appear rapidly

$$T_1 + F_1 = \text{const}.$$

Thus the theorem includes all of the special cases.

The main cosmological conclusion drawn from this is: if there was a uniform state with $T = 0$, $F = 0$, and then some small influences disturbed the uniformity, the only state that can arise under the action of gravitation alone is one in which the kinetic energy does not exceed a definite limit: $T_1 < -F_1$. If an analysis of the observations shows that this inequality is not satisfied, and it is shown that the large kinetic energy did not come from the nuclear energy of reactions in stars—then and only then will it be necessary to renounce the Friedman model for the early stage of the evolution of the Universe.

3. Proof

The proof given below is based on the consideration of a large ($r \gg L$) spherical volume in the framework of classical mechanics and the Newtonian theory of gravitation.

As is well known, this statement of the problem gives correctly all of the local properties of the Friedman model; the results for $\bar{\rho}(t)$ and $H(t)$ are exactly equal to the results of the general theory of relativity. As Bonnor has shown [4], this is also true for the development of perturbations, which in this model agree with the calculations of E. M. Lifshitz. With this statement of the problem there is not trace of the so-called gravitational paradox,[3] and when we let the radius of the sphere become infinite, $r \to \infty$, we get a quite definite, finite answer for all local quantities, and in particular for our quantities F_1 and T_1.

We note that with the classical approach the kinetic energy of the Hubble motion and the potential energy of the entire sphere, referred to unit mass, increase as r^2 and do not approach a definite limit for $r \to \infty$. Nevertheless, against this background we can obtain a relation between the kinetic energy of the random motions and the potential energy of the deviation from uniformity.

Let $\mathbf{r}_i$ and m_i be the radius vectors and the masses of the individual material points which make up the system (galaxies, say), and let $R(t)$ be the characteristic dimension of the system. We introduce relative coordinates ξ_i by means of the formulas

$$\mathbf{r}_i = R(t)\xi_i,$$

$$\frac{d\mathbf{r}_i}{dt} = R\frac{d\xi_i}{dt} + \frac{dR}{dt}\xi_i = R\frac{d\xi_i}{dt} + HR\xi_i = R\frac{d\xi_i}{dt} + H\mathbf{r}_i,$$

so that the random velocity is written

$$\mathbf{v}_i = \frac{d\mathbf{r}_i}{dt} - H\mathbf{r}_i = R\frac{d\xi_i}{dt}.$$

[3]In dealing with this it is precisely necessary and sufficient to consider a sphere; the case of a body of arbitrary shape does not satisfy the requirement that the tensor of the second derivatives of the gravitational potential be isotropic, $\partial^2\phi/\partial x^i \partial x^k = \text{const} \cdot \delta_{ik}$ (in rectangular coordinates).

The equations of motion are

$$\frac{d^2\mathbf{r}_i}{dt^2} = R\frac{d^2\xi_i}{dt^2} + 2\frac{dR}{dt}\frac{d\xi_i}{dt} + \frac{d^2R}{dt^2}\xi_i = \mathrm{grad}_{r_i}\sum_{i\neq j}\frac{\kappa m_j}{|\mathbf{r}_i - \mathbf{r}_j|}.$$

The last term on the left side is equal to the acceleration of the point in the pure Friedman motion, i.e., under the action of the attraction of a sphere of radius R and the uniform density $\bar{\rho}(t)$. Transposing it to the right side, we have

$$R\frac{d^2\xi_i}{dt^2} + 2\frac{dR}{dt}\frac{d\xi_i}{dt} = \mathrm{grad}_{r_i}\left(\sum_{j\neq i}\frac{\kappa m_j}{|\mathbf{r}_i - \mathbf{r}_j|} - \int_0^R \frac{\kappa\bar{\rho}d^3\bar{\mathbf{r}}}{|\mathbf{r}_i - \bar{\mathbf{r}}|}\right).$$

When we now introduce the relative coordinates on the right side, we get

$$\frac{1}{R^2}\mathrm{grad}_{\xi_i}\left(\sum_{j\neq i}\frac{\kappa m_j}{|\xi_i - \xi_j|} - \kappa\bar{\rho}R^3\int_0^1\frac{d^3\xi}{|\xi_i - \xi|}\right),$$

an expression in which the time appears explicitly only in the factor $1/R^2$, since $\bar{\rho}R^3$ is a constant. Multiplying the equation by R^2 and by $m_i d\xi_i/dt$ and summing over i, we get our theorem, in analogy with the derivation of the ordinary energy integral:

$$\begin{aligned} &R^3\frac{d}{dt}\left[\sum_i \frac{1}{2}m_i\left(\frac{d\xi_i}{dt}\right)^2\right] + 4R^2\frac{dR}{dt}\sum_i\frac{1}{2}m_i\left(\frac{d\xi_i}{dt}\right)^2 \\ &= R\frac{d}{dt}\left[\sum_i\frac{1}{2}m_iR^2\left(\frac{d\xi_i}{dt}\right)^2\right] + 2\frac{dR}{dt}\sum_i\frac{1}{2}m_iR^2\left(\frac{d\xi_i}{dt}\right)^2 \\ &= R\frac{dT}{dt} + 2\frac{dR}{dt}T = \frac{d}{dt}\left[\sum_{\substack{(i,j)\\ i\neq j}}\frac{\kappa m_i m_j}{|\mathbf{r}_i - \mathbf{r}_j|} - \kappa\bar{\rho}R^3\sum_i m_i\int\frac{d^3\xi_i}{|\xi_i - \xi|}\right], \end{aligned}$$

where the double sum is taken over all possible pairs (i,j).

When we add to the quantity whose derivative is taken to be a constant—the potential from the action of the sphere of radius R on itself—this quantity becomes $-F$, calculated in the relative coordinates. In fact, when we go over to the continuous way of writing the expressions, we get

$$\sum_{(i,j)}\frac{\kappa m_i m_j}{|\xi_i - \xi_j|} = \frac{1}{2}\kappa\int\int\frac{\rho_1\rho_2}{r_{12}/R}dV_1dV_2 = \frac{\kappa R}{2}\int\int\frac{\rho_1\rho_2}{r_{12}}dV_1dV_2,$$

$$\kappa\bar{\rho}R^3\sum m_i\int\frac{d^3\xi}{|\xi_i - \xi|} = \kappa R\int\int\frac{\rho_1\bar{\rho}dV_2dV_1}{r_{12}}.$$

Finally, the constant we have mentioned as added to the quantity whose derivative is taken is

$$\frac{\kappa R}{2}\int\int\frac{\bar{\rho}^2}{r_{12}}dV_1dV_2.$$

Collecting, we get on the right hand side

$$\frac{d}{dt}\frac{\kappa R}{2}\int\int\frac{\rho_1\rho_2 - 2\rho_1\bar{\rho} + \bar{\rho}^2}{r_{12}}dV_1 dV_2.$$

We finally have

$$R\frac{dT}{dt} + 2\frac{dR}{dt}T = -\frac{d}{dt}RF,$$

that is

$$\frac{d}{dt}R(T+F) + \frac{dR}{dt}T = 0,$$

as was to be proved.

Received
April 15, 1963

REFERENCES

1. *Hoyle F.*—In: *La Structure de l'Universe (11 Conseil de Solvay)*, Bruxelles, p. 66 (1958).
2. *Lifshitz E. M.*—ZhETF **16**, 587 (1946).
3. *Zeldovich Ya. B.*—ZhETF **43**, 1982 (1962).
4. *Bonnor W. B.*—Month. Not. RAS **117**, 104 (1957).

Commentary

This paper obtains the relation between the kinetic energy of peculiar (random, different from the overall expansion) motion and the potential energy of the inhomogeneous distribution of matter (less the divergent gravitational energy of homogeneous matter).

In the literature this result is known as the Irvine-Dmitriev theorem. W. Irvine's name appeared in connection with the fact that D. Layzer, in a later work,[1] cited the unpublished dissertation of his student, W. Irvine.[2] Zeldovich did not enter the name, perhaps, because it is associated with a number of other theorems.

Today many attempts are being made to measure peculiar velocities to determine the mean density of matter in the Universe and to solve the problem of hidden mass (see, for example, J. Peebles' review).[3]

In principle it should also be possible to establish the role of the energy of collective supernova bursts. The general difficulty in applying the Irvine-Dmitriev theorem is related to the fact that the velocities of radiant objects—stars—are measured. However, the density of stars, probably, is not only not equal to, but is not even proportional to the density of all forms of matter in each given volume (see in this connection the article in this book on black regions in the Universe).

What has been said above does not diminish the clarity of the posing of the problem and the mathematical elegance of its solution in the present article.

[1] *Layzer D.*—Ann. Rev. Astron. Astrophys. **2**, 341 (1964).
[2] *Irvine W. M.*—Doctoral Dissertation. Harvard University (1961).
[3] *Peebles P. J. E.*—Ann. Rev. Astron. Astrophys. **21**, 1–32 (1983).

53

The Development of Perturbations of Arbitrary Form in a Homogeneous Medium at Low Pressure*

With A. G. Doroshkevich

The gravitational development of perturbations on the background of matter expanding according to Hubble's law is considered. Similar investigations have been made by Lifshitz [3] and Bonnor [4]. The present study deals with some interesting cases ($p = 0$ and $k \to 0$), where the law of development of perturbations of density and velocity does not depend on the form of the initial disturbances. It is shown that rotational perturbations of velocity are not connected with density perturbations and are damped during the expansion of the Universe. The results are compared with those of Jeans.

The question of the gravitational instability of matter is very important for the clarification of the history of the Universe. Gravitational instability can be a reason for the formation of nebulae and stars during the expansion of a uniform Friedman Universe, as well as one of the reasons for the formation of stars from dust nebulae.

The problem was first considered by Jeans [1, 2] who investigated the evolution of perturbations on the background of stationary matter. His investigation, however, is incorrect, since uniform stationary matter does not satisfy the equations of classical theory. The correct law describing the evolution of perturbations in an expanding Friedman Universe was obtained by E. M. Lifshitz within the framework of the general theory of relativity [3]. He showed that there exist perturbations of density and velocity which grow proportionally to the radius of the Universe. Analogous results in classical theory have been obtained by Bonnor [4]. A detailed analysis of the results of these authors, as well as a review of the current status of the subject, has been given by Ya. B. Zeldovich [5].

The aim of the present work is to show that with the help of Bonnor's results it is possible to obtain the law governing the evolution of arbitrary density and velocity perturbations in a uniform medium at low pressure (dust), as well as the long-wave part of the perturbation in a uniform medium with

*Astronomicheskiĭ zhurnal **40** (5), 807–811 (1963).

an arbitrary equation of state. Both the growing and decaying parts of the density and velocity perturbations are determined by linear combinations of the initial values of the corresponding quantities. Rotational perturbations of the velocity are not linked with density perturbations and decay as the expansion proceeds, whatever the equation of state of the medium.

Let us consider Bonnor's results in greater detail. The solution of the unperturbed problem (uniform matter) for an arbitrary equation of state is given by

$$\rho = \frac{\rho_0}{\varphi^3}; \quad \mathbf{u} = \frac{\dot{\varphi}\mathbf{r}}{\varphi}; \quad \mathbf{r} = \varphi R; \quad \dot{\varphi}^2 = \frac{8\pi G\rho_0}{3\varphi} + \varepsilon, \tag{1}$$

where ρ and $\mathbf{u}$ are the density and velocity of matter, G is the gravitational constant, $\mathbf{R}$ is the initial coordinate of the particle, i.e., the Lagrangian variable, $\varphi(t)$ describes the expansion law of the system, and ε characterizes the energy of the system. For a sphere of finite mass, the total energy (the sum of the kinetic and gravitational energies) is given by $E = \frac{2}{5}\pi\rho_0 R_0^5\varepsilon$. (The possible cases are $\varepsilon > 0$, $\varepsilon = 0$, and $\varepsilon < 0$.)

Bonnor investigates the special case of spherically symmetric perturbations given by

$$\frac{\delta\rho}{\rho} = s(t;\mathbf{r}) = h(t)\frac{\sin kR}{kR} = h(t)\frac{\sin(2\pi r/\lambda)}{2\pi r/\lambda}, \tag{2}$$

where k is a constant and

$$\lambda = \frac{2\pi\varphi}{k}.$$

The conditions that the perturbation will grow with time is given by

$$\frac{k^2}{\varphi^2}\frac{dp}{d\rho} - 4\pi G\rho < 0. \tag{3}$$

If inequality (3) is not satisfied, then the perturbation propagates as an acoustic wave.

Of particular interest are density perturbations of long wavelength ($k \to 0$) and perturbations in a medium at low pressure (dust). In this case, inequality (3) is always satisfied and in the limit, the equation for $h(t)$ is independent of the wavelength of the perturbation. The amplitude of the density perturbation $h(t)$ is determined by two linearly independent solutions with different velocity fields corresponding to one and the same density distribution (2):[1]

$$h_1 = \frac{\dot{\varphi}}{\varphi}, \quad h_2 = \frac{1}{\varepsilon}\left(\frac{3t\dot{\varphi}}{\varphi} - 2\right). \tag{4}$$

The vortical component of the velocity varies as $1/\varphi$ (Appendix A).

The arbitrary density and velocity perturbations at the initial time t_0 can be decomposed into three terms: two linearly independent combinations of s_0 and $\operatorname{div}\mathbf{v}_0$ corresponding to the two solutions h_1 and h_2 and the vortical

[1] As $\varepsilon \to 0$, $h_1 \to t^{-1}$, while an expansion of the terms in (4) gives $h_2 = t^{2/3}$.

component of the velocity $\mathbf{v}_{r0}$, independent of the density variations. Each of these terms has its own form of time variation. Since h_1, h_2, and h_3 are independent of the wave vector $\mathbf{k}$, the Fourier analysis at time $t = t_0$ and the subsequent Fourier synthesis at time t preserve the form of the perturbation. This allows us to write down the result immediately for an arbitrary perturbation in closed form.

Since the velocity and density perturbations are connected by the equation of continuity

$$\frac{ds}{dt} + \operatorname{div} \mathbf{v} = 0, \tag{5}$$

we thus have

$$s = h_1 A + h_2 B, \quad \operatorname{div} \mathbf{v} = -\dot{h}_1 A - \dot{h}_2 B, \quad \mathbf{v_r} = h_3 \mathbf{C}. \tag{6}$$

It should be recalled that h_1, h_2 and h_3 are functions of time only, while A, B, and $\mathbf{C}$ are functions of the Lagrange coordinates only. In order to determine A, B and $\mathbf{C}$ we use the obvious initial conditions

$$s_0 = h_{10} A + h_{20} B, \quad \operatorname{div} \mathbf{v}_0 = \operatorname{div} \mathbf{v}_{p0} = -\dot{h}_{10} A - \dot{h}_{20} B, \quad \mathbf{v_{r0}} = \mathbf{C} h_{30}, \tag{7}$$

where $s_0 = s(t_0; \mathbf{r})$, $\mathbf{v}_0 = \mathbf{v}(t_0; \mathbf{r}) = \mathbf{v}_{p0} + \mathbf{v}_{r0}$; moreover, $\operatorname{div} \mathbf{v}_{r0} = 0$ and $\operatorname{rot} \mathbf{v}_{p0} = 0$. Specifying $\operatorname{div} \mathbf{v}_p$ completely determines the velocity perturbation $\mathbf{v}_p$.

In the simplest case $\varepsilon = 0$, we have

$$\begin{aligned} &h_1 = t^{-1}, \quad h_2 = t^{2/3}, \\ &s(t; \mathbf{r}) = \frac{3}{5} \frac{t_0}{t} \left(\frac{2}{3} s_0 + t_0 \operatorname{div} \mathbf{v}_0 \right) + \frac{3}{5} \left(\frac{t}{t_0} \right)^{2/3} (s_0 - t_0 \operatorname{div} \mathbf{v}_0), \\ &t \operatorname{div} \mathbf{v}_p = \frac{3}{5} \frac{t_0}{t} \left(\frac{2}{3} s_0 + t_0 \operatorname{div} \mathbf{v}_0 \right) - \frac{2}{5} \left(\frac{t}{t_0} \right)^{2/3} (s_0 - t_0 \operatorname{div} \mathbf{v}_0), \\ &\mathbf{v}_r = \left(\frac{t_0}{t} \right)^{2/3} \mathbf{v}_{r0}. \end{aligned} \tag{8}$$

Hence, in particular if $v_0 = 0$ initially, then at any later time $s = f(t) s(t_0; \mathbf{r})$ whatever the value of ε.[2] If at some place $s_0 = 0$, than at the same point $s(t; \mathbf{r}) = 0$. Why is it that the force due to gravity which perturbs the motion of matter everywhere and not only at the points $s \neq 0$ does not change the form of s? To understand this, it should be noted that when $s_0 \sim \delta(\mathbf{r} - \mathbf{r}_0)$, we also have $\operatorname{div} \delta \mathbf{F} \sim \delta(\mathbf{r} - \mathbf{r}_0)$, where $\mathbf{F}$ is the force due to gravity, i.e., $\operatorname{div} d\mathbf{v}/dt \sim \delta(\mathbf{r} - \mathbf{r}_0)$ and, consequently, $\operatorname{div} \mathbf{v} \sim \delta(\mathbf{r} - \mathbf{r}_0)$.

Thus, the velocity perturbation is indeed not restricted to the region $s \neq 0$, but the inverse influence of the velocity perturbation on the density is such that the density distribution remains a delta function.

[2]In particular, if $\varepsilon = 0$, then $f(t) = 0.4 t_0/t + 0.6 (t/t_0)^{2/3}$.

Hence, for the long-wave part of the perturbation as well as for perturbations in a uniform medium with $p = 0$, we can draw the following conclusions:

1) The law of evolution of density and velocity perturbations is independent of their initial form.

2) The growing part of the density and velocity perturbations is governed by a linear combination of the initial perturbations of these quantities. In the special case $\varepsilon = 0$, this combination is $(s_0 - t_0 \mathrm{div}\, \mathbf{v}_0)$.

3) The rotational velocity perturbations are independent of the density perturbations and decay with time.

Appendix A

E. M. Lifshitz has shown that the rotational velocity perturbations are independent of the density perturbations and are damped out whatever the equation of state. An analogous result can be obtained within the framework of classical theory.

Let us write down Euler's equation for velocity perturbations in a uniform medium with the equation of state $p = p(\rho)$:

$$\frac{\partial \mathbf{v}_r}{\partial t} + \mathrm{grad}\,(\mathbf{u}\mathbf{v}_r) - \mathbf{u}_r \,\mathrm{curl}\, \mathbf{v}_r = \delta\mathbf{F} - \mathrm{grad}\, \frac{\delta p}{\rho},$$

$$\mathrm{div}\, \delta\mathbf{F} = -4\pi G \delta\rho. \tag{A.1}$$

Operating on this equation with curl and transforming to the Lagrangian coordinates $\mathbf{R}$ [by means of (1)], we obtain

$$\frac{\partial}{\partial t} \mathrm{curl}_R \varphi \mathbf{v}_r = 0, \tag{A.2}$$

i.e., $\mathrm{curl}_R \varphi \mathbf{v}_r = \mathrm{curl}_R \varphi_0 \mathbf{v}_{\mathbf{r}0}$ and $\mathbf{v}_\mathbf{r} = \varphi_0 \mathbf{v}_{\mathbf{r}0} / \varphi$.

Appendix B

Let us consider, following Bonnor, the case $\varepsilon = 0$, $p = a\rho^{4/3}$. In this case the criterion for the growth of the perturbation (3) becomes

$$\frac{\pi c^2}{G\rho\lambda^2} - 1 = \alpha^2 - 1 < 0, \tag{B.1}$$

where $\alpha^2 = \pi c^2 / G\rho\lambda^2$ does not depend on the time, since $\rho \sim t^{-2}$, $\lambda = 2\pi\varphi/k \sim L \sim t^{2/3}$ and $c^2 = dp/d\rho \sim \rho^{1/3} \sim t^{-2/3}$. For arbitrary λ, the solution for h is of the form

$$h = At^{n_1} + Bt^{n_2},$$

$$n_{1,2} = -\frac{1}{6} \pm \sqrt{\frac{25}{36} - \frac{2}{3}\alpha^2}. \tag{B.2}$$

When $\alpha^2 > 25/24$, we have the case of acoustic waves. The influence of the gravitational force manifests itself in the variation of the frequency for a given wavelength, i.e., in dispersion. Writing the equation for the sound wave in the form $t^{-1/6}\exp\left[ikR + i\sqrt{\frac{2}{3}\alpha^2 - \frac{25}{36}}\ln t\right]$, we find for the frequency

$$\omega = \frac{d}{dt}\left[\sqrt{\frac{2}{3}\alpha^2 - \frac{25}{36}}\ln t\right] = \frac{1}{t}\sqrt{\frac{2}{3}\alpha^2 - \frac{25}{36}} \tag{B.3}$$

We will now show that the decrease in the amplitude of the sound wave according to t^{-1} is a consequence of the expansion of the medium. To do this we make use of the fact that the ratio of the energy of wave packet to the frequency is just the adiabatic invariant. For sound waves $E \sim L^3\rho c^2 s^2 \sim c^2 s^2$, since $L^3\rho \sim M = \text{const}$. Therefore, we see that $c^2s^2 \sim \omega \sim 1/t$, and $s \sim t^{-1/6}$.

The Jeans criterion is now the condition that the expression under the square-root sign becomes zero, which indicates that oscillations cannot occur. We find that $\alpha^2 = 25/24$, while according to Jeans, $\alpha^2 = 1$.

The formula $h_{1,2} = \exp(\pm\int m dt$, where $m^2 = 4\pi G\rho - (4\pi^2/\lambda^2)(dp/d\rho) \simeq 4\pi G\rho$, is a natural generalization of the expression obtained by Jeans for the variation of density in the long-wave region: $h_{1,2} = e^{\pm mt}$. Since $\rho = 1/6\pi G t^2$, we have $h_{1,2} = t^{\pm\sqrt{2/3}}$, which is already close to $h_1 = t^{-1}$, $h_2 = t^{2/3}$ in the exact solution. Taking into account the decrease in the amplitude because of the expansion of the medium, we can generalize Jeans' result as follows:

$$h_{1,2} = t^{-1/6}e^{\pm\int m\,dt}, \tag{B.4}$$

which leads to the following expressions for the amplitudes:

$$h_1 = t^{-1/6-\sqrt{2/3}} = t^{-0.98}, \quad h_2 = t^{-1/6+\sqrt{2/3}} = t^{0.65}.$$

The exponents of t differ form the exact values by 2%.

For the case $p = \alpha\rho^{4/3}$ considered above, we now have

$$h = At^{n_1} + Bt^{n_2}, \quad n_{1,2} = -\frac{1}{6} \pm i\sqrt{\frac{2}{3}(\alpha^2 - 1)}, \tag{B.5}$$

which is also close to the exact solution.

The above analysis shows that the physical picture proposed by Jeans is essentially correct.

Note added in proof. After this paper had gone to press, we became acquainted with [6], [7] and [8], in which perturbations in a contracting mass of gas are considered. The results of these papers are similar to ours, first of all because all of them are based on the work of Bonnor and Lifshitz. As noted by Lifshitz, the growth of density perturbations in a contracting mass of gas is faster than in an expanding medium. However, we deem it important to emphasize the over-all correctness of Jeans' theory and not that it is formally incorrect. In particular, the adiabatic increase in the amplitude of sound waves should be distinguished from gravitational instability and in this sense the Jeans criterion retains its meaning in contrast to the opinion expressed in [7].

Received
December 7, 1962

REFERENCES

1. *Jeans J. H.*—In. *Astronomy and Cosmology.* Cambridge, p. 345 (1929).
2. *Jeans J. H.*—Philos. Trans. Roy. Soc. London A **199**, 44 (1902).
3. *Lifshitz E. M.*—ZhETF **16**, 587 (1946).
4. *Bonnor W. B.*—Month. Not. RAS **117**, 104 (1957).
5. *Zeldovich Ya. B.*—Vopr. kosmogonii **9** (1963).
6. *Hunter C.*—Astrophys. J. **136**, 594 (1962).
7. *Layzer D.*—Astrophys. J. **137**, 351 (1963).
8. *Savedoff P.*—Astrophys. J. **136**, 609 (1962).

Commentary

This article is almost trivial, since in the linear theory the problem reduces to Fourier analysis and subsequent Fourier synthesis under the known limiting law of variation of individual Fourier components.

The significance of the paper, however, lies in the fact that it prepared for the nontrivial analysis of the nonlinear problem, i.e., the construction of the "pancake" theory—see the next article in this book.

Within the frame of the linear problem the continuation of the present work was to take into account the finite wavelength of perturbations as a correction in the coordinate representation.[1]

[1] *Zeldovich Ya. B.*—Astron. Zh. **59**, 636–638 (1982).

54

The Origin of Galaxies in an Expanding Universe*

With A. G. Doroshkevich, I. D. Novikov

The hypothesis outlined in this paper is being developed in the attempt to clarify the sequence of events leading to the formation in the evolving metagalaxy of galaxies, clusters of galaxies, and other cosmic objects.

This hypothesis is predicated on the hot model of the metagalaxy described in Friedman's solution of the problem. It is assumed that the present matter density is close to the critical value $\bar{\rho} \simeq 2 \cdot 10^{-29}$ g/cm^3 and that the present residual-emission temperature is $T_\gamma \simeq 3°$K [1, 2]. Metagalactic evolution according to this hypothesis reduces essentially to the following basic stages. (For arguments in support of the various points, see the sections following this outline.)

1. The time $t \leq t_1 \simeq 10^{13}$ sec (the moment taken as $t = 0$ is $\bar{\rho} = \infty$); $\bar{\rho} \geq \rho_1 \simeq 10^{-20}$ g/cm^3; $T_m = T_\gamma \gtrsim 4 \cdot 10^3$ K. The metagalactic matter is a plasma in equilibrium with radiation and consisting of 70% ionized hydrogen and 30% helium (by weight). The helium is ionized at $T \gtrsim 8 \cdot 10^4$ K and neutral at lower temperatures. The spatial distribution of radiation and plasma is extremely homogeneous and density perturbations are small, $\delta\rho/\rho \ll 1$.

2. The time $t_1 \leq t \leq t_2 \simeq (2-5) \cdot 10^{16}$ sec; the density $\rho_1 \leq \bar{\rho} \leq \rho_2 \simeq (1-10) \cdot 10^{-27}$ g/cm^3. At $t = t_2$ the hydrogen recombines, and matter practically ceases to interact with radiation. During adiabatic expansion there is a drop in the radiation temperature T_γ and the matter temperature $T_m : T_\gamma \sim \bar{\rho}^{1/3}$, $T_m \sim \bar{\rho}^{2/3}$, and gravitational accretion of density perturbations takes place. It is assumed that at $t = t_2$ density perturbations increase so sharply that in some areas, according to the law of chance, there arise gas condensations (protostars) of primary mass $M_3 \simeq (0.5 - 1) \cdot 10^6\, M_\odot$ (cf. Sec. 2 below).

3. The protostars evolve rapidly, releasing a large part of their nuclear energy. As this occurs the remaining matter is heated to a temperature $T_3 \simeq (0.5-1) \cdot 10^6$ K which is many times the residual-emission temperature at that moment.

4. As heating proceeds, inhomogeneities $\delta\rho/\rho \gtrsim 1$ are formed with mass range $M_4 \simeq 10^9 - 10^{10}\, M_\odot$ (the mass associated with one protostar). The

*Astronomicheskiĭ zhurnal **44** (2), 295–303 (1967).

heating of the gas halts the formation of protostars, through which passes about 10^{-4} of the total metagalactic matter. The heated metagalactic gas rapidly cools over a period $\Delta t \lesssim 0.5t_2$. Cooling is accompanied by the birth of second-generation protostars of characteristic mass $M \simeq M_4$. It is possible that some of these protostars evolve into ordinary and radio galaxies and others into quasar-type objects. During the evolution of protoquasars and proto-radio galaxies matter is again heated, over a period of the order of $\Delta t \sim (0.1 - 0.3)t_2$—a quasar lifetime $\Delta t_q \sim 10^{13} - 10^{14}$ sec $\sim 0.01t_2$—to a temperature $T_4 \simeq 10^6$ K. Thus $\sim 10^{-4}$ of all metagalactic matter is concentrated in protoquasars.

5. During the second heating the characteristic mass of the inhomogeneities increases to the value $M_5 \simeq M_4/10^{-4} \sim (10^{13} - 10^{14})M_\odot$, which roughly corresponds to the Jeans wavelength when $T \simeq 10^6$ K, $\rho \simeq 10^{-27}$ g/cm^3. Thus at this moment clouds separate out from which clusters of galaxies of mass $M \simeq M_5$ are formed.

6. During the evolution of clusters of galaxies, thermal instability produces fragmentation, i.e., the breakup of a cluster into individual galaxies. Stars of the ordinary type appear during subsequent galactic evolution.

The above sequence has not been fully demonstrated and may undergo considerable emendation. In particular, the present paper does not deal with the magnitude and function of the magnetic field at all stages. The basic assumptions of the hypothesis should also be more closely defined, but a more precise evaluation of ρ on the basis of the hot model does not alter the qualitative picture. Note, finally, that some parts of the proposed hypothesis are covered in the literature. The function of radiation in the development of inhomogeneities in the first stage is mentioned in [3, 4], the necessity of plasma heating and ionization before the formation of galaxies and galactic clusters and the important part played here by quasars and radio galaxies in [4], and galactic formation by fragmentation of gas clouds constituting protoclusters in [4, 5].

The following basic assumptions are new:

1) The formation of the first generation of protostars and the estimate of their mass, their evolution time, their energy release, and their creation of heated zones;

2) The development with time of the birth of protostars based on the statistical nature of the spectrum of the initial perturbations, account being taken of the exponentially small probability that small rms perturbations will be intense enough to produce condensations;

3) Consideration of the cooling of the gas after the birth of the first protostar generation, the function of the protostars as sources of density perturbations leading to the formation of the second generation, and the parallel relation between the second generation and the galactic clusters;

4) The entire process, from the appearance of the first metagalactic ob-

jects to the separation of clouds which then break up into galaxies, takes place rapidly at $\rho \approx 10^{-26} - 10^{-27}$ g/cm^3. The mean density during this process changes by approximately one order of magnitude. The opposite assumption of lengthy processes [4] seems likely to produce serious difficulties (cf. Sec. 3 below).

In Sec. 1 below we review metagalactic development in the first stage; in Sec. 2 we take a qualitative look at protostar formation and evolution; in Sec. 3 we briefly discuss thermal conditions in the metagalaxy at the various stages of expansion; and in Sec. 4 we consider the processes leading to galactic-cluster formation (cf. stages 4 and 5 in the preceding outline).

1. Development of perturbations in the hot model. Taking the present temperature of equilibrium radiation $T_\gamma \simeq 3$ K [1, 2], which corresponds to a density $\rho_\gamma \simeq 0.7 \cdot 10^{-33}$ g/cm^3, we find, given the above assumptions as to matter density ($\bar{\rho} = \rho_m = 2 \cdot 10^{-29}$ g/cm^3), that when $\bar{\rho} = \rho_0 \simeq 5 \cdot 10^{-16}$ g/cm^3 the radiation density equals the matter density $\rho_m \simeq \rho_\gamma \simeq \rho_0$. Expansion in the earlier stages is due primarily to radiation density and in later stages to matter density [6].

The development of perturbations in the hot model is studied in [3–5, 7].

In the plasma-radiation mixture two types of instability may arise: (1) adiabatic perturbations accompanied by the simultaneous compression of plasma and radiation, and (2) isothermal perturbations with compression of the plasma but continued homogeneity of radiation. The adiabatic perturbations increase when the scale is large, $l \geq \lambda_1$, where $\lambda_1 \simeq (c/\sqrt{3})t$ is the Jeans wavelength for the radiation pressure. Smaller-scale adiabatic perturbations are attenuated. Isothermal density perturbations show practically no increase on any scale, because of the strong radiation drag on the plasma. There is a second Jeans wavelength λ_2, related to the isothermal velocity of sound in a plasma. On the scale $l \leq \lambda_2$ plasma pressure produces attenuation of both adiabatic and isothermal density fluctuations. The relationships are shown in Fig. 1.

When $\bar{\rho} = \rho_0$, $M_1 \simeq 0.5 \cdot 10^{15}\, M_\odot$, $M_2 \simeq 5 \cdot 10^5\, M_\odot$, where M_1 and M_2 are the masses for λ_1 and λ_2 respectively. But when $\bar{\rho} < \rho_0$ radiation-matter interaction suppresses the growth of adiabatic perturbations on the scale $l < \lambda_1$, since radiation pressure still far outweighs plasma pressure and there is still strong radiation drag on the plasma. Only after recombination, which occurs at a radiation temperature $T_\gamma \simeq 3600 - 4000$ K, corresponding to $\bar{\rho} = \rho_1 \simeq 5 \cdot 10^{-20}$ g/cm^3, does matter cease to interact with radiation and perturbations begin to grow on a scale $l > \lambda_2$. When $\bar{\rho} = \rho_1$, $M_1 \simeq 10^{18}\, M_\odot$, and $M_2 \simeq 5 \cdot 10^5\, M_\odot$. As adiabatic expansion proceeds beyond this point the radiation temperature falls in accordance with the law $T_\gamma = T_1(\rho/\rho_1)^{1/3}$, and the matter temperature in accordance with the law $T_m = T_1(\rho/\rho_1)^{2/3}$.

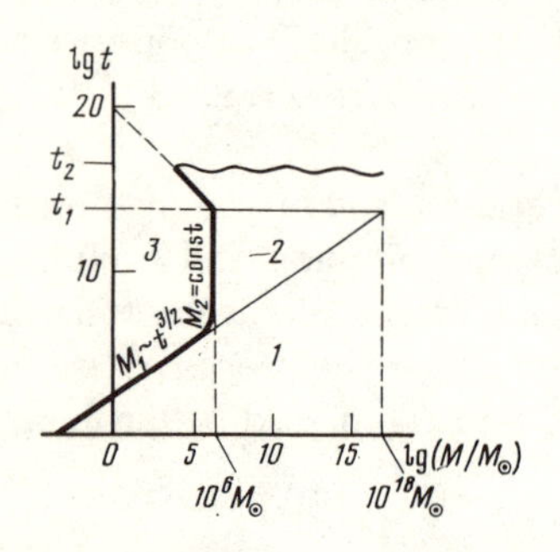

Fig. 1

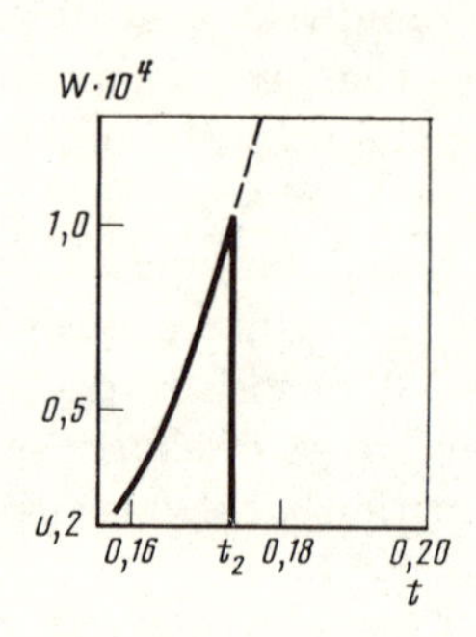

Fig. 2

Fig. 1. Evolution of perturbations in the "hot" model of a metagalaxy. The axes represent $\log(M/M_\odot)$ (abscissa) and $\log t$ (in seconds). The curves $M_1 = M_1(t)$, $M_2 = M_2(t)$ show the time dependence of masses corresponding to the first and second Jeans wavelengths. t_1 is the time of hydrogen recombination, t_2 is the time of heating. Adiabatic perturbations grow in region 1 and decay in region 2. In region 3 both adiabatic and isothermic perturbations decay. At the time of heating a mass corresponding to the critical Jeans wavelength grows sharply (wavy line which for $M \sim 10^{13} M_\odot$ should be bent up to the left).

Fig. 2. Probability W of formation of protostars as a function of time t. t is in relative units: $W \sim t^{-2/3} \exp(t^{-4/3})$. At the time $t = t_2$ heating of the gas occurs, the critical mass quickly increases and the formation of protostars ceases.

We will concentrate here on the later epoch, $\bar{\rho} < \rho_1$, assuming a particular density perturbation distribution at a time $\bar{\rho} = \rho_1$. From now on it will be more convenient to denote the inhomogeneity of matter distribution by the function $\sigma_0^2(M) = (\overline{\Delta M/M})^2$, which characterizes the mass deviation from the mean in a volume of mean mass M. The relation between density perturbations and the function $\sigma_0^2(M)$ is discussed in detail in [7]. It will also be convenient to replace the Jeans wavelength λ_J by the mass corresponding to λ_J:

$$M_\mathrm{J} = \frac{\pi}{6} \bar{\rho}_m \lambda_\mathrm{J}^3. \tag{1}$$

When $\bar{\rho} \leq \rho_1$, as gravitational amplification proceeds, perturbations grow in accordance with the law $\sigma_0^2(M) \sim (t/t_1)^{4/3} = (\bar{\rho}/\rho_1)^{-2/3}$ in a mass range $M > M_2$, where M_2 is the critical mass, related to the adiabatic velocity of sound in neutral hydrogen.

2. Heating of the neutral gas. Following the recombination of matter ($\bar{\rho} \leq \rho_1$), perturbations begin to grow rapidly: $\sigma_0^2(M) \sim (\bar{\rho}/\rho_1)^{-2/3}$. Because of the statistical nature of the fluctuations, it is probable that condensations with $\sigma = \Delta M/M \simeq 1$ (protostars) are formed at the same time.

We will henceforth assume that the condition in which $\sigma \simeq 1$ is the condition of formation of a condensation which over a mean hydrodynamic time $t^* = (6\pi G\bar{\rho})^{-1/2}$ separates out from the surrounding matter and ceases to participate in the overall Hubble expansion.

We assume that the probability of finding the relative perturbation $\sigma = \Delta M/M$ in a volume of mean mass M is determined by the Gaussian distribution

$$W(\sigma, M)d\sigma dM = \frac{1}{\sqrt{2\pi\sigma_0^2(M)}} e^{-\sigma^2/2\sigma_0^2(M)} d\sigma dM; \qquad (2)$$

then, as soon as $\sigma_0^2 \simeq 0.05$, about 10^{-4} of the total matter condenses into protostars. If we assume that $\sigma_0^2(M)$ decreases with the growth of M, then, given a small critical mass M_2 ($M_2 \sim 5 \cdot 10^5 M_\odot$) it is very probable that protostars of mass M_3, close to M_2, will be formed. Using the function $W(\sigma, M)$ we can determine the probability of protostar formation as a function of time: [$W(t, M)$ (cf. Fig. 2)].

These massive protostars evolve rapidly and heat the residual metagalactic gas to a high temperature ($T \simeq 10^5 - 10^6$ K). The Jeans wavelength increases sharply, preventing the formation of more protostars. The reasoning is illustrated in Fig. 2.

Let us take a closer look at protostar evolution.

The theory of stars of mass $\sim 10^5 - 10^6 M_\odot$ has not yet been developed in adequate detail, but the fundamental regularities of the evolution of such stars are known [6, 8]. A star evolves rapidly to the critical state where nuclear energy sources are inadequate to maintain equilibrium luminosity, which is followed by a phase of rapid compression. But for stars of mass $\lesssim 10^6 M_\odot$ close to the critical state, nuclear reactions—$3He^4 \to C^{12}$, the carbon-nitrogen cycle, and others—are initiated, bringing about the rapid release of energy and probably the explosion of protostars [6]. Explosions are described in [16].

The heating of the gas to a temperature $T \simeq 10^5 - 10^6$ requires energy of $\sim 2.5 \cdot (10^{13} - 10^{14}$ ergs/g. But the burning of hydrogen releases $6 \cdot 10^{18}$ ergs/g and the burning of helium (30% of the total matter) releases $1.8 \cdot 10^{17}$ ergs/g. Consequently, even if helium alone is burned, $10^{-3} - 10^{-4}$ of all remaining matter will be heated to a temperature $T \simeq 10^5 - 10^6$ K.[1]

We take the conversion efficiency in the protostars to be approximately $0.1 - 0.01\%$ Mc^2. Then the energy release of protostars of mass $\sim 5 \cdot 10^5 M_\odot$ will be $E_0 \simeq 10^{56} - 10^{57}$ ergs.

If the evolution of a protostar produces a nuclear explosion, a strong shock wave will pass through the surrounding matter. Until the wave velocity $D \geq D_0 \sim 100 - 200$ km/sec this shock wave can be regarded as adiabatic [5]. Over a time $t \sim 0.1t_2$ at the requisite density, it will heat gas of mass $M_4 \sim E_0 D^2 \sim 10^9 - 10^{10} M_\odot$ to a temperature $T \sim (3-8) \cdot 10^5$ K. Protostar

[1]This and all subsequent numerical estimates are very rough approximations.

formation will come to a stop when each exploding protostar has accounted for $\sim 10^9 - 10^{10}\, M_\odot$ of gas, that is, when $\sim 0.5(10^{-3} - 10^{-4})$ of the total matter has passed through the protostars. Under the hypothesis outlined above $\sigma_0^2 \sim 0.05$. The density ρ_2 at which protostars form and explode depends on the fluctuation magnitude σ_0^2 at a recombination time $(\bar{\rho} = \rho_1)$. Further metagalactic evolution is highly dependent on this density value.

3. Thermal conditions in the expanding metagalaxy. Metagalactic thermal conditions are examined by Ginzburg and Ozernoi [9]; here we will merely pause briefly to consider certain features relevant to our problem. The most interesting area, $\bar{\rho} \gtrsim 10^{-27}$ g/cm^3, is not discussed in [9]. We regard the metagalactic medium as an ideal gas with a specific heat ratio $c_p/c_V = \gamma = 5/3$, expanding according to Friedman's model.

The thermal balance equation can be written [10]

$$\frac{1}{\gamma - 1}\frac{dP}{dt} - \frac{\gamma}{\gamma - 1}\frac{P}{\rho}\frac{d\rho}{dt} + \rho L = 0, \tag{3}$$

where $P = \rho RT/\mu$ (usual symbols) and $L = L^- - L^+$; L^- and L^+ denote the energy outflow and inflow per unit time in 1 g of matter. The function L^- is computed from the radiation energy losses in free-free and free-bound transitions and in the hydrogen and helium lines—the part played by helium is examined in detail in [11]—which can be taken into account by changing the term denoting free-bound transitions in the coefficient T_0,

$$L^-_{(1)} = 0.5 \cdot 10^{21} T^{1/2} \rho \left[a_1 + \frac{T_0}{T}\right]. \tag{4}$$

We must also add to the L^- function cooling due to the inverse Compton effect on the residual radiation. The required formula has been obtained by Kompaneets [12] and Weynmann [13] and is

$$L^-_{(2)} = 4\sigma_0 \frac{k(T_e - T_\gamma)}{m_e m_p c} \varepsilon_\gamma,$$

where σ_0 is the Thomson scattering cross section, ε_γ is the energy density of the residual emission, m_e and m_p are the electron and proton masses respectively, and T_e and T_γ are the electron temperature and residual-emission temperature respectively. Considering that $T = T_e \gg T_\gamma$ and that $\varepsilon_\gamma \sim \rho^{4/3}$ we obtain $(a_3 = 2 \cdot 10^4)$

$$L^- = 0.5 \cdot 10^{21} T^{1/2} \rho \left[a_1 + \frac{T_0}{T} + a_3 T^{1/2} \rho^{1/3}\right]. \tag{5}$$

If we disregard helium- and hydrogen-line radiation, then $a_1 = 1$ and $T_0 = 0.385 \cdot 10^6$ K. Assuming the presence of 30% (by weight) ionized helium in the medium, we get $a_1 \simeq 1.4$ and $T_0 \simeq 10^6$ K. If we allow for the line radiation of hydrogen and especially of helium, $T_0 \simeq 1.5 \cdot 10^6$ K $\leq 10^7$ K. Changing to a dimensionless variable $x = \rho/2 \cdot 10^{-29}$ g/cm^3, $T = T_6 \cdot 10^6$ K, we obtain:

$$\frac{\partial T_6}{\partial x} = F^- - F^+,$$

where $F^+ = 0.4 \cdot 10^3 x^{-3/2} L^+$ (L^+ in cgs units),

$$F^- = \frac{2}{3}\frac{T_6}{x} + 0.58 \cdot 10^{-2}\left(\frac{T_6}{x}\right)^{1/2} + \frac{b_1}{T_6}\left(\frac{T_6}{x}\right)^{1/2} + 0.2 \cdot 10^{-2} T_6 x^{-1/6}, \quad (6)$$

and $b_1 = \{1.6 \cdot 10^{-3}, 4.3 \cdot 10^{-3}, 6.5 \cdot 10^{-3} \leq 4 \cdot 10^{-2}\}$ given our assumptions as to the influence of the hydrogen and helium lines.

It is convenient to introduce the thermal relaxation time $t_T = T_6 t^*/2xF^-$ and to consider the ratio t_T/t^*, which characterizes the cooling speed as a function of x and T_6 (ρ and T).

Given a density $\bar{\rho} \gtrsim 10^{-25}$ g/cm^3 ($x \geq 10^4$) we can disregard cooling due to recombination and line radiation, since the medium is optically thick. But even if we neglect this radiation, in the area we are studying $x = 10^4$ and $T_6 = 1 - 0.1$, $t_T/t^* \sim 0.1$, that is, cooling of the medium occurs practically instantaneously. The main influence here is that of Compton cooling on residual emission, the temperature of which ~ 40 K when $x = 10^4$. When $x = 10^3$ ($\bar{\rho} \simeq 10^{-26}$ g/cm^3) the cooling time is highly dependent on temperature. When $T_6 = 1$, $t_T/t^* \simeq 0.6$, but when $T_6 \simeq 0.1$, $t_T/t^* \simeq 0.1$ and is again determined by recombination and line radiation.

Thus, at high densities ($x \gtrsim 10^4$), and also at low temperatures ($T_6 \sim 0.1$) and at all densities, in the interval ($x > 10^2$) under study cooling occurs practically instantaneously if the heat sources are insufficient.[2] Only at high temperatures ($T_6 \simeq 1$) and sufficiently low densities ($x \lesssim 10^3$) does cooling proceed over a period of time comparable to the hydrodynamic time.

4. Formation of galaxies and galactic clusters. The explosion of protostars when $\bar{\rho} = \rho_2$ heats the metagalactic matter to $T \sim (10^5 - 10^6)$ K, where density perturbations greater than ($\delta\rho/\rho \gtrsim 1$) occur in the mass range $M \simeq M_4 \simeq (10^9 - 10^{10})\, M_\odot$.

It is doubtful that this process can take place when $\bar{\rho} \gtrsim 10^{-25}$ g/cm^3, since the cooling time under these conditions is of the same order as the heating time, i.e., the main mass of gas remains cold and the characteristic scale of inhomogeneities due to explosions increases. This would lead to the creation of massive condensations and possibly of collapsed bodies.[3] No

[2] As shown by Weynmann [13], these processes slightly distort the residual-emission spectrum.

[3] A high temperature ($T > 10^5$ K) cannot persist at densities of $\rho > 10^{-25}$ g/cm^3 and more, if only for energy considerations (cf. [3]).

such objects are at present observed, and the possibility of their existence remains an open question.

Heating is effective only when $\bar{\rho} \lesssim 10^{-26}$ g/cm^3, since then the cooling time (from $T \simeq 10^6$ K) considerably exceeds the heating time. For this reason we will henceforth assume that $\rho_2 \sim (1-10) \cdot 10^{-27}$ g/cm^3.

To estimate inhomogeneities in the mass range $M \gg M_4$, we can take $\sigma_1^2(M) = (\Delta \bar{M}/M)^2 \approx M_4/M$, which satisfies the assumption of the statistical independence of explosions.

Since under the above conditions $t_T/t^* \simeq 0.6$, the medium will cool rapidly compared to the hydrodynamic time (but slowly compared to the preceding heating time), and most objects formed have mass $M \simeq M_4$. Thermal instability [4, 10], developing over a period of the same order as the cooling time, should have a considerable effect. This is especially important because the first heating produces large temperature and density inhomogeneities ($\delta T/T \gtrsim 1$ and $\delta\rho/\rho \gtrsim 1$) in the mass range $M \lesssim M_4$. Using for our estimate the results obtained in Field's paper [10] for a stationary medium, and taking $L^- = 0.5 \cdot 10^{21} T^{1/2} \rho [1 + T_0/T]$, we find the critical mass for thermal instability related to electron thermal conductivity (for photons the medium is practically transparent on the scale considered):

$$M_c \simeq 0.4 \cdot 10^{-71} \frac{T^{9/2}}{\rho^2 (1 + 3T_0/T)^{3/2}} M_\odot, \tag{7}$$

which, when $T \simeq 10^6$ K, $\rho \simeq 10^{-26}$ g/cm^3, and $T_0 \simeq 1.6 \cdot 10^6$ K, gives $M_c \simeq 0.3 \cdot 10^7 \, M_\odot$.

To sum up, it appears that under the conditions considered here most objects formed will be of mass $M \simeq M_4$. But this in no way excludes the possibility that a certain number of objects will form throughout the mass range from $M \simeq 10^4 - 10^5 \, M_\odot$ to $M \simeq 10^{11} - 10^{12} \, M_\odot$. Since metagalactic conditions at this stage are similar to present conditions—the gas is ionized, random magnetic fields may be generated—the objects created should be similar to contemporary quasars, galaxies, and radio galaxies.

Note that the typical mass $M_4 \sim 10^9 - 10^{10} \, M_\odot$ agrees with estimates of the mass of the quasar 3C 273, $(3 \cdot 10^9 \, M_\odot)$ [6].

However, the formation at this stage of cosmic objects—protoquasars and protogalaxies—cannot use up an appreciable part of the metagalactic mass, as occurs at the time of first heating. No theory of quasar evolution has yet been developed, and the conditions of quasar formation have not been elucidated. But a quasar lifetime is estimated [14] at $t_q \lesssim 10^{14}$ sec, which amounts to $\sim 0.01 t^*$, that is, a quasar evolves practically instantaneously.

Given a mass $M \sim 10^9 - 10^{10} \, M_\odot$, an evolving protoquasar can release energy $E \simeq 10^{59} - 10^{60}$ ergs only through the consumption of fissionable material at a conversion efficiency of $\sim 0.1\%$. This is enough to heat to $T \simeq 10^6$ K a mass $M_5 \simeq 0.4(10^{12} - 10^{13}) \, M_\odot$. The basic heating mechanism may be the ejection of masses of gas and subsequent shock-wave formation. In this case the estimates used are those in Sec. 2.

When protoquasars have taken up $\sim 10^{-4}$ of the metagalactic mass, the high degree of heating halts their formation. It can be expected that objects on other scales (galaxies and radio galaxies) use up comparable proportions of the metagalactic mass.

We do not consider the contribution to the metagalactic energy balance made by galaxies and radio galaxies, chiefly because their evolution time is notably longer than a quasar lifetime. Some galaxies and radio galaxies formed at this stage may participate in cluster formation and some may remain in the form of background galaxies, but most protoquasars should be members of clusters.

At a density $\bar{\rho} = 10^{-27}$ g/cm^3 and temperature $T = 10^6$ K, the Jeans mass $M_J = 10^{13} M_\odot$, which corresponds to the new scale of inhomogeneities $M_J \sim M_5 \sim 10^{13} M_\odot$. Under these conditions, clouds separate out and produce clusters of characteristic mass $M \sim M_5$. The protocluster mass distribution then will be determined by the spectrum of perturbations resulting from subsequent explosions as well as by processes in existing clusters, and may be very complex. Nevertheless, on a large scale $M \gg M_5$, we can expect that $\sigma_2^2(M) \simeq M_5/M$ and it determines the rate of formation of massive protoclusters. After a mass of gas which will later form galactic clusters has separated out, its heat relaxation time when $T \sim 10^6$ K increases considerably, so that the gaseous mass ceases to take part in the overall expansion.

When this is the case the prime determinant of galactic formation during the evolution of clusters will be thermal instability. The processes leading to the separation of galaxies out of clusters are discussed by Ozernoi [4].

In order to estimate the mass fraction entering into the protoclusters, we use the formula (2) with $\sigma_2^2(M) = M_5/M$ down to $M = M_5$, and find that protoclusters take up a notable proportion (~ 0.3) of the total mass.[4] This is explained largely by the fact that the characteristic scale of an inhomogeneity (M_5) is close to the critical Jeans mass. It is possible, however, that not all separated gas clouds lead to the formation of galactic clusters.

Some protoclusters may evolve more slowly, because (1) these clouds, if formed later, have a lower mean density and hence a longer thermal evolution time, and (2) they may be heated by energy released in the evolution of developed clusters, radio galaxies, cosmic rays, etc.

This hypothesis requires both the first heating of neutral hydrogen and the formation of protoclusters to take place at a density $\rho_2 \sim 10^{-26} - 10^{-27}$ g/cm^3, i.e., in the zone $z = \Delta\lambda/\lambda \sim 5 - 10$. In principle, as shown in [15], this zone is still available for direct observation. According to our hypothesis, when $z \gtrsim 5$ the number of quasars should rise steeply. Experi-

[4]Of course this conclusion is anything but categorical, and wide variations are possible here.

mental confirmation of this point would be of great interest.

The critical mass $M_1 = 10^{18}\ M_\odot$ plays no part at all in this version of the hypothesis, since it is assumed that initial fluctuations on such a scale would be too small to have reached an observable value at the present time, and no new fluctuations arise on so large a scale. An observational test of this assumption by investigating metagalactic homogeneity on maximum scales would be extremely desirable.

In conclusion, we should like to emphasize once more that the proposed hypothesis has not been rigorously proved throughout. A great deal of work remains to be done along these lines. The present paper is an attempt to offer a possible picture of the formation of cosmic objects from the homogeneous metagalaxy.

The authors are indebted to L. M. Ozernoi for making available to them his unpublished paper [4], which encouraged them to write the present essay, and to S. B. Pikelner for his valuable comments.

Note added in proof (March 1, 1967): A study of radio source counts at various flux density levels indicate that density peaks sharply at $z \sim 3-4$ cm—M. S. Longair, Monthly Notices Roy. Astron. Soc. **133**, 421 (1966).

Received
April 22, 1966

REFERENCES

1. *Penzias A. A., Wilson R. W.*—Astrophys. J. **142**, 419 (1965).
2. *Roll P. G., Wilkinson D. T.*—Report at a conference in Miami, 1965.
3. *Peebles P. J. E.*—Astrophys. J. **142**, 1317 (1965).
4. *Ozernoi L. M.*—Collection of works of the symposium "Variable Stars and Stellar Evolution," Moscow: Nauka (1966).
5. *Kaplan S. A., Pikelner S. B.* Mezhzvezdnaya sreda [The Interstellar Medium]. Moscow: Fizmatgiz (1963).
6. *Zeldovich Ya. B., Novikov I. D.*—Uspekhi fiz. nauk **86**, 447 (1965).
7. *Zeldovich Ya. B.*—Adv. Astron. and Astrophys. **3**, 241 (1965).
8. *Fowler W. A.*—Rev. Mod. Phys. **36**, 545, 1104 (1964).
9. *Ginzburg V. L., Ozernoi L. M.*—Astron. zhurn. **42**, 943 (1965).
10. *Field G.*—Astrophys. J. **142**, 531 (1965).
11. *Syunyaev R. A.*—Astron. zhurn. **43**, 1237 (1966).
12. *Kompaneets A. S.*—ZhETF **31**, 876 (1956).
13. *Weynmann.* Preprint USA (1966).
14. *Hoyle F., Fowler W., Burbidge G., Burbidge E. M.*—Astrophys. J. **139**, 909 (1964).
15. *Kardashev N. S., Sholomitskii G. B.*—Astron. tsirkulyar **336** (1965).
16. *Bisnovatyi-Kogan G. S.*—Astron. zhurn. **44** (1967).

Commentary

This article played an enormous role in the development of one of the variants of the modern theory of galaxy formation, in particular, of the assumption of initial entropic density perturbations. The article shows that the Jeans mass in the stage following recombination is equal to $\sim 10^6 M_\odot$. A scenario is proposed in which explosions of "primordial stars" with masses of order $10^6 M_\odot$ as a result of the growth of perturbations ensure the heating of the gas to high temperatures and the growth of the Jeans mass to the scale of galaxy clusters. The picture of the formation of the large-scale structure of the universe as a result of explosions of primordial stars amazingly presaged the popular modern scenario worked out by J. Ostriker[1] related to shock waves formed by explosions of numerous supernova in young galaxies. The problem of the thermal regime of intergalactic gas in the expanding universe, first posed in the pioneering work by V. L. Ginzburg and L. M. Ozernoi (Ref. [9] in the current article) is analyzed in detail here.

It should be noted that in 1968 J. Peebles and R. Dicke proposed their scenario of the evolution of isothermal perturbations, according to which growing perturbations with a mass of $10^6 M_\odot$ correspond to spherical clusters. The question of which scenario is valid—into what are the first clouds of neutral hydrogen and helium with $M \sim 10^6 M_\odot$ transformed—is still unsolved. The solution to this problem (spherical clusters or exploding primordial stars) requires detailed numerical calculations which take into account the formation of molecules, cooling processes, energy removal, and fragmentation. At the same time, the very existence of entropic perturbations is in question (cf. **51**).

[1] *Ostriker J. P., Cowie L. L.*—Astrophys. J. Lett. **243**, L127 (1981).

55

Gravitational Instability: An Approximate Theory for Large Density Perturbations*

An approximate solution is given for the problem of the growth of perturbations during the expansion of matter without pressure. The solution is qualitatively correct even when the perturbations are not small. Infinite density is first obtained on disc-like surfaces by unilateral compression. The following layers are compressed first adiabatically and then by a shock wave. Physical conditions in the compressed matter are analyzed.

1. The Approximate Solution

The linear theory of perturbations, applied to the uniform isotropic cosmological solution, is now well understood. It is generally admitted that its predictions are limited by $\delta\rho/\rho < 1$, and that further events must be followed by numerical calculations. Such calculations, in three dimensions and with random initial conditions, promise to be tedious. Therefore, an approximate method, which gives the right answer at least qualitatively, is of interest.

In this article the linear theory is taken to formulate the answer in terms of lagrangian coordinates: the actual position $\mathbf{r}$ of a particle is given as a function of its lagrangian coordinate $\mathbf{q}$ (i.e., its initial position) and the time t, $\mathbf{r} = \mathbf{r}(t, \mathbf{q})$. The linear theory is applied to the simplest case of pressure $\mathcal{P} = 0$ ("dust") in the Newtonian approximation. Only the growing perturbations are considered. The answer is of the form

$$\mathbf{r} = a(t)\mathbf{q} + b(t)\mathbf{p}(\mathbf{q}). \tag{1}$$

The first term $a(t)\mathbf{q}$ describes the cosmological expansion; the second term describes the perturbations. The functions $a(t)$ and $b(t)$ are known; $b(t)$ is growing more rapidly than $a(t)$, as a result of gravitational instability. The vector function $\mathbf{p}(\mathbf{q})$ depends on the initial perturbation. With given $\mathbf{r}(t, \mathbf{q})$, it is possible to calculate the distribution of velocity and density in space; $\mathbf{r}(t, \mathbf{q})$ contains the whole picture of the motion.

*Astronomy and astrophysics **5**, 84–89 (1970).

The approximation proposed in this article consists in the extrapolation of formula (1) into the region where the perturbations of density $\delta\rho/\rho$ are not small.

Let us first investigate the consequence of the approximation; this will help us to analyze its plausibility. In order to follow the behavior of a small group of particles centered on some definite $\mathbf{q}$, we calculate the tensor of deformation

$$\mathcal{D}_{ik} = \frac{\partial r_i}{\partial q_k} = a(t)\delta_{ik} + b(t)\frac{\partial p_i}{\partial q_k}.$$

The derivatives $\partial p_i/\partial q_k$ define a set of fundamental axes. After choosing the coordinate system along the axes, one obtains[1] for a given $\mathbf{q}$

$$D = \left\| \begin{array}{ccc} a(t) - \alpha b(t) & 0 & 0 \\ 0 & a(t) - \beta b(t) & 0 \\ 0 & 0 & a(t) - \gamma b(t) \end{array} \right\|.$$

A volume which was initially a cube (at $t \to 0$) and which would be a cube in the unperturbed motion, is transformed into a parallelepiped. One can always choose the axis of the cube so that it is transformed into a rectangular parallelepiped; the axes are not rotating in solution (1). The density near a particle with given $\mathbf{q}$ is given by the conservation of mass

$$\rho(a - \alpha b)(a - \beta b)(a - \gamma b) = \bar{\rho} a^3. \tag{2}$$

We recall that α, β and γ are functions of the point $\mathbf{q}$; $a(t)$ and $b(t)$ are the same for all particles. If $\alpha(\mathbf{q}) > 0$, one can specify the moment when $\rho \to \infty$ by $a(t) - \alpha b(t) = 0$. In a given $\mathbf{q}$ volume, we find the point where α has its highest value α_m; this locates the particle in which the density first goes into infinity, at some time t_c for which $a(t_c) - \alpha_m b(t_c) = 0$.

The most important point to be emphasized is that infinite density results form unilateral compression in the direction of the α-axis. The probability of the coincidence of α and β, or of a triple coincidence $\alpha = \beta = \gamma$, is zero.

The picture is very different from a spherically symmetric (SS) compression. The SS case was considered due to its simplicity; I think that it is degenerate and not typical of the general case of random initial perturbations. Later, at $t > t_c$, formula (2) is not applicable to particles which have gone through $\rho = \infty$; the matter stays compressed. But we feel that it is still possible to apply (1) and (2) to other particles. By continuity, particles with high α_1, $\alpha_m - \alpha_1 \ll \alpha_m$, surround the "maximal" particle, lying on some triaxial ellipsoid. The direction of the fundamental axis of $\partial p_i/\partial q_k$ also varies slowly, so that the α direction is nearly the same as long as α_1 is near α_m.

The unilateral compression makes the three-dimensional ellipsoid in $\mathbf{q}$-space into a flat two-dimensional ellipse in the real $\mathbf{r}$-space. The volume

[1]It can be shown that $\partial p_i/\partial q_k = \partial p_k/\partial q_i$ in the growing mode of perturbations. Here $\alpha = \xi_1$, $\beta = \xi_2$, and $\gamma = \xi_3$ are the three roots of $|(\partial p_k/\partial q_i) + \xi\delta_{ik}| = 0$. The sign of α, β, γ is not defined in the usual manner, for the sake of subsequent convenience.

density ρ is infinite, but the product of ρ times the width l (equal to the density per unit of surface σ) is finite on this ellipse.

2. Comparison of Various Approximations

Let us return to the motivation of (1), on which the picture proposed is based.

In the linear approximation, there are other formulations, for example

$$\rho(\mathbf{r}, t) = \bar{\rho}(t)\left(1 + f(t)\delta\left(\frac{\mathbf{r}}{a(t)}\right)\right), \tag{3}$$

with known $f(t)$, $a(t)$ and δ given by the initial conditions. Why should one prefer (1) to (3)? The proposed solution (1) always gives finite velocity and finite acceleration for a particle, up to the moment when this particle is splashed against other particles, giving $\rho = \infty$. Matter with infinite density ρ forms discs with finite density σ. But by the properties of gravitational potential, $\rho = \infty$, $\sigma \neq \infty$ gives a finite potential and a finite acceleration. The approximation proposed is exact at one extreme, when the perturbations are small. But the approximation gives only a *finite* error at the other extreme, when $\rho = \infty$.

This is in contrast with (3): if one attempts to extrapolate (3), meaningless negative densities are predicted for some parts of the volume $\delta < -f^{-1}$, while in other parts the density is only doubled.

The better performance of (1) as compared with (3) can be traced back to the fundamental equations. The solution (3) corresponds to equations written in Eulerian form. During linearization, terms $\operatorname{div}(\mathbf{v}\delta)$ and $(\mathbf{v}\nabla)\mathbf{v}$ are neglected, where $\delta = \delta\rho/\rho$, and $\mathbf{v} = \mathbf{u} - H\mathbf{r}$ is the peculiar velocity.

By adopting solution (1), we adopt a law of motion for every particle:

$$\mathbf{u} = \dot{a}\mathbf{q} + \dot{b}\mathbf{p}(\mathbf{q}); \quad \frac{d\mathbf{u}}{dt} = \ddot{a}\mathbf{q} + \ddot{b}\mathbf{p}(\mathbf{q}). \tag{4}$$

Given formula (1), the density is calculated exactly.

The only error is in the use of the perturbation of gravitational force δF, acting on a particle $\mathbf{q}$, where $\langle \delta F \rangle$ is linearized as a function of the perturbation of position of the particle considered $b\mathbf{p}(\mathbf{q})$, and of all other particles with differing $b\mathbf{p}(\mathbf{q}')$. The analytic evaluation of the error is extremely difficult. Djachenko proposes to make a numerical estimate of the error by taking solution (1) with a definite $\mathbf{p}(\mathbf{q})$ at a definite moment t, calculating the actual distribution of $\rho(\mathbf{r}, t)$, the gravitational potential $\phi(\mathbf{r}, t)$ and the force $F(\mathbf{r}, t)$, and comparing this force with the acceleration given by (4) for the approximate solution. As long as this trial is not made, the real accuracy of the solution for perturbations of different amplitude is unknown. Still, the main qualitative conclusion about the unilateral type of compression seems to be inescapable.

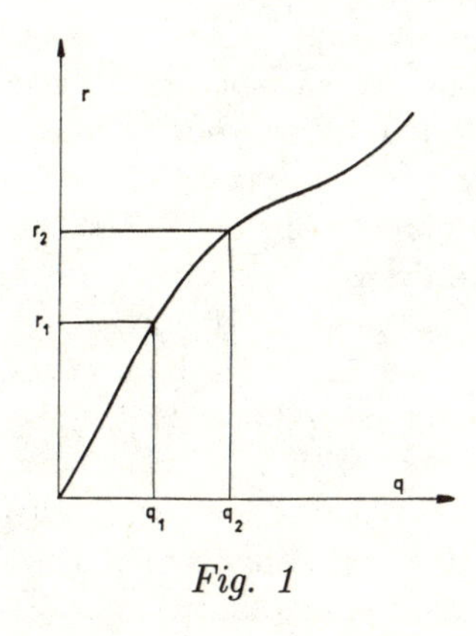

Fig. 1

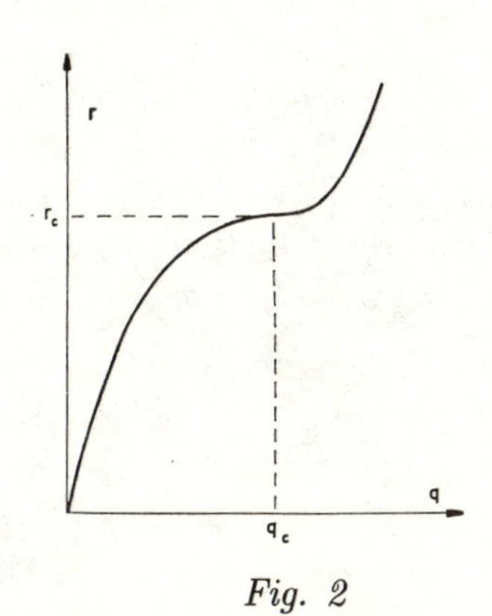

Fig. 2

The compression is due to gravitational interaction: it is due to gravitation that $b(t)$ grows faster than $a(t)$ in (1). The initial excess of density near some particle $\mathbf{q}_0$ increases because other particles are attracted to $\mathbf{q}_0$. But solution (1) and expression (2) for the density also take into account the tidal forces from neighboring perturbations which destroy the spherical symmetry of compression. At small t and small $\delta\rho/\rho$ formula (2) coincides with (3):

$$\rho(\mathbf{q}, t) = \bar{\rho}\left(1 + (\alpha + \beta + \gamma)\frac{b(t)}{a(t)}\right) = \bar{\rho}(1 + S(\mathbf{q})f(t)). \tag{5}$$

The initial growth of the perturbation depends only on the sum $\alpha+\beta+\gamma = S(\mathbf{q})$. But later, as shown by (2), all three parameters α, β, and γ are of importance. Therefore (1) and (2) contain more information than (3). The later nonlinear behavior of the density is not uniquely determined by the initial density amplitude in the linear period; it also depends on the spatial distribution of velocity and density in neighboring regions, on which α, β, and γ depend. This important point is overlooked if one takes the spherically symmetric case as a model for the nonlinear situation.

3. Astronomical Implications

Assuming that the approximate solution is qualitatively true, we must discuss a) whether the necessary conditions ($\mathcal{P} = 0$, newtonian approximation) are fulfilled, b) what physical processes occur in the compressed regions, and c) the place of the solution in the problem of the structure of the Universe. The answer to a) depends on the type of initial perturbation. Widely discussed are adiabatic perturbations, characterized by $\delta T/T = \frac{1}{3}\delta\rho/\rho$ before recombination. As shown by Silk [5], the photon viscosity eliminates perturbations of small scales ($M < 10^{12}\, M_\odot$). On the other hand, it is plausible that the spectrum is decreasing for greater scale. Therefore the perturbations with $M \sim 10^{12}\, M_\odot$ are the most important. We apply the approximate theory to the period after recombination, assuming that the perturbations are small at $Z = 1400$ (just after recombination) and grow so that galaxies, etc., are formed before $Z = 0$ (the present day).

After recombination, the Jeans' mass (depending on the neutral gas pressure) is of the order of $M_J = 10^5 - 10^6\, M_\odot$ [2, 3]. The situation $M \simeq 10^{12}\, M_\odot \gg M_J$ means that pressure can safely be neglected. At the moment of recombination the event horizon (the sphere with $r = ct$) contains $M_h = 10^{19}\, M_\odot$, so that $M \ll \tilde{M}_h$. Therefore newtonian theory is applicable. The inequalities are even better fulfilled later on at $Z < 1400$, because M_J diminishes and M_h grows during further expansion.

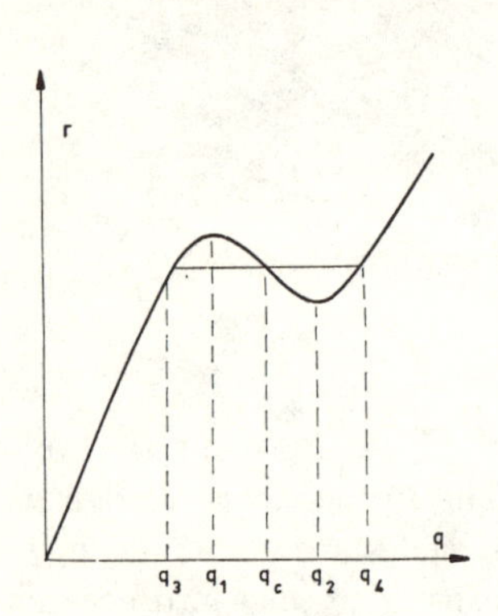

Fig. 3. The ordinate of the maximum of the curve is $t_{\max}$, that of the horizontal segment is t_c.

b) Now we turn from the premises to the consequences of the approximate solution. To study the characteristic unilateral compression, we neglect the motion in other directions and consider the unidimensional problem. The subsequent evolution is shown in three figures: 1—for small perturbations $(t < t_c)$, 2—for the moment when ρ just attains infinity on the line considered $(t = t_c)$, and 3—for $t > t_c$. On each figure the curve $r = f(q)$ is given. To be precise it is r_x as a function of g_x, the x-axis being chosen in the direction of maximal deformation.

The quantity of matter between a pair of points is proportional to $q_2 - q_1$, because q is the lagrangian coordinate. In the physical (r) space it is contained in the strip $r_2 - r_1$; therefore the density ρ is proportional to $\frac{q_2-q_1}{r_2-r_1} \to \left(\frac{dr}{dq}\right)^{-1}$; this is the unidimensional simplification of formula (2). In Fig. 2 the curve has a point with a horizontal tangent. When such a point occurs for the first time, we must have $dr/dq = 0$, $d^2r/dq^2 = 0$ at $t = t_c$; $q = q_c$.

Expanding $r(q)$ near q_c it is easy to obtain

$$\rho \sim \left(\frac{d^3r}{dq^3}\right)(q - q_c)^{-2}, \quad r = r_c + \frac{1}{6}\left(\frac{d^3r}{dq^3}\right)(q - q_c)^3.$$

Thus $\rho \sim (r - r_c)^{-2/3}$ when $t = t_c$.

Going to Fig. 3 one would obtain $\rho = \infty$ at points of maximum (q_1) and minimum (q_2) by formal application of the formulas. But the particle q_1 cannot reach the $r_{\max}$ shown on the curve; to do this, it would have to jump over the particle q_c. Suppose that the density of the disc is infinite: then all particles reaching $r = r_c$ will abruptly come to a standstill by encounter with the disc. But this produces a receding shock-wave. In the encounter, the kinetic energy of relative motion is transformed into heat. The matter is no longer cold; its density is just 4 times the density before the shock. As $t-t_c$ grows, the velocity of impact grows $\sqrt{t-t_c}$, but the density of matter going into the shock decreases as $(t-t_c)^{-1}$. The pressure $\sim \rho u^2$ remains constant in the first approximation (this means that there is no power dependence on $t-t_c$).

Some physical quantities are evaluated below for the case $\Omega = 1$. The linear scale of the perturbations is given by the corresponding characteristic mass M and the amplitude of the perturbations by the moment t_c or the corresponding redshift Z_c. Only orders of magnitude are given; detailed calculations are postponed to a comprehensive to be published in *Astrophysica*. The surface matter density $\sigma\ g/\mathrm{cm}^2$ is given by

$$\sigma = 10^{-4}\left(\frac{M}{10^{12}\,M_\odot}\right)^{1/3}(1+Z_c)^{3/2}(Z_c-Z)^{1/2}.$$

The impact velocity v (cm/s) is

$$v = 10^7(M/10^{12}\,M_\odot)^{1/3}\sqrt{Z_c-Z}.$$

The pressure in the compressed matter is given by

$$\mathcal{P} = 2.5\cdot 10^{-16}(1+Z_c)^4(M/10^{12}\,M_\odot)^{2/3}.$$

The maximum density (in the adiabatically compressed matter) is

$$\rho_{\max} = 4\cdot 10^{-25}(1+Z_c)^{12/5}(M/10^{12}\,M_\odot)^{2/5}.$$

Minimum density corresponds to the pressure given above and a temperature of the order of 5000°K—at higher temperatures the radiation energy losses are great and the temperature drops. This gives

$$\rho_{\min} \simeq 5\cdot 10^{-28}(1+Z_c)^4(M/10^{12}\,M_\odot)^{2/3}.$$

For

$$M = 10^{12}\,M_\odot, \quad Z_c = 4, \quad t_c = 1.5\cdot 10^{16}\,\mathrm{s},$$

we have

$$\rho_{\max} = 2\cdot 10^{-23}, \qquad \rho_{\min} = 3\cdot 10^{-25}.$$

Evaluating the formation of a star with the Schmidt formula [4] one obtains a negative result, the time needed being greater than t_c.

The stars are formed at a later stage and not simultaneously with the disc.

c) In the most favorable case, the approximate solution describes a definite part of the evolution. But it is certainly not intended to cover the whole theory of formation of galaxies. The discs are not in equilibrium. The deformation in the plane of the discs, given by the parameters β and γ (2), occurs in both directions. Nearly half of the matter, compressed in the discs, is not gravitationally bound.

It seems that the formation of discs is an unavoidable result of a definite set of assumptions about the initial perturbations. But even if this hypothesis is confirmed, it is difficult to predict how much of the discs would remain in the structure of nearby contemporary galaxies.

Another problem as yet unsolved is the possible application of the hypothesis to fluctuations of entropy, i.e., fluctuations of matter density at constant temperature. It is assumed that the spectrum is decreasing, so that after recombination the most important fluctuations are those on the scale of the Jeans' mass ($10^5 - 10^6\, M_\odot$). On this scale the gas pressure is important and the approximate solution is inapplicable. Pressure works against the unilateral compression; the formation of protostars [2] or globular clusters [3] is plausible. But nevertheless the fluctuations of greater scale remain, and for them the gas pressure is negligible.

On a greater scale the fluctuations are smaller; therefore they lead to condensation later, after the globular clusters are formed. The question is, what part of the matter has gone into globular clusters? Do they now heat the remaining gas? And finally: should one apply the approximate solution to the "gas" whose atoms are globular clusters or protostars or small gas clouds?

4. Mathematical Appendix

Explicit formulas can be given for the functions $a(t)$ and $b(t)$ of (1). It is convenient to take as independent variable the redshift Z instead of the time. These are connected by

$$t = \frac{1}{H_0} \int_Z^\infty \frac{dZ}{(1+Z)^2\sqrt{1+\Omega Z}}, \tag{A.1}$$

where H_0 is the present value of the Hubble parameter $\sim 100\,\mathrm{km/s}$ megaparsec.

It is easy to make the integration

$$H_0 t = \frac{\sqrt{1+\Omega Z}}{(1-\Omega)(1+Z)} \frac{\Omega}{2(1-\Omega)^{3/2}} \ln \frac{\sqrt{1+\Omega Z}+\sqrt{1-\Omega}}{\sqrt{1+\Omega Z}-\sqrt{1-\Omega}}, \tag{A.2}$$

but to extract the limiting cases ($\Omega = 0$, or $\Omega = 1$, or $\Omega \ll 1$, $\Omega Z > 1$, etc.) it is better to use the integral directly.

The function $a(t)$ [see (1)] is replaced by

$$a(Z) = \frac{1}{1+Z} \tag{A.3}$$

The Lagrangian variable $\mathbf{q}$ is defined by (A.3) so that it coincides with $\mathbf{r}$ in the case of unperturbed motion at the present time ($Z = 0$). The growth of perturbations is given by

$$b(Z) = \sqrt{1+\Omega Z} \int_Z^\infty \frac{dZ}{(1+Z)^2\sqrt{(1+\Omega Z)^3}}. \tag{A.4}$$

We recall that Z decreases with increasing time. For $\Omega = 1$ (flat universe)

$$t = \frac{2}{3} H_0^{-1} (1+Z)^{-3/2}; \quad a(Z) = \frac{1}{1+Z} = \left(\frac{3}{2} H_0 t\right)^{2/3} \tag{A.5}$$

and

$$b(Z) = \frac{2}{5}(1+Z)^{-2} = \frac{2}{5}\left(\frac{3}{2}H_0 t\right)^{4/3}.$$

For $\Omega \ll 1$, the growth of perturbations effectively stops at $Z \sim 1/\Omega$; for the ratio $b(t)/a(t)$ or $b(Z)/a(Z)$ increases more slowly than a power of $(1+Z)^{-1}$ when Z decreases after $Z = 1/\Omega$. For the sake of completeness, the second (damped) mode of perturbations is given:

$$b_d(Z) = \sqrt{1+\Omega Z} \equiv \frac{da}{dt} \equiv \frac{da}{dZ} : \frac{dt}{dZ}. \tag{A.6}$$

The perturbations are given in terms of displacements ($\mathbf{r}$), but not the usual $\delta\rho/\rho$; from (2), it is clear that $\delta\rho/\rho \sim b/a \sim f(t)$ in the linear stage. All expressions given above are written for matter without pressure after recombination, $Z < 1400$. If $\Omega \leq 0.03$, the density of radiation remains greater than the matter density for some time after recombination and the gravitational action of radiation changes the expansion law $a(t)$ or $a(Z)$ and the perturbation law $b(Z)$; $b_d(Z)$ is also changed in the interval $1400 > Z > 40\,000\Omega$.

Suppose that at the moment of recombination, $Z = Z_r = 1400$, the perturbations are given by $\delta\rho/\rho = \delta(\mathbf{r})$, $\mathbf{v} = \mathbf{v}(\mathbf{r})$. They are assumed to be small; for simplicity the case $\Omega > 0.03$ is considered.

To obtain the function $\mathbf{p}(\mathbf{q})$ of (1), it is advisable to work with Fourier decomposition. First we transform from $\mathbf{r}$ to $\mathbf{q}$ by $\mathbf{r} = a\mathbf{q} = \mathbf{q}/(1+Z)$ (here at Z_r perturbations can be neglected).

Having constructed $\delta(\mathbf{q})$ and $\mathbf{v}(\mathbf{q})$ (all for $t = t_r$, $Z = Z_r$ at the moment of recombination), we decompose them:

$$\delta(\mathbf{q}) = \int \delta(\mathbf{k}) e^{i\mathbf{q}\cdot\mathbf{k}} d^3\mathbf{k},$$

$$\mathbf{v}(\mathbf{q}) = \int (\mathbf{n} v_l(\mathbf{k}) + l_1 v_t'(\mathbf{k}) + l_2 v_t''(\mathbf{k})) e^{i\mathbf{q}\cdot\mathbf{k}} d^3\mathbf{k}, \tag{A.7}$$

where

$$\mathbf{n} = \frac{\mathbf{k}}{|\mathbf{k}|}, \quad l_1 \perp \mathbf{n}, \quad l_2 \perp l_1 \quad \text{and} \quad l_2 \perp \mathbf{n}.$$

v_l is the longitudinal component and v_t' and v_t'' the two transverse components of the velocity. The growing perturbation is given at Z_r by Doroshkevich and Zeldovich [1] as:

$$\begin{aligned} \delta_g(\mathbf{k}) &= \frac{3}{5}(\delta(\mathbf{k}) + i|\mathbf{k}| v_l(\mathbf{k})(1+Z_r)), \\ v_g(\mathbf{k}) &= \mathbf{n}\frac{3}{5}\left(v_l(\mathbf{k}) + \frac{i\delta(\mathbf{k})}{|\mathbf{k}|(1+Z_r)}\right). \end{aligned} \tag{A.8}$$

The displacement corresponding to the growing mode (taking into account the fact that $\Omega Z_r \gg 1$) is given by

$$\mathbf{r} - a\mathbf{q} = \int v(\mathbf{q}, t) dt = \frac{3}{4}\mathbf{v}_1(\mathbf{q}, t) t = b\mathbf{p}(\mathbf{q}). \tag{A.9}$$

Because $v_g \sim t^{1/3}$ and applying (1) to $t = t_r$, we obtain

$$\mathbf{p}(\mathbf{q}) = \int \mathbf{p}(\mathbf{k}) e^{i\mathbf{q}\cdot\mathbf{k}} d^3\mathbf{k}, \tag{A.10}$$

with

$$\mathbf{p}(\mathbf{k}) = \frac{t_r}{b(Z_r)} \frac{9}{20} \mathbf{n} v_l(\mathbf{k}) + \frac{i\delta(\mathbf{k})}{|\mathbf{k}|(1+Z_r)} = \mathbf{n}\eta(\mathbf{k}).$$

The $v_l(\mathbf{k})$ and $\delta(\mathbf{k})$ without indices are taken at $t = t_r$. After neglecting the transverse displacement (even if $v_l \sim v_t' \sim v_t''$ at $t = t_c$, thereafter v_l grows while v_t' and v_t'' decrease), $\mathbf{p}(\mathbf{q})$ is vortex-free:

$$\frac{\partial p_x}{\partial q_y} = \int n_x k_y \eta(\mathbf{k}) e^{i\mathbf{q}\cdot\mathbf{k}} d^3\mathbf{k} = \int \frac{k_x k_y}{|\mathbf{k}|} \eta(\mathbf{k}) e^{i\mathbf{q}\cdot\mathbf{k}} d^3 k = \frac{\partial p_y}{\partial q_x}. \tag{A.11}$$

Therefore $\mathbf{p}(\mathbf{q})$ can be written as the gradient of a scalar function $\xi(\mathbf{q})$:

$$\mathbf{p}(\mathbf{q}) = \mathrm{grad}_q\, \xi(\mathbf{q}), \tag{A.12}$$

with the Fourier image $\xi(\mathbf{q}) = \frac{1}{|\mathbf{k}|}\eta(\mathbf{k})$.

It can be shown that the peculiar velocity given by the approximate solution is also vortex-free in physical space, as a function of the Eulerian ($\mathbf{r}$) coordinates. So is the Hubble velocity, and therefore the total velocity also. The exact solution has this property because the motion occurs under the action of gravitation—a force with potential. The approximate solution has this property even when the perturbations are not small ($\mathbf{r} - a\mathbf{q}$ not neglected). This is one more argument in favor of the approximate solution. With a gaussian probability distribution of $\delta(\mathbf{q})$, $v(\mathbf{q})$, and the Fourier-components of δ, v and ψ, the distribution of α, β and γ defined by (2) is not gaussian. As found by Doroshkevich,

$$W(\alpha,\beta,\gamma) \sim (\alpha-\beta)(\alpha-\gamma)(\beta-\gamma)\, \exp\{-m^2[\alpha^2+\beta^2+\gamma^2-\frac{1}{2}(\alpha\beta+\alpha\gamma+\beta\gamma)]\}. \tag{A.13}$$

The probability that all of them are positive ($\alpha > \beta > \gamma > 0$) is 8%, and the probability that $\alpha > \beta > 0 > \gamma$ is 42%; by symmetry, there is a 42% probability that $\alpha > 0 > \beta > \gamma$, and 8% that $0 > \alpha > \beta > \gamma$.

I am grateful to V. F. Djachenko and A. G. Doroshkevich for valuable ideas and discussion and to V. Chechetkin for help.

Institute of Applied Mathematics
USSR Academy of Sciences. Moscow

Received
September 19, 1969

REFERENCES

1. *Doroshkevich A. G., Zeldovich Ya. B.*—Astron. Zh. **40**, 807 (1963).
2. *Doroshkevich A. G., Zeldovich Ya. B., Novikov I. D.*—Astron. Zh. **44**, 295 (1967).

3. *Peebles P. J. E., Dicke R. H.*—Ap. J. **154**, 891 (1968).
4. *Schmidt M.*—Mem. Soc. Roy. Liége **3**, 130 (1959).
5. *Silk, J. I.*—Ap. J. **151**, 459 (1967).

*Commentary**

The importance of this paper, which today is considered one of the foundations of the theory of the large-scale structure of the Universe, only gradually came to be recognized over a period of fifteen years.

Formally the article simply proposes an approximate solution to the nonlinear problem under certain simplifying assumptions (neglect of the pressure). It was not clear how correct the initial assumptions were and what errors the approximation introduced. Since then answers have been found to both questions. Modern cosmology leads to the conclusion that the primordial perturbations were adiabatic and that there were no vortical or entropic perturbations. Ya. B. foresaw this (cf. his report at the Tallin symposium on the large-scale structure of the Universe in 1978—Article **51** in the present volume). Most recently these conclusions have been confirmed by, for example, A. Guth,[1] A. D. Linde[2] and A. A. Starobinskiĭ.[3] The decay of short-wavelength adiabatic perturbations implies that neglecting the pressure for the surviving long-wavelength perturbations is justified.

It should, however, be noted that very recently (1984) there have been attempts to revive the theory of entropic perturbations. In theories with a "cold" hidden mass of heavy particles, the scale of the surviving perturbations may be so small that the pressure of neutral hydrogen and helium can be significant. These hypotheses do not detract from the value of Ya. B.'s results on the behavior of matter under the usual assumptions about its properties.

The accuracy of the approximation has been investigated by A. G. Doroshkevich, V. S. Ryabenkiĭ and S. F. Shandarin.[4] They showed that the error remains finite up to the moment that the infinite density characteristic of this problem and the proposed solution is attained. The solution has gained the interest of mathematicians and served as a starting point for a systematic investigation of Lagrangian maps (see books by V. I. Arnold[5] and by V. I. Arnold *et al.*[6] and Section 12, *Mathematics in the Work of Ya. B. Zeldovich*, of the Introduction to the first volume of these Selected Works).

The approximate solution correctly describes the first singularities (caustics, catastrophes) which arise in the problem. Later Arnold, Shandarin and Ya. B.[7]

*Article 55 in the original Soviet edition of this collection was a somewhat longer and later paper in Russian on the same topic. It is to that paper [*Zeldovich Ya. B.* Astrofizika **6** (2), 319–335 (1970)] that the commentary is directed specifically, however, the issues covered are equally pertinent to the version reprinted here.

[1] *Guth A.*—Phys. Rev. D **23**, 347–356 (1981).

[2] *Linde A. D.*—Phys. Lett. B **180**, 389 (1982).

[3] *A. A. Starobinskiĭ*—Phys. Lett. B **91**, 99 (1980).

[4] *Doroshkevich A. G., Ryabenkiĭ V. S., Shandarin S. F.*—Astrofizika **9**, 257–272 (1973).

[5] *Arnold V. I.*—Catastrophe Theory. Springer (1986).

[6] *Arnold V. I., Varchenko A. N., Gusein-Zade S. M.*—Singularities of Differentiable Mappings. Birkhäuser. V. 1 (1985), V. 2 (1988).

also investigated higher order singularities. The astrophysical aspects of the process of formation of the structure of the Universe, based on the approximate solution, are considered in the work by Ya. B. and R. A. Sunyaev (Article **56**) and in later works by Ya. B.'s students—A. G. Doroshkevich, S. F. Shandarin, A. A. Klypin, by the Estonian astronomers J. Einasto and E. Saar, by the groups of J. Silk and S. Szalai, and by others.

Using the approximate solution obtained for the analysis of later stages, Ya. B. and S. Shandarin predicted the appearance of "black" empty regions in the distribution of galaxies in space, and Ya. B. further predicted a cellular structure for the distribution of matter. Subsequently Shandarin[8] proposed the percolation method for objectively determining the tendency of galaxies to lie along lines on surfaces. The application of this method appears to confirm the entire theoretical concept which is based on the present article.

We note finally that the assumption of a "hidden mass" consisting of weakly interacting particles does not change this picture: Ya. B. and his students considered the scenario of neutrinos with non-zero rest mass in papers published in the present collection (**59–61**). For much heavier particles (gravitinos, axions) the characteristic scale may be less than the Jeans wavelength of neutral hydrogen. In this case the heavy particles will form "pancakes," in accord with the solution of the present paper, but the distribution of gas and stars will be more complex.

In the last few years the solution presented here has been checked directly by numerical simulation.[9]

In sum, it may be safely said that this article has withstood the test of time and has had much fruitful influence both on the development of cosmology and on the development of a number of areas in mathematics.

This paper clearly asserts that the spherically symmetrical growth of perturbations considered in many papers due to its simplicity, is actually a degenerate and atypical case. It is further pointed out that the region bounded by a caustic is a triaxial ellipsoid in Lagrangian space with $\alpha = \text{const}$ and with all three axes of the same order of magnitude, proportional to $\sqrt{t - t_c}$, where t_c is the moment of birth of the caustic. However, one-sided compression of matter in a single direction (which need not coincide with an axis of the ellipsoid caustic) transforms the ellipsoid into a "pancake" whose thickness $\approx (t - t_c)^{3/2}$.

The present state of the theory is laid out in a detailed review article by Shandarin, Doroshkevich and Ya. B.[10]

One of the most authoritative proponents of the theory outside the USSR is J. Oort.[11]

[7] *Arnold V. I., Shandarin S. F., Zeldovich Ya. B.*—Geophys. and Astrophys. Fluid Dan. **20**, 111 (1982).

[8] *Shandarin S. F.*—Pisma v Astron. Zh. **9**, 195 (1983); *Zeldovich Ya. B.*—In: Highlights of Astronomy. Dordreht **6**, p. 29 (1982).

[9] *Klypin A. A., Shandarin S. F.*—Preprint of the Inst. of Appl. Math. AS USSR **136**, Moscow (1982); *Klypin A. A., Shandarin S. F.*—Month. Not. RAS **204**, 891–907 (1983); *Frenk C. S., White S. D. M., Davis M.*—Astrophys. J. **271**, 417–430 (1983); *Delek A. Aarseth S. J.*—Astrophy. J. **283**, 1–23 (1984).

[10] *Shandarin S. F., Doroshkevich A. G., Zeldovich Ya. B.*—UFN **139**, 83–134 (1983).

[11] *Oort J. H.*—Ann. Rev. Astron. Astrophys. **21**, 373 (1983); *Oort J. H.*—Astron. and Astrophys. **139**, 211, 214 (1984).

56

Formation of Clusters of Galaxies: Protocluster Fragmentation and Intergalactic Gas Heating*

With R. A. Sunyaev

Shock waves appearing in the nonlinear stage of growth of the initial density perturbations give a distinctive distribution of temperature and density for a condensed gas. The scale of the perturbations $M \sim 10^{12} - 10^{14}\, M_{\odot}$ is typical of the theory of adiabatic density perturbations and is associated with protoclusters of galaxies. The epoch of formation of protoclusters of galaxies is chosen to correspond to the redshift $z \sim 3 - 5$, in agreement with results for quasars and powerful radio sources counts.

Part of the gas (1%) condenses adiabatically and has a low temperature and high density. A hypothesis on the origin of dense bodies (quasars and galaxies nuclei) from such gas is advanced. After heating, about 20% of the gas cools to 10^4 K, and it was assumed that this gas goes into galaxies. The remaining gas is heated up to 10^6 K, which explains the observed complete ionization of the intergalactic gas.

The paper deals with dynamics of shock waves, heat processes in the gas, its radiation and contribution to the X-ray background as well as the process of fragmentation of the cold gas to form the above mentioned objects, starting from the cosmological model of the hot Universe with small density perturbations.

We are trying to give a consistent theoretical picture based on modern cosmology of galaxy formation. The unperturbed Universe is undergoing expansion, accompanied by decrease of radiation temperature and matter density. At some epoch nuclei (p, α) and electrons associate into neutral atoms. The idealized picture would give cold neutral structureless gas.

We ask for smallest possible perturbations needed to build the observed structure of the Universe. The adiabatic perturbations are chosen ($\delta\rho/\rho = 3\delta T/T$) of initial amplitude of the order of $\delta\rho/\rho \sim 10^{-4}$ before recombination in the scale[1] corresponding to masses $10^{14}\, M_{\odot}$.

*Astronomy and Astrophysics **20** (2), 189–200 (1972).

[1]Perturbations of smaller scale are irrelevant because they are damped out before recombination; observations suggest that perturbations of greater scale are even much weaker.

Small density and velocity perturbations in the initial neutral gas after the recombination of matter in the Universe are increased because of gravitational instability. During the stage of large perturbations, the theory predicts the formation of plane concentrations ("pancakes") and compression of most of the primeval gas by shock waves. Detailed consideration of the hydrodynamic and, mainly, thermal picture lead to conclusions essential to the understanding of the contemporary structure of the Universe. These conclusions generally agree with observations.

Gas heating by shock waves explains the absence of neutral hydrogen in intergalactic space: as is known, observations of quasar spectra having redshifts $z \gtrsim 2$ lead to such conclusions. Inside the first protoclusters, a smooth distribution of temperature from ~ 300 K, in the plane of symmetry, to 10^6 K at the end arises initially due to the fact that the velocity of in-falling gas increases during the process of protocluster formation. In the process of further cooling, the gas is divided sharply into an internal layer having a temperature $T \lesssim 10^4$ K and two external layers with a temperature $T \sim 10^4$ K surrounding it.

Apparently, fragmentation of the internal layer ($T \lesssim 10^4$ K) leads to protogalaxy formation. Estimations give reasonable values for the expected masses.

In the first protoclusters one can distinguish the "most internal" layer with $T \sim 300$ K and a high density of 30 atoms per cm^3 subjected only to adiabatic contraction. One can assume that this layer becomes the galactic nuclei and quasars. Later, soft X-ray radiation of the initial protoclusters heats the total mass of the weakly-perturbed matter in the Universe and a cold "most internal" layer is absent as a rule, occurring only rarely in later plane concentrations. From this fact, one can perhaps explain the observed evolutionary effect whereby the density of radiosources and quasars has decreased a thousand times from $z \sim 2 - 2.5$ to the present time.

In general, the detailed picture of the development of initial small adiabatic perturbations is complicated, intriguing and colorful. It is hoped that this picture can explain the large-scale structure of the Universe up to galactic nuclei and quasars. There has existed an unspoken opinion for a long time that the theory of adiabatic perturbations cannot explain galactic rotation because the growing adiabatic perturbations give a potential velocity field. If this point of view represented the complete truth, then the observed rotation would prove unequivocally the occurrence of initial vorticity: the basic justification of the curl (turbulent) theories of the evolution of the Universe is probably of the same nature [1–3].

However, a series of papers [4–6] showed that, taking into account all nonlinearities, adiabatic perturbations lead to the appearance of vortical motions and to the increase of the rotational moment of protogalaxies and protoclusters of galaxies; thus an essential difficulty of the adiabatic theory is removed. Special attention should be given to the paper [6], where the creation of vorticity is explained by the presence of shock waves which are characteristic of the nonlinear picture. On the other hand the difficulties of the theory of initial vorticity are quite pronounced [7, 8]. On the whole, the results of recent years, including the results discussed below, confirm our confidence that the theory of adiabatic perturbations is "necessary and sufficient" to describe the structure of the Universe.

The scheme of evolution of adiabatic density perturbations at the nonlinear stage is discussed below. The calculations are made without taking into account in detail the statistical character of the problem. Therefore the numerical values given below must be considered as order of magnitude estimates.

I. Thermal Picture

1. Evolution of Density Perturbations on a Nonlinear Stage. The paper ([9], quoted below as A) gives the approximate solution of matter density perturbation behavior in the Universe at the nonlinear stage when the perturbations of density are no longer small compared with the mean density of matter. At this stage, importance is attached to the absence of spherical symmetry of the perturbations and to differences in the amplitudes (and sometimes in the sign!) of velocities (connected with the perturbations) in different directions. Insofar as the density perturbations are small, these velocities have only a weak influence on the universal Hubble expansion of matter. Perturbations of density grow in accordance with a linear theory $\Delta\rho/\rho \sim t^{2/3} \sim (1+Z)^{-1}$, and together with them grow also perturbations of velocity proportional to $t^{1/3}$ ($\Delta\rho/\rho$ and the velocities are connected with the equation of continuity). When, in some part of the matter, $\Delta\rho/\rho$ is positive and ~ 1 a contraction exists, at least on one axis, and this matter separates from the residual expanding background and, moreover, it is compressed, forming a disc: the matter falls onto the disc along the axis where initial velocity of contraction was the greatest. There may occur at that time contraction as well as expansion along two other axes lying in the plane of the disc. As was shown by Doroshkevich [10] (in the reasonable approximation of a Gaussian distribution of perturbation amplitudes) a fraction of about 8% of the total amount of matter contracts along all three main axes of the local tensor of deformation; 42% of the matter contracts along two axes and expands along the third one; 42% contracts along one axis and expands along the other two; and only 8% of the matter expands along all three axes. Therefore, the matter forming a disc in some cases may form a gravitationally bound system and in other cases it may expand and then dissipate. A small fraction of the matter in the disc contracts only adiabatically and reaches a high density. The next gas layers run across the gas already compressed and a shock wave running along the falling matter is formed. A similar qualitative treatment was given by Oort [24], where earlier references are also given.

2. Parameters of a Shock Wave. The velocity at which the matter runs against a disc (see A and the appendix) is equal to

$$U = \frac{\lambda}{2\pi} H_0 \left(\frac{M}{10^{13} M_\odot}\right)^{1/3} (1+z_c)^{1/2} (\pi\mu)^{1/2} (\sin\pi\mu)^{1/2}$$

$$= 9.3 \cdot 10^6 (1+z_c)^{1/2} (\pi\mu)^{1/2} (\sin\pi\mu)^{1/2} \left(\frac{M}{10^{13} M_\odot}\right)^{1/2} \text{ cm/sec} \quad (1)$$

at small $\mu \ll 1$

$$U = 2.9 \cdot 10^7 (1+z_c)^{1/2} \left(\frac{M}{10^{13} M_\odot}\right)^{1/3} \mu \text{ cm/sec.}$$

Here μ is the fraction of the matter within the given scale which has been subjected to contraction; redshift z_c corresponds to the beginning of condensation.

Formula (1) is obtained by considering the perturbations and velocity as plane sinusoidal waves. Wave length λ is connected with the typical scale of mass M by the formula $M = \rho_0(\lambda/2)^3$, $\lambda = 2(M/\rho_0)^{1/3}$. The matter density in the Universe is supposed to be critical $\rho_0 = \rho_{\text{crit}} = 3H_0^2/8\pi G = 10^{-29}\,\text{g/cm}^3$ at the Hubble constant $H_0 = 75\,\text{km/sec} \cdot \text{Mpc}$. (A) and the appendix shows that the fraction of the matter which has passed through a shock wave up to the given moment of time can be derived from the formula $\sin\pi\mu/\pi\mu = (1+z)/(1+z_c)$, $1+z = (t_0/t)^{2/3}$, $t_0 = 2/3H_0$.

The temperature of the matter at the front of a shock wave is equal to

$$T_{fr} = \frac{u^2 m_p}{6k} = 1.7 \cdot 10^5 (1+z_c) \pi\mu \sin\pi\mu \left(\frac{M}{10^{13} M_\odot}\right)^{2/3} \text{K}, \quad (2)$$

at small $\mu \ll 1$

$$T_{fr} = 1.7 \cdot 10^6 (1+z_c) \mu^2 \left(\frac{M}{10^{13} M_\odot}\right)^{2/3} \text{K}, \quad (2')$$

at $\mu \sim 1/2$, when $1+z = 2(1+z_c)/\pi$ and the temperature at the front of the shock wave is close to a maximum value

$$T_{fr} = 2.7 \cdot 10^5 (1+z_c) \left(\frac{M}{10^{13} M_\odot}\right)^{2/3} \text{K}.$$

All formulas are given for full ionization, neglecting the ionization potential $J \ll kT$, or for neutral gas. The matter density in front of the shock wave is

$$\rho_1 = \rho_0 \frac{(1+z_c)^3 (\sin\pi\mu/\pi\mu)^3}{(1 - \pi\mu/\tan\pi\mu)} \underset{\mu \leq 1}{=} 3\rho_0 \frac{(1+z_c)^3}{\pi^2\mu^2}. \quad (3)$$

Behind the shock, the density is equal to $\rho_2 = \frac{\gamma+1}{\gamma-1}\rho_1 = 4\rho_1$. Here ρ_0 is the present mean density of the matter in the Universe; $\bar{\rho}$ is the current mean density of the matter in the Universe, $\rho_1 = \bar{\rho}$ at $\mu = 1/2$ and $1 + z = 2(1 + z_c)/\pi$; $\gamma = 5/3$ is an isentropic exponent. Knowing ρ_1 and u we find the pressure which is attained in the shock wave and which stops the motion of matter.

$$p_{fr} = \rho_1 u^2 = 8.6 \cdot 10^{-16} \frac{(1 + z_c)^4 \sin^4 \mu\pi}{(\pi\mu)^2(1 - \pi\mu/\tan\pi)} \left(\frac{M}{10^{13} M_\odot}\right)^{2/3}$$

$$\underset{\mu \ll 1}{=} 2.5 \cdot 10^{-15}(1 + z_c)^4 \left(\frac{M}{10^{13} M_\odot}\right)^{2/3} \text{dyne/cm}^2. \tag{4}$$

Pressure on the disc surface does not depend (as long as μ is small) on μ—the fraction of the matter which has entered the disc—but depends only on the scale of the perturbations. As μ increases the pressure drops, decreasing 7.5 times for $\mu = 1/2$. The relations given above are obtained in a very rough approximation considering one plane wave. At best (if systematic errors are small) they give mean values of the parameters which, in reality, have a large dispersion of values.

3. Spectrum of Adiabatic Density Perturbations. In spite of the fact that a nonlinear theory of galaxy formation (if $M \gg M_j$, where M_j is a Jeans mass) is applicable to the density perturbations of any nature—entropy, turbulence, etc.—below we shall discuss adiabatic perturbations only. The existing theory gives a typical scale for these perturbations in good agreement with observations.[2]

No matter what is the initial (in the radiation dominated epoch before recombination, when $\rho_r > \rho_m$, $z > 2 \cdot 10^4 \Omega$, where $\rho_r = \sigma T^4/c^2$ is the density of radiation, $\rho_m = \rho_0(1 + z)^3$ is the matter density) spectrum of adiabatic density perturbations, in the stage after recombination the perturbations are small on small scales. In the radiation dominated stage, when the perturbations are acoustical waves, they are damped on the small scales due to the effects of radiative heat conductivity, viscosity and friction of matter against radiation. Silk (1968) has shown that, assuming recombination to be instantaneous, the perturbations are damped on a scale less than

$$M_d = 7 \cdot 10^{10} M_\odot, \text{ if } \Omega = 1, \text{ and} M_d = 5 \cdot 10^{11}, \text{ if } \Omega = 1/30.$$

Here $\Omega = \rho/\rho_{\text{crit}}$ is the dimensionless mean density of matter in the Universe. The most accurate and complete computations, carried out recently by Peebles and Yu (1970), give $M_d = 10^{12} M_\odot$ at $\Omega = 1$ and $M_d = 10^{14} M_\odot$ at $\Omega = 1/30$. Chibisov (1972) obtains that at

$$M < M_d = 5 \cdot 10^{13} \Omega^{-1/2} M_\odot$$

[2]A turbulent theory of galaxy formation (Ozernoi and Chernin, 1968) also considers shock wave heating of matter. However, in this case the main energy release occurs rather early after recombination with redshifts ranging from $z \sim 100$ (Ozernoi and Chibisov, 1970) to $z \sim 1000$ (Peebles, 1971).

The adiabatic density perturbations are damped. As the spectrum of the initial density perturbations cannot be growing in the direction of very large masses—objects with $M \gg 10^{16} M_\odot$ are not observed and small scale fluctuations of the relic radiation are small at the angular scales corresponding to $M \gg 10^{16} M_\odot$ [13]—then the maximum of the perturbation spectrum after recombination should correspond to $M \sim 10^{12} - 10^{14} M_\odot$, the scale typical for clusters of galaxies.

4. The Epoch of Formation of Clusters of Galaxies. Assume that, at $z_c \sim 3 - 5$, in the Universe there takes place an intense condensation of clusters of galaxies. Displacing this process to large redshifts leads to contradiction with observations. The mean distance between clusters of galaxies exceeds by only 5–7 times their sizes and a rough estimate shows that at $z \sim 4 - 6$ they should have closed up and filled the whole volume. The counts of quasars and powerful radiosources are also in accordance with a maximum rate of objects formation at $z \sim 2 - 4$ [14, 15]. (See also more detailed discussion by Sunyaev [16], referred to below as B. In B it is also noted that the objects condensation at $z_c \sim 2 - 4$ corresponds to the assumption of $\Omega > 0.2$ because, according to the linear theory an effective growth of density perturbations occurs only at redshifts $1 + z > 1/\Omega$.)

5. Distribution of Temperature and Density in a Gas Compressed by a Shock Wave. The choice of a typical value of the mass $M \sim 10^{13} M_\odot$ and the time of condensation $z_c \sim 4$ allows us to determine the physical properties of the gas which has passed through the shock wave. We should point out the weak dependence of such properties (first of all of the density) on the mass of the condensing gas and the strong dependence on the chosen value of z_c.

A small fraction of gas condenses adiabatically. Nonperturbed matter had $T \sim 4000$ K at the moment of recombination of hydrogen in the Universe, $z_{\rm rec} \sim 1400$. Its temperature changes as the temperature of radiation according to adiabatic curve with $\gamma = 4/3$ up to $z \sim 200$ and $T \sim 550$ K [17, 18]. Later on, exchange of energy between matter and radiation becomes deficient and the cooling of the matter is given by an adiabatic curve with $\gamma = 5/3$. As a result, when $z_c = 4$ the unperturbed matter had density $\tilde{n} \sim 10^{-3}$ cm^{-3} and temperature $\tilde{T} \sim 0.3$ K.

In an approximation in which the initial temperature of the gas is set equal to zero the shock wave arises immediately after the first two neighboring trajectories have intersected. In this case, the density is adjusted to the definite value P_{fr} at once, when the gas density before the front is proportional to μ^{-2}, $T_{fr} \sim \mu^2$, so that $T_{fr}(\mu = 0) = 0$. Let us take the initial temperature and pressure of the gas different from zero. They do not influence the shock wave propagation and do not change P_{fr} when the wave moves a significant distance and its amplitude is large. However, the initial stage will essentially change; the initial pressure will prevent the appearance of an infinite value of density and the instantaneous formation of a shock wave at $\mu = 0$. An adiabatic contraction in the range from $\tilde{n}$, $\tilde{T}$, p up to P_{fr} will give $n_{\max} = \tilde{n}(p_{fr}/\tilde{p})^{3/5} = 30((1+z_c)/5)^{12/5} \cdot (M/10^{13} m_\odot)^{2/5}$ cm^{-3} and the corresponding temperature $T_{\max}$ of the order of $300((1+z_c)/5)^{8/5}(M/10^{13}M_\odot)^{4/15}$ K. The shock wave will arise only for μ such that

$$T_{fr} = \text{const}\mu^2 \geq T_{\max}. \tag{5}$$

Therefore, the central thin layer having $\mu \leq 1.3 \cdot 10^{-2}((1+z_c)/5)^{3/10} \cdot (M/10^{13}M_\odot)^{1/5}$ must contract adiabatically.

Compressed matter is like a sandwich with the central fraction being cool and dense. It was never heated by the shock wave. The external layers are heated by the shock wave. Thus, when $\mu \sim 10^{-2}$ two shock waves are formed moving towards the falling matter from the central plane of the disc. Near $\mu_1 \sim 6 \cdot 10^{-2}(5/(1+z_c))^{1/2}(M/10^{13}M_\odot)^{-1/3}$ the temperature approaches 10^4 K (a significant level, characterized by strong ionization of hydrogen by electron collisions and rapidly increasing energy losses due to radiation) and the density is $n \sim 1$ cm^{-3} $((1+z_c)/5)^4(M/10^{13}M_\odot)^{2/3}$. After complete ionization, at $\mu > \mu_2 = 3\mu_1$, if we ignore the cooling, the temperature must grow with increasing μ and, when $\mu \sim 1/2$, it approaches $T \sim 1.35 \cdot 10^6((1+z_c)/5) \cdot (M/10^{13}M_\odot)^{2/3}$ K. The ionization relaxation time as well as that of energy transfer between electrons, atoms and ions in the situation considered are small compared with the cosmological one. At $\mu \ll 0.1$, the existence of a small fraction of electrons $e/H \sim 10^{-5} - 10^{-4}$ in the primeval plasma [17] is important. At $\mu > 0.1$ the ionizing radiation from the shock front is important for the appearance of free electrons in gas in front of the shock. Their existence strongly reduced the relaxation time. In fact, in the compressed matter behind the shock wave the temperature will decrease rapidly to 10^4 K because of radiative cooling. The ratio of the cooling time t_c to the hydrodynamic time t_h is

$$\frac{t_c}{t_h} = 3kT_e\frac{\sqrt{4\pi G\rho}}{L^-} = 4 \cdot 10^{-4}\frac{T_e^{1/2}}{n_e^{1/2}} \underset{\mu \ll 1}{\approx} 40\mu^2 \tag{6}$$

at $M = 10^{13}M_\odot$ and $z_c = 4$. The cooling rate due to free-free radiation $L^- \approx 10^{-27}n_eT_e^{1/2}$ erg/sec is chosen for estimation. From (6) follows that $\mu \sim 150$, when $\sim 2 \cdot 10^5$ K, is a critical value (an account of other mechanisms of cooling[3] may increase this critical temperature to $5 \cdot 10^5$ K). During the cosmological time

$$t_{\text{cosm}} \approx \frac{2}{3}(1+z)^{-3/2}H_0^{-1} = \frac{2}{3}H_0^{-1}(1+z_c)^{-3/2}\left(\frac{\pi\mu}{\sin\pi\mu}\right)^{3/2}$$

only a small fraction of the matter $\mu < 1/3$ passing through the shock

[3]This free-free process is most important at high temperatures $T > 10^5$ K; the initial temperature leading to cooling depends on free-free radiation. At lower temperature ($T \sim 10^4 - 10^5$ K) the free-bound and bound-bound radiation are important, but after recombination at $T \sim 10^4$ K, they are switched off. The overall picture (Fig. 2) is not sensitive to radiative processes at low temperatures.

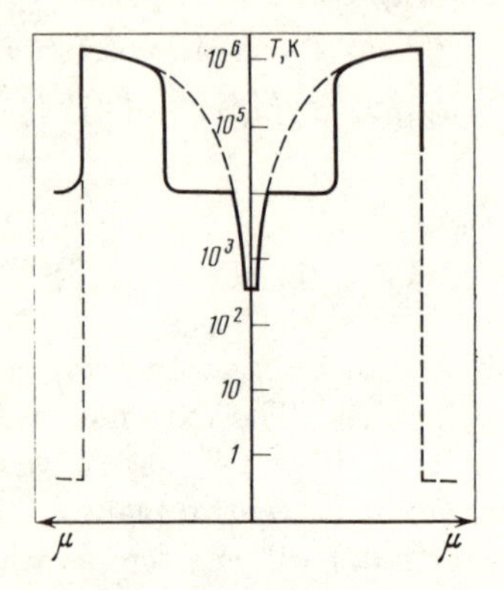

Fig. 1. Temperature distribution in a contracting protogalaxy with $M = 10^{13} M_\odot$ over a time $\mu = 1/2$ and redshift $z_c = 4$. The dotted line is the same distribution without accounting for the effects of radiation cooling.

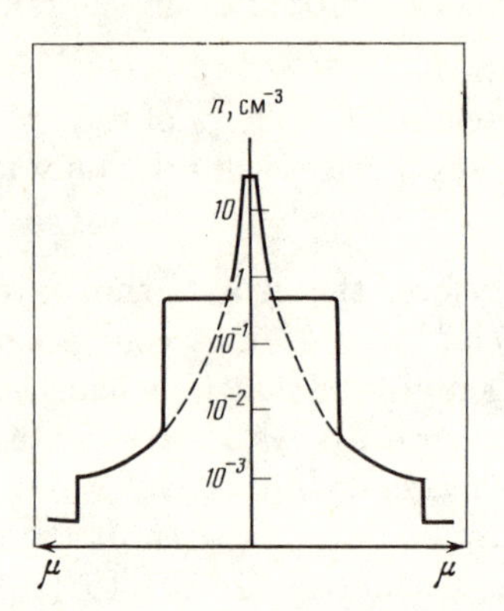

Fig. 2. Density distribution in a gas compressed by a shock wave.

wave becomes cold and the hottest gas is not able to become cold during the lifetime of the Universe. At the same time, the dense matter gets cold rapidly from $T \sim 5 \cdot 10^5$ K to $T \sim 10^4$ K. The time of fragmentation of the cold matter to protogalaxies several times exceeds the hydrodynamic one. It can be shown that only about $\mu \sim 0.25$ fraction of the gas cools to $T \sim 10^4$ K and forms galaxies. Let us recall that the pressure on the shock wave front is constant only when $\mu \ll 1$, and then it begins to decrease. If $\mu = 0.25$ it is 1.5 times less than the initial pressure. The mean density of the matter in a zone having $T \sim 10^4$ K must be constrained *to this value of the pressure and* have[4] $n \sim 0.6\text{cm}^{-3}$. As a result, the profile of density and temperature given in Figs. 1 and 2 is adjusted along the axis where contraction takes place.

Here the difficulties associated with the statistical character of the problem should be noted. Simple formulae have been derived for the initial perturbations in the form of the plane sinusoidal wave $\sin kx$ in which rapid contraction begins simultaneously in all planes $x = 0$, $x = \pm 2\pi/k$, $x = \pm 4\pi/k$, etc. In fact it is evident that some protoclusters are formed earlier than others. This is why the parameter μ should be chosen to be the fraction of the gas contracted by the wave in the neighborhood of the given protoclusters and not an average for the Universe.

[4]Once more we emphasize that all numerical values are rough and depend on the parameters (Ω, z_c) chosen; their comparative values are perhaps more reliable.

According to the formulas given in A and at the beginning of this paper, the gas in-falling to the disc must continue to do so up to $\mu = 1$, i.e., when $t \to \infty$. In fact if we obtain $\mu = 1/2$ the velocity of the in-falling matter begins to decrease and the pressure of the shock wave falls. After that it is impossible to consider the pressure to be constant in the space filled with the condensed gas. The shock wave "tears off" from the matter of high density and runs along the nonperturbed gas heating it to a high temperature (due to the low density, a high pressure is not necessary). A detailed picture requires a gas dynamical calculation, but, taking into account the statistical character of the problem as a whole, a non-idealized calculation is very difficult.

It is very important that the ultraviolet and soft X-ray radiation of the external layers of the disc heated by the shock wave is not able to influence significantly the heat balance of the central part: ionizing radiation is absorbed completely within the thin layer of matter flowing away from the slowly moving ionization front [19]. Hard X-ray radiation is unable to heat this zone on the hydrodynamic time scale which is necessary for fragmentation. At the same time, the existence of hard radiation can greatly influence the degree of ionization of the gas and decrease abruptly the energy losses due to radiation in the resonance lines of hydrogen and He II. Effects of electron heat transfer are negligibly small.

II. Fragmentation of Protoclusters of Galaxies

In the cold, dense fraction of the disc, conditions are favorable for its fragmentation to objects of small mass. In the gas which cools behind the shock wave, thermal instability certainly plays a part but the dominant process is most probably that of gravitational instability.

1. Gravitational Instability in a Medium Subject to an External Pressure. In the cold, thin disc inside the object being compressed, gravitational instability develops in specific conditions [38]: the external pressure—equal to the pressure of the shock wave front—considerably exceeds gravitational forces $({p_{fr}/p_{\text{grav}}}_{\mu \ll 1} = 4/\pi^2\mu^2)$. It is well known that, for a plane layer and the case of long wavelength waves (amplitude e^{ikx} with the wave vector $k \to 0$), the pressure is of no importance and the rate of growth of a perturbation is proportional to $e^{i\omega t}$, where $\omega = \sqrt{2\pi G\sigma k}$ and $\sigma[g/\text{cm}^2]$ is the surface density of matter in the layer.

As $\omega \sim \sqrt{k}$, then the "most dangerous" (causing fragmentation) short waves are always those for which the formula $\omega \sim \sqrt{k}$ proves to be at the limit of validity. This limit of validity is connected with the fact that the gas pressure counteracts the gravitational instability. In the case where there is no external pressure, the pressure is expressed in terms of σ; $p = p_{\text{grav}} = \pi G\sigma^2/2$ and it does not introduce any new parameters as compared to the Jeans theory for a continuous layer. The maximum of k, $k_{\max}$ evidently depends on the finite thickness of layer $2d$: with $\sigma = 2\rho d$, the Jeans wavelength is equal to $\lambda_J \approx 2d$, the Jeans mass $M_J \sim 4\pi\rho d^3$ and the maximum rate of growth of a perturbation $\omega_{\max} \approx \sqrt{2\pi G\sigma/d} \approx \sqrt{4\pi g\rho}$ is defined by the mean density of matter in the layer as in the Jeans theory for an infinite volume.

However, if a heavy layer is compressed from both sides by a light gas and the surface pressure is larger than its own gravitational pressure, gravitation can be of importance only in the radial direction. In this case, the critical wavelength increases in proportion to the pressure: the acceleration due to the pressure is of the order of $p\Delta\sigma/\rho/\lambda\sigma = r\Delta\sigma d/\lambda\sigma^2$ and depends on the wavelength whereas the acceleration due to gravity $G\Delta\sigma$ does not depend on the wavelength. Thus, $\lambda_{\max} \sim dp/g\sigma^2 \sim dp/p_{\text{grav}}$ and the corresponding mass of the perturbations which grow the most rapidly is of the order of

$$M_{\max} \approx 2\pi\lambda_{\max}^2\sigma = 4\pi d^3\rho\left(\frac{p_{fr}}{p_{\text{grav}}}\right)^2.$$

If we compare this mass with the Jeans mass

$$M_J = \pi\lambda_J^3\rho/2 \approx 5\cdot 10^2 T^{3/2} M_\odot/n^{1/2}$$

for a self-gravitating disc with the given density, we obtain

$$\frac{M_{\max}}{M_J} \sim \left(\frac{d}{\lambda_J}\right)^3\left(\frac{p_{fr}}{p_{\text{grav}}}\right)^2 \sim \left(\frac{p_{fr}}{p_{\text{grav}}}\right)^{1/2} \sim \frac{2}{\pi\mu},$$

$$M_{\max} \approx M_J\left(\frac{p_{fr}}{p_{\text{grav}}}\right)^{1/2} \approx \frac{\frac{3}{\mu}\cdot 10^2 T^{3/2} M_\odot}{n^{1/2}}. \tag{7}$$

A body formed due to such an instability originally has the form of a finite disc with thickness $2d \sim \lambda_3\mu$ and characteristic size in its plane $\lambda_{\max} \sim \lambda_J\mu^{-1} \sim d\mu^{-2}$. The characteristic time of growth of such a perturbation is

$$t \sim \frac{1}{\omega_{\max}} = \frac{\sqrt{p_f r}/p_{\text{grav}}}{\sqrt{4\pi g\rho}} = \frac{1}{\mu\sqrt{4\pi g\rho}}. \tag{8}$$

2. *Perturbation Sources.* To obtain the fragmentation of the disc we meet perturbations of appropriate scale $M \ll 10^{13} M_\odot$. The corresponding adiabatic perturbations are exceedingly small due to damping and we feel that another type of perturbation is needed. In the gas which has passed through a shock wave there should exist small perturbations of the matter density. However, the problem of initial small scale perturbations is intriguing. They could be, for example, entropy perturbations. Two types are possible: primeval and secondary entropy perturbations. The dissipation of adiabatic density perturbations produces small (quadratic and cubic with respect to $\Delta\rho/\rho$) entropy density perturbations [20]. They are sleeping during the radiation dominated phase, later they increase slowly in the neutral gas after recombination. In the process of adiabatic compression in condensing protoclusters of galaxies, the amplitude of these perturbations should have

increased significantly and the presence of a gravitational instability would result in the appearance of objects with mass $M_{\text{max}} \sim M \ll 10^{13} M_{\odot}$. In the fragmentation of a plane layer, the initial perturbation spectrum is less important than in the three-dimensional case, due to the maximum of ω ($e^{\omega t}$ growth) at a definite k in the plane case.

3. Formation of Galaxies. According to formula (7), in the region with temperature $T \sim 10^4$ K conditions ($n \sim 0.5$ cm^{-3}) are favorable for the appearance of discrete objects with $M \sim M_{\text{max}} \sim 3 \cdot 10^9 M_{\odot}$. The objects of such mass are naturally associated with protogalaxies. Note that the protogalaxies in this scheme are also weakly ($\lambda_{\text{max}}/d \sim 4\pi^2/\mu^2 \sim 6$) oblate discs being compressed originally only in the plane of symmetry at the constant pressure, i.e., preserving the volume. The final form depends strongly on the rotation obtained by the protogalaxy due to the shock wave and gravitational interaction. This important part of the picture has not been sufficiently worked out to date. A picture of protogalaxy compression should differ greatly from the situation in protoclusters of galaxies (and apparently, it does not result in shock wave formation). It is also indirectly supported by the value of the mean density of matter in the region with $T \sim 10^4$ K, which is close to the observed density of matter in galaxies $n \sim 10$cm^{-3}. The dense $n \sim 30$ cm^{-3}, cold $T \sim 300$ K fractions of discs should, in accordance with (7), break up into objects with characteristic mass $M \sim 3 \cdot 10^7 M_{\odot}$ having initially a disc form with flatness of the order of $\lambda_{\text{max}}/d \sim \mu^{-2} \sim 10^4$. Of the observed astronomical objects, only quasars, galactic nuclei and massive star clusters can be compared with them. In addition, dense, sufficiently massive objects might serve as natural centers of condensation of the galaxies formed afterwards.

In such a scheme, it is possible to explain the observed cosmological evolution of quasars and powerful radiosources [14, 15]. Quasars should be formed in all the first protoclusters of galaxies. Their active life span of $t \sim 10^5 - 10^6$ years proves to be essentially less than the time of fragmentation (8) and the time of evolution of a dense fraction of thin concentrations—"pancakes" ($t \sim 3 \cdot 10^8$ years according to (8)) and especially the time for galaxy formation ($t \sim 10^8 - 10^9$ years); therefore, in such a scheme, clusters of quasars and quasars in clusters of galaxies should be observed rather infrequently).[5]

After condensation of a considerable part of the "pancakes" in the Universe, there appears a noticeable background of soft X-ray radiation which should cause heating of the whole of the unperturbed matter in the Universe up to a temperature of the order of 10^4 K [22–23]. Such a low gas temperature can have no influence on the rate of condensation of objects with $M \sim 10^{13} M_{\odot}$, however, it results in the disappearance of the cold ($T \ll 10^4$ K) dense fraction of "pancakes" and removes the conditions fa-

[5]Komberg and Sunyaev [21] have pointed out that the presence of a condensing protogalaxy around a quasar practically has no influence on its spectrum.

vorable for quasar formation. Thus, the rate of growth and the observed number of quasars should be a strongly decreasing function of time, in good agreement with the observed picture (see also *B*).

4. Discussion. As was stated earlier, in only a small fraction of protoclusters, compression of which takes place along all three directions, should the whole of the matter be a part of the observed objects (similar considerations were advanced by Oort [24]). A large fraction of the discs could prove to be gravitationally unbound and fully disperse forming field galaxies; the other part could partially dissipate trailing a group of galaxies after itself. Thus, in spite of a mass scale $M \sim 10^{12} - 10^{14} M_\odot$ deduced from the theory, the scheme of "pancakes" evolution under consideration leads to the formation of a wide spectrum of bound objects, i.e., clusters, groups of galaxies and single field galaxies; all the galaxies being formed in the dense fractions of the discs in similar conditions. Although galaxies are formed after protoclusters, the density of matter in galaxies is variable over a much smaller range compared with the density of matter in clusters of galaxies).[6] The ultimate density of matter in clusters depends strongly upon the curvature of the "pancake", the possibility that several "pancakes" form a bound system, and other subtle points, which do not effect the density of matter in galaxies. The picture discussed is indisputably idealized in many aspects. First of all, the disc formed is not ideally flat; existing curvature (connected, for example, with non-homogeneity in velocity distribution) together with velocities stimulated by the motion in directions perpendicular to the axis of compression can sphericize originating clusters or groups of galaxies.

The formation of galaxies out of the rather dense gas in the central fraction of a disc and their long-time ($\sim 4t_h$) presence in a close group of bodies simplify the problem of origin of galactic rotational momentum due to tidal interactions [25]. An interesting approach to the question of the origin of galactic rotational momentum is introduced by Doroshkevich [6]. The theorem of vorticity conservation is violated in compression of gas by a shock wave (see also [5]): therefore, the presence of velocity inhomogeneities in directions perpendicular to the shock wave direction can result in the birth of vorticity and galactic rotation. From this point of view, an interesting feature of a central cold layer which is adiabatically compressed is the absence of vorticity in this gas. This remark will probably be essential to the theory of the formation of dense bodies (nuclei, quasars).

III. Heating of the Intergalactic Gas

Hot, noncondensing gas on the periphery of "pancakes" may be gravitationally connected with the clusters of galaxies, but most of it should be distributed more or less homogeneously in the intergalactic space outside the clusters of galaxies.

[6]This behavior was deduced from observational data by Abell and Ozernoi.

Almost all the matter of the Universe should pass through a shock wave, as a shock wave which arises at $\mu \ll 1$ does not stop when $\mu \sim 1/2$ is reached, but continues to spread with decaying amplitude over the nonperturbed gas involved in the Hubble expansion of the Universe.

1. Energy Release in Shock Waves. The total energy released in the shock waves is

$$\begin{aligned} W &= \int \frac{U^2 d\mu}{2} \\ &= 8.6 \cdot 10^{13}(1+z_c)\left(\frac{M}{10^{13}M_\odot}\right)^{2/3} \int_0^M \pi\mu \sin \pi\mu \, d\mu \\ &< 8.6 \cdot 10^{13}(1+z_c)\left(\frac{M}{10^{13}M_\odot}\right)^{2/3} \text{erg/g} \qquad (9) \\ w = W\rho &\leq 8.6 \cdot 10^{13}(1+z_c)\left(\frac{M}{10^{13}M_\odot}\right)^{2/3} \text{erg/cm}^3 \\ &\approx 5.4 \cdot 10^{-4}\Omega(1+z_c)\left(\frac{M}{10^{13}M_\odot}\right)^{2/3} \text{eV/cm}^3. \end{aligned}$$

When $\mu \sim 1/2$, $M = 10^{13}M_\odot$ and $z_c = 4$ we have

$$w \approx 10^{-3}\Omega \,\text{eV/cm}^3 \quad \text{and} \quad W \approx 10^{14}\,\text{erg/g},$$

which is much less than the full reserve of rest energy of the matter $W \sim c^2 \sim 10^{21}$ erg/g. However this energy is released in a large fraction of the matter of the Universe, rather than in the small fraction which is bound into discrete objects (quasars, etc.). No exotic mechanisms of energy transfer from the discrete objects to the intergalactic gas are required for its heating. Shock wave energy release is sufficient to heat the intergalactic gas up to high temperatures.

2. The Observed Properties of the Intergalactic Gas. Observation of quasars having the redshift $z \sim 2$ show that the intergalactic gas (if it exists) is ionized strongly, $n_{\rm H}/n_p \leq 3 \cdot 10^{-8}\sqrt{1+2\Omega}/\Omega$ [26, 27]. The existence of an upper limit $J_\nu < (1.5) \cdot 10^{-21}$ erg/cm^2 sec sterad Hz to the ultraviolet background radiation with $\lambda \sim 2500$ Å suggests that there are difficulties in producing the observed $n_{\rm H}/n_p$ at the expense of photoionization of the cold gas with $\Omega > 0.1$. The measurements of background radiation at $\lambda \sim 2500$ Å give information about the radiation near the Lyman continuum $\lambda \sim 912$ Å at $z \sim 2$. The complete absence of the ionized or neutral gas in the space between galaxies is improbable: it is difficult to conceive a mechanism which is able to lead to condensation in galaxies of more than 99% of the total matter of the Universe.

3. Discussion of the Other Mechanisms of Heating. Different mechanisms of intergalactic gas heating have been suggested; ionization losses of subcosmic rays [28], dissipation of the turbulent energy and the energy of peculiar motion of the galaxies [29], shock waves arising due to explosions of quasars and radiogalaxies [30], etc. Most of the above mentioned mechanisms assume that the main source of energy needed to heat the gas comes from the discrete objects known from observations, i.e., quasars, radiogalaxies, etc. Yet we note that the heating of the intergalactic gas, even not taking into account the energy losses due to radiation and the expansion of the Universe, requires an energy input as great as the complete energy release (in the form of radiation, cosmic rays, shock waves, etc.) of the observed astronomical objects, which is assumed to be close to the energy release in the electromagnetic radiation. The density of heat energy in the intergalactic gas with $T \sim 10^6$ K and $\Omega \sim 1$ is equal to $w = 3kT_e n_e \approx 2 \cdot 10^{-4}\,\mathrm{eV/cm^3}$, which exceeds the energy density of the background nonthermal radio radiation $w \sim 10^{-7}\,\mathrm{eV/cm^3}$, γ- and X-ray $w = 10^{-4}\,\mathrm{eV/cm^3}$ and ultraviolet $w \leq 10^{-4}10^{-4}\,\mathrm{eV/cm^3}$ background radiation. They may be comparable only in optical range (where the main contribution to the background $w \sim 3 \cdot 10^{-3}\,\mathrm{eV/cm^3}$ is from the normal stars in galaxies) and in the infrared range $w \sim 10^{-2}\,\mathrm{eV/cm^3}$ [31]. Earlier heating $z_c \gg 0$ only makes the energy difficulties more acute as in the expanding Universe the energy density of radiation decreases in proportion to $(1+z)^4$, and the density of matter heat energy decreases more rapidly than $(1+z)^5$. The suggested mechanism of intergalactic gas heating by shock waves at the stage of clusters of galaxies formation does not meet energy difficulties (see (9)) as the heating occurs at the expense of gravitational energy released at the moment of the origin of inhomogeneities in the space distribution of the intergalactic gas and of condensation of the observed objects.

4. Discussion. Only the gas condensed by a shock wave when it falls on the "pancake" is heated to extremely high temperatures $T \sim 10^6$ K. We again note that this heating takes place at $1 + z \approx 2(1 + z_c)/\pi \approx 3.2$. It is possible that a significant fraction of the intergalactic gas (10–50%) was not subjected to compression in the "pancakes" and was heated only by the damped shock waves moving away from them (see I.4). The temperature of such gas may prove to be considerably less than $T \sim 10^6$ K required to explain the observed lack of absorption in the Lyα line in the spectra of quasars with $z \gtrsim 2$. In addition, such gas should cool rapidly because of energy losses due to radiation.

Agreement between the discussed scheme of intergalactic gas heating and observations is obtained by taking into account ionization of the intergalactic gas by means of the ultraviolet and soft X-ray radiation of "pancakes" and the objects being formed in them, i.e., quasars, etc. It is difficult to ionize the cold gas by radiation when its temperature $T \approx 10^4$ K because at low temperature the coefficient of radiative recombination is large. This coefficient decreases rapidly with increasing electron temperature. At the same time, the existence of ionizing background radiation should essentially manifest itself in the ionization balance of the hot gas having $T \sim 10^5 - 10^6$ K. Photoionization also manifests itself in the heat balance of the gas: the decrease in the abundance of neutral atoms as compared with equilibrium between ionization by electron collisions and radiative recombination, leads to a decrease in the radiation in lines connected with the excitation of the resonance levels by electron collisions (Doroshkevich and Sunyaev, 1972). We note that radiation in resonance lines determines the balance of the intergalactic gas calculated neglecting photoionization [29, 32, 33] in the temperature range $10^4 < T < 3 \cdot 10^5$ K.

5. The Present Condition of the Intergalactic Gas. Thus, the intergalactic gas having $T \sim 10^5 - 10^6$ K was hot at $z \sim 2-3$. The energy for heating was given by the shock waves and photoionization was due to ultraviolet radiation. The energy losses in line radiation were small: the gas got colder adiabatically while the Universe was expanding. Such evolution leads to the low temperatures $T \sim 10^4 - 10^5$ K at $z \sim 0$. The high degree of ionization $n_{\rm H}/n_p \sim 10^{-2} - 10^{-4}$ at $z \sim 0$ is easily maintained by ultraviolet and soft X-ray background radiation [22, 23, (Rees and Setti, 1970)]. The density and temperature distribution of the gas should be inhomogeneous because a large part of the gas takes part in the compression of the protoclusters of galaxies and only then is included in the Hubble expansion of the Universe. The intergalactic gas which is gravitationally bound in clusters of galaxies should be hotter, i.e., $T \sim 10^6$ K. Its density may be less than that given by formula (3) as the evolution of protoclusters of galaxies allows a certain expansion in the direction perpendicular to the axis of compression.

IV. The Observational Appearance of Intergalactic Gas

The hot gas in intergalactic space might be traced by its X-ray and ultraviolet radiation [34]. In our case the gas which has passed through the shock wave must be strongly inhomogeneous, undoubtedly increasing its contribution to the background radiation. The gas with $T \gtrsim 10^6$ K remained in massive clusters of galaxies which were initially compressed along all three axes: it is possibly observed in the Coma cluster of galaxies [35]. Not knowing in detail the radiation spectrum, it is difficult to distinguish the radiation of hot intergalactic gas and the combined radiation of discrete sources in the same cluster. We recall that we indicated earlier another possible diagnostic method of the hot gas: the decrease in temperature of the relic radiation in the Rayleigh-Jeans part of the spectrum in the direction to the zone containing the hot gas

$$\frac{\Delta T}{T} = -2\sigma_T N_e l \frac{kT_e}{m_e c^2}$$

This decrease is connected with the Compton interaction of the hot electrons with radiation having $T_r \ll T_e$. The high temperature of inhomogeneously distributed intergalactic gas should lead to small scale fluctuation of the relic radiation [19].

The radiation of the "pancake" region which cools to $T \sim 10^4$ K must be transformed mainly to the Lyα line as, at the time of condensation of the "pancake," its external parts have not passed through the shock wave and have a great optical thickness to the ionizing radiation. This radiation comes from regions having $z_c \sim 4$ and is shifted into the optical region of the spectrum. Information on the stage when the Universe was opaque to Lyα radiation is given by the background radiation. After ionization of the entire mass of the intergalactic gas, the light from the sources propagates along a straight line but, at this stage, a smaller portion of the radiation is in the form of Lyα—most is in the continuous spectrum and in the He II $\lambda\, 304$ Å line. Nevertheless, it would have been interesting to discover objects radiating mainly in the narrow redshifted lines of hydrogen and helium.

Another and perhaps more profitable possibility of "pancakes" detection is the observation of redshifted $\lambda\, 21$ cm radiation by the compressed neutral hydrogen in the zone with $T \sim 10^4$ K. Its spin temperature is much higher than the $T_r = 2.7(1+z)$ K relic radiation temperature. The mass of this prestellar matter $M \sim 3 \cdot 10^{12} M_\odot$ is much greater than the total mass of neutral hydrogen in all galaxies of any cluster of galaxies where most of the matter is concentrated in stars. The angular dimensions of "pancakes" at $z_c \sim 4$ are of order $3'$.

A large fraction of the gravitational energy released at the time of condensation of the protoclusters of galaxies is used to heat the intergalactic gas and is radiated in the form of ultraviolet and X-ray quanta. In this case, according to (9), the contribution of the radiation losses of the intergalactic gas to the background radiation is less than

$$w \leq 3 \cdot 10^{-4} \Omega \left(\frac{M}{10^{13} M_\odot} \right)^{2/3} \ \mathrm{eV/cm^3}$$

which may be of the order of total energy density of the background radiation at $\lambda < 1300$Å [Longair and Sunyaev, 1971]. However, the main part of the thermal energy of the intergalactic gas is not able to be radiated away during the lifetime of the Universe. At the mean mass of the "pancakes" $M < 10^{14} M_\odot$ and $0.2 < \Omega < 1$ the radiation of gas heated by the shock waves is not in conflict with the observed X-ray background. This follows in particular from the existence of the upper limit on the temperature of intergalactic gas $T < 2 \cdot 10^6 (1+z)$ K obtained on the basis of observations of the X-ray background [36, 37]. At the same time, by virtue of the fact that the problem is a statistical one, there may exist "pancakes" having anomalously large mass $M \gg 10^{14} M_\odot$ which heat the gas in the shock waves up to very high temperatures $T \sim 10^7 - 10^8$ K. Such objects should be powerful sources of hard X-ray radiation. The observation of fluctuations in the relic radiation may help to solve the problem of their nature. We feel that the overall picture is given in a very definite style, as if we witnessed the processes described. We have done it on purpose, to be better understood by the reader: if all the difficulties and arbitrariness would be mentioned at all places, they would prevent grasping the underlying ideas (as in the Russian proverb "the trees prevent seeing the forest"). We have done it at the expense of the danger of being criticized—but constructive, even severe, critique would be welcome!

Appendix

Calculations of the temperature regime were carried out for a perturbation given by a sinusoidal plane wave on the background of the flat Universe with $\Omega = 1$. The motion of the matter which is not yet compressed by the shock wave is given in a Lagrangian form

$$x = t^{2/3}\xi - bt^{4/3}\sin k\xi$$

$$y = t^{2/3}\eta, \quad z = t^{2/3}\xi.$$

It is usually assumed that k is connected with the characteristic mass M by means of the relationship

$$M = \bar{\rho}(\lambda/2)^3, \quad \bar{\rho} = (6\pi G t^2)^{-1}, \quad \lambda = 2\pi t^{2/3}/k, \quad M = (\pi/k)^3 \,/\, 6\pi G.$$

The given solution is exact if the initial pressure is neglected and for the stated type of perturbation because of the gravitational field of the plane layer $\rho = \rho(x,t)$ does not depend on the distribution of density inside the layer. In the matter which is not yet compressed by the shock wave, the solution remains exact even when the remaining fraction has been compressed by the wave. The density is given by

$$\rho = \frac{1}{6\pi G}\left(\frac{dx}{d\xi}\frac{dy}{d\eta}\frac{dz}{d\xi}\right)^{-1} = \frac{1}{6\pi G t^2}\frac{1}{1 - bt^{2/3}k\cos k\xi},$$

from which the time t_c of appearance of infinite density in the plane $x = 0$ is expressed in terms of b by the relation

$$kbt_c^{2/3} = 1.$$

Let us introduce the variable $\mu = k\xi/\pi$ instead of ξ so that the interval $0 < \xi < \pi/k$ corresponds to $0 < \mu < 1$. We introduce also $\tau = (t/t_c)^{2/3} = (1 + z_c)/(1 + z)$ and $r = 6\pi G t^2 p = \rho/\bar{\rho}$. In terms of these variables

$$r = (1 - \tau\cos\pi\mu)^{-1}.$$

First, we consider shock wave propagation in the limiting case of instantaneous cooling. In this case, the density of the matter compressed in the wave and already cooled is $\rho = \infty$ as $p \neq 0$, $T = 0$; this matter exists in a thin layer at $x = 0$. Hence we find the Lagrangian coordinate μ_1 of the wave front

$$x = 0 = \pi\mu_1 - \sin\pi\mu_1.$$

We express τ in terms of μ_1: $\tau = \pi\mu_1/\sin\pi\mu_1$, when $\tau - 1 \ll 1$, we find that

$$\tau = 1 + \frac{1}{6}(\pi\mu_1)^2, \quad \mu_1 = \sqrt{\frac{6}{\pi^2}(\tau - 1)}$$

and the nondimensional density before the front is given by

$$r_1 = r(\mu_1) = \left(1 - \frac{\pi\mu_1}{\tan\pi\mu_1}\right)^{-1}.$$

At $\tau - 1 \ll 1$ we have asymptotically

$$r(\mu_1) = \frac{1}{2}(\tau - 1) = \frac{3}{\pi^2\mu_1^2}.$$

In the limiting case mentioned pressure is given by

$$p_{fr} = \rho_1 u_1^2 = \rho_c \frac{\bar{\rho}(t)}{\rho_c}\frac{\rho_1}{\bar{\rho}}\left(\frac{dx}{dt}\right)^2 = \lambda^2 H_2^0 \rho_c(1 + z_c)\left(1 - \frac{\pi\mu_1}{\tan\pi\mu_1}\right)^{-1},$$

because

$$u_1 = \frac{dx}{dt} = -\frac{2}{3}\frac{\xi}{t^{1/3}} = -\frac{2\pi}{3}\frac{\mu}{t^{1/3}} = \lambda H_0(1 + z_c)^{1/2}(\pi\mu)^{1/2}(\sin\pi\mu)^{1/2} \cdot \frac{1}{2\pi}.$$

Taking into account that such compression takes place from two sides of the plane $x = 0$ and taking into account the Hubble expansion in all 3 dimensions, the surface density of the compressed layer σ is equal to

$$\sigma = \frac{2\mu \cdot 2\pi\lambda\bar{\rho}}{1 + z} = \frac{4\pi\lambda\rho_c}{1 + z_c}\mu\left(\frac{\sin\pi\mu}{\pi\mu}\right)^2,$$

$$p_{\text{grav}} = \frac{\pi G\sigma^2}{2}, \quad \beta = \frac{p_{\text{grav}}}{p_{fr}} \underset{\mu \ll 1}{=} \frac{\pi^2\mu^2}{4}.$$

To express p and σ in terms of τ it is necessary to solve a transcendental equation for $\mu_1(\tau)$. Let us note that p falls monotonically $p(\mu_1 = 1/2) = (1/7.5)p(\mu = 0)$ but σ passed through the maximum $\sigma = \sigma_{\text{max}}$ when $\tan\pi\mu_1 = 2\pi\mu_1$. It is characteristic that μ approaches 1 (i.e., all the matter is compressed by the shock wave) only as $t \to \infty$. In this case the wave pressure tends to zero. However, these properties depend on the assumption of strong heat losses. An opposite limiting case corresponds to the absence of heat losses, i.e., adiabatic temperature and density variations in the compressed matter after a shock wave has passed through. In this case, the complete calculation requires the solution of partial differential equations for describing the motion of the matter compressed by a shock wave. The motion of the wave front $\mu_1(\tau)$ depends on the solution of such equations. The motion of the matter $\mu_1(\tau) < \mu < 1$ which is not compressed by wave remains the same.

We can find only asymptotic results $\tau \ll 1$ by calculating quantities proportional to $\sqrt{\tau - 1}$ and neglecting $(\tau - 1)$ by comparison. In this approximation, we should neglect the Hubble expansion and consider the matter to be compressed by a wave at rest.

The compressed matter is considered to be completely ionized hydrogen, i.e., $n_e = n_p$, $n_H = 0$; the energy of ionization is neglected. The isentropic index is $\gamma = 5/3$, compression at the front of the wave is $\rho_2 = 4\rho_1$, where index 2 relates to $x_1 - 0$, index 1 to $x_1 + 0$. Hence (taking into account the rest behind the front) the equations of conservation give

$$\frac{dx_1}{dt} = -\frac{u_1}{3}; \qquad p_2 = \frac{4\rho_1 u_1^2}{3}, \quad c_V T_2 = \frac{u_1^2}{2},$$

where c_V is heat capacity at constant volume.

Finally, there exists an approximate solution (not used in the present paper) giving the algorithm for the adiabatic problem in the entire region $0 < \mu < 1$. We suggest that the dependence $p(\tau)/p(\tau_c = 1)$ should be taken from solutions with strong heat losses. It is also suggested to assume an exact solution $p_2(\tau_c)$ of the adiabatic case and, taking into account $p(\tau)$ found in this way, to solve numerically the equation of the wave distribution

$$\frac{d\mu_2}{d\tau} = a\sqrt{\frac{p_2(\tau)}{r(\mu_2, \tau)}}.$$

Institute of Applied Mathematics
USSR Academy of Sciences. Moscow

Received
February 25, 1972

REFERENCES

1. *von Weizsacker C. F.*—Astrophys. J. **114**, 165 (1951).
2. *Gamov G.*—Phys. Rev. **86**, 251 (1952).
3. *Ozernoi L. M., Chernin A. D.*—Astron. Zhurn. **45**, 1137 (1968).
4. *Peebles P. J. E.*—Astrophys. J. **155**, 393 (1969).
5. *Chernin A. D.*—Pisma v ZhETF **11**, 317 (1970).
6. *Doroshkevich A. G.*—Astron. Zhurn. **49**, 74 (1972).
7. *Peebles P. J. E.*—Astroph. and Space Sci. **11**, 443 (1971).
8. *Zeldovich Ya. B., Illarionov A. F., Sunyaev R. A.*—ZhETF **62** 4 (1972).
9. *Zeldovich Ya. B.*—Astrofizika **6**, 319 (1970).
10. *Doroshkevich A. G.*—Astrofizika **6**, 581 (1971).
11. *Ozernoi L. M, Chibisov G. V.*—Astron. Zhurn. **47**, 769 (1970).
12. *Chibisov G. V.*—Astron. Zhurn. **49** 1 (1972).
13. *Sachs R. K., Wolfe A. M.*—Astrophys. J. **147**, 73 (1967).
14. *Longair M. S.*—Uspekhi Fiz. Nauk **99**, 229 (1969).
15. *Schmidt M.*—Astrophys. J. **151**, 393 (1968).
16. *Sunyaev R. A.*—Astron. and Astrophys. **12**, 190 (1971).
17. *Zeldovich Ya. B., Kurt V. G., Sunyaev R. A.*—ZhETF **55**, 278 (1968).
18. *Peebles P. J. E.*—Astrophys. J. **153**, 1 (1968).
19. *Zeldovich Ya. B., Sunyaev R. A.*—ZhETF **56**, 2078 (1969).
20. *Zeldovich Ya. B., Sunyaev R. A.*—Astrofizika (1972).
21. *Komberg B. V., Sunyaev R. A.*—Astron. Zhurn. **48**, 235 (1971).
22. *Sunyaev R. A.*—Astron. Zhurn **49**, 929 (1969).

23. *Silk J., Werner M.*—Astrophys. J. **158**, 185 (1969).
24. *Oort J. H.*—Astron. and Astrophys. **7**, 381 (1970).
25. *Peebles P. J. E., Yu J. T.*—Astrophys. J. **162**, 815 (1970).
26. *Gunn J., Peterson B.*—Astrophys. J. **142**, 1633 (1965).
27. *Burbidge G., Burbidge M.* Quasi-Stellar Objects. San Francisco: Freeman (1967).
28. *Ginzburg V. L., Ozernoi L. M.*—Astron. Zhurn. **42**, 943 (1965).
29. *Gould R., Ramsay W.*—Astrophys. J. **144**, 587 (1966).
30. *Christiansen W.*—Month. Notic. Roy. Astron. Soc. **145**, 327 (1969).
31. *Longair M. S., Sunyaev R. A.*—Uspekhi Fiz. Nauk **105**, 41 (1971).
32. *Weymann R.*—Astrophys. J. **147**, 887 (1967).
33. *Doroshkevich A. G., Sunyaev R. A.*—Astron. Zhurn. **46**, 20 (1969).
34. *Kurt V. G., Sunyaev R. A.*—Kosmich. issled. **5**, 573 (1967).
35. *Gursky H., Kellogg E., Murray S., et al.*—Astrophys. J. **167**, L81 (1971).
36. *Field G., Henry R. C.*—Astrophys. J. **140**, 1002 (1964).
37. *Vainshtein L. A., Sunyaev R. A.*—Kosmich. issled. **6**, 635 (1968).
38. *McCrea D. A.*—Month. Notic. Roy. Astron. Soc. **132**, 641 (1957).

Commentary

This article embodies the foundations of the astrophysical aspect of the theory of the formation of the large-scale structure of the universe. It is a natural extension of the previous paper (**55**), which presents the theory which predicted a one-dimensional picture of compression of galaxy protoclusters—"pancakes."

In the present article the parameters of a gas compressed by a shock wave in a "pancake" are calculated for the first time. It indicates the existence of

a) zones of strong cooling of the plasma (to 10^4 K), which is where galaxies should in fact form;
b) zones of adiabatically compressed dense gas—it is possible that precisely its evolution leads to the formation of nuclei of galaxies and quasars;
c) zones of hot gas which does not cool in cosmological time and which is the source of X-ray radiation.

It is remarkable that subsequent exact calculations changed little in this picture: parameters were refined, but the very same ideas were discussed and developed in more detail. Even in the currently popular scenario of the evolution of the universe, in which the dominant contribution to the mean density of matter is made by weakly interacting particles (neutrinos?), the thermal picture for ordinary matter has changed little, preserving the primary features outlined in the present article (see recent reviews by S. F. Shandarin, A. G. Doroshkevich and Ya. B.,[1] by Ya. B. in Soviet scientific reviews,[2] and in the original paper by J. Silk and G. Efstathiou.[3])

[1] *Doroshkevich A. G., Zeldovich Ya. B., Shandarin S. F.*—Uspekhi fiz. nauk **139**, 83–134 (1983).

[2] *Zeldovich Ya. B.*—In: Itogi nauki i tekhniki. Astronomiĭa [Summary of science and technology. Astronomy] **22**, 4–32. Moscow: VINITI (1983); Astrophys. and Space Phys. Soviet Sci. Rev. **3**, 1 (1984).

[3] *Efstathiou G., Silk J.*—Fundamentals of Cosmic Physics **9**, 1 (1983).

The article notes for the first time that the formation of galaxies in the "pancake" picture occurs at constant pressure, which leads to an increase of the scale at which the rate of growth of density perturbations is at a maximum. The picture considered in the article of formation of galaxies reduces to the problem of gravitational instability of a flat layer which experiences a pressure effected by hot gas with negligibly small density. Without taking account of the external pressure, the gravitational instability of a flat layer was earlier considered by A. Pakholchik[4] and A. Toomre.[5] A review of early results is given in a book by V. L. Poliachenko and A. M. Friedman.[6] Later the behavior of a flat layer in the presence of an external pressure was studied in detail by A. G. Doroshkevich.[7]

It is precisely in the present paper that a method for observing protoclusters in the stage of formation (in the $\lambda = 21$ line of neutral hydrogen shifted by the red cosmological shift to the meter range of wavelengths). Later this topic was developed in a paper by the authors.[8] The search for such lines is being conducted now by R. Davis' group at the Jowdrell Bank observatory,[9] and at the world's largest aperature synthesis radiotelescope, the VLA in New Mexico, by J. Bond, J. Silk and S. Szalay. This method of searching for protoclusters is discussed (and promoted) by the eminent Dutch astronomer and patriarch of radioastronomy, Jan Oort.[10] Under his direction observations of these lines on the aperature-synthesis radiotelescope at Vesterbork are beginning.

[4] *Pakholchik A.*—Astron. zhurn. **39**, 953 (1962).

[5] *Toomre A.*—Astrophys. J. **139**, 1217 (1964).

[6] *Poliachenko V. L., Friedman A. M.* Ravnovesie i ustoichivost' gravitiruiushchikh sistem [Equilibrium and stability of gravitating systems]. Moscow: Nauka, p. 448 (1976).

[7] *Doroshkevich A. G.*—Astron. zhurn. **57**, 259–267 (1980).

[8] *Sunyaev R. A., Zeldovich Ya. B.*—Month. Not. Roy. Astron. Soc. **174**, 375–379 (1974).

[9] *Davis R. D.*—In: Krupnomasshtabnaia struktura Vselennoi [Large-scale structure of the Universe]. Moscow: Mir, p. 489 (1981).

[10] *Oort J. H.*—Astron. and Astrophys. **139**, 211–214 (1984).

57
Growth of Perturbations in an Expanding Universe of Free Particles*

With G. S. Bisnovatyĭ-Kogan

A study is made of the growth of perturbations in a collision-free medium on the background of the exact solution for an expanding universe with the critical density. The Newtonian theory of gravitation is used. The solution of the linearized transport equation is obtained by the method of "integration along trajectories." The results obtained for long waves coincide with the hydrodynamic treatment. For short waves, aperiodic damping is obtained, in contrast to the hydrodynamic result. It is shown that the presence of streams in the distribution function exerts a stabilizing influence on the growth of perturbations; the gravitational interaction does not lead to two-stream instability.

1. Introduction

During the early stages of expansion, the matter in the Universe was in thermodynamic equilibrium as a result of collisions and could be described hydrodynamically.

The growth of perturbations in an expanding Friedman universe was considered in [1, 2] for different equations of state. The growth of perturbations in a Newtonian universe was considered in [3, 4]. During the late stages of expansion, different classes of condensation are formed, namely, stars, galaxies, and clusters of galaxies. The time between collisions of such condensations is very large and during this stage one may therefore consider a universe of free particles. Layzer [5] has proposed that the galaxies and clusters were formed as a result of clustering.

If one neglects the motion of the particles, the corresponding kinetic treatment is identical with the well-known hydrodynamic solution for dust [6]; the same results are therefore obtained for the growth of perturbations. Allowance for the motion of the free particles calls for a kinetic treatment. In

*Astronomicheskiĭ Zhurnal **47** (6), 942–947 (1970).

a Friedman model, the distribution function of the particles is statistically isotropic. However, situations are possible in cosmology when this function is anisotropic; in addition, the study of anisotropic distribution functions is of interest from the point of view of a comparison of the growth of perturbations in a collision-free plasma with an electromagnetic interaction and in a gravitating medium of free particles.

The stability of a uniform collision-free medium at rest was considered in [7]. The result obtained in this investigation was virtually the same as the classical result of Jeans concerning the existence of a critical wavelength.

For smooth distributions of a Maxwellian type the value of the critical wavelength differs only slightly from the hydrodynamic result of Jeans.

A characteristic new feature in a system of free particles arises if a group of particles with a small spread of velocities is present. The growth of perturbations in such a group occurs with an increment that corresponds to the density of these particles. It is, however, well known that the Jeans treatment is incorrect although certain exact results can be interpreted as a generalization of the Jeans solution [4]. In order to investigate the finer points, one must construct an exact solution and investigate the growth of perturbations on this background.

A marked difference from hydrodynamics arises for short waves, for which there is a rapid damping of the perturbations in contrast to the undamped oscillations in the hydrodynamics of an ideal gas. This is connected with the fact that an ideal gas with a vanishing viscosity requires a vanishing mean free path, which is the exact opposite of a collision-free medium.

It should be noted that the expression for the critical wavelength obtained in [7] is incorrect since the resulting integral diverges and integration by parts in the derivation is inadmissible. The correct result $k_{\text{crit}}^2 = -4\pi Gm \int_{-\infty}^{\infty} \frac{\partial f_0/\partial v}{v} dv$ is given in the same paper. In this expression, k is the wavenumber, G is the gravitational constant, m is the mass, v is the velocity, f_0 is the unperturbed distribution function, and the integral is to be understood in the sense of the principal value.

The problem of the instability of a homogeneous medium of free particles must be considered on the background of an exact nonstationary expanding or contracting solution. In this connection, interest attaches to an investigation of the stability of different distribution functions, including those when streams are present that give rise to instability in a plasma. A rigorous investigation of the stability of stationary but inhomogeneous or anisotropic systems of point masses was made in [8–10].

In the present paper, we construct an expanding model of free particles and consider the growth of perturbations in such a model. We shall restrict ourselves to the Newtonian problem.

A relativistic kinetic model of an expanding universe was constructed in [11]; however, the growth of perturbations was not considered. The difficulty of treating the growth of perturbations on the background of an expanding universe derives from the fact that the system of free particles during expansion is equivalent to a hydrodynamic system with an adiabatic exponent $\gamma = 5/3$, for which there are no exact analytic solutions. Every Lagrangian volume becomes unstable with time and the critical mass decreases with time.

We have considered the cases of very short and very long waves. For short waves, we obtain an aperiodic damping of the perturbations, in contrast to hydrodynamics, in which adiabatically damped oscillations occur in the stability region. The results for long waves are analogous to those obtained in hydrodynamics.

The problem considered here describes the clustering of globular clusters, if they were formed first of all [12], galaxies, or clusters of galaxies in larger complexes.

2. *Solution of the Kinetic Equation in an Unperturbed Expanding Universe*

As is well known, the solution of the transport equation in a homogeneous isotropically expanding Newtonian universe is an arbitrary function f_0 of the velocities v, which decrease in accordance with an adiabatic law.

If we use an Eulerian system of coordinates $\mathbf{x}$, we must also take into account the Hubble rate of expansion. We have (m is the mass):

$$f_0 = (m, \eta), \quad \eta = a(t)(\mathbf{v} - H\mathbf{x}). \tag{1}$$

In a homogeneous universe with critical density, the Hubble constant depends inversely on the time, i.e., $H \sim 1/t$, and the scale factor satisfies $a(t) \sim t^{2/3}$, although (1) also remains valid in the case of an arbitrary density. In a nonrelativistic theory the function f is arbitrary; for example, it may be anisotropic with respect to the velocities and determine an anisotropic pressure tensor $P_{ik} = \int f p_i v_k \, d\mathbf{v}$ (p_i is a momentum). In a relativistic theory, the arbitrariness in the choice of the function (1) is reduced. In a homogeneous, isotropically expanding universe the pressure tensor P_{ik} must also be isotropic, i.e., it must reduce to a scalar. The presence of the time-dependent scale factor $a(t)$ in the argument of the distribution function (1) means that the average random velocity decreases during the expansion as $\sim 1/a$.

We shall now show that (1) is an exact solution of the problem. We shall solve the kinetic equation

$$\frac{\partial f}{\partial t} + \mathbf{v}\frac{\partial f}{\partial \mathbf{x}} - \frac{\partial \Phi}{\partial \mathbf{x}}\frac{\partial f}{\partial \mathbf{v}} = 0 \tag{2a}$$

and Poisson's equation

$$\Delta\Phi = 4\pi G\rho \tag{2b}$$

by the method of characteristics. Here Φ is the gravitational potential. The solution for a Newtonian universe with critical density is [6]

$$\rho = \frac{1}{6\pi G t^2}, \quad \Phi = \frac{2}{3}\pi G \rho(t) r^2, \quad \frac{\partial \Phi}{\partial \mathbf{x}} = \frac{2}{9}\frac{\mathbf{x}}{t^2}. \tag{3}$$

The instant $t = 0$ corresponds to the start of expansion from an infinite density. The characteristics of the equations (2a) are determined by the system

$$dt = \frac{dx_i}{v_i} = -\frac{dv_i}{\frac{2}{9}x_i/t^2} \tag{4}$$

Integrals of the system (4) are

$$C_{1i} = 3v_i t^{1/3} - \frac{x_i}{t^{2/3}}, \quad C_{2i} = v_i t^{2/3} - \frac{2}{3}\frac{x_i}{t^{1/3}}. \tag{5}$$

The solution must be such that $\int f\, dv\, dm = 1/6\pi G t^2$. It is readily seen that an arbitrary function $\alpha f(C_{2i})$ satisfies this condition. The constant α is determined by the normalization condition

$$\alpha \int f(u_i)\, d\mathbf{u} dm = \frac{1}{6}\pi G t^2, \quad u_i = v_i t^{2/3} - \frac{2}{3}\frac{x_i}{t^{1/3}} \tag{6}$$

In what follows, we shall consider a Maxwellian distribution function for particles with the same mass m:

$$f_0 = \frac{1}{6\pi G m}\frac{1}{(\pi\theta)^{3/2}} \exp\left(-\sum_i \frac{u_i^2}{\theta}\right) \tag{7}$$

and a distribution with colliding streams along the x_1 axis:

$$f_0 = \frac{1}{6\pi G m}\frac{1}{(\pi\theta)^{3/2}} \exp\left(-\frac{u_2^2 + u_3^2}{\theta}\right)\frac{1}{2}\left(e^{-(u_1-v)^2/\theta} + e^{-(u_1+v)^2/\theta}\right). \tag{8}$$

3. *Growth of Perturbations*

We shall solve the linearized transport equation by the method of integration along trajectories [13]. The equations of the trajectories coincide with the equations of the characteristics of the transport equation.

The solution of this system, which expresses the velocity v' and the coordinate x' at the time t' in terms of their values at t, is

$$\begin{aligned} x' &= \left(3v_x t^{1/3} - \frac{x}{t^{2/3}}\right) t'^{2/3} + \left(\frac{2x}{t^{1/3}} - 3v_x t^{2/3}\right) t'^{1/3}, \\ v'_x &= \frac{2}{3}\left(3v_x t^{1/3} - \frac{x}{t^{2/3}}\right)(t')^{-1/3} + \frac{1}{3}\left(\frac{2x}{t^{1/3}} - 3v_x t^{2/3}\right)(t')^{-2/3}, \end{aligned} \tag{9}$$

In accordance with [13], the solution of the linearized transport equation is

$$f = \int_0^t \frac{\partial \Phi}{\partial \mathbf{x}'}\frac{\partial f_0}{\partial \mathbf{v}'} dt, \quad f(0) = 0. \tag{10}$$

In the unperturbed solution (3) every scale increases as $\sim t^{2/3}$; it follows that the wavelength of a perturbation also increase as $\sim t^{2/3}$.

We shall seek the solution in the form of an expansion in a series in eigenfunctions for $\mathbf{k} = (k, 0, 0)$:

$$\Phi = e^{ik\xi}\varphi(t), \quad f = e^{ik\xi}, \quad \xi = \frac{x}{t^{2/3}}. \tag{11}$$

We now substitute (11) into (10) and take into account (5), (6), and (9). Integrating by parts, we obtain

$$f = ik\frac{\partial f_0}{\partial u_1}\int_0^t \varphi(t')\exp[3iku_1(t^{-1/3} - t'^{-1/3})]\,dt'. \tag{12}$$

Using Poisson's equation and (11), we obtain

$$\Delta\Phi = -\frac{k^2\varphi(t)}{t^{4/3}}e^{ik\xi} = e^{ik\xi}\int f\,dv. \tag{13}$$

Integrating over the velocities for the function (8), we obtain an integral equation for $\varphi(t)$:

$$\varphi(t) + \frac{2}{t^{2/3}}\int_0^t \varphi(t')\tau\exp\left(-\frac{9k^2\theta}{4}\tau^2\right)\cos 3kv\tau\,dt' = 0, \quad \tau = t^{-1/3} - (t')^{1/3}. \tag{14}$$

For the Maxwellian function (7), an equation can be obtained from (14) for $v = 0$. In the case of long waves or large times, (14) reduces to a differential equation. In this case, $9k^2\theta\tau^2/4 \ll 1$ and the exponential function in (14) can be replaced by unity. Differentiating (14), we obtain an equation for φ:

$$t^4\varphi^{\mathrm{IV}} + \frac{32t^3\varphi'''}{3} + \left[\frac{254t^2}{9} + 2t^{4/3}(kv)^2\right]\varphi'' + \left[\frac{448t}{27} + \frac{16t^{1/3}(kv)^2}{3}\right]\varphi' +$$
$$+ 2[(kv)^2t^{-2/3} + (kv)^4t^{-4/3}]\varphi = 0. \tag{15}$$

A small value of $kv/t^{1/3}$ does not affect the growth of perturbations. With the passage of time, the importance of the streams continually decreases and the solution invariably goes over into the well-known hydrodynamic solution for dust [3, 4]. An equation for this solution can be obtained from (14) for $v = 0$:

$$t^2\varphi'' + \frac{8\varphi'}{3} = 0. \tag{16}$$

The solution of equation (16) is $\varphi_1 = c_1$, $\varphi_2 = c_2t^{-5/3}$; this gives $\delta\rho/\rho \sim t^{2/3}$ for the first solution and $\delta\rho/\rho \sim t^{-1}$ for the second [3, 4]. Formally, the second (damped) solution gives a divergence at $t = 0$ when substituted into the original equation (14). As in a plasma, it is to be regarded as the analytic continuation of the converging (growing) solution with a corresponding path around a Landau type pole [14].

In the case of short waves, the growth of perturbations differs significantly from the hydrodynamic case. One can find a solution if

$$\frac{9k^2\theta t^{-2/3}}{4} \gg 1; \quad \frac{9k^2\theta}{4} = \lambda; \quad \lambda \gg k^2v^2. \tag{17}$$

In this case, the integral in (14) can be calculated by the method of steepest descents [15]. Assuming that $\varphi(t')$ is a slowly varying function and $\tau e^{9k^2\theta\tau^2/4}$ is a function that decreases rapidly for large λ, we obtain the following equation from (14):

$$\varphi + 9\sqrt{\frac{\pi}{2e}}\frac{t^2}{\lambda^{3/2}}\varphi' \cos\frac{3kv}{\sqrt{2\lambda}} = 0, \tag{18}$$

and, hence,

$$\varphi = C \exp\left[\frac{1}{9}\sqrt{\frac{2e\lambda^3}{\pi}}\frac{1}{\cos(3kv/\sqrt{2\lambda})}\frac{1}{t}\right]. \tag{19}$$

The characteristic time of damping from the time t_0 is T:

$$T = 9\sqrt{\pi/(2e\lambda^3)}t_0^2 \cos(3kv/\sqrt{2\lambda}), \tag{20}$$

where, by virtue of the criterion (17), the argument of the cosine is less than $\pi/2$, $3kv/\sqrt{2\lambda} < \pi/2$. The damping time is of the order of

$$T = 9\sqrt{\pi/(2e\lambda^3)}t_0^2 \sim t_\lambda(t_\lambda/t)^2,$$

where t_λ is the time taken by the particles to traverse a distance equal to the wavelength at the time t_0.

Thus, the presence of streams on the velocity distribution function does not lead to an instability, but exerts a stabilizing influence.

The rate of damping of perturbations depends mainly on the random energy of a unit mass.

For short waves, one can simplify (14). Going over to the variable $x = t^{-1/3}$, we obtain from (14)

$$\varphi(x) + 6x^2\int_x^\infty \varphi(x')(x-x')e^{-a(x-x')^2}\cos b(x-x')\frac{dx'}{x'^4} = 0 \tag{21}$$

$$a = \frac{9k^2\theta}{4}, \qquad b = 3kv.$$

For $\alpha \gg 1$, we can set $x'^4 = x^4$, and we then obtain

$$\varphi(x) + \frac{6}{x^2}\int_x^\infty \varphi(x')(x-x')e^{-a(x-x')^2}\cos b(x-x')dx' = 0. \tag{22}$$

Equation (22) admits solutions of the type $\varphi \sim e^{\omega x}$, where ω satisfies the equation

$$1 - \frac{6}{x^2}\int_0^\infty \tau e^{-a\tau^2+\omega\tau}\cos b\tau\, d\tau = 0. \tag{23}$$

Unfortunately, for arbitrary b, $a \gg 1$, one cannot solve (23); for small b one obtains a result that is identical with (19). A noteworthy feature of the solution (23) is that one obtains a local growth increment; the integral (true) increment can be obtained approximately in the form [4] $\varphi \sim e^{\int \omega\, dx}$.

For the local law of growth $\varphi \sim e^{\alpha/t} \sim e^{\alpha x^3}$ the integral law $\sim e^{\alpha \int x^2\, dx}$ has the same functional form as the local law.

For two streams moving with a high relative velocity, the growth increment corresponds to the density of each stream.

The growth is monotonic for an observer moving with a stream; for an observer at rest $\varphi \sim e^{\int \omega\, dt} e^{ik(x - \int v\, dt)}$ and there also arise oscillations with frequency kv.

Similar results were obtained in [10], in which a study was made of the instability of a distribution function with two streams of arbitrary relative velocity in an axisymmetric model.

As the wavelength increases there is a transition from stability to instability at a certain k_{crit} and k_{crit} is an increasing function of the time.

4. Conclusions

The results obtained in this paper can be summarized as follows.

1. The growth of long-wave perturbations in an expanding universe of free particles occurs in the same way as in the hydrodynamic treatment.

2. Short-wave perturbations are damped aperiodically because of a pure kinetic mechanism.

3. In contrast to a plasma with an electromagnetic interaction, streams on the distribution function do not lead to an additional instability; rather the kinetic energy of the motion of the streams tends to decrease the growth increment of the perturbations.

Institute of Applied Mathematics
USSR Academy of Sciences. Moscow

Received
April 9, 1970

REFERENCES

1. *Lifshitz E. M.*—ZhETF **16**, 587 (1946).
2. *Lifshitz E. M., Khalatnikov I. M.*—UFN **30**, 391 (1963).
3. *Bonnor W. B.*—In: Appl. Math., Vol. 8, Relativity Theory and Astrophys., Part 1 Relativity and Cosmology, 263 (1967).
4. *Doroshkevich A. G., Zeldovich Ya. B.*—Astron. Zh. **40**, 807 (1963).
5. *Layzer D.*—Ann. Rev. Astron. and Astrophys. **2**, 341 (1964).
6. *Zeldovich Ya. B., Novikov I. D.* Relativistic Astrophysics. Univ. of Chicago (1971).

7. *Maksumov M. N., Marochnik L. S.*—Dokl. AN SSSR **164**, 1019 (1965).
8. *Lee E. P.*—Astrophys. J. **148**, 185 (1967).
9. *Bisnovatyĭ-Kogan G. S., Zeldovich Ya. B., Sagdeev R. Z., Friedman A. M.*—Zh. Prikl. Mekh. i Tekhn. Fiz. **3** (1969).
10. *Bisnovatyĭ-Kogan G. S.* Preprint of the Institute of Applied Mathematics **8** Moscow (1970).
11. *Bel L.*—Astrophys. J. **155**, 83 (1969).
12. *Peebles P. J. E., Dicke R. H.*—Astrophys. J. **154**, 898 (1968).
13. *Shafranov V. D.*—Vopr. Teorii Plazmy **3**, 3 (1963).
14. *Landau L. D.*—ZhETF **16**, 574 (1946); UFN **93**, 527 (1967).
15. *Lavrentiev M. A., Shabat B. V.* Metody teorii funktsiĭ kompleksnogo peremennogo [Methods of the Theory of Functions of a Complex Variable]. Moscow: Fizmatgiz (1958).

Commentary

The study of multiparticle systems in the hydrodynamic approximation was a typical feature of classical physics at the turn of the century. The problem of a collisionless gas with collective long-distance action appeared in connection with plasma theory. The present article followed the pioneering, but incomplete work by M. N. Maksumov and L. S. Marochnik, cited in the paper.

The problem of gravitational instability of a collisionless gas was solved by G. S. Bisnovatyĭ-Kogan and Ya. B. rigorously, i.e., not in the Jeans approximation, but accounting for the cosmological expansion.

As is well known, the problem of the evolution of the Friedman adiabatic perturbations in the hydrodynamic approximation (including the case of zero matter pressure) was completely solved in E. M. Lifshitz' classic work. However, very important for applications is the evolution of adiabatic perturbations in a collisionless gravitating gas (depending on the particular problem, either heavy neutrinos and other hypothetical massive particles, or stars and entire galaxies, may be considered to be "particles"). This problem was first solved, not in the Jeans approximation, but with account of the cosmological expansion, by G. S. Bisnovatyĭ-Kogan and Ya. B. in 1970. It turned out that in the long-wave region ($\lambda \to \infty$) the classic Jeans instability is reproduced and the result does not differ from the case of a continuous medium. For $\lambda < \lambda_{\text{cr}}$, where λ_{cr} differs little from the Jeans wavelength, perturbations decay rapidly (in the case of a continuous medium there would be sound waves in this region). Lately this paper has found an important new application in connection with the hypothesis of dominance of massive weakly-interacting particles (massive neutrinos and others) in the Universe.

Another important result consists in the absence of the two stream instability which is characteristic of a plasma with electromagnetic interaction. This group of questions was studied in great detail in the monograph by A. M. Friedman and V. L. Polyachenko.[1]

[1] *Friedman A. M., Polyachenko V. L.* Physics of Gravitating Systems. Vol. 1. Equilibrium and Stability. Springer-Verlag (1984).

58

Peculiar Velocities of Galaxy Clusters and the Mean Density of Matter in the Universe*

With R. A. Sunyaev

1. Introduction

In a recent paper the authors have shown [10] that by observing the microwave background radiation in directions toward clusters and remote superclusters of galaxies, one might, in principle, be able to measure both the radial and the tangential component of the peculiar velocity of the member galaxies relative to the background radiation.

At present all we know is the peculiar velocity of the Sun and of our Galaxy [7]. What information could we acquire if we were successful in tying rich clusters of galaxies to a reference frame coupled to the background radiation, that is, to a co-moving reference frame?

It will be shown in this letter that such observations might yield information of great value on the deceleration parameter q_0, or equivalently, on the density parameter $\Omega = 2q_0 = \rho_0/\rho_{cr}$, where ρ_0 is the mean density of matter in the Universe at the present epoch and $\rho_{cr} = 3H_0^2/8\pi G = 5 \cdot 10^{-30}[H_0/(50\,\mathrm{km \cdot s^{-1} \cdot Mpc^{-1}})]^2\,\mathrm{g/cm^3}$ is the critical density. Furthermore, observations of this sort could furnish new information on the spectrum of large-scale density irregularities in the Universe whose characteristic size $\lambda \gg 100\,\mathrm{Mpc}$. On such scales, according to evidence regarding the space distribution of galaxies [5], any density irregularities are of low amplitude, and the linear theory for the evolution of perturbations in the density and velocity of matter applies.

Earlier Sunyaev and Zeldovich [9] pointed out the importance of measuring the peculiar velocities of galaxies and clusters. Such observations, it was maintained, could yield highly valuable information on the spectrum of density irregularities. Sandage *et al.* [11] remarked that the small scatter of points in the Hubble diagram (apparent magnitude of galaxies against their redshift z) implies that galaxies have a small peculiar velocity ($V \approx 50$km/s) and militates in favor of an open cosmological model. In 1974 Silk [6] sought to correct for the influence of nonlinear effects upon the amplitude of the

*Pisma v Astronomicheskiĭ zhurnal **6** (10), 738–741 (1980).

peculiar velocity of galaxies in world models with $\Omega < 1$. Nonlinear effects would arise because of the existence of irregularities with $\delta\rho/\rho \gtrsim 1$ on certain spatial scales.

We now return to this theme in the context of our new method for measuring the peculiar velocities of clusters, and with the growing interest of observational astrophysicists in cosmology. One would hope that the specialized Einstein X-ray Observatory will discover some new X-ray galaxy clusters having $z \geq 1$. Detailed radio and submillimeter observations can convey information on the peculiar velocity of these objects. If such observations are successful, one might well be able to make a start in applying the cosmological test discussed below: peculiar velocity as a measure of the mean density of matter in the Universe.

2. *Peculiar Velocity in Linear Theory for Growth of Small Density Perturbations*

Following standard procedures [1–3], let us linearize the equations of hydrodynamics, writing the simple Newtonian equations that describe, against the background of the expanding Universe, the time evolution of the amplitude $\delta = \delta\rho/\rho$ of small density perturbations and of the peculiar velocity $\mathbf{v}$ of matter in a medium with zero pressure:

$$\ddot{\delta} + 2\frac{\dot{R}}{R}\dot{\delta} - 4\pi G\rho\delta = 0; \quad \dot{\delta} = \frac{i(\mathbf{kV})}{R}. \tag{1}$$

In deriving Eqs. (1) we have assumed that the irregularities are irrotational and take the form of plane waves:

$$\delta = \frac{1}{(2\pi)^3}\int a_k \exp(i\mathbf{kr}/R(t))\, d^3k, \quad \mathbf{V} = \frac{1}{(2\pi)^3}\int \mathbf{b}_k \exp(i\mathbf{kr}/R(t))\, d^3k. \tag{2}$$

Since the scale factor

$$R(t) = \frac{R_0}{1+z}, \quad \rho = \Omega\frac{3H_0^2}{8\pi G}(1+z)^3 \quad \text{and} \quad \frac{dz}{dt} = -H_0(1+z)^2\sqrt{1+\Omega z},$$

we obtain an equation for the z-dependence of δ ($\delta_z \equiv \partial\delta/\partial z$):

$$\ddot{\delta}_{zz}(1+z)(1+\Omega z) + \frac{1}{2}\Omega\delta_z(1+z) - \frac{3}{2}\Omega\delta = 0. \tag{3}$$

Bonnor [1] showed that the Newtonian treatment yields the same result as an analysis in terms of general relativity theory [4].

On scales $\lambda = 2\pi/k \ll ct$, this statement is quite evident. On large scales $\lambda \gg ct$, the evolution of density irregularities should be handled as follows. The expansion of the Universe depends on the mean density of matter. If density irregularities occur on scales $\lambda \gg ct$, then regions of differing density will evolve differently, and as a result the density contrast will grow with time; that is, large-scale density perturbations will tend to

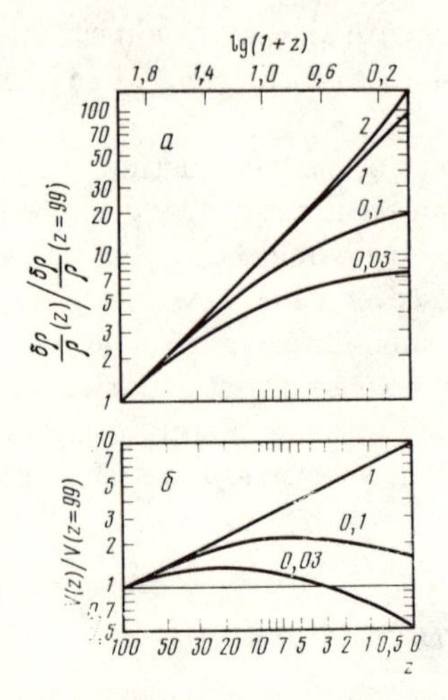

Fig. 1. The evolution of small perturbations of density (a) and velocity (b) of matter during expansion of the Universe depends strongly on the mean density of matter, i.e., on Ω. Values of Ω are indicated on the curves. The curve $\Omega = 1$ corresponds to the laws: (a) $\delta\rho/\rho \sim 1/(1+z) \sim t^{2/3}$, (b) $V \sim 1/\sqrt{1+z} \sim t^{1/3}$.

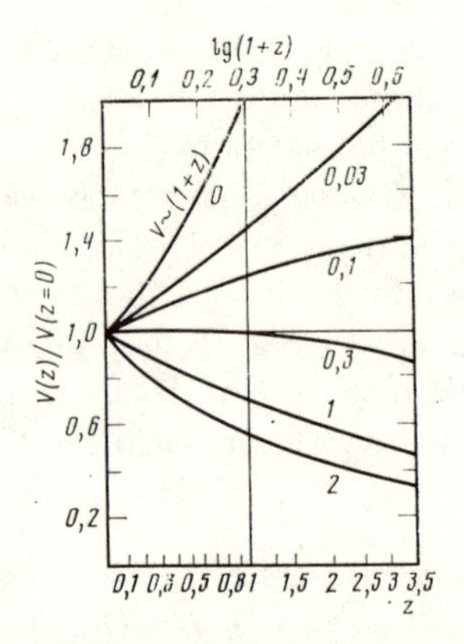

Fig. 2. Change of the mean-square peculiar velocity of matter with a redshift (from the linear theory) at different Ω. Values of Ω are indicated on the curves. The curve $\Omega = 0$ corresponds to the law $v \sim (1+z)$, the curve $\Omega = 1$ to the law $v \sim (1+z)^{-11/2}$.

grow. The growth rate will be described by the same equations as the growth of irregularities on scales $\lambda \ll ct$. As the Universe proceeds to expand, the wavelength $\lambda = \lambda_0/(1+z)$ of an irregularity will grow more slowly than the horizon ct. Hence any scale which today is small compared with the horizon would at one time have surpassed the horizon. The fact that Eqs. (1) are applicable to irregularities of any scale enables us to follow their evolution with time, and it furthermore shows that at the state when pressure may be neglected, small perturbations of all scales will evolve according to the same law.

The solutions of (3) are $\delta_1 = (1+z)\sqrt{1+\Omega z}$ (a decaying mode) and

$$\delta_2 = (1+z)\sqrt{1+\Omega z}\int_z^\infty \frac{dz'}{(1+z')^2(1+\Omega z')^{3/2}} \tag{4}$$

(a growing mode). The integral in (4) is easily evaluated, and we finally obtain [8]

$$\delta = A\left[\frac{1+2\Omega+3\Omega z}{(1-\Omega)^2} + \frac{3}{2}\frac{\Omega(1+z)\sqrt{1+\Omega z}}{(1-\Omega)^{5/2}}\ln\frac{\sqrt{1+\Omega z}-\sqrt{1-\Omega}}{\sqrt{1+\Omega z}+\sqrt{1-\Omega}}\right] \quad \text{for } \Omega < 1, \tag{5}$$

$$\delta = A\left[\frac{1+2\Omega+3\Omega z}{(\Omega-1)^2} + \frac{3\Omega(1+z)\sqrt{1+\Omega z}}{(\Omega-1)^{5/2}}\left[\arctan\sqrt{\frac{1+\Omega z}{\Omega-1}} - \frac{\pi}{2}\right]\right]$$

$$\text{for } \Omega > 1,$$

$$\delta = \frac{A}{1+z} \text{ for } \Omega = 1,$$

where A is a normalizing factor. Using the second of Eqs. (1), we find

$$\mathbf{V} = \frac{1}{(1+z)}\frac{\mathbf{k}}{k^2}\dot{\delta} = \frac{1}{1+z}\frac{\mathbf{k}}{k^2}\frac{d\delta}{dz}\frac{dz}{dt} = H_0(1+z)\sqrt{1+\Omega z}\frac{\mathbf{k}}{k^2}\frac{d\delta}{dz},$$

$$\mathbf{V} = \frac{3H_0\mathbf{k}}{4k^2}A\frac{\Omega(1+z)}{(1-\Omega)^{5/2}}\left[6\sqrt{1+\Omega z}\sqrt{1-\Omega} + \right. \tag{6}$$

$$\left. + (2+\Omega+3\Omega z)\ln\frac{\sqrt{1+\Omega z}-\sqrt{1-\Omega}}{\sqrt{1+\Omega z}+\sqrt{1-\Omega}}\right] \quad \text{for} \quad \Omega < 1,$$

$$\mathbf{V} = \frac{3}{2}\frac{H_0 k}{k^2}A\frac{\Omega(1+z)}{(\Omega-1)^{5/2}}\left[3\sqrt{1+\Omega z}\sqrt{\Omega-1} + \right.$$

$$\left. + (2+\Omega+3\Omega z)\left(\arctan\sqrt{\frac{1+\Omega z}{\Omega-1}} - \frac{\pi}{2}\right)\right] \quad \text{for} \quad \Omega > 1, \tag{7}$$

$$\mathbf{V} = A\frac{H_0\mathbf{k}}{k^2\sqrt{1+z}}, \quad \text{for} \quad \Omega = 1.$$

Density perturbations in a world with $\Omega \ll 1$ will grow as long as $\Omega \gg 1/z$; afterward the growth will slow down markedly (Fig. 1a). Peculiar velocities in a world with $\Omega \ll 1$ also will tend to rise at first, but then they will begin to diminish (Fig. 1b). Table 1 shows how $z_{\max}$ (the redshift at which maximum peculiar velocity is reached) depends on the parameter Ω.

Table 1

Ω	$z_{\max}$	Ω	$z_{\max}$	Ω	$z_{\max}$
0.375	$2.3 \cdot 10^{-3}$	0.225	1.1	0.07	7
0.37	0.02	0.2	1.4	0.06	8.4
9.36	0.067	0.175	1.8	0.05	10
0.35	0.12	0.15	2.4	0.04	13
0.325	0.25	0.125	3.2	0.03	18
0.3	0.4	0.1	4.4	0.02	28
0.275	0.59	0.09	5.1	0.01	59
0.25	0.81	0.08	5.9		

Some interesting information is presented in Fig. 2. The dependence of peculiar velocity on redshift differs fundamentally for different Ω. If $\Omega = 1$, the peculiar velocity diminishes with regression into the past, while if $\Omega = 0.03$ or 0.1, it rises. If $\Omega = 0.3$ the peculiar velocity hardly depends on z at all. Notice how greatly the $V(z)$ curve for $\Omega = 0.03$ differs from the curve for $\Omega = 0$ [which represents the standard solution $V \sim (1+z)$]. The bare fact that the behavior of $V(z)$ depends on the value of Ω has long been recognized. But now, thanks to the Einstein Observatory, it has become feasible to search systematically for clusters of galaxies with $z > 0.5$, and by examining the microwave background in the directions of such clusters one might be able to measure their peculiar velocity. It is important to understand that the change in the microwave background intensity in the direction of a cluster moving with a peculiar velocity will be independent of the redshift at which the cluster is located [10]. The only quantity depending on z is the angular size of the cluster. Note also that even for z as small as 0.2, the difference between the $V(z)$ curves for $\Omega = 2$ and $\Omega = 0.03$ is quite appreciable.

If the peculiar velocities of galaxy clusters depend on low-amplitude density irregularities of large scales, then observations of the microwave background in the directions of remote clusters would pave the way to a new determination of the parameter Ω, for one might be able to apply a novel cosmological criterion: peculiar velocity against redshift.

Table 2

Ω	$B(\Omega)$		Ω	$B(\Omega)$		Ω	$B(\Omega)$
0.01	$4.9 \cdot 10^{-2}$		0.15	0.32		0.8	0.88
0.02	$8.2 \cdot 10^{-2}$		0.2	0.38		0.9	0.94
9.03	0.1		0.3	0.49		1	1
0.04	0.13		0.4	0.58		1.1	1.1
0.05	0.16		0.5	0.67		1.5	1.3
0.06	0.18		0.6	0.74		2	1.5
0.08	0.21		0.7	0.81		2.5	1.7
0.1	0.25						

Some interesting information on the parameter Ω might also be gained by measuring the peculiar velocities of clusters in our neighborhood ($z < 0.1$), provided the amplitude δ on large scales is known. Equations (1), (5) and (7) imply that

$$V(z=0) = \frac{\lambda_0 H_0}{2\pi} B(\Omega)\delta(z=0). \tag{8}$$

Furthermore, from the equation of continuity $\partial\delta/\partial t + \operatorname{div} \mathbf{V} = 0$ we at once infer that $B = 1$ for $\Omega = 1$. If $\Omega \neq 1$ the $B(\Omega)$ dependence is more complicated; it is given in Table 2. In the range of interest to us, $0.03 < \Omega < 2$, the power law $B = \Omega^{0.615}$ provides a good fit. The smaller Ω is, the smaller the peculiar velocity would be expected to be. This result is quite natural; in a world with small Ω ($\Omega \ll 1$), linear density irregularities will become "frozen in", but velocity perturbations will diminish with the expansion. We have given (8) in order to illustrate the relationship between δ and V. To evaluate the peculiar velocity, one would have to make some assumptions about the density irregularity spectrum.

3. Spectrum of Density Irregularities

In the range of small k (large λ), the spectrum of linear density irregularities is unknown. Suppose, as an illustration, that it is given by a power law: $a_k \sim k^n$ in (2). Then (5), (7) imply that $b_k \sim k^{n-1}$. Hence the mean-square peculiar velocity (the quantity measured),

$$\mathbf{V}^2 = \frac{1}{(2\pi)^3} \int b_k^2 \, d^3k \sim \int k^{2n} \, dk$$

will be determined by longer waves than the density perturbations

$$\frac{\delta\rho}{\rho} = \frac{1}{(2\pi)^3} \int a_k^2 \, d^3k \sim \int k^{2n+2} \, dk.$$

For n in the interval $-\frac{3}{2} < n < -\frac{1}{2}$, the main contribution to the peculiar velocity will come from scales $\lambda \sim ct$, whereas $(\delta\rho/\rho)^2$ will be determined by small scales. Evidently in order to estimate theoretically the peculiar velocity of clusters one needs further information on the spectrum and amplitude of the density irregularities on large scales. That information could, in principle, be furnished by measurements of angular fluctuations in the microwave background.

Our discussion above neglects nonlinear effects, although they actually might contribute as much to V as the linear effects we have considered.

Institute for Space Research
USSR Academy of Sciences. Moscow

Received
August 4, 1980

REFERENCES

1. *Bonnor W. B.*—Month. Not. RAS **117**, 104 (1957).
2. *Weinberg S.* Gravitation and Cosmology. N.Y.: Wiley (1972).
3. *Zeldovich Ya. B., Novikov I. D.* Stroenie i evolyutsiĭa Vselennoĭ [Structure and Evolution of the Universe]. Moscow: Nauka, 735 p. (1975).
4. *Lifshitz E. M.*—ZhETF **16**, 587 (1946).
5. *Davis M., Groth E. J., Peebles P. J. E.*—Astrophys. J. Lett. **212**, L107 (1977).
6. *Silk J.*—Astrophys. J. **193**, 529 (1974).
7. *Smoot G. F., Gorenstein M. V., Muller R. A.*—Phys. Rev. Lett. **39**, 898 (1977).
8. *Sunyaev R. A.*—Astron. and Astrophys. **12**, 190 (1971).
9. *Sunyaev R. A., Zeldovich Ya. B.*—Astrophys. and Space Sci. **7**, 3 (1970).
10. *Sunyaev R. A., Zeldovich Ya. B.*—Month. Not. RAS **190**, 413 (1980).
11. *Sandage A., Tammann G., Hardy E.*—Astrophys. J. **172**, 253 (1972).

Commentary

This article is closely related to article **68**. It gives fairly simple and convenient formulas describing the growth of density perturbations and the evolution of peculiar velocities of probe points (galaxy clusters) in the Universe. Similar results were independently obtained in Peeble's group.[1]

Also of interest is the possibility of determining the model of the world (open, flat, closed) from data on the dependence of the peculiar velocities of clusters on the red shift z. Presently the authors have carried out calculations also for the flat model of the Universe in the presence of a Λ-term, $\Omega = 1$, $\Omega_B + \Omega_r < 1$, where Ω_B, Ω_r characterize the contributions of matter (including hidden mass) and radiation, respectively, to the average density of matter in the Universe.

We note that for a very flat density perturbation spectrum, when the chief contribution to the peculiar velocity is given by long-wave density perturbations, radiation is also brought into motion. In this case the relic radiation may no longer be considered as a "new ether."

In the meantime, precisely the velocity of clusters with respect to radiation may be, apparently, measured with the least error (see **68** by Ya. B. and Sunyaev in this volume).

At small scales the difficulty of interpreting measurements of peculiar velocities of galaxies is related to the fact that the distribution of galaxies does not mimic the average distribution of gravitating matter, especially in the presence of hidden mass (cf. the commentary to article **52** by N. A. Dmitriev and Ya. B.).

[1] *Peebles P. J. E.* The Large-Scale Structure of the Universe. Princeton Univ. Press (1980).

59

Astrophysical Implications of the Neutrino Rest Mass I. The Universe*

With R. A. Sunyaev

Recent measurements of the spectrum of tritium decay electrons in the USSR suggest that electronic neutrinos may have a rest energy $m_\nu c^2 \approx 30\,\text{eV}$. Primordial neutrinos would then make a major contribution to the mean density of matter in the Universe, enough for the Universe to be closed. If future reactor and accelerator experiments should reveal that muonic and tau neutrinos have a mass of the same order, then one would infer an age of less than 10^{10} years for the Friedman Universe, contrary to the age observed for the oldest stars in the Galaxy. The contradiction can be removed by introducing the cosmological constant Λ. Conversely, measuring the rest mass of all neutrino species as well as the age of the Universe would afford a unique opportunity to evaluate Λ.

Our understanding of the Universe as a whole would be fundamentally altered if it should be proved that neutrinos have a nonzero rest mass. The latest measurements announced by Lyubimov *et al.* [7] suggest that electronic antineutrinos do have a rest energy $m_\nu c^2 \approx 30\,\text{eV}$, corresponding to a rest mass $m_\nu \approx 5 \cdot 10^{-32}\text{g}$.

Now if the combined rest energy of all three kinds of neutrinos known to elementary-particle physics (ν_e, ν_μ, ν_τ) should exceed 20 eV, then the Universe would be closed.[1] And if their total rest energy were greater than 60 eV, the age of the Universe, on the usual assumptions, would turn out to be less than 10^{10} years, contrary to other age estimates. In the event that direct laboratory measurement yield such a result, it would inescapably follow that the cosmological constant is different from zero.

The value of the cosmological constant required to resolve this conflict corresponds to a positive energy density of order $10^{-8}\,\text{erg/cm}^3$ and a negative pressure of the flat vacuum. Along with the condition that the Universe must have passed through a state of high density, the age arguments serve

*Pisma v Astronomicheskiĭ zhurnal **6** (8), 451–456 (1980).

[1]If $\sum m_\nu c^2 = 17.6\,(H_0/50\,\text{km/(s}\cdot\text{Mpc))}^2$ eV, we would have the case of a flat universe, despite the low density of ordinary forms of matter.

to place the possible value of the cosmological constant within a narrow range. If such a cosmological constant exists, then unbounded expansion would be predicted as the future course of the Universe, even though three-dimensional space may be closed.

It has been remarked in the literature that a neutrino rest mass would change our ideas about the origin of structure in the Universe at the level of clusters of galaxies—that it would explain the hidden-mass paradox for galaxies and clusters. These points will be developed more fully in the next Letters of this series.

We would finally mention that the existence of a neutrino rest mass would permit (though it would not prove) the occurrence of the neutrino oscillations that Pontecorvo [9] has predicted. The result of R. Davis's attempts to measure the flux of neutrinos from the Sun might perhaps be interpreted in this way. From this brief inventory we see that a measurement of the neutrino mass would be of prime significance for astronomy.

In this first letter we shall confine our attention to the evolution of the Universe as a whole.

1. Neutrino Density in the Universe

The Universe is permeated with primordial background radiation at a temperature $T_\gamma = 3(1+z)$ K. In 1cm^3 there are $N_\gamma = 20T_\gamma^3 = 540(1+z)^3$ cosmic background photons (z is the redshift). At early phases of the expansion the radiation temperature exceeded $m_e c^2/k \approx 6 \cdot 10^9$ K; electron-positron pairs and all kinds of neutrinos in equilibrium with the radiation. Let us consider neutrinos of mass $m_\nu \ll m_e$, which are stable against spontaneous decay.

One will recall [6] that the density of neutrinos in thermodynamic equilibrium is equal to $N_\nu = 7.5 g_\nu T_\nu^3$ cm^{-3}, where T_ν is the temperature of the neutrinos in kelvins and g_ν is the statistical weight. Weak interactions involve neutrinos of one helicity and antineutrinos of the other one of all sorts (ν_e, ν_μ, ν_τ and $\tilde{\nu}_e$, $\tilde{\nu}_\mu$, $\tilde{\nu}_\tau$), so that $\sum g_\nu = 6$ for all kinds together and $g_\nu = 2$ for each kind separately. If $kT_\gamma \gg m_e c^2$, the neutrinos were in equilibrium with the radiation (that is, $T_\nu = T_\gamma$), and as the expansion proceeded the neutrinos became collisionless. Annihilation of electron-positron pairs at $kT_\gamma \approx 0.2 m_e c^2$ converted the energy and entropy of the pairs to radiation, and T_γ rose relative to T_ν. According to statistical physics, throughout the wide temperature range $m_e c^2 > kT_\gamma > m_\nu c^2$ the relation $T_\nu = (4/11)^{1/3} T_\gamma = 0.7 T_\gamma$ holds, implying that $N_\nu = 2.7 g_\nu T_\gamma^3$ cm^{-3}. This expression for N_ν remains true both if $kT_\gamma \ll m_e c^2$, for which neutrinos will no longer interact with radiation, and if $kT_\gamma \ll m_\nu c^2$, for which they will become nonrelativistic. The 3θ K temperature of the microwave background

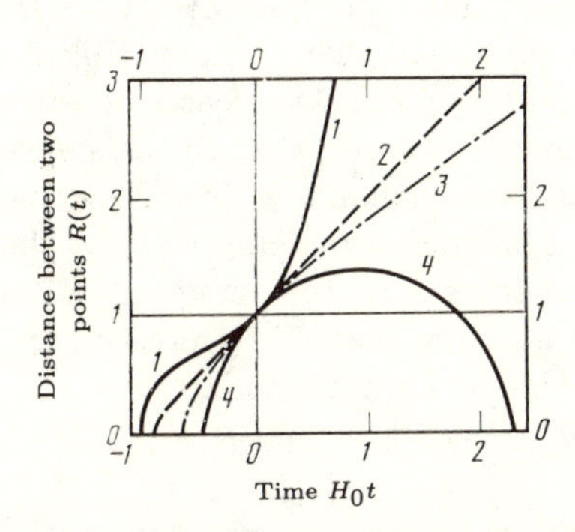

Fig. 1. Past and future of the Universe in different world models: 1–closed world with a cosmological constant, $\Omega = 3.5$, $\lambda = 4.5$; 2–open world, $\Omega = 0.03$, $\lambda = 0$; 3–flat world, $\Omega = 1$, $\lambda = 0$; 4–closed world, $\Omega = 3.5$, $\lambda = 0$.

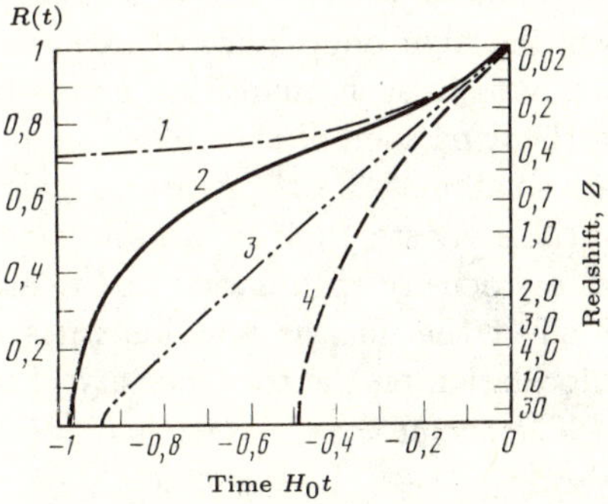

Fig. 2. Age of the Universe according to different world models: 1–$\lambda = 4.9$, $\Omega = 3.5$; 2–$\lambda = 4.5$, $\Omega = 3.5$; 3–$\lambda = 0$, $\Omega = 0.03$; 4–$\lambda = 0$, $\Omega = 3.5$.

radiation (the factor θ takes into account the uncertainty in the measured values of the background radiation temperature) corresponds to a density today of $N_\nu = 75\theta^3 g_\nu \,\mathrm{cm}^{-3}$, or $150\theta^3\,\mathrm{cm}^{-3}$ for ν_e and $\tilde{\nu}_e$, and $450\theta^3\,\mathrm{cm}^{-3}$ for all three kinds of neutrinos together. Setting $\sum m_\nu c^2$ equal to the combined rest energy of the three kinds of neutrinos, expressed in electron volts, we may write the cosmological neutrino density as

$$\mu_\nu = \sum m_\nu N_\nu = 2.7 \cdot 10^{-31}\theta^3 \sum m_\nu c^2 \,\mathrm{g/cm}^3 .$$

From their analysis of the tritium decay $\mathrm{T} \to \mathrm{He}^3 + e^- + \tilde{\nu}_e$, Lyubimov *et al.* [8] conclude that with 99% probability $m_{\nu_e}c^2$ is in the range $25\,\mathrm{eV} < m_{\nu_e}c^2 < 47\,\mathrm{eV}$. Taking $m_{\nu_e}c^2 = m_{\tilde{\nu}_e}c^2 = 30\,\mathrm{eV}$ and $m_{\nu_\mu} = m_{\nu_\tau} = 0$, we find $\rho_\nu = 8 \cdot 10^{-30}\theta^3\,\mathrm{g/cm}^3$, while if $m_{\nu_\mu} = m_{\nu_\tau} = m_{\nu_e}$ we obtain $\rho_\nu = 2.4 \cdot 10^{-29}\theta^3\,\mathrm{g/cm}^3$.

2. Cosmology and Neutrinos

A Hubble constant $H_0 = 50h\,\mathrm{km/s \cdot Mpc}$ corresponds to a critical density $\rho_{\mathrm{cr}} = 4.7 \cdot 10^{-30}h^2\,\mathrm{g/cm}^3$. (The factor $h \sim 1$ describes the uncertainty in the experimental values for the Hubble constant.) Hence the existence of a rest energy $m_{\nu_e}c^2 = 30\,\mathrm{eV}$ itself implies that the Universe is spatially closed, with $\Omega = \rho_\nu/\rho_{\mathrm{cr}} = 1.7\theta^3h^{-2}$; and after a certain time span of order $(2-3)\cdot 10^{10}$ years the expansion observed today will give way to a contraction (see Fig. 1).

This is a far different result from that conjectured earlier, based on a determination of the density of matter concentrated within galaxies: $\Omega_m = 0.03$. One of the chief arguments for an open model universe with $\Omega = 0.03$ was the high abundance of deuterium in the interstellar medium [10]; its synthesis in such quantities is possible only through nuclear reactions early in the expansion of the Universe. It is evident, however, that the existence of a small mass for the neutrino does not affect the dynamics of the Universe in the early stages of expansion (since the dynamics were determined by the energy density of radiation and relativistic neutrinos at that time), nor does it affect the nuclear reaction rates. The neutrino mass therefore conveys information on the total density of matter in the Universe today.

The age of the Universe is expressed by

$$t = H_0^{-1} \int_0^\infty \frac{dz}{(1+z)^2\sqrt{1+\Omega z}} \approx \frac{2.1 \cdot 10^{10}\,\mathrm{yr}}{\left(h + \sqrt{\theta^2 \sum m_\nu c^2/40\,\mathrm{eV}}\right)}. \tag{1}$$

As $\Omega \to 0$ the age becomes $t = H_0^{-1} = 1.96 \cdot 10^{10} h^{-1}$ yr, while if $\Omega = 1$ we have $t = (2/3)H_0^{-1} = 13 \cdot 10^9$ yr. If $\sum m_\nu c^2 = 30$ eV, then $\Omega_\nu = 1.7$ and $t = 0.59H_0^{-1} = 1.17 \cdot 10^9$ yr; if $\sum m_\nu c^2 = 90$ eV, then $\Omega_\nu = 5.1$ and $t = 0.44H_0^{-1} = 8.63 \cdot 10^9$ yr; and if $\sum m_\nu c^2 = 120$ eV, then $\Omega_\nu = 6.8$ and the age becomes $t = 0.385H_0^{-1} = 7.57 \cdot 10^9$ yr [12].

In a recent review, Tammann *et al.* [12] quote estimates of $t \geq (14-16) \cdot 10^9$ yr for the oldest stars in star clusters [2], and $t = (11-18) \cdot 10^9$ yr for the age of the Universe derived by the rhenium-osmium method of nuclear cosmochronology [11]. They conclude that t probably exceeds $15 \cdot 10^9$ yr, and in any event is greater than $12 \cdot 10^9$ yr. We arrive, then, at a contradiction, which would become intolerable if new experimental data should raise $\sum m_\nu c^2$ appreciably above 30 eV.

It is hard to imagine that the masses of muonic and tau neutrinos can be measured in the foreseeable future in experiments analogous to that [8] performed for electronic neutrinos. But it seems entirely reasonable that oscillations might be detected, that is, mutual transformations of the type $\nu_e \rightleftharpoons \nu_\mu \rightleftharpoons \nu_\tau$, by means of experiments in which neutrinos from reactors and accelerators are recorded. In such an experiment, strictly speaking, it is superpositions of the form $\alpha\nu_e + \beta\nu_\mu + \gamma\nu_\tau$ that have definite masses. These masses can be determined through accelerator and reactor experiments, and these are the very masses that would enter into cosmological calculations.

If the existence of neutrino oscillations should be demonstrated, then in accord with Pontecorvo's ideas the expected number of reactions in a solar-neutrino detector should be cut, and the conflict between Davis's measurements and the electronic neutrino flux from the Sun calculated from very simple assumptions would be eliminated. This drop is explained in a natural way by the conversion of electronic neutrinos emitted through nuclear reactions in the solar interior: they would undergo oscillations along their way to the Earth and change into a mixture of all three types of neutrino, of which only one form can be detected with Davis's equipment. Oscillations are feasible only if at least two of the three kinds of neutrino differ in mass.

3. The Cosmological Constant

If measurements should reliably yield $\sum m_\nu c^2 > 30\,\text{eV}$ and if the estimates of the background radiation temperature, the Hubble constant, and the age of the Universe should remain unchanged, then one may assert that the cosmological constant is non-zero. The cosmological constant may be interpreted as the non-zero energy density ε_v of a vacuum for small space-time curvature [3, 4]. The vacuum would then have a pressure $p_v = -\varepsilon_v$ and a mass density $\rho_v = \varepsilon_v/c^2$. When these quantities are introduced into the equations describing the dynamics of the Universe for $\varepsilon_v > 0$, the expansion time scale of the Universe corresponding to a given Hubble constant and a given density Ω of matter increases. Denoting[2] $\lambda = \rho_v/\rho_{\text{cr}} = 8\pi G\varepsilon_v/c^2 H_0^2$, we obtain the expression

$$t = \frac{1}{H_0}\int_0^\infty \frac{dz}{(1+z)\sqrt{(1+z)^2(1+\Omega z) - \lambda z(2+z)}}, \tag{2}$$

for the age, which shows that for arbitrary H_0 and Ω we can get as great an age as desired by permitting a finite increase in λ, approaching the critical value λ_{cr} such that the expression in the square root vanishes at the point of tangency of the curves $(1+z)^2(1+\Omega z)$ and $\lambda z(2+z)$.

One can graphically interpret the increase in the age t for positive λ, ρ_v, ε_v by noting that when $\lambda > 0$ is introduced and allowed to increase, the present radius a of the closed Universe will simultaneously decrease, as given in terms of H_0, Ω, and λ by the relation (c is the velocity of light)

$$\left(\left.\frac{d\ln(1+z)}{dt}\right|_{z=0}\right)^2 = H_0^2 = H_0^2\Omega + H_0^2\lambda - \frac{c^2}{a^2}; \quad a = \frac{c}{H_0}(\Omega+\lambda-1)^{-1/2} \tag{3}$$

The inequality $t > t_1$ gives for λ and ε_v inequalities of the same sign: $\lambda > \lambda_1$, $\varepsilon_v > \varepsilon_{v_1}$. On the other hand, we must at the same time have $\lambda < \lambda_{\text{cr}}$; otherwise the Universe could not have evolved from its extremely dense state (as required to create the cosmic background radiation) to a state with the present values of H_0 and ρ_m. Thus $\sum m_\nu c^2$ should turn out to be of order 10–100 eV and if the other premises are correct, the cosmological constant would be known to within 10–40 percent, even though only a lower limit might be known for the age of the Universe. In particular, for $\sum m_\nu c^2 = 60$ eV we would have $\Omega_\nu = 3.5$ and $4.4 < \lambda < 4.9$ (see Fig.

[2]The usual way of writing the Einstein equations with the cosmological constant is $R_{ik} - \frac{1}{2}g_{ik}R = \Lambda g_{ik} + 8\pi G T_{ik}/c^4$; the dimensions of Λ are then [cm^{-2}], and the dimensionless quantity introduced above is defined as $\lambda = \Lambda c^2/H_0^2$.

2). The acceleration parameter $q_0 = \Omega/2 - \lambda$ would then be negative and confined to the interval $-2.65 < q_0 < -3.15$.

If $\sum m_\nu c^2 > 200$ eV, a plateau will develop on the evolution curve $z(t)$ at $z \approx 0.35$, contravening available evidence on the distribution of galaxies and quasars with respect to z. Hence the condition $\sum m_\nu c^2 < 200$ eV is a rigid (even allowing for the cosmological constant) upper limit on the sum of the masses of the stable neutrinos. This value agrees with a cautious estimate by Gershtein and one of us [1] (cf. [5]).

By applying some of the observational tests of modern cosmology one might be able to impose further constraints on $\sum m_\nu c^2$. In a model with the Λ-term, a significant departure from the Friedman models in terms of the number of objects, their properties, and the age of the Universe would be expected even for redshifts as small as $z \sim 0.2-0.3$, which telescopes can now reach for detailed inspection of many galaxies and clusters of galaxies.

4. Flat Universe with Cosmological Constant

At early stages of its expansion ($z \gg 1$, $\Omega z \gg 1$), our Universe had a practically flat geometry. Since our present era has virtually no distinguishing feature, it is natural to suppose (although we certainly have no proof) that our world is flat now. For values of Ω different from unity, this situation will be possible only if the cosmological constant exists, with $\lambda = 1 - \Omega$ [see (3)]. In this event the age of the Universe will be given by

$$t = H_0^{-1} \int_0^\infty \frac{dz}{(1+z)\sqrt{\Omega(1+z)^3 + 1 - \Omega}}.$$

For $h = 1$ and $\Omega = 5.1$ we have $t = 7.2 \cdot 10^9$ yr; if $\Omega = 1.7$, then $t = 10.9 \cdot 10^9$ yr; and if $\Omega = 1$ we will of course have $t = (2/3)H_0^{-1}$. In the low-density case we have discussed, $\Omega = 0.03$, the age $t = 1.65 H_0^{-1} = 32.4 \cdot 10^9$ yr. Clearly if the energy $\sum m_\nu c^2 > 20$ eV the age of the flat world with heavy neutrinos will conflict with the age of the oldest objects observed in the Universe. The condition becomes less stringent if the temperature correction $\theta < 1$, for $\sum m_\nu c^2 < 20\theta^{-3}$ eV.

5. Concluding Remarks

The solution containing the cosmological constant seems artificial, expressly because only a narrow range of mutually consistent values of $\sum m_\nu c^2$, Λ, and a will give an admissible evolution scenario. Yet early in the evolution there was nothing to distinguish these particular values.

However, one ought to remember that: a) fundamental theory does not demand $\Lambda \equiv 0$ (indeed, the prevailing view among physicists at the present time runs, "Everything that needn't equal zero doesn't equal zero"; one example is the finite probability of proton decay, and another, more pertinent, is the neutrino mass); b) physics and astronomy are experimental and observational sciences. If experiment and observation should tell us that $\Lambda \neq 0$, then fundamental theory will simply have to explain why!

Institute for Space Research
USSR Academy of Sciences. Moscow

Received
June 28, 1980

REFERENCES

1. *Gershtein S. S., Zeldovich Ya. B.*—Pisma v ZhETF **4**, 174 (1966).
2. *Demarque P., McClure R. D.*—In: The Evolution of Galaxies and Stellar Populations, Ed. B. M. Tinsley, R. B. Larson. Yale: Univ. Observ., p. 199.
3. *Zeldovich Ya. B.*—Pisma v ZhETF **6**, 1050 (1967).
4. *Zeldovich Ya. B.*—UFN **115**, 169 (1975).
5. *Cowsik R., McClelland J.*—Phys. Rev. Lett. **29**, 669 (1972).
6. *Landau L. D., Lifshitz E. M.* Statistical Physics. Reading, MA: Addison-Wesley (1969).
7. *Lyubimov V. A., Novikov E. G., Nozik V. Z. et al.*—Preprint Inst. of Theor. and Exper. Physics **62**. Moscow (1980).
8. *Lyubimov V. A., Novikov E. G., Nozik V. Z. et al.*—Yader. Fiz. **32**, 301 (1980).
9. *Pontecorvo B. M.*—ZhETF **53**, 1717 (1967).
10. *Rogerson Y. B., York D. C.*—Astrophys. Lett. **186**, 95 (1973).
11. *Steigman G., Schramm D. N., Gunn J. E.*—Phys. Lett. B **66**, 202 (1977).
12. *Tammann G., Sandage A., Yahil A.*—Preprint Basel Univ. N 28 (1980).

60

Astrophysical Implications of the Neutrino Rest Mass II. The Density-Perturbation Spectrum and Small-Scale Fluctuations in the Microwave Background*

With A. G. Doroshkevich, R. A. Sunyaev, M. Yu. Khlopov

If the neutrino has a rest mass, a scale comparable with the distance between clusters of galaxies should be singled out in the Universe. The amplitude of neutrino density perturbations on smaller scales should diminish in proportion to a high power of the perturbation wavelength. The evolution of adiabatic and entropic (isothermal) density perturbations is considered, and it is shown that the existence of a neutrino rest mass implies smaller fine-scale fluctuations in the microwave background radiation than models in which $m_\nu = 0$.

As Szalay and Marx [18] have pointed out, if neutrinos have a rest mass such that $m_\nu c^2 > 5\,\mathrm{eV}$, then they would have become nonrelativistic long before the era of hydrogen recombination in the Universe. Neutrinos would have governed not only the mean density of matter in the Universe and the expansion dynamics but also the growth of density perturbations in the Universe during the linear phase.

A neutrino rest energy $m_\nu c^2 \approx 20-30\,\mathrm{eV}$ would serve to single out a characteristic length in the Universe comparable with the distance between clusters of galaxies. Any primordial neutrino density perturbations on smaller scales would have decayed. Neutrino density perturbations on the scale of galaxy clusters would have grown even during the stage prior to hydrogen recombination in the Universe, at an era when perturbations of the mixture of radiation, baryons, and electrons would not have intensified. Afterward, regions of enhanced neutrino density would have attracted neutral matter to themselves, inducing the formation of dense clouds of ordinary gas, which could then have turned into clusters of galaxies.

*Pisma v Astronomicheskiĭ zhurnal **6** (8), 457–464 (1980).

This picture is an appealing one (compared with the situation for massless neutrinos), because the formation of clusters here calls for a lower level of initial perturbations of the metric, and a lower level of fine-scale fluctuations is predicted for the angular distribution of the microwave background radiation.

1. Spectrum of Primordial Neutrinos

Tremaine and Gunn [17] have rightly remarked that at low temperatures ($kT_\nu < m_\nu c^2$) neutrinos of finite mass would have a nonequilibrium spectrum.

At temperatures $kT_\nu \gg m_\nu c^2$, neutrinos would have been in thermodynamic equilibrium with other particles: photons, electrons, and positrons. One would be entitled at the same time to neglect the mass of the neutrino (since $kT_\nu \gg m_e c^2 \gg m_\nu c^2$) and to take $E_\nu = cp_\nu$, where E_ν denotes the energy and p_ν the momentum of a neutrino.

According to Fermi statistics, in phase space ultra-relativistic neutrinos would have had an occupation number of the form

$$n = (\exp(cp_\nu/kT_\nu) + 1)^{-1} \tag{1}$$

with a chemical potential $\mu = 0$. Moreover, $n_\nu = n_{\tilde{\nu}}$. When $kT_\nu \approx m_e c^2$ the neutrinos would have become collisionless. For collisionless neutrinos the momentum of each given group of particles would fall off as $(1+z)$, and their density (occupation number) in phase space would have been preserved during the expansion of the Universe. According to (1), this conservation condition is equivalent to a decrease in T_ν proportional to $(1+z)$.

Later, when $kT \approx 0.1 m_e c^2$, annihilation of electron-positron pairs would have taken place in the Universe, raising the radiation temperature above the neutrino temperature:

$$T_\nu = \left(\frac{4}{11}\right)^{1/3} T_r = 2.14(1+z)K, \qquad T_r = 3(1+z)K \quad \text{for} \quad kT_r < 0.1 m_e c^2.$$

The distribution (1) implies a neutrino number density $N_\nu = h^2 \int n d^3 p = 75(1+z)^3\,\mathrm{cm}^{-3}$.

Note that at temperatures $kT_\nu \ll m_\nu c^2$ the expression (1) does not represent an equilibrium distribution because in the nonrelativistic domain one may write $_\nu = cp_\nu$ and $p_\nu c = (E_\nu^2 - m_\nu^2 c^4)^{1/2}$; i.e.,

$$n = [\exp((E_\nu^2 - m_\nu^2 c^4)^{1/2}/kT_\nu) + 1]^{-1} \simeq [\exp(\sqrt{2m_\nu c^2 E_k}/kT_\nu) + 1]^{-1},$$

since $E_\nu - m_\nu c^2 \simeq p_\nu^2/2m_\nu = E_k \ll m_\nu c^2$ and $E_\nu + m_\nu c^2 \simeq 2m_\nu c^2$ when $kT_\nu \ll m_\nu c^2$. Adopting (1) and taking an average over phase volume, we obtain

$$\overline{E}_k = \frac{\overline{p_\nu^2}}{2m_\nu} = 6.47\frac{(kT_\nu)^2}{m_\nu c^2},$$

$$\overline{v^{-2}} = 0.385 \frac{m_\nu^2 c^2}{k^2 T_\nu^2},$$

$$\sqrt{\overline{v^2}} = 3.6 \frac{kT_\nu}{m_\nu c} = 6.64(1+z)\,\text{km/s}.$$

For a Maxwellian gas, such a velocity would correspond to a temperature $T_{\text{eq}} = m_\nu v^2/3k = 5.5 \cdot 10^{-5}(1+z)^2$ K. Neutrinos would have become non-relativistic when $z \approx 4.5 \cdot 10^4 (m_\nu c^2/30\,\text{eV})$, long before the hydrogen recombination epoch $z_{\text{rec}} \approx 1500$.

2. Evolution of Primordial Adiabatic Density Perturbations

First let us recall the primordial picture of adiabatic perturbations evolving in the case where neutrinos have zero rest mass, as outlined by us on previous occasions [6, 7].

The deuterium abundance observed in the interstellar medium suggests a low baryon density in the Universe: $\Omega_m = 8\pi G\rho_m/3H_0^2 \simeq 0.03$. In this event the radiation energy density and the energy density of massless neutrinos would have exceeded $\rho_m c^2$ prior to the epoch $z_{\text{rec}} \approx 1500$ of hydrogen recombination in the Universe. When $z > 1500$ the Jeans wavelength would have practically coincided with the horizon:

$$\lambda_{mJ} = \frac{2\pi}{k_J} \simeq \frac{c}{\sqrt{3}} \sqrt{\frac{\pi}{G(\rho_r + \rho_m)}} = 1.3 \cdot 10^6 (1+z)^{-2}\,\text{Mpc}. \tag{2}$$

Here and subsequently all lengths are given in co-moving space. In order to find the corresponding scale today, the value above has to be multiplied by $(1+z)$.

All perturbations on scales $\lambda > \lambda_{mJ}$ will tend to grow [with $\delta\rho/\rho \sim (1+z)^{-2}$]; those on smaller scales will behave like acoustic waves. Small-scale acoustic waves will decay through radiative viscosity and heat conduction [14]. The decay process will serve to isolate a scale comparable with the distance separating clusters of galaxies. After recombination, the Universe would have become transparent to radiation; the Jeans mass would have dropped to $M_J \approx 10^5 M_\odot$, and perturbations on all larger scales would have continued to grow. In an empty universe perturbations would not have grown, while in an open world with $\Omega = 0.03$ they would have been able to build up from epoch $z = 1000$ to $z = 0$ only by a factor of 15.3 [5, 13].

If the initial perturbation spectrum when $z \gg z_r$ is taken in the form $\delta\rho/\rho = \int a_k d^3k$, then after recombination it would have developed to

$$\frac{\delta\rho}{\rho} = \int a_k c_k \, d^3k, \tag{3}$$

where $c_k = \frac{\sin(kR_J)}{kR_J} e^{-kR_c/2}$ for $k > \frac{2\pi}{R_J}$ and $c_k = 1$ for $k < \frac{2\pi}{R_J}$, $k = \frac{2\pi}{\lambda}$ is the wavenumber, R_c specifies the scale of viscous damping at the epoch of recombinations, and R_J is the Jeans wavelength at the same epoch.

How will this picture change if neutrinos have a rest energy of around 30 eV? When $z > 5 \cdot 10^4$ the neutrinos would have been relativistic, and a finite value for their mass would not affect the situation described above. But as soon as $z < 5 \cdot 10^4 (m_\nu c^2/30\,\text{eV})$ the neutrinos would have become nonrelativistic, and we can apply to them the standard solution of the problem of instability in a collisionless gravitating gas [2, 10, 15]. In such a gas one may introduce (in co-moving space), by analogy to the Jeans length, the characteristic length

$$\lambda_J = \frac{2\pi}{k_J} = \sqrt{\frac{\pi}{G\rho_\nu \overline{v^{-2}}}} = \frac{kT_\nu}{m_\nu}\sqrt{\frac{2.6\pi}{G\rho_\nu}}$$
$$= 0.135(1+z)^{-1/2}\theta^{1/2}\left(\frac{30\,\text{eV}}{m_\nu c^2}\right)^{3/2}\ \text{Mpc},$$
$$M_j = \rho_\nu\left(\frac{\lambda_J}{2}\right)^3 = 8 \cdot 10^8 (1+z)^{3/2}\theta^{3/2}\left(\frac{30\,\text{eV}}{m_\nu c^2}\right)^{3.5} M_\odot.$$

The numerical estimates assume that $m_\nu c^2 = 30\,\text{eV}$ and that all three kinds of neutrinos have the same mass. Perturbations will decay on scales $\lambda > \lambda_J$ (see Appendix II) and grow on scales $\lambda > \lambda_J$.

One very important parameter turns out to be the scale equal to the Jeans wavelength at the epoch when neutrinos become nonrelativistic. Today this parameter has the value

$$R_\nu = \lambda_J(z_1)(1+z_1) = 30\,\text{Mpc}$$

and it is of the same order as the distance between clusters of galaxies. On smaller scales, neutrino perturbations would have decayed rather than continue to grow (see Appendix II). This behavior alters the initial spectrum of neutrino density perturbations:

$$c_k = \left(1 + \frac{k^2 R_\nu^2}{4\pi^2}\right)^{-4},$$

i.e., $c_k \approx 1$ for $kR_\nu < 2\pi$ and $c_k \sim k^{-8}$ for $kR_\nu > 2\pi$. Neutrinos would make the Universe either closed or flat; hence when $z < z_1$, neutrino density perturbations on scales $R > R_\nu$ would have grown according to the law

$$\frac{\delta\rho}{\rho} \sim (1+z)^{-1}.$$

Perturbations of the baryon and radiation density on scales smaller than the horizon would not have grown when $z > z_r$ (Appendix I), and by the epoch of recombination they would have become smaller than the neutrino density perturbations by a factor $3\lambda^2/(2\pi R_J)^2$, or a factor $z_1/z_r \approx 40$ for the scale R_ν. After recombination the neutrino perturbations would have pulled perturbations of matter up to their own level (Appendix II). As a result, primordial density perturbations in matter on scale R_ν could have grown by a factor of $5 \cdot 10^4$ from z_1 to $z = 0$ (apart from the dependence on $\Omega_M < \Omega_\nu$), whereas in a universe with $\Omega_m = 0.03$ and $m_\nu = 0$ they would have grown only by a factor of 150 (by a factor of 15 after recombination and another factor of $kR_J \approx 10$ due to the jump in the perturbation amplitude during the recombination process [16]). In order for the observed structure of the Universe to have developed in a world model with $m_\nu c^2 \approx 30\,\text{eV}$, the primordial perturbations in the metric should have been 300 times smaller.[1]

It is reasonable to suppose that metric perturbations δh near the singularity [defined as $(h_k^2 k^2)^{1/2}$, where h_k is the Fourier amplitude of the metric perturbation] would have been independent of k. This spectrum, corresponding to $a_k \propto k^{1/2}$ in (3), is the only possible spectrum that does not single out some characteristic scale. It is described by the single dimensionless quantity δh. The world metric near the singularity is similar to the fractals of Mandelbrot [11].

3. Fine-Scale Fluctuations in Microwave Background Radiation

The existence of perturbations in the density and velocity of matter at the time of recombination would have produced angular fluctuations in the background radiation [7, 12, 14, 16]. On scales R_ν or R_c, the chief mechanism responsible for inducing fluctuations should have been scattering of radiation by electrons in motion due to the presence of velocity perturbations [16].

At the epoch of recombination the amplitude of the mass density perturbations in a world with $m_\nu c^2 = 30\,\text{eV}$ would have been smaller than in the case with $m_\nu = 0$, so the peculiar velocities of matter should also have been smaller. Peculiar velocities in matter could have increased for either of two reasons: the presence of small ($\delta_m < \delta_\nu$) perturbations δ_m in the density of matter prior to recombination, and their growth due to the action of large perturbations δ_ν in the neutrino density. On an angular scale $\theta < 10'$, which would correspond to $\lambda \sim R_{\nu'}$, the expected fluctuations are about 20 times smaller, even in comparison with a world model having $\Omega_m = 1$ and $m + \nu = 0$. In the theory of adiabatic density perturbations in a world with $m_\nu c^2 = 30\,\text{eV}$, the expected level of fluctuations in the primordial background radiation is $\delta T/T \approx 3 \cdot 10^{-6}$ on scales $\theta \leq 10'$.

4. Evolution of Entropic Density Perturbations

As we demonstrate in Appendix I, in a universe with heavy neutrinos and a low baryon density entropic (isothermal) perturbations on scales $\lambda < R_\nu$ cannot play any appreciable role if $\delta_m = \delta\rho_m/\rho_m \ll \Omega_\nu/\Omega_m$ at epoch z_1.

[1]Compared with a model having $\Omega_m = 1$ and $m_\mu = 0$, this increment would not be nearly so great, only a factor of 5–6.

On scales $\lambda > R_\nu$, perturbations of the baryon density during the era $z < z_1$ will induce perturbations in the neutrino density, which will grow and after recombination will pull the matter density perturbations up to their own level. In order for objects (clusters of galaxies) to develop when $z \approx 5$, an initial perturbation $\delta_m > 1/40$ on scale R_ν would be needed.

Appendix I
Growth of Small Fluctuations in Density of Matter

Linearized equations describing the growth of small perturbations can readily be obtained in the context of Newtonian mechanics [3, 4, 8]. For a universe which consisted, prior to recombination, of two noninteracting neutrino components having a rest mass and a mixture of baryons, electrons, atoms, and radiation, these equations take the form

$$\ddot{\delta}_\nu + \frac{4}{3}\frac{\dot{\delta}_\nu}{t} - \frac{2}{3}\frac{\delta_\nu}{t^2} = 0 \tag{A1}$$

$$\ddot{\delta}_m + \frac{4}{3}\frac{\dot{\delta}_m}{t} + \frac{c^2}{3}k_1^2\left(\frac{t_1}{t}\right)^{4/3}\delta_m = \frac{2}{3}\frac{\delta_\nu}{t^2}. \tag{A2}$$

The derivation of these equations assumes that when z was in the range $z_r < z < z_1$ the dynamics of the Universe was determined by nonrelativistic neutrinos, and that the Universe was practically flat; scales smaller than the horizon are considered, and the attraction of baryons and radiation is neglected.

This system of equations implies that when $z_r < z < z_1$ the neutrino perturbations will grow:

$$\delta_\nu = A(t/t_1)^{2/3},$$

and

$$\delta_m = A + t^{-1/3}\{c_1 \sin[\sqrt{3}ck_1(t/t_1)^{1/3}] + C_2\cos[\sqrt{3}ck_1(t/t_1)^{1/3}]\}.$$

The constants C_1, C_2 can easily be evaluated from the initial conditions at $t = t_1$. We see that the oscillating component will decay gradually, and because of the presence of the growing neutrino mode a constant component $\delta_m = 2A/(k_1^2c^2t_1^2)$ will appear, growing with wavelength and becoming comparable to δ_ν near R_J. One will recall that k_1 represents the wave vector at the epoch t_1 when neutrinos became nonrelativistic (A detailed discussion of this and other points raised in this Appendix is given in a separate letter by Rozgacheva and one of us [13]). In the post-recombination era, matter and radiation will no longer interact with each other, and we may neglect the third term in the left-hand member of (A.2). Specifying initial conditions $\delta_m(t_r) = 0$, $\delta_{m'}(t_r) = 0$, we find that $\delta_\nu = A(t/t_1)^{2/3}$ and

$$\delta_m = \delta_\nu \frac{1 + 2t_r}{t - 3(t_r/t)^{2/3}}.$$

If $t = t_r(1+\Delta)$, then $\delta_m = 8\delta_\nu \Delta^2/3$; $\delta_m = \delta_\nu/2$ only when $z \approx 350$.

When $z > z_1$, entropic perturbations would have taken the form of fluctuations in the baryon density relative to a homogeneous background of radiation and neutrinos. When $z < z_1$, in order for perturbations in scale smaller than the horizon and greater than R_ν to have occurred, fluctuations in the baryon density would have had to produce perturbations in the neutrino density growing according to the equation

$$\ddot{\delta}_\nu + \frac{4}{3}\frac{\dot{\delta}_\nu}{t} - \frac{2}{3}\frac{\delta_\nu}{t^2} = \frac{2}{3}\frac{\Omega_m}{\Omega_\nu}\frac{\delta_m}{t^2},$$

which has the solution

$$\delta_\nu = \frac{\Omega_m}{\Omega_\nu}\delta_m[-1 + \frac{3}{5}(t/t_1)^{2/3} + \frac{2}{5}(t_1/t)].$$

For $\Omega_\nu/\Omega_m = 150$, perturbations δ_m, δ_ν on scale R_ν would have been able by epoch $z = 0$ to grow by a factor of 180, or 250 times less than adiabatic perturbations of the same amplitude by epoch z_1. Neutrino perturbations with $R < R_\nu$ could not have grown, because of strong damping. After recombination the growth of δ_m would have depended entirely on the level of δ_ν. IF $\delta_\nu = 0$ (that is, in the range $\lambda \ll R_\nu$),

$$\ddot{\delta}_m = \frac{4}{3}\frac{\dot{\delta}_m}{t} - \frac{2}{3}\frac{\Omega_m}{\Omega_\nu}\frac{\delta_m}{t^2} = 0$$

so that there would be a decaying mode $\delta_m \propto t^{-1/3}$ and a weakly growing mode $\delta_m \propto t^n$, where

$$n = \frac{1}{6}\left(\sqrt{1 + 24\frac{\Omega_m}{\Omega_\nu}} - 1\right) \approx 2\frac{\Omega_m}{\Omega_\nu} \approx \frac{1}{75}.$$

On scales $\lambda > R_\nu$ the perturbations of matter would be pulled up to the level of the neutrino perturbations.

Appendix II

The problem of the damping of waves in a collisionless gas has been examined in detail, both in the literature on plasma theory [1] and for a neutral gas. Our problem differs from the cases discussed earlier, in that the departure of the neutrinos from a Maxwellian spectrum causes the decay to be not exponential but power-law. Waves whose length is substantially shorter than R_ν will decay primarily during the period (t_1, z_1), when neutrinos change from relativistic to nonrelativistic (as will be shown below). During this period the condition $\lambda < R_\nu$ means that the gravitation may be neglected. In this event the problem reduces to the trivial one of particles traveling at constant velocity.

If at initial time t_1 the density in phase space is specified by a function $n = n(\mathbf{p}, \mathbf{x})$, $t = t_1$, then at arbitrary time we will have $n = n[\mathbf{p}, \mathbf{x} - \mathbf{v}(t - t_1)]$. If we take $n = n_0(\mathbf{p}) + \delta n_1(\mathbf{p}) e^{i\mathbf{kx}}$ at $t = 0$, then later on $\delta n = \delta n_0(\mathbf{p}) e^{i\mathbf{k}(\mathbf{x} - \mathbf{v}t)}$, and the density perturbations $\delta\rho(\mathbf{x}, t) = \int \delta n(\mathbf{p}, \mathbf{x}, t)\, d\mathbf{p} = e^{i\mathbf{kx}} \int \delta n_0(\mathbf{p}) e^{-i\mathbf{kv}t}\, d\mathbf{p}$. Here $\mathbf{v}$ should be regarded as a function of momentum:

$$\mathbf{v} = \frac{c^2\mathbf{p}}{E} = \frac{c^2\mathbf{p}}{\sqrt{(m_\nu c^2)^2 + p^2c^2}}.$$

In an expanding universe the product $\mathbf{kv}t$ should evidently be replaced by $\int \mathbf{k}\frac{d\mathbf{x}}{dt}\, dt$. If $\mathbf{k}$ and $\mathbf{x}$ are here written in co-moving coordinates, the $d\mathbf{l} = a d\mathbf{x}$, where a denotes the radius (scale factor) of the Universe, so that $d\mathbf{x}/dt = \mathbf{v}/a$. Furthermore, the momentum of a freely moving (test) particle will decay in inverse proportion to a; that is, $p = p_0 a_0/a$. Thus the integral takes the form

$$J = \text{const} \cdot \int k_\nu \frac{\mu c^2 p_0 a_0\, dt}{a^2\sqrt{p_0 c^2 a_0^2/a^2 + m_\nu^2 c^4}},$$

where $\mu = \cos(k_\nu, p)$.

During the relativistic era this integral simplifies to $\int k_\nu \frac{c}{a}\, dt$ and it converges as $t \to 0$ (since $a \propto \sqrt{t}$). In the nonrelativistic stage we will have $J = \text{const} \int \frac{k_\nu p_0 a_0}{ma^2}\, dt$. For $a \propto t^{2/3}$ this integral converges as $t \to \infty$. These two estimates show that the integral reaches a maximum in just the period of transition from one law to the other; and to order of magnitude the value at the maximum is given by

$$J = k_\nu \frac{ct_1}{a(t_1)} \frac{p_0}{\overline{p}_0} = \frac{k_\nu}{k_{\max}} \frac{p_0}{\overline{p}_0},$$

where k_ν denotes the "maximum" wave vector corresponding to the "minimum" wavelength R_ν introduced above, p_0 is the momentum of a given particle at some early ultra-relativistic stage when $t_0 \ll t_1$, and $\overline{p}_0$ denotes the momentum at the same early epoch averaged over the neutrino spectrum (incidentally, the ratio $p/\overline{p}$ remains constant with time). It is the quantity $\overline{p}_0$ that enters into the calculation of t_1 and k_ν.

The law for the decay of perturbation depends significantly on the form of the function $\delta n_0(\mathbf{p})$. For example, if this function is Maxwellian in form the perturbations in nonrelativistic gas will obey the law $\delta n \propto e^{-\mathbf{v}^2/\theta}$, and over a fixed time interval we would obtain

$$\int e^{-v^2/\theta - ikvt} dv \approx e^{-k^2t^2\theta/4}.$$

But if only particles at rest are perturbed, that is, $\delta n(p) = \delta(p)$ (here the second δ denotes the Dirac delta-function), then such a perturbation will not decay at all.

It is vital, then, to ascertain the character of the initial perturbation that arose prior to the onset of damping, in a "non-primordial" era when the wavelength was longer than the horizon.

One will recall that in the Lifshitz theory [9] there is no density perturbation near the singularity; it arises later, in the course of non-uniform expansion. Thus it is natural to suppose that the perturbation consists of a differing amount of expansion at different points of space leading to differences in temperature and a perturbation of the type $(\partial n/\partial T)e^{ikx}$. A one-dimensional calculation of the perturbation arising in relativistic gas through the action of a gravitational field confirms this idea. Assume that at the beginning $n = n(p) = n(cp/kT)$. The influence of a metric perturbation may be considered equivalent to the action of a gravitational field. The change in the momentum of a relativistic particle in a given metric or in a given field will be proportional to the momentum itself; hence

$$\frac{\partial n}{\partial t} = f(x)p\frac{\partial n(cp/kT)}{\partial p},$$

where $f(x) \propto e^{ikx}$. It is important here that

$$p\frac{\partial n(cp/kT)}{\partial p} = -T\frac{\partial}{\partial T}n\left(\frac{cp}{kT}\right).$$

But if $\delta n \sim \partial n/\partial T$, then $n = n_0 + \delta n = n(p, T + e^{\mathbf{ikx}}\delta T)$, which is what we wanted to show. The treatment of the three-dimensional problem is necessarily approximate.

Thus the decay of perturbations is determined by the convolution

$$\delta\rho = e^{ikx}\int\frac{d}{dt}\frac{1}{e^{cp/kT}+1}e^{-ikpR_\nu/\overline{p}}p^2\,dp\,d\mu\,d\varphi.$$

We may write the expansion

$$\frac{1}{e^{cp/kT}+1} = e^{-cp/kT} - e^{-2cp/kT} + \ldots$$

One readily finds that the contribution of the first term predominates. Evaluating the integrals for $R_\nu k \gg 1$, we obtain

$$\left(\frac{\delta\rho}{\rho}\right)_\nu\bigg|_{t\gg t_\nu} = \left(\frac{\delta\rho}{\rho}\right)_\nu\bigg|_{kct/a=1}\left(\frac{k_{\max}}{k}\right)^4.$$

In comparing the expected amplitude of short-wavelength perturbations at the present epoch, we must also recognize that the shorter the wavelength, the later (already in the nonrelativistic era, when $t > t_\nu$) a given perturbation will enter the growing mode, and the less will be its growth from that time to the present epoch. This circumstance introduces a factor $(kR_\nu)^2 = (\lambda_J/\lambda)^2$. On the other hand, during the relativistic era $z > z_1$, only perturbations with $\lambda > ct$ will grow, thereby introducing another factor $(kR_\nu)^2$. The law for the decline in the amplitude of the perturbation toward short wavelengths is finally given, then, by the expression

$$\left(\frac{\delta\rho}{\rho}\right)_\nu = \frac{\delta\rho}{\rho}(R_\nu)(kR_\nu)^{-8}.$$

This decrease in amplitude proportional to the eighth power of the wavenumber is fast enough for R_ν to be considered the controlling scale in further analysis of the nonlinear growth of perturbations.

Institute for Space Research
USSR Academy of Sciences. Moscow
M. V. Keldysh Institute of Applied Mathematics
USSR Academy of Sciences. Moscow

Received
May 28, 1980

REFERENCES

1. *Artsimovich L. A., Sagdeev R. Z.* Fizika plazmy dlya fizikov [Plasma Physics for Physicists]. Moscow: Atomizdat (1979).
2. *Bisnovatyĭ-Kogan G. S., Zeldovich Ya. B.*—Astron. Zh. **47**, 942 (1970).
3. *Bonnor W. B.*—Month. Not. RAS **117**, 104 (1957).
4. *Weinberg S.* Gravitation and Cosmology. N.Y.: Wiley (1972).
5. *Guyot M., Zeldovich Ya. B.*—Astron. and Astrophys. **9**, 227 (1970).
6. *Doroshkevich A. G., Zeldovich Ya. B., Sunyaev R. A.*—In: Confrontation of Cosmological Theories and Observational Data. Ed. M. S. Longair. Dordrecht (1974).
7. *Doroshkevich A. G., Zeldovich Ya. B., Sunyaev R. A.*—Astron. Zh. **55**, 913 (1978).
8. *Zeldovich Ya. B., Novikov I. D.* Stroenie i evolyutsiĭa Vselennoĭ [Structure and Evolution of the Universe]. Moscow: Nauka (1975).
9. *Lifshitz E. M.*—ZhETF **16**, 587 (1946).
10. *McCone A. I.*—Nucl. Sci. Abstrs. **25** 15 (1971).
11. *Mandelbrot J.* Fractals. San Francisco: Freeman (1975).
12. *Peebles P. J. E., Yu I. T.*—Astrophys. J. **162**, 815 (1970).
13. *Rozgacheva I. G., Sunyaev R. A.*—Pisma v Astron. Zh. **21**, 370 (1980).
14. *Silk J.*—Astrophys. J. **151**, 459 (1968).
15. *Silvestrov V. V.*—Astron. Zh. **51**, 293 (1974).
16. *Sunyaev R. A., Zeldovich Ya. B.*—Astrophys. and Space Sci. **7**, 3 (1970).
17. *Tremaine S., Gunn J. E.*—Phys. Rev. Lett. **42**, 407 (1979).
18. *Szalay A. S., Marx G.*—Astron. and Astrophys. **49**, 437 (1976).

61

Astrophysical Implications of the Neutrino Rest Mass III. Nonlinear Growth of Perturbations and the Missing Mass*

With A. G. Doroshkevich, R. A. Sunyaev, M. Yu. Khlopov

In two previous Letters we have discussed how neutrinos of finite mass would influence the structure and evolution of the Universe as a whole, as well as the growth of irregularities during the linear phase and the amplitude of fine-scale fluctuations in the temperature of the primordial background radiation. We shall now examine the behavior of the nonlinear stage in the evolution of irregularities and the formation of galaxies and clusters of galaxies, and we shall consider what bearing this behavior has on the problem of the missing mass.

In particular, we shall demonstrate that: 1) if neutrinos have a rest energy of $\approx 30\,\mathrm{eV}$, then a hidden-mass effect will inevitably arise, and the hidden mass will be distributed very differently from ordinary visible matter; 2) the nonlinear evolution stage of irregularities in the "gas" of finite-mass neutrinos will result in the formation of highly flattened, extended condensations of neutrinos and matter, like gas-dynamical "pancake" structures; 3) the enhanced ultraviolet and X-ray luminosity of the matter compressed into a "pancake" would probably serve to explain the low H 1 content in the space between clusters of galaxies.

The missing-mass problem [16, 22] arises from the serious disparity between the value of the mass inferred from the virial theorem and the mass given by the mass-luminosity relation for pairs of galaxies as well as groups and clusters of galaxies. The first authors to inquire whether this phenomenon might be attributable to the existence of a small ($\approx 10\,\mathrm{eV}$) rest energy for neutrinos seem to have been Marx and Kovesi [14], and more recently Marx and Szalay [15]. Gunn and Tremaine [4] have pointed out several difficulties with such an explanation if $\Omega_\nu \ll 1$, that is, if $m_\nu c^2 < 1.2\,\mathrm{eV}$. However, if $m_\nu c^2 \approx 30\,\mathrm{eV}$ and $\Omega_\nu \gg \Omega_b$, then the evolution of irregularities

*Pisma v Astronomicheskiĭ zhurnal **6** (6), 465–469 (1980).

in the neutrino gas will in fact control the large-scale structure of the Universe. Accordingly, the question of the ultimate distribution of the missing mass might well be answered if the evolution of these irregularities could be studied in detail.

We would point out here that efforts to explain the high density of hidden mass by appealing to some type of low-mass stars face difficulties in explaining the primordial deuterium abundance [2,10,23]. The observed abundance of deuterium in the interstellar gas would be consistent with calculations of cosmological nucleosynthesis on the big-bang model only for a very small baryon density in the Universe: $\Omega_b \approx 0.03$.

The presence of a neutrino rest energy $m_\nu c^2 \approx 30\,\mathrm{eV}$ during the era of nuclear reactions (when $kT \geq 100\,\mathrm{keV}$) would have had no effect on either the expansion dynamics of the Universe or the rate of nuclear reactions. We thereby arrive at a picture wherein the neutrino mass density today is approximately 100 times the density of matter.

The thermal velocities of the neutrinos in the expanding Universe would have diminished in proportion to $(1 + z)$, and by epoch $z = 0$ they should have dropped to about 5 km/s, corresponding to a neutrino "temperature $T \approx 0.75 \cdot 10^{-4}$ K (one may speak rather conventionally of the neutrino temperature, since the distribution function of neutrinos should be far from Maxwellian; see Letter II). If one primitively regards the formation of clusters of galaxies as a process whereby neutrinos pile up in the potential well of the cluster gravitational potential, then with neutrino velocities $v \approx 10^3$ km/s in such a potential, and with density being conserved in phase space, the density of neutrinos in galaxy clusters ought to be roughly 10^3 times the values "observed." But the cluster formation process would actually have been more complicated.

1. Following the recombination of hydrogen, there would have been negligible interaction between matter and radiation, and the gravitational influence of perturbations in the neutrino background would soon have given rise to perturbations in matter as well (see Letter II). Intrinsic matter perturbations would also have been preserved, but the differences between the matter and neutrino perturbations would not have become magnified with time.

According to the nonlinear theory of gravitational instability [9], the perturbations in the neutrino background would have evolved into neutrino "pancakes" with a characteristic scale $R_\nu \approx 12\beta_\nu(1 + z)^{-1}(\Omega_\nu h^2)^{-1}$ Mpc determined by the scale for decay of neutrino background perturbations (here $\beta_\nu \approx 0.5$). Since the growing modes of perturbations in the neutrino background and in matter are identical, the matter in the neutrino pancake should also become flattened. And if the neutrinos can travel freely through the pancake, then matter will be decelerated, compressed, and heated in the shock wave bounding the gaseous pancake.

Calculations performed in the one-dimensional approximation show that as a pancake consisting of collisionless particles evolves, finite-mass neutrinos should form a gravitationally bound pancake of their own which is two to three times as thick as the gaseous pancake. Hence the mean neutrino density in the pancake will, as before, exceed the mean density of the gaseous component, but not by so great an amount as in the Universe as a whole. This circumstance could weaken Sandage and Tammann's restrictions [19] on the maximum value of Ω.

If we compare the parameters of the gas flattened into a pancake in the model with finite-mass neutrinos and in models with $m_\nu = 0$, we find that in the former case the density of the compressed matter, and thereby the luminosity of the pancake in the ultraviolet and X-ray regions, should be several times as great. The principal reason is the change in the perturbation spectrum (see letter II). We know that the gas temperature depends only weakly on the form of the spectrum, and is determined by the characteristic scale of irregularity. Since the characteristic scale $R_\nu \approx 12\beta_\nu(1+z)^{-1}(\Omega_\nu h^2)^{-1}$ Mpc is approximately the same as the scale of damping used [20, 21] in models with $m_\nu = 0$, namely $R_c = 2\beta_c(1+z)^{-1}(\Omega_b h^2)^{-1}$ Mpc, $\beta_c \approx 0.5$ (allowing for the difference in the values of Ω_ν and Ω_b), the compressed gas should change little in temperature as we pass to models with $m_\nu \neq 0$. As m_ν decreases the scale R_ν will become longer and there will be a corresponding rise in the gas temperature. Current observational limits on the intensity of the ultraviolet and X-ray background, as well as information on the large-scale structure of the Universe, impose restrictions on the scale: $R_\nu \lesssim 100 - 200$ Mpc, so that we can evidently set the lower limit $m_\nu c^2 \gtrsim 5$ eV on the neutrino rest energy, with the additional condition $\Omega_\nu \gg \Omega_b$.

Thus in models with finite-mass neutrinos the flattened gas would have three to five times the density, pressure, and thermal energy, and with the corresponding increase in emission measure the pancake would become more luminous. In earlier estimates we have shown [5, 8] that in models with $m_\nu = 0$ the radiation of the flattened gas would not alone be adequate to produce strong photoionization of the hydrogen between clusters and thereby to explain the Gunn-Peterson effect [3]. But with the enhanced luminosity of the pancake, this radiation could probably ensure adequate ionization of the intercluster hydrogen.

2. Radiative cooling of the gas heated in the shock wave, along with thermal and hydrodynamic instability, should have caused the cooled gas to break up into separate clouds and gas-star complexes [7], which then would have undergone successive clumping to form galaxies of various sizes and galaxy clusters. In addition to the gas-star complexes, finite-mass neutrinos would have taken part in this amalgamation process. It is important to recognize that the participating neutrinos would have taken part in this amalgamation process and that the participating neutrinos would have belonged to the pancake, rather than to the uniformly distributed background.

When neutrinos condense into a pancake, a multistream distribution of particles will arise, with a sharply anisotropic distribution function [12]. The caustic surface separating the one- and three-stream distributions may be regarded as the boundary of the neutrino pancake. Any small departures from one-dimensional flattening will induce large neutrino velocities in the plane of the pancake as well. The effective entropy of the neutrino gas in the pancake, due to the development of the multistream distribution and determined by the averaged random motions of the neutrinos in the macroscopic volume, will be about the same as the entropy of the hot gas and of the "gas" composed of cool gaseous clouds and gas-star complexes, although both the density and the random velocities of the neutrino "gas" will be several times as great as the same quantities for the cloud "gas" and for the ordinary hot gas belonging to the pancake. We would therefore expect that further agglomeration of neutrinos, clouds, and hot gas into separate galaxies and clusters of galaxies would proceed along the lines developed [7] for models with $m_\nu = 0$. Recent estimates show [6] that the entropy of the matter in a pancake should be comparable with the entropy of the matter belonging to clusters of galaxies.

Rapid relaxation processes should play an important role in the clumping dynamics, for they should cause a quasi-steady state to be established in the system [13, 17]. Probably this is how clusters of galaxies have formed, but rapid relaxation processes alone cannot account for the formation of individual galaxies [1, 6]. An important contribution to individual galaxy formation probably is made by dissipative processes such as collisions among agglomerating clouds, which would weld them together, causing them to slow down and settle toward the center. Only by taking processes like this into consideration can one hope to explain the very high densities encountered in the central regions of galaxies (as high as $10^{-20}\,\mathrm{g/cm^3}$).

These effects and their possible influence on the agglomeration of finite-mass neutrinos during galaxy formation are discussed in a separate paper by two of us and colleagues [11]. We have shown that the rise in the density ρ_b of colliding particles will cause the gravitational field to depend on time, leading in turn to a rise—although a more gradual one—in the density ρ_ν of collisionless particles. The density ratio ρ_b/ρ_ν will vary in the course of collapse according to the law

$$\frac{\rho_b}{\rho_\nu} \propto \rho_b^{1/4},$$

and ρ_ν will grow with time. The density of collisionless particles will fall off more slowly with radius than will the density of matter; and if in the central regions the density ratio can be $\rho_b/\rho_\nu \approx 3 - 10$, then near the periphery of galaxies and beyond we can expect that $\rho_\nu > \rho_b$, while at distances of several galaxy radii the neutrino density should be similar to the density of the neutrino background in the pancake, or two to three times the density of visible matter in the same region of space.

If the missing mass is attributable to neutrinos, it should be manifested in very different ways in rich clusters, such as in hypergalaxies and groups of galaxies. This difference will result from the differing gravitational potentials of rich clusters, superclusters, and galaxies. Clusters have a large gravitational potential, corresponding to velocities of order 10^8 cm/s; tidal forces are also large, serving to merge the extended envelopes of the member galaxies. Hence the neutrinos in clusters probably would not be associated with individual galaxies. In clusters it would be more natural for a general neutrino background to form, by analogy to the hidden background mass consisting of dwarf stars, black holes, and so on. Probably the massive neutrino background in clusters would not exhibit any distinctive properties.

On the contrary, in individual galaxies, hypergalaxies, and groups of galaxies the contribution of the neutrino background to the missing mass might be several times smaller than in clusters and in supergiant galaxies. Upon condensing into a pancake, finite-mass neutrinos would acquire velocities of order 10^7 cm/s, and these velocities might impede gravitational capture of the neutrinos by systems having a small gravitational potential. But if the missing mass is associated with ordinary matter in the form of some sort of stars or black holes, then possible collisions and dissipative processes during the era when those stars were being formed would facilitate their gravitational amalgamation with systems of galaxies. Thus the development of missing mass in the form of neutrinos will depend strongly on the mass of the individual galaxy or, properly speaking, the group of galaxies, and it will diminish rapidly as that mass diminishes.

Since our Galaxy has a gravitational potential corresponding to virial velocities $v \approx 300$ km/s, the neutrinos in the Galaxy should have velocities of the same order. For the Earth, the gravitational potential implies velocities smaller by an order of magnitude, and for the Sun they would be similar, so the density of neutrinos in the Sun and in the Earth should differ insignificantly from their mean density in the Galaxy. Thus neutrinos will make a negligible contribution to the mass of the Earth and the Sun.

We would also expect the hidden neutrino mass to be more prominent at the periphery of large superclusters or in sparse superclusters of galaxies formed from primordial pancakes of low mass, because in such cases the neutrinos would, on the average, have acquired lower velocities and a smaller effective entropy when they condensed into their pancake. As a result, one might find a correlation between the position of pairs or groups of galaxies and the occurrence of hidden neutrino mass. The system Galaxy + Andromeda Nebula might well harbor hidden mass to a striking degree: each galaxy is quite massive and both are felicitously located at the outer edge of the local supercluster. To see whether this predicted behavior agrees with the observations would make a very interesting test.

Measurements by Roberts and Whitehurst [18] of the rotational velocity of M31 in the 21-cm line as a function of radius have shown that the velocity $v \sim (GM/R)^{1/2}$ does not fall off with radius. The familiar result follows, that $M \propto R$ and that the missing (neutrino) mass has a density $\rho_\nu \propto R^{-2}$, whereas the density of visible matter drops more rapidly with radius.

In a short letter it is hard to touch on all the manifold implications that the existence of a background of finite-mass neutrinos would have for cosmology and for the theory of galaxy formation. To sum up our discussion in a general way, we would merely point out that finite-mass neutrinos do not alter the fundamental principles and conclusions of Friedman cosmological models or the adiabatic theory of the formation of large-scale structures in the Universe—galaxies and clusters of galaxies. Only the numerical predictions of the theory change—and these for the better. An interesting (because of the possible observational consequences) qualitative effect of the existence of a mass for the neutrino would be the sharp rise in the density of primordial neutrinos in the Galaxy: by a factor of $10^4 - 10^5$ compared with the case of massless neutrinos distributed uniformly throughout the Universe.

Institute of Applied Mathematics
USSR Academy of Sciences. Moscow
Institute for Space Research
USSR Academy of Sciences. Moscow

Received
May 28, 1980

REFERENCES

1. *Aarseth S. J., Binney J.*—Month. Not. RAS **185**, 227 (1978).
2. *Wagoner R.*—Astrophys. J. **179**, 343 (1973).
3. *Gunn J. E., Peterson B. A.*—Astrophys. J. **142**, 1633 (1965).
4. *Gunn J. E., Tremaine S.*—Phys. Rev. Lett. **42**, 407 (1979).
5. *Doroshkevich A. G., Zeldovich Ya. B., Sunyaev R. A.*—In: Confrontation of Cosmological Theories and Observational Data. Ed. M. S. Longair. Dordrecht (1974).
6. *Doroshkevich A. G., Klypin A. A.* Preprint of the M. V. Keldysh Inst. for Appl. Math. AS USSR **2**, Moscow (1980).
7. *Doroshkevich A. G., Saar E. M., Shandarin S. F.*—Month. Not. RAS **184**, 648 (1978).
8. *Doroshkevich A. G., Shandarin S. F.*—Astron. Zh. **52**, 643 (1975).
9. *Zeldovich Ya. B.*—Astrofizika **6**, 319 (1970).
10. *Zeldovich Ya. B.*—Pisma v Astron. Zh. **1** 1, 10 (1975).
11. *Zeldovich Ya. B., Klypin A. A., Khlopov M. Yu., Chechetkin V. M.*—Yader. Fiz. **31**, 1286 (1980).
12. *Zeldovich Ya. B., Myshkis A. D.* Elementy matematicheskoĭ fiziki [Elements of Mathematical Physics]. Moscow: Nauka (1973).

13. *Lynden-Bell D.*—Month. Not. RAS **136**, 101 (1967).
14. *Marx G., Kovesi-Domoros F. I.*—Acta Phys. Acad. Sci. Hung. **17**, 171 (1964).
15. *Marx G., Szalay S.*—Astron. and Astrophys. **49**, 437 (1976).
16. *Ostriker J. R., Peebles P. J. E., Yahil A.*—Astrophys. J. Lett. **193**, L1 (1974).
17. *Peebles P. J. E.*—Astron. J. **75**, 13 (1970).
18. *Roberts M. S., Whitehurst R. V.*—Astrophys. J. **201**, 327 (1975).
19. *Sandage A., Tammann G. A.*—Astrophys. J. **196**, 313 (1975).
20. *Silk J.*—Astrophys. J. **151**, 459 (1968).
21. *Chibisov G. V.*—Astron. Zh. **49**, 74 (1972).
22. *Einasto J. E., Kaasik A., Saar E. M.*—Nature **250**, 309 (1974).
23. *Reeves H.*—Ann. Rev. Astron. Astrophys. **12**, 437–469 (1974).

Commentary

The theory of the hot Universe predicts a large concentration of neutrinos. The cosmological consequences of this fundamental fact were first considered in an article by Ya. B. and S. S. Gershtein (see **26** and accompanying commentary).

The report of laboratory measurements of the neutrino mass coincided with the accumulation of difficulties in cosmology (the problem of hidden mass and the slowed growth of perturbations in the open model of the Universe with relic radiation—article **58**). The present article heralded the beginning of a new wave in cosmology. It was followed by a huge flow of papers on the role of heavy neutrinos and other not-yet-observed particles in cosmology. The final clarification of the physical situation will not be achieved soon; it is unthinkable without extraordinarily difficult breakthroughs in elementary particle physics.

At the same time, the pioneering role of these three papers will remain forever.

62

The "Black" Regions of the Universe*

With S. F. Shandarin

The fragmentation model, corresponding to the scenario whereby primordial adiabatic density fluctuations produced the large-scale structure of the Universe, implies that vast regions should exist nearly devoid of galaxies. A discussion is given of the physical properties of the gas that should occupy these "black" regions and the prospects for detecting it.

Radial velocities have lately been measured in large quantity for galaxies in several parts of the celestial sphere and have furnished distances for all the galaxies brighter than some limiting magnitude. The distribution of these galaxies in space suggests that there exist vast regions, containing up to a million cubic megaparsecs, in which galaxies are almost or totally absent [4, 14, 15, 16, 19].

We have remarked on previous occasions, although perhaps not loudly and persistently enough, that regions of this kind would arise in a natural way from the "pancake" scenario that we proposed more than a decade ago [12] for the formation of structure in the Universe. Without pretending to regard the choice of scenario or the quantitative theory as finalized, we should like to concentrate here on the question of the "black" regions of the Universe and the problems involved in studying them.

Deliberately, consciously, we shall limit our arguments to the latest stages in the evolution of the Universe, bypassing the earlier period of fully ionized plasma. The question of how irregularities would have evolved during the post-recombination phase reduces to the following alternatives: 1) comparatively small structural units ($\sim 10^6 M_\odot$) formed at first, later collecting into larger structures, as large as superclusters [3, 5] (the "clumping" process); 2) irregularities initially developed on the largest scale ($\sim 10^{15} M_\odot$), with smaller-sized objects being formed in places of enhanced density—that is to say, the denser gas clouds broke up into separate galaxies [6, 13, 18] (the fragmentation process).

In the clumping version, during the late phase one has to consider the motion of a gas whose molecules are represented by globular clusters or galaxies. In the fragmentation version the global structure would be determined by the motion of an atomic hydrogen and helium gas. In either case finite-mass neutrinos may participate; we shall not dwell on their effects here [10].

Setting aside the question of the high-density regions, let us turn now specifically to the regions of diminished density. It is in fact for the low-density regions that the clumping and fragmentation pictures differ most strikingly.

*Pisma v Astronomicheskiĭ Zhurnal **8** (3), 131–135 (1982).

In the clumping case, the galaxies that originally formed would have been distributed more or less uniformly. Through their mutual gravitation they would then have collected into more massive aggregates: clusters and superclusters of galaxies. Accordingly, in the clumping version the places of subnormal density would be regions containing more or less ordinary galaxies; the galaxies would merely be less highly concentrated than on the average for the Universe as a whole.

Now let us take the fragmentation case. In this situation, on entering the regions of high density (the pancakes) the gas would be compressed by a shock wave and heated to $T \sim 10^6 - 10^7$ K, becoming ionized. Part of the heated gas would then radiate and cool off to $T \sim 10^4$ K. As a result the gas would recombine and its density would rise further. All in all the pancakes would acquire the conditions needed for galaxy formation. The galaxy formation process would take place quite rapidly, since the perturbation growth rate in gravitational instability theory is determined primarily by the local density.

If this picture is correct, the formation of dense-gas zones would naturally have been accompanied by the emergence of regions of diminished gas density. In these latter regions, however, no fragmentation would take place at all: no galaxies would develop. The main reason is that in the low-density regions the time scale for irregularities to grow to galactic scale would exceed the cosmological time.

One can easily set a lower limit on the *average* density of the gas remaining in the "black" regions. In the scenario of fragmentation during the stage prior to formation of the first pancake object, every gas element would belong to one of four groups. The particles in the first group would undergo a contraction (with respect to a reference frame co-moving with the mean expansion of the Universe) along all three principal axes; those of the second group would contract along two axes, expanding along the third; the particles in the third group would contract along only one axis; and the fourth group of particles would expand along all three axes. According to Doroshkevich [7], the first and fourth groups each contain 8% of all matter; the second and third groups, 42% each. Over a prolonged time span the gas expanding along all three axes would be unable to fragment into objects of smaller size, as its density would be lower than the mean density in the Universe. Thus at least 8% of all matter would persist in the "black" regions as gas. If this gas were to occupy the *whole* volume, its mean density would amount to about 10% of the average density in the Universe. Actually, however, its density is higher.

Even at low density, neutral gas would be detectable from its absorption of the $L\alpha$ line. In continuous gas caught up in the cosmological expansion, the $L\alpha$ line would be converted into a full band. Realizing that the spectra of distant quasars showed no continuous absorption of radiation at wavelengths shortward of $L\alpha$, Gunn and Peterson [2] concluded that the neutral hydrogen density is $10^5 - 10^6$ times lower than the mean density of matter. Believing that the total gas density could not be excessively low, they inferred that the gas in the space between clusters of galaxies is almost fully ionized.

Now such a picture would arise in a natural way in the fragmentation model [8]. Ultraviolet radiation and soft X-rays from the dense regions (both the emission of the shock-compressed primordial gas and the emission of individual objects born in the dense regions) would penetrate the low-density regions, ionizing and heating the residual gas there. In the black regions the gas temperature would rise to a few tens of thousands of degrees. This temperature is set by the condition that a balance be established between the radiation flux external to this gas and its own radiation at longer wavelengths. The gas temperature would in fact be high enough to inhibit galaxy formation. For a density of $0.1\rho_c$ and $T = 2 \cdot 10^4$ K the Jeans mass $M_J \sim 10^{11} M_\odot$.

We have explained that galaxies cannot be formed inside the black voids, but that does not mean that the gas there will necessarily be distributed uniformly. Indeed, the process whereby the dense pancake zones are created will stretch out in time, in accord with the statistical distribution of the initial fluctuations. Thus along with the denser regions created early on from neutral gas, condensations (pancakes) formed later should also be present. By that time the first pancakes to develop would have heated all the gas not yet incorporated into the later pancakes.

The density would have been less greatly enhanced in the latter pancakes for two reasons: a) those pancakes developed from weaker fluctuations, and thus were formed at a later stage; b) they developed in hot ionized gas, whose elasticity impeded contraction. These late pancakes would never be able to turn into galaxies. They would not spoil the black region, not populate it with optically luminous matter (galaxies, stars). Nevertheless, the late pancakes might conceivably be detectable if small amounts of neutral hydrogen there were to produce absorption lines in the spectra of distant quasars [9, 17].

If in that fashion one were to prove that hot gas exists in a black void that contains no galaxies, then the whole pancake concept would receive major support; our interpretation of the origin of the black regions would be strengthened, and the quantitative theory would take a step forward.

Let us turn to the process by which the black regions were formed. Intuitively it is hard to conceive how any region of space might be emptied by gravitational instability. Hence the existence of large zones containing no bright galaxies (the black regions) probably means that the physical conditions in these regions inhibited galaxy formation when the density of matter was only *moderately* low. The hypothesis that galaxies once occurred everywhere with approximately the same number density, and that every last one later departed from a whole big region, seems very unlikely. It would appear, then, that the discovery of black regions supports the fragmentation theory and the concomitant theory of adiabatic perturbations, and it militates against the theory of successive agglomeration based on the idea of entropic (isothermal) perturbations.

In a certain sense black regions are even more characteristic of the fragmentation theory than are cellular and lattice structures. In fact, in the pancake theory as the fluctuations grow the first features to develop will be high-density zones, typified by contraction along a single axis. As time passes other types of features will appear; we have given elsewhere [1] a general classification of them for the motion of cold matter. A well-defined porous and filamentary structure will arise, on the whole.

This structure will, however, break down during the late evolutionary stage. The galaxies originally formed in the pancakes will undergo secondary clustering [11]. In the process the distinct cellular and lattice format will be lost. As before, motion will take place most intensively in the high-density regions, and the rarefied regions will be affected to a far less extent. Conclusions drawn from analysis of the black regions therefore seem more dependable.

However, as emphasized at the outset, the question is by no means definitely answered. Our aim in this letter is not to arrive at any hasty inference based on incomplete data. It is the prospects for investigating the black regions that we wish to stress. At the present stage it is important to recognize that detailed study of the black regions, both observational and theoretical, might before too long give us the answer we seek. What ought to be done along these lines?

The first and most obvious need is to enlarge the field of study by detecting and measuring a large number of black regions, enough of them for their statistical properties to be considered.

Progress is required in measuring the spectra of fainter galaxies, in order to reinforce the claim that none occur in the black regions. Such a claim will always have the character of an upper limit, referring to a number of galaxies smaller than some specified percentage of the expected average number; and this limit has to be lowered in the same way as the limiting luminosity of any faint galaxies which might happen to turn up in a black region.

It would be worthwhile learning how to distinguish a galaxy of high peculiar velocity in a dense region from an isolated galaxy that is advancing by Hubble's law into a black region. There is an opposite side to the coin: one must be able to discriminate galaxies born in a black region from galaxies that arrived there after having been ejected from dense regions. A peculiar velocity of ≈ 1000 km/s would introduce into the estimated distance of the galaxy an error of $10h_{100}^{-1}$ Mpc. Traveling at that speed, a galaxy would cover a distance of $\approx 10h_{100}^{-1}$ Mpc in the Hubble time$t_0 \sim H_0^{-1}$.

A direct test for deciding between the fragmentation and clumping alternatives (or equivalently, whether the adiabatic or the entropic scenario prevailed when structure was developing in the Universe) would be made by a positive discovery of gas inside the black regions. Determining its chemical composition would practically solve the problem. In the fragmentation scheme the black regions ought to be filled with gas having the pregalactic abundances: about 75% hydrogen and 25% helium. In the clumping model there should not be any regions devoid of heavy elements.

The observational results should be compared against the pictures given by numerical calculations, and also against theoretical estimates. But such estimates are difficult to make. Above all, the theory calls for definite input data: the density of matter, or if the neutrino does have a rest mass, the ordinary matter density and the neutrino density. The spectrum and mean amplitude of the density fluctuations are required.

The matter density, neutrino mass, and perturbation amplitude will together determine the epoch when the large density irregularities were formed. If these parameters are not known, one must resort to considering various cases. The same holds for numerical calculations. We would mention that until recently in presenting the results of numerical analyses (see, for example, the diagrams in the proceedings of the Tallin symposium [13]) we customarily depicted the density distribution of matter, rather than that of galaxies; in giving illustrations we ignored the fact that galaxies would have been formed in dense regions but not in rarefied regions. The distribution of *galaxies* has higher contrast than the distribution of matter.

Summing up our arguments, we can only reemphasize the importance of investigating the "black" regions. The available evidence on these regions already tends to favor the fragmentation picture, which would come about if the adiabatic scenario for development of structure in the Universe is correct. New data on the black regions should at last help us decide whether fragmentation or agglomeration was mainly responsible for producing the structure in the Universe today.

Note added in proof. A new model for the development of structure in the Universe has recently been suggested [*Ostriker J., Cowie L.*—Astrophys. J. Lett. **243**, L127 (1981)] in which the dominant role would be played by explosions of fully evolved objects. The model predicts that the distribution of heavy elements in the Universe should be highly uniform, and, like the entropic model, it differs in this respect from the adiabatic model.

M. V. Keldysh Institute of Applied Mathematics
USSR Academy of Sciences. Moscow

Received
December 3, 1981

REFERENCES

1. *Arnold V. I., Zeldovich Ya. B., Shandarin S. F.*—Preprint of the M. V. Keldysh Inst. of Appl. Math. AS USSR **100**, Moscow (1981); Geophys. and Astrophys. Fluid Dyn. **20**, 111 (1982).
2. *Gunn J., Peterson S.*—Astrophys. J. **142**, 1633 (1965).
3. *Gott J. R., Rees M. J.*—Astron. and Astrophys. **45**, 365 (1975).
4. *Gregory, S. A., Thompson L. A.*—Astrophys. J. **222**, 784 (1978).
5. *Dicke R. H., Peebles P. J. E.*—Astrophys. J. **194**, 838 (1968).
6. *Doroshkevich A. G., Shandarin S. F.*—Astron. Zh. **51**, 41 (1974).
7. *Doroshkevich A. G.*—Astrofizika **6**, 581 (1970).
8. *Doroshkevich A. G., Shandarin S. F.*—Astron. Zh. **52**, 9 (1975).
9. *Doroshkevich A. G., Shandarin S. F.*—Month. Not. RAS **179**, 95 (1977).
10. *Doroshkevich A. G., Khlopov M. Yu., Sunyaev R. A., et al.*—In: Proc. Xth Tex. Symp. on Relativ. Astrophys. N.Y., 1981, p. 32.
11. *Doroshkevich A. G., Kotok E. V., Novikov I. D., et al.*—Month. Not. RAS **192**, 321 (1980).
12. *Zeldovich Ya. B.*—Astrofizika **9**, 319 (1970).
13. *Zeldovich Ya. B.*—In: The Large Scale Structure of the Universe. 1978. (IAU Symp. N79).
14. *Joeveer M., Einasto J., Tago E.*—Preprint Tartu Astron. Observ. N 1 (1977).
15. *Chincarini G., Rood H¿ J.*—Astrophys. J. **230**, 648 (1978).
16. *Kirshner R., Oemler A., Schechter P., Schectman S.*—Astrophys. J. Lett. **248**, L57 (1981).
17. *Oort J. H.*—Astron. and Astrophys. **94**, 359 (1981).
18. *Sunyaev R. A., Zeldovich Ya. B.*—Astron. and Astrophys. **20**, 189 (1972).
19. *Tarenghi M., Tifft W. C., Chincarini G., Rood H. J.*—Astrophys. J. **234**, 793 (1979).

Commentary

This paper investigates one of the most important consequences of the theory of large-scale perturbations of the Universe. It predicts the existence of regions in which galaxies never formed. It is obvious that the alternative theory, in which galaxies formed everywhere, but later grouped together, leads to completely different conclusions about the properties of matter in the black regions. The detailed investigation of the black regions has become a most important problem of cosmology. We note a more detailed article by Ya. B., Ya. E. Einasto and S. F. Shandarin in *Nature.*[1]

The interest generated by these works may be see in the review by J. Oort.[2]

[1] *Einasto J., Shandarin S. F., Zeldovich Ya. B.*—Nature **5891**, 407–412 (1982).
[2] *Oort J.*—Astron. Astrophys. **139**, 211–214 (1984).

63

The Origin of Large-Scale Cell Structure in the Universe*

A qualitative explanation is offered for the characteristic global structure of the Universe, wherein "black" regions devoid of galaxies are surrounded on all sides by closed, comparatively thin, "bright" layers populated by galaxies. The interpretation rests on some very general arguments regarding the growth of large-scale perturbations in a cold gas.

Both observations and theory indicate that great regions devoid of galaxies—"black" regions—exist in the Universe, separated by comparatively thin layers where the galaxies congregate. Such structure may be termed "cellular." The number density of galaxies rises along the lines of intersection of layers, and for this reason one sometimes speaks of a cellular-lattice structure. As explained by Shandarin and the author in another letter in this issue [4], it appears that the black regions are filled with ionized gas whose density is several times lower than the average density of matter in the Universe. Galaxies were never born in this gas. The combined volume of the black regions apparently surpasses by a large margin the "bright" volume occupied by galaxies.

In this letter I consider a topological question: Why should the bright volume comprise an unbroken ensemble of cell walls, whereas the black regions are disjoint, separated from one another by the bright walls?

Clearly if the two types of regions were distributed at random, one would expect the regions encompassing the smaller combined volume to be strewn like blobs within the continuous bulk of the region covering the larger volume. An everyday example: milk consists of 6% fat and 94% water by volume, the separate droplets of fat being surrounded by a continuous mass of water.

Roughly speaking, we may suppose that (60–80)% of the matter in the Universe by weight belongs to the bright regions, whose density is about 10 times the average. Thus the bright regions occupy (6–8)% of the whole volume. The remaining (20–40)% of matter extends throughout (92–94)% of the volume, so that the density of the matter in the black regions is 0.43–0.22 of the mean value. It would seem, then, that the bright regions ought to be disconnected and surrounded by dark space. But the cellular structure means that, remarkably enough, the situation is actually just the opposite.

*Pisma v Astronomicheskiĭ zhurnal **8** (4), 195–197 (1982).

By bright regions I refer not to the individual galaxies themselves but to the entire domain outlining the distribution of galaxies in space. As Oort [5] points out, the cell structure is not very firmly established, despite a variety of evidence that clusters of galaxies may be arrayed in thin layers along definite surfaces (to a large extent the observations perhaps tend to substantiate a lattice structure). The belief that cellular structure exists is based mainly on theoretical results. These calculations have examined the motion of cold gas in its own gravitational field. The initial conditions determine the general (Friedman, Hubble) expansion of matter and the growth of small irregularities, which are also specified by the initial conditions. The theory of such motion [3] shows that at first small, thin, relatively dense clumps will form—"pancake" structures. The character of these objects has been analyzed rigorously and in some detail by Arnold, Shandarin, and the author [1]. During the early stages there will be no cellular structure. Later in the numerical calculations, cell (in two-dimensional calculations, lattice) structure does develop, but its topology has not yet been validated theoretically. I propose here to use the Lagrangian description of the motion, as in my 1970 papers, but without insisting on a solution of any particular form.

Let the position $\mathbf{r}$ of a particle be a function of its initial position ξ and of time t:

$$\mathbf{r} = \boldsymbol{\psi}(\xi, t).$$

We take the initial position at an early epoch t_0, so that the density ρ of matter has not yet been appreciably perturbed but is everywhere equal to ρ_0. We shall consider the "bright" region (one in which matter will contract and galaxies will be born) in terms of the coordinates ξ. In these coordinates the fraction of the volume occupied by bright regions will be equal to their fraction by mass; that is, $dm = \rho_0 d^3\xi$. Hence if the bright regions contain more than half of all matter [say the (60-80%) suggested above], it would be natural for them to link up into a single connected medium. The dark regions, occupying (20–40)% of ξ-space, would form separate blobs in that space.[1]

The central point of this whole argument is the continuous dependence of $\mathbf{r}$ upon ξ: the vector function $\boldsymbol{\psi}(\xi, t)$ has no discontinuities. Physically this means that neighboring points will always remain so.

In ξ-space let us take some closed, continuous surface confined to a bright region, with a blob of dark space located inside. This connected surface will remain so even after transformation from ξ- to $\mathbf{r}$-space. But that implies the dark regions will be isolated from each other at later stages as well, which is what we wanted to show.

Several additional remarks may now be made.

[1]In speaking of blobs I am emphasizing the isolation of the individual black regions and do not have in mind shapes that are nearly round, there being no basis for this.

The transformation function has no singularities as a function of ξ. This property does not prevent the function from being non-monotonic[2] at a later stage; in that event the singularity would appear in the inverse function [in the expression $\xi = \psi^{-1}(\mathbf{r}, t)$]. The physical significance of the singularity is that two or more particles coming from different initial positions will pass through the same point in **r**-space. The development of a situation of this kind (analogous to the caustic for light rays) will be accompanied by a strong contraction of matter. Nevertheless, the property invoked above, that a closed bright surface must remain closed, will not be violated. Folds can develop in the surface, but not holes: the black region enclosed by the bright surface cannot slip through it and join up with some other black region.

As we are regarding motion as taking place under the action of gravitational forces, we may assert that it is vortex-free, with a potential transformation. The transformation can be written down by expressing the general expansion in the form

$$r_1 = a(t)\xi_i - \frac{\partial}{\partial \xi_i}\varphi(\xi_k, t).$$

The approximate theory assumes that

$$r_i = a(t)\xi - b(t)\frac{\partial}{\partial \xi_i}\Phi(\xi_k), \quad \text{i.e., } \varphi(\xi_k, t) \approx b(t)\Phi(\xi_k).$$

High density (formally, infinite density) will arise when

$$\frac{b(t)}{a(t)}\frac{\partial^2 \Phi}{\partial \xi_k^2} = 1,$$

where ξ_k is taken along the most "hazardous" direction whereby this derivative will be maximized at the point in question. The region in which $\partial^2\Phi/\partial\xi_k^2 < 0$ and where infinite density is never reached will occupy only 8% of the volume in ξ-space [2].

A supplementary physical analysis should, however, be made to establish the conditions under which galaxies can form in places where high density is reached at a late stage.

In regions where, on the contrary, galaxies developed early, their subsequent motion would lead to a concentration along the lines and points where two or three regions intersect. Formally, as time passes a continuous bright region may become so depleted in galaxies as to appear unobservable between the intersection lines.

The cellular structure represents, in a sense, an intermediate, asymptotic solution to the problem of the growth of long-wave perturbations in cold gas. Evidently the cell structure (the bright surfaces) will be followed by lattice structure (bright lines), and as a concluding stage in the relaxation process, bright blobs will develop. Because the problem is statistical in character, the

[2]For a function of a single variable, $\partial r/\partial\xi$ vanishes and then changes sign. For a function of several variables, the Jacobian $|\partial r_i/\partial\xi_k|$ plays the role of the derivative.

replacement of one structure by the next will not take place in precise fashion: at any given time different stages may coexist in different parts of space.

Observations of the cellular structure might yield highly valuable information on the era of formation and the age of the structure, as well as the amplitude of the initial fluctuations.

In any event, the cell structure does not at all resemble an aggregate of individual, more or less spherical, huge superclusters of galaxies. One should recognize here that after the first pancakes have formed, the star formation process will set in without delay. Evolving on a time scale short compared with the cosmological time, massive stars will end up as supernovae, releasing an energy far in excess of their gravitational energy. These outbursts might in fact exert a decisive influence on the global structure [6] out to distances of 100 Mpc. During these late stages analysis of the motion of cold gas might turn out to be an overidealization, even in the theory of primordial adiabatic perturbations.

Cell structure represents a highly specific consequence of the distinctive initial irregularities and course of evolution of the Universe, so that its investigation will be of enormous significance for cosmology.

I should like to thank A. G. Doroshkevich and particularly S. F. Shandarin for discussion and assistance.

M. V. Keldysh Institute of Applied Mathematics
USSR Academy of Sciences. Moscow

Received
January 26, 1982

REFERENCES

1. *Arnold V. I., Zeldovich Ya. B., Shandarin S. F.*—Preprint M. V. Keldysh Inst. for Appl. Math. AS USSR **100**, Moscow (1981).; Geophys. and Astrophys. Fluid Dyn. **20**, 111 (1982).
2. *Doroshkevich A. g.*—Astrofizika **6**, 581 (1970).
3. *Zeldovich Ya. B.*—Astrofizika **6**, 319 (1970); Astron. and Astrophys. **5**, 84 (1970).
4. *Zeldovich Ya. B., Shandarin S. F.*—Pisma v Astron. Zh. **8**, 131 (1982).
5. *Oort J. N.*—Preprint Sterrewaht. Leiden (1981).
6. *Ostriker J. P., Cowie L. L.*—Astrophys. J. Lett. **243**, L127 (1981).

Commentary

This article was the first to pose the problem of the global topological properties of the distribution of matter in the Universe. At the same time it poses the question of the properties of the mathematical mapping describing the motion of matter. Thus, this paper opened a fundamentally new direction in the theory of mappings and the associated structures which arise. This aspect of the topic was further developed in papers by Ya. B. (**64**) and by Ya. B. and S. F. Shandarin.[1]

[1] *Zeldovich Ya. B., Shandarin S. F.*—Phys. Rev. Lett. **25**, 1488–1491 (1984).

64
Topological and Percolation Properties of Potential Mapping with Glueing*

The potential movement of particles forms a continuous mapping of the Lagrangian space on the space of Eulerian coordinates of particles. The force-free inertial movement with random smooth initial velocity field is considered. The condition of particles on caustics glued together to form singularities leads to a peculiar structure with localized regions of rarefied matter surrounded by glued surfaces with higher-order singularities on them.

Continuous mapping is a common subject of mechanics and catastrophe theory. Let us consider the simplest example

$$x_i(t,\xi) = \xi_i + t\frac{\partial\varphi(\xi)}{\partial\xi_i}, \tag{1}$$

which corresponds to free inertial motion of particles with the initial velocity being potential

$$\mathbf{v} = \operatorname{grad}\varphi. \tag{2}$$

We assume the scalar function φ to be random but smooth.

The intersection of particle trajectories leads to local singularities equivalent to caustics made by intersection of rays in geometrical optics [1].

These local singularities now have been well investigated and classified [2–5]. I am changing the problem from one of collisionless particles by using the condition of glueing particles whose trajectories are intersecting. For those particles after glueing the simple (1) is no longer valid. Instead of the simplest type of singularities, "lips," one obtains first small elliptical two-dimensional surfaces in three-dimensional space. On these surfaces the density is the Dirac δ-function of the coordinate normal to the surface. The classification of subsequent higher-order singularities [4] will change due to glueing. But in this article I am interested in the global, topological properties of the solution, in particular in the late stage.

It is clear with a smooth $\varphi(\xi)$ that at small t the amount of glued particles is small. The glued surfaces are disconnected and the not-yet-glued matter (NYGM) prevails and occupies a simply connected region.

My point is that, in due time at $t \sim t_c$, the intersecting and merging glued surfaces will form a connected network, making the region occupied by NYGM nonconnected. The final era at $t > \kappa t_c$, $\kappa > 1$ but yet unknown,

*Proceedings of National Academy of Sciences, USA **80**, 2410–2411 (1983).

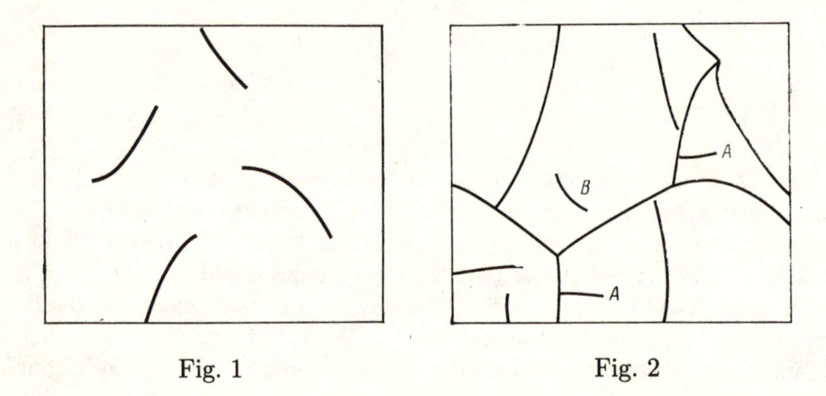

Fig. 1 Fig. 2

is characterized by NYGM being confined in volumes totally enclosed by the glued surfaces, so that the region occupied by NYGM is disconnected and one obtains cell structure.

In the two-dimensional case the situation is even simpler. The first lines of glued matter make the initial plane nonconnected (Fig. 1). Later, the growing segments of glued lines make a net dividing the plane into disconnected parts containing the NYGM (Fig. 2).

I observed the two-dimensional situation in the swimming pool of Debrecen, Hungary. The random waves on the surface of the water, due to many swimming people, refracted the sun's rays. The bright caustics on the ceramic bottom formed a pronounced net structure in a definite range of depth (cf. [6] and references therein).

To evaluate the characteristic time one uses the dimensional analysis, $\varphi\,(\mathrm{cm}^2\cdot\mathrm{s}^{-1})$. Therefore,

$$t_c \approx \left(\overline{\left(\frac{\partial^2\varphi}{\partial\xi_i\partial\xi_k}\right)}\right)^{-1/2} . \tag{3}$$

The proof of the statement made above about cell structure consists of two parts. First, we evaluate the division of matter between glued phase and NYGM. It gives us also the division of volume in Lagrangian (ξ) space between these two phases.

We apply the results of percolation theory to decide about the topology of the phases in Lagrangian space.

The second step consists of the statement that the mapping (1), being continuous and done with a smooth $\varphi(\xi)$ does not change the topology: every two points initially being at an infinitesimal distance remain close;

every closed line in Lagrangian space remains closed in Eulerian (x) space, and the same is true for closed surfaces.

In connection with the first step, one can be more specific. Taking the initial density in ξ-space $\rho_0 = 1$, it is easy to obtain the density around the same particle at an arbitrary time t:

$$\rho = \left(\frac{\partial^3 x}{\partial \xi^3}\right)^{-1} = \left|\delta_{ik} + t\frac{\partial^2 \varphi}{\partial \xi_i \partial \xi_k}\right|^{-1}. \tag{4}$$

Let us denote by α, β, and γ the negative three principal values of the deformation tensor—i.e., the solution of the cubic equation

$$\left|\frac{\partial^2 \varphi}{\partial \xi_i \partial \xi_k} - \lambda \delta_{ik}\right| = 0, \tag{5}$$

$$\alpha = -\lambda_1, \quad \beta = -\lambda_2, \quad \gamma = -\lambda_3, \tag{6}$$

with the additional condition

$$\alpha > \beta > \gamma. \tag{7}$$

In terms of α, β, and γ, the density is given by

$$\rho = [(1 - \alpha t)(1 - \beta t)(1 - \gamma t)]^{-1}. \tag{8}$$

This is true as no one denominator equals zero

$$1 - \alpha t > 0, \quad t < \frac{1}{\alpha}. \tag{9}$$

Thereafter, the particles are glued and the density is a one-dimensional δ-function, as explained above. This is the description of the simplest type of singularities.

In a simple-minded approach, ignoring condition (7) and the way of calculating α, β, and γ from (5) and (6), one would think of α, β, and γ as three random independent functions of ξ. This would give the probability P to every one of them being positive $P = 0.5$. Therefore, the probability that a given particle remains not glued would be $N = P^3 = 0.125$.

Actually calculation [1] shows that the probability N of $\alpha < 0$, which implies $\beta < 0$, $\gamma < 0$ by condition (7), is $N = 0.08$. It is smaller than the simple $2^{-3} = 0.125$ due to the well-known phenomenon of repulsion of eigenvalues.

Therefore the glued matter is $1 - 0.08 = 92\%$ of the total in the three-dimensional case and more than 75% in the two-dimensional case for large enough times.

Obviously the NYGM which occupies 8% in Lagrangian three dimensions and less than 25% in two dimensions consists of isolated islands in the ocean of glued matter. By the reasoning of the second step it remains in isolated islands in the Eulerian space—although now the glued matter is thin, and the islands are divided by narrow interconnected channels rather than by the wide ocean.

There are additional remarks. There is matter which is glued by sticking with finite velocity on the already built surface. Asymptotically, as $t \to \infty$ the NYGM is going to 0, not to 0.08 or 0.25.

But it is important that our statement about the isolated islands does not need so deep depletion of NYGM. It is true when NYGM is less than 20% in three dimensions or less than 50% in two dimensions. So the isolated islands appear at some finite t, rather early. The exact calculation of this t is of course very difficult; perhaps it is dependent on the spectrum of φ. The specification of φ as random, Gaussian, isotropic, and uniform on the average is enough for calculation of α, β, and γ probabilities, but it is possible that more information is needed to obtain the time of isolation.

The glueing of particles changes their velocities abruptly. It is reasonable to assume that the velocity of every point of singular (glued) surface is the average velocity of all particles glued at a given spot of the surface.

For the line or surface drawn in Lagrangian space, the change of velocity makes an angle in Eulerian space, but it does not disrupt the line or the surface. Our statements remain valid.

The structure obtained was compared by Biermann with that of champagne foam: a small volume of liquid divides much larger volumes of gas into isolated bubbles. But there is also a difference. The foam structure depends on the surface tension and specific properties of the champagne stabilizing the foam.

In our case there is no surface tension. Therefore protruding one-sided singular surfaces (A) or isolated surfaces (B) (lines in two dimensions) are possible (see Fig. 2). There is one subtle point that is important for the quantitative theory. It is assumed that the distribution of α values in Lagrangian ξ-space is rather similar to a random function (but with $\overline{\alpha} > 0$), although α is obtained by a nonlinear operation. Therefore, we apply the criteria of percolation theory to the Lagrangian space. In the Eulerian space the volume occupied by NYGM is 100% by definition, but the distribution of density is quite different from random, so direct application of percolation theory would be totally wrong.

Last but not least, the connection with real physics should be mentioned. In the case of light [7] propagating in one dimension and being smoothly refracted in the other two dimensions there is no glueing. The rays make caustics and are propagating thereafter on straight lines. The caustics of the first generation are rather bright and thin; our picture is a good approximation, but it is an intermediate asymptotics theory only. In the long run a statistical picture emerges without sharp caustics.

If the glueing is replaced by shock-wave compression with high density of a gas with low but finite initial temperature, then all infinite densities disappear. Still the boundaries of NYGM are well defined by shock waves and the topological consequences remain. In this case due to finite densities, the mapping remains forever, but it loses its potential character due to vortex generation in the shocks.

In the case of the large-scale structure of the universe due to gravitational instability [5] the shock wave in compressed gas and the gravitational binding of massive neutrinos (if any) play the role of glueing. It was in connection with observations of giant voids in the distribution of galaxies that the considerations were done [6].

Thanks are due to V. I. Arnold, A. G. Doroshkevich, and S. F. Shandarin.

Space Research Institute
USSR Academy of Sciences. Moscow

Submitted
September 29, 1982

REFERENCES

1. *Doroshkevich A. G.*—Astrofizika **6**, 581 (1970).
2. *Arnold V. I., Varchenko A. N., Gusein-Zade S. M.*—Singularities of Differentiable Mappings. Birkhäuser. V. 1 (1985), V. 2 (1988).
3. *Arnold V. I.* Mathematical Methods of Classical Mechanics. Springer (1978).
4. *Arnold V. I., Shandarin S. F., Zeldovich Ya. B.*—Geophys. and Astrophys. Fluid Dyn. **20**, 111 (1982).
5. *Zeldovich Ya. B.*—Astr. Astrophys. **5**, 84 (1970).
6. *Zeldovich Ya. B.*—Pisma v Astron. Zh. **8**, 195 (1982).
7. *Berry M. V., Upstill C.*—In: Progress in Optics, ed. Wolf E., V. 18. Amsterdam: North-Holland (1980).

Commentary

In this paper, just as in the preceding article (**63**) and in a paper by S. F. Shandarin and Ya. B. in *Phys. Rev. Lett.*,[1] a new chapter is opened in research on the global properties of mappings. The problem posed is of interest both for optics and for mathematics. Until recently, the object of investigation were only local properties (caustics, catastrophes). There is not doubt that this new area will interest many and work will follow (in particular, using numerical modeling methods) to establish the relation between global properties of the mapping and the spectrum of the initial random function. In line now is the question of global properties of vector fields of various origins (see, for example, Ya. B. article "Percolation properties of a two-dimensional random stationary magnetic field"[2] and references therein). The percolation analysis of the structure was suggested by S. F. Shandarin.[3]

[1] *Zeldovich Ya. B., Shandarin S. F.*—Phys. Rev. Lett. **25**, 1488–1491 (1984).
[2] *Zeldovich Ya. B.*—Pisma v ZhETF **38**, 51–54 (1983).
[3] *Shandarin S. F.*—Pisma v Astron. Zh. **9**, 195 (1984).

IX

Observational Effects in Cosmology

65

Observations in a Universe Homogeneous in the Mean*

A local nonuniformity of density due to the concentration of matter of the Universe into separate galaxies produces a significant change in the angular dimensions and luminosity of distant objects as compared to the formulas for the Friedman model.

The propagation of light in a homogeneous and isotropic model of the expanding Universe (first studied by A. A. Friedman) has been investigated in a number of papers [1–3].

In these papers expressions were obtained for the observed angular diameter θ and the observed brightness of an object with a known absolute diameter and absolute brightness as a function of the distance or, strictly speaking, the red shift of the object $\Delta = (\omega_0 - \omega)/\omega_0 = z/(1+z)$.

In particular, there is a remarkable feature in the function $\theta(\Delta)$, namely, the presence of a minimum when Δ is approximately equal to 1/2. Formula (10) and Fig. 6 in the appendix show the variation of the function $f(\Delta) = rH/c\theta$ which is inversely proportional to θ for a given density of matter. Here r is the radius of the object, H is Hubble's constant, c the velocity of light; the maximum of the curve $f(\Delta)$ corresponds to a minimum of the angle θ. Such behavior of θ is caused by the curvature of space due to the matter filling the Universe.

*Astronomicheskiĭ zhurnal **41** (1), 19–24 (1964).

In the Milne model where the density of matter is zero, θ decreases monotonically to zero to the horizon where $\Delta = 1$, $\omega = 0$ [see Fig. 6 of the appendix, curve $f_0(\Delta)$]. In principle, in a homogeneous universe, observations of $\theta(\Delta)$ can provide an independent determination of the average density of matter ρ.

In fact, the distribution of matter is obviously not that of the Friedman model since matter is concentrated in individual stars which form galaxies.

What is the effect of the nonuniformity on density on the function $\theta(\Delta)$? The angular diameter θ subtended by the object AB is determined by the two rays AO and BO emitted by the outer points of the object (Figs. 1, 2).

A mass situated between these rays bends the latter in such a way that θ is increased (Fig. 2). What we have in mind is the bending of light rays by the gravitational field, predicted by Einstein; this bending amounts to 1.75" for a light ray passing near the limb of the solar disc and has been confirmed by observation.

This type of bending of light rays by matter situated inside the cone defined by the rays is the physical reason for the features of the variation of θ with Δ obtained in the Friedman solution. It can be seen from this that there are very strict constraints on the uniformity of the distribution of matter: it is necessary that the density of matter within the cone of light rays be the same as the mean density of matter in the Universe; only in this case will the angle θ be the same as that obtained in the uniform case.

Let us make an extreme assumption, namely, we will neglect the mass of the intergalactic medium (which includes such forms of matter as neutrinos and gravitons) in comparison with the mass of matter contained in galaxies.

In this case the angular diameter of those objects which do not have galaxies within the cone subtended by them at the observer is given by $\theta_1(\Delta)$ which differs substantially from the function $\theta(\Delta)$ of the uniform model (see derivation in the appendix).

In particular, $\theta_1(\Delta)$ does not have a minimum in all of the region open to observation, $0 < \Delta < 1$. For a given density we give in the appendix the formula (21a) for the quantity f_1, inversely proportional to θ (a curve of f_1 is shown in Fig. 6).

For objects AB which have a galaxy C within their cone we have the formation of a ring as was pointed out by Zwicky [4] (see also [5]), while if C does not lie on the axis of the cone, there is a complex pattern of refraction recently investigated by Klimov [6].

It is not possible to derive any definite expression for $\theta_2(\Delta)$, since θ_2 depends on the mass of C and the exact disposition of AB, O, and C. The value of $\theta(\Delta)$ calculated on the basis of the uniform model represents, in a nonhomogeneous universe, a weighted mean of $\theta_1(\Delta)$ and all possible values of $\theta_2(\Delta)$.

Consequently, in a nonhomogeneous universe it is not possible to study

$\theta(\Delta)$ directly, even if the difficulties of selecting objects with known linear dimensions are ignored. It is clear that as a rule objects not screened by others are observed, i.e., the observations refer to $\theta_1(\Delta)$.

The above remarks are also pertinent in the case of the relationship between absolute and apparent magnitudes of distant objects, including those which are not resolved, for example, individual stars.

The amount of light received by an observer from a distant star is proportional to the solid angle subtended by it. Therefore, the well-known formulas should be corrected by a factor of $[\theta_1(\Delta)/\theta(\Delta)]^2$ if the star is not screened by other stars and if the intergalactic density is negligibly small.

In conclusion, it should be noted that the matter which fills the Universe has an influence on the red shift of photons, as well as on the total number of objects of a given type having a red shift less than a given value $N(\Delta)$.

In principle, the nonuniform distribution of matter could also bring about changes in the expressions derived on the basis of the uniform model.

However, the latter model involves the average density in a sphere whose radius is of the order of the distance from the observer to the object under observation. For very distant objects the volume of such a sphere is large and in spite of local fluctuations, density differs little from the average value, so that the usual formulas are fully valid.

Appendix

Let us consider the case of a flat Friedman model with pressure equal to zero. The metric can be conveniently written as

$$ds^2 = c^2dt^2 - (at^{2/3})^2[dx^2 + dy^2 + dz^2], \tag{1}$$

where x, y, z are co-moving coordinates.

In this solution the density of matter is given by

$$\rho = \frac{1}{6\pi\kappa t^2}, \tag{2}$$

where κ is the Newtonian gravitational constant and the Hubble constant at the current time t_0 is

$$H = \frac{2}{3}\frac{1}{t_0} \tag{3}$$

The light ray traveling along the x axis ($y = 0$, $z = 0$) and reaching the point of observation (the coordinate origin) at time t_0 satisfies the equation

$$dx = -\frac{cdt}{at^{2/3}}; \quad x = 3c\frac{t_0^{1/3} - t^{1/3}}{a}. \tag{4}$$

Let $3(c/a)t_0^{1/3} = x_0$ be the distance to the horizon and let us rewrite the above expression as

$$t^{1/3} = \frac{a(x_0 - x)}{3c}. \tag{4a}$$

Let us consider a neighboring light ray which lies in the x, y-plane and which passes through the origin at a small angle θ. It is obvious that in the co-moving reference system the light also propagates along straight lines, so that

$$y = \theta x. \tag{5}$$

The absolute linear dimensions of an object included between two rays is

$$r = at^{2/3}y = \theta 3ct^{2/3}(t_0^{1/3} - t^{1/3}). \tag{6}$$

Let us express the relationship between r and θ through the Hubble constant and the red shift.

As is known,

$$\frac{\omega_1}{\omega_0} = \frac{t^{2/3}}{t_0^{2/3}}, \tag{7}$$

where ω_1 is the frequency of light received by the observer at time t_0 and ω_0 the frequency of light emitted by the object at time t. Consequently, we have

$$t^{1/3} = t_0^{1/3}\sqrt{1-\Delta}, \quad \Delta = \frac{\omega_0 - \omega_1}{\omega_0}, \tag{8}$$

$$\frac{r}{\theta} = 2c(1-\Delta)\frac{1-\sqrt{1-\Delta}}{H}, \tag{9}$$

$$\frac{Hr}{c\theta} = f(\Delta) = 2(1-\Delta)(1-\sqrt{1-\Delta}), \tag{10}$$

where $f(\Delta)$ is a function characterizing the change in the metric in the uniform model. For small Δ, $f(\Delta) = \Delta$, i.e., the distance is equal to $(c/H)\Delta$ as should be the case. $f(\Delta)$ has a maximum of $8/27$ when $\Delta = 5/9$.

It can be stated that the maximum distance determined as the ratio of the linear to the angular distance (in Euclidean geometry $\theta = r/R$) is given by

$$R = \frac{8c}{27H} = 1200\,\mathrm{Mpc} \tag{11}$$

with $H = 75\,\mathrm{km}/(\mathrm{s}\cdot\mathrm{Mpc})$.

The above considerations refer to the uniform model and, in addition, to the flat case $\rho = \rho_c = 3H^2/8\pi\kappa$ [cf. (2) and (3)].

Let us now consider the following problem: the density distribution is not uniform in that within the cone of opening angle α (where $\alpha > \theta$) and axis along the x-axis there is no matter, while everywhere else ρ is given by (2) and is independent of position. It is assumed that both θ and α are much less than unity so that the amount of matter removed is small and the general motion is not affected. As is known, a ray passing near a concentrated mass m, the impact parameter being b, will be deflected through an angle

$$\Delta\beta = \frac{4\kappa m}{c^2 b}. \tag{12}$$

Since the density perturbation consists of the removal of matter, i.e., in effect the addition of negative mass, the deflection caused by such a perturbation will be towards the outside (see Figs. 4, 5).

The function $E = 2\kappa m/b$ represents a field which corresponds to the Newtonian potential of the two-dimensional problem

$$\rho = 2\kappa m \ln b + \text{const.} \tag{13}$$

The total deviation of the light ray over all of its path due to the mass m is equal to $\beta = 2E/c^2$.

In the case of uniform density distribution, the potential φ_1 due to the mass along unit path length is given by

$$\Delta\varphi_1 = \frac{1}{r}\frac{d}{dr}r\frac{d\varphi_1}{dr} = 4\pi\kappa\rho, \tag{14}$$

where r is the distance of the point from the x-axis; thus, inside the cavity we have

$$\varphi_1 = \pi r^2 \rho, \quad E_1 = 2\pi r\rho. \tag{15}$$

Generalizing Einstein's expression for the bending of the light ray, we find the deflection per unit path length to be

$$\frac{d\beta}{dl} = \frac{2E_1}{c^2} = \frac{4\pi r\rho}{c^2}, \tag{16}$$

$$\frac{d\beta}{dx} = \frac{4\pi r\rho}{c^2}\frac{dl}{dx}. \tag{17}$$

Making use of expression (2) for the density, expression (4a), expression (6), and the fact that $dl/dx = at^{1/3}$, we find that

$$\frac{d^2y}{dx^2} = \frac{d\beta}{dx} = \frac{6y}{(x_0 - x)^2}. \tag{18}$$

This homogeneous differential equation can be easily solved

$$y = k_1(x_0 - x)^3 + k_2(x_0 - x)^{-2}. \tag{19}$$

Imposing the condition that at $x = 0$, $y = 0$ (the rays meet at the point of observation), $\frac{dy}{dx}\big|_{x=0} = \theta_1$ by definition, we find that

$$y = \frac{2}{5}\theta_1 x_0[x_0^2(x_0 - x)^{-2} - x_0^{-3}(x_0 - x)^3]. \tag{20}$$

From this we can find the relation between the absolute diameter of the object r [Eq. (6)], its angular diameter θ_1, and the distance which can be characterized by the magnitude of the red shift Δ with the help of formulas (8) and (4), (4a), connecting Δ, t, and x. We thus find that

$$\frac{r}{\theta_1} = \frac{c}{H}\frac{2}{5}[1 - (1 - \Delta)^{5/2}], \tag{21}$$

$$f_1(\Delta) = \frac{2}{5}[1 - (1 - \Delta)^{5/2}]. \tag{21a}$$

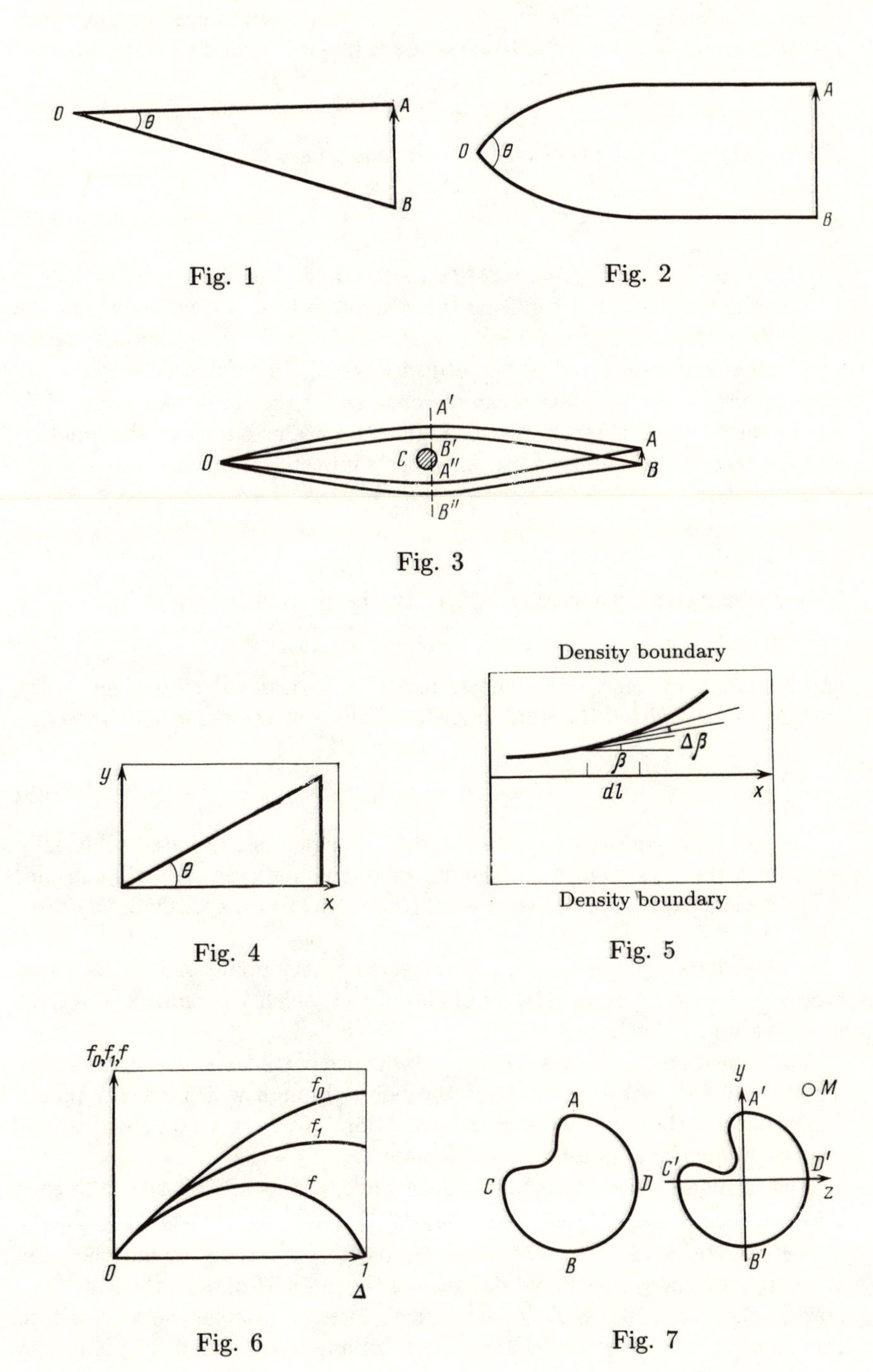
O
θ
A
B
Fig. 1
O
θ
A
B
Fig. 2
A'
C
B'
A''
O
A
B
B''
Fig. 3
y
θ
x
Fig. 4
Density boundary
Δβ
β
dl
x
Density boundary
Fig. 5
f0,f1,f
f0
f1
f
0
1
Δ
Fig. 6
A
C
D
B
y
A'
M
C'
D'
z
B'
Fig. 7

By contrast with f, the function f_1 increases monotonically right up to the horizon ($\Delta = 1$) where it reaches the value 2/5, which corresponds to

$$R_1 = \frac{2}{5}\frac{c}{H} = 1600\,\text{Mpc}.$$

The correction to the brightness of a distant object is

$$L_1 = L\left(\frac{f(\Delta)}{f_1(\Delta)}\right)^2 \tag{22}$$

for $\Delta = 5/9$, $f = 8/27$, $f_1 = 418/1215$, $L_1 = 0.7L$. For $\Delta \to 1$, $f/f_1 \to 0$.

The above calculation confirms the statement that the maximum of the ratio r/θ or the minimum of θ for a given r occurs only when there is matter within the cone subtended by the object at the point of observation.

In principle, the calculation can be repeated for the case when $\rho \neq \rho_c$, i.e., for a hyperbolic or closed universe. Figure 6 contains curves of the function $f(\Delta) = rH/c\theta$ for three cases: an empty universe (limiting case $\rho \to 0$, Milne model)

$$f_0 = \Delta\left(1 - \frac{\Delta}{2}\right), \tag{23}$$

for a flat universe with a uniform density distribution $\overline{\rho} = \rho_c$

$$f = 2(1-\Delta)(1-\sqrt{1-\Delta}), \tag{24}$$

and for a flat universe with a mean density $\overline{\rho} = \rho_c$ in the case of an object with no matter within the solid angle it subtends at the point of observation (21a)

$$f_1 = \frac{2}{5}[1-(1-\Delta)^{5/2}]. \tag{25}$$

Let us briefly consider the effect due to a single mass situated arbitrarily relative to the rays going from the object to the observer. We will consider the y, z-plane to be normal to the light rays and to pass through the mass (Fig. 7).

In this plane the light rays are represented by the points $A'B'C'D'$ which correspond to the points $ABCD$ of the object which has emitted the light rays (see Fig. 7, left).

The deflection of the rays under the influence of the mass M occurs in the direction of the vector which joins the point through which the ray passes (A, etc.) and the point M and is proportional to $F = \operatorname{grad}\varphi$, where φ is the two-dimensional potential of the mass M.

The change in the angle between A' and B' depends on the difference between the values of F_y at these points, i.e., on $\partial^2\varphi/\partial y^2$, while the change in the angle between C' and D' depends on $\partial^2\varphi/\partial z^2$. Since outside the mass $\Delta\varphi = 0$, then correct to higher derivatives the mass M outside the cone (i.e., outside the curve $A'B'C'D$ in the figure) does not change the area within this curve, i.e., it does not change the solid angle subtended by the object at the point of observation and, consequently, it does not change the total amount of light received by the observer.

The mass M deforms the observed shape of the object, so that the latter becomes contracted along the axis joining it to M and elongated in the perpendicular direction.

REFERENCES

1. *Tolman R. C.* Relativity, Thermodynamics and Cosmology, Oxford (1934).
2. *Robertson H. P.*—Astrophys. J. **15**, 69 (1938).
3. *McVittie G.* General Relativity and Cosmology. Urbana: Univ. of Illinois Press (1965).
4. *Zwicky F.* Morphological Astronomy. Springer (1957); Helv. Phys. Acta **6**, 1 (1933); Astrophys. J. **86**, 217 (1937); Phys. Rev. **51**, 290, 679 (1937).
5. *Tikhov A. G.*—Dokl. AN SSSR **16** 4 (1938); Izv. Gl. Astron. Observatorii v Pulkovo **16**, 1 (1937).
6. *Klimov Yu. G.*—Dokl. AN SSSR **148**, 789 (1963).

Commentary

Any investigation of distant sources, both for the purpose of determining parameters of the Universe (in particular, the mean density of matter and the acceleration parameter) and for determining properties of the sources themselves, requires knowledge of the relations between the angular dimension of the image, the absolute dimension of the source, and the magnitude of the red shift z. This dependence is significant even for objects which may be resolved by a telescope since it enters in the relation between the absolute luminescence and the received radiation.

Only at small distances and small z can one use a Euclidean geometry; the angle is inversely proportional to z, and the solid angle is inversely proportional to z^2.

One of the most important means of determining modern cosmological parameters (the ratio of the density of matter to critical value Ω, the acceleration parameter) is the study of the dependence of the angular dimension of distant galaxies on their red shift. A calculation of this dependence, done for a completely homogeneous universe with $\Omega = 1$, shows that the angular dimension passes through a minimum at $z = 1$ and increases for $z \to \infty$. Ya. B. noticed that this calculation is only valid when the density of matter inside the cone in which the distant object is visible differs little from the mean density in the Universe. Since the view angle is small, the volume of the cone is small and as a result there is a high probability that the density of matter in the cone will prove different from the mean and, in particular, that the cone will prove empty. In the paper "Observations in a Universe Homogeneous in the Mean" Ya. B. has calculated the dependence of the angular dimension on the red shift in the extreme case of an empty cone. The dependence found differs significantly from that calculated according to a completely homogeneous universe with $\Omega = 1$; it has no minimum and is more like the dependence in the case of a completely homogeneous universe with $\Omega = 0$ (Milne's model). In subsequent papers by V. M. Dashevskiĭ and Ya. B.[1] and by V. M.

Dashevskiĭ and V. I. Slysh[2] this result was generalized to the case of arbitrary Ω and the presence of intergalactic gas.

The present paper is also interesting in that there appears a graphic physical interpretation of the effect of an increase of the angular dimension as gravitational deflection of rays by matter inside the cone (the "gravitational lens").

It is interesting that after publication of Ya. B.'s paper it was noted in responses by U.S. scientists that analogous considerations were simultaneously and independently developed by the well-known theoretical physicist R. Feynman. Later the problem of the effect of an inhomogeneous distribution of matter outside the light cone on the luminescence was studied by Refsdal.[3]

[1] *Dashevskiĭ V. M., Zeldovich Ya. B.*—Astron. Zh. **41**, 1071–1074 (1965).
[2] *Dashevskiĭ V. M., Slysh V. I.*—Astron. Zh. **42**, 863–864 (1965).
[3] *Refsdal S.*—Month. Not. RAS **128**, 295–306 (1964).

66
Small-Scale Fluctuations of Relic Radiation*

With R. A. Sunyaev

Perturbations of the matter density in a homogeneous and isotropic cosmological model which leads to the formation of galaxies should, at later stages of evolution, cause spatial fluctuations of relic radiation. Silk assumed that an adiabatic connection existed between the density perturbations at the moment of recombination of the initial plasma and fluctuations of the observed temperature of radiation $\delta T/T = \delta\rho_m/3\rho_m$. It is shown in this article that such a simple connection is not applicable due to

1) The long time of recombination;
2) The fact that when regions with $M < 10^{15} M_\odot$ become transparent for radiation, the optical depth to the observer is still large due to Thompson scattering;
3) The spasmodic increase of $\delta\rho_m/\rho_m$ in recombination.

As a result the expected temperature fluctuations of relic radiation should be smaller than adiabatic fluctuations. In this article the value of $\delta T/T$ arising from scattering of radiation on moving electrons is calculated; the velocity field is generated by adiabatic or entropy density perturbations. Fluctuations of the relic radiation due to secondary heating of the intergalactic gas are also estimated. A detailed investigation of the spectrum of fluctuations may, in principle, lead to an understanding of the nature of initial density perturbations since a distinct periodic dependence of the spectral density of perturbations on wavelength (mass) is peculiar to adiabatic perturbations. Practical observations are quite difficult due to the smallness of the effects and the presence of fluctuations connected with discrete sources of radio emission.

1. *Introduction*

In the contemporary "big-bang" model of the Universe it is hypothesized that in the distant past, before recombination of the initial plasma at times corresponding to a red shift $z \sim 1000$, there were no galaxies and the origin

*Astrophysics and space science **7** (1), 1–20 (1970).

of galaxies is connected with insignificant deviations from strict homogeneity existing in that period. In the first approximation it can be considered that after recombination of protons and electrons "matter"—neutral atoms—do not interact with radiation and relic radiation (having at present an average temperature of 2.7 K) immediately gives us information about conditions for $z \sim 1000$. In particular, the dependence of deviations of temperature on the direction of observation now being performed from the Earth characterizes the dependence of physical values, i.e., deviations of density, on the spatial coordinates at an earlier stage. These deviations grow in the future (after recombination) due to gravitational instability. For the moment of formation of separate objects it is reasonable to take the time of origin of regions with densities at least twice the average density, i.e., $\delta\rho/\rho \sim 1$. It is assumed that this occurred relatively recently (on a logarithmic time scale) at $z \sim 2-10$. In this case an estimate for the perturbation at the moment of recombination gives $\delta\rho/\rho \sim 10^{-2} - 10^{-3}$, i.e., it is possible to speak about small perturbations. Not only the perturbations which lead to formation of galaxies, but the whole spectrum of perturbations is of interest for the characteristics of the initial inhomogeneity.

The natures of the perturbations may be qualitatively divided:

1) Density perturbations of nuclei and electrons ρ_m on a background of a constant density of quanta ρ_r (so-called entropy perturbations).
2) Compression and rarefaction waves of the plasma as a whole with simultaneous changes of ρ_m and ρ_r (adiabatic perturbations).
3) Turbulent motions of the plasma.
4) Chaotic magnetic fields and perhaps other types of perturbations.

Different types of perturbations evolve differently at plasma periods ($z > 1400$) and give different predictions concerning the formation of galaxies and fluctuations of relic radiation. Silk [14–16] first made quantitative predictions concerning adiabatic perturbations. His results were obtained on the assumption that recombination of the initial plasma occurs quite suddenly for a definite z_r. Earlier in time, for $z > z_r$, perfect adiabaticity is assumed so that

$$\frac{\delta\rho_m}{\rho_m} = 3\frac{\delta T}{T}, \quad \frac{\delta\rho_r}{\rho_r} = 4\frac{\delta T}{T}.$$

Later, for $z < z_r$, matter is completely transparent and an observer measures $\delta T/T$ reached at the moment $z = z_r$ directly. On the other hand, due to gravitational instability $\delta\rho/\rho$ subsequently grows proportional to $(1+z)^{-1}$, so that

$$\left(\frac{\delta\rho}{\rho}\right)_{z_r} = \frac{1+z_0}{z_r}\left(\frac{\delta\rho}{\rho}\right)_0 = \frac{1+z_0}{z_r}$$

(for a definite z_0 at the moment $\delta\rho/\rho = 1$).

It is clear that measurements of the fluctuations of relic radiation allow a judgment of the time of galaxy formation only if radiation does not interact with matter after recombination. In fact, recombination may not occur instantaneously. Even if it occurred according to the equilibrium Saha equation, the results of Silk would require substantial corrections. Moreover, as shown by detailed calculations [4, 11], recombination of hydrogen does not occur according to the equilibrium Saha equation, but much more slowly. When a protogalaxy or proto-cluster of galaxies becomes transparent to radiation, the optical depth for Thompson scattering by matter between the proto-objects and the observer is still very large, as a result of which temperature fluctuations of relic radiation are smoothed out. Here by proto-object we mean a small perturbation of density at the moment of recombination encompassing a mass which is later transformed into a currently existing object.

The thesis about the small values of adiabatic fluctuations considered by Silk is confirmed by the spasmodic increases of the amplitudes of density perturbations during recombination.

Consideration of processes occurring before recombination leads to the following picture: at a later stage of expansion the amplitude of density perturbations turns out to be a periodic function of wavelength (mass). Such a picture was previously obtained by Sakharov [6] for a cold model of the Universe. Together with density perturbations, peculiar plasma motions (connected in particular with these perturbations) occur superimposed on the general cosmological expansion. The value $\delta T/T$ mentioned above is the change of temperature measured by an observer moving together with the plasma: an observer on Earth also measures a change of intensity (fluctuation) due to the Doppler effect which equals $\delta T/T = (u/c)\cos\theta$, where u is the velocity of the plasma and θ is the angle between the velocity and the direction to the observer.

Throughout the entire article $\delta\rho/\rho$ will be the density perturbations and $\delta T/T$ the temperature fluctuations. Due to the long time of recombination adiabatic temperature perturbations are smoothed out and effects connected with the presence of peculiar velocities determine the observed temperature fluctuations of relic radiation connected with initial perturbations. Since the growth of adiabatic and entropy density perturbations follow the same law in the stage after recombination, their velocities of peculiar motions and temperature fluctuations of the radiation, which are related to various types of perturbations, turn out to be of the same order. Entropy perturbations played no role in Silk's approximation. It is quite probable that at the time of formation of present objects, i.e., when density perturbations became large $\delta\rho/\rho \geq 1$, a secondary heating and ionization of the gas occurred. Due to Thompson scattering the ionized gas decreases the amplitude of fluctuations of radiation arriving from the period $z \sim 1000$. On the other hand for small z the ionized gas itself creates fluctuations of radiation due to both its macroscopic motion, and the simultaneous presence of inhomogeneous heating and thermal motion of electrons. In conclusion we compare the effects described above with the gravitational effects calculated by Sachs and Wolfe [13]. Gravitational influence on radiation for the same values of $\delta\rho/\rho$ is much smaller than the effects described above, but it continues after matter becomes completely transparent to radiation. It is important for large masses.

2. Recombination and its Influence on Density Perturbations and Temperature Fluctuations

In this section and future sections we will use the following notation: z = red shift; $\Omega = \rho/\rho_c = n/n_c$ = present dimensionless average density of matter in the Universe, $\rho_c = 3H_0^2/8\pi G = 2\cdot 10^{-29}\,\mathrm{g/cm^3}$, $n_c = 10^{-5}\,\mathrm{cm^{-3}}$ = critical density, $H_0 = 10^{-10}\,\mathrm{yrs^{-1}}$ = Hubble constant, $T_0 = 2.7$ K = present temperature of relic radiation, $I = 13.6$ eV = ionization potential of atomic hydrogen, and $w = 8\,\mathrm{s^{-1}}$ = probability of a two quantum transition in a hydrogen atom from the $2s$ state to the ground state $\mathrm{H}(2s) \to \mathrm{H}(1s)+\gamma_1+\gamma_2$. The concentrations of corresponding particles are designated by the symbols p, e, H, and ρ_r and ρ_m are the energy densities of radiation and matter, respectively. For the present temperature of radiation $T_0 = 2.7$ K, ρ_m and ρ_r are comparable in the process of expansion of the Universe for $z = 4\cdot 10^4\Omega = z_1\Omega$. At the stage before recombination of the initial plasma the speed of sound depended on the relation between ρ_r and ρ_m,

$$a_s = \frac{c}{\sqrt{3}}\left(\frac{4\rho_r}{4\rho_r + 3\rho_m}\right)^{1/2} = \frac{c}{\sqrt{3}}\left(1 + \frac{3}{4}\frac{\rho_m}{\rho_r}\right)^{-1/2} = \frac{c}{\sqrt{3(1+\frac{3}{4}\Omega(z_1/z))}}.$$

1. Recombination rate. Solving the equation of recombination (4) from the work [4], it is easy to find that in the region $900 < z < 1500$,

$$x(z) = \frac{p}{p+H} = \frac{e}{e+H} = \frac{A}{z\sqrt{\Omega}}e^{-B/z},$$

where

$$A = \frac{(2\pi m_e kT_0)^{3/2} H_0 I}{4wn_c kT_0 h^3} = 6\cdot 10^6,$$

and

$$B = \frac{I_{23}}{kT_0} = \frac{I}{4kT_0} = 1.458\cdot 10^4.$$

this analytic solution is obtained from the following physical picture. For $z_r \sim 1500$ ($T_r = T_0 z = 4000$ K) in agreement with the Saha equation $p = e = H$. In the process of further temperature decreases in the expansion of the Universe the Saha equation ceases to describe the process of recombination. Free electrons and radiation are only found in equilibrium with the excited levels of atomic hydrogen; the recombination rate is determined by two-quantum decays of the $2s$ level and is described by the equation

$$\frac{dp}{dt} = -3\mathrm{H}p - w\mathrm{H}_{23}; \quad \mathrm{H}_{23} = \mathrm{const}\, p^2 e^{-(B/z)} \tag{2}$$

the solution of which is given by equation (1). For $z < 900$ thermodynamic equilibrium is destroyed between H_{23} and free protons and electrons, the ionization time from the $2s$ levels by collisions with quanta and electrons becomes greater than w^{-1} and therefore the solution (1) is correct only in the interval $900 < z < 1500$. The optical depth of the Universe due to Thompson scattering equals

$$\tau(z) = \Omega^{1/2}\sigma_T n_c H_0^{-1} \int_0^z \eta^{1/2} x(\eta)\, d\eta = \tau_0 + a z^{3/2} e^{-B/z}, \tag{3}$$

where $a = \sqrt{\Omega}\sigma_T n_c c H_0^{-1}(A/B) = 27.3$. The constant $\tau_0 = \sqrt{\Omega}\sigma_T n_c c H_0^{-1} \cdot \int_0^{900} x(z) z^{1/2}\, dz = 0.4$ is easy to find with the equations for $x(z)$ for $z < 900$, given in [4].

2. Weakening of fluctuations. We will find the moment z_2 when the proto-object which interests us with mass $M = \frac{4}{3}\pi\rho(z_2) l^3(z_2)$ becomes transparent (the conditions of transparency: $\Omega\sigma_T l(z_2) n_c z_2^3 \sim 1$)

$$z_2 = \frac{4.4 \cdot 10^4}{7.5 + \ln(M\Omega^{1/2}/M_\odot)}. \tag{4}$$

Equation (4) is correct only for $M > 10^9 M_\odot$ when $z < 1500$ and (3) is applicable.

It is clear that the adiabatic relation $\delta T/T = \delta\rho_m/3\rho_m$ will not be satisfied and the temperature is smoothed out somewhat earlier than the moment z_2. Thus z_2 is larger, the smaller the mass M. On the other hand, at this moment the optical depth to a present observer is still large and adiabatic fluctuations of radiation connected with masses smaller than $2\cdot 10^{15}\Omega^{-1/2} M_\odot$ will be weakened strongly only after scattering by $e^{-\tau}$ times, where

$$\tau = 0.4 + 8.3 \cdot 10^4 (M\Omega^{1/2}/M_\odot) = 0.4 + (6 \cdot 10^{14} M_\odot / M\Omega^{1/2}). \tag{5}$$

For example, if the density of the Universe $\Omega = 0.1$, the effect from a single galaxy like ours with $M = 10^{11} M_\odot$ amounts to a 10^{11}-fold decrease. Below we will need the function

$$e^{-\tau}\frac{d\tau}{dz} = \sigma_T n_c c H_0^{-1} A z^{-1/2} \exp\left\{-a z^{3/2} e^{-B/z} - \frac{B}{z} - \tau_0\right\}, \tag{6}$$

which in agreement with (3) has a sharp maximum for $z_{\max} = 1055$ ($e^{-\tau} \cdot (d\tau/dz)_{z=z[\max]} = 3.32\cdot 10^{-3}$) and exponentially decreases in both directions, the value of the function decreasing to half its maximum value for $z_3 = 960$ and $z_4 = 1135$. It will be convenient in what follows to approximate this function by a Gaussian function with variance $\sigma_z = 75$ whose integral equals 1.

3. Amplitude jump of the adiabatic perturbations. In the process of recombination the amplitudes of the adiabatic density perturbations grow. Before recombination the Jeans wavelength was equal to

$$\lambda_J = a_s \left(\frac{\pi}{G(\rho_r + \rho_m)} \right)^{1/2} \approx \frac{c}{\sqrt{G}} \frac{\sqrt{\rho_r}}{\rho_r + \rho_m} \approx \frac{c/\sqrt{G\rho_c}}{z\sqrt{z_1}\Omega(1 + (z/z_1\Omega))}.$$

We note that in a period $z > z_1\Omega$ ($\rho_r > \rho_m$) it will be of order ct. The Jeans mass of matter equaled

$$M_J = \frac{\pi}{6}\rho_m \lambda_J^3 = \frac{10^{16}\Omega^{-2}}{(1 + (z/z_1\Omega))^3} M_\odot, \tag{7}$$

i.e., it grew with decreasing z right up to $z_1\Omega$, but afterwards practically did not change. On a smaller scale density perturbations were manifested by sound waves of amplitude $\delta\rho/\rho$ which were connected with the velocity u of motion of matter by the relation

$$\left.\frac{\delta\rho}{\rho}\right|_{\text{ion}} = \frac{K}{\omega}\bar{u}, \tag{8}$$

where $\mathbf{K}$ is the wave vector, $\omega = a_s K$, and the average is carried out over a half-cycle. Furthermore, $\delta\rho/\rho|_{\text{ion}}$ and $\delta\rho/\rho|_{\text{neutral}}$ are the corresponding density perturbations before and after time of recombination, respectively, and only close to the time of recombination. After recombination the Jeans wavelength which is determined only by the plasma pressure decreases ($m_J = 10^5 - 10^6 M_\odot$) and perturbations corresponding to masses greater than m_J grow as $t^{2/3}$. Using the equation of continuity (see below) it can be shown that

$$\left.\frac{\delta\rho}{\rho}\right|_{\text{neutral}} \approx \frac{3}{5} K t \bar{u}, \tag{9}$$

where t corresponds to the hydrodynamic time during recombination and the wave number

$$K = 10^3 z \frac{H_0}{c} \left(\frac{10^{15}\Omega M_\odot}{m} \right)^{1/3},$$

$\delta\rho/\rho$ goes asymptotically to $\delta\rho/\rho|_{\text{neutral}}$ in the hydrodynamic time. Comparing (8) and (9) we see that the recombination of hydrogen in the Universe leads to an increased amplitude of density perturbations since the velocity of matter may not strongly change, as given by the equation

$$\frac{\delta\rho/\rho|_{\text{neutral}}}{\delta\rho/\rho|_{\text{ion}}} = \frac{3}{5} a_s K t \approx \left\{ \frac{M_J(z_r)}{M} \right\}^{1/3} = \frac{1}{1 + 27\Omega} \left(\frac{2 \cdot 10^{20}\Omega}{M} \right)^{1/3}. \tag{10}$$

For $\Omega = 0.1$ and $M = 10^{11} M_\odot$ this relation is close to $2 \cdot 10^2$, which supports the thesis about the smallness of initial adiabatic temperature fluctuations of radiations $\delta T/T < (\delta\rho/3\rho)_{\text{neutral}}$. Equation (10) is true only for masses $M < M_J(z_r)$.

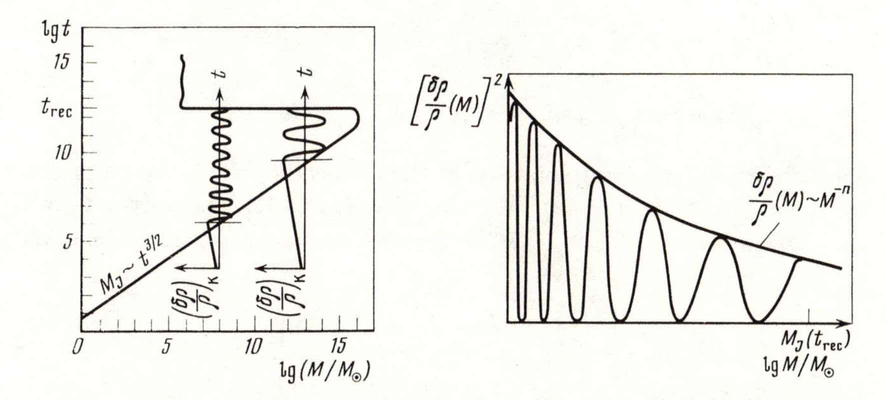

Fig. 1. Stability diagram in the hot model. To the right of the line $M_J(t)$ is the region of instability; the region of stability lies to its left. Two additional graphs demonstrate the evolution of density perturbations with time: growth as long as the mass remains smaller than the Jeans mass, and oscillations after it is larger. It is clear that perturbations corresponding to different masses arrive at the time of recombination with different phases.

4. The periodic amplitude dependence of adiabatic density perturbations and temperature fluctuations. The picture presented above is only a rough approximation since the phase relations between density and velocity perturbations in standing waves in an ionized plasma were not considered. As mentioned in the introduction, Sakharov [6] showed that the amplitude of perturbations of matter at a later stage when pressure does not play a role (in our case after recombination) turns out to be a periodic function of wavelength. This characteristic dependence is superimposed on the usually assumed power law dependence. From the expression for the dependence of the Jeans mass on the red shift we find z_J, the time when a given mass M is equal to the Jeans mass

$$z_J = z_1 \left[\left(\frac{10^{16}\Omega^{-2}}{M} \right)^{1/3} - 1 \right] \Omega.$$

In the interval between z_J and z_r the adiabatic density perturbations of a given scale are manifested by standing sound waves. Waves corresponding to different scales have different frequencies and different periods of existence (see Figure 1). As a result, at the moment of recombination sound waves should have occurred with phases depending on the scale, i.e., density perturbations (and corresponding velocity perturbations and temperature fluctuations of the radiation) should depend on the mass of the proto-object. From the previous result of the presence of growing modes, ion-acoustic waves at the stage $z_r < z < z_J$ should only be standing waves

$$\begin{aligned} \left(\frac{\delta\rho}{\rho}\right)_k &\sim \sin\left(\int_{z_J}^{z} \omega\, dt + \varphi_k\right), \\ u_k(z) &\sim \cos\left(\int_{z_J}^{z} \omega\, dt + \varphi_k\right) = \sin\left(\int_{z_J}^{z} \omega\, dt + \varphi_k + \frac{\pi}{2}\right). \end{aligned} \tag{11}$$

The conditions for joining solutions at the moment z_J require equal phases for sound waves of all scales. We note that $(\omega t)_{z_J}$ does not depend on k since $\omega = a_s k$ and $a_s t \sim \lambda_J \sim k_J^{-1}$. Then at the moment of recombination z_r,

$$\left(\frac{\delta\rho}{\rho}\right)_k \sim \sin\varphi_r, \quad u_k \sim \cos\varphi_r.$$

Considering that $z_r \lesssim z_1\Omega$ for all possible Ω's, we find[1]

$$\varphi_r = \int_{z_J}^{z_r} \omega\, dt + \varphi_0 \approx \frac{ck}{\sqrt{3z_1\Omega z_r} H_0} + \varphi_1 = bk + \varphi_1,$$

where

$$\varphi_1 = \varphi_0 - (\omega t)_{z_J}.$$

Now it is easy to find zeros of the functions $\sin\varphi_r$ and $\delta\rho/\rho(M)$,

$$\begin{aligned} \varphi_r &= n\pi, \quad n = 0, 1, 2, 3, \dots \\ bk_0 + \varphi_1 &= n\pi, \\ k_0 &= \frac{n\pi - \varphi_1}{b}, \end{aligned}$$

which corresponds to the equations

$$\begin{aligned} \left(\frac{M_J}{M_0}\right)^{1/3} &= n\pi - \varphi_1, \\ \frac{M_0}{M_J} &= \frac{1}{(n\pi - \varphi_1)^3} \underset{n\gg 1}{\approx} \frac{1}{(n\pi)^3}. \end{aligned} \tag{12}$$

The zeros of the function $u(M)$ will be moved by

$$\frac{M_0}{M_J} = \frac{1}{\left\{\left(n + \frac{1}{2}\right)\pi - \varphi_1\right\}^3}. \tag{13}$$

It was shown in the first paragraph that the amplitude jump of density perturbations at the recombination time is connected with the presence of the velocity u so that the actual mass distribution of objects should be described by the last equation. It is barely possible to observe such a distribution since the mass of a galaxy $M \sim 10^{11} M_\odot$ for $\Omega = 1$ corresponds to $n \sim 15$, and for $\Omega = 0.1$ to $n \sim 70$ and $\Delta M/M \ll 1$. In the case $\Omega = 1$, this effect gives

[1]We replace $\int_{z_J}^{z_r} \omega\, dt$ by $\omega t|_{z_r} - \omega t|_{z_J}$ which considerably simplifies the calculation, but does not significantly change the qualitative results.

$\Delta M \sim M$ for a cluster of galaxies, but it is possible that it is masked by processes at a later stage of expansion of the Universe.

The dependence on the scale of both the amplitude of density perturbations and the speed of material motion should be reflected in the fluctuations of relic radiation. We are preparing to investigate this question elsewhere. We note that only observations of the small-scale fluctuations of relic radiation with a periodic dependence on scale may give information on the large-scale density perturbations which are small at the present time. The effect considered in the paragraph is applicable only to adiabatic density perturbations and will allow a choice to be made between density perturbations of various types.

3. The Observational Picture for Initial Fluctuations

In the first section of this article it was shown that adiabatic temperature fluctuations of small scale are strongly damped in the process of recombination. Below we shall consider fluctuations connected with the motion of matter at the time of recombination.

1. Scattering on Moving Matter. With the aid of the equation of continuity it is possible to find the velocity of matter for any given density perturbation

$$\frac{\partial(\delta\rho/\rho)}{\partial t} = \operatorname{div}\mathbf{u}. \tag{14}$$

The scattering of photons on moving electrons leads to a frequency shift (depending on the direction of motion) due to the Doppler effect. In the first approximation this effect is proportional to the optical depth of moving matter and decreases, as already mentioned in the first section, due to the subsequent scattering of radiation propagating from the proto-object to the observer, i.e.,

$$\frac{\delta T}{T} = \int_0^\infty \frac{u_1(z)}{c} e^{-\tau(z)} \frac{d\tau}{dz} dz, \tag{15}$$

where u_1 is the projection of velocity along the direction of the ray and the integral is carried out along the ray $r = r(z)$, $t = t(z)$, $\theta = \text{const}$, and $\varphi = \text{const}$. The function $e^{-\tau}(d\tau/dz)$ has already been found above [see (6)]. All the effect (15) is concentrated in the region of the maximum of the function of (6). The maximum effect produces objects for which the sign of u_1 does not change within the limits of a Gaussian curve. This result is applicable for perturbations containing a mass $M > M_{\text{max}}$, where

$$M_{\text{max}} = \frac{\Omega\rho_c z_{\text{max}}^3}{2}\left[cH_0^{-1}\int_{z_3}^{z_4}\frac{dz}{z^{2.5}\Omega^{0.5}}\right]^3 = 7\cdot 10^{14}\Omega^{-1/2}M_\odot. \tag{16}$$

For smaller objects it is necessary to take into account the change in sign of u_1.

The mean square temperature fluctuations which interest us will be calculated below. We represent the density fluctuations and the velocity of matter connected with them in the form of Fourier integrals

$$\frac{\delta\rho}{\rho} = \frac{1}{(2\pi)^3}\int a_{\mathbf{p}} e^{i\mathbf{pr}} d^3p, \qquad \frac{\mathbf{u}}{c} = \frac{1}{(2\pi)^3}\int \mathbf{b}_{\mathbf{p}} e^{i\mathbf{pr}} d^3p, \tag{17}$$

for which the dimensionless variable r is defined by

$$r = 1 - \frac{RH_0}{2c} = (1+\Omega z)^{-1/2}, \tag{18}$$

where

$$R = \int_t^{t_0} \frac{dx}{a(x)} = \frac{2c}{H_0}\left\{1 - \left(\frac{t}{t_0}\right)^{1/3}\right\} = \frac{2c}{H_0}\left\{1-(1+\Omega z)^{-1/2}\right\}$$

is the co-moving horizon coordinate and $a(t)$ is the radius of the Universe. With this definition of r,

$$p = \frac{K}{1+z}\frac{2c}{H_0} = 2\cdot 10^3\left(\frac{M\Omega^{-1}}{10^{15}M_\odot}\right)^{-1/3} \tag{19}$$

does not depend on z. In (19),

$$M = \frac{4}{3}\pi\left(\frac{\pi}{p}\right)^3\left(\frac{2c}{H_0}\right)^3 \rho_c\Omega$$

is the mass within a half-wavelength radius sphere. In agreement with [2] after recombination perturbations vary with time according to the law

$$\frac{\delta\rho}{\rho}(t) = F\left[\left(\frac{t}{t_r}\right)^{2/3} - \left(\frac{t}{t_r}\right)^{-1}\right] \tag{20}$$

for any $z < z_r$ if $\Omega = 1$, and for $z > \Omega^{-1}$ if $\Omega < 1$. In (20) it is assumed that the density perturbations $\delta\rho/\rho$ were small, for $t = t_r$, $(\delta\rho/\rho)(t_r) = 0$ before recombination and the perturbations were excited due to the presence of a velocity. The interesting mode for us is the growing mode for $t \gg t_r$,

$$\frac{\delta\rho}{\rho}(t) \sim t^{2/3} \sim (1+z)^{-1}. \tag{21}$$

If the observed objects were formed when z_0 and $\delta\rho/\rho(z_0) \sim 1$, then it follows from (21) that

$$\frac{\delta\rho}{\rho}(z) = \frac{1+z_0}{1+z}.$$

Since p does not depend on z, $a_{\mathbf{p}}$ varies in the same way, and

$$\left(\frac{\delta\rho}{\rho}\right)^2 = \frac{1}{(2\pi)^3}\int |a_{\mathbf{p}}|^2\, d^3p \quad\text{and}\quad a_{\mathbf{p}} \sim \frac{1+z_0}{1+z}. \tag{22}$$

Using (20) with the equation of continuity (14) for $t = t_r$

$$\frac{\partial(\delta\rho/\rho)}{\partial t} = \frac{5}{3}F\frac{1}{t_r} = \frac{5}{3}FH_0 z_r^{3/2}\Omega^{1/2}$$

and taking into account the relation between the physical coordinate $x = ct$ and the variable r

$$\operatorname{div}\mathbf{u} = \frac{1}{(2\pi)^3}\int i\mathbf{p}\mathbf{b}_{\mathbf{p}}e^{i\mathbf{p}\mathbf{r}}d^3p\frac{1}{2}\frac{H_0}{c}z,$$

we find the desired relation between $a_{\mathbf{p}}$ and $b_{\mathbf{p}}$

$$\mathbf{b}_{\mathbf{p}} = -5i\Omega^{1/2}z^{1/2}\frac{\mathbf{p}}{p^2}a_{\mathbf{p}}. \tag{23}$$

In this calculation it is assumed that the transition from adiabatic plasma oscillations connected with the radiation to the growth of perturbations occurs instantaneously: the corresponding time is small in comparison to the hydrodynamic time.

Temperature fluctuations are evidence by velocity perturbations in agreement with (15) from which we find with the aid of (17) for $\mathbf{u}$ the equation

$$\frac{\delta T}{T} = \frac{1}{(2\pi)^3}\int d^3p\int\frac{\mathbf{b}_{\mathbf{p}}\mathbf{r}}{r}e^{-\tau+i\mathbf{p}\mathbf{r}}d\tau = \frac{1}{(2\pi)^3}\int c_{\mathbf{p}}d^3p. \tag{24}$$

The coefficients $c_{\mathbf{p}}$ are calculated in the following manner: substituting $e^{-\tau}(d\tau/dz)$ in the form of a Gaussian function with maximum at $r_{\max} = \Omega^{-1/2}z_{\max}^{-1/2}$ and variance σ_r, and taking into account $\sigma_r = \frac{1}{2}\sigma_z z_{\max}\Omega_{\max}^{-3/2}$, we obtain by standard methods from (22), (23), and (24)

$$\begin{aligned} c_{\mathbf{p}} &= 0.54\sqrt{3}\pi^{3/2}\Omega^{1/2}\sigma_z\cos\theta p^{-5/2}z_{\max}^{-1}(1+z_0)\exp\left[-\frac{p^2\sigma_z^2\cos^2\theta}{8z_{\max}^3\Omega} + i\mathbf{p}\mathbf{r}_{\max}\right]a_{\mathbf{p}}(z_0) \\ &= Ca_{\mathbf{p}}(z_0)\cos\theta p^{-3/2}\exp\left[-\frac{\alpha}{2}\cos^2\theta + i\mathbf{p}\mathbf{r}_{\max}\right], \end{aligned} \tag{25}$$

where

$$\alpha = \frac{p^2\sigma_z^2}{4z_{\max}^3\Omega}.$$

The mean square temperature fluctuations of the radiation are equal

$$\left(\frac{\delta T}{T}\right)^2 = \frac{1}{(2\pi)^6}\int c_{\mathbf{p}'}d^3p'\int c_{\mathbf{p}}d^3p. \tag{26}$$

Due to the orthogonality of waves with different $\mathbf{p}$ and $\mathbf{p}'$ and using (25), we obtain

$$\left(\frac{\delta T}{T}\right)^2 = \frac{C^2}{(2\pi)^2}\int dpp^{-3}\int_0^1\cos^2\theta e^{-\alpha\cos^2\theta}d(\cos\theta). \tag{27}$$

The integral over angles is easy to calculate: namely,

$$I = \int_0^1 \cos^2\theta e^{-\alpha\cos^2\theta} d(\cos\theta)$$
$$= \int_0^1 y^2 e^{-\alpha y^2} dy = \frac{1}{2\alpha}\left\{\sqrt{\frac{\pi}{4\alpha}}\Phi(\alpha\sqrt{2}) - e^{-\alpha}\right\},$$

where

$$\Phi(x) = \frac{2}{\sqrt{2\pi}} \int_0^x e^{-t^2/2} dt$$

is the probability integral. For large $\alpha > 1$, $I \approx \sqrt{\pi}/4\alpha\sqrt{\alpha}$; for small $\alpha \ll 1$, $I \simeq \frac{1}{3}$; and for $\alpha = 1$, $I = \frac{1}{2}$. We adopt the approximation

$$I^{1/2} = \frac{2\pi^{1/4}}{\sqrt{3}\pi^{1/4} + 2\alpha^{3/4}} \tag{28}$$

(in which we increase by a factor of 2 the asymptotic value of $\delta T/T$ for small and large masses) and note that $\alpha = 1$ for $p = 10^3\Omega^{1/2}$ and $M = 8 \cdot 10^{15}\Omega^{-1/2} M_\odot$. Supposing that $a_\mathbf{p}$ has a maximum corresponding to mass M, we finally obtain from (25), (27), and (28)

$$\sqrt{\left(\frac{\delta T}{T}\right)^2} = 2 \cdot 10^{-5} \frac{(M\Omega^{1/2}/10^{15} M_\odot)^{1/3}}{1 + 2.5(10^{15} M_\odot / M\Omega^{1/2})^{1/2}} (1 + z_0). \tag{29}$$

The characteristic angular scale of fluctuation is [5]

$$\theta = \frac{l H_0}{2c\psi(z, \Omega)},$$

where

$$\psi = \Omega^{-2}(1+z)^{-2}[\Omega z + (\Omega - 2)(\sqrt{1 + \Omega z} - 1)].$$

For $z \gg \Omega^{-1}$ which applies in the case of interest to us we obtain, taking into account that $l = 2\pi c H_0^{-1} p^{-1} (1+z)^{-1}$,

$$\theta = \frac{2\pi\Omega}{p} \approx 10' \left(\frac{M\Omega^2}{10^{15} M_\odot}\right)^{1/3}. \tag{30}$$

2. Discussion. We note especially that perturbations corresponding to small masses in comparison with $10^{15} M_\odot$ give quite a small contribution to $\delta T/T$; for example, for a single object with mass $M = 10^{11} M_\odot$, in the case $\Omega = 1$ and $(\delta\rho/\rho) = 1$ for $z_0 = 2$ for a wave vector inclined at an angle of 45° to the direction of the observer, we obtain $\delta T/T = 3 \cdot 10^{-8}$ with equations (6), (9), and (15). After integrating over all angles, $\delta T/T = 10^{-8}$. Perturbations effectively arise from a region whose absolute size (for $z_{\max} = 1055$ and $\sigma_n = 75$) is of order $L \sim 10^{23}$ cm. A proto-object with mass $M = 10^{11} M_\odot$ had a size $l \sim 10^{21}$ cm at that time. If in the length L there are $N = L/l$ characteristic lengths and one object produces an effect $(\delta T/T)_1$, then it would turn out that the integrated effect should be simply $\delta T/T = \sqrt{N}(\delta T/T)_1$

according to the combination rule for random values. Why is the effect much smaller in the proposed calculations and why is the dependence on the sizes and mass of an object different? Physically, this signifies an assumption about the specific strength of anticorrelations between velocities of matter in neighboring regions of space. It is most probable that at a distance of the order of the size of an object the velocity changes sign. This result is formally connected with the fact that at a quite early stage, long before recombination, initial adiabatic perturbations are established by the spectral function $a_{\mathbf{p}}$. There is no basis for considering that this function leads to a constant value for all or even for small $\mathbf{p}$ (see a discussion about this point in [5]). The so-called "natural" or "random" distribution, for which the elementary law $\delta\rho/\rho \sim n^{-1/2}$ and $\Delta n = n^{1/2}$ (where n is the number of objects in a given volume) is correct, corresponds to $a_{\mathbf{p}}$ = const. The power law $a(p) \sim p^m$ for $m > 0$ corresponds to a large degree or order: for example, for $m = 2$, $\Delta n \sim n^{1/6}$ [18], i.e., an anticorrelation between fluctuations in neighboring volumes exists. Apparently, this type of dependence of $a(\mathbf{p})$ as $p \to 0$ actually occurs in nature.

4. Fluctuations of Radiation Connected with a Secondary Heating of Matter in the Universe

After recombination the hydrogen in the Universe should remain neutral. However, in our neighborhood neutral intergalactic hydrogen is not detected right up to $z \sim 2$. The possible variants are:

(a) The average density of matter in the Universe considerably exceeds the average density which went into galaxies which corresponds to $\Omega = 1/45$. Sometime for z_h between $z_r = 1500$ and $z = 2$ a secondary nonequilibrium heating and ionization of the intergalactic plasma occurred; the main part of the plasma remained distributed homogeneously and did not enter into the matter of galaxies. In this case the optical depth of the ionized gas for Thompson scattering,

$$\tau = \Omega\sigma_T n_c c H_0^{-1} \int_0^{z_h} \frac{1+z}{\sqrt{1+\Omega z}} dz = 6.65 \cdot 10^{-2}\Omega \int_0^{z_h} \frac{1+z}{\sqrt{1+\Omega z}} dz \qquad (31)$$

may turn out to be sufficient for a strong decrease of the initial fluctuations of radiation. Due to the large energy losses of the ionized plasma by the inverse Compton effect of electrons on quanta of relic radiation, there is a small probability that heating occurred much earlier than $z \sim 10$ [20, 3] which results in $\tau < 2$, i.e., this heating may not decrease the fluctuations of relic radiation resulting from perturbations for $z \sim 1000$ by more than an order of magnitude.

(b) Matter is mostly concentrated in galaxies whose average density is extremely small $\Omega \sim 1/45$. Further, if matter was ionized before the formation of galaxies, τ should be extremely small $\tau \ll 1$.

(c) Intergalactic gas in clusters of galaxies determines the average density of matter in the Universe. Since neutral hydrogen is not founded in clusters of galaxies, its density $< 10^{-7}\,\mathrm{cm}^{-3}$ [7], so that this gas must have a high temperature leading to $\tau < 2$, as in the case (a).

We return to fluctuations excited by the interaction with electrons at the stage $z < 10$. In [19] it was shown that the spectrum of relic radiation is changed by Compton scattering of hot electrons on the photons of this radiation leading to an intensity increase for $h\nu > 3.83kT$, but to an intensity decrease in the Rayleigh-Jeans region which is most convenient for observations. The effective temperature for $h\nu \ll kT_0$ is

$$T = T_0 e^{-2y}, \tag{32}$$

where the parameter y is determined by

$$y = \int \frac{kT_e}{m_e c^2} d\tau. \tag{33}$$

It is obvious that in variant (a) the inhomogeneity of electron temperature and the inhomogeneity of plasma heating in various directions should lead to temperature fluctuations of the radiation with possible observational effects even for $\tau < 1$; the equations for the Rayleigh-Jeans region are true even for this condition. In variant (c) the determination of $\delta T/T$ in the direction of the nearest cluster of galaxies presents considerable interest. In agreement with (32) variations of the small-scale fluctuations of relic radiation are sensitive to extended high-temperature objects with small electron densities; with an increase of the plasma temperature the optical depth due to bremsstrahlung decreases, but the parameter γ and $\delta T/T$ grow. The intensity of X-ray and radio emission are proportional to the square of the electron density, but y is proportional to the first power. A decrease in intensity of relic radiation in the Rayleigh-Jeans part of the spectrum close to X-ray sources would resolve the question about the radiation mechanism and allow a precise determination of the size of sources. Thus, for example, the presence of intergalactic gas with a temperature of $3 \cdot 10^8$ K and $n_e \approx 10^{-3}\,\mathrm{cm}^{-3}$ in the Coma cluster of galaxies whose diameter is 10^{25} cm [9] would lead to $\delta T/T = -2y = -10^{-3}$. The sign of the effect (a decrease of T) and large cluster sizes which allow the elimination of the contribution of the brightest radio sources gives rise to the possibility of verifying the interpretation of X-ray data.

Limits on the temperature of the intergalactic gas were obtained from X-ray measurements of the background emission for $\Omega \sim 1$; $T_e < 10^6(1+z)$ K [12, 1]. On the other hand, the existing data on the fluctuations of relic radiation which gives $\delta T/T < 10^{-3}$ on a scale $10' - 15°$ [17, 8] indicate an absence of extended objects with temperatures exceeding 10^8 K in the investigated areas of the sky. The final estimate does not depend on z, since y is a function of the ratio $kT_e/m_e c^2$.

5. Conclusions

Due to a decrease of adiabatic fluctuations of radiation by subsequent scattering for $M < 10^{15}\Omega^{-1/2}M_\odot$ the initial fluctuations will be determined by the effect of scattering on moving matter and in the case of adiabatic and entropy perturbations will be equal to

$$\sqrt{\left(\frac{\delta T}{T}\right)^2} = 10^{-5}\left(\frac{M\Omega^{1/2}}{10^{15}M_\odot}\right)^{5/6}(1+z_0), \tag{34}$$

where z_0 corresponds to the moment when $\delta\rho/\rho \sim 1$. However, for $M > 10^{15}\Omega^{1/2}M_\odot$ and large Ω, even taking into account the jump in amplitude of density perturbations due to recombination, adiabatic fluctuations become of the same order as fluctuations connected with the motion of matter. This is easy to see from a comparison of equations (9), (10), and (15). We introduce for comparison two equations for $M > 10^{15}\Omega^{-1/2}M_\odot$.

(a) effects connected with motion:

$$\sqrt{\left(\frac{\delta T}{T}\right)^2} = 2\cdot 10^{-5}\left(\frac{M\Omega^{1/2}}{10^{15}M_\odot}\right)^{1/3}(1+z_0); \tag{35}$$

(b) adiabatic fluctuations:

$$\sqrt{\left(\frac{\delta T}{T}\right)^2} = 10^{-6}(1+27\Omega)\left(\frac{M\Omega^{-1}}{10^{15}M_\odot}\right)^{1/3}(1+z_0). \tag{36}$$

The second equation is true only for masses less than $M_J(z_r)$ (i.e., less than the Jeans mass in the period before recombination). For $M > M_J$ again the effects of scattering on moving matter become most important; the velocity and this effect which is proportional to the velocity grow as $M^{1/3}$ at a time when adiabatic temperature fluctuations in this region no longer depend upon mass, as

$$\sqrt{\left(\frac{\delta T}{T}\right)^2} = 2\cdot 10^{-5}\left(\frac{M\Omega^{1/2}}{10^{15}M_\odot}\right)^{1/3}\frac{\delta_0\rho}{\rho}; \tag{37}$$

where $\delta\rho/\rho < 1$ is the density perturbation corresponding to the present time (if $\Omega = 1$ or for $\Omega < 1$ at $z \sim \Omega^{-1}$) to objects with mass $M > M_J(z_r) = 10^{16}\Omega^{-2}M_\odot$. After recombination entropy density perturbations evolve in the same way as adiabatic perturbations. Temperature fluctuations connected with them are excited by scattering on moving matter and are described by (22). Using the functional form of $e^{-\tau}(d\tau/dz)$ from (6), it is easy to calculate temperature fluctuations of the relic radiation for any given distribution of velocities. Sachs and Wolfe [13] estimated the temperature fluctuations of the relic radiation for $\Omega = 1$ excited by the gravitational

Table 1. Temperature Fluctuations of the Relic Radiation

Mass of Object, $M_\odot$	Exciting cause of fluctuations				Density Perturbation of Matter
	Adiabatic Connection	Scattering on Moving Matter	Gravitational Influence	Doppler Velocity of Observer	
1	2	3	4	5	6
			$\Omega = 1$		
10^{11}	10^{-13}	10^{-8}	—	—	$\delta\rho/\rho = 1$
10^{13}	$5 \cdot 10^{-7}$	10^{-6}	—	—	for $z_0 = 2$
10^{15}	10^{-4}	$3 \cdot 10^{-5}$	—	—	
10^{15}	$3 \cdot 10^{-6}$	$2 \cdot 10^{-6}$	$2 \cdot 10^{-7}$	10^{-4}	$\delta_0\rho/\rho = 0.1$
10^{17}	$2 \cdot 10^{-5}$	10^{-5}	$4 \cdot 10^{-7}$	10^{-4}	for $z = 0$
10^{19}	$2 \cdot 10^{-5}$	$5 \cdot 10^{-5}$	10^{-4}	$2 \cdot 10^{-3}$	
			$\Omega = 1/45$		
10^{11}	10^{-13}	$4 \cdot 10^{-8}$	—	—	$\delta\rho/\rho = 1$
10^{13}	$5 \cdot 10^{-7}$	$2 \cdot 10^{-6}$	—	—	for $z_0 = 45$
10^{15}	10^{-4}	10^{-4}	—	—	
10^{15}	$3 \cdot 10^{-7}$	10^{-6}	—	—	$\delta_0\rho/\rho = 0.1$
10^{17}	10^{-6}	$5 \cdot 10^{-6}$	—	—	for $z = 0$
10^{19}	$4 \cdot 10^{-6}$	$2 \cdot 10^{-5}$	—	—	

action of the nearest objects (gravitational effects on the early stage of expansion have little effect because of the growth of density perturbations)

$$\sqrt{\left(\frac{\delta T}{T}\right)^2} = 2 \cdot 10^{-6} \left(\frac{M}{10^{15} M_\odot}\right)^{2/3} \frac{\delta_0 \rho}{\rho}; \tag{38}$$

Comparing (35), (36), and (38) we see that for $\Omega = 1$ fluctuations of the radiation formed during recombination exceed gravitational effects up to masses of the order of $10^{18} M_\odot$. For the case $\Omega = 1$ the existing experimental data $\delta T/T < 10^{-3}$ on a scale $\theta \sim 10' - 15°$ [17, 8] contradict $\delta\rho/\rho \sim 1$ for masses $M > 10^{19} M_\odot$.

In Table 1 fluctuations of the radiation corresponding to the same density perturbation of matter, but due to various physical effects, are compared. In the fifth column the probable values of the temperature perturbations connected with the peculiar motion of the observer (i.e., our own Galaxy) relative to the relic radiation are given. This motion should manifest itself within a 24-hour period. Investigation of one plane [17] gave for this value $\delta T/T|_{24\,\mathrm{h}} \lesssim 10^{-3}$. It is clear from the table that the absence of observed motion and a 24-hour asymmetry provide the most stringent bounds for large-scale perturbations. It is necessary, however, to emphasize the specific character of this variation: we can in principle only obtain data on the velocity relative to the relic radiation for one object—our solar system. The possibility remains that in the average of the peculiar velocity $\overline{u}$ over all galaxies, randomly the velocity of our Galaxy is less than u_0. The *a priori* probability of this is of the order $\alpha < (u_0/\overline{u})^2$ for the observations in one plane as it was carried out at the present time (two components of u_0). If the velocity remains less than u_0 as before for the measurements in two perpendicular planes (all three components) we obtain $\alpha < (u_0/\overline{u})^3$. For $u_0 = 200\,\mathrm{km \cdot s^{-1}}$ which gives $\alpha > 0.01$ we find $\overline{u} < 1000\,\mathrm{km \cdot s^{-1}} = 3 \cdot 10^{-3}c$ which corresponds to the fifth column in the table.

The scattering of quanta of relic radiation on hot electrons in clusters of galaxies may lead to temperature fluctuations of relic radiation $\delta T/T \lesssim 10^{-4} - 10^{-5}$ since the observed background X-ray radiation in galactic clusters is contradictory to the anomalously high temperatures of the gas in galactic clusters for any $z < 4$. At the same time fluctuations of the background radiation due to the presence of discrete sources of radio emission exceed $10^{-4} - 10^{-5}$ and complicate the detection of effects connected with the formation of observed objects [10].

The authors would like to thank A. F. Illarionov for discussions.

Institute of Applied Mathematics
USSR Academy of Sciences. Moscow

Received
September 11, 1969

REFERENCES

1. *Vainshtein L. A., Sunyaev R. A.*—Kosmich. Issled. **6**, 635 (1968).
2. *Doroshkevich A. G., Zeldovich Ya. B.*—Astron. Zh. **40**, 807 (1963).
3. *Doroshkevich A. G., Sunyaev R. A.*—Astron. Zh. **46**, 20 (1969).
4. *Zeldovich Ya. B., Kurt V. G., Sunyaev R. A.*—ZhETF **55**, 278 (1968).
5. *Zeldovich Ya. B., Novikov I. D.* Relativistic Astrophysics. Univ. of Chicago (1971).
6. *Sakharov A. D.*—ZhETF **49**, 345 (1965).
7. *Allen R. J.*—Nature **220**, 147 (1968).
8. *Conklin E. K., Bracewell R. N.*—Nature **216**, 777 (1967).
9. *Felten J. E., Gould R. G., Stein W. A.*—Astrophys. J. **146**, 955 (1967).
10. *Longair M. S., Sunyaev R. A.*—Nature **223**, 719 (1969).
11. *Peebles P. J. E.*—Astrophys. J. **153**, 1 (1968).
12. *Rees M., Sciama D., Setti G.*—Nature **217**, 326 (1968).
13. *Sachs R. K., Wolfe A. M.*—Astrophys. J. **147**, 73 (1967).
14. *Silk J.*—Astrophys. J. **151**, 459 (1967).
15. *Silk J.*—Nature **215**, 1155 (1967).
16. *Silk J.*—Nature **218**, 453 (1968).
17. *Wilkinson D. T., Partridge R. B.*—Nature **215**, 719 (1967).
18. *Zeldovich Ya. B.*—Adv. Astron. and Astrophys. **3**, 241 (1965).
19. *Zeldovich Ya. B., Sunyaev R. A.*—Astrophys. and Space Sci. **4**, 302 (1969).
20. *Sunyaev R. A.*—Dokl. AN SSSR **179**, 45 (1968).

Commentary

The relation between the amplitude of angular fluctuations of the relic radiation and the structure of the observed Universe is a central problem of cosmology at the present stage. Fluctuations of the relic radiation on a scale less than one arc degree give us unique information on perturbations of density and velocity of matter in the distant past—in the period of recombination of hydrogen in the Universe. It is precisely observations of fluctuations that proved that in that period the Universe was homogeneous to a high degree. Data on observations of the dependence of $\Delta T/T$ on the angle should, in principle, give us initial data for a full description of the evolution of the structure of the Universe up to its present state. However, if one does not take into account all of the quantitative complexity in this endeavor, mistakes are possible which lead to fundamentally invalid conclusions. The present article laid the foundation for a detailed investigation of the problem.

The paper obtains an asymptotic solution which describes the rate of hydrogen recombination in its most interesting stage where it determines the appearance of small-scale fluctuations of the relic radiation and the maximum wavelength of adiabatic density perturbations decaying due to radiation viscosity and heat conduction.

In the literature one still finds widespread discussion of three effects which were first noted in this paper (see, for example, the work by Peebles[1] and the paper by Jones and Wyse[2]):

1) Smearing of the initial adiabatic fluctuations of the relic radiation due to the non-instantaneity of recombination scales smaller than the horizon at the moment of recombination;
2) Periodic dependence of the amplitude of perturbations on their scale at scales smaller than the horizon at the moment of recombination. It was independently found by Peebles and Yu[3] from numerical calculations;
3) A jump in the amplitude of adiabatic density perturbations resulting from recombination which is related to a sharp change in the speed of sound.

The present article gives a method, now widely accepted, for calculating the small-scale fluctuations of relic radiation. It is shown that the initial adiabatic fluctuations of $\Delta T/T$ at small scales are smeared in the course of recombination.

It is commonly accepted that this article is one of the ground-laying papers on the theory of small-scale generation of relic radiation fluctuation. This is noted, for example, in one of the best observational papers.[4]

In addition, the article was the first to point out the effect of reduction of brightness of relic radiation in directions toward galaxy clusters. It is shown here for the first time that inhomogeneities of heating of the hot intergalactic gas lead to fluctuations of the relic radiation. This topic is presently being intensively studied by Hogan.[5]

Intensive theoretical and experimental investigations are continuing. Experiments are being carried out both on Earth-based radiotelescopes and high-altitude balloons and from space instruments. The successful experiment "Relict" on the satellite "Prognoz-9" should be noted here.[6]

[1] *Peebles P. J. E.* The Large-Scale Structure of the Universe. Princeton Univ. Press (1980).

[2] *Jones B. J. T., Wyse R. F. G.*—Month. Not. RAS **123**, 171–182 (1983).

[3] *Peebles P. J. E., Yu I.*—Astrophys. J. **162**, 815–836 (1970).
[4] *Uson J., Wilkinson D.*—Astrophys. J. **283**, 471–478 (1984).
[5] *Hogan C. J.*—Astrophys. J. Lett. **284**, L1–L5 (1984).
[6] *Strukov I. A., Skulachev I. P.*—Pisma v Astron. Zh. **10**, 3–13 (1984).

67

The Observation of Relic Radiation as a Test of the Nature of X-Ray Radiation from the Clusters of Galaxies*

With R. A. Sunyaev

Introduction

The X-ray radiation from a number of clusters of galaxies (Coma, Virgo, Perseus) was discovered recently [1]. It is assumed that clusters of galaxies form an important class of powerful X-ray sources, possibly giving the main contribution to the X-ray background radiation of the Universe [2]. What is the nature of these sources? What physical mechanisms give the observed X-ray radiation?

Most likely this is either the bremsstrahlung radiation of hot intergalactic gas or inverse Compton scattering on the relativistic electrons. Again the question arises—what kind of radiation and where is it scattered? The relic photons in the intergalactic space [3], or in haloes of massive elliptical galaxies [4], or infrared radiation in the vicinity of nuclei of galaxies [5]? The observations of a small perturbation in angular distribution of relic radiation can give an answer to these questions. These observations enable us to distinguish between hot nonrelativistic electron gas (being a bremsstrahlung source) and a less numerous group of relativistic electrons. The hole in the relic radiation—the decreasing of its brightness temperature in the Rayleigh-Jeans spectral region,

$$\frac{\Delta T_r}{T_r} = -2\frac{kT_e}{m_e c^2}\sigma_T N_e l,$$

must be observed in the directions to the source with hot electrons [6], if the latter are optically thin on bremsstrahlung absorption and Thomson scattering. At large optical depth on Thomson scattering the effect increases strongly. The basic physical mechanism is the transfer of soft photons ($h\nu \ll kT_r$) into short wave regions ($h\nu \gg kT_r$).

*Comments on Astrophysics and Space Physics **46**, 173–178 (1972).

The Properties of Relic Radiation

We propose to exploit the 2.7 black-body relic radiation. It is undoubtedly the best source of information on any deviations from the isotropy and homogeneity of the Universe and its characteristic features are astonishing: (1) Excellent isotropy—independence in the direction $\delta F_\nu / F_\nu$ less than 10^{-3} in all angular scales from 10 angular minutes to 24 hours; (2) its dominance over the radiation of discrete radio sources (at observations with broad angles) in the centimeter and millimeter wavelength bands of the radio background spectrum; (3) the spectrum follows the Rayleigh-Jeans formula $F_\nu = 2kT_r/\lambda^2$ with great precision (better than 5% or 10%) in the 70 cm–0.3 cm wavelength interval.

These properties are quite enough for the use of relic radiation in the proposed test. And, what is more, the realization of tests on different wavelengths can promote elucidation of the exact spectrum of relic radiation.[1]

The Distortion of Relic Radiation Spectrum via Thomson Scattering on Thermal Electrons

As is well known, Thomson scattering of radiation on the thermal electrons leads not only to a change of the direction of photon propagation (isotropization) but also to the broadening of spectral lines due to Doppler effect. Our problem is the investigation of thermal electrons interacting with spectrally broad radiation having Planckian distribution over frequencies. It is assumed that in the isotropically expanding Universe there is a nonexpanding cloud of Maxwellian electrons with $T_e \gg T_r$.

Neglecting the frequency shift by scattering, we would obtain no effect. Obviously elastic scattering of photons cannot change the equilibrium isotropic radiation. Photons 1 scattered into 1′ do not reach the observer but, instead, photons 2 are scattered into the observer 2′ and replace the lost 1. The compensation is exact and valid even in multiple scattering situations (Fig. 1).

Owing to thermal motion (velocity V) of electrons the frequency is shifted randomly by

$$\frac{\Delta v}{v} = f\frac{V}{c},$$

where f depends on the angles between the electron velocity and photon propagation before and after scattering; soft photons $h\nu \ll kT_e$ are considered—an excellent approximation in the radio frequency band. Assume electrons moving isotropically (with equal probability in all directions; the Maxwellian distribution in the partial case) in the coordinate system in which the unperturbed radiation is isotropic. At cosmological distance

[1]This remark is due to Yu. N. Pariĭskiĭ.

this means that the peculiar velocity of the plasma investigated is neglected. This means that the number of photons whose frequency has been raised ($\Delta\nu > 0$) and lowered ($\Delta\nu < 0$) owing to scattering are equal. In a broad spectrum the first order effect vanishes. However, a second-order effect, proportional to $\overline{(\Delta\nu)^2}$, i.e., to v^2/c^2, remains. In turn v^2/c^2 for the electrons is proportional to kT_e/m_ec^2. A rather long calculation gives the answer for an arbitrary radiation spectrum in the form[2] [7, 8]

$$F(\nu)_{\text{before}} - F(\nu)_{\text{after}} = y\nu\frac{d}{d\nu}\nu^4\frac{d}{d\nu}\nu^{-3}F(\nu)_{\text{before}},$$

where $y = (kT_e/m_ec^2)\tau_T \ll 1$; $\tau_T = \int N_e\sigma_T dl$, $\sigma_T = (8\pi/3)r_0^2 = 6.65 \cdot 10^{-25}\,\text{cm}^2$ is the Thomson cross section.

Inserting the Rayleigh-Jeans initial spectrum $F_\nu = A\nu^2$ into the right-hand side, we obtain

$$\Delta F_\nu = F(\nu)_{\text{before}} - F(\nu)_{\text{after}} = -2yF(\nu)_{\text{before}}.$$

The Planckian spectrum of radiation as a whole is not preserved: its Rayleigh-Jeans part is reduced and its Wien region $h\nu \gg kT_r$ is increased.

$$\Delta F_\nu = y\left(\frac{h\nu}{kT_r} - 2\right)F(\nu)_{\text{before}}.$$

The energy of radiation integrated over all the spectrum increases,

$$\int F(\nu)_{\text{after}}d\nu = e^{4y}\int F(\nu)_{\text{before}}d\nu, \quad \Delta\mathcal{E} = (e^{4y} - 1)\mathcal{E} = 4y\mathcal{E},$$

but the number of photons is constant,

$$\int \frac{1}{h\nu}F(\nu)_{\text{after}}d\nu \equiv \int \frac{1}{h\nu}F(\nu)_{\text{before}}d\nu.$$

Qualitatively the results are obvious: scattering has preserved the number of photons. The electron temperature is assumed much higher than the radiation temperature, therefore the general trend is a cooling of electrons and an increase in the radiation energy, with conservation of photon number. This is possible only by means of photon transfer from the low-frequency region $h\nu \ll kT_r$ to the high-frequency region $h\nu \gg kT_r$.

[2]This is the appropriate approximation of the general kinetic equation [9, 10]

$$\frac{\partial n}{\partial t} = \frac{\sigma_T N_e h}{m_e c}\nu^{-2}\frac{\partial}{\partial\nu}\nu^4\left(n^2 + n + \frac{kT_e}{h}\frac{\partial n}{\partial\nu}\right)$$

where $n = (c^2/8\pi h\nu^3)F_\nu$ is the occupation number in photons phase space. In the situation considered, the terms n^2 and n can be neglected as compared with $(kT_e/h)\partial n/\partial\nu$, because $T_e \gg T_r$. Inserting the initial spectrum into the right-hand side of the general equation, we easily obtain the form given above for small distortions of the initial spectrum.

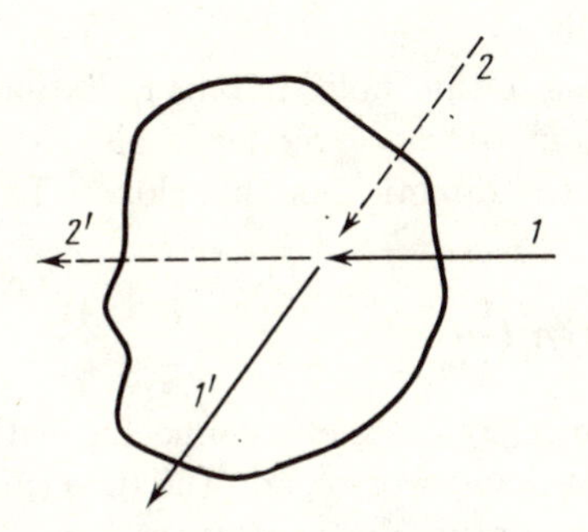

Fig. 1. Scattering of an isotropic radiation field by a cloud of electrons.

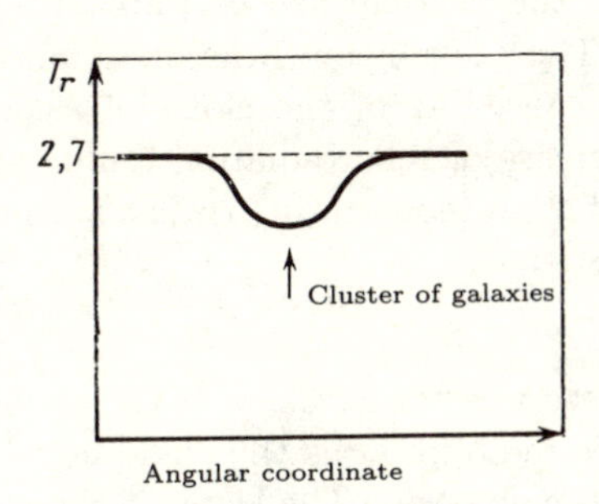

Fig. 2. Decrease of brightness in the relic radiation.

Clusters of Galaxies—The Best Candidate for Observation of the Proposed Effect

In the rarefied intergalactic or interstellar gas the Compton optical depth is small, $kT_e/m_e c^2 \ll 1$, and therefore the effect is small. However, the properties of relic radiation mentioned and the great precision of comparative measurements allow us to measure the small-scale fluctuations with a sensitivity unequaled by absolute measurements.

If hot intergalactic gas really exists in clusters of galaxies, there are all the conditions for action of the proposed effect. Thus, for example, the X-ray radiation of the Coma cluster of galaxies is interpreted as the bremsstrahlung radiation of a hot intergalactic gas having $T_e \sim 7 \cdot 10^7$ K and the density $N_e \sim 10^{-3}\,\text{cm}^{-3}$. The linear dimension of the source is estimated to be $l \sim 10^{25}\,\text{cm}^{-3}$. Multiplying these figures, we find

$$\frac{\Delta T_r}{T_r} = -2y = -2\sigma_T n_e l \frac{kT_e}{m_e c^2} \sim 2 \cdot 10^{-4},$$

the value accessible for observations. Namely, such effect was recently discovered in Coma by Pariĭskiĭ [11].

The deficit of brightness (hole) in the Coma (Fig. 2) is difficult to understand by any other mechanism! The radiation temperature is 2.7 K; an arbitrary absorption of radiation is always accompanied by spontaneous emission. If the temperature of the absorber is higher than 2.7 K, the net result would be an increase of radiation, instead of the observed deficit.

The common radiation of galaxies and radiogalaxies of the cluster must also lead to an increase of radiation intensity in the direction to the cluster. However, in the centimeter band the contribution of galaxies to the background is small, as follows from direct observations (see, for instance, [12]) and this is confirmed by Pariĭskiĭ's observations.

There is only one other mechanism leading to the "hole" in relic radiation. The receding of the cloud of electrons from the observer leads also to a decrease of relic radiation temperature in the direction of this cloud. The radiation temperature deficit is equal to

$$\frac{\Delta T_r}{T_r} \sim \tau_T \frac{v}{c} \cos\theta = \sigma_T n_e l \frac{v}{c} \cos\theta,$$

where v is the velocity of the cloud in the reference frame connected with relic radiation, θ is the angle between the velocity vector and the direction to the observer. The Doppler change of temperature does not depend on the frequency, contrary to the statement above (diffusion of the photons on the frequency axis due to scattering on hot electrons). Therefore it is possible to distinguish these effects observing the "hole" in the low-frequency, $h\nu < kT_r$, and high-frequency, $h\nu > kT_r$, parts of the spectrum.

Both effects are equal at

$$\frac{2kT_e}{m_e c^2} = \frac{v}{c}.$$

If the observed "hole" in Coma results from its receding, even at $\cos\theta = -1$, the velocity v of its motion in the reference frame connected with the relic radiation must be of the order of 7000 km/sec. This value is of the order of the velocity found from redshifts of the lines in the spectra of galaxies in Coma. However, this velocity is most likely connected with the expansion of the Universe, and the value of v given above contradicts observations.

Conclusion

Thus, the decrease of brightness temperature of relic radiation in the direction of the cluster of galaxies uniquely testifies to the existence of hot intergalactic gas in clusters. The effect is proportional to the product $\int N_e T_e dl$. Knowing the angular dimensions of the "hole" and the distance to the cluster of galaxies, we can identically evaluate $\int \Delta T_r \, d\Omega$ in $\int N_e T_e \, dV$, which in turn gives the full thermal energy of all free electrons in a given object. Most interesting data may be obtained by comparing this information with the data of X-ray observations. The X-ray radiation intensity of the hot gas is proportional to $\int N_e^2 T_e^{-1/2} \, dV$: the X-ray measurements allow us to determine the temperature of the gas. Using all these data it is possible to define the lower bound of the mass of intergalactic gas in the cluster of galaxies and to find its spatial distribution. To confirm our explanation, it would be important to make measurements on different wavelengths. In principle, by refining the measurements, one could extract information on possible deviations of the spectrum from the simple Rayleigh-Jeans formula, which are of extreme cosmological importance [8].

We are grateful to Yu. N. Pariĭskiĭ for stimulating discussions and for early communication of his observations.

REFERENCES

1. *Gursky H, Solinger A., Kellogg E. M., et al.*—Astrophys. J. **173**, L99 (1972).
2. *Fabian A. C.*—Nature. **237**, 19 (1972).
3. *Brecher K., Morrison P.*—Astrophys. J. **150**, L61 (1967).
4. *Kurilchik V. N.*—Astron. Zh. **49**, 89–93 (1972).
5. *Longair M. S., Sunyaev R. A.*—Astrophys. Lett. **4**, 65 (1969).
6. *Sunyaev R. A., Zeldovich Ya. B.*—Astrophys. and Space Sci. **7**, 3 (1970).
7. *Zeldovich Ya. B., Sunyaev R. A.*—Astrophys. and Space Sci. **4**, 302 (1969).
8. *Sunyaev R. A., Zeldovich Ya. B.*—Comments Astrophys. and Space Phys. **2**, 66 (1970).
9. *Kompaneets A. S.*—ZhETF **31**, 876–885 (1956).
10. *Weymann R.*—Phys. Fluids **8**, 2112 (1965).
11. *Pariĭskiĭ Yu. N.*—Astron. Zh. **50**, 453 (1972).
12. *Welch G. A., Sastry G. N.*—Astrophys. J. **175**, 323 (1972).

Commentary

This article considers the effect which, in the foreign literature, is known as the Sunyaev–Zeldovich effect. This effect predicts a decrease in the luminescence of the background in directions toward galaxy clusters in the region $h\nu < 3.83kT_r$ (this makes them "negative" radiosources) and an increase in the luminescence of the background in the submillimeter range at $h\nu > 3.83kT_r$, $\lambda < 1.25$ mm (transforming clusters into very bright sources of submillimeter radiation). We note that the first to carry out observations of the effect in the direction of the galaxy cluster Coma was Yu. N Pariĭskiĭ using the large Pulkovo antenna.[1]

Now the effect is entering into the observational program of many of the largest radiotelescopes in the world; it has been discovered in the cluster $A\,2218$ (red shift $z = 0.17$),[2] and in the cluster $0016 + 16$ (red shift $z = 0.56$).[3]

Recently Birkinshaw, Gull and Hardebeck[4] announced that as a result of many years of observations on the 40–meter antenna at Cal-Tech in Owens-Valley they had definitely discovered the effect (at a level exceeding seven mean-square deviations) in three rich clusters at a wavelength of 1.5 cm.

The effect attracts the attention of researchers because, together with data on X-ray observations of clusters, it gives rise to the possibility of determining the absolute dimensions of the cluster and, consequently, the distance to it and the Hubble constant.[5,6,7,8,9] Detailed observations of the effect at several frequencies also allows one to determine the peculiar velocity of a cluster with respect to the relic radiation, i.e., to use the relic radiation as a "new ether" (see article **68** by Ya. B. and R. A. Sunyaev in the present volume in which this possibility is first pointed out). The very fact of discovery of the effect in the direction of a cluster with red shift $z = 0.56$ (cluster 0016 + 16) is of immense importance—it experimentally proves that the relic radiation indeed has a cosmological, not local, origin. It existed at $z = 0.56$, when the Universe was almost half as old.

[1] *Pariĭskiĭ Yu. N.*—Astron. Zh. **50**, 453–458 (1972).
[2] *Schallwich D., Wielebinski R.*—Astron. and Astrophys. **71**, L15–L16 (1979).
[3] *Birkinshaw M., Gull S. F., Moffet A. T.*—Astrophys. J. **251**, L73 (1981).
[4] *Birkinshaw M., Gull S. F., Hardebeck H.*—Nature **309**, 34 (1984).
[5] *Birkinshaw M.*—Month. Not. RAS **187**, 847–862 (1979).
[6] *Cavaliere A., Danese L., De Zotti G.*—Astron. and Astrophys. **75**, 322–325 (1979).
[7] *Silk J., White S.*—Astrophys. J. Lett. **226**, L103–L106 (1978).
[8] *Sunyaev R. A., Zeldovich Ya. B.*—Ann. Rev. Astron. and Astrophys. **18**, 537–560 (1980).
[9] *Zeldovich Ya. B., Sunyaev R. A.*—In: Astrofizika i Kosmich. Fizika [Astrophysics and Cosmic Physics]. Moscow: Nauka, 9–65 (1982).

67a

Long-Wavelength Perturbations of a Friedman Universe, and Anisotropy of the Microwave Background Radiation*

With L. P. Grishchuk

Observational evidence, which is necessarily confined to a region of the Universe limited in space (within the observer's horizon), implies a high degree of homogeneity and isotropy for the large-scale structure of the Universe. In principle, substantial deviations of the properties of the real Universe from the parameters of an idealized Friedman cosmological model could have prevailed on scales exceeding that of the horizon. Constraints on the amplitude of perturbations with such long wavelengths are imposed by the virtual isotropy ($\delta T/T < 10^{-4}$) of the observed background radiation. This information on $\delta T/T$ together with the natural hypothesis that the perturbations are statistically independent implies that on spatial scales exceeding the horizon there exist no significant (with amplitude of order greater than $\delta T/T$) perturbations in density. For certain types of perturbations in the metric (in the gravitational field), the amplitude could be appreciable without contradicting the empirical limits on $\delta T/T$.

Studies of the 3 K microwave background radiation are a most valuable tool for probing the large-scale structure of the Universe [1]. Empirical evidence concerning the spectrum and angular distribution of this radiation can set limits on the perturbations that may exist—the departures of the properties of the real world from the parameters of an idealized Friedman cosmological model. But the observations are necessarily confined to a region restricted in space. If we nevertheless take advantage of the measurements that have been made and invoke a few general hypotheses, what can we say about the state of the Universe beyond the region which is in principle accessible to observation today? To answer this question, we should at the outset make clear what is meant by the observable region, perturbations and their Fourier spectrum, and statistical independence.

We are presently observing photons emitted by the primordial plasma in the remote past, at an epoch when the plasma had become transparent as a result of recombination. Thus the photons of the background radiation

*Astronomicheskiĭ zhurnal **55**, 209–215 (1978).

have been traveling freely, without being scattered, for an extremely long, although finite, time. The time of free photon propagation determines the spatial scale that we call the horizon of recombination.

Small perturbations of the Friedman model can be expanded in Fourier harmonics, each characterized by the corresponding wavelength λ_n. As time passes the wavelength changes in proportion to the scale factor of the isotropic Universe.

Among the perturbations of various wavelengths, there are some for which λ_n exceeds not only the horizon of recombination but also the observer's horizon (or simply the horizon), as specified by the finite time of expansion from a superdense (singular) state to the present epoch. (Actually the relative difference between the recombination horizon and the observer's horizon is small, amounting to just a few percent.) Such waves are said to be long.

Now we should ascertain what quantitative constraints on the amplitude of long-wavelength perturbations can be imposed in view of the observed high degree of isotropy of the background radiation: $\delta T/T < 10^{-4}$. Might it not turn out that at our present epoch, on scales greater than the horizon, substantial perturbations in the density and in the metric exist (say with an amplitude $\delta\rho/\rho \approx 10^{-1}$, or with a dimensionless amplitude in the metric of the same order) which we do not even suspect, because we cannot yet observe them directly? Such effects would become accessible only to astronomers of the very remote future, $t \gg 2 \cdot 10^{10}$ yr, when the observer's horizon and the recombination horizon become comparable with the corresponding wavelength.

We would emphasize that we are here interested in smooth, long-wavelength perturbations that encompass the region of space inside the horizon as well, rather than perturbations that originate beyond the horizon, remaining always unchanged within it. In other words, we shall make the natural assumption that harmonic perturbations of differing wavelength are not specially correlated. If this were not the case, they could be selected in such a way that within the horizon the deviations from the Friedman model would be particularly small (and hence $\delta T/T$ would be small), but beyond the horizon the perturbations would be large. But such a choice of different harmonics is most unlikely, for it would imply that an observer on the Earth is in a singular position. Our starting assumption is, on the contrary, that all observers are equivalent and perceive approximately the same picture; thus for all observers, even for those causally unrelated, $\delta T/T < 10^{-4}$. The question nevertheless remains of whether small deviations, not noticeable with measurements of the present accuracy within each observer's horizon, might add up to a substantial large-scale perturbation. Graphically expressed, might not observers living on the slopes and humps and in the valleys of a long density wave or gravitational wave be able to recognize this fact by examining, with limited accuracy, only their immediate neighborhood?

The conclusion we shall reach may be stated as follows. Measurements of $\delta T/T$ indicate that the density perturbations cannot be appreciable; on any scale they are limited to an amplitude which is an infinitesimal of at least the same order as $\delta T/T$. As for the perturbations in the metric, for the same empirical constraints on $\delta T/T$ these can be larger than $\delta T/T$, and indeed in the range of long wavelengths the metric perturbations can reach values approaching unity, so that they are at the limit of applicability of the theory of small perturbations.

In this respect, with a restriction of order $\delta T/T$ imposed on the dimensionless amplitude of the perturbations, the result may appear paradoxical. At first glance it may seem obvious that for a fixed recombination horizon, a transition to increasingly long wavelengths would be equivalent to smoothing out the perturbations within the horizon, diminishing their contribution to $\delta T/T$, and thereby permitting a rise in the admissible amplitude of the perturbations without coming into conflict with the observations. Actually, however, it is not only the spatial dependence of a perturbation that is important, but also the speed of its time variation. A long-wavelength perturbation will manifest itself not as a wave but as an anisotropy in the deformation. It will make some contribution to $\delta T/T$ even as the wavelength characterizing the spatial periodicity increases without bound.

However, perturbations of the metric also exist which vary slowly with time and induce a particularly small deformation anisotropy. The constraints on the amplitude of such perturbations resulting from their connection with $\delta T/T$ will be relaxed. This is the reason why perturbations of the metric with a growing density mode, as well as metric perturbations such as the nonsingular mode of gravitational waves, may exceed $\delta T/T$ and reach substantial values for waves of long wavelength.

We shall examine here a homogeneous and isotropic model with flat three-dimensional space, the density being equal to the critical density. Assume that "smeared-out" matter has the equation of state $P = 0$, which is a good approximation for the postrecombination era. Then the metric of the Universe (the gravitational field), including small perturbations, can be described by the line element

$$ds^2 = a^2(\eta)(\eta_{ab} + h_{ab})dx^a dx^b = a^2(\eta)$$

$$(d\eta^2 - dx^2 - dy^2 - dz^2) + a^2(\eta)h_{ab}dx^a dx^b.$$

The corrections h_{ab} are determined by solutions of the Einstein equations. As Lifshitz demonstrated [2], three independent types of corrections can be identified: 1) density perturbations, 2) eddy perturbations, 3) gravitational waves.

One should keep in mind that prior to recombination matter was hot, possessing entropy, so that density perturbations could have been either adiabatic or nonisentropic [3]. After recombination the adiabatic perturbations could have had a growing and decaying mode, whereas the entropy perturbations in the long-wavelength range would have been transformed only into the decaying mode of density perturbations. In the discussion below we shall consider the entropy perturbations separately.

The metric h_{ab} will be represented in the form proposed by Sachs and Wolfe [4], except for trivial changes of notation. If $P = 0$, the equations given by Lifshitz [2] and Sachs and Wolfe [4] coincide for all perturbations except for the eddy (rotational) perturbations.

Suppose that the photons become free instantaneously at the epoch η_E of emission. Let η_R denote the epoch of reception today. If the last scattering of photons occurred at the epoch of recombination, when $z \approx 1400$, then[1] $\eta_E/\eta_R \approx 1/40$. The direction of arrival of a light ray is defined by a vector e^α having the components $e^\alpha = \{\sin\theta\cos\varphi,\ \sin\theta\sin\varphi,\ \cos\theta\}$. If at the epoch η_E of photon emission the temperature of the primordial background radiation was everywhere the same and equal to T_E, then at the time of photon arrival it would possess variations depending on the angle of arrival and on the observer's position. The temperature will take the value

$$T_R = T_E \frac{\eta_E^2}{\eta_R^2}\left(1 + \frac{\delta T}{T}\right),$$

where[2]

$$\frac{\delta T}{T} = \frac{1}{2}\int_0^{\eta_R - \eta_E}\left(\frac{\partial h_{\mu\beta}}{\partial \eta} e^\mu e^\beta - 2\frac{\partial_{0\beta}}{\partial \eta} e^\beta\right) dw \tag{1}$$

The integration extends along the light geodesic $\eta = \eta_R - 2$, $x^\alpha = e^\alpha w$, with a parameter w varying from zero to $\eta_R - \eta_E$.

We shall consider the three types of perturbations individually, representing each of them by a single plane wave.

1. Density Perturbations

The growing mode of density perturbations is described by the equations

$$\frac{\delta\rho}{\rho} = \frac{1}{2} h (n\eta)^2 e^{i(nx+\xi)},$$

$$h_{\mu\beta} = (10 h \eta_{\mu\beta} + h\mu^2 n_\mu n_\beta) e^{i(nx+\xi)}, \quad h_{0\mu} = 0, \tag{2}$$

where h depends, in general, on n. The number n, the modulus of the wave vector, characterizes the spatial periodicity of the perturbation. For long waves the condition $n\eta_R < 1$ is satisfied. Here and subsequently we shall regard the z axis as directed along the vector $\mathbf{n}$, with the observer having coordinates $x = y = z = 0$.

[1]This statement presupposes that secondary ionization occurred quite late, with the ionized gas having an optical depth less than unity.

[2]Greek indices take the values 1, 2, 3.

Substituting the expressions (2) into (1) and integrating, we find that

$$\frac{\delta T}{T} = he^{i\xi}[in\eta_R \cos\theta - in\eta_E \cos\theta e^{in(\eta_R-\eta_E)\cos\theta} + 1 - e^{in(\eta_R-\eta_E)\cos\theta}].$$

In the long-wavelength limit, with $n\eta_R \ll 1$ we have

$$\frac{\delta T}{T} = \frac{1}{2}h(n\eta_R)^2 \cos^2\theta \Bigg\{\left(1 - \frac{\eta_E^2}{\eta_R^2}\right)\cos\xi - \frac{1}{3}n\eta_R \cos\theta \left(1 - 3\frac{\eta_E^2}{\eta_R^2} + 2\frac{\eta_E^3}{\eta_R^3}\right)\sin\xi\Bigg\}. \tag{3}$$

Since $\frac{1}{2}h(n\eta_R)^2 = (\delta\rho/\rho)_R$, it is evident from (3) that at the present epoch large $\delta\rho/\rho$ values are precluded in the long-wavelength range by the high degree of isotropy of the background radiation. As for the amplitude of the metric, these restrictions become considerably weaker because of the small value of the factor $(n\eta_R)^2$.

The parameter ξ describes the relative position of the observer. For $h > 0$ and $\xi = 0$ the observer will be at a minimum. The integral of $\delta T/T$ over a sphere will be greater than zero for $\xi = 0$, so that at the density maximum the radiation will appear somewhat hotter than the average over all space. Correspondingly, if $\xi = \pi$ the radiation will seem just as much cooler. Only the angular variations in the temperature are significant for the observer, since he has no opportunity to compare the mean temperature he observes with the temperature averaged over all space.

If the observer is located near the inflection points, that is, at $\pi/2$, then for fixed limits on $\delta T/T$ the quantity $(\delta\rho/\rho)_R$ can be increased because of the small factor $n\eta_R \cos\theta$. However, the probability that the observer will be situated in this range of ξ values is very small, since the departures from $\xi = \pm\pi/2$ should not exceed a small quantity of order $n\eta_R \cos\theta$. As very long wavelengths are approached ($n\eta_R \to 0$), this ξ interval will also tend to zero.

For the decaying mode of density perturbations we have

$$\frac{\delta\rho}{\rho} = \frac{1}{2}hn^2\frac{1}{\eta^3}e^{i(nx+\xi)}, \quad h_{\mu\beta} = \frac{1}{\eta^3}hn_\mu n_\beta e^{i(nx+\xi)}, \quad h_{0\mu} = 0. \tag{4}$$

If $n\eta_R \ll 1$, an evaluation of $\delta T/T$ yields

$$\begin{aligned}\frac{\delta T}{T} &= -\frac{1}{2}hn^2\cos^2\theta\frac{1}{\eta_E^3}\left\{\left(1 - \frac{\eta_E^3}{\eta_R^3}\right)\cos\xi - n\eta_R\cos\theta\left(1 - \frac{3\eta_E}{2\eta_R} + \frac{\eta_E^3}{2\eta_R^3}\right)\sin\xi\right\}\\ &= -\left(\frac{\delta\rho}{\rho}\right)_E \cos^2\theta\,\Phi(\xi,\theta),\end{aligned} \tag{5}$$

where $\Phi(\xi, \theta)$ designates the expression in braces.

Since

$$\frac{1}{2}hn^2\frac{1}{\eta_E^3} = \left(\frac{\delta\rho}{\rho}\right)_E,$$

the density and metric perturbations at epoch η_E could not have exceeded $\delta T/T$, and *a fortiori* they should be small today.

After recombination, entropy perturbations in the long-wavelength range will have gone over primarily to the decaying mode of the density perturbations. In fact, the general equations for entropy perturbations [5] provide us with asymptotic relations in the long-wavelength limit which are applicable both before and after recombination. During the stage when radiation dominates (but after the annihilation of antibaryons has terminated, when $kT \ll m_p c^2$), an entropy perturbation of the total density will grow according to the slower of the two modes characteristic of adiabatic perturbations, that is, according to the law

$$\frac{\delta\varepsilon}{\varepsilon} \sim \left(\frac{\delta S}{S}\right)\frac{\eta}{\eta_e} \quad (\eta \ll \eta_e), \tag{6}$$

whereas the fast mode would give a law of the form $\delta\varepsilon/\varepsilon \sim \eta^2$ [in (6), η_e designates the epoch when the radiation density and the nonrelativistic plasma density are equal; $\delta S/S = \text{const} \cdot e^{inx}$ represents the entropy perturbation]. During the stage when nonrelativistic plasma dominates (when $\eta \gg \eta_e$) the slower mode will be the decaying mode $\delta\rho/\rho \sim \eta^{-3}$ of density perturbations; by contrast, the growing mode will give $\delta\rho/\rho \sim \eta^2$. One finds that after recombination a density perturbation will in fact follow the $\delta\rho/\rho \sim \eta^{-3}$ law (at least until the wavelength of the perturbation exceeds the scale of the horizon). This statement is demonstrated by an actual calculation. If $\eta \gg \eta_e$, one can obtain for perturbations in the metric and in the density the approximate relations

$$\frac{\delta\varepsilon}{\varepsilon} = \frac{\delta\rho}{\rho} = \frac{1}{2}Cn^2\left(\frac{\eta_e}{\eta}\right)^3 e^{i(nx+\xi)}, \quad h_{\mu\beta} = C\left(\frac{\eta_e}{\eta}\right)^3 n_\mu n_\beta e^{i(nx+\xi)}, \tag{7}$$

$$h_{0\mu} = 0 \qquad (n\eta < 1),$$

where C is a constant related to $\delta S/S$. The perturbations in the velocity are negligible. In the long-wavelength range, then, we obtain from entropy perturbations after recombination the decaying mode of density perturbations.

An intuitive explanation can also be given for this fact [1]. An entropy perturbation, by changing the state of primordial matter in various regions of space (within the wavelength of the perturbation) in the initial state, will insignificantly alter the total density and space geometry of the Universe near the singularity. In every part of space, a three-dimensionally flat world will remain practically flat. After recombination, the density will be determined by nonrelativistic gas with a pressure $P = 0$ and, without regard to initial deviations in the entropy, it will fall off according to the law $\rho = [6\pi G(t + \tau)^2]^{-1}$. The quantity τ corrects for the fact that one cannot consider $P = 0$ at the beginning of the expansion. Entropy perturbations will cause τ to differ in different parts of space.

It is readily verified that the difference in density (that is, the perturbation) will diminish with increasing t as $\delta\rho/\rho = -2\delta\tau/t + \tau$, where the time t is measured from the singularity. Thus as time passes the density perturbation will behave in accordance with the decaying mode. This process will continue until the wavelength of the perturbation becomes comparable with the horizon. At that time very insignificant initial deviations in the geometry of various parts of space from a flat geometry and differences in the radiation density will become important. Various parts of the Universe will now increasingly resemble a spatially open or a spatially closed world. The growing mode of density perturbations, negligibly small immediately after recombination, will surpass the decaying mode in amplitude a certain time after equality between the horizon and the wavelength has been achieved, and will thenceforth determine $\delta\rho/\rho$.

With regard to the question of the $\delta T/T$ produced by long-wavelength perturbations, we are interested in the solution (7). This solution coincides with (4) and leads to the same conclusions. We may infer from the limits $\delta T/T < 10^{-4}$ that density perturbations (originating as entropy perturbations) and the corresponding perturbations in the metric are bounded by a quantity of order 10^{-4} in the long-wavelength range.

2. Eddy Perturbations

A rotational perturbation in the metric will have the form

$$\delta\rho = 0, \quad h_{\mu\beta} = -\frac{2}{n}(n_\beta \kappa_\mu + n_\mu \kappa_\beta)\left(\frac{8}{\eta^3} + \frac{n^2}{\eta}\right) e^{i(nx+\xi)},$$

$$h_{0\mu} = -i\frac{2}{\eta^2} n \kappa_\mu e^{i(nx+\xi)}, \quad \kappa^\mu n_\mu = 0.$$

The last condition here means that κ^μ has the components $\kappa^\mu = (C_1, C_2, 0)$. The angular velocity vector Ω^α of matter lies in the x,y-plane. If $C_1 = 0$ the vector Ω^α will be oriented along the x-axis; if $C_2 = 0$, along the y-axis. The observer is located at the maximum absolute value $\Omega_\alpha \Omega^\alpha$ for $\xi = 0$ and $\xi = \pi$, and the orientation of Ω^α is here distinguished by sign.

If $n\eta_R \ll 1$, we obtain for $\delta T/T$ the expression

$$\frac{\delta T}{T} = 8 \sin 2\theta (C_1 \cos\varphi + C_2 \sin\varphi) \frac{1}{\eta_E^3} \left\{\Phi(\xi,\theta) + \left[\frac{n\eta_E}{8\cos\theta}\left(1 - \frac{\eta_E^2}{\eta_R^2}\right)\right]\right\}. \quad (8)$$

The bracketed term owes its origin to the components $h_{0\mu}$. The integral of $\delta T/T$ over a sphere everywhere vanishes.

Equation (8) implies that with the constraints that can now be placed on the quantity $\delta T/T$, the amplitude of eddy-type perturbations in the metric cannot be significant for long wavelengths. It is worth recalling, incidentally, that near the singularity the eddy perturbations in the metric are large and are divergent [6, 7].

3. Gravitational Waves

The perturbation in the metric is

$$\delta\rho = 0, \qquad h_{\mu\beta} = -\frac{d_{\mu\beta}}{\eta^3}(1 - in\eta)e^{i(nx+n\eta+\xi)}, \qquad h_{0\mu} = 0, \qquad d_\beta n^\beta, \qquad d^\mu_\mu = 0. \tag{9}$$

The nonvanishing components of the matrix $d_{\alpha\beta}$ will be written in the form $d_{11} = -d_{22} = C_1$, $d_{23} = d_{32} = C_2$, where C_1, C_2 correspond to the two possible polarizations of the wave.

In the form (9) given above, the real part of the solution describes the decreasing (singular) mode of the perturbations, divergent near the singularity; the imaginary part of the solution corresponds to the nonsingular mode, which, like the growing mode of density perturbations, is compatible with a quasi-isotropic (locally Friedman) solution near the singularity.

If $n\eta_R \ll 1$, we have approximately for the decaying mode

$$h_{\mu\beta} = -\frac{d_{\mu\beta}}{\eta^3}\left(1 + \frac{(n\eta)^2}{2} + \ldots\right)e^{i(nx+\xi)},$$

which gives

$$\frac{\delta T}{T} = \frac{1}{2}\sin^2\theta(C_1\cos 2\varphi + C_2\sin 2\varphi)\frac{1}{\eta_E^3}\Phi(\xi,\theta). \tag{10}$$

Since the components C_1, C_2 directly yield the characteristic amplitude of the wave for the decaying mode (that is, the value of the metric at the epoch when the length of the perturbation becomes comparable with the horizon), (10) implies that the amplitude of the decaying mode cannot be greater than $\delta T/T$.

For the nonsingular mode we have approximately

$$h_{\mu\beta} \approx -\frac{1}{3}d_{\mu\beta}n^3\left(1 - \frac{(n\mu)^2}{10} + \ldots\right)e^{i(nx+\xi)},$$

which leads to the expression

$$\frac{\delta T}{T} = \frac{n^3}{60}\sin^2\theta(C_1\cos 2\varphi + C_2\sin 2\varphi)(n\eta_R)^2 \times\left\{\left(1 - \frac{\eta_E^2}{\eta_R^2}\right)\cos\xi - n\eta_E\cos\theta\left(1 - 2\frac{\eta_E}{\eta_R} + \frac{\eta_E^2}{\eta_R^2}\right)\right\}. \tag{11}$$

In this case the characteristic amplitude of the wave is determined by the quantities $C_1 n^3$, $C_2 n^3$. In the main approximation the perturbation in the metric is constant and does not depend on time, so that the main terms make no contribution to $\delta T/T$. The relation between the quantities $C_1 n^3$, $C_2 n^3$ and $\delta T/T$ contains the small factor $(n\eta_R)^2$. Thus in the limit of long wavelengths the amplitude of the nonsingular mode of gravitational wave perturbations could have been comparatively high, of order $(\delta T/T)(n\eta_R)^{-2}$, without contradicting the observations.

For both modes the integral of $\delta T/T$ over a sphere vanishes. Both $\delta T/T$ itself and its gradient drop to zero for $\theta = 0$.

Equations (3), (5), (8), (10) and (11) give the exact relation between $\delta T/T$ and the amplitude of the perturbations in the long-wavelength limit. As ought to be the case, $\delta T/T \to 0$ as $\eta_E \to \eta_R$. For perturbations growing with time, the value of $\delta T/T$ will be determined by the amplitude of the perturbations at epoch η_R; for decaying perturbations, at epoch η_E. An approximate expression for $\delta T/T$ could have been obtained directly from (1) by replacing the integral by the difference in the values of the expression $h_{\alpha\beta}e^\alpha e^\beta - 2h_{0\alpha}e^\alpha$, taken at epochs η_R and η_E.

Thus empirical evidence on $\delta T/T$ in conjunction with the plausible hypothesis of statistically independent perturbations leads to the conclusion that on scales exceeding the size of the horizon there exist no significant (with an amplitude exceeding $\delta T/T$, in order of magnitude) density perturbations, rotational perturbations, or gravitational waves for the singular mode. As for gravitational waves representing the nonsingular mode, as well as perturbations in the metric associated with the growing modes of density perturbations, these could be appreciable without coming into conflict with the observational limits on $\delta T/T$.

REFERENCES

1. *Zeldovich Ya. B., Novikov I. D.* Struktura i evoliutsiĭa Vselennoĭ [Structure and Evolution of the Universe]. Moscow: Nauka (1975).
2. *Lifshitz E. M.*—ZhETF **16**, 587 (1946).
3. *Doroshkevich A. G., Zeldovich Ya. B., Novikov I. D.*—Astron. Zh. **44**, 295 (1967).
4. *Sachs R. K., Wolfe A. M.*—Astrophys. J. **147**, 73 (1967).
5. *Chibisov L. V.* Ph.D. Dissertation, Moscow Phys. Tech. Inst. (1972).
6. *Ozernoĭ L. M., Chernin A. D.*—Astron. Zh. **44**, 1131 (1967).
7. *Zeldovich Ya. B., Novikov I. D.*—Astrofizika **6**, 379 (1970).

68
The Velocity of Clusters of Galaxies Relative to the Microwave Background. The Possibility of Its Measurement*

With R. A. Sunyaev

Observations of the microwave background intensity and polarization in the direction of clusters of galaxies permit us, in principle, to measure their peculiar velocities relative to the background radiation.

1. *Introduction*

It is well known according to X-ray observations that rich clusters of galaxies contain a large amount of hot intergalactic gas [1,2]. Hence clusters of galaxies may be considered as fully ionized gas clouds with high temperature and finite optical depth with respect to Thomson scattering.

The scattering of 3 K microwave background radiation on these clouds of intergalactic gas opens the possibility of measuring the velocity of each cloud relative to the coordinate frame determined by the background radiation.

The radial motion changes the observed radiation temperature $\Delta T/T \sim -(v_r/c)\tau$ in the direction to the cloud. Here

$$\tau = \int_{-\infty}^{+\infty} \sigma_T N_e \, dr$$

is the optical depth of the cloud with respect to Thomson scattering, v_r is the radial component of the peculiar velocity of the cloud (positive v_r corresponds to a recession velocity exceeding that corresponding to Hubble's law).

In addition there is the possibility, at least in principle, of also finding the tangential component of the peculiar velocity of the cloud v_t. There are effects of the order of $0.1\beta_t^2\tau$ and $\frac{1}{40}\beta_t\tau^2$ (with $\beta_t = v_t/c$) in the microwave background polarization in the leading (with respect to the direction of motion) and trailing parts of the cloud.

*Monthly Notices of the Royal Astronomical Society **190** (2), 413–420 (1980).

Measurements of the large-scale anisotropy of the microwave background gave a motion velocity $V = 390\,\mathrm{km/sec}$ for the solar system. Using data on the rotation of the galaxy, it was also possible to estimate the velocity of the center of the galaxy relative to the microwave background to be $v = 600\,\mathrm{km/sec}$ [2, 3]. This is a rather small velocity. It is of great importance for cosmology to know what are typical peculiar velocities of other galaxies, clusters of galaxies and rich clusters of galaxies.

Earlier we analyzed the distortion of the microwave background spectrum due to scattering on free hot electrons in such clusters of galaxies [4]. The scattering on hot electrons redistributes the photons over the spectrum. In the long-wave (Rayleigh-Jeans, RJ) region of the spectrum, the radiation intensity and its brightness temperature decrease in the same ratio

$$\frac{\Delta T}{T} = \frac{\Delta J_\nu}{J_\nu} = -\frac{2kT_e}{m_e c^2}\tau, \tag{1}$$

where T_e is the electron temperature.

In the shortwave (Wien) region they increase. Below we write (see curves with $v_r = 0$ in Fig. 1) the dependence of the intensity distortions on the dimensionless frequency $x = h\nu/kT_r$ in the direction to the cluster [5] (see also [6, 7]):

$$\left(\frac{\Delta J_\nu}{J_\nu}\right)_1 = 2\frac{kT_e}{m_e c^2}\tau\frac{xe^x}{e^x - 1}\left\{\frac{x}{2tanh(x/2)} - 2\right\}, \tag{2}$$

or

$$\left(\frac{\Delta T}{T}\right)_1 = \frac{2kT_e}{m_e c^2}\tau\left\{\frac{x}{2tanh(x/2)} - 2\right\}.$$

The formulas are independent of the redshift. They are valid even for $z \gtrsim 1$.

The difference between (2) and the simple formula (1) increases with x. It is equal to 0.26 percent with $x = 0.18$, $\lambda = 3\,\mathrm{cm}$; 4.7 percent with $x = 0.67$, $\lambda = 8\,\mathrm{mm}$ and 53 percent with $x = 2.7$, $\lambda = 2\,\mathrm{mm}$.

The experimental study of this effect in the RJ spectral region was begun by Pariĭskiĭ [8] and Gull and Northover [9]. Now the effect has been definitely observed in the direction of the richest clusters of galaxies. For example, Lake and Partridge [11] (the observations were carried out on wavelength $\lambda = 9\,\mathrm{mm}$) and Birkinshaw *et al.* [10] ($\lambda = 2.6\,\mathrm{cm}$) found a brightness diminution in the direction of the cluster of galaxies Abell 2218. According to [11] (see also [12]) this diminution is equal to $\Delta T = (-2.65 \pm 0.23) \cdot 10^{-3}\,\mathrm{K}$. Let us assume that the gas temperature in this cloud of intergalactic gas is equal to $kT_e = 5\,\mathrm{keV}$ or is of the same order as in the X-ray clusters of galaxies Coma, Virgo, etc. [13], having similar velocity dispersion of the galaxies ($\sim 1500\,\mathrm{km/sec}$) as in Abell 2218. Then from (1) it is easy to estimate the optical depth of the cloud:

$$\tau = \frac{1}{2}\frac{\Delta T}{T}\frac{m_e c^2}{kT_e} \approx \frac{1}{20}. \tag{3}$$

2. Radial Component of the Velocity

We turn now to the peculiar motion of the cluster. It was mentioned in our paper [4] that the motion of the cloud as a whole relative to the background must lead (due to the Doppler effect) to an additional change of the radiation intensity $(\Delta J_\nu/J_\nu)_2$ and temperature $(\Delta T/T)_2$ in the direction to the cloud. Small effects superpose linearly,

$$\left(\frac{\Delta T}{T}\right)_{\text{total}} = \left(\frac{\Delta T}{T}\right)_1 + \left(\frac{\Delta T}{T}\right)_2 .$$

If τ is small

$$\left(\frac{\Delta T}{T}\right)_2 = -\frac{v_r \tau}{c}; \quad \left(\frac{\Delta J_\nu}{J_\nu}\right)_2 = \frac{x e^x}{e^x - 1}\frac{v_r}{c}\tau. \tag{4}$$

(In the rest system of the cloud the background radiation has anisotropy of dipole type. After Thomson scattering radiation loses its dipole component, and in first approximation becomes isotropic. Doppler transformation to the observer's frame makes the scattered radiation field anisotropic, of dipole type again, increasing the effective temperature and total flux in the direction of cloud motion.)

The sign of the effect depends on the velocity direction. The amplitude of the temperature perturbation ΔT does not depend on the frequency; the perturbation of the intensity ΔJ_ν depends on x, but does not change its sign, contrary to the case of the thermal effect (2). Therefore, in principle, two measurements at different frequencies are enough to separate two effects of interest for us[1]

$$\frac{\Delta T}{T}(\nu) = \left[\frac{\Delta T}{T}(\nu)\right]_1 + \left(\frac{\Delta T}{T}\right)_2 .$$

The index 1 corresponds to the thermal effect, and index 2 to the effect arising from radial peculiar velocity of the cloud. Fig. 1 shows that at wavelength $\lambda = 2\,\text{mm}$ the thermal effect decreases by a factor of 2 in comparison with the Rayleigh-Jeans region of the spectrum

$$\left(\frac{\Delta T}{T}\right)_1\Bigg|_{2\,\text{mm}} = -\frac{kT_e}{m_e c^2}\tau.$$

Observations at wavelength $\lambda = 2$ are possible from the Earth's surface [14]. The thermal effect changes its sign in the Wien region of the spectrum, where $x = 3.83$ or $\lambda = 1.3\,\text{mm}$. Unfortunately observations in the band $\lambda < 1.3\,\text{mm}$ are possible only from outside the Earth's atmosphere and this becomes a problem for the next decade. The curves with $v_r = \pm 3000\,\text{km/sec}$ illustrate the change of the ΔT dependence on x or λ with simultaneous

[1] We do not take into account here other effects, connected with non-thermal radio sources in the cluster, difference of microwave background spectrum from the black body one, free-free emission of the cloud, etc.

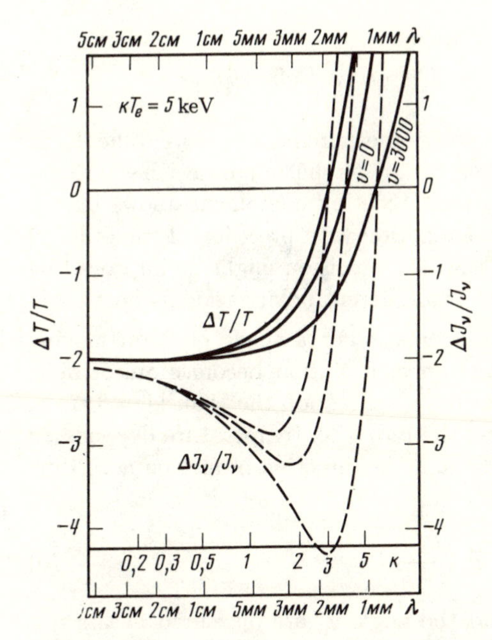

Fig. 1. The change of microwave background radiation temperature in the direction to a cluster of galaxies containing hot intergalactic gas with $kT_e = 5.11$ keV: (a) the cloud moves towards the observer. The peculiar velocity is equal to $v_r = -3000$ km/sec, $\tau = 0.2$. (b) The cluster is at rest relative to the microwave background; $\tau = 0.1$. (c) The peculiar recession velocity $v_r = 3000$ km/sec, $\tau = 1/15$. All cases give equal effect in the Rayleigh-Jeans part of the spectrum.

action of thermal effect and overall radial motion. They are normalized to equal effects in the Rayleigh-Jeans region.

In the Rayleigh-Jeans region the amplitudes of the effects are of the same order when

$$\left|\frac{v_r}{c}\right| = \frac{2kT_e}{m_e c^2} = \frac{1}{50} \quad \text{or} \quad v_r \approx 6000\,\text{km/sec}.$$

The observations of Abell 2218 [10–12] give an estimate of the upper limit to the radial peculiar velocity of the cluster

$$|v_r| < c\frac{\Delta T}{T}\frac{1}{\tau} = 6000\,\text{km/sec}. \tag{5}$$

Let us recall that the Hubble recession velocity for $z = 0.17$ is equal to 50 000km/sec. Precision of the distance and of the Hubble constant[2] estimates are much worse than 10 percent. Therefore the usual methods of velocity estimation using the Doppler effect do not allow us to find a peculiar velocity of the order of 6000 km/sec if the distance to the object corresponds to $z = 0.17$. Even the result (5) is of great interest in connection with the problems of large-scale structure of the Universe and of the spectrum of density perturbations in the Friedmanian model. Measurements of $\Delta T/T$ at $\lambda = 3$ cm and $\lambda = 2$ mm, with precision of the order of several per cent of

[2]Cavaliere, Danese & De Zotti [15] and Silk & White [16] proposed a new method of determining the Hubble constant based on the thermal effect.

the effect itself, could reveal velocities of the order of our Galaxy's motion velocity relative to the microwave background.

3. Tangential Component of the Peculiar Velocity

The problem of the tangential velocity measurement is more difficult. It is possible to find v_t using the observation of the microwave background polarization of the scattered radiation. Existing equipment allows one to measure the microwave radiation polarization with precision of the order of 10^{-4} [17]. However, one can hope that the precision might be improved by two orders of magnitude. There are two different polarizational effects.

3.1 Scattering on a single electron. For a single electron moving relative to the background the microwave radiation becomes anisotropic, transformed by the Doppler effect. In each direction the Planckian form of the spectrum is conserved; however, the radiation temperature depends on the angle θ between the line of sight and the direction of motion according to the well-known formula [18]

$$T_0 = T_r \frac{\sqrt{1-\beta^2}}{1+\beta\mu_0} \tag{6}$$

where $\beta = v/c$; $T_0, \mu_0 = \cos\theta_0$ and the angle θ_0 are measured in the rest frame of the electron. The expansion of the formula for T_0 has, in the second order of β, a quadrupole component in the angular distribution

$$T_0 = T_r[1 - \beta\mu_0 + \beta^2(\mu_0^2 - \frac{1}{3}) + \ldots]. \tag{7}$$

It is well known that the scattering of the unpolarized collimated light beam leads to linear polarization of the scattered light. We turn to the general case of a smooth anisotropic intensity distribution.

Polarization is a tensor quantity. Therefore it is proportional to the quadrupole term but does not depend on the dipole and higher terms in the intensity angular distribution.

After scattering on a free electron the linear polarization must arise in the direction perpendicular to the direction of its motion. The electric vector is perpendicular to the plane defined by the vectors of the motion velocity V and line of sight. If radiation angular distribution has the form

$$J = J_r[1 + \alpha\mu_0 + b(\mu_0^2 - \frac{1}{3} + \ldots],$$

the degree of polarization is equal to

$$p = \frac{J_\parallel - J_\perp}{J_\parallel + J_\perp} = 0.1b\mu_0^2.$$

It does not depend on a and coefficients of higher harmonics. In the Rayleigh-Jeans region $b = \beta_t^2$. Therefore $p = 0.1\beta_t^2$.

In the case of small τ there is unscattered light in the direction to the cloud. The scattered light contributes only a small fraction τ to the total intensity. Therefore,

$$p = 0.1\tau\beta_t^2.$$

The transformation to the observer's reference frame does not change this value. With $v_t = 6000\,\mathrm{km/sec}$, $\beta_t = 0.02$ and $\tau = 1/20$ we have $p = 2\cdot 10^{-6}$. This value must be compared with the observed $\Delta T/T \approx 10^{-3}$. It is difficult to find β_t using this method. With $x \gtrsim 1$ we must take into account the nonlinear dependence of intensity at a given frequency on the temperature.

$$J_\nu = B_\nu(T_0) + \left.\frac{dB_\nu}{dT}\right|_{T_0}(T - T_0) + \left.\frac{1}{2}\frac{d^2B_\nu}{dT^2}\right|_{T_0}(T - T_0)^2. \tag{8}$$

Using this expansion it is easy to find the coefficient b and the degree of polarization for any frequency.

3.2 Effect of finite optical depth. There is another effect connected with two consecutive scatterings, i.e., with finite optical depth of the cloud. Although τ is small, it is possible that effects of the order of τ^2 may become observable. These effects were absent when we considered scattering on a single electron.

In the rest frame of the cloud the unperturbed radiation field is in the first approximation on β,

$$T_0 = T_r(1 - \beta\mu_0), \qquad J_0 = J_r(1 - k\beta\mu_0),$$

where coefficient k depends on x [see (8)].

Due to symmetry of the angular dependence of Thomson scattering relative to the change $\theta_0 \to \pi - \theta_0$, $\mu_0 \to -\mu_0$ the scattered radiation becomes isotropic, and the dipole term $k\beta\mu$ vanishes, in the scattered radiation field.

It is obvious that the isotropic radiation field does not change under the action of Thomson scattering (the scattered photons exactly compensate infalling photons).

Therefore we must take into account only the uncompensated decrease due to scattering of the dipole component (measured in a frame moving with the cloud) of the radiation.

For example, for a plasma sphere with constant electron density within its boundary, we obtain in the first approximation

$$J_0 = J_r(1 - k\beta\mu_0) \quad \text{for} \quad -1 < \mu_0 < 0,$$

and

$$J_0 = J_r(1 - k\beta\mu_0 e^{-\tau\mu_0}) \quad \text{for} \quad 0 < \mu_0 < 1,$$

where

$$\tau = 2\sigma_T N_e R.$$

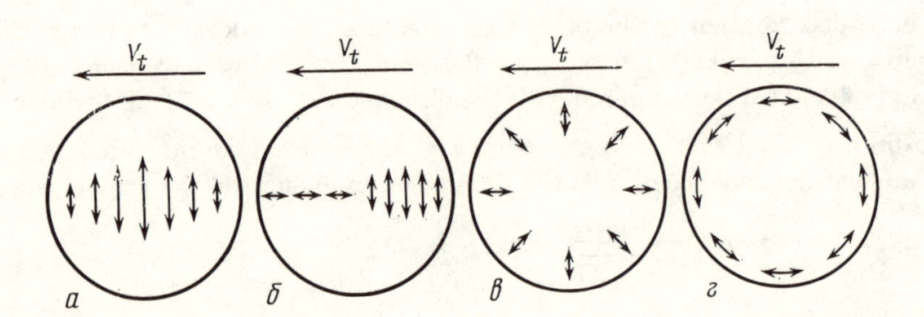

Figure 2. Predicted polarization of the microwave background in the direction to the cluster of galaxies. (a) Effect connected with finite optical depth of the cloud and proportional to $\beta_t \tau^2$. (b) Effect connected with scattering on a single electron and proportional to $\beta_t^2 \tau$. (c) Polarization connected with the thermal effect. (d) Polarization due to existence of the unpolarized radio radiation sources inside the cloud.

Such a distribution contains the second harmonics with coefficient $\sim k\beta\tau$. The scattered light becomes polarized, the polarization having opposite signs on different sides of the cloud.

Taking into account that only a small fraction τ of the detected radiation is scattered in the cloud we obtain finally the maximal degree of polarization

$$p_{\max} = \pm \frac{k}{40} \tau^2 \beta_t.$$

In the Rayleigh-Jeans region $k = 1$ and with $\tau = 0.1$, $v_t = 3000\,\text{km/sec}$, $\beta_t = 0.01$ we find

$$p_{\max} = \pm 2.5 \cdot 10^{-6}.$$

3.3 Other sources of polarization. The existence of non-thermal radio sources in clusters and radiation from intergalactic gas (of any nature) also contribute to the detected radiation flux and its polarization. Scattering of such radiation in the cloud must lead to its polarization. The sign and degree of polarization must depend on the angle between the line of sight and the direction to the center of the cloud. This effect is well known for optically thick objects [19, 20]. Obviously it occurs also for $\tau < 1$. In this case it is proportional to τ and to the intensity of the source radiation. The coefficient is smaller than 1/40.

The thermal effect decreases the intensity in the Rayleigh-Jeans region. Its influence can be compared with a "negative radiation source" inside the cloud. Therefore even in the case $V = 0$ it can lead to the polarization of the microwave background in the direction to the cloud. In this case

$$p_{\max} = \gamma \frac{kT_e}{m_e c^2} \tau^2$$

and γ is smaller than 1/40.

The polarization distribution over the cloud projection for different cases is given in Fig. 2.

4. Concluding Remarks

Observations of the polarization distribution over the cloud image in several wavelengths must give information of extraordinary importance for cosmology. Possibly there are clusters of galaxies, superclusters and even intergalactic gas bridges between them, with $1/20 < \tau < 1$. Second-order effects become observable in this case. They do not depend on the electron temperature.

Even with $\tau = 1/20$ the cloud must scatter several percent of the radiation of the galaxies and must appear as a diffuse radiation source in all spectral bands (in optical light, in particular). Narrow spectral lines must be absent in this scattered light due to the Doppler effect.

$$\frac{\Delta\nu}{\nu} \sim \sqrt{\frac{2kT_e}{m_e c^2}} \sim \frac{1}{7}.$$

Let us remark also that formulas (2) are derived using the Kompaneets [21] differential equation, which was obtained assuming multiple scatterings. In reality, τ is small and we consider the contribution of the small fraction of single-scattered photons. However, the formulas (2) are exact for $x < \sqrt{m_e c^2/2kT_e} \approx 7$ [22]. Important deviations from the formulas (2) occur only for $x > 7$ or $\lambda < 0.7$ mm, where the observations are very difficult.

Institute of Applied Mathematics
USSR Academy of Sciences. Moscow

Received
May 31, 1979

REFERENCES

1. *Forman W., Jones C., Cominsky L., et al.*—Astrophys. J. Suppl. **38**, 357 (1978).
2. *Cooke B. A., Rickers M. J., Maccacaro T. et al.*—Month. Not. RAS **182**, 489 (1978).
3. *Smoot G. F., Gorenstein M. V., Muller R. A.*—Phys. Rev. Lett. **39**, 14, 898 (1977).
4. *Sunyaev R. A., Zeldovich Ya. B.*—Astrophys. and Space Sci. **7**, 3 (1970).
5. *Sunyaev R. A., Zeldovich Ya. B.*—Comments Astrophys. and Space Phys. **4**, 173 (1972).
6. *Zeldovich Ya. B., Sunyaev R. A.*—Astrophys. and Space Sci. **4**, 301 (1969).

7. *Gould R. J., Raphaeli Y.*—Astrophys. J. **219**, 12 (1978).
8. *Pariĭskiĭ Yu. N.*—Astron. Zh. **40**, 453 (1973).
9. *Gull S. F., Northover K. J. E.*—Month. Not. **173**, 535 (1975).
10. *Birkinshaw M., Gull S. F., Northover K. J. E.*—Month. Not. **185**, 245 (1978).
11. *Lake G., Partridge R. B.*—Nature **270**, 502 (1977).
12. *Shallwich D., Wielebinsky R.*—Astron. and Astrophys. **71**, L15 (1979).
13. *Mitchell R. J., Culhane J. L.*—Month. Not. **178**, 75 (1977).
14. *Fabbri R., Melchiorri F., Natale V.*—Astrophys. and Space Sci. **59**, 223 (1978).
15. *Cavaliere A., Danese L., De Zotti G.*—Astr. Astrophys. **217**, 6 (1977).
16. *Silk J., White D.*—Astrophys. J. **226**, L103 (1978).
17. *Lubin P. M., Smoot G. F.*—Phys. Rev. Lett. **42**, 2, 129 (1979).
18. *Landau L. D., Lifshitz E. M.* The Classical Theory of Fields. New York: Pergamon (1971).
19. *Chandrasekhar S.* Radiative Transfer. Oxford (1950).
20. *Sobolev V. V.* A Course of Theoretical Astrophysics. NASA (1969).
21. *Kompaneets A. S.*—ZhETF **31**, 876 (1956).
22. *Sunyaev R. A., Zeldovich Ya. B.*—In: *Astrofizika i kosmicheskaĭa fizika [Astrophysics and Space Physics]*. Moscow: Nauka, 9–65 (1982).

Commentary

The discovery of dipole anisotropy of the background relic radiation[1] proved that the solar system moves with respect to the background with a velocity of about 350 km/sec. However, it long seemed that this measurement would remain unique, that observers in other distant galaxies and exchange of information with them would be needed to obtain further information.

This article proposes methods which allow measurement of both radial and tangential components of the velocity of distant galaxy clusters contained ionized intergalactic gas with respect to a system of coordinates in which the background radiation is isotropic, i.e., with respect to the so-called "new ether."

Massive observations of peculiar velocities should give very important information on the spectrum and amplitude of large-scale density perturbations of matter. This same method, in principle, allows one to study the dynamics of galaxy clusters in superclusters and, in particular, to understand the presence or absence of rotation of clusters and superclusters of galaxies.

In order to single out in the change of brightness of the relic radiation in the direction of a cluster the component related to motion of the cluster with respect to the relic radiation, measurements are needed at various wavelengths and especially at $\lambda = 2$ mm, where the temperature effect becomes zero.

[1] *Smoot G. F., Gorenstein M. V., Muller R. A.*—Phys. Rev. Lett. **39**, 898–901 (1977).

PART THREE

THE HISTORY OF PHYSICS

PERSONALIA

69

The Creative Work of a Great Physicist and Modern Science. Albert Einstein*

Einstein was a modest and self-critical man. He used to say that the theory of Brownian motion, the idea of photons, and the special theory of relativity—all were in the mainstream of scientific development—and that if he had not, other physicists would have arrived at the same results no more that 2 or 3 years later. But for the general theory of relativity he made an exception: it, perhaps, might not have been discovered for another 50 years. This estimate found remarkable confirmation after Einstein's death: indeed, it was only in the sixties and seventies of the twentieth century that methods appeared which naturally, without brilliant guesses, led to the formulations of the general theory of relativity. Even the interval of 50 years was correctly predicted. Einstein's son, when he was 10 years old, asked his father why he was so famous throughout the world. His father answered: "Mankind is like a caterpillar crawling along a branch. I have discovered that this branch is curved."

Above we spoke of the science of the past, of the indisputable achievements associated with the works and name of Einstein. But Einstein's ideas live even in today's and tomorrow's science, they inspire those walking on uncharted paths. I will concentrate on two directions: unified field theory and modern astrophysics. It is well known that Einstein worked tirelessly for 40 years on a unified field theory, to his last days he hoped for success, but did not achieve his desired result. Failure? Tragedy? An optimistic tragedy because now, 25 years after Einstein's death, a new unified field is emerging. Its unity is different—it is not a single field, but many fields constructed according to a single plan, the unity of symmetry. This unity is achieved when we break ourselves away from the diversity of masses of particles. Modern physicists act as described by Alexander Blok:

"Erase erratic features
And you will see—the world is splendid"

*Voprosy filosofii **6**, 32–45 (1980) (abbreviated version). The article is based on a report read on 12 December 1979 at a meeting of the USSR Academy of Sciences and M. V. Lomonosov Moscow State University dedicated to the 100th birthday of Albert Einstein.

Had Einstein lived to our times, he would have appreciated the papers in which—even though by other methods—his dream is being realized and his intuitive faith in a unified plan for the structure of nature and the world is being justified.

We spoke above** about the law of baryon conservation and its vital importance. Modern theory admits the possibility that this law is not absolutely correct! It is possible that protons too—the nuclei of ordinary hydrogen—are capable of decaying, that hydrogen and other atoms are able to transform completely into energy, but over a time of the order of 10^{32} years. Imagine a 1, followed by 32 zeros! In practical terms this means that in one ton of water such decay occurs once in a hundred years, and in a cubic kilometer in the ocean—once per second. Such a rate of decay is negligible even on geological scales, and we see no ways to accelerate it in either Earth or stellar conditions. The decay and generation of protons could play a role only in the densest and hottest pre-stellar phase of the Universe's existence, but this discussion would carry us too far afield.

If the subject of the unified theory is the microworld, comprising the first section of the natural sciences, then a second very important section of natural science is the macroworld, the Universe. Einstein's general theory of relativity plays an enormous role in modern astronomy. Initially the solar system was the laboratory where the barely detectable differences between the predictions of the general theory of relativity and those of Newtonian theory were sought. Astronomy nurtured physics and the general theory of relativity, but received little in return. The general theory of relativity (GTR, for short) explained anomalies of the motion of Mercury which found no explanation in Newtonian celestial mechanics. GTR pointed out the correct significance of bending of light and radio waves by the Sun. Photons also have mass, and their trajectories are bent by gravitational forces. GTR predicts a change in the frequency of light: decrease, i.e., a red shift for photons moving away from a heavy body, and a blue shift for falling particles. All of these predictions of the theory are confirmed by observations and experiments, with all presently possible accuracy. Yet still we can better say that astronomy confirmed GTR, strengthened it with observations. This situation has changed radically in the last 15–20 years. The theory of the expanding Universe has already been discussed in P. L. Kapitsa's speech. We must add here a discovery, made later—the discovery of the microwave radiation which fills the Universe. The theory of a hot universe proposed by Gamow has triumphed. The microwave radiation is living evidence coming to us from the Universe's distant past, evidence of the fact that the Universe consisted of a hot, amorphous plasma from which galaxies and stars crystallized only later. Each point of the expanding Universe has its own specific system of coordinates. This coordinate system is determined by the

**This refers to a part of the speech which has not been included in this volume.

microwave radiation. Only in this coordinate system is the microwave radiation identical from all directions. A body (for example, a man) which moves forward with respect to a specific system, will encounter in front a greater number of hotter, more energetic photons, while his back will freeze. We may say that the microwave radiation is a new ether which exists at every point of the Universe.

If all cosmic matter does not move with respect to the new ether, i.e., with respect to the radiation, this means that matter is expanding together with the expansion of the radiation. In such an expansion the homogeneous distribution of matter is preserved. On the contrary, motion with respect to the new "ether" is the motion which leads to the formation of the structure of the Universe—clusters of galaxies and individual galaxies and on down to stars. On the background of the general expansion the solar system is moving with a velocity of 400 km/sec. Recently a means was found to determine the motion of very distant galaxy clusters. There thus appears a new method in radioastronomy to study the depths of the Universe. We have now the certainty that soon measurements will give us the possibility of choosing between the closed world, in which today's expansion will be replaced by contraction, and the open world, i.e., an indefinitely expanding Universe.

The general theory of relativity gives wonderful concrete predictions. Two very dense stars are found which rotate with respect to one another. At the same time they radiate gravitational waves, lose energy, and gradually fall in on one another and, approaching, spin still faster; this has been confirmed by observations. Einstein's formulas for radiation of gravitational waves are confirmed. Another example: on the line of rays towards us emitted by a distant quasar there is a galaxy. The mass of the galaxy bends the light rays. We receive two beams, we see images of one and the same quasar from two directions, separated by six angular seconds.

The most remarkable prediction is related to the evolution of stars. When the mass is large, the gravitational force grows such that neither electrons nor nuclei can resist the gravity. A puzzle appears: what is the final stage of the evolution of stars. Strange though it may be, to the end of the fifties this question did not much interest astronomers. At the congress of the International Astronomical Society in Moscow in 1958 I compared the position of the astronomer with the world-view of a healthy man who does not think about his own death (which is, of course, as it should be), although, if one were to ask him, he understands the inevitability of death. The justification for this position is that in the active period of their lives stars are luminous, they may be observed and studied using a telescope. It was presumed that after the store of energy is used up the star in one way or another is transformed into something, a passive, unobservable, uninteresting body. In reality, under specific conditions a black hole arises—the matter of the star ends up inside the black hole and is surrounded by such a gravitational field that even light and neutrinos are unable to get out of the boundaries of the black hole. At first glance this is a dead body. But let some other piece of matter approach the black hole, and the gravitation accelerates it almost to the speed of light. Woe to obstacles in the path of such matter. By lowering a weight on a thread toward a black hole, we could extract all the rest energy—any kind of matter would give 1000 times more energy than uranium in a reactor. There is no such thread. Under natural conditions, falling gradually and colliding with one another, particles of the matter give 0.1–0.2 c^2 per gram, carrying the remaining 80 or 90% of the rest energy into the black hole. But even an efficiency of 10–20% exceeds by dozens of times the energy of nuclear reactions! This explains the source of energy of the X-ray star in Cygnus. It also explains the enormous luminosity of quasars. Recently R. A. Sunyaev has found convincing details in the spectra of quasars which support these views. In a number of cases we see that from quasars and nuclei of the galaxy streams of matter are ejected, but each kilogram of luminescent matter thrown out is preceded by the fall of 100 kilograms into the gravity field. Matter is not born, just as energy is not born; only a redistribution of energy is occurring.

For stars and for the Sun the source of luminosity and activity is nuclear energy. For larger-scale phenomena gravity turns out to be the primary, still stronger source. We emphasize that we do not have to think up new, unknown laws of physics.

Astronomical phenomena find natural explanation in the existing laws of physics, in particular in the general theory of relativity. This theory dominates modern astronomy; it is impossible to ignore it. The scientific-technical revolution in astronomy has not ended—it is taking new forms and becoming deeper. We face questions not only of the area about us, but also the most difficult questions of superdense states and of the earliest stages of the evolution of the Universe. Today, with regard to the earliest stages of the Universe, there is a persistent search for a synthesis of GTR and quantum theory—science moves forward, *per aspera ad astra*: through the thorns of difficulties and mistakes to the stars of knowledge of the truth. Finally, we note one of the predictions of GTR—the existence of gravitational waves, in some measure similar to electromagnetic waves, but extraordinarily weakly interacting with matter. On one hand, this ensures the arrival of gravitational waves from such depths of the stars and galaxy nuclei whence electromagnetic waves cannot come. On the other hand, the weakness of the interaction makes it singularly difficult to find gravitational waves. And yet we may hope that before the end of our century there will appear an observational gravitation-wave astronomy, and laboratory instruments will thus reveal collisions and explosions of stars in far-away galaxies. Perhaps it will be possible to obtain information on the earliest, pre-stellar stage of the

Universe. All of this will be a tremendous contribution to astronomy and the search for knowledge of the world.

Let us return to Einstein, to his work and his fate after the creation of GTR and completion of his work on radiation (1917–1924). He lived more than 30 years more in the complex circumstances of the rise of Nazism in Germany, the second world war, and the first post-war decade. In Germany, and even in neighboring Holland, he did not live in safety. From 1934 until his death in 1955 he lived in the USA; he was well-off, and famous as no other scientist. And at the same time he worked only with a few colleagues; he began to be alienated from the main trends of modern physics. Here begins the part that is most difficult for the author of an anniversary paper. Perhaps, out of respect to Einstein, it would be best to leave out the story of the less productive period of his life and creative work? No, even the mistakes of a man of his stature are interesting and instructive.

Einstein worked with extraordinary persistence, trying different variations of the theory; several times he wrote to friends that he was close to success, but would be again forced to admit that a unified theory did not emerge. A psychological question arises: Why did Einstein work on the unified field theory with such self-denying tenacity, in spite of all his disappointments? There were not a few people who advised him to turn to quantum theory, which was undergoing rapid development at the time. With great humor Landau told of the time in 1930, when he was quite young (22), that he visited Einstein in order to put him on the true path ...

I can only reconstruct Einstein's mentality in an unsubstantiated way. The enormous success of the general theory of relativity, undoubtedly, affected him. His own statement, given above, is evidence of this. But for the creation of the GTR Einstein did not need the latest, freshest experimental data. What every schoolboy had known for 200 years turned out to be sufficient. It was so natural and so tempting to repeat this world record, once more to construct a theory on the basis of a minimum of information. But "that which is broken accidentally brings luck, but that which is broken on purpose doesn't count" (Yevtushenko). In constructing the theory of gravitation, Einstein knew all the physics of 1917: it turned out that for the foundation of the theory only a small part of it was required. It appears that in the 20's, 30's, 40's and 50's Einstein purposely limited the initial foundations of the theory he sought, and this work of his turned out "not to count."

Today, at the end of the seventies, a unified theory is emerging, but in a completely different guise. This is a huge, difficult task, moving forward step by step, controlled and directed by experiments on all the most powerful accelerators. One may hope that in the future (the first shoots are already appearing) gravitation too will be brought into the circle of unified interactions; however, even now it is clear that many intermediate stages will have to be passed on this route, each of which will inject qualitatively new ideas.

And still, perhaps, not we but our descendants, leaving aside all the details, will give due to the main thing. They will appreciate Einstein's idea that there must exist a unified theory—an idea based on a deep conviction of the unity of nature.

I am completing the scientific part and wish to speak of human qualities and their relation to the scientific and technical specifics of creativity. How can Einstein's successes be explained? How can we understand the extraordinary role of one man in modern science? Least of all do I wish to deify him, or to speak of divine inspiration bestowed upon him. You recall the words of P. L. Kapitsa about Einstein's humility and self-criticism? But an explanation is needed, and I see two chief factors. One of these is the factor of time. Einstein's completion of his studies and his blossoming when he was 20–25 years old, coincided with a moment of crisis when changes in physics had become ripe, had become necessary. As Tyuchev (a Russian poet, 1803–1873) said: "Blessed is he who visits this world in its moments of destiny." But this is not all. The young Einstein's contemporaries numbered not less than 100 young and not-so-young theoretical physicists. Three outstanding discoveries (molecules, quanta, relativity theory) made by one man, without doubt, mean that this man is exceptional. But he was exceptional not, I think, in quickness or brilliance or formal education. He stood out in his depth, his aspiration to clarity, and perseverance of thought, and this is the second factor in his success. In one of his letters Einstein wrote: "God gave me the stubbornness of a mule, but he didn't give me a mule's thick skin." How far this is from pomposity, but note the emphasis which the self-critical author put precisely on his stubbornness.

In his autobiography he writes that when he was four he was struck by the miracle of a magnetic arrow rotating without being touched, and at 15 he thought about what a man moving at the speed of light would see. His perseverance of thought did not allow him to forget or avoid this question, and he returned to it, creating the special theory of relativity. Education had not suppressed the natural curiosity of the youth. Einstein was led by a deep feeling for the unity of nature, of the harmony of the laws of nature. He did not think that he was organizing his perceptions or optimizing his thought. He discovered objectively existing laws of nature, without being diverted from what was most important and fundamental. Recall Pasternak's words: "In everything I want to get to the very heart of the matter: in work, in the search for a path ...", and further "to the foundations, the roots, to the core."

I would note here for physicists one characteristic technical detail: Einstein understood extraordinarily clearly that new theories do not destroy, but rather enrich old ones, he understood the "correspondence principle." In particular, he consciously and purposefully used the fact that the motions of dust particles and large molecules are always the same and do not change in quantum theory. He understood that a new theory of gravitation must be constructed such that in the limit it coincides with the old theory, that of Newton. In other words, in constructing something new, he leaned on what had been achieved in physics, with great intuition chose in the old that which would remain forever. Finally, after 1905 he had that soaring feeling of success which is independent of material well-being or job situation. Some part of that success was in the recognition by colleagues and in the fact that his articles were published and recognized. But still more important was the feeling that nature's secrets were being discovered. The feeling of success gave Einstein the strength to work on constructing the theory of gravitation and the general theory of relativity for 12 years.

I cannot give prescriptions to leadership on how to grow Einsteins. I can only repeat that which is generally accepted: we must develop the sciences, we must take care of the people. My wish for the young: carry your thoughts to their end, to complete clarity.

Returning to the subject of the article, I must say that perhaps most important is the fact that Einstein was a truly good person with pure intentions, caught up in the essence of his work. This is not so trivial. There is the eternal problem: "genius and evil are incompatible." Let us not speak of criminal evil. Many good and necessary scientific studies are done for mundane reasons, for example, to obtain a degree. My point is this: that extra-class work can only be done on the highest moral level of self-denial—and Einstein achieved this level.

We are celebrating the hundredth birthday of a great physicist. But we also honor a great man, one of the organizers of Society of Friends of Soviet Russia, an Honorary Foreign Member of the USSR Academy of Sciences. When after the cruel war in 1919 English astronomers began to verify the theory of the German physicist, people on the street who knew nothing of his theory knew that there was a great scientist, and letters addressed to "Planet Earth, Einstein" made it to the addressee. When in 1923 Einstein accepted the theory which originated in our country of an expanding Universe, it was a visible sign of peace among nations that it was thus accepted all over the world. When in 1933 Einstein was exiled from Germany, this meant that a new war was approaching. Here are the words from the manifesto of Einstein, Russell, Joliot-Curie: "... Therefore the question which we set before you is a stern, terrible and unavoidable one—should we destroy mankind or should mankind abolish war." Further, "... Before us lies the path of uninterrupted progress, happiness, knowledge and wisdom. Will we instead choose death?" Einstein signed this document several days before his death in 1955; this was his legacy to mankind. We honor a great scientist and we honor a man who consistently defended peace on Earth, fought for justice, for social progress, and for peoples' happiness.

Eternal glory to the physicist and the man—Einstein!

70

The Main Book of D. A. Frank-Kamenetskiĭ and Its Author

Presently there is a third, posthumous edition of the remarkable monograph "Diffusion and Heat Transfer" in preparation.

Published in the USSR in 1947 and a second edition in 1967, the book was translated in the FRG in 1959 and in the USA in 1969.

The monograph played an enormous role in the development of post-war science and technology, and in the education of more than one generation of chemists, engineers and chemical physicists.

To this day precisely this monograph has remained the best known and most cited in papers relating to the foundations of chemical machine design and to the theory of combustion.

The book has become a bibliographic rarity and in this connection, of course, the question arose of a new edition. To our great regret, the third edition will be posthumous. David Albertovich Frank-Kamenetskiĭ prematurely died on June 2, 1970, before he reached 60.

In planning a new edition without the participation of its author, it seems necessary to speak in detail in the foreword both about the book and about its author. One must explain the reason for the unique longevity and importance of the book. Its distinctive feature is the abundance of results of the author himself—one of those who constructed the foundations of a new branch of science.

In the foreword to the first edition, Academician N. N. Semenov writes: "The author also managed to work out new, fruitful methods, to single our important limiting areas, to introduce a number of new physical concepts and to obtain valuable physical results. One may say that this work heralds the beginning of the transformation of macroscopic kinetics into an independent branch of science. The author has succeeded in showing that the problems at hand have not only particular and practical value, but also general scientific value and are of general theoretical interest."

The development of science over the last 35–40 years has completely confirmed this evaluation by N. N. Semenov. Macroscopic kinetics, under the new names "synergetics," "theory of dissipative structures" and "catastrophe theory," is rapidly developing. In this, naturally, new problems and new results arise. However, for a deep understanding of the new branches of science with their historical roots it is necessary to turn to the foundations, to the pioneering works in the area of macroscopic kinetics—and here the book by Frank-Kamenetskiĭ is the best resource.

But there is still another task related to the passing of the book's author. Fewer and fewer people remain who knew him personally. Not all readers of the book are aware of the creative path and personality of the author, not all know of the breadth of David Albertovich's scientific interests and, in particular, of his results in other areas of science—in plasma physics, astrophysics and cosmology. The present sketch makes the attempt to fill this gap. I think that such a story will also help to evaluate the present monograph "Diffusion and Heat Transfer in Chemical Kinetics."

David Albertovich began his scientific career under somewhat unusual circumstances. He spent his childhood and youth in Siberia; there too he received his engineering-metallurgy degree (he graduated from the Tomsk Technological Institute) and began to work at an ore-enrichment factory. In early 1935 David Albertovich wrote a letter to N. N. Semenov in which he discussed problems of chemical thermodynamics. The author's talent was so obvious that he was invited to the Institute of Chemical Physics. At that time, at the Leningrad Physico-Technical Institute and associated institutes, in particular the Institute of Chemical Physics, there was broad, conscious and successful effort made to find and attract capable young people.

As part of a large collective David Albertovich took part in work on the problem of oxidation and fixation of atmospheric nitrogen in combustion and explosions. Mentions of this problem, for example by Cavendish, appeared immediately after the discovery of nitrogen and after the composition of air was established. This problem was addressed by such outstanding chemists as F. Gaber, V. Nernst (Germany) and R. Bone (England). In connection with the development of the theory of chain reactions there arose the problem of the possibility of direct use of combustion energy to transform nitrogen into nitrogen oxide. Research carried with David Albertovich's participation showed that the process was related to the mechanism of chain reaction with the participation of N and O atoms, but that the yield of nitrogen oxides was limited by conditions of thermodynamic equilibrium. The directions of David Albertovich were outlined quite naturally: on the one hand—the theory of combustion and explosion, and on the other—the general foundations of chemical engineering. David Albertovich was close to these questions through his engineering education and experience as well.

In a fundamental paper in 1939, David Albertovich posed the problem of thermal explosion taking into account the spatial temperature distribution in a medium in which a chemical reaction is occurring. The solution of this problem concluded almost a century's research. The possibility appeared of exactly precalculating the conditions of explosion. Numerous experiments fully confirmed D. A. Frank-Kamenetskiĭ's theory. Thanks to this theory important results were obtained in chemical kinetics.

Hidden in this work, however, was a deeper content, one which went beyond the bounds of the problem of explosion. Only after many years was the fruitfulness of the way the problem of critical conditions was posed as the limit of existence of a solution understood. Using the example of thermal explosion David Albertovich developed a similarity theory of processes of energy release and removal. He proposed an asymptotic expression, $k_1 \exp(\alpha(T - T_1))$, replacing the exponential dependence $k_2 \exp(-A/RT)$, in which the solution relating to some temperature is obtained by a similarity transformation of the solution relating to some other temperature. In modern terminology, David Albertovich used group properties of the equations and consciously chose the approximation needed for the appearance of a group which is additive in temperature and multiplicative in coordinates.

These general physical and mathematical ideas were used widely in works on the theory of combustion, both those carried out with the direct participation of D. A. Frank-Kamenetskiĭ and as continuations of his research.

David Albertovich's research on the foundations of chemical engineering are summarized in his remarkable monograph.

Heat transfer, diffusion and hydrodynamics are all branches of classical physics. The monograph wonderfully combines analytic solutions, similarity theory and a semi-empirical approach to phenomena and processes of varying degrees of complexity. The broad scientific program of a physico-mathematical approach to technology presently being carried out uses in many ways the deep ideas and methods presented in this work.

The scientific interests of David Albertovich soon after the war and after writing the book shifted to the area of astrophysics. Problems of the calm evolution of stars are masterfully presented in the monograph "Physical Processes in Stars." D. A. Frank-Kamenetskiĭ also solved the problem of how a shock wave intensifies in the outer layers in the explosion of a star. This phenomenon is significantly related to laws of variation of luminosity of supernova stars and also, perhaps, with the process of the initial acceleration of cosmic rays. David Albertovich was among the first to understand, as evidenced by works on "epiplasm," the role in astrophysics and in particular in cosmology played by the process of generation of particle-antiparticle pairs in extreme conditions. In 1956, on an invitation by I. V. Kurchatov, D. A. Frank-Kamenetskiĭ moved to the Institute of Atomic Energy where he led a new direction—research on the interaction of waves with plasma. Here he was the first to clearly formulate the problem of plasma heating due to dissipation of waves excited within it by an external source of oscillations.

David Albertovich theoretically predicted the important phenomenon of magneto-sonic resonance which was later found experimentally with his direct participation. Subsequently he and his students studied this phenomenon in detail both theoretically and experimentally. The dispersional properties of a wide class of oscillations were studied: normal and oblique magneto-sonic waves. The appearance of resonant oscillation of electromagnetic fields in plasma with magneto-sonic resonance (the effect of "spatial amplification of a magnetic field"). Finally, an anomalous dissipation of magneto-sonic oscillations was found and, as a result, the possibility of heating plasma to high temperatures by magneto-sonic resonance was demonstrated experimentally. In particular, using this method heating of dense hydrogen and helium plasma to a temperature of $5 \cdot (10^6 - 10^7)$ K was accomplished.

In recent years together with this study David Albertovich undertook an investigation of the instability of plasma in electron cyclotron heating, and also began work on plasma phenomena in a solid body.

I should note the immense influence of David Albertovich on creative young people. As early as 1956–1957 he called for studies of collective processes in plasma and the phenomena in which they should appear: nonlinear and shock waves without collisions, pinch, etc. His ideas without doubt played a fundamental role in the development of plasma physics.

The breadth of erudition, encyclopedic knowledge, unusual literary talent and the ability to speak simply and clearly about the most complex things made him one of the greatest populizers in the natural sciences. He is the author of a series of popular books on various areas of physics: *Energy in Nature and Technology*, *The Formation of Chemical Elements in the Depths of Stars*, *Nuclear Astrophysics*. The most popular and well-known book was that written on the suggestion of I. V. Kurchatov, *Plasma—The Fourth State of Matter*, which has been reprinted many times and translated into Bulgarian, Polish, Czech, German and Japanese.

David Albertovich was one of the most competent specialists on a huge spectrum of questions in physics, chemistry, astrophysics and biophysics. His honest and completely selfless desire to help everyone who came to him for advice, his colossal erudition, and constant readiness to enthusiastically delve into the solution of scientific problems, even those lying outside his current interests, made David Albertovich an irreplaceable creative consultant. A vivid picture of the unique breadth of his scientific interests, and also of his lively style of presentation and sense of humor, is given in reviews printed over many years in the bulletin "New Books Abroad," and by articles and notes on the latest accomplishments of Soviet and world science and on a number of fundamental philosophical problems in the journal "Nature." D. A. Frank-Kamenetskiĭ edited many Russian editions of scientific and popular science books by famous foreign scientists. His public lectures on the most important problems of modern natural science and his brilliant appearances on radio and television have become well-known.

D. A. Frank-Kamenetskiĭ's teaching activity, begun as early as the early thirties in the Chitin Mining-Metallurgy Technical School and at Irkutsk University, continued almost without interruption all of his life. In his last years he headed the department of plasma physics, organized by him, at the Moscow Physico-Technical Institute. David Albertovich is the author of the textbook *Lectures on Plasma Physics.*

The Soviet government held the work of D. A. Frank-Kamenetskiĭ in high regard: he was awarded the Lenin Prize and the Red Banner of Labor, and three times received the State Prize of the USSR.

A simple enumeration of David Albertovich's scientific works does not give a full idea of the authority and scientific influence of D. A. Frank-Kamenetskiĭ; these in many ways depended on his personal qualities and the generosity with which he related his ideas not only to his own students and colleagues, but to anyone who turned to him for advice.

Enormous optimism and good-will, nobility and attentiveness—these qualities of David Albertovich are well known not only to his colleagues and students, but also to a wide circle of physicists and chemists. He was distinguished by a complete lack of jealousy or envy toward others' results. The natural active desire of a scientist to open up his abilities and to obtain an important result was harmoniously combined in David Albertovich with an interest in the essence of the matter, an interest in nature.

David Albertovich lived a vital and happy life. His memory will remain forever in our hearts.

Time has confirmed the evaluations given above.

Problems of diffusion and heat transfer in chemical kinetics are, as before, at the center of attention of chemical engineering and the theory of combustion. In the last chapter of the book *Diffusion and Heat Transfer in Chemical Kinetics* David Albertovich devotes much attention to periodic chemical reactions. Today in our country a large cycle of studies has been performed on the periodic reaction of Belousov–Zhabotinskiĭ.

At a recent international conference, "Synergetics 1983" (August 1983) at the Biological Center of the USSR Academy of Sciences in Pushchino, near Moscow, it became especially clear that research on periodic reactions and related spatial pictures are growing into an independent branch of science—synergetics. The theory of thermal explosion was the first proverbial swallow of the modern theory of catastrophe. Due to the development of catastrophe theory the results of David Albertovich and his co-workers are being generalized to a multitude of areas, areas which seem at first glance, externally, to be far from combustion and explosion—from polymer mechanics to biological phenomena. David Albertovich's work on the theory of plasma is of great value. His popular book, *The Fourth State of Matter*, remains even today the best introduction for young people to plasma physics. His deep penetration into the physics of dissipative processes allowed him to make a significant contribution to the theory of diffusion and heat transfer of plasma.

He has written a number of methodological manuals on the problems of transport and waves in plasma. He is the author of one of the methods of plasma heating using sonic resonance.

The book *Physical Foundations of the Theory of Stars* (1960) has no equals in clarity and penetration into the essence of problems, and in the rational application of similarity theory. In addition, the investigation of hydrogen stars has proved to be more important than it seemed 20 years ago. Today it is established that heavy elements did not form in the cosmological nucleosynthesis. Consequently, first-generation stars consisted of a hydrogen-helium combination without heavier ions.

David Albertovich developed the ideas on epiplasma, and he is also responsible for the very name plasma, consisting of particles and antiparticles. Today it is established without doubt that this was indeed the state of matter in the early stage of evolution of the Universe.

David Albertovich was the embodiment of broadness of knowledge and of striving toward synthesis of our knowledge.

There are sometimes complaints about the ever growing complexity and differentiation of individual branches of the natural sciences. The life and creative work of David Albertovich are a convincing rebuttal of these complaints. David Albertovich's talent and erudition made it possible for him to overcome barriers dividing different areas and to remain a universal scientist on a high professional level in our difficult second half of the twentieth century.

It is becoming obvious today that in a whole series of questions David Albertovich's ideas were ahead of their time and, perhaps, have for that reason not gotten enough response.

All of us (and the author of this forewarn above all) are belatedly realizing that in the sixties we had insufficient contact with David Albertovich, did not listen enough to his ideas and advice. Maybe these lines will serve as an impetus to intensified contacts and discussions among Soviet physicists, chemists, biologists.

Although it is not customary to write of this, we recall that David Albertovich was not chosen either as an Academician or as a Member-Correspondent of the Academy of Sciences. I am convinced that David Albertovich was completely worthy of these titles. He did not receive them largely because the range of his work was wider than the subjects of the sections (general chemistry, physics and astronomy) into which the Academy is presently divided.

His personal modesty also played a role ... Yet, still, there remains a feeling of injustice and insufficient persistence by David Albertovich's colleagues and friends, including the author. It must be said here that David Albertovich himself was absolutely indifferent to careerist considerations.

He was a person with an exceptionally bright attitude; he loved nature and art. His work, his students and, finally (last but not least), his family provided him unsurpassed joy.

I wish to share one deeply personal reminiscence that is very dear to me. In March 1970 under the influence of some internal impulse, without concrete reason or invitation, my wife and I went to visit David Albertovich. After a discussion of science (mainly about astronomy), we went to the kitchen for dinner. We were joined by David Albertovich's wife, Elena Efimovna, his children, and his children's friends. The topic of poetry came up, and Albert Davidovich inspiredly read the poem "Ravenna" by Blok, beginning with the words, " All that is transient and fleeting was buried by you in the ages ..."

I will remember always the delight of the young people.

David Albertovich was already very ill, but in my memory of this meeting his image remains as it was all his life—talented, inspired, happy.

71
In Memory of a Friend. Boris Pavlovich Konstantinov*

Boris Pavlovich Konstantinov will always remain for me a friend, someone dear, simply Boris, and for my children—kind "Uncle Borya." Those who feel that one must write impartially and without passion—let them not read this essay. I loved Boris, and I am now speaking of the fact that he is no longer with us, that he has left so early. He remained witty, inventive, deep and happy, kind and active up to the last months, when his ailing heart had already become tragically weary. Boris Pavlovich did not live to be sixty. These notes are dedicated to his 75th birthday. And couldn't we dedicate them to a living recipient of honor? How much Boris Pavlovich could have done with his mind and talent in 15 years!

We met in 1932 in the Konstantinov's large apartment on Malaĭa Podĭacheskaĭa street near the Griboedov canal in Leningrad. We were introduced to one another by Varvara Pavlovna, Varya, Boris' sister, my wife.

I learned that Boris was a physicist, that he worked in the Acoustics laboratory and, just as I did, he lacked higher education. I also found out that he had recently been very ill (then the disease was called flaw of the heart), that he had been saved by Nina Nikolaevna Ryabinina, who later became his wife. Among a large family (six brothers and three sisters) Boris was the most talented, the most brilliant and generous, the most musical, with the most singing ability and excitability. But all of the Konstantinov family was extraordinary.

In difficult years the brothers and sisters stayed together, they helped one another to live, to obtain an education. They had too little to eat, saved on tram fares, but by hook or crook they managed to get to the gallery of the Mariinskiĭ Opera Theater, and sometimes to the stage—as extras. They all skied together.

Probably in Boris Pavlovich's ability to work with people, to understand them and to help them, his childhood and youth spent in a large, happy family played a major role. By the time that I entered this family the parents were no longer alive. The father died in 1918. Recently in the thick book *All of Petrograd, 1916* I found the line "Konstantinov Pavel Fedoseevich, hereditary honored citizen, building contracts, Malaĭa Podĭacheskaĭa,

*Voprosy istorii estestvoznaniĭa i tekhniki **2**, 71–75 (1984); Nauka v SSSR **5**, 109–112 (1984).

Number 10."

Boris' father was a self-made man who began his life with labor, and achieved an independent position. To this day I remember the story of how he sat down to work every day before dawn at 5 am to work with documents and bills.

The house in which the Konstantinovs lived, and the six-story building opposite, were built by him. Today we have departed from the primitive idea that only he who lays the bricks is a builder (Pavel Fedoseevich began as a bricklayer). Today we also see the value of the organizer.

In 1929 the mother, Agrippina Fedoseevna, died. The older sister, Ekaterina Pavlovna, married and left the country with her husband. The older brother, the talented radio-engineer Aleksandr Pavlovich Konstantinov, also married and became more distant from the family. He is described as an inventor in the area of television. He worked in the Pulkovo observatory, and was also associated with geophysicists. I remember at an exhibition of the Academy of Sciences his instrument for registering earthquakes according to changes in electrical capacity.

The older brother highly valued the mind, talent and inventiveness of Boris, and helped to get him the job in the Acoustics laboratory. Both were tall, fair, with broad facial features; they were also similar in build, but in childhood Boris had often been ill, and this laid its mark on him. I think that Aleksandr projected more self-confidence, and Boris—more kindness and attention.

I cannot help thinking, should I be writing about this? That which is dear to me and exciting—are they interesting or necessary for the reader? Especially a reader who has never met the living Konstantinov.

His specific scientific works and inventions have long since been published; the theories created by him continue to develop in the works of those who followed. They have become faceless. But no! It only seems thus, that they are faceless. If we try to more deeply understand the creative work itself, questions arise: why did the man take up a particular problem, why did his colleagues come to him and follow him, why did he become a leader? In order to understand this, one must return to the man himself, to his childhood, his family—to the sources of his personality.

During the war Boris and his close ones, like I and my family, moved to Kazan. I knew of the serious work of Boris in the Physico-Technical Institute, of the brilliant doctoral defense in 1943, of the very high regard for him by our elders—Abram Fedorovich Ioffe, Petr Leonidovich Kapitsa, Nikolai Nikolaevich Andreev.

After the war, together with his colleagues of the Institute, Boris Pavlovich returned to Leningrad. I had ended up in Moscow even before the end of the war. In parallel with work on the internal ballistics of powder rockets, I was brought into atomic affairs. For several years Konstantinov and I met rarely. Later work again brought us closer. In the course of research which I was carrying out with Kurchatov and Khariton, a complicated problem arose which allowed for several different technical solutions. It was also set before the Leningrad Physico-Technical Institute (PTI), and before Boris Pavlovich in particular. He exhibited amazing good sense in choosing the approach and immense inventiveness in constructing it. A new period in his life began: research in the Leningrad laboratory, discussion of their results, coordination of plans, provision for materials—in Moscow in government institutions and at the Academy of Sciences, then construction and release of the project.

In most cases, when he was in Moscow, Boris preferred to stay in our apartment; he slept on the couch in our room, sat with me and his sister until late in the evening. In the morning—I am ashamed to recall this now—we woke him with our loud calisthenics, and the children with their merry bustle. Then he would leave for the ministry or the train station.

I remember one evening when Boris was extraordinarily pale, untalkative, serious. A difficulty had arisen, perhaps associated with an earlier mistake. The fate of the approach was in question. Boris drank coffee, did not go to bed—and by morning had found the necessary ideas and saved the job.

Need I say that it was successfully completed! Boris Pavlovich received the Gold Star of the Hero of Socialist Labor, and he was soon chosen as a Member-Correspondent of the Academy of Sciences. Finally,—and this, of course, is the most important point—the process proposed and worked out by Konstantinov today remains unsurpassed, and even now finds wide application.

In 1955 Boris Pavlovich was appointed director of the Leningrad Physico-Technical Institute. Now a bronze bust of him stands before the door. Rather, two busts rise up opposite one another: Abram Fedorovich Ioffe, the founder of PTI and its director from 1918 to 1948, and Boris Pavlovich Konstantinov, who occupied that post from 1955.

The role of Abram Fedorovich Ioffe in the development of Soviet physics and Soviet science is impossible to overestimate. All of us, including Konstantinov, were and are aware of what we owe him, and consider him in the highest sense our teacher. Therefore the question whether the bronze bust of Konstantinov has the right to stand next to the bust of Ioffe is a difficult one. I nevertheless prefer to speak in Konstantinov's defense openly, leaving no room for reservations.

The chief point to note is that Ioffe ceased to be director of PTI not because Konstantinov took over the post. This is clear from a comparison of the dates (1948 and 1955). Boris Pavlovich is not the second, but the third director of PTI. His arrival heralded the beginning of a renaissance of the Institute, which had lost somewhat its authority and traditions over seven years. Together with his energetic support of the old classic directions,

Konstantinov organized new directions: chromatography, technical electrolysis; he paid attention to astrophysics (more on this later), and began work on diagnostics of thermonuclear plasma. Research on atomic collisions also found its place. Finally, under Konstantinov's leadership the theoretical divisions were strengthened. Theoreticians understood that they need not seek another place to work, they understood that they were needed and valued; they saw a future for working themselves and for teaching young people. The division of nuclear physics at PTI in Gatchina grew into an independent institute, rightly bearing the name B. P. Konstantinov.

Probably I have not listed everything, not mentioned dozens of good deeds. I know one thing: he sacrificed his time, his strength, his health, and he died before he reached sixty. Ask any worker at PTI who has been there for 15–20 years and he will tell you what Konstantinov was for the Institute.

Boris Pavlovich became director when Ioffe was already heading the Semiconductor Institute and did not wish to return to the Physico-Technical Institute. Konstantinov visited him, sought his advice and supported him.

Let us recall the day of celebration of the 100th birthday of Ioffe in Leningrad. We still hear the words of Academician Aleksandrov and many others about what the institute created by Abram Fedorovich was and remains for the country, for science, physics and physicists. But do we understand well enough that they would not reverberate thus had not Konstantinov, and later Tuchkevich, saved the institute!

I will not dwell on Konstantinov's activity as vice-president of the USSR Academy of Sciences. Others, undoubtedly, will write of this and write well. I highly value his work in the post of vice-president. At the same time I saw that the excessive load during these years had an effect on his health. A weak heart. I will never forget how Boris came to our dacha to rest on a Saturday evening. He had trouble ascending a single flight of stairs. We carried a chair into the garden, surrounded Boris Pavlovich, laughed and joked. But in our hearts we felt immensely anxious for him, for his life. During this period he did not leave his responsibilities. Attempts to help him in round-about ways did not work. Should a doctor or someone close to him have told him of the fatal danger? Did he himself know that he was burning out? Probably, optimism and hope for recovery are necessary components of a healthy psyche, even if one's heart is sick.

I wish to pause more on one of Boris' works, one which involved him and was very controversial. Once again I must warn you—do not expect of me impartiality!

We are speaking of Konstantinov's astrophysical idea, the idea of "antimatter here at home," in the solar system, in meteors. This, of course, is an incorrect idea. It has been disproved by experiments set up by Konstantinov's laboratories. Moreover, a close analysis of indirect data could show that the idea of antimatter in the solar system (and even in our galaxy) is extremely unlikely.

Now, after 20–30 years, the question stands somewhat differently: should we not consider work on the search for antimatter evidence of bad knowledge of the subject, of unwillingness to listen to the opinion of specialists or, what is even worse, of a desire to gain fame, honor and position by dishonest means?

It is easiest to answer the last of these questions. It is sufficient to note that by the time he took up astrophysics, Boris Pavlovich was already an Academician, the director of a major institute, and a Hero of Socialist Labor. All his positions, titles and rewards he obtained for indisputable achievements in other areas. For him personally astrophysics offered nothing but additional work, hassle, trouble—but also, of course, satisfaction of scientific interest!

I well remember my reaction to the idea of antimatter. I did not believe it, considered it unlikely. But I also thought about the fact that with this low probability the value of such a discovery, if it were to occur, would be enormous.

In physics and in life we distinguish between probability and mathematical expectation. There is a joke that "mathematical expectation is the product of probability and trouble." A probability of one percent is small, small compared to unity. But if this is the probability of a fatal outcome of a disease, then the disease is considered very serious.

So then: if I considered the probability of finding antimatter to be small, I did not consider the mathematical expectation too small. From such a point of view, the contra-arguments become weaker. Yes, for ten or a hundred meteor showers it is proved that they consist of ordinary matter. Yes, the majority, even the vast majority of meteors are "ordinary." Does this prove, strictly speaking, that the eleventh or hundred and first shower is not of antimatter? In this approach the question goes to another plane: can we, should we undertake only projects with guaranteed success, should we reject out-of-hand projects which promise great risk? As an example of research where the risk of a negative result is great, we may point out the search for radio signals from extraterrestrial civilizations. However they have continued for many years and are not considered hopeless or compromising.

Let us return to the question of antimatter. Indeed, it does not exist in the solar system. Konstantinov's co-workers came to this conclusion. For many astronomers this conclusion was considered to be obvious from the very beginning. But one can and should look deeper, pose the question why is there no antimatter? Then the pragmatic point of view "no and that's the end of it" becomes similar to Chekhov's expression "this cannot be because it cannot ever be."

The posing of the problem by Konstantinov was a manifestation of a sacred concern. And if one were to pose the problem especially rigorously:

why is there no antimatter in a hot universe where in the distant past over a very short time there had to be super-high temperatures at which pairs of protons and antiprotons were generated? Then it becomes obvious that there is a question, a difficult and nontrivial question.

Being literally incorrect, Konstantinov's idea, without doubt, stimulated an extremely important direction of modern cosmology—the theoretical investigation of the very early Universe. Boris Pavlovich did not live to see the day when these questions began to be widely discussed, when hypotheses appeared which, perhaps, explain the absence of antimatter.

Let us return in our story to the Leningrad Physico-Technical Institute. Caught up by an idea, Boris Pavlovich created an astrophysics department which even now is one of the leading centers of high-energy astrophysics. In 1981 a group of his co-workers headed by E. P. Mazets obtained remarkable results on cosmic sources of gamma radiation. And Doctor of Physical-Mathematical Sciences E. P. Mazets is still grateful to Boris Pavlovich for bringing him, a qualified nuclear physicist, to astrophysics. There are many such examples.

And so an incorrect idea nevertheless proved fruitful both theoretically and experimentally.

Let us look back in our minds on the total life journey of Boris Pavlovich Konstantinov. A man, a scientist of striking integrity, he lived a short, but bright, fruitful, worthy life.

71a
Remembering a Teacher.
For the Eightieth Birthday of L. D. Landau*

It is very hard to write about Dau, knowing that his eightieth birthday is coming up. For all of us who knew him personally, he remains in our memories lively, active, clever, young (both in mind and body)—a unique and dear man. It seems as if, had the irreversible accident of 1962 not occurred, Dau would remain as he was—fifty years old and in his prime.

Now that I am over seventy, I want to qualify "fifty" with the word "young." But I knew Dau when he was truly young, twenty or so, in a fashionable jacket with gold-plated buttons, in pre-war Leningrad. I remember isolated episodes (about them, later), but beyond them I remember something bigger—the Teacher role, with a capital "T."

Dau created a particular style in theoretical physics, insisting on accuracy and unassailable logic, demanding investigation of all qualitative aspects of the phenomenon being studied. At the same time, after this investigation, he considered it all right to take absolute quantities from experiment. "It saves time, and life is short"—a saying which was tragically confirmed in his case.

I recall one episode: Dau had come to Leningrad from Moscow to hear and consult with the theory group at the Institute of Chemical Physics. Yakov Il'ich Frenkel and the still active Lev Emmanuilovich Gurevich headed the group. I occupied the position of youngest: they were teaching me physics, but were quite as likely to send me for beer ... Naturally, it was I who drove Dau to the consultations in the Institute car. On the last day, I took him to the Institute accounting office and was astounded to see Dau recount the money he was given: "But, Dau, you taught us that one must count only in order of magnitude; they obviously haven't given you ten times less than they should." Dau lost his composure for an instant only, then said: "Money is in the exponent ..."

It was great luck for our science that Dau worked with Pyotr Leonidovich Kapitsa. During the difficult year of 1938, Kapitsa resolutely, taking the full responsibility on himself, helped Landau out, providing him with everything. As Kapitsa had promised the government, Dau brilliantly developed a macroscopic theory of superfluidity. I am sure that others will write,

*Prepared for a collection of articles celebrating Landau's eightieth birthday.

and write well, about the Landau and Lifshitz course of theoretical physics. This is our common treasure; no other textbook is so all-encompassing, while managing at the same time to emphasize what is most important.

Dau never refused discussion with anyone. Sometimes, of course, the discussion would be quite short. We all remember Dau's wondering: "Why is *XX* mad at me?! I never said he was a fool, I just said the work was idiotic ..."

All modern theoretical physicists throughout the world are to some extent Dau's pupils, thanks to the famous *Course.* But there is also a small group who form Landau's own school, his direct students and colleagues. I, unfortunately, belong only in part to this group.

Most unforgettable were the discussions with Landau. Actually, the word "discussion" seems inappropriate: Dau briefly, often with annoyance, gave his own opinion, and I then left to analyze the situation as "homework." Sometimes I even went back. Dau was neither cold-blooded nor impassive. He watched the creation of quantum mechanics take place before his very eyes in Copenhagen. The first applications of quantum mechanics to molecules, to electrons in metal and to nuclear physics belong to him.

He had an amazingly keen ability to see, many steps ahead, attempts to corrupt the principles of quantum mechanics. In discussions he was merciless.

Dau was also generous: I still remember with shame how he suggested the problem of flame stability to me. I made an error in the calculations; Dau never admonished me, he simply did the work correctly himself, work which lives on to this day.

Did Dau have any faults and should one write of them? Today I think that Dau and his students tended to isolate themselves from other, equally strong scientific schools. One may see two sides to this. The negative one needs no explanation. However, there is another, no less important positive aspect: the absolute refusal to recognize the authority of anyone's name or title. In this sense, Dau's position was unassailable since he never intended his own words or claims to be taken on faith. Dau did not use his authority to dominate—he "dominated" with his logic, talent and keen mind.

What is written above, however, relates more to individual papers. There is another, more subtle question—that of the choice of scientific directions—where it is not the correctness of individual transformations, formulas or numbers that is at issue. It seems to me that the car accident of 1962 occurred precisely in a period when Dau was making an internal shift from macroscopic physics to the theory of elementary particles and fields. There is no doubt that, had he remained "on duty," he would soon have renounced such grandiose pronouncements as "The lagrangian is dead, we have only to bury it with full honors," and he would be in the vanguard of researchers of the microworld and cosmology.

Flame Instability

To characterize Dau is, above all, to tell of his works, of how they came about and of their subsequent fate. Dau took up the theory of combustion, a subject very far from his most important works, in part because of his relationship with the late D. A. Frank-Kamenetzkiĭ and me. I hope that the reader will not take this as boasting.

The one-dimensional theory of flame propagation was developed by D. A. Frank-Kamenetzkiĭ and myself in 1938; our basic assumption was of a strong temperature dependence of the chemical reaction. The result consisted in an expression for the normal flame speed and the realization that the reaction zone—the actual flame—was quite thin. Dau posed the question of the possible instability, in the hydrodynamic sense, of the one-dimensional solution in three-dimensional space as the boundary of a discontinuity of density and the normal flow speed. It turned out that a flat flame surface, i.e., the one-dimensional solution, is unstable! Dau found a dispersion relation: the growth rate is equal to the flame speed multiplied by the perturbation wave vector lying in the flame plane and by a dimensionless coefficient of order one. The criterion for correctness of the formulation was that the Reynolds number be greater than one, $\mathrm{Re} > 1$. In the expression for Re the perturbation wavevector plays the role of length, and the flame speed is used for velocity.

In industry, as a rule, flames are turbulent simply because the flows of hot mixture are turbulized in advance, before combustion begins. But even under laboratory conditions (Bunsen burners, slow layered burning of a mixture in a pipe or vessel) the Reynolds number is much larger than 1. For this reason, Dau was certain that laminar flames are practically nonexistent. Nevertheless, it is sufficient to look at a Bunsen burner to convince oneself of the stationary and laminar character of the process. Special experiments with central ignition showed that spontaneous auto-turbulization of combustion such as Landau proposed occurs only for $\mathrm{Re} \geq 10^4 - 10^6$! Several decades were required, after Dau's accident and his death, to understand the reason for such a large critical Reynolds number.

A. G. Istratov and V. B. Librovich were the first to show that in the propagation of a spherical flame from an ignition point in the center, the perturbations grow slowly. Indeed, the wave-length of each mode of the perturbation grows proportionally to the radius; the growth rate is inversely proportional to the wavelength. As a result, one obtains algebraic, not exponential growth of perturbations. Later, in the seventies, this author (together with Kidin and Librovich) considered the instability of a stretched parabolic flame. In this case, at the same time that perturbations are growing, they slip off the portion protruding into the hot mixture toward the edge, i.e., toward the walls. Perturbations leave the scene before they even have a chance to grow! All of this explains the results of observations in laboratory conditions. At the same time, there is no doubt that under industrial conditions in the combustion of a turbulent gas the auto-turbulization discovered by L. D. Landau plays a role in maintaining the turbulence.

Detonation

During and shortly after the war, Dau worked actively with Kirill Petrovich Staniukovich on the problem of detonation of explosive materials and related problems of explosion shock waves. In detonation, hot explosion products appear with densities greater than 2 g/cm^3 and temperatures of several thousand degrees. Formally, one may say that it is a hot gas. Previous researchers, by inertia mimicking the descriptions of low- and medium-density gases, used the Van der Waals equation. This formulation corresponds to a picture of inelastic molecules with a specific minimum volume $\delta_m = b$. Landau immediately took into account the smooth law of repulsion of the molecules, leading to an algebraic pressure dependence, $P = K\rho^n = K/\delta^n$. Subsequent terms were also accounted for in the work with Staniukovich. What was important, however, was to get past the flaws in the old ideas.

The works above still retain their importance in a number of scientific and technological areas in which explosives are used. Of some of his hydrodynamics work, Dau recalled during and after the war that they had been done in 1938 and concluded: hydrodynamics is, after all, simpler than quantum theory and the theory of elementary particles.

How grateful we all are to Pyotr Leonidovich Kapitsa that Dau's hydrodynamics period did not extend too long.

The Theory of Metals

I probably had the most chance to talk with Dau on the theory of metals. In these discussions, every time, our differing understanding of the essence of the problem manifested itself. For Dau, and to a great extent for his entire school and the related school of Ilya Mikhaĭlovich Lifshitz, the chief problem was of the *properties* of metals. As examples, we may mention electrical conductivity and its temperature dependence, the specific heat of electrons, etc. I, on the other hand, was concerned with a perhaps more primitive question—*will* a given substance or element be a metal or a dielectric?! How does the answer depend on the density? Here it should be mentioned that under given conditions (pressure and temperature) a substance generally has only a single, quite concrete density: it is either a gas, a liquid or a solid. Thus, if we wish to change the density, we must change the conditions. But in this case, not all situations are possible, i.e., not all densities are possible at low temperature. Therefore we have an idealized problem of atoms (nuclei) "nailed down" in specific places at particular distances from one another. Such a formulation of the problem was foreign to Dau; he considered it unphysical and uninteresting.

Dau's position was to some extent influenced by the revolution in physics which occurred when quantum mechanics was established. The desire to work only with observables was very characteristic for this period. With respect to the present problem, the results are well known: the properties of metals were excellently established by Ilya Lifshitz, Dau and his school. The solution of the problem of creation of metals brought glory to Neville Mott and others. Dau and I published a small joint Letter on the transformation of a dielectric to a metal with a simultaneous gas-liquid transition.

On the Reasons for Superfluidity

It is interesting that the approach to the theory of metals is echoed in the theory of superfluidity. Dau's theory of superfluidity—one of his best creations, rewarded with a Nobel Prize—is in essence a theory of the *properties* of a liquid whose superfluidity was established by experiment. Dau was fairly critical of the works of Tissa and others who related superfluidity to Bose-condensation of helium atoms. The brilliant work of N. N. Bogoliubov on the theory of superfluidity had to do with the case of a gas whose atoms only repelled one another, i.e., it was not directly applicable to the case of liquid helium. All of this led Dau to develop a phenomenological theory (as opposed to a microscopic one) of the properties of helium. To this day I remember a seminar in Kapichnik—probably during the fifties—where Dau was particularly insistent that there was only one liquid which God made superfluid. I do not guarantee the accuracy of the quotation, perhaps he spoke of nature rather than God. However, I clearly recall the gist: the situation is that we must study the *properties* of such a liquid, rather than ask whence the superfluidity itself comes. But I also remember the comment by the late Academician Obreimov: there are two stable isotopes of helium: ^{4}He and ^{3}He. Thus, there arises the question whether both are superfluid, and the comparison allows the characterization of the role of Bose-condensation. I will not write here about what happened in science (in particular, in superfluidity and superconductivity) after the accident which tragically cut off Dau's work.

Dau on Dau

The evaluation of his own activity (and even talent, personality) has always been and remains one of the most important features of a human. In his youth, Dau often spoke of the classification of scientists. He sharply distinguished "Zeroeth" and "First" class scientists—the latter restricted to 10–12 men, from Newton and Einstein to Schroedinger, Heisenberg and Dirac. He self-critically said that no single individual work of his achieved the level of creation of the theory of relativity or quantum mechanics. For this reason, he did not include himself in the first class.

It seems to me now that Dau's opinion was probably too humble. If we take all of Dau's work together—as an "integral"—and if we take into account his influence on physics as a whole through the *Course of Theoretical Physics* and through personal interactions—Dau belongs to the highest class. In any case, in his ability to instantly—and sometimes in advance—find a mistake, Dau had no equal. It is interesting that Dau did not like to talk about why this or that physicist—his partner in a discussion, or he himself, or someone else—did not manage to do something that had been published by another physicist. He felt that one must evaluate *what had been done*, not worry about what might have been.

Dau was self-critical, but his self-criticism never went overboard into fruitless hand-wringing. Dau's life was honorable and in harmony—until the horrible catastrophe which cut off his work and practically cut off his interaction with us.

The memories of Landau as his eightieth birthday approaches, of all the years talking with him, are surprisingly disturbing. A huge portion of life comes to mind which is closely tied to the life and work of Landau.

I remember the time when I could call Dau on the phone and stop by 10 minutes later to see the teacher and to receive a bit of clarity and understanding.

I understand now better than then how happy were life and work next to this great physicist and Man.

Note: This article was prepared for a collection devoted to Landau's eightieth birthday. Landau's role in creating the theory of superfluidity, phase transitions and many other problems have been illuminated in detail in other articles. Therefore, in this article only the lesser-known works of Landau are discussed—and, of course, my own impressions of his personality.

71b

The Creation of Particles and Antiparticles in Electric and Gravitational Fields*

John Archibald Wheeler, who is now completing his 60th year, is remarkable for his youthful enthusiasm and his ability to find what is new and unexpected in nature. Wheeler's enthusiasm and eloquence come through in his scientific writings, and in the attractive force that his name exerts on young people. J. A. Wheeler may look with pride not only on his achievements, but also on his students, who have acquired the depth and style of their teacher.

The subject of this paper would be worthy of the pen of Wheeler; having experienced the charm of his personality and of his works, I am aware of my imperfection, my limited imagination, the stinginess of my tongue. Wheeler may be admired, but he cannot be imitated. Still, the problem of the creation of particles and antiparticles is imposing—one could say, in Russian, that the problem is "worthy of the brush of Aivazovskiĭ." [1]

Newtonian mechanics gave us laws of motion for bodies whose existence and properties are postulated. Quantum mechanics changed those laws; the changes have especially strong effects on the motion of elementary particles, but the very existence and the properties of the particles remain, as before, postulated from without.

Only the theory of Dirac has predicted the existence and properties of new particles, and the possibility of particle creation and annihilation! In retrospect, we can see prototypes of particle theory in acoustics and gas dynamics, where normal-mode equations and the theory of wave packets lead us to the concept of quasiparticles. The photon with its energy $\hbar\omega$ and its momentum $\hbar\omega/c$ was a "particle" glimpsed by Einstein in the chaos of the electromagnetic field. However, profound understanding did not come until Dirac, transcending semiclassical ideas based on the correspondence

*In: "Magic without Magic," Ed. Klauder. N.Y., 277–288 (1972).

[1] I. K. Aivazovskiĭ (1817–1900), the celebrated seascape painter, who recorded the grandeur of the tempest. The fascinating properties of a stormy sea are due to the nonlinear interaction of waves on the surface of a heavy liquid; in this sense, they are connected with the creation of quasiparticles, which is similar to the creation of pairs. The designation "gravitons" has already acquired a different meaning. Why not use the designation "aivasons" for the quanta of waves on water? I note especially the use of hydrodynamic analogies by Wheeler [1].

principle, presented the theory of radiation strictly as a theory of the birth of new particles. Fock's treatment of second quantization should be mentioned.

A major landmark in the development of particle theory was the formulation of the Pauli principle, with its division of particles into bosons and fermions, and its explanation of the symmetry properties of wave functions. The idea of identical particles emerged—and simultaneously the idea that particles have statistical internal structures was buried. To prevent the resurrection of the deceased, in heathen times aspen stakes were driven into graves; this was the role of the Pauli principle.

Until now, particle creation in various circumstances has not always found a sufficiently clear and picturesque description. In the spirit of Wheeler, I will try to clear up some problems about the birth of particles.

Is There a Limiting Charge for an Atomic Nucleus; Does the Mendeleyev Table Have a Limit?

It is known that, in the Coulomb field of a point charge, the solution of the Dirac equation for the electron becomes singular and the energy becomes zero when $Z = \alpha^{-1} = \hbar c/e^2 = 137$.

Taking into account the finite dimensions of the nucleus, Pomeranchuk and Smorodinski [2] find that the critical charge is $Z_c \sim 170$, with the energy approaching $E = -mc^2$. The question arises: does such a Z represent a natural limit of Mendeleyev's periodic system; will any larger charge be screened back to the value Z_c? (We ignore here the problem of the stability of such a nucleus with respect to nuclear forces and the Coulomb repulsion of protons.)

Gerstein and I [3] have answered this question negatively. For $Z > Z_c$, the $1s$ level (K shell) of the atom has energy below $-mc^2$. As a result, the K shell in a certain sense is dissolved in the continuum. However, the sum rules lead to the conclusion that (in the language of the Dirac sea), besides the usual infinite charge density, $\rho_0 = \infty$, which is made zero by charge symmetry of the vacuum, and besides the infinite additional density, $\int \rho_1\, dV = \infty \cdot Z$, which is taken into account by a renormalization of the nuclear charge Z in the heavy nucleus, when $Z > Z_c$ there arises an additional term, ρ_2, with

$$\int (\rho_1 + \rho_2)\, dV = \infty \cdot Z - 2e,$$

the same term as would arise if two electrons were present.

Popov [4] has shown that for $Z > Z_c$, there arises a resonance in the wave functions of the electron, with a finite width for $E < -m_e c^2$.

In other words, nonradiative scattering of a positron is possible on an atom with $Z > Z_c$. Described graphically: the positron annihilates with one K electron, and then another electron settles into the resultant hole in the K shell with a consequent creation of a positron.

It is crucial that the repulsion between nucleus and positron makes the process "sub-barrier." Consequently, for small $Z - Z_c$, the width is exponentially small. On this point, Popov has corrected erroneous statements by Gerstein and myself concerning the polarization of the vacuum for $Z > Z_c$.

It is important that the continuum $E < m_e c^2$ for $Z > Z_c$ is orthogonal to the $2s$, $2p$, and other higher levels. Thus, the dropping of the $1s$ state into the continuum does not affect the further building up of electron shells. The chemical properties of Mendeleyev's table must extend into the region $Z > Z_c$.

We note that the renormalization of charge is set forth clearly in the linear approximation in connection with the self-energy diagram of the photon. It would be important to show, equally clearly, that the renormalization is linear, independent of the charge of the nucleus.

By analogy with the physics of dielectrics, it is possible to state that the renormalization depends on the polarization far from the charge, that is, where the field is always weak. There it is strictly linear.

Pair Production in a Constant Electric Field

This process has been studied in a series of papers; the most recent, in which previous references are quoted, is Nikishov and Narozhny [5].

I. D. Novikov has pointed out a paradox in the classical description of the phenomenon: the total momentum $\mathcal{P}$ of the pair must be equal to zero, because their charges are equal and opposite; but the total energy $\mathcal{E}_p$ is not zero. Query: how can this result be Lorentz-invariant?

The electric field E_x, assumed constant in space and time, is obviously invariant with respect to motion in the x-direction; but the result $\mathcal{P} = 0$, $\mathcal{E}_p \neq 0$, would not seem to be.

The solution of the paradox, it turns out, is tied to the time-dependence of the individual momenta of e^- and e^+:

$$\frac{dp_+}{dt} = eE, \qquad \frac{dp_-}{dt} = -eE, \qquad \mathcal{P} = p_+ + p_- = p_+(t) + p_-(t) = 0.$$

In the last equation, simultaneity of the measurements of p_+ and p_- is essential. In a Lorentz transformation not only must $\mathcal{P}$ be transformed as a component of the 4-vector $(\mathcal{P}, \mathcal{E}_p)$, where $\mathcal{E}_p = \mathcal{E}_+ + \mathcal{E}_-$, but time must also be transformed as a component of (x, t), with a resulting change in the identification of simultaneous events on the e_+ and e_- world lines. Thanks to this, it is possible to construct trajectories of the electron and the positron for which $\mathcal{P} = 0$, $\mathcal{E} \neq 0$, in any reference frame.

To be specific, the equation for the trajectories

$$x_\pm = x_0 \pm \sqrt{x_m^2 + c^2(t - t_0)^2}, \qquad \text{where} \quad x_m = mc^2/eE_x,$$

resolves Novikov's paradox: one computes the momentum in the usual elementary way, $p = Mc\beta/\sqrt{1 - \beta^2}$, with $\beta = c^{-1}dx/dt$ and with x_0 and t_0

arbitrary. The Lorentz-invariance is obvious. It can be shown easily that the appropriate selection of x_m given above guarantees that the equations of motion are satisfied.

By resolving one paradox, we fall under the influence of others.

The trajectories are such that between any point on $x_+(t)$ and any point on $x_-(t)$, the interval is spacelike.

Any pair of points on the two world lines can be chosen as the points of particle creation and, whatever points are chosen, their spacelike separation guarantees the existence of a reference frame in which the particles are created simultaneously. Hence, in the classical description the births of the e^+ and e^- are a pair of causally independent events. Moreover, at each point separately charge cannot be conserved; charge conservation holds only for the whole space pierced by both world lines, and even there only in reference frames where the pair creation is simultaneous.

These paradoxes, obviously, are the price paid for adopting a classical description of the phenomenon.

In an exact quantum-mechanical solution of this problem with finite time $t_1 < t < t_2$ and region for the field activity, the vacuum 4-current j^μ is continuous in space and time; that is, it satisfies the conservation law $\partial j^\mu/\partial x^\mu = \partial p/\partial t + \mathrm{div}\,\mathbf{j} = 0$. And, in the limit of a small and slowly changing external electric field, j^μ is proportional to $\partial F^\mu/\partial x^\mu$; that is, it is proportional to the external 4-current J^μ that creates the field. Such a vacuum current is a renormalization of the charge. In the general case, for a finite field and/or a finite frequency, this proportionality is violated and, in particular, although both external 4-currents and vacuum 4-currents may be absent in the past, at $t < t_1$, and future, $t > t_2$, a vacuum 4-current $j^\mu \neq 0$ may arise at $t > t_2$ in the future. That is precisely the birth of charged pairs.

It is especially interesting and simple to follow this when the external field is spatially homogeneous (in some particular reference frame): $E_x = E_x(t)$, $A_x = A_x(t)$, $A_y = A_z = \phi = 0$. In this reference frame, the external current creating the field is

$$\frac{4\pi}{c} J^x = \frac{1}{c^2}\frac{\partial^2 A_x}{\partial t^2} = -\frac{1}{c}\frac{\partial E_x}{\partial t}, \qquad J^0 = J^y = J^z = 0.$$

By symmetry the vacuum current will also be in the x-direction, j^x. In this problem one can trace the contributions to j^x from particles with definite initial momenta, since (see Nikishov and Narozhny) the solutions of the Dirac equation have the form $e^{ipx} f(t)$ with a conserved momentum p.

The renormalization of the vacuum, that is, the current j for $J \neq 0$, gives a divergent integral, but the pair creation described by the residual current $j \neq 0$ for $t > t_2$, $J = 0$ is finite. For a slowly changing or static field, the residual vacuum current is $dj/dt \sim e^{-E_c/E}$ if $E < E_c$, and $dj/dt \sim e^3 E^2$ if $E > E_c$, where $E_c = m^2c^3/\hbar c$. Reference [5] also analyzes pair creation in a rapidly changing ($\hbar\omega > 2mc^2$) weak field and shows it is proportional to E^2.

The impossibility, or incompleteness, of any classical description of pair creation can be traced in still another manner. The quantum mechanical equations, as is well known, contain the law of energy conservation. In a spatially homogeneous problem we have

$$\frac{d\mathcal{E}}{dt} = jE,$$

where $\mathcal{E}$ is the energy density and j is the current density. Let us apply this equation to charges originating in a vacuum.

For classical particles already hatched from the eggshell of the vacuum, the inequality $|j| < (e/mc)\mathcal{E}$ exists, because

$$j = \sum e_i v_i, \quad \mathcal{E} = \sum M_i c^2 / \sqrt{1 - \beta_i^2}, \quad |v_i| < c.$$

Suppose that this inequality were always valid. Then it would be easy to prove that pair creation is impossible. In particular, in this case there exists the inequality

$$\frac{d\mathcal{E}}{dt} < \frac{eE\mathcal{E}}{mc}, \qquad \int E\,dt > \frac{mc}{e} \int \frac{d\mathcal{E}}{\mathcal{E}}.$$

Notice that the integral on the right diverges at $\mathcal{E} = 0$. Now, it is known that a finite density of pairs can appear in an initial vacuum ($\mathcal{E} = 0$ initially) as a result of the action of a finite field E in the course of a finite time; that is, for a limited $\int E\,dt$. But this violates the above inequality, which was based on the single assumption that all velocities are smaller than c. The connection of this result with the super-light separation of e^+ and e^- on classical trajectories should be obvious (see above).

According to a remark by L. P. Pitayevski, the case of a small field is fully described by a quantity called [15] the "photon self-energy term," $\prod_{\mu\nu}^{(2)}(k)$, in which regularization [the subtraction of $\prod_{(0)}^{(2)} + k^2(\partial \prod^{(2)} / \partial k^2)(0)$] has already been carried out. There remains a finite quantity with imaginary terms for $k^2 > 4m^2$ ($\hbar = c = 1$). In this approximation it is easy to verify that, in the course of the process, $j \sim E$ and $\mathcal{E} \sim E^2$, so the classical inequality is violated.

On the other hand, after the external field is turned off, the remaining number of pairs is $\sim \int E^2\,dt$, and the current created by them is $\sim \int E^3\,dt$, so there remain only "real" particles and antiparticles satisfying the classical inequality.

In a paper by Hawking [16] this type of paradox was met with full formal rigor. Hawking introduces the condition of "energy dominance," $T_0^0 > |T_a^b|$, and on the basis of it he proves that the creation of particles by a gravitational field in a theory with non-quantum metrics is impossible.

The condition of "energy dominance" is similar to the inequality $\mathcal{E} > (mc/e)|j|$, which forbade pair creation in an electric field. Hawking's objection to pair creation must therefore be rejected.

Very recently, L. P. Pitayevski and I have shown concretely that Hawking's condition is violated under the influence of a gravitational field, if the particles are described by quantum field theory [18].

The Birth of Charged Pairs in Higher Order

The annihilation cross section for a semi-relativistic ($E - mc^2 \sim mc^2$) pair e^+e^- is of the order of πr_0^2 where r_0 is the classical electron radius. Obviously the cross section for the reverse process, pair creation, must be of the same order.

Let us consider high-frequency electromagnetic radiation so intense that each state in phase space in the domain $\hbar\omega \sim mc^2$ contains more than one photon. For this it is necessary that the radiation temperature be $kT > mc^2$. In such a case one can speak of the birth of pairs in a high-frequency, classical, free ($J^\mu = 0$), electromagnetic field. We shall systematically express all relevant quantities [the photon density, N cm^{-3}; the photon flux, Q cm^{-2} sec^{-1}; and the rates of pair creation, dn/dt and of energy pumping $d\mathcal{E}/dt$] in terms of the classical field parameters [the spectral density of the radiation, $\mathcal{F}$(erg $\cdot$ sec/cm^3); the width of the spectrum, $\Delta\omega$; and the mean frequency, ω]. Let us consider the case

$$\Delta\omega \sim \omega \sim mc^2/\hbar.$$

We obtain

$$N = \mathcal{F}\Delta\omega/\hbar\omega, \qquad Q \sim cN \sim c\mathcal{F}\Delta\omega/\hbar\omega,$$

$$\frac{dn}{dt} \sim \pi r_0^2 NQ, \qquad \frac{d\mathcal{E}}{dt} \sim \hbar\omega\frac{dn}{dt} \sim \frac{e^4\mathcal{F}^2}{\hbar^2 mc}.$$

This free-field situation is remarkable because the whole vacuum current corresponds to the birth of real pairs; there is no contribution to vacuum polarization since the external field is free.

We note that the above formula for the energy pumping rate is identical to the expression for plasma heating by the induced scattering of the classical field on free electrons,

$$\frac{d\mathcal{E}}{dt} = \text{const } n_0 r_0^2 c^2 m^{-1} \int F^2\omega^{-2}\, d\omega$$

(see [6]). This identification requires, however, that the vacuum—under the action of radiation of frequency $\hbar\omega \sim 2mc^2$—be assumed similar to plasma with a characteristic density $n_0 \sim (mc/\hbar)^3$, in agreement with the naive ideas of the Dirac sea. For $\omega > 2mc^2/\hbar$, the gap between $E_0 = \pm mc^2$ is negligible.

If we raise the brightness temperature still higher, to $T \sim 137mc^2$ (keeping, however, only part of the spectrum; for example, $mc^2 < \hbar\omega < 2mc^2$), then processes involving any number of photons become equally probable. This is not surprising, since then the mean square field corresponds to the critical value E_c^2 for which the barrier disappears. In this connection, notice that a field E that is constant in space and in time can be considered either as the limit of a homogeneous field $E(t)$ produced by an external current, or as the limit (for a region smaller than a wave length) of a standing wave of low frequency. For $E \sim E_c$, when intensive pair creation occurs, it is impossible to determine the order of the process. Pair creation in optically thin plasmas also diverges at $T \sim \alpha^{-1}mc^2$, but for different reasons [19].

Back Action of the Pairs; Energy and Collisions

An exceedingly complete discussion of the system of quantum-field-theory equations for particles interacting with an electromagnetic field is contained in the well-known dissertation by E. S. Fradkin, and in his subsequent papers [14]; however, his general expressions lack visual appeal. We consider in what follows the simple case of a sharply time-confined large field $E > E_c$, created by a single impulse of external current that lasts for a time of the order of $\hbar/mc^2$. If there were no creation of real pairs, the electric field would subsequently remain constant in time.

The created pairs rapidly (in a time $\tau \sim mc/eE < \hbar/mc^2$) accelerate up to $v \sim c$. Considering all the e^- and e^+ as moving relativistically along the x-axis, we obtain $j = 2enc$. According to Nikishov and Narozhny,

$$\frac{dn}{dt} = \frac{1}{18\pi}\frac{(eE)^2}{\hbar^2 c}, \qquad \frac{dj}{dt} = \frac{1}{9\pi}\frac{e^3E^2}{\hbar^2};$$

from this follows an equation for the current. After termination of the external current, which created the initial field, the field satisfies

$$\frac{dE}{dt} = -4\pi j.$$

From this system of equations it is easy to calculate how long it takes for the field to return to zero:

$$t_0 = \sqrt{\frac{27}{8}}\frac{\hbar}{\sqrt{E_0 e^3}}\int_0^1 \frac{dx}{\sqrt{1-x^3}} = \frac{3\hbar}{\sqrt{E_0 e^3}}.$$

It is readily seen that for $E > E_c$ this time is longer than the time for accelerating the pairs up to relativistic velocity; this justifies the approximations made.

After t_0, the $e^\pm$ pairs continue to move by inertia; recombination is negligible, the field changes sign, and the system subsequently approaches equilibrium ($E = 0$, $j = 0$, $dE/dt = 0$) by way of damped oscillations.

In a more precise treatment, the field E due to the external current produces a particular coherent electron-positron state. Its entropy is zero and, in principle, there can arise echo-type plasma phenomena, which distinguish a coherent state from a thermodynamically balanced plasma.

Collisions of particles destroy the coherence, create entropy, and drive the system toward equilibrium.

If the probability of collision during the characteristic time $\hbar/mc^2$ is not small, then collisions may substantially change the rate of pair creation. These remarks—requiring, incidentally, further elaboration—refer to pair creation in both electromagnetic fields (discussed above) and gravitational fields (discussed below).

Pair Creation in a Gravitational Field

The possibility, in principle, of pair creation in vacuum under the action of a gravitational field becomes obvious if we follow, step by step, the following sequences:

(1) The cross section for pair annihilation with emission of two gravitons is known and has been computed [7,12].

(2) From this follows the possibility of pair creation when gravitons collide.

(3) An aggregate of gravitons with large occupation number represents a classical gravitational wave.

Thus, pair creation should take place in the approximation where one considers quantized fields of electrons and neutrinos, in a nonquantized, classical gravitational field; that is, in a well-determined metric of curved space-time. The above numbered sequence corresponds to pair creation in a high-frequency ($\omega > mc^2/\hbar$), free ($T_{ik \text{ extern}} = 0$) gravitational field. By analogy with electrodynamics, one can foresee the possibility of pair creation also under the action of a high-frequency, longitudinal (not-free, $T_{ik \text{ extern}} \neq 0$) field ($0 - 0$ type conversion), and pair creation by quasi-static tunneling under the barrier. In the latter case, the Newtonian potential must be sufficiently great: $|\phi_1 - \phi_2| > c^2$. This threshold condition does not depend on the mass of the particles about to be born, but the penetrability of the barrier does depend on the mass; the probability of birth is exponentially small if the distance L between points with a potential difference of c^2 is larger than the Compton wave length, $L > \hbar/Mc$. For massless particles, L limits the momentum at the moment of birth, $|p| < \hbar/L$ and, consequently, it limits the available volume in phase space. Dimensionality arguments yield the following expression for the rate of pair creation: $dn/dt \sim cL^{-4}$ for $L < \hbar/Mc$ and for any L if $M = 0$. In a spatially homogeneous situation the potential becomes a quadratic function of the coordinates,[2] $\phi = -kx^2$; so the work done by gravity on a pair which move apart from the origin to points 1 and 2, where they are "created", is

$$\phi_1 + \phi_2 - 2\phi_0 = 2kL^2 = 2c^2.$$

Thus, the separation required for pair creation is $L = c/\sqrt{k}$, and the pair creation rate is $dn/dt \sim k^2c^{-3}$ for $k > (Mc^2/\hbar)^2$. The connection between gravity and metric of spacetime is known to cause an incompatibility between a static situation and a superstrong gravitational field. It is known, for example, that inside the gravitational radius, $r < r_g$ of the Schwarzschild field, t is no longer a time coordinate, and the situation is dynamical, not static [20]. In the cosmological problem, the coefficient k in the Newtonian potential depends on the rate of expansion, and hence on the age of the Universe: $k \sim t^{-2}$. Hence, the particles are born at distances exceeding ct from each other—but after the analysis of pair creation in an electric field, this should not frighten us. It is obvious that here too a systematic quantum-mechanical investigation will remove the paradox. The nonstatic character of the gravitational field and metric makes it impossible to delineate, as separate processes in the cosmological problem, barrier penetration and high-frequency excitation. L. Parker [8], in his well-known paper, considers nonadiabatic changes of wave functions in a variable metric as the cause of pair creation.[3] This procedure is analogous to the analysis of nonadiabatic excitations of an oscillator in the paper by Popov and Perelomov [9]. Since the harmonic-oscillator equation describes the elementary excitations of a Bose field (for instance, of the electromagnetic field), the excitation of an oscillator is equivalent to the birth of bosons.

In connection with pair creation in a gravitational field there arises a deep question of principle: What is the relationship between the birth of particles and quantum mechanical corrections to the equations of general relativity?

At first sight the equations $T^k_{i;k} = 0$, which result from the Bianchi identities and the Einstein field equations, do not permit the birth of particles: these equations are linear in T^k_i and, therefore, if $T^k_i \equiv 0$ for $t < t_0$, then a continuation of $T^k_i \equiv 0$ for $t > t_0$ seems to be necessary. For this reason, in a letter [10] by this writer (Zeldovich, 1970) devoted to the influence of pair creation on cosmology, the erroneous suggestion was put forth that pair creation is impossible unless the equations of general relativity are changed.

In the course of discussions with Professors Treder and Novikov at the Astrophysical Institute in Potsdam, Professor Dautcourt has remarked that $T^k_{k;i} = 0$ is a set of only 4 equations for 10 unknowns and that, therefore, it does not determine unambiguously the total behavior of T_{ik} in the future.

[2]The constant gravitational field corresponding to linear $\phi = ax$ is unobservable; so it is clear that the second derivative of the potential must come into play.

[3]Pair creation in the Friedman metric has been considered recently by Grib and Mamayev [17]. See also Zeldovich and Starobinskiĭ [21].

In particular, the reaction of a nonzero T_{ik} from vacuum (in a region where in the past one had $T_{ik} \equiv 0$) does not contradict the equations $T^k_{i;k} = 0$.

But the problem reappears when one tries to make more explicit the form of the stress-energy tensor T_{ik}. For a system of particles all the components of T_{ik} are proportional to a common factor, the number density of particles n (at least for small values of n). Therefore, the system of equations $T^k_{i;k} = 0$ has the form $\partial n/\partial t = f(x,t)n$. For finite f such a system does not permit one to join an initial $n = 0$ state to a solution $n \neq 0$. We come once more to Hawking's paradox [16], which should be answered more explicitly here.

This problem arises from the use of an invalid classical description for the essentially quantum mechanical phenomenon of pair creation.[4]

In a proper quantum-field-theory description of the particles by a superposition of the vacuum state (ψ_0, with $T_{ik} = 0$) and of filled states (ψ_n, with $T_{ik} \neq 0$), $\psi = c_0\psi_0 + \sum c_n\psi_n$, the expectation value of the T_{ik} components is expressed in terms of the coefficients c_n.

One may assume, for small c_n, that $T^0_0 \sim \sum |c_n|^2$, while $T^\alpha_\beta \sim \sum |c_0 c_n|$, so that $T^0_0 \sim |T^\alpha_\beta|^2$. In this case, a smooth join of the birth of particles with the vacuum is possible.

For example,

$$\frac{\partial T^0_0}{\partial t} \sim \Gamma^{0\beta}_\alpha T^\alpha_\beta$$

with

$$\Gamma^{0\beta}_\alpha \sim t^m, \quad t > 0, \quad m > 0, \quad \Gamma^{0\beta}_\alpha \equiv 0, \quad t < 0$$

is compatible with the solution

$$T^\alpha_\beta \sim t^{m+1}, \quad T^0_0 \sim t^{2(m+1)}, \quad t > 0$$

(compare with the situation for pair creation in an electric field: $j \sim E$, $\mathcal{E} \sim E^2$, $\mathcal{E} \sim j^2$).

One can give a rough analogy. Consider an atom in a non-degenerate ground state; it has a definite energy and zero dipole moment. Under the action of an electric field the atom is deformed and passes to a superposition of the initial state with an excited state, $\psi_0 + cn\psi_n$; its energy changes, $\Delta E \sim c_n^2$, and a non-zero dipole moment arises, $D \sim c_n$. The different behavior of D and E is due to the fact that the atom was initially in an eigenstate of the energy operator (Hamiltonian), but not in an eigenstate of the dipole moment. The dipole moment was equal to zero only in its expectation value. Whether similar behavior is possible for the different components of T_{ik}, despite the fact that in a Lorentz transformation they mix among each other, remains to be shown. By analogy with pair creation in an electric field, one would expect no contradictions between pair creation in a vacuum and the classical equations of general relativity (including the Bianchi identity and its consequence, $T^k_{i;k} = 0$).

[4]Let us emphasize that we insist here on a quantum description of the fields of electrons, photons, and other particles in a classical, nonquantum metric.

There is another side to the problem. If the creation of real pairs from the vacuum is possible, then there must also exist virtual pair creation, connected to real pair creation by dispersion relations. Virtual pair creation must lead to renormalization and to radiative corrections in the equations of general relativity (Feynman [11], DeWitt [12], Ginzburg *et al.* [13]). These corrections, including real pair creation, may be represented on the right-hand side of the Einstein equations by terms, $T_{ik\,\text{vac}}$, which are nonlinear functionals of the metric in the past light cone. As before, $T^k_{i;k} = 0$ will be satisfied, though the customary relations $T^0_0 > |T^\alpha_\beta|$ are not obligatory. Such a situation should not be called the quantization of gravity or the metric.

Order-of-magnitude estimates lead to the conclusion that pair creation can influence, in a most essential manner, the evolution of the universe near a singularity (Parker [8]; and for an anisotropic universe, Zeldovich [10, 22]). Parker [8] came to the conclusion that pair creation disappears for both $m^2 = \infty$ and $m^2 = 0$. The last result, as shown by Starobinsky and myself [21], is due to the conformal invariance of the wave-equation for massless particles, and therefore the result is specific for the conformally flat isotropic Friedman universe considered by Parker.

But the anisotropic Kasner-type universe $[ds^2 = dt^2 - \sum t^{2\pi i}(x^t)^2]$ that I have considered [10] is not conformally flat. The creation does not vanish in the $m^2 = 0$ limit. Therefore, at $t \to 0$ it must give, by dimensional arguments, infinite $\mathcal{E}$, $T_{\alpha\alpha} \sim \hbar c^{-3} t^{-4}$. Such energy and stresses will be the dominant term in the dynamic general-relativity equations, probably working in the direction of turning the Kasner metric into Friedman isotropic solution. It is possible that the observed energy (or, to be more precise, entropy) of the universe and its observed isotropy are both connected to pair creation (near the universe's "initial singularity"). It is quite probable that the quantization of the metric (near the initial singularity) was unavoidable and affected its subsequent state. But here we encounter also an aspect new in principle: in cosmology—as distinguished from all other fields of science—one requires ideas concerning the choice of the initial state, and not simply ideas concerning the evolution of a given state by given laws. What is needed is a special intuition, of a type unusual for physicist-theoreticians.

I should like to close this paper—dedicated to the 60th birthday of J. A. Wheeler—with an appeal to him: Come and solve this problem in the next 60 years, between 1970 and 2030.

The initial state of Wheeler on 31 December 1970, his force, courage, clearness of mind, and rich imagination permit us to hope for and expect results.

I take this opportunity to express my thanks to G. Dautcourt, I. Kobzarev, I. Novikov, D. Okun, L. Pitayevski, A. Starobinski, G. Treder, and E. Fradkin, for discussions.

REFERENCES

1. *Wheeler J. A.*—In: Geometrodynamics. N.Y.: Academic Press (1962).
2. *Pomeranchuk I. Ya., Smorodinski Ya. A.*—J. Physics (USSR) **9**, 97 (1945).
3. *Gerstein S. S., Zeldovich Ya. B.*—ZhETF **57**, 654 (1969).
4. *Popov V. S.*—ZhETF **59**, 965 (1970).
5. *Narozhny N. B., Nikishov A. I.*—Yader. Fiz. **11**, 1072 (1970).
6. *Peyraud J.*—J. de Physique **29**, 88, 306, 872 (1968);
   Zeldovich Ya. B., Levich E. V.—Pisma v ZhETF **11**, 497 (1970).
7. *Vladimirov Yu. S.*—ZhETF **45**, 251 (1963).
8. *Parker L.*—Phys. Rev. Lett. **21**, 562 (1968); Phys. Review **183**, 1057 (1969); Phys. Review **3**, 346 (1971).
9. *Popov V. S., Perelomov A. M.*—ZhETF **56**, 1375 (1969).
10. *Zeldovich Ya. B.*—Pisma v ZhETF **12**, 443 (1970).
11. *Feynman R. P.*—Acta Phys. Polonica **24**, 697 (1963).
12. *DeWitt B. S.*—Phys. Rev. **160**, 1113 (1967); Phys. Rev. **162**, 1195 (1967); Phys. Rev. **162**, 1239 (1967).
13. *Ginzburg V. L., Kirzhnits D. A., Lyubishin A. A.*—ZhETF **60**, 451 (1971).
14. *Fradkin E. S.*—Dissertation, Inst. Theor. Phys., 1960, Trudy Lebedev Inst. of Physics, AS USSR **29** (1965); Acta Phys. Hungarica **XIX**, 176 (1965); Nucl. Phys. **76**, 588 (1966).
15. *Ahiezer A. I., Berestetski V. B.* Kvantovaĭa Elektrodinamika [Quantum Electrodynamics]. Nauka Press, Moscow (1969).
16. *Hawking S.*—Comm. Math. Phys. **18**, 301 (1970).
17. *Grib A. A., Mamayev S. G.*—Yader. Fiz. **10**, 1276 (1969).
18. *Zeldovich Ya. B., Pitayevski L. P.*—Comm. Math. Phys. **23**, 185 (1971).
19. *Bisnovatyĭ-Kogan G. S., Zeldovich Ya. B., Sunyaev R. A.*—JETP Lett. **12**, 64 (1970).
20. *Zeldovich Ya. B., Novikov I. D.* Relativistic Astrophysics, Vol. I, Theory of Gravitation and Stellar Evolution, Chicago Univ. Press (1971).
21. *Zeldovich Ya. B., Starobinski A. A.*—ZhETF **61**, 2161 (1971).
22. *Zeldovich Ya. B.*—Comments on Astroph. and Space Sci. **3**, 179 (1971).

An Autobiographical Afterword

The last page of the last article has been turned and, naturally, the problem arises of a summing-up of seventy years of life and fifty three years of work, and of the lessons for the future which may be extracted from this summary.

The first question—that of a final summary—is the subject of the introductory essay, compiled by the editorial committee and situated at the beginning of the first volume, although covering the contents of both volumes. In my opinion the introduction contains an exaggerated evaluation of my results and their influence on modern science.

It would not be appropriate, however, to argue whether the significance of this or that article was greater or less. There might be some interest in the qualitative difference between evaluations of my work, as well as of the general state of physics from various angles—from the outside, by specialists, even the most well-disposed toward me, and from within, by myself. Thus the present afterword is written from very subjective positions, without any pretension to objectivity.

I well remember the first, still childish (I was 12 years old) choice of subjects, a conversation with my father. For mathematics exceptional abilities were required (which I did not sense in myself). Physics seemed to be a finished science: this was the influence of a venerable teacher of physics at school who triumphantly read the immutable laws of Newton first in Latin, then in Russian. The rebel spirit of the new physics had not yet penetrated into high schools in 1926. On the other hand in the chemistry course mysteries abounded—what is valence? catalysis? And the chemists did not hide the absence of a fundamental theory. Ya. I. Frenkel's book *The Structure of Matter* made a great impression on me, especially the first part devoted, for the most part, to atomic theory and the kinetic theory of gases, the definition of Avogadro's number and Brownian motion. But atomic theory, just as thermodynamics, relates equally to physics and chemistry. Later fate placed me in the Institute of Chemical Physics (ICP).

In 1930 I was a laboratory assistant at the Institute for Mechanical Processing of Useful Minerals (Mekhanobr); I studied ground ends of mineral ores. I will remember always the richness of the Kola peninsula, and my respect for Academician A. S. Fersman is imprinted in me. In March, 1931 with an excursion of co-workers from Mekhanobr I visited the chemical physics division of the Leningrad Physico-Technical Institute (PTI). In S. Z. Roginskiĭ's laboratory I became interested in the crystallization of nitroglycerin in two modifications. L. A. Sena told me about this (Roginskiĭ was abroad).

After the discussion (in which neither I nor Sena yet knew the truth) I was invited to work in the laboratory in my free time. Soon the question of an official transfer arose. By the time I was hired (May 15, 1931) the division had been turned into the independent Institute of Chemical Physics. In the interim I recall my review presentation on the kinetics of the transformation of para-hydrogen to ortho-hydrogen. Without fully understanding what it was, I nevertheless stood firmly and heatedly for the principle of detailed equilibrium. N. N. Semenov, S. Z. Roginskiĭ and many others of my future colleagues were present.

Many years later I heard three rumors. The first: Mekhanobr traded me to ICP for a fuel pump. Second: Academic A. F. Ioffe wrote to Mekhanobr that I would never be useful for the solution of practical problems. Three: Ioffe could not stand *wunderkinde* and that is why he gave me up to ICP.

To this day I do not know how much truth there is in each of them. I can only testify that I did not see Ioffe until 1932, and I saw him under remarkable circumstances: A general seminar of PTI and its daughter institutes was called. Ioffe read a telegram from G. Chadwick about the discovery of the neutron, commented on it, and in conclusion a resolution was taken and a reply sent saying that we (all?!) too were taking up neutron physics. For me—not immediately—the resolution proved prophetic.

In my interest in chemistry a large role was played by the purely visual perception of bright colors and forms, beginning with the "transformation of water into blood" in the interaction of iron salts and thiocenogen potassium, and from the formation of sediments and crystallization. This was followed by an interest in the sharpness of the transition of the indicator color, and further to the sharpness of phase transitions.

In neighboring laboratories atomic spectra were being studied. I clearly remember that in comparison with the variety of colors and forms of macroscopic phenomena the detailed theory of the atom seemed boring. I write of this today as evidence of my deep misunderstanding then of the theory of physics.

Together with this was the correct and natural feeling that behind the randomness of forms and alternation of smooth and sharp dependencies were hidden general patterns. Today they are called catastrophe theory and synergetics.

In the thirties, in developing the theory of combustion, we in essence were studying specific examples of these new sciences without knowing their names. Remember Moliere's hero who discovered in his old age that all his life he had been speaking prose.

The tremendous, unsurpassed accomplishment of Abram Fedorovich Ioffe and Nikolai Nikolaevich Semenov was the creation of institutes which attracted capable young people from everywhere. There arose a "supercritical" situation of rapid growth in people and in their work. For me a huge role was played by the possibility of learning from the young (but older than me!) theoreticians. I am deeply indebted to my then teachers and now friends—L. E. Gurevich, V. S. Sorokin, O. M. Todes, S. V. Izmailov. For about two years I studied (but did not complete my studies) through the continuing education department at the university. I attended remarkable lectures on electrodynamics by the late M. P. Bronstein. I remember now the words "gradient invariance" which I did not understand at the time.

A great stroke of luck for me was the combination of experimental and theoretical work on one and the same problem. I first observed the Freundlich adsorption isotherm experimentally in studying the system $MnO_2 - CO - O_2 - CO_2$. Only after this was a corresponding theory worked out (see article **1** in my book *Chemical Physics and Hydrodynamics*). Without delay I checked in an experiment the dependence on the temperature of the exponent n in the formula $q = cP^n$. In the experiment there was nothing new; as the name itself shows, the Freundlich isotherm was discovered by Freundlich, not me. However, doing my own experiment amazingly activated my desire to understand the phenomenon and to provide a theory for it. I think that this is a common phenomenon. I highly recommend to theoreticians working in macroscopic physics that they take part in an experiment!

A certain cycle of papers on adsorption and catalysis comprised my doctoral dissertation. Blessed times those were when the Higher Attestation Committee allowed people without a university degree to defend their theses! The defense took place in September 1936.

Even earlier I had begun to swim independently and decided to take up fuel elements. My interest in electrochemistry was encouraged by my respect for Academic A. N. Frumkin, who had looked favorably upon my works on adsorption which were in significant measure parallel to his work with M. I. Temkin. Considerations of the means of transformation of fuel energy into electricity arose naturally under the influence of A. F. Ioffe.

However, practically speaking, in Leningrad at the ICP I turned out to be alone working on the question of fuel elements. The work proceeded very slowly.

In 1935 the extraordinarily energetic and forceful Odessite A. A. Rudoy arrived at the institute, or better said, stormed the institute. He was inspired by the chain theory of chemical reactions. What is to keep one from finding a means of transforming combustion energy into energy of active centers and to use it for the endothermic reaction of nitrogen oxidation? Why not obtain several liters of nitrous acid from a kilogram of fuel and free air? Beyond the mist idyllic pictures were drawn: a tractor plows a field while simultaneously providing it with nitrogen fertilizers, and the classic equipment for synthesizing ammonia would lie unneeded. Semenov took Rudoy into the Institute, but at the same time he created a serious group to investigate the problem. It included the late P. Ya. Sadovnikov, D. A. Frank-Kamenetskiĭ and A. A. Kovalskiĭ. I also entered the group. It turned out that the formation of nitrogen oxides in the combustion of hydrogen in air was observed even by Cavendish, before the composition of air was established.

I will not describe here the results of this large collective work—they are presented in the materials in my book *Chemical Physics and Hydrodynamics.*

I again worked as both an experimenter and a theoretician. The work forced me to learn and apply the theory of dimensionality, similarity and self-similarity; it expanded my horizons and introduced me to problems of turbulence, convection and thermal power engineering. The book by Gukhman, *Similarity Theory*, inspired me. I formed a strong and fruitful friendship with David Frank-Kamenetskiĭ. An engineer by training, he had sent a letter to ICP behind which N. N. Semenov espied talent. He summoned David Albertovich from Siberia to Leningrad and soon put him to work on nitrogen oxidation. From Frank-Kamenetskiĭ with his engineering education I learned about Reynolds number, supersonic flow, the Lavalle nozzle and much more.

Much later, also in connection with nitrogen oxidation, I met Ramzin who had by then received the State Prize; he was still active, but already hopelessly ill. Working at home in the evenings, he completed in two weeks a job which some other scientific research institute would stretch out over years. But qualitatively the problem was solved earlier. In the best case, with heating of the air and fuel and even with the addition of oxygen, relatively low concentrations of nitrogen oxide were obtained. The limiting factor turned out to be the process of transformation of NO to NO_2 according to the classical trimolecular reaction $2NO + O_2 = 2NO_2$. Only NO_2 can be consumed and used, but the technological volumes necessary for its formation are prohibitively large. The dream did not come true, and only in recent decades has the theory of nitrogen oxidation found a new, ecological significance. The theory of nitrogen oxidation was the topic of my doctoral dissertation, defended at the end of 1939. It is pleasant for me to note that among the opponents was Aleksandr Naumovich Frumkin. A natural continuation of the work in which combustion was the source of high temperature, was the investigation of the combustion process itself.

Combustion appears in many forms: the combustion of explosive mixtures, combustion of unmixed gases, detonation, etc. All of these processes had been studied before, but without penetration into the chemical kinetics of the reactions. The previous generation of researchers had approached from the standpoint of thermal power engineering and gasdynamics. A brilliant exception was the Frenchman Taffanel, who published papers in 1913–1914 which anticipated much. In 1914 he fell silent. Only in April of 1985 did I learn that Taffanel had lived until 1946, successfully working on engineering problems.

We faced a wide field of activity and the period 1938–1941 was a productive one. The vital interest of N. N. Semenov made its mark. As a rule, within 10 minutes after I had arrived home in the evening, Nikolai Nikolaevich would call and dinner would be delayed for an hour. The discussion centered on individual parts of a well-known review article by Semenov in *Uspekhi Fizicheskikh Nauk* [**23**, 251 (1940); **24**, 433 (1940)].

A combustion laboratory was organized at the institute where we systematically studied the kinetics of the reaction $2CO + O_2 = 2CO_2$ up to the very highest temperatures. Perhaps more important was the fact that an internal combustion engine laboratory, where K. I. Shchelkin was studying detonation, had long existed nearby in the institute. The neighboring explosives laboratory had the greatest influence on me. My peers A. F. Belyaev and A. A. Appin where there. Yuliĭ Borisovich Khariton had organized this laboratory and directed it. This is my friend and teacher to the present day. I will say much more about joint work with Yuliĭ Borisovich below.

As a theoretical physicist I consider myself to be a student of Lev Davidovich Landau. There is no need here to explain Landau's role in the creation and development of Soviet theoretical physics. At the same time, without belittling this role, I want to add that as I matured with the years—grew older, alas!—I began to better understand and value more the roles of other people and schools. This includes above all Ya. I. Frenkel with his immense intuition, optimism and breadth. It includes V. A. Fok with his deep and brilliant mathematical technique. It includes I. E. Tamm and his students and the school of perturbation theory originating with L. I. Mandelshtam. Finally, this includes many, including mathematicians alive and well today, who have successfully worked in theoretical physics.

I beg you not to read the above paragraph as ill-intentioned. If I write that Frenkel had intuition, while Fok was a good mathematician, do not draw the conclusion that Landau had neither intuition nor knowledge of mathematics—this is not what I have in mind! Landau's talent was harmonious, his judgment was severe, but almost always just. What is said about schools of theoretical physics may be applied to schools of physics as a whole.

In my youth my horizon was limited by ICP and PTI. There is no doubt that PTI produced a brilliant constellation of physicists, it raised Igor Vasilevich Kurchatov and his colleagues who carried out a very important state assignment. This has been described in many excellent books and articles. But in the pre-war years, and even in the first post-war years it seemed to me, for example, that optics was a science in which the fundamental questions had been exhausted. Today it is enough to name Cherenkov radiation and lasers to refute this incorrect, superficial judgment of mine. The line running from Lebedev through Rozhdestvenskiĭ and Vavilov, Mandelshtam and Tamm, Cherenkov, Frank, Ginzburg, Prokhorov and Basov, proved infinitely more fruitful that it seemed to me in the thirties.

It is difficult now for me to determine whether it was my own nearsightedness or in some measure an underestimation of the other school (other schools) that divided my colleagues. In any case, from the very open reminiscences of Gamow and certain remarks by Scobeltzyn I can now definitely judge the views of representatives of the other approach. Lebedev's school very definitely felt its existence separate from the school of Ioffe. But let us leave this topic to the historians of science. Today, fortunately, there is no such antagonism; a sufficiently close mixing of those schools which could earlier be distinguished has occurred.

Returning to my work at the end of the thirties, I see one significant defect: insufficient attention to the promotion of my results abroad. I knew foreign work well, and published some papers in Soviet journals in English. However, it never even occurred to me to send out my reprints to foreign scientists. Trips abroad were out of the question. The time was at fault, but perhaps, in some measure, our older comrades were also at fault; they should have cared more about personal connections.

Let us move on. The discovery of uranium fission and the possibility in principle of a chain fission reaction predetermined the fate of the century—and mine as well. The appropriate papers by Yu. B. Khariton and me are published at the beginning of this book and I have nothing (and no reason) to add to the commentaries regarding the scientific essence. I want only to note the leading role of my teacher—Khariton—in the understanding of the significance of the problem to all of mankind. I think I was more interested in specific problems of calculation method, etc. It is no accident that precisely Yuliĭ Borisovich became in 1940 a member of the Uranium commission (see UFN, March 1983). The subsequent development of the work is well-known from the many reminiscences of participants.

Yuliĭ Borisovich notes a curious detail: we considered the work on the theory of uranium fission to be apart from the official plan of the Institute and we worked on it in the evenings, sometimes until very late. In fact, the administration of the Institute, apparently, held the same view—a capable, but more practical colleague asked 500 rubles for a review on the theory of fission of isotopes, but this sum was not found.

In speaking of subsequent work, I want to emphasize the role of the theory of detonation and explosions.

The surprise of U.S. scientists in August, 1949, when air samples showed that their nuclear monopoly had ended is well known. August 1949—the test of a Soviet atomic weapon—was the natural result of the huge, directed effort of an entire people; the scientific potential of the country, developed in the prewar years, also played a role. The surprise in the U.S. would have been less had they read our pre-war articles published in Russian. I speak not only of papers on the chain reaction of uranium. The science of explosion and the theory of detonation are also a necessary part of that knowledge without which it would not have been possible to solve the problem. We recall that Khariton formulated the condition of the detonation limit as early as 1938. A complete one-dimensional theory of detonation was formulated by me in 1940. In the U.S. the same problem was solved by John von Neumann—the great mathematician—only in 1943. We note that von Neumann took up detonation precisely in connection with the atomic bomb.[1]

Soon after the beginning of the war the Institute was evacuated to Kazan. The problem arose of detailed analysis of processes associated with rocket weapons—"katyushas." The theory of powder combustion sufficient for the internal ballistics of cannons needed correction. A delicate balance between the entry of powder gases in combustion and their exit from the barrel is characteristic for the combustion chamber of a reactive charge. New conceptions of powder combustion, the phenomenon of blowing-out discovered in our laboratory by O. I. Leipunskiĭ, the role of a heated layer of powder—all of this was unusual for artillery specialists and received differing evaluations from powder and internal ballistics specialists.

I wish to note the interest and support in this work on the part of General Professor I. P. Grave, the well-known rocket-builder Yu. A. Pobedonostsev (both are gone ...) and the still-living and well G. K. Klimenko. But we did not always meet such support; there were also sharp arguments, attempts at administrative pressure, and substitutions of arguments by orders.

In connection with the papers on powder combustion, our group relocated to Moscow. We turned out to be the forward division, in whose path the entire Institute of Chemical Physics came to Moscow (and not back to Leningrad) at the end of the war. Work on combustion and detonation, as also work on powder combustion, continued at ICP even after a group of theoreticians shifted (together with me) to a new subject. I wish here to express my deep gratitude for this to A. G. Merzhanov and his group, to B. V. Novozhilov, G. G. Manelis, A. I. Dremin and many others (Institute of Chemical Physics of the USSR Academy of Sciences). In the course of their work they have not forgotten my work—and do not allow others to forget. Without this continuity there undoubtedly would have been much rediscovered abroad. There is no more thankless task than a belated struggle for priority.

One does not forget one's first love—and thus in 1977 a scientific commission on the theoretical foundations of combustion processes was formed. To this day I continue to work on combustion problems, although not full force. In connection with problems of combustion, in close interaction with G. I.

[1]For a detailed history of the development of the theory of detonation, see the articles and commentaries in *Chemical Physics and Hydrodynamics.*

Barenblatt in the fifties, the concept of "intermediate asymptote" was formulated, with general significance for mathematical physics. He and I also found in the perturbation theory of self-oscillatory processes (for example, flame propagation) a very general solution which corresponds to a shift and has an identically zero increment. Physicists who work with field theory will recognize here an analogy with the so-called Goldstone particle.

Together with A. P. Aldushin and S. I. Khudyaev (ICP) we studied the transition from the theory of Kolmogorov, Petrovskiĭ, Piskunov and the Englishman Fisher to the theory of Frank-Kamenetskiĭ and myself. In the most general case of reaction kinetics and arbitrary initial conditions the correct approach to the problem of propagation again turned out to be related to the idea of an intermediate asymptote.

The problem of the hydrodynamic flame instability discovered by L. D. Landau turned out to be far from simple: here, after the very fundamental work of A. G. Istratov and V. B. Librovich, only in the eighties has it been possible to move forward together with V. B. Librovich and N. I. Kidin.

Ideas borrowed from field theory allow us to take a new approach to the nonlinear theory of spin combustion. Recently, within the framework of the Commission, we have been forced to direct much attention to organizational work related to the large power technology of coal burning.

Let us return to the atomic problem and the forties and fifties.

A huge collective was headed by Igor Vasilievich Kurchatov. The most important part of the work was directed by Yuliĭ Borisovich Khariton. So this problem completely took over me as well. During very difficult years the country spared nothing to create the best possible working conditions. For me these were happy years. A great new technology was being constructed in the best traditions of great science. Attention to new proposals and to criticism completely without regard to the ranks and titles of the authors, the absence of secretiveness and suspicion—this was the style of our work.

The country was living through the difficult post-war years. However, the immense authority of Kurchatov created a healthy atmosphere. In addition, our work had a beneficial effect on Soviet physics as a whole. Once, when I was in Kurchatov's office, there was a phone call from Moscow: "Well, should we print in *Pravda* an article by a philosopher which refutes the theory of relativity?" Igor Vasilievich, without thinking for a moment, answered: "Then you may shut down our whole operation." The article was not published.

Toward the middle of the fifties some of the first-priority problems had already been solved. New trends were appearing: the Geneva Conference on the peaceful use of nuclear energy and the famous presentation by Kurchatov at Harwell (England) on thermonuclear reactions became milestones of detente.

Part of the work which related to applications were of general scientific interest and were published. These include papers on strong shock waves, their structure and their optical qualities.

Interest in phenomena occurring at high temperatures also led to the fundamental posing of the problem of establishing thermodynamic equilibrium between photons and electrons. The peculiarity of the problem lay in the fact that at a sufficiently high temperature, scattering becomes predominant over radiation and absorption. A. S. Kompaneets performed brilliant work in this area. It was published in 1965 and proved extraordinarily important for cosmology and astrophysics, for the plasma of a hot universe and for the radiation of matter falling into the gravitational field of a black hole.

Work in the theory of explosions prepared us psychologically for research on explosions of stars and the very greatest explosion—the Universe as a whole.

Simultaneously industrial activity stimulated interest in nuclear physics and neutron physics. In the fifties it was but a short jump to elementary particle physics. I was especially impressed and stimulated by Enrico Fermi's little book, *Theory of Elementary Particles.* In the English edition which I used, but not in the Russian translation, on the slip-cover the editor (not Fermi!) gave the following introduction: "The publication of this book is made possible by a grant from a certain rich lady who bequeathed it for proof of the existence of God. The discovery of the laws of nature and their harmony proves God's existence better than any theological treatise."

If we take the existence of God to mean objectivity of the laws of nature, existing independently of our knowledge and desires, then any Marxist philosophy can sign this thesis.

As a kind of self-education I worked through the best presentation of the general theory of relativity—the second part of *Field Theory*, the two-volume course of theoretical physics by Landau and Lifshitz.

I want to again emphasize the tremendous role that interaction with Lev Davidovich Landau played for me. In Kazan, and then in Moscow, we lived nearby to one another, and were closely associated at work. The possibility of going to him, to seek advice, to take my proposals, ideas, papers to him for judgment—I felt all of this as a great blessing. I found out about the tragedy of January 1962, when Landau ceased to be a theoretical physicist (although he remained alive), while I was far from Moscow. The disturbing days, weeks and months of struggle to save his life, the unity of physicists which transcended national boundaries—are unforgettable. The school created by Landau has been preserved! It lives in those who are continuing the monumental *Course of Theoretical Physics*—E. M. Lifshitz and L. P. Pitaevskiĭ. It lives in the AS USSR L. D. Landau Institute of Theoretical Physics. Its organization, the choice of people, the maintenance of the highest professional level of theoreticians—this is the enormous accomplishment of I. M. Khalatnikov and his colleagues. To the Landau school in the narrow sense one may also include the theoretical division of the AS USSR Institute of Theoretical and Experimental Physics—the brain-child of I. Ya. Pomeranchuk, currently headed by L. B. Okun. In the broad sense the ideas and methods of Landau together with the ideas and methods of other outstanding Soviet theoreticians (I listed them briefly above) have entered organically into all of Soviet theoretical physics.

Returning to memoir style, I wish to say that my work with Kurchatov and Khariton has given me very much. The main thing was and remains the internal feeling that my duty is done before my country and people. This gave me the definite moral right to investigate in the subsequent period such questions as particles and astronomy, without second thoughts about their practical value. Above I wrote about how my scientific interest in these questions matured. I must at the same time self-critically tell of the weaknesses and difficulties which I encountered in the new turn of my scientific activity. Recall that in 1964 I officially transferred to the AS USSR Institute of Applied Mathematics (IAM), organized by M. V. Keldysh as early as 1953. After his death, A. N. Tikhonov has directed this institute. In this institute I have worked 19 years (up to my transfer to the Institute of Physical Problems in 1983).

Before my transfer to IAM my works on particles and astronomy were outside of my normal work, in some sense optional—and now I see that this is reflected in their quality. Up until recently I was proud of the fact that I obtained a maximum of physical results with a certain fairly elementary supply of mathematical knowledge; but now, and particularly in connection with the theory of elementary particles, I face the reverse side of the statement. Why, in fact, should one limit oneself to a certain modest set of mathematical knowledge? Thus arose the book *Higher Mathematics for Beginning Physicists and Engineers.*

I cite a part of my letter, published in the American journal *Physics Today* (Sept., p. 95), in connection with the discussion in this journal of the reasons for the decline in the level of physics teaching in the U.S.

"In connection with the discussion of how to teach physics to the young generation, I would like to mention one common difficulty.

The laws of physics are formulated in the form of differential equations: such, for example, are Newton's laws of the motion of a material point, a solid body, or even a gyroscope. Maxwell's laws of an electromagnetic field—these are partial differential equations; the laws of gasdynamics are written in the same way.

Teenagers are capable of understanding all of this material.

However it would be more accurate to say that they are not capable of understanding deeply and loving physics, if they lack the supply of mathematical terms necessary for this. Here is my main point: in the majority of cases the teaching of mathematical analysis is begun late and includes difficult elements of the theory of sets and limits.

So-called 'rigorous' proofs and theorems of existence are much more complex than an intuitive approach to derivatives and integrals.

As a result the mathematical ideas needed to understand physics reach children too late. It is the same as serving the salt and pepper not at lunch, but a little later—at five-o-clock tea."

But let us return to the mathematics that works and is used in modern theoretical physics.

The theory of particles to a very large degree develops under the influence of advanced mathematical ideas and in directions which are indicated by mathematical elegance. I will not invoke the chrestomatic example of the Dirac theory of a relativistic electron which leads to the concept of antiparticles. Let us consider isotopic invariance. Experimentally a discrete symmetry is observed: replacing a proton by a neutron (or the reverse) in the same quantum state does not change the energy of the nucleus. However Heisenberg considered it necessary to introduce a continuous group of rotations in isotopic space which smoothly translates a neutron to a proton through a rotation of 180° through mystic intermediate states! Not the simplest, but the more complex and more elegant formulation proved the more fruitful. The depth of Heisenberg's formulation manifested itself in the shift from nuclei to mesons. The concepts built in analogy with isotopic rotation perform especially clearly in connection with the theory of quark colors, gradient invariance, and the Yang-Mills theory.

I will not describe in detail my work on particles. They are given in the book and are followed by quite well-qualified commentary. From the commentaries, washing them of celebratory politeness, it is clear how many mistakes I made. The mistakes in published works which are not included in the present collection are even greater.

This book presents, with commentaries, my papers on astrophysics. It is not reasonable to argue with these commentaries. Today the most important individual work seems to me to be the nonlinear theory of formation of the structure of the Universe or, as it is now called for short, the "pancake" theory. The structure of the Universe, its evolution and the properties of the matter which forms the hidden mass, have to this day not been fully established. A major role in this work was played by A. G. Doroshkevich, R. A. Sunyaev, S. F. Shandarin and Ya. E. Einasto. The work continues. However the "pancake" theory is "beautiful" in and of itself; if the initial assumptions hold, then the theory gives a correct and nontrivial answer. The "pancake" theory is a contribution to synergetics. It was especially pleasant for me to learn that this work in some measure initiated mathematical investigations by V. I. Arnold and others. A large volume of work on the spectrum of relic radiation in the presence of perturbations was "left hanging"—the Universe turned out to be very smooth, and the perturbations too small.

A method that has survived and is of great interest is the diagnostics of hot plasma by scattering of relic radiation with distortion of the spectrum, proposed by me together with R. A. Sunyaev.

In significant measure my work (together with my closest co-workers, above all R. A. Sunyaev, A. G. Doroshkevich, S. F. Shandarin and—until 1978—I. D. Novikov) in astrophysics turned out to be promotional, popularizing and pedagogical. All of this is necessary and useful, however it is to be judged on a different scale than obtaining original results.

At the beginning of my activity in astrophysics I was hindered by habits picked up in the course of practical work. The astrophysicist should ask questions like: how is nature structured? what observations will make it possible to find this out? I, meanwhile, was more inclined to define the problem thus: how can we best construct the Universe, or how can we construct a pulsar in order to satisfy given technical conditions—excuse me, I meant to say: first observations. That is how the idea of a cold universe came up, that is how we got the idea of a pulsar as a white dwarf in a state of strong radial oscillations. In my own defense I can only say that I did not insist on my mistakes. Apparently, nevertheless, as a whole my activity—scientific and promotional—was useful. Astronomers have taken me into their ranks. It is my works on astronomy that are associated with my election to the National Academy in the U.S. and the Royal Society, the gold medals of the Pacific Astronomy Society and the Royal Astronomy Society. It was a great honor for me to be asked to read the report on modern cosmology at the XIII General Assembly of the International Astronomical Union. Greece, the colonnade of an ancient theatre, above me the black starry sky, the audience on marble benches, my nervousness before the report and during the report and the happy conclusion. Life continues, and cosmology is entering an area where physics has long since broken away from experimental verification. A new generation of theoreticians will speak not of the first three minutes or seconds, and not of nuclear reactions and plasma. Processes are being discussed at a "planckian" length 10^{-33} cm, over a "planckian" time 10^{-43} s with a "planckian" energy 10^{19} GeV. S. Hawking, A. D. Linde, A. A. Starobinskiĭ, A. Guth and others are leading. In field theory 5-, 11- and 26-dimensional spaces are being discussed. In laboratory conditions they will certainly imitate our customary $(3+1)$ space-time, the extra dimensions will hide and collapse, leaving traces only in the systematization of particles and fields. Twenty-year-old kids come in and immediately, without the burden of earlier work and traditions, take up the new topics. Do I not look among them like a mastodon or a archeopterix?

I am consoled by the restructuring of mentality with age. Today (several days before my seventieth birthday) I am already less interested in competitive motives, whether it be I who say that "Aha" that Dobchinskiĭ and Bobchinskiĭ argued about. The final result, the physical truth is what interests me, almost independently of who finds it first. If only I have the strength to understand it!

Mankind, as never before, is at the threshold of marvelous discoveries. The idea of an all-encompassing unified physical theory seems brighter, and geometry is playing an increasingly great role. Perhaps, in the highest sense, not literally, Einstein will prove to be right, and his theory, which reduces the forces of gravity to geometry, will turn out to be the model of the unified theory.

It is possible that precisely cosmology will turn out to be the testing ground for verification of new theories. Then I will recall the works of S. S. Gershtein, V. F. Shvartsman, S. B. Pikelner, L. B. Okun, I. Yu. Kobzarev, M. Yu. Khlopov and myself as the first timid applications of cosmological arguments for the solution of problems in particle theory which are inaccessible to today's experiments. We, I together with L. P. Grishchuk and A. A. Starobinskiĭ, are trying to move forward in the analysis of the birth of the Universe. In the middle of the eighties the most difficult and fundamental problems of natural science are bound together into a tight knot. I have no greater wish than the wish to learn the answer and to understand it.

Moscow, 3 March, 1984